T0314030

The Map Reader

The Map Reader

Theories of Mapping Practice and Cartographic Representation

Edited by

Martin Dodge

Department of Geography, School of Environment and Development, University of Manchester, Oxford Road, Manchester, M13 9PL, UK

Rob Kitchin

National Institute for Regional and Spatial Analysis, National University of Ireland, Maynooth, Co. Kildare, Ireland

Chris Perkins

Department of Geography, School of Environment and Development, University of Manchester, Oxford Road, Manchester, M13 9PL, UK

A John Wiley & Sons, Ltd., Publication

Wiley-Blackwell is an imprint of John Wiley & Sons, formed by the merger of Wiley's global Scientific, Technical and Medical business with Blackwell Publishing.

Registered office: John Wiley & Sons, Ltd, The Atrium, Southern Gate, Chichester, West Sussex, PO19 8SQ, UK

Editorial offices: 9600 Garsington Road, Oxford, OX4 2DQ, UK
The Atrium, Southern Gate, Chichester, West Sussex, PO19 8SQ, UK
111 River Street, Hoboken, NJ 07030-5774, USA

For details of our global editorial offices, for customer services and for information about how to apply for permission to reuse the copyright material in this book please see our website at www.wiley.com/wiley-blackwell.

Library of Congress Cataloging-in-Publication Data

The map reader : theories of mapping practice and cartographic representation / edited by Martin Dodge, Rob Kitchin and Chris Perkins.
p. cm.
Includes index.
ISBN 978-0-470-74283-9 (cloth)
1. Cartography. 2. Maps. I. Dodge, Martin, 1971- II. Kitchin, Rob. III. Perkins, Chris.
GA101.5.M38 2011
912–dc22

2010049397

A catalogue record for this book is available from the British Library.

This book is published in the following electronic format: ePDF 9780470979594; Wiley Online Library 9780470979587; ePub 9780470980071

Set in 9.75/11.25 pt, Minion by Thomson Digital, Noida, India.

First Impression 2011

Copyright Notice

Contents

The Editors

Martin Dodge
Department of Geography, School of Environment and Development, University of Manchester, UK

Martin is Senior Lecturer in Human Geography in Manchester where his research focuses on conceptualising the socio-spatial power of digital technologies and urban infrastructures, virtual geographies, and the theorisation of visual representations, cartographic knowledge and novel methods of geographic visualisation. He curated the well known Web-based *Atlas of Cyberspaces* and has co-authored three books covering aspects of spatiality of computer technology: *Mapping Cyberspace* (Routledge, 2000), *Atlas of Cyberspace* (Addison-Wesley, 2001) and *Code/Space* (MIT Press, 2011). He has also co-edited two books, *Geographic Visualization* (John Wiley & Sons, 2008) and *Rethinking Maps* (Routledge, 2009), focused on the social and cultural meanings of new kinds of mapping practice.

Rob Kitchin
National Institute for Regional and Spatial Analysis and Department of Geography, National University of Ireland, Maynooth, Co. Kildare, Ireland

Rob is Professor of Human Geography and Director of the National Institute of Regional and Spatial Analysis (NIRSA) at the National University of Ireland, Maynooth, and Chair of the Management Board of the Irish Social Sciences Platform (ISSP). He has published sixteen books, is editor of the international journal *Progress in Human Geography*, and co-editor-in-chief of the *International Encyclopedia of Human Geography* (Elsevier, 2009).

Chris Perkins
Department of Geography, School of Environment and Development, University of Manchester, UK

Chris is Senior Lecturer in Geography and emeritus University Map Curator. He is the author of four books, including standard texts documenting the changing contexts of map availability (*World Mapping Today* with R.B. Parry; Bowker-Saur, 2000), and has co-edited the second edition of the *Companion Encyclopaedia to Geography* (Routledge, 2006) and *Rethinking Maps* (Routledge 2009). His research interests are centred on the different ways in which mapping may be employed and he is the first Chair of the International Cartographic Association's Commission on Maps in Society.

Preface

Introducing *The Map Reader*

Martin Dodge, Rob Kitchin and Chris Perkins

Delineating maps and mapping

> A map is, in its primary conception, a conventionalised picture of the Earth's pattern as seen from above.
>
> Erwin Raisz, *General Cartography*, 1938.

Mapping provides a uniquely powerful visual means to classify, represent and communicate information about places that are too large and too complex to be seen directly, and cartography is the practice of map making. Importantly, the places that maps are able to represent need not be limited to physical, geographical spaces like continents, rivers, mountain ranges and such like: maps can be used to represent human activities, cultural patterns and economic exchanges, and indeed to construct worlds of the imagination. In this Preface we delineate the nature of maps and mapping, and outline the aims of *The Map Reader* and the practicalities of its making.

The ability to create and use maps is one of the most basic means of human communication, at least as old as the invention of language and, arguably, as significant as the development of mathematics. The recorded history of cartography clearly demonstrates the practical utility of maps in all aspects of Western society, being most important for organising spatial knowledges, facilitating navigation and controlling territory. They are instrumental in the work of the state, in aiding governance and administration, and in assisting trade and the accumulation of capital. Some have gone further to argue that mapping processes are culturally universal, an innate human activity, evident across all societies (Blaut *et al.* 2003), although the visual forms of the resulting cartographic representations are very diverse. At the same time, maps are rhetorically powerful graphic images that frame our understanding of the human and physical world, shaping our mental image of places and constructing our sense of spatial relations. So, in a very real sense, maps make our world.

Conventionally, maps are material artefacts that visually represent a geographical landscape using the cartographic norms of a planar view – looking straight down from above – and a consistently applied reduction in scale. However, it makes little sense to neatly define maps according to the type of phenomena mapped or the particular mode of presentation, or their medium of dissemination. Maps have traditionally been used as static paper repositories for spatial data, but now they are much more likely to be interactive tools displayed on a computer screen. Indeed, many national mapping agencies are discontinuing their printed topographic map products as customers increasingly use digital geospatial data. Today, we live in a map-saturated world (Wood 1992), continually exposed to conventional maps, along with many other map-like spatial images and media (e.g. animated satellite views on television news, three-dimensional city models in video games, medical scans in hospitals and clinics), along with myriad artistically deployed maps, and pictorial and cartoon cartographies meant to amuse and persuade.

Maps have long been used in scholarly research into social and physical phenomena. They are a primary technique of analysis in geography but they are also used widely in other disciplines, such as anthropology, archaeology, history and epidemiology, to store spatial data, to analyse information and generate ideas, to test hypotheses and to present results in compelling, visual forms. Mapping as a method of enquiry and knowledge creation also plays a growing role in the natural sciences, in disciplines such as astronomy and particle physics, and in the life sciences, as exemplified by the metaphorical and literal mapping of DNA by the Human Genome Project. This mapping work is not limited to cartography; many other spatial visualisation techniques, often using multidimensional displays, have been developed for handling very large, complex spatial datasets without gross simplification or opaque statistical output (e.g. volumetric visualisation in atmospheric modelling, three-dimensional body imaging in medical sciences or huge fractal graphs – see Colour Plate Two). At the start of the 1990s, Hall (1992: 22) claimed that 'more mapping of more domains by more nations will probably occur in the next decade than has occurred at any time since Alexander von Humboldt "rediscovered" the earth in the eighteenth century, and more *terra incognita* will be charted than ever before in history'; two decades on not only has this happened, but the trend shows no signs of slowing.

Mapping processes

The production of cartography and other spatial visualisations involves a whole series of mapping processes, from the initial selection of what is to be measured to the choice of the most appropriate scale of representation and projection, and the best visual symbology to use. The concept of 'map as process' is useful methodologically because it encourages particular ways of organised thinking about how to generalise reality, how to distil inherent, meaningful spatial structure from the data, and how to show significant relationships between entities in a legible fashion. Mapping provides a means of organising large amounts of, often multidimensional, information about a place in such a fashion as to facilitate human exploration and understanding. Yet, mapping practices are not just a set of techniques for information 'management', they also encompass important social processes of knowledge construction. As scholars have come to realise, maps and culture are intimately entwined and inseparable.

Mapping not only represents reality, it has an active role in the social construction of that reality. Mapmakers do not so much represent space, as create it. As Winichakul (1994; excerpted here as Chapter 5.4) and Pickles (2004) persuasively argue, maps precede and make the territory they seek to portray. So, for example, the first maps of Siam delineated the nation providing the model for an imagined community, rather than depicting it. Maps then are a key resource of states in the formation of national identity (Anderson 1991). It is rarely the case, however, that people are conscious of this constructive role when they make or use maps. Sparke (1998: 466, excerpted here as Chapter 5.7) calls this the 'recursive proleptic effect' of mapping, 'the way maps contribute to the construction of spaces that later they seem only to represent'. The power of maps comes from the fact that they are both a practical form of information processing and also a compelling form of rhetorical communication.

Maps work, essentially, by helping people to visualise the unseeable. This is achieved through the act of visualisation, premised on the common notion that humans can reason and learn more effectively in a visual environment than when using textual or numerical descriptions. Maps provide graphical displays which renders a place, a phenomenon or a process visible, enabling one of our most powerful information processing abilities – that of spatial cognition associated with the human eye–brain vision system – to be brought to bear. Visualisation is thus a cognitive process of learning through interactions with the multiple visual signs that make up the map. Effective cartographic visualisation can reveal novel insights about spatial relations, patterns and trends that are not apparent with other methods. In an instrumental sense, then, map use is a powerful prosthetic enhancement for the human body: '[l]ike the telescope or microscope, it allows us to see at scales impossible for the naked eye and without moving the physical body over space' (Cosgrove 2003: 137). The ideal of obtaining a reliable capacity to see the unseen is particularly applicable to much of thematic cartography, because it renders statistical information about people, places and geographical processes tangible by revealing their spatial pattern.

Their ability to communicate effectively means that maps are widely deployed as devices to present ideas, themes and concepts that are difficult to express verbally and to persuade people to their message. Most of the maps encountered on a daily basis (often with little conscious thought given to them) are used in the service of persuasion, ranging from marketing maps and city-centre tourist maps to the more subtle displays such as states' claims to sovereign power over territory, implicitly displayed in daily weather maps. Maps work because they are able to *sell* a particular vision of the world *and* because people are willing to *buy* into this vision: people believe in the authority of the image as a trustworthy representation of reality.

Objectives of *The Map Reader*

The map is one of the key components of visual culture and has proved to be a vital representational technology in many fields for hundreds of years. Maps enjoy widespread functional use for a range of tasks. In recent years, maps have started to gain more significance in the wider academy given the visual and spatial turns across the social sciences. As a consequence, there is an increasing interest is spatial representations and mapping practices in disciplines such as anthropology, literary studies, sociology, history and communications (Elkins 2007; Warf and Arias 2008). Similarly, mapping approaches are proving useful in the information sciences, bio-informatics and human–computer studies as the basis for novel knowledge discovery strategies (Börner 2010). In addition, there is also a lively engagement with cartography beyond academia with growing artistic interest (see Wood 2010 for a recent overview), numerous exciting participatory mapping projects and mass consumer enrolment of interactive spatial media on the Web, on mobile phones and with in-car satnavs to solve myriad daily tasks.

However, despite this attention and their widespread production and use, at a theoretical and analytical level, maps are still somewhat taken for granted: they are spatial representations that portray the spatial relations of the world. As such, analysis of the rhetorical power and technical complexity of how maps work has largely been confined to the small field of cartography, with some contributions from across the social sciences and humanities. Compared with other visual cultures, such as art and film, this literature is relatively small and, we feel, often overlooked. In compiling *The Map Reader* we wanted to draw together into a single source some of the most influential articles from the last half century to provide an intellectually-driven and interpretative anthology of cartographic research which could act as a primer for students, academics and lay readers interested in understanding the appeal and power of maps.

In that sense, the book cuts through the 'information overload' generated by bibliographic databases and ready online access to e-journals and digital books by providing direct access to a careful selection of the most influential texts. The materials selected for inclusion in *The Map Reader* are diverse in their agendas and approaches, drawn from leading scholars and researchers from a range of cognate fields, including cartography, geography, architecture, anthropology, literature, political science, graphic design and geomatic engineering. Each reading provides a thought provoking analysis, and collectively they demonstrate the diverse philosophy, history, praxis and technologies of mapping. They thus provide an insight into how influential cartographic ideas arise and how they circulate as catalysts that can codify and instigate important areas of research within cartography. While the focus on past 'classics' might seem rather backward looking in an era of such rapid change in mapping techniques and technologies, there is nonetheless real intellectual value understanding the roots and routes of cartographic thinking because it places current developments in context and provides a basis on which to build and extent contemporary analysis.

To aid the reader, we have structured the readings around five broad themes: (1) concepts, (2) technologies, (3) aesthetics and design, (4) cognition and culture, (5) politics and power. Each theme is set into context by an original interpretative essay from the editors. A series of full-page colour plates between sections present distinctive map exemplars that we hope will serve as provocative visual 'think-pieces' that counterpoise the surrounding texts.

Making our selection

The task of drawing up a limited, yet definitive, list of significant work for inclusion in a 'reader' text that would achieve widespread agreement is, for any academic discipline, an almost impossible one. We therefore acknowledge our final selection for *The Map Reader* is subjective, reflecting our personal biases, partial knowledge and political agendas. To guide our selection we used a number of parameters. Firstly, we decided to focus on the post Second World War period. This period has seen a diverse range of new theoretical ideas and technological developments, when cartography emerged as a distinct scholarly discipline with its own peer-review journals. Secondly, our remit centred upon pieces that were concerned in the main with contemporary mapping. There is only a limited consideration here of the history of cartography. Thirdly, we limited our selection to the English language given our own language limitations and the prospective readership of the book. As a consequence, the book is unavoidably reflective of Anglo-American scholarship. Fourthly, we sought to select material that speaks in a scholarly fashion to trends and concepts, rather than include more applied, technical and practical papers. Fifthly, nearly all of the readings were published in peer-review journals and scholarly monographs.

We did not, however, use quantitative metrics to guide the assessment of what counts as 'significant'. There is a panoply of projects that seeks to 'scientifically' assess the most significant scholarly work using citations counts, impact factors, h-scores and an assortment of other quantitatively derived metrics. While such calculative approaches seem to offer objectivity, this is very much a veneer that masks a whole host of messy realities, fallacies and contingencies with citation data, particularly relating to relative comparability through time and across subject areas. The material we have selected for *The Map Reader* has a range of citation counts from over one thousand to more recent articles which have so far attracted little attention. For example, according to Google Scholar (July 2010), the foundational semiological work of Jacques Bertin has been cited 566 times in the original French language version, and 1341 times in the 1983 translation (excerpted here as Chapter 1.2). Another well-cited 'classic' article in this collection is *Deconstructing the Map* by Brian Harley (excerpted here as Chapter 1.8), with well over 500 citations since its publication in 1989. A few of the pieces we have included have, as yet, negligible numbers of citations (e.g. Aitken and Craine's 2006 article in *Directions Magazine* cited only seven times so far; excerpted here as

Table P.1 Count of excerpts in *The Map Reader* by decade in which they were first published

Decade	Count
1940–1949	1
1950–1959	2
1960–1969	3
1970–1979	4
1980–1989	6
1990–1999	23
2000–2010	15
Total	54

Chapter 3.10). We have included such pieces because we think they have something important to contribute and are worthy of a wider audience.

While the material in *The Map Reader* covers a wide time span – nearly 60 years – running from 1942 up to 2010, there is an uneven spread of material selected and we are perhaps guilty of overlooking earlier significant work. Styles and conventions of academic discourse evolve and the pace of change in mapping tends to focus attention on the more recent past. Looking at the dates of the pieces included grouped by decade (Table P.1) it is somewhat evident that there is a bulge of material post 1990. This is reflective of the notable upsurge in philosophical engagement given the influx of social theory into cartographic debates and the explosion of new mapping technologies underpinning the growth of digital cartography. Given that we wanted to cover a wide range of topics it has meant that none could be covered in depth; not all the issues are as well represented as perhaps needed and, consequently, many important topics are represented by a single piece of work (in some cases this is unavoidably a placeholder for larger subfield). Clearly in these cases, these articles cannot encompass the full complexity and nuances of on-going debates. We hope that our introductory essays will help provide some additional context.

Editorial practicalities

In terms of the editorial process we have employed in *The Map Reader*, working within practical constraints of an affordable and commercially viable book, has meant that the pieces included are mostly reprinted as excerpts rather than verbatim. For monographs, we have generally excerpted from a single, most pertinent chapter. Where material has been deleted from the original this is indicated in the text by [. . .]. In some cases sizable edits have been made, but we have endeavoured to preserve the core intellectual argument as well as the narrative flow of the original, whilst removing extraneous examples or more elliptical context.

Each entry has been reformatted for consistency and to remove variability in the layout and referencing style evident in the original versions. The degree of standardisation, particularly the switch from footnote citations to Harvard style referencing in some excerpts, has necessitated some very minor changes to the texts themselves involving the insertion of references. Bibliographies have been edited from the originals to include only the references used in the excerpted text. Spelling has generally been standardised to British English. Some tables and figures have been omitted (to save space and for copyright reasons), so the numbering of these sometimes differs from the original. Many of the original illustrations included have been faithfully redrawn for this book by Graham Bowden (Cartographic Unit, University of Manchester) to ensure higher quality reproduction than scans of the originals.

Conclusion

Over the past fifty years there has been a sustained scholarly engagement in thinking about the ontological bases of cartographic representation and an exploration of new epistemologies of mapping. Moreover, there is burgeoning interest from many scientists, social scientists and humanities disciplines in theorising the nature of cartography and productively applying mapping and geographic visualisation to solve research problems. This coupled with tremendous socio-technical developments in the production of cartographic representations has led to a widening and more vibrant array of different kinds of mapping being employed by scholars. We hope *The Map Reader* will further advance understandings of cartography by illustrating the ways in which maps have been thought about and researched and that it will encourage a wider appreciation of where mapping has come from, and perhaps where it might go.

References

Aitken, S. and Craine, J. (2006) Guest editorial: Affective geovisualizations. *Directions Magazine*, 7 February. www.directionsmag.com. (Excerpted as Chapter 3.10.)

Anderson, B. (1991) *Imagined Communities: Reflections on the Origins and Spread of Nationalism*, 2nd edn, Verso, New York.

Bertin, J. (1983) *Semiology of Graphics* (trans. W.J. Berg), University of Wisconsin Press, Madison, WI. (Excerpted as Chapter 1.2.)

Blaut, J.M., Stea, D., Spencer, C. and Blades, M. (2003) Mapping as a cultural and cognitive universal. *Annals of the Association of American Geographer*, **93**(1), 165–185.

Börner, K. (2010) *Atlas of Science*, MIT Press, Cambridge, MA.

Cosgrove, D. (2003) Historical perspectives on representing and transferring spatial knowledge, in *Mapping in the Age of Digital Media*, (eds. M. Silver and D. Balmori), John Wiley & Sons Ltd, Chichester, UK, pp. 128–137.

Elkins, J. (2007) *Visual Practices Across the University*, Wilhelm Fink, Munich.

Hall, S.S. (1992) *Mapping the Next Millennium*, Vintage Books, New York.

Harley, J.B. (1989) Deconstructing the map. *Cartographica*, **26**(2), 1–20. (Excerpted as Chapter 1.8.)

Pickles, J. (2004) *A History of Spaces: Cartographic Reason, Mapping and the Geo-Coded World*, Routledge, London.

Raisz, E. (1938) *General Cartography*, McGraw-Hill, New York.

Sparke, M. (1998) A map that roared and an original atlas: Canada, cartography, and narration of nation. *Annals of the Association of American Cartographer*, **88**(3), 463–495. (Excerpted as Chapter 5.7.)

Warf, B. and Arias, S. (2008) *The Spatial Turn: Interdisciplinary Perspectives*, Routledge, London.

Winichakul, T. (1994) *Siam Mapped: A History of the Geo-Body of a Nation*, University of Hawai Press, Honolulu, HI. (Excerpted as Chapter 5.4.)

Wood, D. (1992) *The Power of Maps*, Guilford, New York.

Wood, D. (2010) *Rethinking the Power of Maps*, Guilford, New York.

Acknowledgements

We would like to give thanks to Rachael Ballard, Fiona Woods, Izzy Canning and colleagues at Wiley-Blackwell for all their hard work in the development and production of the book.

We are also very grateful to Graham Bowden and Nick Scarle in the Cartographic Unit at the University of Manchester for all their help with the slog of scanning and expertise with the illustrations and to Justina Senkus for formatting a number of the entries. Many thanks to Chris Fleet and Karla Baker at the National Library of Scotland for their help in finding us a really nice map example from the Bartholomew Archive. Thanks also to Jason Dykes and Mark Harrower in sourcing images for us.

We would also like to acknowledge the tremendous book and journal collections available to us in the John Rylands University Library at the University of Manchester; these provided such a vital infrastructure to support the production of *The Map Reader*.

SECTION ONE

Conceptualising Mapping

Chapter 1.1

Introductory Essay: Conceptualising Mapping

Rob Kitchin, Martin Dodge and Chris Perkins

It is all too easy to think of maps and cartography in a naïve, commonsense way – a map is a two-dimensional, spatial representation of the Earth, and cartography is the creation of such maps. If only it were so simple! The history of cartography reveals a rich engagement with different philosophies of science. As a result, the scholarly understanding of what maps are and the processes, procedures and protocols through which they are created and deployed has changed enormously over time. This has never been more so than over the past fifty years as academics from a broad range of disciplines have focused on conceptualising mapping.

In this section, a broad range of readings are excerpted; they span more than 60 years and have sought to advance how maps and cartography are conceptualised. What unites the thirteen chapters is the common pursuit of rethinking the ontological and epistemological bases of cartography. That is, they each put forward a novel way to conceptualise maps as artefacts and mapping practices, each moving beyond commonsense and naïve understandings to set out a viewpoint that they believe provides a more robust and useful theoretical underpinning. At the time of writing, none of the approaches detailed in the readings is considered hegemonic amongst academics. For some, this conceptual plurality is considered a hindrance because it means that there is no generally accepted way to understand maps, thus introducing uncertainty and undermining the credibility of cartography as a 'science', with well-grounded theory and prevailing methods and an established canon. For others, it is a sign of intellectual fervent that has reinvigorated what was arguably becoming an increasingly technical discipline that was progressing largely through technological advances and methodological refinement rather than more philosophical ideas (Crampton 2003).

According to Harvey (1989, excerpted as Chapter 5.2) the first major change in how maps were conceptualised, in a Western context, occurred in the Renaissance through the application of Enlightenment thought and technologies to cartography. Prior to this, knowledge of the geographical world was parochial and documented from multiple perspectives to no formal, universal standards. Areas that were unknown were literally off the map, filled with religious cosmology and figures of myth and imagination. Maps were understood more as reminders – as spatial stories – than as scientific representations of the world based on surveyed data (Ingold 2000). Replacing the piecemeal frameworks of medieval cartography was the adoption of a single, universal system of measuring and representing the world that used perspectivism and Cartesian rationality, underpinned by notions of objectivity, functionality and ordering. This perspective understood space and time in quite different ways to the medieval period, and the resulting transformation in cartographic thinking made the world knowable, navigable and claimable, for a privileged and powerful few, through a shared framework of scientific endeavour that was translatable across peoples and places (see Latour 1992,

The Map Reader: Theories of Mapping Practice and Cartographic Representation, First Edition. Edited by Martin Dodge, Rob Kitchin and Chris Perkins.

excerpted as Chapter 1.9). In the centuries that followed, the science of cartography – wherein maps provided objective, truthful representations of the spatial relations of the world – was refined through improvements in surveying and mapping techniques and the development of a set of established principles of design and aesthetics.

Attempts to historicise the nature of (Western) cartography through categorisations of map forms and taxonomies based on purpose, often implicitly use the notion of evolutionary advancement driven by technological development. The end result narrates cartography as a beneficent pursuit, characterised by improving accuracy and comprehensiveness with each new generation of map. Examples of this conceptualisation are quite common in the literature, such that '[t]he normative history of cartography is a ceaseless massaging of this theme of noble progress' (Harley 1989: 4). For example, Crone (1953: xi) notes, '[t]he history of cartography is largely that of the increase of accuracy with which . . . elements of distance and direction are determined and the comprehensiveness of the maps' content.' Histories of cartography in this tradition were histories of techniques, with an underlying assumption that rational decision making leads to the adoption of improved technologies and institutional practices when they become available. The result is that cartographic development can be conceptualised as a 'tree' with evolving complexity of mapping (Figure 1.1.1).

The apparent 'naturalness' of this account belies the politics behind the progressive conceptualisation of the development of cartography from a primitive past to the sophisticated present (Edney 1993, excerpted as Chapter 1.10). The underlying goal of this kind of construction of cartographic history – achievable only through a carefully selective reading of extant map artefacts – is to 'prove' that the objectivity of *current* scientific methods is predestined. It grants an important legitimisation to the positivist notion of contemporary professional cartography as the 'best' and provides a discursive mechanism to dismiss maps that do not fit 'acceptable' scientific standards. Scientific worldviews see technological progress almost like a force of nature that somehow operates outside society and beyond the political concerns of money, power and ego. The way one approaches cartographic history is therefore worthy of consideration, as it is at the heart of the recent political theorisation of cartography and directly informs our understanding of the nature of the map and contemporary positivistic epistemological foundations of cartography (including much of the work on online mapping and GIScience).

This Cartesian rationality still predominates the general understanding of maps. However, over the last half century or so there has been a fresh engagement between cartography and philosophy that has either sought to refine and advance scientific cartography, or has sought to challenge and reconfigure its ontological and epistemological underpinnings. The first of these engagements by Ernest Raisz (1938) and Arthur Robinson and colleagues (Robinson 1952; Robinson and Petchenik 1976; excerpted as Chapters 1.3 and 3.3) sought to provide formalised rules and principles of map design, drawing on a range of disciplines such as mathematics and

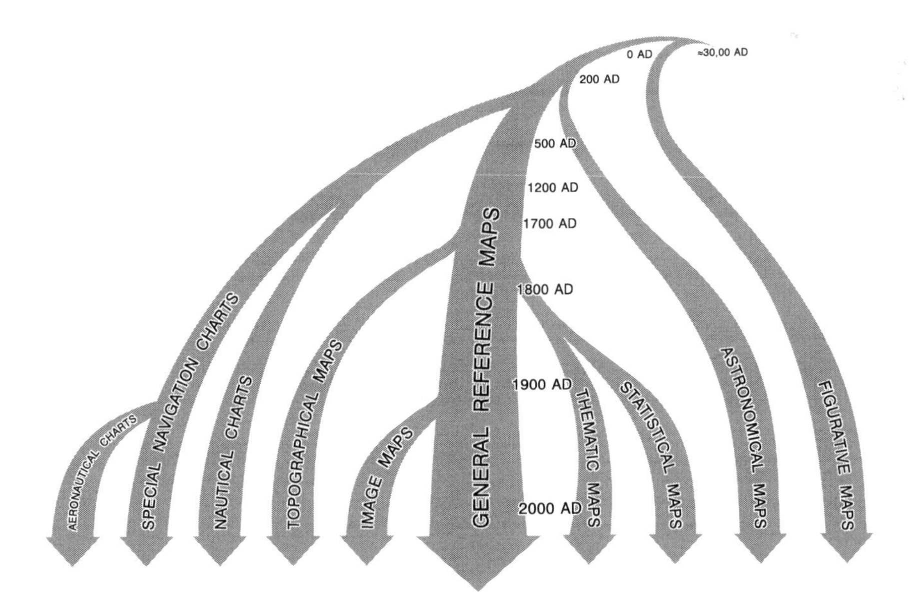

Figure 1.1.1 Cartography explained as a 'story of progress'. Mapping is shown to evolve over time with the development of increasingly complex forms. (Source: Redrawn from Robinson *et al.* 1995: 22.)

psychology. These approaches tended to see cartography as a blend of art and science, but where the aesthetic elements could be formalised through colour and visual theory and thus made more effective. Robinson, in particular, sought to advance a communications model approach that drew inspiration from psychology and information theory, and which sought to foreground the fact that maps serve as communication devices. As such, cartographic research needed to be framed around the goal of effective communication, wherein maps capture and portray information in a way that an idealised map reader could easily and intuitively analyse and interpret. Here, the aims of the cartographer were framed normatively to reduce error in the representation and to increase map effectiveness through appropriate design. Research thus sought to improve map designs by carefully controlled scientific experimentation that focused on issues such as: how to represent location, direction and distance; how to select information; how best to symbolise these data; how to combine these symbols together; and what kind of map to publish.

Robinson's ideas were extended and developed by others such as Joel Morrison (1976, excerpted as Chapter 1.4) and Christopher Board (1981, excerpted as Chapter 1.6). These scholars sought to forward the communication model as the new dominant paradigm for academic cartographic research, producing increasingly sophisticated conceptual models of how maps worked. Links were forged with cognitive scientists and behavioural geographers interested in cognitive mapping and how maps were learnt and people used and interpreted maps (Downs and Stea 1973; Lloyd 2000; excerpted are Chapters 4.3 and 4.9). Morrison, for example, envisaged cartography developing as 'communications science' with researchers working to understand the structural transmission of mapped information from data collection through to map use – including the science of data classification, generalisation, symbolisation and so on – in order to develop more effective cartographic syntax and grammar suitable for a given situation. By the early 1980s, Board was able to provide an overview of different information flow models, which by that stage had started to engage with the ideas of semiology.

Whilst Anglo-American cartographic researchers were examining the communicative properties of maps from a functional and pragmatic perspective, French academics were examining the utility of semiology – the study of signs and symbols – for map design. This work was based principally on the influential work of Jacques Bertin (1967) on graphic design (the 1983 English translation is excerpted as Chapter 1.2). Bertin set out what he saw as key properties of graphic systems and a set of rules for their presentation founded on a semiological analysis of the presentation of information in graphic form. These rules were influential in informing map design, and the science of semiology became an important touchstone for Anglo-American researchers in the late 1980s and early 1990s seeking to move beyond the limitations of the communications model. A semiological approach elided the divide between mapmaker and reader that underpinned communications theory. Technological advances were already exposing this divide as a fiction by the late 1980s and communication theory also failed to recognise the social and cultural aspects of mapping. A representational theory of cartography offered a more useful and practical grounding for scientific research. For example, MacEachren (1995) sought to blend cognitive and semiotic approaches, along with visualisation theory, to provide a coherent picture of how maps worked. Such work became influential amongst those working in GIScience and geovisualisation seeking ways to scientifically conceptualise and improve mapping within increasingly exploratory and interactive media. This new representational orthodoxy is borne out in research agendas of the geovisualisation community (MacEachren and Kraak 1997, excepted as Chapter 1.11) and in emerging work in multimedia cartography (Cartwright 1999, excerpted as Chapter 2.11).

A different challenge to cartographic theory emerged at the end of the 1980s, however. The communications model and its subsequent offshoots are still framed within a scientific rationality that sought to produce objective, 'value free', accurate representations. In a landmark paper, Brian Harley (1989, excerpted as Chapter 1.8) argued that, far from presenting the truth of the world, maps were social constructions presenting subjective versions of reality. Harley was by then a well-established scholar in the history of cartography, able to draw on a wealth of empirical material. He built on an emerging critical tradition, dating back to research from John Kirkland Wright in the 1940s (Wright 1942, excerpted as Chapter 4.2). Although there is a long history of analysis examining the role of maps in society, and the part they have played as cultural artefacts in political and economic development of nations and empires, including the 'persuasive cartography' of propaganda maps (Tyner 1974), Harley changed the tenor of such analysis by focusing on the power of maps and the power invested in maps. Drawing on the ideas of Foucault and Derrida, Harley argued that maps are a product of the society that creates them, and regardless of how much they seek to represent 'the truth', they inherently capture the interests of those that produced them and work to further those interests. Such a position recognises that in the production of maps many subjective decisions are made about what to include, how the map will look, and what the map is seeking to communicate. As such, maps are the product of power and they exert power, and therefore in any theory or history of cartography it is necessary to be mindful of the historical and social context in which mapping has been employed and to deconstruct the power relations inherent within its production. His goal was to 'subvert the apparent naturalness and innocence of the world shown in maps both past and present' (Harley 1992: 232).

Harley's ideas opened the floodgates for a re-imagining of cartography and maps, and a re-examination of works that had been suggesting such a reorientation but had, at that time, received little attention. Shortly after its publication other significant pieces by Pickles (1991, excerpted as Chapter 5.3) and Wood (1992) were published. Wood's book, *The Power of Maps*, drew together and extended a number of his works published over the previous decade (see the 1986 Wood and Fels' article 'Designs on Signs', excerpted as Chapter 1.7). Wood, drawing on linguistic structural thought and Barthean semiotics sought to detail how maps worked as a complex sign system, through their design and structuring, to produce certain versions of truth in order further the interests of those that created them. 'No sooner are maps acknowledged as social constructions than their contingent, their conditional, their . . . arbitrary character is unveiled. Suddenly the things represented by these lines are open to discussion and debate, the interest in them of owner, state, insurance company is made apparent' (Wood 1992: 19). Maps from this perspective are always political, working to (re)produce certain ways of thinking about the world. Rather than drawing on semiology, Pickles (1991, excerpted as Chapter 5.3) argued for a hermeneutic reading of maps that understood them as texts. As with Harley and Wood, Pickles argued that maps are inherently political and to fully understand them requires a method of deconstruction that seeks to provide a multifaceted, contextual and interpretive reading of their meaning and purpose. This includes examining the context in which a map is made, how the map is framed by other texts, the situation into which it is projected, and the world of the reader.

These new ways of thinking opened up fresh opportunities for historians of cartography. For example, Edney (1993, excerpted as Chapter 1.10) argued that historical accounts of cartography and the role of maps in society up to that point were largely empiricist and teleological, charting out in linear ways the progression of cartographic knowledge and method. He points out that the development of cartography has not followed such a well-defined path, but rather has been contingent on the social, cultural and technical relations at particular times and places. In so doing, Edney argues for a non-progressivist and non-presentist history; that is, a history that is sensitive to the science, politics and technologies of map production at the time of their creation and which does not judge maps *vis-à-vis* the standards and norms of the present day. In other words, maps from the past are seen as no better or worse than now, but rather are simply different and can be thought of a rhizomatic tangle of mapping modes rather than a family tree of cartographic progress.

While this body of critical writing on cartography has been forceful (and sometimes polemical), it is not without its problems, inconsistencies and critics (Andrews 2001; Belyea 1992; Godlewska 1989). Keates (1996: 194), for example, undermines the methodological agenda of Harley and 'critical cartography' paradigm more broadly, commenting: 'The question of how the production and publication of maps is controlled in any society is an interesting and important issue, but it is not illuminated by uttering clichés about hidden agendas.'

Ideologically-driven cartographic deconstruction can also be seen as unproductive in that it offers little in the way of an agenda for changing or improving mapmaking practice other than exalting cartographers to be aware of the power of the maps they create (Crampton 2001; Kitchin and Dodge 2007, excerpted as Chapter 1.14). Indeed, the influence of new critical theoretical approaches within academic discourse is in marked contrast to the work of the large majority of cartographers in practitioner communities, in university drawing offices, in government departments and in commercial design firms. The profession has largely ignored this new epistemological line as it offers little of value for those tasked with real world demands of making effective maps and they have little reason to contribute to wider theoretical debates; as Petchenik (1985, quoted in Keates 1996: 190) wryly notes: 'Practising cartographers tend to be so busy earning their living by making and selling maps that there isn't 'free' time or energy left to be expended on research and writing projects: as a consequence, their point of view is not accurately reflected in the literature.' Equally disappointing in terms of effecting progressive change in the nature of cartography is the failure of human geographers and other social scientists to make critical use of maps in their research. Accordingly, Perkins (2004: 385) laments: '[d]espite arguments for a social cartography employing visualisations to destabilise accepted categories most geographers prefer to write theory rather than employ critical visualisation'.

Other accusations levelled at critical cartography include: a misreading and superficial misuse of social theories; of simply jumping on the cultural 'bandwagon' of deconstruction; and the foisting of a false 'conspiracy' view of cartography through biased sampling of empirical evidence (Black 1997). 'In contrast to Harley's experience of cartographers', Godlewska (1989: 97) notes, 'I have found that most have a subtle and critical sense of the nature of their work and do not perceive cartography as an objective form of knowledge'. Of course, the critical scholars themselves had an agenda in their attacks on mainstream cartography, being 'propelled by an odd mixture of cynicism and idealism' (Lemann 2001).

As Crampton (2003) has noted, these new theorisations – maps as communicative models, sign systems, social constructions – whilst significantly advancing the conceptual ideas for understanding and interrogating maps, are still rooted in representational ways of thinking. As the new millennium turned, a small number of cartographic theorists started to rethink maps from a post-representational perspective. In particular, they drew on post-structural theory that was becoming popular across the social sciences and humanities at the time. From a post-representational perspective, the questions applied to cartography change from what maps represent and mean, to focus more on how maps work and their effects on the world (Corner 1999, excerpted in Chapter 1.12). Further, the separation of map and territory – a fundamental tenet of representational cartography – becomes problematised.

For example, the landscape architect, James Corner, following Baudrillard, argued that a territory does not precede a map, but that space becomes territory through bounding practices that include mapping (see also Winichakul's 1994 work on Siam, excerpted as Chapter 5.4). And since places are planned and built on the basis of maps, so space is itself a representation of the map; maps and territories are co-constructed. In other words, he demonstrates that space is constituted through mapping practices, amongst many others, so that maps are not a reflection of the world, but a re-creation of it; mapping activates territory. Corner thus develops an understanding of maps as unfolding potential; as conduits of possibilities. He thus argues for a processual understanding of maps, wherein mapping is seen to consist of multiple processes of action that have effects in the world. In so doing, maps endlessly remake territory through their employment. The power of maps then is not simply in their capturing and presentation of the world, but in their use and suggestion of new possibilities. For him, cartographic research thus needs to focus on mapping actions and mapping effects and not solely on map design, map meaning and the reading of maps.

Del Casino and Hanna (2005, excerpted as Chapter 1.13) similarly draw on poststructural theory, and in particular the idea of performativity, to argue that maps, far from being fixed, immutable objects, are in a constant state of becoming; that they are 'mobile subjects' whose meaning emerges through socio-spatial practices of use that mutate with context and is contested and intertextual. They argue that the map is not fixed at the moment of initial construction but is in constant flux, where each encounter with the map produces new meanings and engagements with the world, the product of the map as representation and material object, the knowledges the subject brings to bear on it and the space it represents, and the context of its use. Maps are produced and used through practices, and maps and space co-produce each other through their creation and use. They thus argue that maps can only be fully understood by examining the complex, recursive interplay between map and the world.

Likewise, Kitchin and Dodge (2007, excerpted as Chapter 1.14) have argued for a shift in cartographic theory from seeking to understand the nature of maps (an ontological project) to examining the practices of mapping (an ontogenetic project). This move denies maps any ontological security as representations of reality and instead posits that they are always in the state of becoming, bought into being through embodied, social and technical practices to solve relational problems such as plotting, planning, navigating and so on. Maps then emerge through a mix of creative, reflexive, playful, tactile and habitual practices; affected by the knowledge, experience and skill of the individual to perform mappings and apply them in the world, and shaped by the context of its reproduction. The map does not re-present the world or make the world, it is a co-constitutive production between inscription, individual and world; a production that is constantly in motion, always seeking to appear ontologically secure. Of course, this process very often succeeds – hence the real utility of cartography in all kinds of contexts for all manner of pragmatic tasks. Conceiving of maps in this way reveals that they are never fully formed but emerge in process and are mutable. Such a re-imagining of maps changes in quite fundamental ways the focus of cartography, moving it away from notions of accuracy, design, aesthetics and power, to emphasising the complex, contingent interactions between cartographers, users, maps and the world.

As is clear from this discussion, how maps are presently conceptualised varies substantially between scholars. Understanding maps and conceiving of how to undertake cartographic research is anything but straightforward. Mapping is a lot more complex than it at first seems; the theory, history and principles of map creation and use are contested. And so it should be. The engagement between cartography and philosophy is enormously important because it sets the parameters through which maps are thought about, produced and used; it shapes our assumptions about how we can know and measure the world, how maps work, their techniques, technologies, aesthetics, ethics, ideology, what they tell us about the world, the work they do in the world, and our capacity as humans to engage in mapping (Kitchin, Perkins and Dodge 2009). There are many fundamental ontological, epistemological, ideological and methodological questions that need further examination and debate, and yet more questions that have not yet received sufficient attention. This is the challenge for cartographers going forward, to continue to debate, refine and extend our theories during the search for a conceptual framework that adequately accounts for the nature of maps and the work that they do in the world.

References

Andrews, J.H. (2001) Meaning, knowledge and power in the map philosophy of J.B. Harley, in *The New Nature of Maps: Essays in the History of Cartography* (ed. P. Laxton), The Johns Hopkins University Press, Baltimore, MD, pp. 1–32.

Belyea, B. (1992) Images of power: Derrida/Foucault/Harley. *Cartographica*, **29** (2), 1–9.

Bertin, J. (1967) *Sémiologie Graphique*, Gauthier-Villars, Paris.

Black, J. (1997) *Maps and Politics*, Reaktion Books, London.

Board, C. (1981) Cartographic communication. *Cartographica*, **18** (2), 42–78. (Excerpted as Chapter 1.6.)

Cartwright, W. (1999) Extending the map metaphor using web-delivered multimedia. *International Journal of Geographical Information Science*, **13** (4), 335–353. (Excerpted as Chapter 2.11.)

Corner, J. (1999) The agency of mapping: speculation, critique and invention, in *Mappings* (ed. D. Cosgrove), Reaktion Books, London, pp. 213–252. (Excerpted as Chapter 112.)

Crampton, J.W. (2001) Maps as social constructions: power, communication and visualization. *Progress in Human Geography*, **25** (2), 235–252.

Crampton, J. (2003) *The Political Mapping of Cyberspace*, Edinburgh University Press, Edinburgh.

Crone, G.R. (1953) *Maps and Their Makers: An Introduction to the History of Cartography*, Hutchinson's University Library, London.

Del Casino, V.J. and Hanna, S.P. (2005) Beyond the 'binaries': a methodological intervention for interrogating maps as representational practices. *ACME: An International E-Journal for Critical Geographies*, **4** (1), 34–56. (Excerpted as Chapter 1.13.)

Downs, R.M. and Stea, D. (1973) Cognitive maps and spatial behavior: process and products, in *Image and Environment: Cognitive Mapping and Spatial Behavior* (eds R.M. Downs and D. Stea), Aldine Press, Chicago, pp. 8–26. (Excerpted as Chapter 4.3.)

Edney, M.H. (1993) Cartography without 'progress': reinterpreting the nature and historical development of mapmaking. *Cartographica*, **30** (2/3), 54–68. (Excerpted as Chapter 1.10.)

Godlewska, A. (1989) To surf or to swim? Responses to J.B. Harley's article 'Deconstructing the map'. *Cartographica*, **26** (3/4), 96–98.

Harley, J.B. (1989) Deconstructing the map. *Cartographica*, **26** (2), 1–20. (Excerpted as Chapter 1.8.)

Harley, J.B. (1992) Deconstructing the map, in *Writing Worlds: Discourse, Text and Metaphor in the Representation of Landscape* (eds T.J. Barnes and S. Duncan), Routledge, London, pp. 231–247.

Harvey, D. (1989) *The Condition of Postmodernity*, Blackwell, London. (Excerpted as Chapter 5.2.)

Ingold, T. (2000) *The Perception of the Environment: Essays in Livelihood, Dwelling and Skill*, Routledge, London.

Keates, J.S. (1996) *Understand Maps*, 2nd edn, Addison Wesley, Harlow, England.

Kitchin, R. and Dodge, M. (2007) Rethinking maps. *Progress in Human Geography*, **31** (3), 331–344. (Excerpted as Chapter 1.14.)

Kitchin, R., Perkins, C. and Dodge, M. (2009) Thinking about maps, in *Rethinking Maps: New Frontiers in Cartographic Theory* (eds M. Dodge, R. Kitchin and C. Perkins), Routledge, London, pp. 1–25.

Latour, B. (1992) Drawing things together, in *Representation in Scientific Practice* (eds M. Lynch and S. Woolgar), MIT Press, Cambridge, MA, pp. 19–68. (Excerpted as Chapter 1.9.)

Lemann, N. (2001) Atlas shrugs: the new geography argues that maps have shaped the world, *The New Yorker*, 9 April.

Lloyd, R. (2000) Cognitive maps: encoding and decoding information, in *Cognitive Mapping: Past, Present and Future* (eds R. Kitchin and S. Freundschuh), Routledge, London, pp. 84–107. (Excerpted as Chapter 4.9.)

MacEachren, A.M. (1995) *How Maps Work: Representation, Visualization and Design*, Guilford, New York.

MacEachren, A.M. and Kraak, M.J. (1997) Exploratory cartographic visualization: advancing the agenda. *Computers & Geosciences*, **23** (4), 335–343. (Excerpted as Chapter 1.11.)

Morrison, J.L. (1976) The science of cartography and its essential processes. *International Yearbook of Cartography*, **16**, 84–97. (Excerpted as Chapter 1.4.)

Perkins, C. (2004) Cartography: cultures of mapping, power in practice. *Progress in Human Geography*, **28**, 381–339.

Pickles, J. (1991) Texts, hermeneutics and propaganda maps, in *Writing Worlds: Discourse, Text and Metaphor in the Representation of Landscape* (eds T.J. Barnes and J.T. Duncan), Routledge, London, pp. 193–230. (Excerpted as Chapter 5.3.)

Raisz, E. (1938) *General Cartography*, McGraw-Hill, New York.

Robinson, A.H. (1952) *The Look of Maps*, University of Wisconsin Press, Madison, WI. (Excerpted as Chapter 3.3.)

Robinson, A.H. and Petchenik, B.B. (1976) *The Nature of Maps*, University of Chicago Press, Chicago. (Excerpted as Chapter 1.3.)

Robinson, A.H., Morrison, J.L., Muehrcke, P.C. *et al.* (1995) *Elements of Cartography*, 6th edn, John Wiley & Sons, Inc., New York.

Tyner, J. (1974) *Persuasive Cartography*, University of California Press, Los Angeles.

Winichakul, T. (1994) *Siam Mapped: A History of the Geo-Body of a Nation*, University of Hawai Press, Honolulu, HI. (Excerpted as Chapter 5.4.)

Wood, D. (1992) *The Power of Maps*, Guilford, New York.

Wood, D. and Fels, J. (1986) Designs on signs / myth and meaning in maps. *Cartographica*, **23** (3), 54–103. (Excerpted as Chapter 1.7.)

Wright, J.K. (1942) Map makers are human: comments on the subjective in maps. *Geographical Review*, **32** (4), 527–544. (Excerpted as Chapter 4.2.)

Chapter 1.2

General Theory, from *Semiology of Graphics*

Jacques Bertin

Editors' overview

Bertin's *Sémiologie Graphique* is one of the foundational texts for graphic design and has enjoyed significant influence on research in cartographic design. Based upon years of practical design experience, it is the first formal specification of rules for what might be achieved when representing different kinds of information in a graphic form. A number of translations of the original text have been made into English – excerpted here is the introductory material summarising the book from Berg's 1983 translation of the 1973 French language revision of the original book. Bertin first defines the semiological basis for his work, before exploring the kinds of analytical approach to data that are required. He then defines key properties of the graphic system and the rules that follow. Amongst the influential aspects presented are notions of visual hierarchy and, in particular, the definition of visual variables. A more detailed exposition and visualisation from later in the book is also included, to highlight the significance of these visual variables.

Originally published in English translation in 1983: Jacques Bertin, *Semiology of Graphics* (translated by William J. Berg), University of Wisconsin Press, Madison, WI, pp. 2–13, 42–43.

General theory

Graphic representation constitutes one of the basic sign systems conceived by the human mind for the purposes of storing, understanding and communicating essential information. As a 'language' for the eye, graphics benefits from the ubiquitous properties of visual perception. As a monosemic system, it forms the rational part of the world of images.

To analyse graphic representation precisely, it is helpful to distinguish it from musical, verbal and mathematical notations, all of which are perceived in a linear or temporal sequence. The graphic image also differs from figurative representation, essentially polysemic, and from the animated image, governed by the laws of cinematographic time. Within the boundaries of graphics fall the fields of networks, diagrams and maps. The domain of graphic imagery ranges from the depiction of atomic structures to the representation of galaxies and extends into the spheres of topography and cartography.

Graphics owes its special significance to its double function as a storage mechanism and a research instrument. A rational and efficient tool when the properties of visual perception are competently utilised, graphics is one of the major 'languages' applicable to information processing. Electronic displays, such as the cathode ray tube, open up an unlimited future to graphics.

Definition of graphics

Based on rational imagery, graphics differs from both figurative representation and mathematics. In order to define it rigorously in relation to these and other sign systems, we shall adopt a semiological approach and begin with two rather obvious statements: (a) the eye and the ear have two distinct systems of perception; (b) the meanings which we attribute to signs can be monosemic, polysemic or pansemic (Figure 1.2.1).

The Map Reader: Theories of Mapping Practice and Cartographic Representation, First Edition. Edited by Martin Dodge, Rob Kitchin and Chris Perkins.

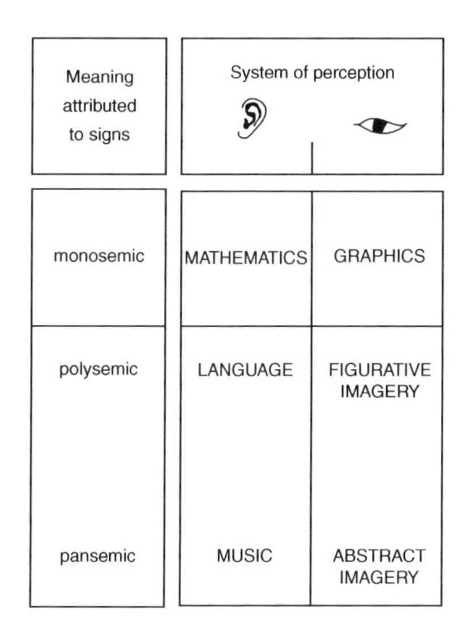

Figure 1.2.1 Graphics in relation to other basic sign systems.

A monosemic system

A system is monosemic when the meaning of each sign is known prior to observation of the collection of signs. An equation can be comprehended only when the *unique* meaning of each term has been specified. A graphic can be comprehended only when the unique meaning of each sign has been specified (by the legend). Conversely, a system is polysemic when the meaning of the individual signs follows, and is deduced from, consideration of the collection of signs. Signification becomes subjective and thus *debatable.*

Indeed, a figurative image, and for that matter an ordinary photograph or an aerial photograph, is always accompanied by a certain amount of ambiguity: 'Who is this person?', 'What does this black mark or that shape represent?' To these questions each person will respond individually, since interpretation is linked to the repertoire of analogies and structures characterising each 'receiver'. And this repertoire varies from one individual to another, according to personality, surroundings, period and culture. Faced with the polysemic image, the perceptual process translates into the question: 'What does such an element or collection of elements signify?', and perception consists of decoding the image. The reading operation takes place between the sign and its meaning.

The abstract painting represents an extreme form of polysemy. In its attempt to signify 'everything' it no longer signifies anything precise and so becomes 'pansemic'.

On the other hand, in graphics, with a diagram or map, for example, each element is defined beforehand. The perceptual process is very different and translates into the question: 'Given that such a sign signifies such a thing, what are the relationships among all the signs, among all the things represented?' Perception consists of defining the relationships established within the image or among images, or between the image and the real world. The reading operation takes place among the given meanings. This distinction is fundamental because it suggests the true purpose of 'graphics' in relation to other forms of visualisation.

What does it actually mean to employ a monosemic system? It is to dedicate a moment for reflection during which one seeks a maximum reduction of confusion; when, for a certain domain and during a certain time, *all the participants* come to agree on certain meanings expressed by certain signs, and *agree to discuss them further.*

This convention enables us to *discuss the collection of signs* and to link propositions in a sequence which can then become 'undebatable', that is, 'logical'[...]. This is the object of mathematics, which deals with problems involving a temporal sequence. It is the object of graphics, which operates in areas linked to the tri-dimensionality of spatial perception. On this point, graphics and mathematics are similar and construct the 'rational moment'.

A visual system

But graphics and mathematics differ in the perceptual structure which characterises each of them. It would take at least 20 000 successive instants of perception to compare two data tables of 100 rows by 100 columns. If the data are transcribed graphically, comparison becomes easy; it can even be instantaneous.

As we see in Figure 1.2.2, auditory perception has only *two* sensory variables at its disposal: sound and time. All the sign systems intended for the ear are linear and temporal. Remember that written transcriptions of music, language and mathematics are merely formulae for setting down systems which are fundamentally auditory, and that these formulae do not escape from the linear and temporal character of the systems themselves.

On the other hand, visual perception has at its disposal *three* sensory variables which do not involve time: the variation of marks and the two dimensions of the plane. The sign systems intended for the eye are, above all, spatial and atemporal. Hence their essential property: in an instant of perception, linear systems communicate only a *single sound or sign*, whereas spatial systems, graphics among them, communicate in the same instant the *relationships among three variables.* Maximum utilisation of this considerable perceptual power within the framework of logical reasoning is the true purpose of graphics, the monosemic domain of spatial perception.

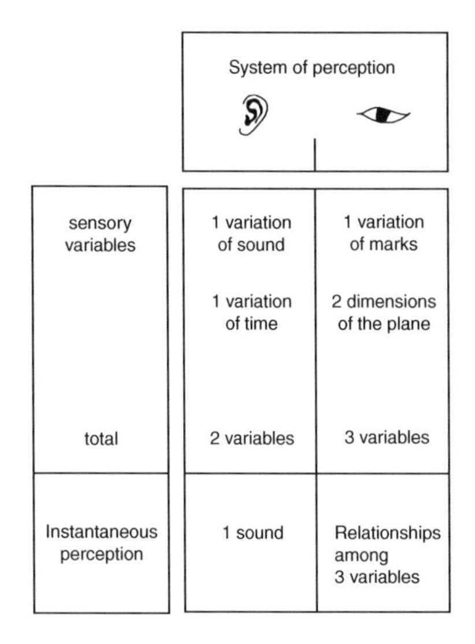

Figure 1.2.2 Perceptual properties of linear and spatial systems.

Evolution of graphics

The effectiveness of graphics has long been recognised. The most ancient graphic representations which have been discovered are geographic maps engraved on clay, and they probably date from the third millennium before Christ. Graphic images were first conceived, and are still usefully conceived, as reproductions of the visible world, which benefit from only one degree of freedom, that of the scale. In a molecular model, a geometric figure, an assembly diagram, an industrial drawing, a geologic section or a map, the two dimensions of the coordinate plane are identical to those of the visible space, adjusted for the scale of the drawing.

One had to wait until the fourteenth century to suspect, at Oxford, and until the eighteenth century to confirm, with Charles de Fourcroy (de Dainville 1958), that the two dimensions of a sheet of paper could usefully represent *something other than visible space.* This amounts to a transition from a simple representation to a 'sign system' that is complete, independent and possesses its own laws, thus falling within the scope of semiology.

And now, at the end of the twentieth century, with the pressure of modern information and the advances of data processing, graphics is passing through a new and fundamental stage. The great difference between the graphic representation of yesterday, which was poorly dissociated from the figurative image, and that of tomorrow, is the disappearance of the congenital fixity of the image.

When one can superimpose, juxtapose, transpose and permute graphic images in ways that lead to groupings and classings, the graphic image passes from the *dead image*, the 'illustration', to the *living image*, the widely accessible research instrument it is now becoming. The graphic is no longer only the 'representation' of a final simplification, it is a point of departure for the discovery of these simplifications and the means for their justification. The graphic has become, by its manageability, an instrument for information processing. Its study must begin, then, with the analysis of the information to be transcribed.

Analysis of the information

Thought can only be expressed within a system of signs. Mimicry is a natural form of coding; verbal language is a code of auditory signs (which must be learned in order to communicate with others); the written language is another code; graphic representation yet another. Memory storage on disks, tapes or in computers necessitates appropriate new codifications. Graphic representation is the transcription, into the graphic sign system, of 'information' known through the intermediary of any given sign system. Graphic representation can thus be approached by semiology, a science which deals with all sign systems.

Information and representation

Any transcription leads necessarily to a separation of content from form. The 'content', those elements of the thought which can remain constant, regardless of the sign system into which they are translated, must be distinguished from the 'container', that is, the means available in a given system and the laws which govern their use. These elements are constant, whatever the thought to be transcribed.

Whether we are studying the means, properties and limits of the graphic system, or planning a design, it is first necessary to strictly separate the content (the information to be transmitted) from the container (the properties of the graphic system). [. . .] What matters to us is the quality and the efficacy of its graphic transcription. Incidentally, it is the singular characteristic of a good graphic transcription that it alone permits us to evaluate fully the quality of the content of the information.

Knowing that each sign system has its properties, its style, its aesthetic, what constant can be isolated in a thought, throughout its diverse translations? A thought is a relationship among various concepts which have been recognised and isolated at a given moment from among the multitude of imaginable concepts. Consider the following example: 'On July 8, 1964, stock X on the Paris exchange is quoted at 128 francs; on July 9, it is quoted at 135 francs'. Whatever form the phrase may take, the content will always involve the pertinent correspondence among certain elements:

(1) the concept 'quantity of francs', or a VARIATION in the number of francs;

(2) the concept 'time', or a VARIATION of the date;

(3) an element – stock X – of the concept 'different stocks quoted on the Paris exchange'. This element is, by definition, invariant since it constitutes the common ground which relates the francs to the dates.

In graphic representation the translatable content of a thought will be called the information. It is constituted essentially by one or several pertinent correspondences between a finite set of variational concepts and an invariant. The information to be transcribed can be furnished in any given sign system, so long as it is known by the transcriber, that is, the graphic designer. Let us stress once and for all that we will never use the term 'information' in the very limited and precise sense which it has in 'information theory', but as a synonym for 'data to be transcribed'.

The invariant and the component

Whenever we discuss information to be transcribed, the central notion common to all the pertinent correspondences will be called the invariant; the variational concepts involved will be called the components. Thus, the preceding example can be said to have an invariant – a given stock – which relates two components: a variation in the number of francs and a variation in time. Whatever sign system is employed, at least two components will be needed to translate the information. In the graphic system, two visual components are normally utilised: the two dimensions of the plane. The wording of the titles and legends is the first application of these notions [. . .]

In order to facilitate explanation, the components of the graphic sign system will be called visual variables (or simply 'variables'), and the two variables furnished by the plane itself will be called planar dimensions. Information will, therefore, be formed by pertinent correspondences among given components and its graphic representation by correspondences among given variables. Visual perception will admit only a small number of variables. [. . .]

The number of components

The determination of the number of components is the first stage in the analysis of the information. Components and variables are, by definition, divisible.

The different identifiable parts of a component or of a variable will be called elements or categories (or 'classes' or 'steps' [. . .]). We can talk, for example about categories of 'departments' [major administrative subdivisions of France] in a geographic component, about the categories 'bovine', 'ovine', 'caprine', of the component 'different domestic animals', about steps of grey in the variable 'value', about annual classes in the component 'time', or about elements of the component 'different persons'. The complexity of a figure is linked to the number of categories in each component.

The length of the components

We will use the term length to describe the number of elements or categories which we are able to identify in a given component or variable. This is the second stage in the analysis of the information.

Thus the binary component 'sex' can be said to have a length of two; the geographic component 'departments of France' a length of ninety. In a quantitative component we must not confuse the 'length', the number of useful steps, with the range of the series, the ratio between the largest and smallest numbers in the statistical series.

The level of organisation of the components

What is properly called graphics depicts only the relationships established among components or elements. These relationships define three levels of organisation, and each component, each visual variable, is located on one of these levels:

The *qualitative level* (or nominal level) [. . .] includes all the concepts of simple differentiation (trades, products, religions, colours, etc.). It always involves two perceptual approaches: This is similar to that, and I can combine them into a single group (association). This is different from that and belongs to another group (differentiation).

The *ordered level* [. . .] involves all the concepts that permit a ranking of the elements in a universally acknowledged manner (e.g. a temporal order; an order of sensory valuations: cold–warm–hot, black–grey–white, small–medium–large; an order of moral valuations: good–mediocre–bad etc.). This level includes all the concepts which allow one to say: this is more than that and less than the other.

The *quantitative level* (interval ratio level) [. . .] is attained when one makes use of a countable unit (this is quarter, triple, or four times that).

These levels are overlapping: What is quantitative is likewise ordered and qualitative. What is ordered is also qualitative. What is qualitative is neither quantitative nor ordered but is arbitrarily reorderable.

The *levels of organisation* form the domain of universal meanings, of fundamental analogies, in which graphic representation can stake a claim. This is the third stage in the analysis of the information.

All forms of signification other than the above relationships are in fact exterior to graphics and merely serve to link the graphic system to the world of exterior concepts. They must rely either on an explanation coded in another system (legends) or on a *figurative analogy* of shape or colour (symbols), which is based on acquired habits or learned conventions and can never claim to be universal. Each visual variable has its particular properties of length and level. It is important that each component be transcribed by a variable having at least a corresponding length and level.

Graphics is concerned with the representation of these three levels of organisation. However, the relationships of similarity and order, based on metrics, are those which constitute the foundation for information processing and analysis.

The properties of the graphic system

The scope of the system

What variables does the graphic sign system have at its disposal? The eye is the intermediary in a great number of perceptions. Not all of these concern the system we are studying; the intervention of real movement, for example, although perceptible by vision, would make us pass from the graphic system (atemporal) into film, whose laws are very different. We will only consider that which can be represented by readily available graphic means, on a flat sheet of white paper of standard size and under normal lighting.

Within these limits, we will consider that the graphic system has at its disposal eight variables. A visible mark expressing a pertinent correspondence can vary in relation to the two dimensions of the plane. It can further vary in *size, value, texture, colour, orientation and shape.* Within the plane, this mark can represent a point (a position without area), a line (a linear position without area) or an area.

The plane

A given French department can be represented either by a point or a line, as in a diagram, or by an area, as in a map. These 'implantations' are the three moments of the sensory continuum applied to the plane. They constitute the three elementary figures of geometry.

The level of organisation of the plane is maximum [...]. Its two dimensions furnish the only variables which can correctly represent any component of the information, whatever its level of organisation.

The utilisation of the two dimensions of the plane [...] depends upon the nature of the pertinent correspondences expressed on the plane and enables us to divide graphic representation into four groups. In effect, the correspondences on the plane can be established:

- among all the elements of one component and all the elements of another component. The construction is a diagram. [For] example: variation in the quotations for stock X on the Paris exchange. Any date (component: time) can correspond a priori to any price (component: quantity of francs), and there are no grounds for predicting a correspondence between two dates or between two prices.

- among all the elements of the same component. The construction is a network. [For] example: the relationships of conversations among individuals situated around a table. Any individual [...] is capable of communicating with any other individual (of the same component).

- among all the elements of the same geographic component, inscribed on the plane according to the observed geographic distribution. The network traces out a geographic map.

- between a single element and the reader (road signs, various codes based on shape, industrial colour code etc.). The correspondence is exterior to the graphic image. This is a problem involving symbolism, which relies upon figurative analogies.

In diagrams and networks [...], the free disposition of the dimensions of the plane leads us to distinguish arrangements dispersed over the entire plane from those which structure it in some manner (rectilinear, circular orthogonal, polar) and to define types of construction which can be characterised by schemas of construction.

The retinal variables

We will term 'elevation' the utilisation of the six variables other than those of the plane, that is, the retinal variables. A qualitative variation between two cities can be represented on a map by a variation in size, value, texture, colour, orientation, shape, or by a combination of several of these variables.

Retinal variables must be utilised whenever a third component appears in the information. But none of these variables has the capability of the plane to represent any component of the information, whatever its level of organisation. We must, therefore, determine the level of organisation [...] for each variable, its properties of length [...] and its applicability.

The rules of the graphic system

The basic graphic problem

The great diversity of graphic constructions, within a group or even from one group to another, is due to the designer's apparent freedom to represent a given component by using any one of the eight visual variables or a combination of several of them. Faced with such a choice, the graphic designer can, for example, represent a geographic component by a single dimension of the plane, thereby constructing a diagram; or by both dimensions of the plane, constructing a map. A variation in colour or value could also be used. In fact, to construct 100 different figures from the same information requires less imagination than patience. However, certain choices become compelling due to their greater 'efficiency'.

Image theory

Efficiency is defined by the following proposition: if, in order to obtain a correct and complete answer to a given question, all other things being equal, one construction requires a shorter period of perception than another construction, we can say that it is more efficient for this question. This is Zipf's notion of 'mental cost' applied to visual perception [...]. In most cases the difference in perception time between an efficient construction and an inefficient one is considerable. The rules of construction enable us to choose the variables which will construct the most efficient representation.

Efficiency is linked to the degree of facility characterising each stage in the reading of a graphic. [...] Five aspects of image theory are discussed here.

Stages in the reading process

To read a drawing is to proceed more or less rapidly in successive operations:

- external identification: what components are involved? It is necessary to clearly define and situate the concepts proposed for examination.

- internal identification: by what variables are the components expressed? For example, quantities by the vertical dimension of the plane, time by the horizontal dimension; or, alternatively, quantities by the length of the radius, time by the length of the arc subtended by the angle about the point of origin.

These operations link the graphic system to other systems through the use of written notations (titles and legends) or figurative analogies (shape and colour). The two stages of identification are indispensable and must precede any study of the information itself: [...] perception is always the result of a question, conscious or not. [...]

Possible questions – levels of reading

In the preceding example two types of question are possible:

- On a given date, what is the price of stock X?

- For a given price, on what date(s) was it attained?

There are as many TYPES OF QUESTIONS as components in the information, but within each type there are numerous possible questions:

(a) Questions introduced by a single element of a component, for example: 'On a given date...' and resulting in a single correspondence. This is the elementary level of reading. Here, questions tend to lead outside of the graphic system.

(b) Questions introduced by a group of elements in the component, for example: 'In the first three days what was the trend of the price?' Answer: 'The price rose'. Such questions are quite numerous, since we can form highly diverse groups. This is the intermediate level of reading. Here, questions tend to reduce the length of the components.

(c) A question introduced by the whole component: 'During the entire period, what was the trend of the price?' Answer: 'General upward movement'. This is the overall level or 'global' reading. Such a question tends to reduce all the information to a single ordered relationship among the components. We can say that:
- there are as many types of questions as components in the information;
- for each type there are three levels of reading: the elementary level, the intermediate level and the overall level;
- any question can be defined by its type and level.

These reading levels are comparable to the integrative levels of the mind. Their analysis permits knowing in advance the *totality of the questions* which any given information can generate; as a result it permits studying the probability of their occurrence and, if appropriate, of taking them into account in the graphic construction.

Definition of an image

Answering a given question involves: (a) an input identification: 'On a given date... ?'; (b) perception of a correspondence between the components; (c) an output identification: the answer 'so many francs'. This implies that the eye can isolate the input date from all the other dates and, during an instant of perception, see only such correspondences as are determined by the input identification, but see all of these. During this instant, the eye must disregard all other correspondences. This is *visual selection.* [...] In certain graphic constructions, the eye is capable of including all the correspondences determined by an input identification within a single 'glance', within a single instant of perception. The correspondences can be seen in a single visual form.

The meaningful visual form perceptible in the minimum instant of vision will be called the *image.*

In this sense, *image* corresponds to 'form' in 'form theory', to 'pattern' and to 'Gestalt'. A synonym would be 'outline'. Other constructions do not permit the inclusion of all the correspondences within a single instant of perception; the entire set of correspondences could only be constructed in the memory of the reader, by the addition of the various images perceived in succession. It is therefore obvious that:

- The most efficient constructions are those in which any question, whatever its type or level, can be answered in a single instant of perception, in a single image.
- The image, the temporal unit of meaningful visual perception, must not be confused with the figure, which is the apparent and illusory unit defined by the sheet of paper, by a linear frame or by a geographic border.

Construction of an image

The image is built upon three homogeneous and ordered variables: the two dimensions of the plane and a retinal variable. Rules of construction [...] thus lead the designer to utilise the two dimensions of the plane in a homogeneous, rectilinear and orthogonal manner and also to employ an ordered retinal variable, such as size, value, or texture. Consequently, all information with three components or less can be represented as a single image[...]. It is necessary and sufficient that the rules of construction be respected. In this case, whatever the type or level of the question, the answer will be seen in a single instant of perception. We can say that the graphic representation is an image.

In any construction not respecting these rules, certain types and levels of questions will necessitate the perception of several images in succession; there will be a high mental cost. The formulation of the answer will be very difficult and often impossible. We will call these graphic constructions figurations. They are obviously less efficient than constructions involving a single image.

Limits of an image

An image will not accommodate more than three meaningful variables. As a result, all information with more than three components cannot be constructed as an image. This means that for certain questions identification will necessitate several instants of perception, several images and information with more than three components; it is necessary to choose preferred questions and construct the graphic so that they can be answered in a single instant of perception. At the same time one reserves input identifications that necessitate several instants of perception for questions which are less useful or less likely to be posed.

Visual efficiency is inversely proportional to the number of images necessary for the perception of the data; it is this rule which, in the final analysis, governs the choice of preferred questions and leads to identifying the three functions of graphic representation.

Three functions of graphic representation

(1) Recording information: creating a storage mechanism which avoids the effort of memorisation. The graphic utilised for this purpose must be comprehensive and may be non-memorisable in its totality.

(2) Communicating information [...]: creating a memorisable image which will inscribe the information in the viewer's mind. The graphic used here must be memorisable and may be non-comprehensive. The image should be a simple one.

(3) Processing information [...]: furnishing the drawings which permit a simplification and its justification. The graphic should be memorisable (for comparisons) and comprehensive (for choices).

Information with three components or less, constructed as a single image, can fulfil all three functions of graphic representation. But information involving more than three components must be constructed differently, according to the intended function, that is, according to the nature of the preferred questions.

Rules of construction

The rules of construction, represented by standard schemas, define the most efficient construction for a given case.

Rules of legibility (or rules of separation)

The rules of construction govern the choice of visual variables. Once chosen, however, the variables can still be utilised well or poorly. Efficiency also depends on the sensory differentiation which we can obtain from each variable or combination of variables; this differentiation will increase or reduce the capacity for 'separation' within the variables. For example, sensory differentiation is greater between blue and red than between blue and green, between black and white than between black and grey.

The observations which permit us to best accomplish sensory differentiation will be called rules of legibility. Linked to the faculties of human perception, these rules apply to each variable as well as each combination of variables and are related to their length. But length, in turn, varies according to the level of organisation involved; selective perception, for example, calls for the greatest amount of differentiation.

The scope of the graphic system

[...] A sign system cannot be analysed without a strict demarcation of its limits. This study does not include all types of visual perception, and real movement is specifically excluded from it. An incursion into cinematographic expression very quickly reveals that most of its laws are substantially different from the laws of atemporal drawing. Although movement introduces only one additional variable, it is an overwhelming one; it so dominates perception that it severely limits the attention which can be given to the meaning of the other variables. Furthermore, it is almost certain that real time is not quantitative; it is 'elastic'. The temporal unit seems to lengthen during immobility and contract during activity, though we are not yet able to determine all the factors of this variation.

Actual relief representation (the physical third dimension) has no place here either and will be referred to only for purposes of comparison.

In this study, we will consider only that which is:

- representable or printable
- on a sheet of white paper
- of a standard size, visible at a 'glance'
- at a distance of vision corresponding to the reading of a book or an atlas
- under normal and constant lighting (but taking into account, when applicable, the difference between daylight and artificial light)
- utilising readily available graphic means.

Consequently, we will exclude:

- variations of distance and illumination
- actual relief (thicknesses, anaglyphs, stereoscopics)
- actual movement (flickering of the image, animated drawings, film).

In order to be visible a mark must have a power to reflect light which is different from that of the paper. The larger the mark, the less pronounced the difference need be. A black mark of minimum visibility and discriminability must have a diameter of 2110 mm. But this is not absolute, since a constellation of smaller marks is perfectly visible.

[...] A visible mark can vary in position on a sheet of paper. In Figure 1.2.3, for example, the black rectangle is at the *bottom* and toward the *right* of the white square. It could

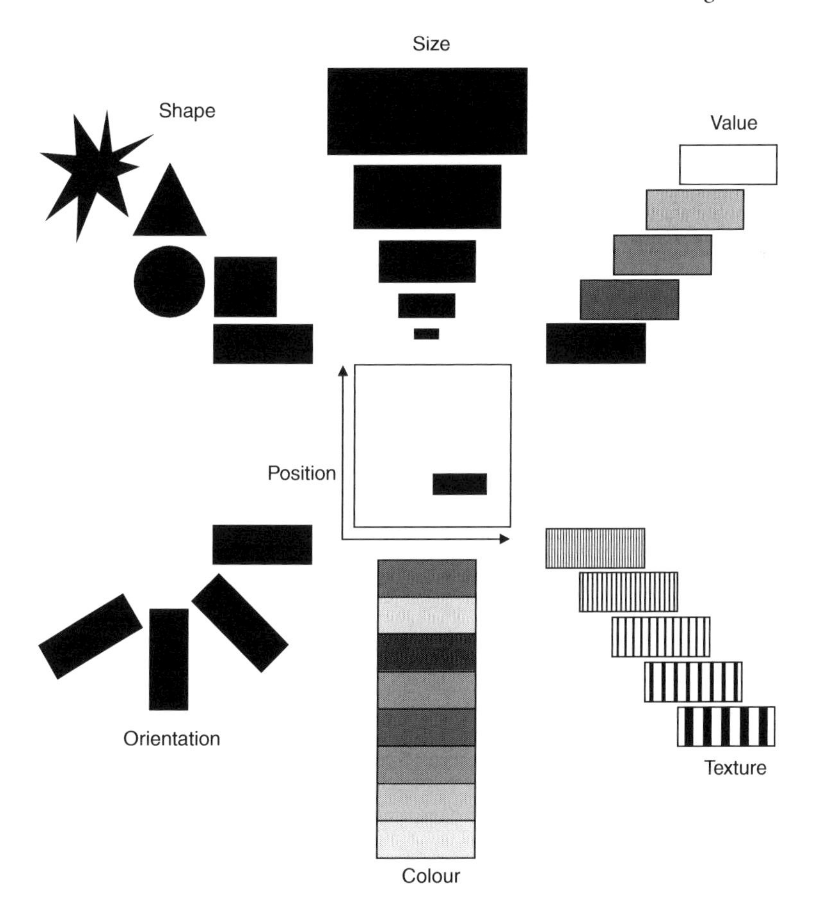

Figure 1.2.3 The visual variables.

just as well be at the bottom and towards the left, or at the top and toward the right.

A mark can thus express a correspondence between the two series constituted by the two planar dimensions. Fixed at a given point on the plane, the mark, provided it has a certain dimension, can be drawn in different modes. It can vary in:

SIZE
VALUE
TEXTURE
COLOR
ORIENTATION
SHAPE

and can also express a correspondence between its planar position and its position in the series constituting each variable.

The designer thus has eight variables to work with. They are the components of the graphic system and will be called the '*visual variables*'. They form the world of images. With them the designer suggests perspective, the painter reality, the graphic draftsman ordered relationships, and the cartographer space.

This analysis of atemporal visual perception in eight factors does not exclude other approaches. But, combined with the notion of 'implantation', it has the advantage of being more systematic, while remaining applicable to the practical problems encountered in graphic construction.

These variables have different properties and different capacities for portraying given types of information. As with all components, each variable is characterised by its level of organisation and its length. [. . .]

References

de Dainville, F. (1958) Grandeur et population des villes au XVIIIe siécle. *Population*, **13** (3), 459–480.

Further reading

Bertin, J. (1983) *Graphics and Graphic Information-Processing*, (trans. W.J. Berg and P. Scott), de Gruyter, Berlin. [The English language translation of Bertin's subsequent research monograph.]

Daru, M. (2001) Jacques Bertin and the graphic essence of data. *Information Design Journal*, **10** (1), 20–25. [Establishes the significance of Bertin's work through a contextual analysis of the influences and working practices that led to the book's publication.]

Garlandini, S. and Fabrikant, S. (2009) Evaluating the effectiveness and efficiency of visual variables for geographic information visualization, in *Lecture Notes in Computer Science 5736* (eds K. Stewart-Hornsby, C. Claramunt, M. Denis and G. Ligozat), Springer, Berlin, pp. 195–211. [An empirical, perception-based evaluation of Bertin's and subsequent reworkings of the visual variables concept.]

Koch, W.G. (2001) Jacques Bertin's theory of graphics and its development and influence on multimedia cartography. *Information Design Journal*, **10** (1), 37–43. [Argues Bertin's formal theoretical grounding is flexible enough to be extended into the multimedia age with interactive map design.]

MacEachren, A.M. (2001) An evolving cognitive-semiotic approach to geographic visualization and knowledge construction. *Information Design Journal*, **10** (1), 26–36. [Explores the impact of Bertin's work and extends his conception of visual variables, in the context of geovisualisation and the re-evaluation of theoretical approaches offered by MacEachren in his own monograph *How Maps Work*.]

Muller, J.C. (1981) Bertin's theory of graphics: a challenge to North American thematic cartography. *Cartographica*, **18** (3), 1–8. [This article argues that Bertin's conceptual design schema might profitably have been applied to challenge orthodoxies of cartography design.]

See also

- Chapter 1.7: Design on Signs / Myth and Meaning in Maps
- Chapter 1.11: Exploratory Cartographic Visualisation: Advancing the Agenda
- Chapter 3.3: Cartography as a Visual Technique
- Chapter 3.6: The Roles of Maps
- Chapter 4.9: Cognitive Maps: Encoding and Decoding Information

Chapter 1.3

On Maps and Mapping, from *The Nature of Maps: Essays Toward Understanding Maps and Mapping*

Arthur H. Robinson and Barbara B. Petchenik

Editors' overview

Robinson and Petchenik's book of essays provided the intellectual justification for the communications paradigm and its deployment by academics writing about cartography in the 1970s. Excerpted here is the introductory chapter, which situates the status of the map and mapping by seeking to define and justify terms. It starts by examining the philosophical literature relating to maps from cartography, geography, language and semantics, psychology and the philosophy of science, and from this suggests that the map is central to our cognitive make-up. Definitions of maps, mappers and mapping are derived, anchored in a representational view of the world, and the notion of the percipient, as a knowledgeable map reader is distilled. They conclude with a call for more cognitive and perceptual research to inform map design and use.

Originally published in 1976: Chapter 1 in Arthur H. Robinson and Barbara B. Petchenik, *The Nature of Maps: Essays Toward Understanding Maps and Mapping*, University of Chicago Press, Chicago.

[...] The apparent simplicity of an ordinary sketch map is deceptive; in fact, even the simplest map is a remarkably complicated instrument for understanding and communicating about the environment. It is quite reasonable to suppose that the map, as a communicative device, has been around as long as written language: like writing, a map is a way of graphically expressing mental concepts and images. On the other hand, whereas language has had the benefit of numerous penetrating studies of its form and function, the map has not. Consequently, for something both venerable and common, the map seems to be surprisingly mysterious [...].

Initially we are particularly concerned with the questions *What is a map?* and *Where does the map fit into the larger system of cognition in general?* It seemed appropriate, therefore, firstly to examine the philosophical literature representative of the thinking in such fields as cartography, geography, language and semantics, psychology and the philosophy of science, to determine how others have looked at maps. While some cartographers and geographers have cast about for things to which they can liken the map, in order that they may gain additional insight into it and the cartographic process in general, scholars in other fields tend to use the map as the fundamental analogy.

The literature of geography and cartography is replete with statements extolling the virtue of the map, sometimes with an almost religious fervour [...]. Only in recent years has the study of the map by geographers and cartographers begun to probe beyond the physical object into its theoretical foundations. The increasing concern of geographers with cognitive maps and spatial behaviour (Downs and

The Map Reader: Theories of Mapping Practice and Cartographic Representation, First Edition. Edited by Martin Dodge, Rob Kitchin and Chris Perkins.

Stea 1973) and the analogy of maps and language (Harvey 1969: 369–376) is encouraging, to say the least, but it does serve to emphasise the paucity of our previous understanding.

In a strangely general/specific explanation of the map as an example of the static class of models in science, Chorley and Haggett (1967: 48–49) point out that 'it is characteristic that maps should be likened to languages and scientific theories ... we sometimes think of maps as models for languages and scientific theories.' [...] What can there be about the map which is so profoundly fundamental? Why should a representational system for space be so basic? Further, assuming that the concept 'map' provides some sort of universally accepted, fundamental datum, what simpler analogies or definitions can we use to clarify our understanding of it? These are difficult questions.

In most areas of knowledge, the attempt to explain that which is more complex proceeds by analogising what is less familiar with what is more familiar – that is, employing simple concepts to illuminate those which are more complicated. Unfortunately, in writing or thinking about mapping, this is not possible. In the first place, we cannot explain mapping by mapping or in any kind of mapping terms. Instead we need to use a language, composed of discrete words which follow each other either in a rigorously arranged temporal sequence of sounds or, when written, in a linear sequence of marks. This puts us in a difficult situation: we are forced to use words to clarify to ourselves what mapping is all about, and yet the language we must use is itself a different medium and is often taken to be the more complex of these two distinct communication systems! [...]

It has turned out to be both astonishing and frustrating for the introspective cartographer to sample the large literature that has sprung up in the last few decades on the general topic of communication. [...] The frustration comes with the discovery that the universal metaphor turns out to be *map* itself! Hence, when the map is the 'atom' of investigation, how can the mapper contemplate his own activity? [...] Clearly, something very fundamental is assumed about maps and the process of mapping. When non-cartographic writers use the term 'map' [...] they seem to mean that it is possible to take isolated incidents, experiences and so on and arrange them intellectually so that there is some coherence, some total relation, instead of individual isolation. Since plotting the information they have [...] gives them the knowledge they already have *plus* the sort of automatic potential organisation provided by the continuous cognitive field (cf. the surface of the paper) [...], mapping is considered to be the most fundamental way of converting personal knowledge to transmittable knowledge. The basic significance of maps, then, seems to lie particularly in the fact that maps are surrogates of space.

As we experience space, and construct representations of it, we know that it will be continuous. Everything is somewhere, and no matter what other characteristics objects do not share, they *always* share relative location, that is, spatiality; hence the desirability of equating knowledge with space, an intellectual space. This assures an organisation and a basis for predictability, which are shared by absolutely everyone. This proposition appears to be so fundamental that apparently it is simply adopted a priori.

Students in fields concerned with language systems, signs and meanings extend the assumption of the knowledge–space–map relationship to include an assumed isomorphism between the map and cognitive spatial territory. Perhaps the prime examples of this come from students of what is called 'general semantics.' Without going into detail, general semantics is the study of the relation of signs (in this case spoken and written language) to referents and the consequences in human behaviour. According to its founder, Korzybski (1941), a great many human problems stem from the use of a language system which can be misleading. The form of the language may not be properly matched to 'reality', but since language tends to be accepted as reality, people are misled by it, function less well than they should, communicate inadequately and so on. Maps appear to enter into this analysis in a very fundamental way, because the relationship between the language system and reality is analogised as that between a map and the territory it represents. This fundamental analogy is considered to be a universally shared known [...].

What Korzybski finds that maps retain of the territory is structure. He goes on to say, 'Two important characteristics of maps should be noticed. A map is not the territory it represents, but, if correct, it has a similar structure to the territory, which accounts for its usefulness' (Korzybski 1941: 58). Structure, unfortunately, is not further defined. Apparently it is thought to be sufficiently obvious from the example of maps so that the term need not, and perhaps cannot, be defined more simply. [...]

Korzybski assumed that the map–territory image was simple and needed no analysis or explanation. In his examples, 'structure' seems to mean certain kinds of spatial relationships, particularly topological ones. He equates language with a map because the former is assumed to preserve certain relationships which exist in an intellectual sort of space. Further scrutiny of this whole analogy, quite central to Korzybski's thinking, reveals that cartographically it is so elementary as to be of questionable utility. While stressing that the map is not *the* territory, he seems to forget that a map is *a* territory, and that it may be meaningful to employ a variety of transformations to retain particular relationships from one territory to the other. Reality and language must, therefore, *both* be converted into some kinds of spaces before one can be mapped on the other [....].

Students of perception and philosophy also frequently employ the map metaphor and make much of the fundamental significance of the spatial conception. Polanyi (1963) makes considerable use of maps and mapping in his discussions of knowledge and knowing. Maps are one form of what he calls explicit knowledge, a kind of factual reality, and as such they are subject to critical reflection, both by the person who made the map and by those who may later use it. He finds maps to be a particularly useful example, for they lend themselves easily to checking; truth may be more easily agreed upon with maps than with certain other forms of knowledge, such as writing. Such an assertion is true, of course, only of the simplest notion of getting from one place to another with the use of a map; there are other aspects of mapping – cartographic generalisation, for example – which can be as elusive as anything in print. Polanyi points out that there is also a personal knowledge, or tacit knowledge as he calls it, which is also present whenever an individual encounters some form of explicit knowledge, such as a map. He points out that this tacit knowledge is the fundamental understanding process, and it is not available for critical surveillance [. . . .]. Hence the impossibility of using the map to explain the map.

Cassirer (1957) is concerned with the fundamental nature of the spatial experience. Of basic importance is the 'symbolic concept', man's chief activity being symbol formation. This is thought to be true whether the symbols are those of language, art, history, or science. The world which man experiences is not one of raw sensations, merging perceptions, but is one mediated by a constructive, active intellect which organises these perceptions into symbol systems. As Cassirer sees it the most fundamental organisation is spatial in nature [. . . .]. The articulation of things in space is therefore preceded by an articulation in judgment. Differences in spatial articulation (in terms of position, size and distance) can be grasped and assigned, since the separate sensory impressions are differentiated by judgment because different significances are imputed to them. These concepts are not a priori in any way, but are constructed because human activity finds them significant. There is great similarity between this view and that of Piaget and Inhelder, on the development of representational space (Harvey 1969: 192–193).

Cassirer finds that the problem of space enters all fields of knowledge, indicating to him that its pervasiveness is somehow related to the pervasiveness of symbolism. He decides that representational space is a construction, a result of symbolic activity, and [. . .] concludes that the transition from experience in real space to the ability to represent space internally is a crucial one in man's intellectual development. In Cassirer's view, spatiality is also the basis for all language development [. . .]. Labelling signifies an articulation of phenomena, and for any things or impressions to be separated from the flow of ongoing existence, there must be the separation which we call spatial. The sundering of the 'I' from 'the world' is, Cassirer feels, based on a spatial intuition [. . .]. Cassirer outlines a sequence of development for spatial representation which is identical to that of Piaget and Inhelder. From elementary relation, such as near and far, in topological space, with no particularly consistent point of view, the human gains flexibility as he develops a conception of space which includes other points of view, such as projective and Euclidean space.

The truly great achievement which the ability to map represents in the history of human intellectual development is pointed out in a number of ways by Cassirer. For example, he describes the impairment of spatial representational ability in aphasic disorders, where [. . .] a patient may operate with reasonable facility in concrete space, but be unable to perform the integrative activity necessary to create a coherent representational space. Cassirer observes that certain patients could not draw sketches of their rooms or routes, but could orient themselves reasonably well when presented with a sketch in which the basic schema was already laid down [. . .].

The ultimate significance of the spatial concept has led to the map metaphor being widely employed in other fields. In the field of psychology, for example, the term 'cognitive map' is currently much in vogue. Apparently the first psychologist to use the term was Edward Tolman (1948). [. . .] Much attention in the field then was being devoted to stimulus–response associations. [. . .] Tolman's conception of learning as consisting of more than just a string of simple associations was something of a major departure in the field. [. . .] In his article on cognitive maps, Tolman (1951) is attempting to develop theory about the behaviour of rats as they learn to run mazes [. . .]. He distinguishes between those cognitive maps which are narrow and strip-like and those which are broad and comprehensive. This distinction is surprisingly like the development of representational space in the child as outlined by Piaget and Inhelder (1967), which involves a transition from restricted topological connections to the greater flexibility of a comprehensive Euclidean space. Tolman, in summary, believes that learning consists not so much of building a series of stimulus–response connections as of developing sets in the nervous system which function like cognitive maps. This view has become far more popular in recent times than it was when he proposed it [. . .].

Students of scientific methodology also employ the map as a metaphor. The cartographer Board (1967: 719) develops the analogy of maps as models and concludes, 'By recognising maps as models of the real world and by employing them as conceptual models in order better to understand the real world, their central importance in geographic methodology is assured'. Important for our purposes here are the instances where models, languages, understanding, theories of reality and so on are likened to maps. Kuhn (1962: 108) provides a clear example of the a priori use of the map metaphor in relation to the functioning of the paradigm in science:

> . . . [The paradigm] functions by telling the scientist about the entities that nature does and does not contain and about the ways in which those entities behave. That information provides a map whose details are elucidated by mature scientific research. And since nature is too complex and varied to be explored at random, that map is as essential as observation and experiment to science's continuing development. Through the theories they embody, paradigms prove to be constitutive of the research activity. [. . .] our most recent examples show that paradigms provide scientists not only with a map but also with some of the directions essential for mapmaking.

Perhaps the most penetrating analysis of this is provided by Toulmin (1960) [. . .], who stresses the efficiency of maps, in providing more information than the set of point observations which went into their making, in order to show that theories can behave in similar ways. Toulmin finds that the selectivity which any mapmaker exercises is similar to the degree of completeness which exists in any particular theory in physics. In either case, there is no absolute standard of ultimate completeness. He emphasises that what is recorded or theorised about is what is significant to man in a particular situation [. . .]. He also makes a significant distinction between itineraries, or route maps, and more general maps – a distinction which nicely parallels the stages of the development in human consciousness from purely topological space notions (and the restrictions they impose) to an all-points-of-view, projective, Euclidean space. [. . .].

The foregoing cursory review of the 'map idea' in several disparate fields of learning provides clear evidence that the map is something fundamental to man's cognitive make-up. While a map itself exists and occupies space, it derives its meaning and usefulness from the fact that it represents some other space. There is fairly widespread philosophical agreement, which certainly accords with common sense, that the spatial aspects of all existence are fundamental. Before an awareness of time, there is an awareness of relations in space, and space seems to be that aspect of existence to which most other things can be analogised or with which they can be equated.

There also appears to be something appealingly simple about the 'map idea', but because it seems so often to have been taken for granted, its fundamental character remains obscure. The hazy notions of structure, of topological and Euclidean relations, of the importance of an organisation of space, all seem to be parts of the central but elusive concept called 'map'. In contrast, the 'real' map – that produced by the cartographer – is something that is a great deal more concrete: one can look at it and touch it. Nevertheless, because it obviously involves the concept of space in the ways in which we have seen that others use that term, the material map is fundamentally just as elusive. [. . .] We shall turn next to the definitions of a few limited terms needed to explore this subject.

It is obvious that maps have tremendous utility, but it seems all but impossible to make an exhaustive list of the uses to which maps may be put. Many people have tried. Such inventories usually point out that a map can be a popular and scholarly tool which aids in perceiving and understanding geographical relationships; that it is an efficient means for storing spatially-anchored data; and that it is a technical device permitting easy recovery of distance, direction and areal measurements. As a scholarly tool it serves a multitude of purposes, such as making possible inferences about the occurrence of unobserved or unobservable data, and aiding in the development and testing of spatial hypotheses. The list can go on almost indefinitely. Maps can serve these functions in less sophisticated ways, as in the form of the common road map, or in highly complex ways, as in an isarithmic map of population potential. In the past, maps have sometimes functioned as graphic, spatial-allegorical, didactic tools, as for example in the religious cartography of the Middle Ages. Maps even serve as ornaments. In the map, then, we seem to be dealing with something clearly fundamental, something which has nearly unrestricted potential utility.

Thus, since the map is so basic and has such a multiplicity of uses, the variety of its occurrences is vast. There are specific maps and general maps, maps for the historian, for the meteorologist, for the sociologist, and so on without limit. Anything that can be spatially conceived can be mapped – and probably has been. Maps range in size from those on billboards or projection screens to postage stamps, and they may be monochrome or multicoloured, simple or complex. They need not be flat – a globe is a map; they need not be of earth – there are maps of Mars and the moon; or for that matter, they need not be of anyplace real – there have been numerous maps made of imaginary 'places'. [. . .] The more one contemplates the variety of the map in its forms, its versatility and its fundamental nature, the more one is impressed with the fact that a general definition of a map must be based on its being simply a representation of things in space; *representation* and *space* are the two critical elements [. . .]. Here we need to consider briefly what kind of space is the accepted domain of the tangible map. Similarly, we need to look into the nature of the possible representations of that space in order to limit our concern to that which is appropriately called a map.

'Space' is a word with a large range of meanings. [. . .] Clearly, the space represented by a tangible map normally refers to the three-dimensional field of our experience; this is referred to in various ways by using such terms as 'area', 'territory', 'region', 'section of the earth', and so on. These terms imply a limited extent of land, but often something more. For example, 'territory' properly applies to a segment of the earth, but in lay language it also connotes some kind of proprietorship; but maps are neither limited necessarily to a particular segment nor even to the earth. To a geographical sophisticate the term 'region' has a specialised meaning. Since maps are not limited to earth, one cannot use such terms as 'section of the earth'. The word 'environment' could be employed because a map of a 'real' area always includes the conception of the maker in one way or another, and hence the mapped section is a part of his environment. Unfortunately, for our purposes, the term 'environment' has come to have such aspatial ecological overtones that it too seems inappropriate. The most general terms are probably 'place' and 'area' – simply a portion of space. They are quite impersonal, though, and because of the involvement of the cartographer in a map, they leave something to be desired. It is our view that the word 'milieu' best connotes one's surroundings or environment in addition to its meaning of place, and thus involves the cartographer. Our definition of a map, then, would be 'a representation of the milieu'. [...]

To represent is to stand for, symbolise, depict, portray, present clearly to the mind, describe, and seems to occasion no problem in meaning; but what of the form that the representation takes? [...] It seems necessary for our purposes here to put forward the proposition that a map is a graphic thing made of marks of various kinds. Traditionally, a map itself is a space in which marks that have been assigned meanings are placed in positions relative to one another in such a way that not only the marks, but also the positions and the spatial relationships among the elements, have meaning. It is thus a graphic or visual construct, and it follows that one must be able to *see* a map. [...] For something to be a real map it must be a graphic thing that is visible, such as paper with marks on it or the fluorescence of a cathode ray tube [...].

Our definition of a map turns out to be deceptively simple. 'A map is a graphic representation of the milieu' does no violence to the English-language definition of a map that appears in Meynen (1973): 'A map is a representation normally to scale and on a flat medium, of a selection of material or abstract features on, or in relation to, the surface of the Earth or of a celestial body.' [...] This ICA definition includes one element that is specifically and intentionally missing from ours, namely an indication of the medium employed or the form the representation usually takes.

Having defined the term 'map', it would appear logical to turn next to the concept expressed by the words 'mapmaker' or 'cartographer'. But a more fundamental label takes priority: the term 'mapper'. The considerable analytical work in what is called 'cognitive mapping' makes it clear that humans and some other creatures appear to process some forms of sensory input such that information obtained from the milieu is arranged or converted so that they can operate as if there were an internal space like a map (Downs and Stea 1973). In our view, whatever it is that actually occurs, this is the phenomenon that makes one a mapper. Creatures that have an elevated eye level and the mental capacity to arrange what they see into some sort of spatial framework are all potential mappers. Such an assertion covers a wide range, from the high-flying eagle to the darting dragonfly to no longer earthbound man. The eyes may range from the binocular equipment of the whole man to the 'vision' in the fingertips of the blind, but the essential *sine qua non* of the mapper is the ability to operate in a spatial mode. At the minimum this is two dimensional (x, y) and at the maximum it is three dimensional (x, y, z), with the possibility of some sort of integration of time as a fourth dimension, but the spatial framework must exist as part of the cognitive endowment of the mapper. Nothing else is really essential.

[...] The term 'mapper' [...] refers simply to anyone who actively conceives of spatial relationships in the milieu [...]. The conception of things in spatial relationship is the critical operation, and he who does it is a mapper. [...] It really makes no difference whether he obtains his conception of the milieu by simply observing nature at scale, by looking at a map, by plotting observed items at reduced scale on some handy medium, or by compiling items from a variety of other such plottings. The mapmaker and the map viewer are, therefore, both mappers in this broad sense of the term.

Questions naturally arise regarding the nature of the spatial conception of the milieu which forms in the mind of the mapper. [...] What does the term 'map' mean with respect to the cognitive map? Are there lines and points or is it restricted to shapes defined by surfaces and edges? Are things in colour? Can the mapper transform in rotation and scale at will? How sophisticated can he become with practice? Are people, as potential mappers, largely alike or quite dissimilar? And so on [....]. We really know very little about the cognitive map of the mapper.

We [...] can assert here that the cognitive map is best termed an image. Naturally the mapper's image [...] will be a function of his past experience and his ability to involve himself in a spatial framework. Therefore, it will vary from person to person; one can confidently assert that the images of no two mappers are alike, and that the same milieu can be mapped in different ways by the same mapper.

It would be easy to confuse the concept of mapper, as here outlined, and mapmaker, because in a very real sense a mapmaker is a mapper. But in our definition of mapper we have specifically restricted the map he develops to an image which is not tangible, that is, it does not materially exist to be touched and seen by another. On the other hand, the map made by a mapmaker we define as being corporeal, and in common parlance, the terms mapmaker and cartographer are essentially synonymous.

Among those who are concerned with the preparation and production of maps (charts) a rather careful distinction is generally drawn between mapping in the sense of mapmaking on the one hand and cartography on the other. 'Mapping' refers to all the operations involved in the production of a tangible map. Because the terms 'map' and 'chart' refer to all kinds of maps [. . .] the contributory activities range along a spectrum from geodetic survey and position finding, through various kinds of compilation, generalisation and decision making, to the construction of the artwork for the printer at the far end, with myriad technical procedures and executions in between. The term 'cartography' is generally restricted to that portion of the operation often termed 'creative', that is, concerned with the design of the map, 'design' being used here in a broad sense to involve all the major decision making having to do with specification of scale, projection, symbology, typography, colour and so on. [. . .] Usually when we employ the terms 'cartographer' or 'mapmaker' we will be using them in this sense of 'author'. Naturally, a cartographer must be a mapper in the sense of our definition of that term, but it is evident that a mapper need not be a mapmaker.

So 'mapper' is the all-inclusive term encompassing all who increase their spatial knowledge of the milieu by any sort of sensory input. [. . .] A [. . .] particular mapper who augments his spatial knowledge of the milieu as a consequence of looking at a map can be termed a [. . .] 'percipient'. The map percipient obtains information about the milieu by looking at a map, is coordinate with the cartographer, one who attempts to communicate spatial information about the milieu by making a map.

The use of the term 'percipient' makes it possible to distinguish those who, by looking at a map, add to their fund of spatial knowledge or acquire additional meaning, from those we designate by the more restricted terms 'map reader' and 'map user'. The term 'map reading' implies a rather specific and limited action, such as looking up the name of a city or country, or finding out how high a particular hill is. Similarly, the term 'map user' connotes the employment of a map for a specific purpose, such as that of the farmer who obtains the data needed for contour ploughing, or the engineer who lays out a road with the help of soil and topographic maps. Neither the map reader nor the map user is necessarily adding to his spatial knowledge. Both terms suggest operations similar to that of using a dictionary to find out simply how to spell or pronounce a word, which adds little if anything to one's understanding of the meaning of the word and, therefore, of the language to which it belongs.

[. . .] Although the concept of the map is thought by many students to be of fundamental significance, the nature of the map as an image and the manner in which it functions as a communication device between the cartographer and the percipient need much deeper consideration and analysis than they have yet received. [. . .] Until recently the emphasis in cartography has been restricted largely to the map itself. Cartographers are frequently engaged – on deadline – in producing maps for immediate consumption. [. . .] The maps they make work well enough, most of the time. But there is a need in cartography for another point of view, an analysis from someplace other than the production line. It seems that, too often, meaning can be obtained from the map only with great effort and annoyance, or with an amazing loss of information between cartographer and percipient. There has not been, in the field of cartography, a thorough delineation of the methodological and philosophical bases on which an analysis of the acquisition and transmission of spatial knowledge via the map could be conducted.

Such an analysis clearly must deal with more than just the physical characteristics of the map. It must probe the characteristics of human beings as they see and know, spatially, and how they use maps to understand and communicate this knowledge. The emphasis must shift from the map as a static graphic display to the cognitive and perceptual activities of the individuals who interact with maps, namely, those mappers who are mapmakers or map percipients. Such a study leads us inevitably into the broad field of psychology, but also into aspects of philosophy, epistemology and other fields as well. This is a large order [. . .] and so far a broad research paradigm specific to the field of cartography has not emerged from fragmentary research activity.

For the most part, up to this time, the psychological research which has been conducted in connection with cartography and percipients has been carried on with the methodological assumptions of the field of behavioural psychology, as, for example, in psychophysical studies which attempt to measure and compare stimulus and response magnitudes. Some of this material has proved quite useful in some practical mapmaking situations, but its applicability has been restricted. Furthermore, its capacity for generalisation is severely limited, and thus it has furnished little basis for prediction in new applications in cartography.

[. . .] We find it necessary to turn to the ideas and terminology of other academic fields for assistance in understanding our own. We must come to terms with what constitutes 'meaning' in our own field of cartography–geography. We must draw on the work of others for some insight into the nature of the human being as learner and knower, particularly about things spatial. For example, we find that we become very heavily indebted to Piaget and his coworkers for the understanding of the conception of space and articulation within it. Moreover, we owe much to the writings of Polanyi for a grasp of the difference between tacit and explicit knowledge, and for his very lucid analysis of part-whole relationships and the nature of meaning [. . .].

We hope that in the various comparisons of maps with other things, this point does not become obscured: maps are both *unique* and *fundamental.* [. . .]

References

Board, C. (1967) Maps as models, in *Models in Geography* (eds R.J. Chorley and P. Haggett), Methuen, London.

Cassirer, E. (1957) *The Philosophy of Symbolic Forms,* Yale University Press, New Haven, CT.

Chorley, D. and Haggett, P. (1967) *Models in Geography,* Methuen, London.

Downs, R.M. and Stea, D. (eds) (1973) *Image and Environments,* Aldine Publishing, Chicago.

Harvey, D. (1969) *Explanation in Geography,* Arnold, London.

Korzybski, A. (1941) *Science and Sanity,* International Non-Aristotelian Library & Publishing Press, New York.

Kuhn, T.S. (1962) *The Structure of Scientific Revolutions,* Chicago University Press, Chicago.

Meynen, E. (1973) *Multilingual Dictionary of Technical Terms in Cartography,* Franz Steiner, Wiesbaden, Germany.

Piaget, J. and Inhelder, B. (1967) *The Child's Conception of Space,* Norton, New York.

Polanyi, K. (1963) Ports of trade in early societies. *The Journal of Economic History,* **23**, 30–45.

Tolman, E.C. (1948) Cognitive maps in rats and men. *Psychological Review,* **55** (4), 189–208.

Tolman, E.C. (1951) *Behavior and Psychological Man,* University of California Press, Berkeley.

Toulmin, S. (1960) *Philosophy of Science,* Harper, New York.

Further reading

Board, C. (1984) New insights in cartographic communication. *Cartographica,* **21** (1). [A relevant theme issue, coming towards the end of the research heyday of cartographic communication theory, devoted to exploring ongoing implications of cognitive scientific approaches to mapping.]

Crampton, J. (2001) Maps as social constructions: power, communication and visualization. *Progress in Human Geography,* **25** (2), 235–252. [Situates the development of the communications paradigm in cartography in the light of subsequent social constructivist challenges and emergence of visualization technologies.]

Olsson, G. (2007) *Abysmal: a Critique of Cartographic Reason,* University of Chicago Press, Chicago. [A philosophical text that explores the deep cultural associations between mapping and ways of thinking about the world.]

Sismondo, S. and Chrisman, N. (2001) Deflationary metaphysics and the natures of maps. *Philosophy of Science,* **68**, 538–549. [This paper develops and explores the metaphor that scientific theories are maps of the natural world in order to examine limits in realist notions of mapping as representation.]

Vasiliev, I., Freundschuh, S., Mark, D.M. *et al.* (1990) What is a map? *The Cartographic Journal,* **27**, 119–123. [An insightful experimental investigation into definitions of the map.]

See also

- Chapter 1.4: The Science of Cartography and its Essential Processes
- Chapter 1.6: Cartographic Communication
- Chapter 1.14: Rethinking Maps
- Chapter 3.3: The Look of Maps
- Chapter 4.2: Map Makers are Human: Comments on the Subjective in Maps
- Chapter 4.3: Cognitive Maps and Spatial Behaviour: Process and Products
- Chapter 4.9: Cognitive Maps: Encoding and Decoding Information

Chapter 1.4

The Science of Cartography and its Essential Processes

Joel L. Morrison

Editors' overview

Morrison's article provides a description of the processes he saw as central for the practices of cartography as a 'communications science'. It represents a growing orthodoxy amongst cartographic researchers that reached its peak in the mid 1970s. He argues for a unified paradigm and expands on Ratajski's notion of structural transmission of mapped information from data collection through to map use. Morrison's ideas are strongly influenced by cognitive science and situate the cartographer as working between different cognitive realms; those involved in collecting data about the world and those concerned with decoding these mapped information. He identifies processes that are of direct concern to the cartographer as the selection of data, its classification, simplification and symbolisation, and concludes with a clarion call for the progressive improvement of map design as a scientific practice based upon a properly developed understanding of cartographic syntax and grammar.

Originally published in 1976: *International Yearbook of Cartography*, **16**, 84–97.

The deduction of theories from a structure representing a basis for the cartographic discipline has received the attention of a growing number of cartographers during the past ten years. [...]

To establish a unified body of theory, scientists in a field must agree on a fundamental paradigm. Koláčný's work set forth a reasonable schematic for the cartographic discipline based on the paradigm of cartography as a science of the communication of information (Koláčný 1969). This author is not aware of how Koláčný came to select his paradigm. In 1972 Ratajski outlined three possible paradigms for the discipline of cartography and selected the paradigm used by Koláčný as the one which offered the broadest perspective (Ratajski 1973). Ratajski outlined the cartographic discipline based on this paradigm, and the completeness of this outline is sufficient evidence to support his opinion that the paradigm of cartography as part of the science of communication offers cartographic scientists the broadest perspective.

Cartographic scientists in many nations are now accepting this paradigm, and the impact of it on the discipline is becoming very pervasive. Cartography, under this paradigm, is a science. [...] The Multilingual Dictionary recently published by Commission II of the ICA defines cartography as: 'The art, science and technology of making maps, together with their study as scientific documents and works of art. In this context maps may be regarded as including all types of maps, plans, charts and sections, three dimensional models and globes representing the Earth or any celestial body at any scale' (ICA Commission 1973).

This definition needs revision. A new definition must be formulated which will delete the equating of art and technology with science, but which will explicitly state that the science of cartography, as well as other sciences, relies on art and technology. Ratajski (1973) is correct in saying 'Technological research mainly concerns the adaptation of achievements in other sciences to cartography'. As such, it is not within the province of cartographers to perform purely technological research. The subsidiary and most helpful role of art to the science of cartography is clearly evident in any research structure based on the paradigm of cartography as an information or communication science.

The Map Reader: Theories of Mapping Practice and Cartographic Representation, First Edition. Edited by Martin Dodge, Rob Kitchin and Chris Perkins.

But again, cartographers adapt artistic principles to cartography; pure art is not cartography. Considering that during the past two millennia no unified structure of the cartographic discipline has been successfully outlined based on any alternative paradigm, the acceptance of this new paradigm and the deduction of a unified body of theory is long overdue. [...]

The central theme of this paper is to elaborate and expand Ratajski's structural model of cartographical transmission shown in Figure 1.4.1. The emphasis will be placed on defining the processes which operate within the science of cartography. There are at least two realms of activity in a communication process: initiation of the communication and the reception of the communicated message. In cartography the fact that the initiator of the map can also be the receiver of the communicated message is recognised. Communication takes place from one cognitive realm (that of the initiator) to another cognitive realm (that of the receiver). It does not matter if these cognitive realms are identical prior to the communication. For communication to have occurred, however, the cognitive realm of the receiver must register a change after receipt of the communicated message. If no change occurs, communication has not taken place.

Communication via a map between two cognitive realms utilises a cartographic language. Successful communication results in the addition of correct information about reality to the cognitive realm of the receiver. Unsuccessful communication results in the addition of incorrect information about reality to the cognitive realm of the receiver. If nothing is added to the cognitive realm of the receiver, no communication has taken place. The nature of the cartographic language will be outlined in more detail below.

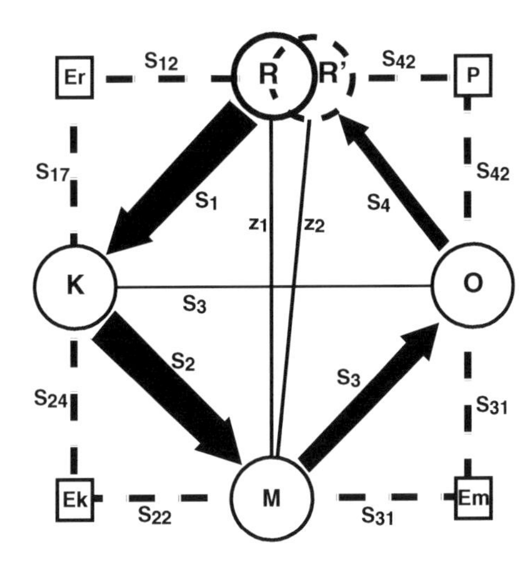

R – reality as source of direct information
Er – informative emission of direct source
S_2 – relation of direct recognition
S_{11} – relation of direct emitting
S_{13} – relation of direct perceiving
K – cartographer, sender of message
E_k – message emission
(informative emission of sender)
S_2 – relation of message creating
s_{31} – relation of message emitting
S_{21} – relation of message perceiving
M – map or message informant
S_2 – relation of indirect recognition
(by means of a map)
S_{31} – relation of indirect emitting
S_{22} – relation of indirect perceiving
O – receive (map user)
P – mental transformation
(mental processes)
S_4 – relation of reality image reconstruction
(relation of recognition)
s_{41} – relation of recalling
s_{42} – relation of imagination
R^1 – imagination of reality
z_1 – message relation
z_2 – relation of identification
(relation of map efficiency)
z_2 – relation of making available
(relation of informing receiver)
α – degree of transmission correctness

Figure 1.4.1 A model of cartographical transmission after Ratajski (1973: 219).

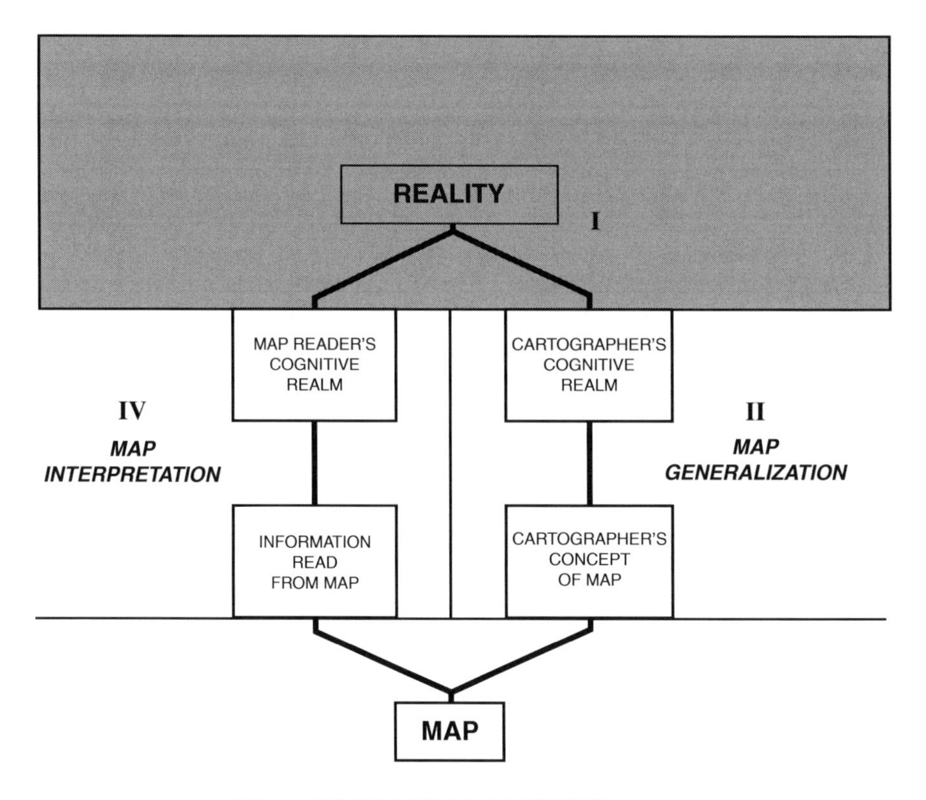

Figure 1.4.2 Outline of the science of cartography (lines denote areas in which processes of concern to cartographers occur).

Figure 1.4.2 illustrates the discipline of cartography as communication between two cognitive realms. To facilitate discussion, Figure 1.4.2 has been divided into four segments: (I) the relationship of the cognitive realms of the initiator and the receiver to reality; (II) the relationship of the cognitive realm of the initiator (cartographer) to the initiator's conception of the map (communication); (III) the cartographic language; and (IV) the relationship of the receiver's (map reader) cognition of the map to the cognitive realm of the receiver. For each segment, the process involved will be outlined and defined.

The first segment, as shown in Figure 1.4.2, is not part of the cartographic discipline. Since it is important, however, in establishing the cognitive realms between which cartographic communication can take place, it is necessary that it be included in our discussion.

Reality to cognitive realms

The cognitive realms of the cartographer (initiator) and the map reader (receiver) are established and enlarged through two related processes. The first of these processes can be characterised as deliberate or direct. The second process can be characterised as indirect or inadvertent.

The first process corresponds closely to Ratajski's relation of direct recognition. It is composed of several tasks which can be considered sequential: detection, discrimination, recognition or cognition, and estimation (Day 1969). Detection is the uncovering or discovery of the existence of a bit of information. Discrimination is the process of gaining distinct information about the detected target. In discrimination, the observer notes or distinguishes one bit of information as being different from another. The third task, recognition, is the identification of information as having been previously seen. If the information is new, the identification of the information alone represents cognition. The final task of estimation allows one to form an approximate judgment or opinion regarding the value, amount or size. Thus, in creating a cognitive realm by a direct process, reality is searched until a target is detected. The detected target is discriminated and recognised or identified. When these three tasks are completed, the cognitive realm is established or enlarged. Estimation takes place after these three tasks are completed. When estimation does occur, it results in a further enlargement of the cognitive realm. If the first three tasks are incorrectly completed, misinformation about reality enters the cognitive realm. Further misinformation can occur by incorrect estimation. This misinformation may cause misconceptions about reality.

The second process (indirect) differs from the first process (direct) in that no search task is employed. Repeated haphazard encounters with reality serve to enlarge the cognitive realm. Still the detection, discrimination, recognition and estimation tasks take place. Again, the tasks through cognition must occur for the enlargement of the cognitive realm to take place. Once a cognitive realm has been established, one more process, verification, can operate. The verification process often is performed to substantiate the validity of information in one's cognitive realm against reality. Three examples can be outlined. Firstly, the cartographer may verify information in his cognitive realm during his formation of a conception of the map he is about to make. Secondly, the map reader, after receiving communicated information via a map which enlarges his cognitive realm, may verify that information against reality. Thirdly, communicated information may conflict with information in the map reader's cognitive realm. To resolve the conflict the map reader verifies his received information against reality. Often this serves to enlarge his cognitive realm even further.

The Venn diagram shown in Figure 1.4.3 illustrates two cognitive realms. Let us call the entire area reality, R. Let area A in the Venn diagram correspond to the cognitive realm of the cartographer. As can be seen, while the cognitive realm of the cartographer contains considerable information about reality, R∩A, it also contains incorrect information or misconceptions about reality, $\tilde{R}$∩A.

Let us consider area B to be the cognitive realm of the map reader. Again, the map reader has some misconceptions about reality, $\tilde{R}$∩B. The stippled cross-hatched area shows that both the cartographer and the map reader share misconceptions of reality $\tilde{R}$(A B) and the cross-hatched area R∩(A∩B) illustrates shared correct cognition about reality. If A∩B was null, it is unlikely that useful, successful, efficient communication could take place, because the cartographer and the map reader share no common cognition of reality. The map might communicate information to the map reader which would segment his cognitive realm. When this happens, the verification process must be used by the map reader to 'build a bridge' so to speak, between the separate areas of his cognitive realm. Otherwise the map reader must accept the information on faith, which he is not likely to do because he shares no common cognition of reality with the cartographer. Communication which causes segmentation of the cognitive realm is termed 'inefficient' communication.

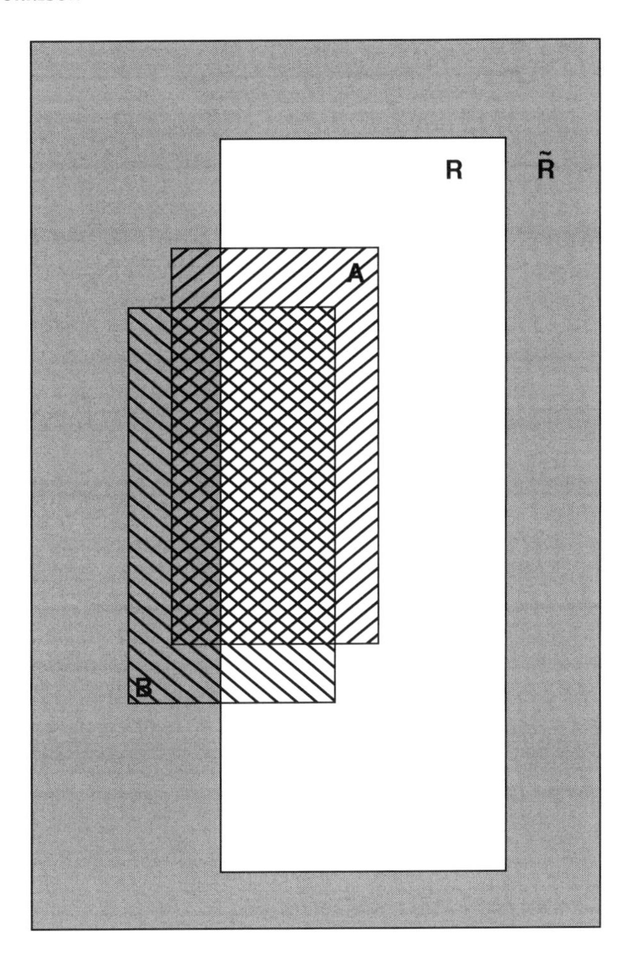

Figure 1.4.3 A Venn diagram illustrating reality, R; misconceptions of reality, Ř; and two cognitive realms A and B.

In transformational terms, all the processes involved in the formation, enlargement and substantiation of cognitive realms are one-to-one transformations. Since a process that results in enlargement or substantiation exhausts neither reality nor the cognitive realm, the transformation is not onto. Thus, these processes in transformational terms resemble the generalisation element symbolisation (Morrison 1974). Symbolisation is restricted, however, by the controls of generalisation, which do not affect the formation, enlargement or substantiation of cognitive realms.

Finally, it must be emphasised that, throughout, everything is dynamic. Reality itself changes with time and cognitive realms are constantly undergoing change in living organisms. However, it is the fact that different cognitive realms exist that makes communication between cognitive realms desirable and necessary. If reality and, especially, cognitive realm formation and enlargement was not dynamic, only a finite amount of communication could take place before all cognitive realms were equal and communication therefore unnecessary.

The dynamic nature of reality and the formation of cognitive realms require communication, and thus gives cartography as well as the other communication sciences their essential raison d'etre. Cartography as one communication science probably plays a rather limited role in the total formation of cognitive realms. However, cartography does possess the ability to enlarge the cognitive realm not only of the receiver of the communication but also of the initiator, the cartographer. In essence, the map allows the cartographer to communicate with himself. This is possible because of the existence of induction as one of the elements of cartographic generalisation. [...]

Formation of the cartographer's conception of a map from his cognitive realm

The science of cartography begins at this point. The previous discussion, which dealt with the formation, enlargement and verification of cognitive realms, is not included in the discipline of cartography. Rather, it is because of the dynamic nature of reality and cognitive realms that cartography is a useful communication science. Hence, cartography begins with the cartographer's desire to communicate a portion of his cognitive realm to someone else. The first step in this communication science is the formation of the cartographer's conception of a map.

Three principle processes operate in the cartographer's formation of his conception of a map. These processes are: (1) selection, (2) classification and (3) simplification. [...]

The first process is that of selection. The cartographer makes a conscious selection of part of his cognitive realm which he wishes to communicate to a map reader. [...] Consciously the cartographer selects a proper subset, C, of his cognitive realm for mapping. To the cartographer this is a one-to-one and onto transformation. In fact, however, due to the dynamic nature of reality, the incompleteness of the cartographer's cognitive realm and any misconceptions the cartographer may have as part of his cognitive realm, the selected set is not set C, but a set $A = C \cup D$ (Figure 1.4.4). Set A is not a proper subset of the cartographer's cognitive realm because it usually contains data elements of which the cartographer is unaware, namely set D. Only when $D = \phi$, the null set, is A a proper subset of the cartographer's cognitive realm. The cartographer himself never knows if $D = \phi$ or not. [...]

The second process which the cartographer uses in forming his conception of a map is classification. Classification is one of four elements of generalisation (Robinson and Sale 1969). In this process the cartographer takes his conscious set C (in fact the set A) and classifies its elements into the categories that he wishes to map. Classification can only be applied to set C, resulting in the set C' in Figure 1.4.4, although set A' will actually be mapped. In transformational terms, the process is not one-to-one, but it can be considered onto. Several elements selected for mapping are often combined in classification. Presumably, however, the consistent cartographer will apply classification equally to the entire set C to form set C'. Classification

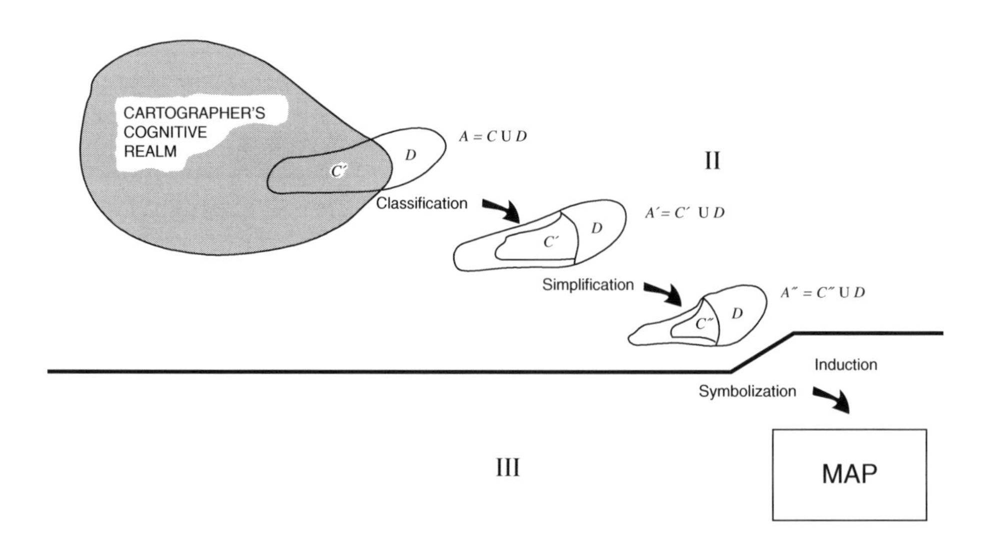

Figure 1.4.4 Processes involved in producing a map from a cartographer's cognitive realm.

can be applied at any scale of measurement: nominal, ordinal, interval or ratio.

The third process that the cartographer employs is simplification, which is also an element of generalisation (Robinson and Sale 1969). Each data element that has already been selected and classified still possesses several characteristics. The cartographer must decide which of these characteristics he will portray on the map and which he will ignore. Consciously the cartographer transforms his set C′ onto the set C″ and then maps the set A″. [...]

The formation of the cartographer's conception of his map from his cognitive realm is, therefore, an onto transformation which is not one-to-one. The three processes of selection, classification and simplification tend not to have a set sequence of application, but a cartographer may iterate through each process several times, even returning to reality to en large or verify his cognitive realm. [...]

The cartographic language

Once the cartographer has formulated his conception of the map, he must prepare the communication channel (map) for the receiver of the message (map reader). A cartographic language must be used. The cartographer must symbolise his conception of the map and the map reader must read the encoded message. Therefore, it is essential that both the cartographer and the map reader understand the same cartographic language. Two basic criteria for cartographic communication can be set: (1) two developed cognitive realms, preferably having some information in common, and (2) knowledge of a given cartographic language by both the cartographer and the map reader.

The information common to both the cognitive realm of the cartographer and of the map reader (areas 1 and 5 in Figure 1.4.5) constitutes a basis that allows for efficient communication. The part of this information that is on the map is defined as base data. In the Venn diagram, area 1 represents base data on the map. It is essential to have base data before efficient communication can take place. In the Venn diagram, area 3 represents data to be communicated from the cognitive realm of the cartographer to the cognitive realm of the map reader. Only areas 3 that border on areas 1 can be efficiently communicated, that it, transferred from the cognitive realm of the cartographer to that of the map reader by the use of a map. If an area 3 is not adjacent to an area 1, the map reader has no frame or base upon which to receive part of an area 3, that is, to enlarge his cognitive realm without segmenting it. If he segments his cognitive realm he then must substantiate the information received from the map by the process of verification. That a map reader must do this connotes inefficient communication.

There are two reasons for inefficient cartographic communication: both are the fault of the cartographer. In one instance the map reader may have an insufficiently developed cognitive realm. When inefficient communication results, the cartographer has failed to become aware of his intended audience and to design his map using base data that the map reader is known to possess. In the second case, where the map reader has a developed cognitive realm and still inefficient communication results, the cartographer has failed to include sufficient base data. The faults respectively are (1) assuming information to be base data when it is not, and (2) failing to use a sufficient quantity of base data.

The second criterion for efficient cartographic communication is a knowledge of the cartographic language. If the map reader is not fluent in the language, he will fail to

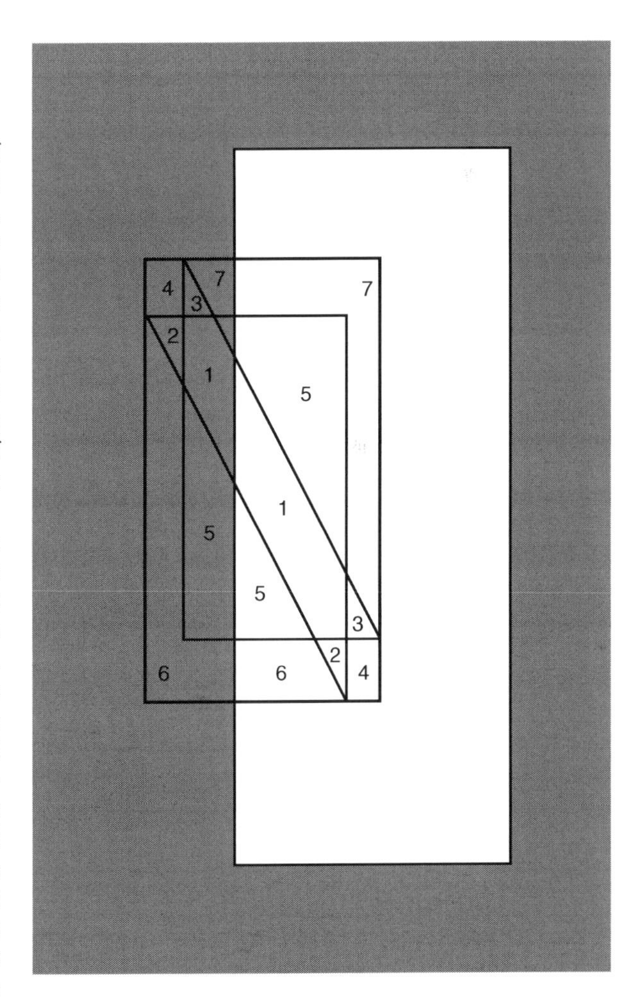

Figure 1.4.5 A Venn diagram illustrating the map (areas 1, 2, 3 and 4) in relation to reality and the two cognitive realms involved in cartographic communication.

receive all of the area 3 that is being communicated. If the cartographer is not fluent, he makes the task of the map reader more difficult and the area 3 that is being communicated is small.

Efficient cartographic communication transfers areas of correct conceptions of reality from the cartographer's cognitive realm to the cognitive realm of the map reader. A good map, therefore, contains a sufficient area 1, in conceptions of reality, a small area 1 of misconceptions of reality, a large area 3 adjacent to area 1, in conception of reality and no area 3 which is not adjacent to an area 1. The size of areas 2 and 4 are relatively unimportant.

The cartographic language, as well as most non-verbal languages, has two aspects: (1) writing and (2) reading. In order to write a language, a syntactical structure and a grammar are necessary. The generalisation element of symbolisation is coincidental with conscious writing in cartography. Symbolisation, in transformational terms, is a one-to-one process from the cartographer's cognitive realm to the map. It is not onto, however, due to the symbolisation (Figure 1.4.4). Induction, the fourth element of generalisation, is neither a one-to-one nor an onto process. The cartographer consciously symbolises the set C'' in Figure 1.4.4. This corresponds to areas 1 and 3 in the Venn diagram of Figure 1.4.5. Induction corresponds to areas 2 and 4 in the Venn diagram and to set D in Figure 1.4.4. Since the cartographer actually maps set A'' in Figure 1.4.4, the map corresponds to areas 1, 2, 3 and 4 in Figure 1.4.5.

The cartographer has no control over induction. Therefore, the best safeguard for a cartographer when writing the cartographic language is to perform his symbolisation as logically as possible. Because of induction, however, the map becomes more than the cartographer envisaged, and hence the cartographer can, by becoming a map reader, enlarge his own cognitive realm. In essence the cartographer, by making a map which has inherent in it induction, can enlarge his own cognitive realm without referring to reality. This is a characteristic of cartographic communication that is either not possessed, or possessed to a lesser degree, by other forms of communication.

Map reading ends with perception and/or cognition of the information on the map. This is a one-to-one transformation from the map into the cognitive realm of the map reader. Map reading requires detection, discrimination, recognition or cognition and estimation. Map analysis or interpretation takes place within the cognitive realm of the map reader and not in the communication channel between the two cognitive realms. [...]

The efficient use of the cartographic language, therefore, requires a syntactical structure and grammar for symbolisation based on the thresholds of a map reader's ability to perform detection, discrimination, recognition and estimation tasks in a spatial framework.

[...] The pragmatics of the symbol dimensions used by cartographers require knowledge of the abilities of map readers to perform detection, discrimination, recognition and estimation tasks on [...] eight symbol dimensions – [size, shape, colour hue, colour value, colour intensity, pattern orientation, pattern arrangement, pattern texture] [...]

Once the map reading tasks are thoroughly researched, the grammar of the cartographic language should become apparent. [...]

Map reading and the cognitive realm of the observer

For the purposes of discussion in this section, it is necessary to assume that cartographic communication has taken place. The map reader has 'read' the map, that is, he has completed the detection, discrimination, recognition and estimation tasks necessary to transfer the areas 3 in the Venn diagram into areas 1.

To avoid confusion, precise definitions for map reading, map analysis and map interpretation must be established. The term map reading refers only to the completion by the map reader of the processes involved in the cartographic language as discussed above. Map interpretation refers to the interaction within the map reader's cognitive realm of the communicated information received from the map, and the information existing previously in the cognitive realm of the map reader. The term map analysis refers to the resolution of any conflicting information or the substantiation of new information that forces the map reader to return to reality. Therefore, the processes resulting in the enlargement of the cognitive realm to combine segmented cognitive realms, or the process of verification to substantiate the information, will be map analysis processes.

Map interpretation is an iterative process within the cognitive realm of the map reader. It includes the processes of simplification, classification and selection. Map interpretation integrates the newly communicated information and the base data with the existing cognitive realm of the map reader. In transformational terms, map interpretation is a composite function that needs be neither one-to-one nor onto.

Cartographers, as information communication scientists, must be concerned that information has been communicated correctly to allow map interpretation and analysis to take place. The specific results of map interpretation and/or analysis may be of far more concern to other scientists than to cartographers. [...]

For map interpretation the important point is that the set A'' is being read for integration with the reader's cognitive realm. Since the cartographer symbolised only the set C″, where A″ = C″∪D, the inductive generalisation accompanying symbolisation allows for extension of the map reader's cognitive realm. This clearly explains the utility of the map for communicating between a cartographer and a map reader where the two are synonymous.

Conclusions

Cartography is the detailed scientific study of a communication channel. In this paper the processes that are of direct and of auxiliary concern to the cartographer have been identified. The processes of direct concern include: selection of data from the cartographer's cognitive realm to form his conception of a map; classification and simplification of selected data to form a conception of the map; symbolisation of a conception of the map to create a physical map; induction or inductive generalisation accompanying symbolisation; detection of symbols on the map; discrimination of differences between symbols on the map; recognition (identification) or cognition of the specific symbol; and estimation of characteristics of the identified symbol.

The auxiliary processes of concern are: detection of reality; discrimination of differences between elements of reality; recognition (identification) or cognition of specific elements of reality; estimation of characteristics of identified elements of reality; simplification, classification and selection of information read from the map during map interpretation; verification of information received by reading the map or through map interpretation during map analysis; and resolution of conflicts between information read from the map and information existing in the map reader's cognitive realm by referring to reality during map analysis. The above processes have been defined in both transformational terms and via a Venn diagram to illustrate their relation to the cartographic science. A syntactical structure for symbolisation and a research structure for the pragmatics of map reading are outlined, which when complete will hopefully lead to a grammar for a cartographic language.

[C]artographic science utilises a language to communicate between individual cognitive realms. The mental processes leading to the conception of the map, the rendering and reading of the map and the mental processes involved in interpreting the map are all legitimate cartographic concerns. The successful use of this communication channel is dependent upon a well established syntactical structure and grammar. This grammar must take into account both the physiological and the psychological thresholds of the map reader's abilities to perform map reading tasks. Once these are established, the cartographic scientist can more fully utilise art and technology, logically, to facilitate communication via his map.

References

Day, R.H. (1969) *Human Perception*, John Wiley & Sons, Inc., New York.

ICA Commission (1973) *Multilingual Dictionary of Technical Terms in Cartography*, Franz Steiner, Wiesbaden, Germany.

Koláčný, A. (1969) Cartographic information – a fundamental concept and term in modern cartography. *The Cartographic Journal*, **6** (1), 47–49.

Morrison, J.L. (1974) A theoretical framework for cartographic generalization with emphasis on the process of symbolization. *International Yearbook of Cartography*, **14**, 115–127.

Ratajski, L. (1973) The research structure of theoretical cartography. *International Yearbook of Cartography*, **13**, 217–228.

Robinson, A.H. and Sale, R.D. (1969) *Elements of Cartography*, 3rd edn, John Wiley & Sons, Inc., New York.

Further reading

Board, C. (1984) New insights in cartographic communication. *Cartographica*, **21** (1). [A relevant theme issue, coming towards the end of the research heyday of cartographic communication theory, devoted to exploring ongoing implications of cognitive scientific approaches to mapping.]

Crampton, J. (2001) Maps as social constructions: power, communication and visualization. *Progress in Human Geography*, **25** (2), 235–252. [Situates the development of the communications paradigm in cartography in the light of subsequent social constructivist challenges and emergence of visualisation technologies.]

Krygier, J.B. (1995) Cartography as an art and a science. *The Cartographic Journal*, **32** (1), 3–10. [This paper provides an insightful critique of the art-science dualism as a means of understanding cartography and argues they both inform all mapping practices in complex ways.]

MacEachren, A.M. (1995) *How Maps Work*, Guilford, New York. [A thorough scientific reworking of cognitive approaches to mapping incorporating semiotic ideas into a representational framework.]

See also

- Chapter 1.2: Semiology of Graphics
- Chapter 1.3: The Nature of Maps
- Chapter 1.6: Cartographic Communication
- Chapter 1.8: Deconstructing the Map
- Chapter 1.10: Cartography Without 'Progress': Reinterpreting the Nature and Historical Development of Mapmaking
- Chapter 1.14: Rethinking Maps
- Chapter 4.3: Cognitive Maps and Spatial Behaviour: Process and Products
- Chapter 4.9: Cognitive Maps: Encoding and Decoding Information

Chapter 1.5
Analytical Cartography
Waldo R. Tobler

Editors' overview

By the mid 1970s an orthodoxy of the cartographic sciences as a means of graphical communication had emerged, and the majority of teaching either focused on the graphical skills needed to design maps as illustrations or regarded the discipline as at best a technical adjunct to geography. Tobler's article was a call for a rethinking of cartography, informed by a much more theoretical consideration, and for concerns with analytical approaches to mapping. Tobler had been one of the pioneers for using computers in mapping and he echoes rising concern for the impact of technological change on standardised procedures and knowledge. His notion of analytical cartography focused on mathematical principles underlying many aspects of mapping, from photogrammetry and geodesy, through to generalisation and map design. The ideas summarised in this article strongly informed shifts towards GIS and away from purely visual approaches to mapping and remain a powerful influence on contemporary practice 35 years later.

Originally published in 1976: *The American Cartographer*, **3** (1), 21–31.

[...] Cartography has been a university subject in the United States since the late [1930s] [...]. Of the 3000 or so colleges in the United States, cartography is taught in most of those which offer geographical instruction, but less than a dozen have developed research specialisations in this area. To these one should add a small number of engineering schools which have professional programs in surveying, photogrammetry, or geodesy. For various reasons there was a gap between official governmental cartography and academic geographical cartography. Geography was once associated with geology, but it has now, at least in North America, moved away from this focus to become a social science. The tradition is thus somewhat different than that of continental Europe in both geography and cartography. Geography in the late 1950s and early 1960s went through the so-called 'quantitative revolution', in which statistical description replaced verbal description, and formal abstract model building was recognised as superior to anecdotal explanation. [...] [Cartography] seemed sometimes to be treated as a 'technique', which could be taught without especial qualifications by any staff member, and this attitude certainly could not lead to any expansion of the theoretical core of the field. At the same time it was clear that computers would play a crucial role in the future of cartography, the first computer maps having been produced in 1951, and that many changes were needed in traditional cartography (Tobler 1959). The world seemed to be changing so fast that 50% of what one learned was obsolete within five years. I had the hope that my lectures would have a half-life of 20 years. The problem was to revise drastically the advanced cartography curriculum. The eventual result was Analytical Cartography. [...] It is perhaps also appropriate to draw especial attention to the contrast between this direction of development of the field of cartography with other developing directions. Much of the current literature, for example, emphasises cartography as geographic illustration, in which communication is paramount (Robinson and Petchenik 1975).

A popular title would have been Computer Cartography. This did not appeal to me because it is not particularly critical which production technology is used. Such a title would also imply the existence of other courses, perhaps titled Handicraft Cartography, or Pen and Ink Cartography. The substance is the theory that is more or less independent of

The Map Reader: Theories of Mapping Practice and Cartographic Representation, First Edition. Edited by Martin Dodge, Rob Kitchin and Chris Perkins.

the particular devices; equipment becomes obsolete rather quickly anyway. Mathematical Cartography could have been used for the title, but this already has a definite meaning (Graur 1956; Solovyev 1969), and I had in mind more than is usually covered under this heading. Cartometry, the study of the accuracy of graphical methods, is another available term, but has a rather narrow meaning. One could also speak of Theoretical Cartography. This did not appeal to me on two grounds. It would frighten students, who are always concerned that they learn something practical. Secondly, the precedent is not very attractive. Max Eckert (1921), for example, wrote a great deal about theoretical cartography but did not solve many problems. I wished to emphasise that mathematical methods are involved, but also that an objective is the solution of concrete problems. [...] All professions must constantly fight their myopia. But clearly the application of mathematical methods to cartography is growing rapidly. Thus a simple introduction to Analytical Cartography is through Photogrammetry and Geodesy. These fields have a long mathematical tradition and a healthy literature to which the student need only be referred. The principal equations of the method of least squares and its newer derivatives, of the theory of errors, and some theorems of projective geometry and of potential theory, and so on, are all easily reviewed. One can also see tendencies, such as the direct production of mosaics from aerial photographs and the computer recognition of objects seen by an imaging system (Duda and Hart 1973). Another trend is the replacement of triangulation by trilateration.

To see how a more general view is useful, consider this last topic, in which one determines locations from measurements. Suppose that we have identified n points on the surface of the earth. Between these there exist $n \times (n-1)/2$ distances, d_{ij}. Let us assume that all of these distances have been measured and that the locations of the points are to be found. This means that 2n coordinates must be determined. In high school one learned that one could compute distances from coordinates by use of the Euclidean formula:

$$\sqrt{[(x_i - x_j)2 + (y_i - y_j)2]} \rightarrow d_{ij}$$

The surveying problem reverses the arrow, and every surveyor knows how to find the coordinates when given the distances. In this problem the value of $n \times (n-1)/2$ grows much faster (essentially quadratically) than does 2n, and there are thus more equations than unknowns. This has three useful consequences:

(1) No solution satisfies all equations, since all empirical measurements have error. This leads naturally to a discussion of least squares methods, error ellipses and iterative solution procedures. But one also obtains internal checks on the accuracy of the measurements.

(2) One only needs to know the ordinal relations of the distances in order to obtain a solution (Shepard 1966), which comes as quite a shock to the cartographer who strongly believes in numbers.

(3) Only 2n measurements are really necessary, and one is then led to consider the 'optimal' positioning of surveying measurements, a rather recent development in the literature (Grafarend and Harland 1973).

A cartographic application of this 'geodetic' technique might be as follows: Suppose one is successful in getting people to complete a questionnaire in which they are required to estimate the distances between, say, prominent buildings in Vienna. The result of this empirical operation yields data similar to the measurements obtained from a geodetic survey in that it is possible to compute coordinates and their standard errors (Tobler 1976). One can thus obtain a type of mental map (Gould and White 1974) and its degree of variance. Comparison with geodetic positions enables one, using Tissot's Theorem, to measure the amount of distortion of these mental maps. Analysing such data over time may reveal the rate of spatial learning, which is geographically a very interesting question. A similar technique, incidentally, is used in psychology when one attempts to portray separations between personality types. In some respects the psychology literature on trilateration, under another name (Shepard 1962) is more advanced than the geodetic literature.

Two other examples of old topics in new guises stem from the subject of map projections. In the 1880s Francis Galton invented the geographical isochrone, a line connecting all points which can be reached in a given time. Isochronic maps are now quite popular, and the concept has been generalised to include travel costs (isotims). These are really geographical circles, recalling that a circle is the locus of points equidistant from a centre point. Measure distance in units of time and you have a geographical circle. But what curious circles. They have holes in them, and disjoint pieces, and the ratio of circumference to radius is hardly 2π. What a curious geometry – it makes Einstein seem simple! Or consider a set of concentric isochrones. Now draw in the orthogonal trajectories and one has the equivalent of a set of polar coordinates. Technically these are the polar geodesic coordinates of Gauss, familiar from differential geometry, and for which the metric takes on a particularly simple form. One can draw maps of this geographical geometry by using the usual ideas from the study of map projections, but the reference object is no longer a sphere or spheroid, rather it is more like a pulsating Swiss cheese. As a second map projection example, consider the problem of dividing the United States into compact cells, each of which contains the same number of people, and how it might be approached by a map projection. [...]

[...] One knows, a priori, that all maps that can be drawn by hand can also be drawn by computer-controlled devices. This follows from Turing's (1936) theorem. Of course *can* does not imply *should.* [...] The user of geographical data is, in principle, indifferent as to whether the data are on a geographical map or on magnetic tape. One of the principal uses of geographical maps is that of a graphical data storage device. But tapes are often more convenient than are drawings. These are simply two alternative methods of storing geographical information. We can assert that, when one has enough information on a magnetic tape to be able to draw a geographical map, one also has enough information on the tape to be able to solve all of the problems which could be solved using that map. But to store geographical information in electronic form in such a manner that a geographical map can be drawn requires that the substantive data be given geographical referencing. The usual procedure is to reference points, lines and areas by coordinates: geodetic latitude and longitude, or Gauss–Kruger coordinates, and so on (Maling 1973). So these must be treated in [...] analytical cartography [...]. But this is really too narrow a point of view. Call the fire department and announce that there is a fire in this room, at 48° 15′22″N, 16°23′l0″E. It would never be done; one would use the street address or the building name. But there are no cartography books that describe the street naming/numbering system. If I can locate a house using the street address then this label must contain exactly the same amount of information as does the latitude, longitude designation (Huffman 1952). The telephone area code number for Vienna, 0222, locates this place to about ±20 km. If I call from the United States to a phone in Vienna, I need to dial 12 digits, and these digits locate an area not much bigger than one square mile. If I know the postal code for Laxenburg, 2361, then I have specified a region to ±5 km.

Equivalently, Gauss–Kruger coordinates can be calculated from latitude and longitude $\varphi, \lambda \rightarrow G, K$ and this is invertible $G, K \rightarrow \varphi, \lambda$. Make a list of as many ways as you can recall of how locations are identified. Some will define points, others will refer to regions (Werner 1974; Clayton 1971). Now form a table by repeating this list in the orthogonal direction and consider this table as a transformer; place name to latitude and longitude, and the inverse, might be an example of two transformations. The concern in the literature with computerised address coding, the DIME system (Corbett 1973), point-in-polygon programs and so on, all relate to these transformations (Barraclough 1971). More exactly considered, coordinates are a way of naming places which, *inter alia*, allow all places to be given a unique name, and which allow relations between places to be deduced from their names. The North American telephone area codes, for example, have the property that if two area codes are similar, the places are most likely widely separated, and the converse. Thus the telephone area code scheme implicitly includes a relation between the places. From this relation one can use the trilateration procedure, already described above, to compute and draw a map of North America. In order to do this simply use the implicit relation and then interpret 'A is near to (or far from) B' as 'adjacent (or non-adjacent)', and then compute 2n coordinates from the set of these n (n – l)/2 adjacencies. For details see the delightful paper by Kendall (1971). The point of these remarks is that there is a great deal of theoretical structure buried in the topic of locational coding, and it has remained almost completely unexplored by theoreticians.

When one puts geographical information into a computer, one finds that it is extremely voluminous. A four-colour map of size 10×10 cm contains perhaps $100 \times 100 \times 4 = 40\,000$ discrete elements, not a great amount for a computer these days, but this is a small map. One can ask whether all of these elements are necessary. It is clear that not all of the $4^{10\,000}$ possible four-colour maps of this size can occur, because of the inherent geographical structure. One could throw away a goodly proportion of the 4×104 elements and still have a very useful map. I use the word 'goodly' for lack of a numerical estimate. Attempts have been made to apply information theory to this type of cartographic situation, but the bi-dimensional frequency statistics which seem needed are not generally available, except in the case of television pictures (Connelly 1968). [...] [T]hese topics lead naturally into the question of geographical map simplification. The condensed book, the overture to a musical work and a graphical caricature are all somewhat similar modifications of an original. A very related topic is also treated in economics and in sociology under the heading of optimal aggregation (Fisher 1969; Hannan 1971). [...]

As a simple example of a problem in the processing of geographical data, one can take the case of map overlays, for example, given a soils map of Michigan and a geological map of Michigan, find the logical intersection of the two. Conceptually this is not a difficult problem, and it can be done using a computer [...]. One has several choices in the way in which the regions are stored in the computer. These options become quite technical, but, for example, an area can be described (Meltzer *et al.* 1967) as $F(x,y) = 1$ if in R, 0 otherwise, or the boundary can be described as an equation $x(s) + iy(s) = z(s)$, or one can store the skeleton (Blum 1967) of the region. From all of these representations (and there are more), one can compute the area of a region, calculate whether or not a point lies inside of the region, find whether two regions overlap, and draw maps. It turns out, however, that some representations are more convenient for particular purposes than others, even though they are algebraically equivalent in the sense that each can be converted into all of the others (Palmer 1975). A comparison to two methods of solving a pair of simultaneous linear equations may be appropriate. If one remembers a bit of algebra and that linear equations have the form:

$$Y1 = A1 + B1X$$

$$Y2 = A2 + B2X,$$

then the intersection point can be found by the simultaneous solution of these equations. But sometimes it is easier to plot the lines and to read the coordinates of the intersection from the drawing. Many uses of maps are of this nomographic nature. Recall that Mercator's projection, for example, provides a graphic solution to the problem of finding the intersection angle between a North–South great circle and a logarithmic spiral on a sphere. Or consider the problem of finding the nearest gas station when the automobile gauge indicates 'nearly empty'. With the entire street system stored in the core of a pocket calculator, and the location of all the gas stations that accept my credit card also stored, and with a minimal path algorithm (Gilsinn and Witzgall 1973), which is efficient for networks of this size, it seems like a trivial computation.

What is easy, convenient, or difficult depends on the technology, circumstances and problem. The teaching of cartography must reflect this dynamism, and the student can only remain flexible if he has command of a theoretical structure as well as specific implementations. The spirit of Analytical Cartography is to try to capture this theory, in anticipation of the many technological innovations which can be expected in the future; wrist watch latitude/longitude indicators, for example, and pocket calculators with maps displayed by coloured light emitting diodes, do not seem impossible. [...].

References

Barraclough, R. (1971) Geographic coding, in *Federal Statistics Report*, US President's Commission on Federal Statistics, Government Printing Office, Washington, DC, pp. 219–296.

Blum, H. (1967) A transformation for extracting new descriptors of shape, in *Models for the Perception of Speech and Visual Form* (ed. W. Wathen-Dunn), MIT Press, Cambridge, MA, pp. 362–380.

Clayton, K. (1971) Geographical reference systems. *Geographical Journal*, **137** (1), 1–13.

Connelly, D. (1968) The coding and storage of terrain height data: an introduction to numerical cartography. Unpublished MSc thesis. Cornell University, Ithaca, NY.

Corbett, J. (1973) The use of the DIME system for identification of places and description of location, in *Land Parcel Identifiers for Information Systems* (eds K. Fisher and D. Moyer), American Bar Foundation, Chicago, pp. 149–155.

Duda, R. and Hart, P. (1973) *Pattern Classification and Scene Analysis*, John Wiley & Sons, Inc., New York.

Eckert, M. (1921) *Die Kartenwissenschaft*, de Gruter, Berlin/Leipzig, Germany.

Fisher, W. (1969) *Clustering and Aggregation in Economics*, The Johns Hopkins University Press, Baltimore, MD.

Gilsinn, J. and Witzgall, C. (1973) *A Performance Comparison of Labeling Algorithms for Calculating Shortest Path Trees*, National Bureau of Standards, Washington, DC.

Gould, P. and White, R. (1974) *Mental Maps*, Pelican Books, Harmondsworth, UK.

Grafarend, B. and Harland, P. (1973) Optimales design geodätiseher netze, in *Höhere Geodäsie*, Heft Nr. 74, Deutsche Geodätische Kommission, Munich.

Graur, A. (1956) *Matenaticheskaya Kartografia*, Leningrad University Press, Leningrad.

Hannan, M. (1971) *Aggregation and Disaggregation in Sociology*, Heath Books, Lexington, KY.

Huffman, D. (1952) A method for the construction of minimum redundancy codes. *Proceedings of the Institute of Radio Engineers*, **40** (10), 1098–1101.

Kendall, D. (1971) Construction of maps from odd bits of information. *Nature*, **231**, 138–159.

Maling, D. (1973) *Coordinate Systems and Map Projections*, G. Philip, London.

Meltzer, B., Searle, N. and Brown, R. (1967) Numerical specification of biological form. *Nature*, **216**, 32–36.

Palmer, J.A.B. (1975) Computer science aspects of the mapping problem, in *Display and Analysis of Spatial Data* (eds J. Davis and M. McCullagh), John Wiley & Sons, Inc., New York, pp. 155–172.

Robinson, A.H. and Petchenik, B.B. (1975) The map as a communication system. *The Cartographic Journal*, **12** (1), 7–15.

Shepard, R. (1962) The analysis of proximities. *Psychometrika*, **27**, 125–140, 219–246.

Shepard, R. (1966) Metric structures in ordinal data. *Journal of Mathematical Psychology*, **3**, 287–315.

Solovyev, M. (1969) *Matematichcskaya Kartografiya*, Nedra, Moscow.

Tobler, W. (1959) Automation and cartography. *Geographical Review*, **49** (4), 526–534.

Tobler, W. (1976) The geometry of mental maps, in *Essays on the Multidimensional Analysis of Perceptions and Preferences* (eds R. Golledge and G. Rushton), Ohio State University, Columbus, OH.

Turing, M. (1936) On computable numbers, with an application to the Entscheidungsproblem. *Proceedings of the London Mathematical Society 2nd Series*, **42**, 230–265.

Werner, P. (1974) National geocoding. *Annals of the Association of American Geographers*, **64** (2), 310–317.

Further reading

Burrough, P.A. (1986) *Principles of Geographical Information Systems for Land Resources Assessment*, Clarendon Press, Oxford. [One of the earliest monographs exploring what was to become GIScience and clearly informed by Tobler's analytical ideas.]

Buttenfield, B.P. and McMaster, R.B. (1991) *Map Generalization: Making Rules for Knowledge Representation*, Longman Scientific, New York. [The standard work on generalisation strongly influenced by Tobler's work on algorithmic processing of cartographic data.]

Clarke, K.C. (1995) *Analytical and Computer Cartography*, 2nd edn, Prentice Hall, Englewood Cliffs, NJ. [A textbook, following Tobler's lead, that provides a useful introduction to standard practices of computer-based cartography of the 1990s.]

Clarke, K. and Cloud, J.G. (2000) On the origins of analytical cartography. *Cartography and Geographic Information Sciences*, **27** (3), 195–204. [This research paper traces the social and intellectual influences underpinning Tobler's work and situates these in the context of cold war North American geopolitics.]

Moellering, H. (2000) The scope and conceptual content of analytical cartography. *Cartography and Geographic Information Science*, **27** (3), 205–224. [An insightful paper which defines the lasting impact of Tobler's ideas and ongoing research interest in the field.]

See also

- Chapter 1.4: The Science of Cartography and its Essential Processes
- Chapter 1.11: Exploratory Cartographic Visualisation: Advancing the Agenda
- Chapter 2.2: A Century of Cartographic Change
- Chapter 2.5: Automation and Cartography
- Chapter 2.6: Cartographic Futures on a Digital Earth
- Chapter 2.7: Cartography and Geographic Information Systems
- Chapter 2.9: Emergence of Map Projections
- Chapter 3.4: Generalisation in Statistical Mapping
- Chapter 3.5: Strategies for the Visualisation of Geographic Time-Series Data

Chapter 1.6

Cartographic Communication

Christopher Board

Editors' overview

Published in the middle of the so-called 'Quantitative Revolution' in geography, Board's paper represents a sustained attempt to establish a set of theoretical principles that might validate the place of mapping in the discipline. This is one of the most complete published discussions of cartographic communication as a paradigm. He traces the development of 'cartography as communication' from Arthur Robinson's formulations onwards, drawing on contributions from North America and various European scholars. Board explores increasingly sophisticated developments of initial information flow models and evaluates the processes underpinning these attempts to understand mapmaking and use, focusing upon semiological research on map language. His paper makes useful comments, in particular, on human factors inflected work before drawing normative conclusions about map use and design in geography.

Originally published in 1972: *Cartographica*, **18** (2), 42–78.

Geography, map use and cartographic communication

[...] Are not all maps expected to communicate geographical information in one way or another? If so, the term cartographic communication would appear to be somewhat tautologous. One might reply that cartographic communication is only one form of communication and that it emphasises not only the medium but both the initiator and receiver of the information being communicated. It emphasises a process rather than a product and it certainly says more than cartography *tout court* does or than most of the current definitions of the field do.

It used to be accepted that whatever else it was and, however one defined it, geography was about maps. For long, Hartshorne's assertion that geography was 'represented in the world of knowledge by its techniques of map use' (1939: 463–464) and that it depended 'first and fundamentally on the comparison of maps' went unchallenged. However, Bunge's influential *Theoretical Geography* argued that despite the advantages of maps as spatial tools 'mathematics is the broader and more flexible medium for geography'. This in the middle of the so-called quantitative revolution may have been just what the 'new geographers' were looking for. To some extent the writings of geographer/cartographers such as J.K. Wright and Arthur H. Robinson had unwittingly helped to undermine the geographer's confidence in his classical technique by stressing the subjective qualities of maps. It was assumed, in line with Wooldridge and East, that the map was 'pre-eminently the geographer's tool both in investigation of his problems and the presentation of his results' (1951: 61). Furthermore, it was generally accepted that 'no one claiming the title of geographer, however humbly, is entitled to be ignorant of how maps are made' (1951: 70). Even when *Maps and Diagrams* was first published, apart from a reference in the preface to the map being the traditional medium of the geographer (Monkhouse and Wilkinson 1952), it was clearly assumed that all geographers were required to include map work in their training and the thematic arrangement of the book was therefore aimed principally at the topics with which geographers would be concerned. By the 1960s this notion was being challenged by the shock troops of the quantitative revolution, or, as Duncan and others put it, 'those who are interested in "statistical geography" and who are dissatisfied with the level of rigour of inferences based

The Map Reader: Theories of Mapping Practice and Cartographic Representation, First Edition. Edited by Martin Dodge, Rob Kitchin and Chris Perkins.

solely on maps – the mainstay of the "classical" geographer' (1960: 15). Nevertheless, although Berry explains that four maps do about the same job as 43 for portraying the basic similarities in the development of countries world-wide by means of factor analysis, the resulting patterns were nevertheless displayed by *maps* (1961: 110). [...]

[...]

In her classic book on geographical map interpretation, Alice Garnett expressed the nature of the relationship between reality and the map in the geographer's mind thus: '... to gain even a slight perception of what the map truly portrays the mind must be enriched by contact with, for example, true mountain, plain or fenland, with rural life and scattered farmstead opposed to the unending maze of streets within large industrial centres. Without this our mental equipment possesses no standard or scale of associations or of space relations whereon to base impressions of other regions seen only through map reading' (1930: 25).

Textbooks and manuals on the geographical interpretation of landscape now abound [...]. Considering their long experience in using maps, particularly topographical ones, it is perhaps rather strange that more geographers have not contributed to the discussion of cartographic communication. Such concerns have been left to the cartographers to take up. [...]

Cartography as a science of communication

One of the first to explore the idea of cartography as a science of communication was Robinson (1960: v) [...] He saw the primary process of cartography as 'the conceptual planning and design of the map as a medium for communication or research'. [...]

Without question the cartographic equivalent of the professional communication expert has become the technically trained and equipped geographer involved in designing thematic maps for special purposes. This was certainly not always the case, for Robinson pointed out that the artist was better able than geographer-cartographers to solve the design problem for many special purpose maps required in the Second World War (1960: viii). Even further back, thematic mapping, at least of census data, appears to have been the job of the graphic-statistician like Levasseur (1885). During the 1960s the emergence of a more self-confident and distinctive cartography, characterised by professional status, research and its own publications, has been attributed to the influence of the view of the thematic map as a medium of communication (Morrison 1974a: 13). However, for the time being, topographic mapping tended to be left out of this new view of cartography as 'a science, or at least part of the science of graphic communication' (Morrison 1972: 22).

Emboldened by the tacit and increasingly explicit support of other workers, Robinson and Petchenik (1975: 14) felt able to write that 'mapping is basically an attempt at communication between the cartographers and the map percipient' (the one who, by viewing a map, augments his understanding of the geographical milieu). Such a view embraces general (including topographic) and thematic maps. Similar views were adopted by Ratajski, for whom cartography was 'a field of human activity comprising the *creation* and *functioning* of all forms of cartographic transmission' (1970: 97–110).

Morrison has more positively and forcefully argued for a science of cartography (1976a: 84–97) [...]. He defines it as 'the science of communicating information between individuals by the use of a map' (1976a: 97).

Salichtchev also regards cartography as a science but in a rather different way from Morrison. For him cartography is much broader than that suggested by the communication science paradigm, which is criticised as narrow and technical. In 1970, Salichtchev reproduced his 1967 definition of cartography as 'the subject of the spatial distribution, combinations, and interdependence of nature and society (and their changes in time) by means of representation through a special symbolic system – cartographic symbols' (1970: 83). [...] Salichtchev (1973) not only considers that the communication science approach to cartography is narrowly technical but is inseparable from the mathematical theory of information. That, he argued, cannot explain the increase in information which occurs in map interpretation, something which Morrison's papers in fact allowed for, but which he set partly outside cartography as he defined it (1974b: 120; 1976a: 89, 93). For Morrison the science of cartography begins with the 'cognitive realm of the cartographer' or with his conception of the geographical space and its contents that he is mapping. It is the detailed study of a communication channel and, as such, strictly excludes all those processes (map reading, interpretation and analysis) taking place in the cognitive realm of the map reader. Whether we adopt this narrow view or not, it is obvious even from what Morrison wrote that there is a constant flux of information passing between the mind of the cartographer and the real world he is mapping or the map reader for whom the map is intended; and, at times, cartographers enlarge their own cognitive realms by the very process of mapmaking.

Thus while Salichtchev's definition of cartography appears to resemble geography through maps, Morrison's more nearly approaches the Austrian views of Arnberger and Kretschmer of cartography as a formal science (Formalwissenschaft) like mathematics, as opposed to geography which is categorised as an object-oriented science (Objektwissenshaft) (Arnberger 1970: 1–28).

[. . .]

In France, although Moles had firmly regarded cartography as a particular case of the theory of messages or theory of information (1964: 11), Bertin preferred to regard it as a method of graphic representation, itself one code employing a system of signs or symbols and, hence, a part of semiology (1966). By the mid 1970s it was possible for Robinson and Petchenik to assert 'cartography has opened wide its arms to welcome the concept that it is a communication system' (1976: 25). [. . .]

Attempts to model cartographic communication

The last fifteen years have witnessed a proliferation of models of the process of cartographic communication. [. . .] Keates was apparently the first specifically to identify and name the concept of cartographic communication (1964: 9), to link it with information theory, referring to coding, decoding and the noise affecting the communication system in a cartographic context. [. . .]

In France, also in 1964, Moles had independently published a paper relating the theory of information to the transmission of cartographic information, thus foreshadowing many of the notions found in subsequent attempts to model systems of cartographic communication. This was apparently quite separate from efforts by cartographers at that time, but Bertin acknowledges criticism and advice received from Moles in the preparation of *Sémiologie Graphique* (1967). Whereas Moles's research appears rooted in information theory, psychology and linguistics, Bertin's approach seems more that of the practical cartographer. Notwithstanding the title of his book, the concept of semiology is employed only as a general context and is never itself the subject of a diagram or graphical model.

The essence of the approach first appeared in 1966 and was abbreviated as the introduction to the book itself. Its main significance derives from recognition that graphic representation, whether by maps or diagrams, was part of the wider study of semiology, the science that dealt with systems of signs. Distributed throughout the profusely illustrated text are various rules of construction and of legibility. However, these are similar to the principles of visual design referred to by Robinson (1960: 14–15). [. . .]

Information flow models

Although De Lucia's review (1978) of models of cartographic communication designated my model of the communication process an engineering-like model, it was put forward as Robinson and Petchenik correctly state as a broad analogy (1976: 27). Both that model, which was derived essentially from communication theory, and the more complex model attempt to portray a flow of information (Figure 1.6.1) (Board 1967; Freitag 1980). The latter model hints at a progressive loss of information in both mapmaking and map analysis. Furthermore, this model described a cyclical process which includes a new view of the real world much the same as Koláčný's (1968) indicated a discrepancy between the cartographer's and map user's reality (Figure 1.6.2). In the Map–Model Cycle, however, the discrepancy was not generalised, being split between noise generated by the mapmaker and map user, and the deliberate pursuit of scientific analysis aimed at explaining something from mapped patterns.

Whereas the map–model cycle was closely related to the new paradigm in geography, Koláčný's was developed quite independently and stems from a decade of research especially with school atlases. Although the focus and top of Koláčný's diagram is 'reality', with the intersecting subsets cartographer's reality and map user's reality, the text makes it plain that one must start with the cartographer who observes and selects either directly the 'geographical medium', the concrete world of the earth's surface and/or indirectly maps and other source materials of the same area. In simplifying Koláčný's diagram to eliminate the reciprocal relationship between cartographer's reality and the contents of the cartographer's mind and by re-labelling cartographer's reality 'mapped part of reality', Salichtchev (1978: 108–109) did less than justice to Koláčný's.

There is no reason why the cartographer's reality should not include both the real world, of which he has direct knowledge, and the maps and other data, which contribute indirectly. Salichtchev's re-interpretation of Koláčný's model is more radically different when it refers to map consumption or map use. Here, Salichtchev's processes of map study and the formation of ideas about reality derived from the information extracted by the map reader resemble the right-hand part of the map–model cycle, which is concerned with testing the model. Salichtchev frequently and correctly stresses the importance of this process and the increase of information which results. 'Scientific mapmaking, understood as a process of modelling, always aims at a deeper understanding of that part of reality being investigated and the obtaining of new knowledge about it, that is, new information' (1978: 109). Koláčný fundamentally differs, since he refers to the map user's reality being widened. This exchange of views demonstrates how useful such models as Koláčný's have been, for they focus attention on relationships and processes rather than merely on the products of those processes. Also, by getting away from the essentially linear form of the standard or generalised communication systems in particular by building in feedback, models such as

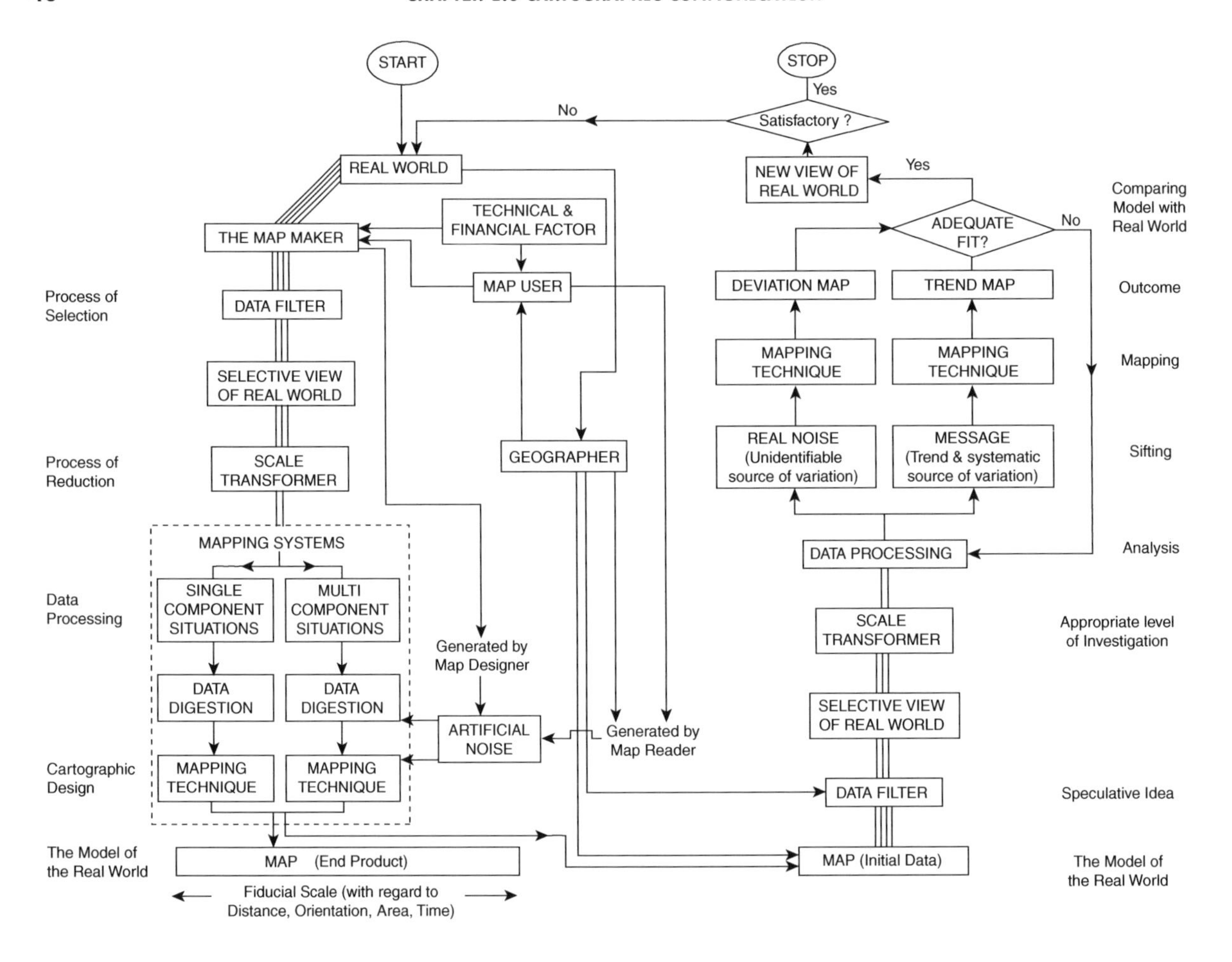

Figure 1.6.1 Board's (1967) Map-Model Cycle.

this have greatly helped both mapmakers and map users to appreciate one another's problems.

Such exercises also provide a very valuable link between the more theoretically minded and the more practically minded cartographers. [. . .]

Two further and completely independent works, each of which generated further model building, occurred in the United States (Dornbach 1969) and Georgia in the Soviet Union (Aslanikashvili 1967, 1968). It is quite remarkable that the mid 1960s yielded so many attempts to bring some order to understanding the process of cartographic representation and map use. It is fashionable to suggest that such a need arose from the very diverse and proliferating literature on psycho-physical experiments to do with maps, but that would be being wise after the event. For those who were connected with the exciting movements in geography, the atmosphere was right and to an extent the temptation to emulate and imitate other geographers must have been significant. [. . .]

In historical terms the second generation of models is well represented by Ratajski's (1970, 1972) model of cartographic transmission, which acknowledges the models of 1967–1968. Ratajski's model incorporates an interesting notion, by the angle between the centres of the circles which represent reality, the source of information and the imagining of reality. He called it 'degree of transmission correctness' (1972: 65) and in effect it measures a loss of information. Ratajski spells out in more detail some of the factors affecting the transmission of information and expresses many of the concepts in mathematical notation. Finally, the diagrammatic model of cartographic transmission provides the basis for a model of what he called cartology and then of cartography itself. Cartology was for him, in effect, a system of theoretical cartography and studied the process of transmitting spatial information by maps and the relationship between cartography and sciences in general. [. . .]

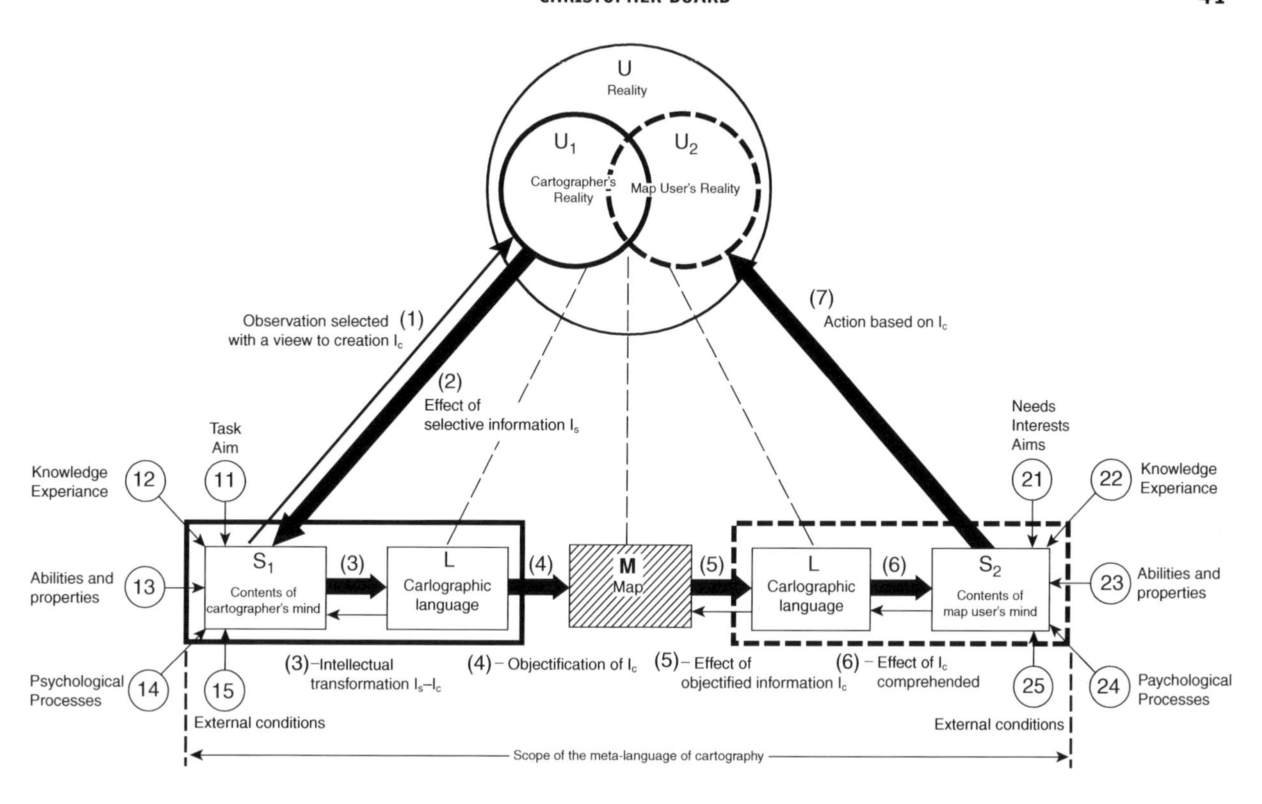

Figure 1.6.2 Board's model of 1967 contrasted with that of Koláčný (1968).

Among mapmakers, models such as Koláčný's were received with some scepticism and were regarded as too complex (Bartholomew and Kinniburgh 1973). But although they may have preferred Koeman's formulations (1971), they nevertheless quote frequently enough from Koláčný's to suggest that he had struck the right chords and might have given two cartographers in one of the world's most famous mapmaking establishments cause to re-evaluate the process of designing maps for different makers. In Germany the first substantive reaction to the first wave of model building was Freitag's paper (1971), which led on to further development by Hake (1973). Freitag's schematic model incorporates time and may therefore be included among models of the process of cartographic communication (1971: 181). It also links such models with the notion of cartography as a language by identifying those parts which may be regarded as syntactics, pragmatics and semantics (Figure 1.6.3). In the accompanying text Freitag carefully explains the relationship between semiology and cartography, going much further than Bertin (1966). Although the model provides for a continuous repetition of mapmaking and map using returning to reality from which new data are obtained for the next cycle, there is no explicit reference to a discrepancy between the mapmaker and map user's views of reality. [...] Hake brought automation into the mapmaking process in his model (1973: 140) but also makes use of semiological concepts in his interesting decision making approach to the achievement of symbol perception through map use (Figure 1.6.4) (1973: 146).

[S]ome explanation of terms is required at this point in order to realise the significance of Freitag's and Hake's contributions. Within semiotics, the theory of signs, three overlapping fields are generally recognised:

Pragmatics is concerned with the relationships between signs and their users. For cartography it is largely the empirical study of the uses and effects of symbols on maps.

Semantics is concerned with the relations between signs and the designations of signs or what is represented by signs. This is to do with the meanings of symbols on maps in terms of geographical and other concepts about the real world, which is the subject of the map.

Syntactics looks at the relations between signs themselves and concerns rules abstracted from users of signs and real world environments. Some map specifications have

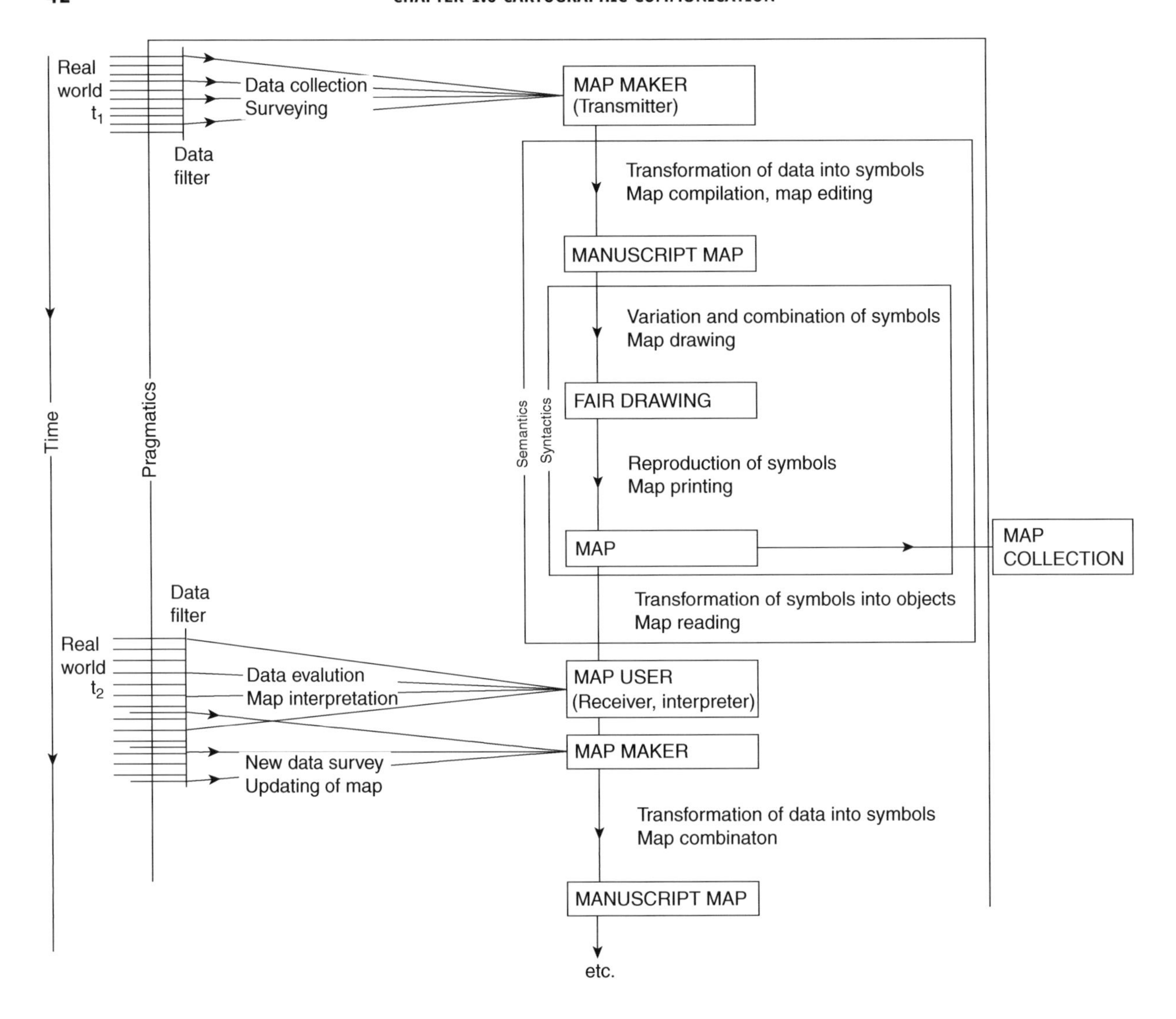

Figure 1.6.3 Freitag's schematic model of the process of cartographic communication. (Translated from the original diagram in Freitag, 1971.)

a strong syntactic component in that relatively arbitrary rules are set down for the relationships between symbols such as placing the smaller of two circular symbols which must 'overlap' 'on top of' the larger. Rules for positioning lettering also come into this category (Cherry 1961; Dacey 1970; Board 1973).

In Hake's model, which is one of the earliest to detail map using activities, the goal of map use has a prominent and crucial place, allowing the map user either to search actively or to browse (passive perception). After the stage of symbol perception, understanding symbol structure is labelled syntax, which is followed in turn by a succession of activities (1) relating symbols with objects, (2) explaining them using the legend, and (3) explaining them using other aids, all leading to comprehending the significance or meaning of the symbols (labelled semantics). Pragmatics is reserved for the stage of information processing in the form of map reading and map measurement (map analysis or cartometry), at which stage the full significance of symbol patterns is related to the features and phenomena of the real world. At the end of the flow diagram if no further search is undertaken, other non-cartographic activities (e.g. way finding) may take place.

A third generation of models of cartographic communication may be identified [...] culminating in Ratajski's paper (1978) which summarised the field as he saw it in 1977.

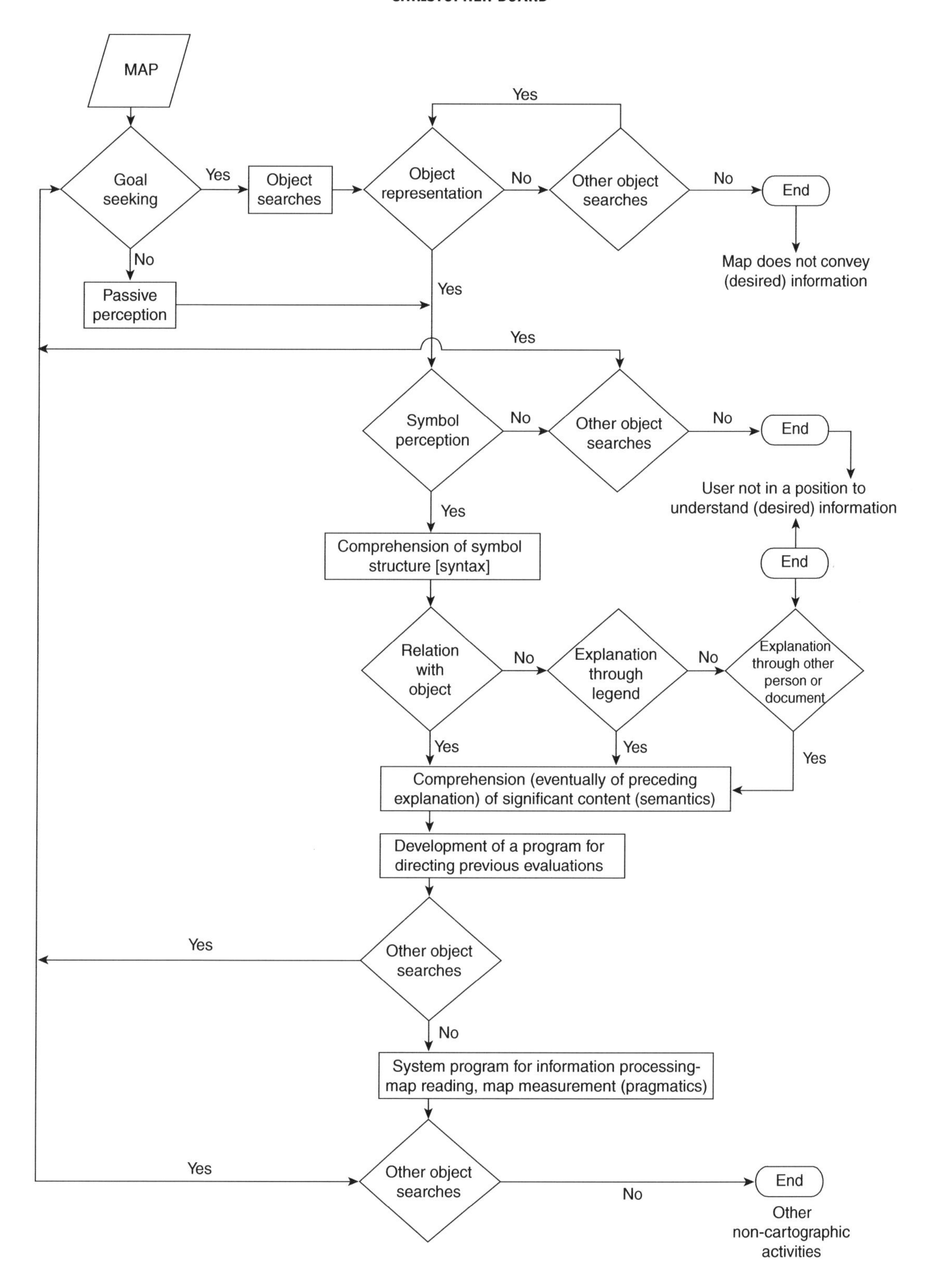

Figure 1.6.4 Process of the recognition of symbols in map evaluation. (Translated from Hake, 1973: Figure 7.)

[...] De Lucia relates the models to three levels of communication problems: A – the technical; B – the semantic; C – effectiveness, and upon these bases a typology of models. Most of those he selects for review belong to level A, concerned with how accurately can the symbols be communicated. [...] The second group, which De Lucia does not review, are those which incorporate the basic ideas of the information flow models, include the notion of discrepancy between what information is disseminated and what information is received but which also go on to generalise about the meaning of the symbols communicated. This last characteristic stems chiefly from attempts to model map reading processes. [...]

Extended information flow models

One further illustration of the development of one group of ideas is provided by American papers, which examine the stages in communicating geographical information by means of maps. Convinced of the necessity to distinguish between *mapmaking* (the practical side of map production) and *cartography*, Muehrcke characterises the latter as 'philosophical and theoretical bases, principles and rules for maps and mapping procedures', and by 'conceptual, problem-oriented research directed at formalising the *science* underlying the *art* of cartography' (1972: 1). Having noted the increasing use of communication jargon in the cartographic literature, Muehrcke prefers to regard cartographic processing as an 'active feedback system' involving three transformations:

(1) Data collection – Real world to Raw Data

(2) Mapping – Data to Map

(3) Map Reading – Map to Map Image

with a feedback from the Map Image to the Raw Data (Muehrcke 1972). This last is a process of evaluating the accuracy of the map image. [...] By 1973 Morrison had developed a critique of the simple model first published by Muehrcke which he elaborated in later years (Morrison 1973, 1976a, 1976b). Building on the essential notion of transformations he shows that those postulated by Muehrcke's model are much more complicated than they first appear, the second being composed of two functions and the third, properly regarded as one function, is not comparable to the former transformation. Morrison expressed these notions in set-theory terms and explored the stages involved in the process of cartographic generalisation which corresponded to the second of the two functions involved in the first transformation – data sensed by the cartographer to be included in the map he is making. The three processes involved in generalisation – classification, simplification and symbolisation – are shown by Venn diagrams clearly implying a progressive loss of information (1974a: 121). A fourth process, induction, is summoned up from Robinson and Sale's text on cartography (1976: 52), but this is almost an accidental by-product of the map-making process. [...]

Identifying some of the processes in models of cartographic communication

For Dury, map reading was like learning a new language, whereas map interpretation was beginning to speak it and involved 'a synthesis in which complex ideas are deduced and combined from simple ones' (1952: 2). He saw map analysis as a preliminary activity, tracing elements of the total pattern appearing on a topographic map. As such, this mirrors the approach adopted by French geographers who, when discussing thematic maps, clearly distinguish between analytical mapping and synthetic mapping (Claval and Wieber 1969; Rimbert 1968). The former isolate elements or components of the physical, cultural or economic landscape and one reads them as one would look at classes of elements in a painting. Synthetic maps demonstrate holistically the relationships between these elements and one interprets these by comprehending the complex *ensemble*, the structure of the pattern, much as one would appreciate the composition of a painting (Claval and Wieber 1969). Morrison's view of map interpretation involves the interaction between information communicated from the map with that previously in the map reader's cognitive realms [...]. However, Morrison has an unconventional view of map analysis which 'refers to the resolution of any conflicting information or the substantiation of new information that forces the map reader to return to reality' (1976a: 95). He thus regards verification as map analysis. [...] Robinson and Petchenik (1975: 45) found [...] that cartographic practice differed, but suggested two new terms, *map viewer* and *map percipient*, to distinguish between the person who merely views a meaningless graphic array from someone who sees a meaningful figure or pattern on a map. The difficulty of drawing the distinction *in practice* nevertheless remains. Few geographers would be able to look at a map without seeing a meaningful pattern somewhere upon it and one would like to think that this capability grew with training and experience. It is, nevertheless, useful to recognise and maintain the distinction and perhaps to name each process. The fact that map viewing and perception are often virtually simultaneous and inseparable merely illustrates the limitations of modelling the process of cartographic communication.

Finally in this section we should bear seriously in mind Guelke's criticism of a communications model which does not concern itself with the *meanings* of the configurations viewed on maps (1976). As Morrison pointed out, map interpretation takes place within the cognitive realm of the map reader and not in the communication channel (1976a: 93). A narrow view of communication might, therefore, exclude the assignment of meaning and significance to what was viewed on a map. Here perhaps Petchenik's characterisation of a thematic map as being one 'whose meaning is experienced as *knowing-about-space*' is useful (1976). This depends upon a locational base which at its most primitive level enables the map reader *to place* the information telling him about a particular segment of geographical space. This may require no more than a numbered grid or graticule, named relative location or a recognisable coastline. Guelke's warning that cartographers should not, by becoming specialists in graphic communication, neglect the geographical context, is well taken. [. . .]

[. . .]

Cartographic style and communication

One comparatively neglected aspect of cartographic communication which can affect the success with which the message is put over is cartographic style. Although the concept is commonly applied to lettering which has a special quality that conveys a distinctive character to a map, there is no question that not only official topographic maps but coloured thematic maps and even black and white illustrations in geographical texts demonstrate different styles. [. . .] Rimbert introduced the subject of cartographic style in the context of the subjectivity of the individual mapmaker but pointed out that imitation by beginners from the examples placed before them was more influential than individual differences (1968). Hence, one can recognise a newspaper or a magazine style, that of an urban planner or advertiser, or a Swiss style. [. . .] Petchenik suggested that the 'look' of entire maps, rather than their elements, be examined through the semantic differential (1974). The scales developed from word pairs with opposite meanings helped her, as a cartographic designer of an atlas, to set out her objectives more clearly. They were also useful in comparing whole maps and defining house styles or the styles of particular periods of the history of cartography. There are instances in Western Europe where one can compare the styles of up to three different agencies for the same piece of terrain. The different appearance of the maps at the same scale is imparted by specification and house style and must convey quantitatively if not qualitatively different information (Meine 1974: 405). As Petchenik pointed out, little attention has been given to the total appearance of entire maps. If one continues to analyse a map's elements separately, there is a risk that the nature of cartographic style and its influence on communication of geographical information will be ignored. [. . .]

[. . .] In J.K. Wright's classic essay (1942) cartographic style is never far from his thoughts and is exemplified best by the parallel he draws with verbal communication. 'The trim, precise and clean-cut appearance that a well drawn map presents lends it an air of scientific authenticity that may or may not be deserved. A map may be like a person who talks clearly and convincingly on a subject of which his knowledge is imperfect'. If Wright was correct to argue that an ugly version were less likely to inspire confidence, despite its being as intrinsically accurate as a beautiful version of the same map, cartographic communicators should pay more attention to the aesthetic aspects embraced in the concept of style.

Some who have investigated the effectiveness of maps have argued forcibly that likes and dislikes, preferences for and attitudes to particular maps influence their acceptability. Hopkin and Taylor express concern that objective measures of performance on maps required by ergonomic studies will overlook holistic, aesthetic assessments in favour of objective measures of, for example, information density (1979: 182–183). They refer to work which suggests that individuals can satisfactorily assess information density subjectively and go on to suggest that maps might be improved by reducing the subjective impression of information density while retaining the original quantity possibly by de-emphasising some of the detail. [. . .]

[. . .]

The human factors approach in cartographic communication

[. . .] '[T]he need for efficient communication of map information to meet the needs of the user was recognised in aviation cartography long before it was recognised in other areas of cartography' (Taylor and Hopkin 1973: 87). [. . .] The extension of ergonomic principles to map design for various aspects of aviation has been prompted by the very practical requirements of highly specialised map users whose technological environment has been changing extremely rapidly. It owes little to the growth of the theory of cartographic communication, yet the pragmatism of the human factors approach provides the best opportunity yet for testing the models and theories which now abound. If nothing else, the growth of communication theory has drawn attention to factors affecting the loss of information in map design and to the ways in which people use maps.

Had geographers in general retained their once very flourishing interest in map use, their experience might have been useful in applying *ergonomic* principles to map design. However, [...] map techniques were exchanged for quantitative techniques and, furthermore, geographers themselves were never very clear or explicit about what information they expected to acquire by using maps.

[...]

Implications for geography

The relevance of cartographic communication for improving map design is, or should be, obvious, but it is less easy to identify the links between cartographic communication and geography. The professionalisation of cartography has occurred roughly simultaneously with the adoption of the scientific paradigm within geography. Cartographers have struggled to establish an identity separate from geographers, explorers or surveyors. Geographers reacting to jibes about colouring maps, the banality and naiveté of cartography can be forgiven for forsaking conventional maps. For their part, cartographers, especially those in the area of cartographic communication, have not made as much as they might have done of mental and cognitive mapping. [...]

[...]

Too often maps are poorly designed for use in textbooks, theses or reports without much consideration for their authors' objectives, if indeed these are ever properly worked out. Correspondingly little is known of the use to which these maps are put (discussion with Jenks 1978). [...] One suspects that too frequently the spatial patterns and relationships there portrayed are transferred from the source to the new document without any adequate mental analysis by its author. If the principles of cartographic communication are to have any real significance for geographers, they should throw light on the nature of map use in geography and lead ultimately to improving map design in order to increase the flow of information on and appreciation of the nature of geographical space.

References

Arnberger, E. (1970) Die kartographie als wissenschaft und ihre beziehungen sur geographie und geodäsie, in *Grundsatzfragen der Kartographie* (ed. E. Arnberger), Osterreischischen Geographishen Gesellschaft, Vienna.

Aslanikashvili, A.F. (1968) *Metakartografiya Osnovniye Problemi*, Metsniereba, Tbilisi.

Bartholomew, J.C. and Kinniburgh, I.A.G. (1973) The factor of awareness. *The Cartographic Journal*, **10**, 59–62.

Berry, B.J.L. (1961) Basic patterns in economic development, in *Atlas of Economic Development* (ed. N. Ginsburg), University of Chicago Press, Chicago.

Bertin, J. (1966) Le langage graphique et la cartographie. *Bulletin du Comité Français de Cartographie*, **28**, 60–69.

Bertin, J. (1967) *Sémiologie Graphique*, Gauthier-Villars, Paris.

Board, C. (1967) Maps as models, in *Models in Geography* (eds R.J. Chorley and P. Haggett), Methuen, London.

Board, C. (1973) Cartographic communication and standardization. *International Yearbook of Cartography*, **13**, 229–38.

Cherry, C. (1961) *On Human Communication: A Review, a Survey, and a Criticism*, John Wiley & Sons, Inc., New York.

Claval, P. and Wieber, J.C. (1969) La cartographie thématique comme méthode de recherche. *Les Belle Lettres, Cahiers de Géographie de Besancon*, **18**, 8–103.

Dacey, M.F. (1970) Linguistic aspects of maps and geographic information. *Ontario Geography*, **5**, 71–80.

De Lucia, A. (1978) Models of cartographic communication: a survey. Paper presented to the Canadian Cartographic Association, Vancouver.

Duncan, O.D., Cozzort, R.O. and Duncan, B. (1960) *Statistical Geography: Problems in Analyzing Areal Data*, The Free Press of Glencoe, Illinois.

Dury, G.H. (1952) *Map Interpretation*, Sir Isaac Pitman and Sons, London.

Freitag, U. (1971) Semiotik und kartographie: über die anwendung kybernetischer disziplinen in der theoretischen kartographie. *Kartographische Nachrichten*, **21**, 171–182.

Freitag, U. (1980) Can communication theory form the basis of a general theory of cartography? *Nachrichten aus dem Karten – und Vermessungswesen*, **38**, 17–35.

Garnett, A. (1930) *The Geographical Interpretation of Topographical Maps*, George G. Harrap & Co, London.

Guelke, L. (1976) Cartographic communication and geographic understanding. *The Canadian Cartographer*, **13**, 107–122.

Hake, G. (1973) Kartographie und Kommunikation. *Kartographische Nachrichten*, **23**, 137–148.

Hartshorne, R. (1939) *The Nature of Geography*, Association of American Geographers, Lancaster, PA.

Hopkin, V.D. and Taylor, R.M. (1979) *Human Factors in the Design and Evaluation of Aviation Maps*, NATO, Neuilly-Sur-Seine, France.

Keates, J.S. (1964) Cartographic communication, in *Abstracts of Papers* (ed. F.E.I. Hamilton), Nelson, London.

Koeman, C. (1971) The principle of communication in cartography. *International Yearbook of Cartography*, **11**, 169–176.

Koláčný, A. (1968) Cartographic information – a fundamental concept and term in modern cartography. *The Cartographic Journal*, **6** (1), 47–49.

Levasseur, E. (1885) La statistique graphique. *Statistical Society of London*, **50**, 218–250.

Meine, K.H. (1974) *Kartographische kommunicationsketten und kartographisches alphabet: ein beitrag zur theorie* der kartographie, Mitteilungen der Österreichischen Geographischen Gesellschaft.

Moles, A.A. (1964) Théorie de l'information et message cartographique. *Science et Enseignement des Sciences*, **5**, 11–16.

Monkhouse, F.J. and Wilkinson, H.R. (1952) *Maps and Diagrams*, Methuen, Boston.

Morrison, J.L. (1973) Towards a schematization of the cartographic process. Paper presented to the NATO Advanced Study Institute on Design and Analysis of Spatial Data, Nottingham, UK.

Morrison, J.L. (1974a) Changing philosophical-technical aspects of thematic cartography. *Proceedings of Sixth International Cartographic Conference*, Ottawa, pp. 13–22.

Morrison, J.L. (1974b) A theoretical framework of cartographic generalization with emphasis on the process of symbolization. *International Yearbook of Cartography*, **14**, 115–127.

Morrison, J.L. (1976a) The science of cartography and its essential processes. *International Yearbook of Cartography*, **16**, 84–97.

Morrison, J.L. (1976b) The relevance of some psychophysical cartographic research to simple map reading tasks. Paper presented at 8th Technical Conference of ICA, Moscow.

Muehrcke, J.P. (1972) *Thematic Cartography*, Association of American Geographers, Washington, DC.

Petchenik, B.B. (1974) A verbal approach to characterizing the look of maps. *The American Cartographer*, **1**, 63–71.

Petchenik, B.B. (1976) From place to space: the psychological achievement of thematic mapping, *The American Cartographer*, **6**, 5–12.

Ratajski, L. (1970) Kartologia. *Polski Przeglad Kartograficzny*, **2**, 97–110.

Ratajski, L. (1972) Cartology. *Geographia Polonica*, **21**, 63–78.

Ratajski, L. (1978) The main characteristics of cartographic communication as a part of theoretical cartography. *International Yearbook of Cartography*, **18**, 21–32.

Rimbert, S. (1968) *Leçons de Cartographie Thématique*, Société d'Edition d'Enseignement Supérieur, Paris, pp. 65–70.

Robinson, A.H. (1960) *Elements of Cartography*, 2nd edn, John Wiley & Sons, Inc., New York.

Robinson, A.H. and Petchenik, B.B. (1975) The map as a communication system. *The Cartographic Journal*, **12**, 7–15.

Robinson, A.H. and Petchenik, B.B. (1976) *The Nature of Maps: Essays Toward Understanding Maps and Mapping*, University of Chicago Press, Chicago.

Salichtchev, K.A. (1970) The subject and method of cartography: contemporary views. *The Canadian Cartographer*, **7**, 77–87.

Salichtchev, K.A. (1973) Some reflections on the subject and method of cartography after the Sixth International Cartographic Conference. *The Canadian Cartographer*, **10**, 106–111.

Salichtchev, K.A. (1978) Cartographic communication/its place in the theory of science. *The Canadian Cartographer*, **15**, 93–99.

Taylor, R.M. and Hopkin, V.D. (1973) Human factors principles in map design. *Revue de Médecine Aeronautique et spaciale*, **49**, 87–91.

Wooldridge, S.W. and East, W.G. (1951) *The Spirit and Purpose of Geography*, Hutchinson, London.

Wright, J.K. (1942) Map makers are human: comments on the subjective in maps. *Geographical Review*, **32**, 527–544.

Further reading

Board, C. (1984) New insights in cartographic communication. *Cartographica*, **21** (1). [A relevant theme issue, coming towards the end of the research heyday of cartographic communication theory, devoted to exploring ongoing implications of cognitive scientific approaches to mapping.]

Crampton, J. (2001) Maps as social constructions: power, communication and visualization. *Progress in Human Geography*, **25** (2), 235-252. [Situates the development of the communications paradigm in cartography in the light of subsequent social constructivist challenges and emergence of visualisation technologies.]

MacEachren, A.M. (1995) *How Maps Work*, Guilford, New York. [A thorough scientific reworking of cognitive approaches to mapping incorporating semiotic ideas into a representational framework.]

Montello, D.R. (2002) Cognitive map-design research in the twentieth century: theoretical and empirical approaches. *Cartography and Geographic Information Science*, **29**, 283-304. [This paper explores the legacy of empirical and theoretical work in the aftermath of Robinson's initial framing of mapping as a cognitive endeavor concerned with cognition.]

See also

- Chapter 1.2: Semiology of Graphics
- Chapter 1.3: The Nature of Maps
- Chapter 1.4: The Science of Cartography and its Essential Processes
- Chapter 1.8: Deconstructing the Map
- Chapter 1.10: Cartography Without 'Progress': Reinterpreting the Nature and Historical Development of Mapmaking
- Chapter 1.14: Rethinking Maps
- Chapter 4.3: Cognitive Maps and Spatial Behaviour: Process and Products
- Chapter 4.9: Cognitive Maps: Encoding and Decoding Information

Chapter 1.7

Designs on Signs / Myth and Meaning in Maps

Denis Wood and John Fels

Editors' overview

Wood and Fels' lengthy article initially made little impact, but has subsequently been influential in its structuralist reading of the coding process though which mapping constructs particular discursive positions. Their analysis starts from the apparently functional map legend to reveal its political role and, drawing on largely contemporary examples, develop a semiotic reading of how maps work. At the core of the piece is a Barthean analysis of how maps function to construct particular myths about the places they represent. Wood and Fels then develop notions of semiotic codes through which these myths becomes naturalised in society. They also explore the design of individual signs in the map, but also the extent to which the map as a whole functions as a sign system, whilst purporting to be neutral and authoritative.

Originally published in 1986: *Cartographica*, **23** (3), 54–103.

Spread out on the table before us is the *Official State Highway Map of North Carolina*. It happens to be the 1978–1979 edition. Not for any special reason: it just came to hand when we were casting about for an example. If you don't know this map, you can well enough imagine it, a sheet of paper – nearly two by four feet – capable of being folded into a handy pocket or glove compartment sized four by seven inches. One side is taken up by an inventory of North Carolina points of interest [. . .] On the other side, North Carolina, hemmed in by the margins of pale yellow South Carolina and Virginia, Georgia and Tennessee, and washed by a pale blue Atlantic, is represented as a meshwork of red, black, blue, green and yellow lines on a white background, thickened at the intersections by roundels of black or blotches of pink. [. . .] Constellated about this image are, *inter alia,* larger scale representations of ten urban places and the Blue Ridge Parkway, an index of cities and towns, a highly selective mileage chart, a few safety tips and [. . .] a legend (Figure 1.7.1).

Legends

[. . .]

Clearly this legend – to say nothing of the rest of the map – carries a heavy burden, one that reflects aggressively the uses to which this map was put. We stress the plural because it is a fact, not so much *overlooked* (cartographers are not *that* naive), but nonetheless ordinarily ignored, denied, suppressed. For certainly in this case the first and primary 'user' was the State of North Carolina, which *used* the map as a promotional device [. . .] It is conventional to pretend, as Robinson and Sale (1969: 270) have put it, that 'legends or keys are naturally indispensable to most maps, since they provide the explanations of the various symbols used', but that this is largely untrue hardly needs belabouring. Legends [. . .] are, nonetheless, still dispensed with more often than not, and never provide explanations of more than a portion of the 'symbols' found on the maps to which they refer. [. . .]

Nor is this dispensability a result of the 'self-explanatory' quality of the map symbols, for, though Robinson and Sale (1969: 270) might insist that, 'no symbol that is not self-explanatory should be used on a map unless it is

The Map Reader: Theories of Mapping Practice and Cartographic Representation, First Edition. Edited by Martin Dodge, Rob Kitchin and Chris Perkins.

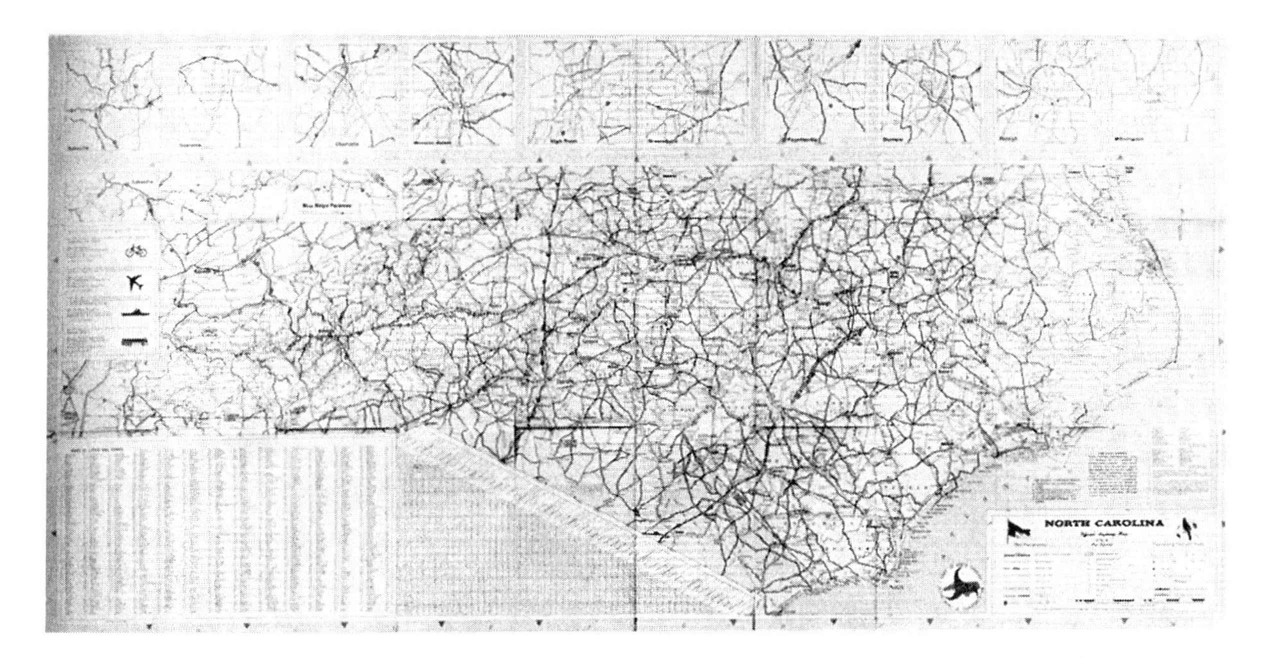

Figure 1.7.1 The '1978–79 North Carolina Transportation Map & Guide To Points of Interest'. That's what it says on the cover fold. The map image proper is titled on the legend block as the 'North Carolina Official Highway Map 1978–79', whereas the headline on the other side of the sheet reads, 'North Carolina Points of Interest'. Unfortunately the distinctions among the pale blue, yellow, pink and white are all but lost in this reproduction. (North Carolina Department of Transportation.)

explained in a legend', the fact is that NO symbol *explains* itself, stands up and says, 'Hi, I'm a lock', or 'We're marsh', anymore than the *words* of an essay bother to explain *themselves* to the reader. Most readers make it through most essays (and maps) because as they grew up through their common culture (and *into* their common culture), they learned the significance of most of the words (and map symbols). Those they don't recognise they puzzle out through context, or simply skip, or ask somebody to explain. [...]

It is not, then, that maps don't need to be decoded; but that they are by and large encoded in signs as readily interpreted by most map readers as the simple prose into which the marks are translated on the legends themselves.

[...]

No matter how many readers are convinced that blue naturally and unambiguously asserts the presence of water, or that little pictograms of lighthouses and mountains explain themselves, signs are *not* signs for, dissolve into marks for, those who don't know the code. [...]

There's no way around it: each of these signs is a perfectly conventional way of saying what is said ('lighthouse', 'mountain', 'water') – which is why the *map seems* so transparent, so easy to read. But *were* the function of the legend to explain such conventions (or at least translate them into words), then these would belong on it as surely as those that are there. [...]

Yet, while all these signs are absent, *on* the legend we find interpretative distinctions made among the shapes and colours of the road signs of the Interstate, federal and state highway systems. Does the person really exist for whom the graticule is self-evident, yet the highway signs obscure? [...] What becomes gradually clear is that if the purpose of the legend ever were 'explanation', everything is backwards: the things least likely to be most widely known are the very things about which the legend is reticent, while with respect to precisely those aspects both natives and travellers are most sure to be familiar, the legend is positively garrulous.

[...] *All* the maps of the genre, and most other genres as well, are characterised by legends (like ours) which in a more or less muddled fashion put into words map signs that are so customary as to be widely understood without the words, while leaving the map images themselves *littered* with conventions it taxes professional cartographers to put into English.

Myths

[...] *There is nothing wrong with the design of these legends: they are supposed to be the way they are.* This will be difficult for many to accept, but once it is understood that the role of the legend is less to elucidate the 'meaning' of this or that map element than to function as a sign in its own right, this conclusion is even more difficult *to evade.* [...] the legend as a whole is itself a signifier. As such, the legend refers not to the map (or at least not directly to the map), but back, through a judicious selection of map elements, to that to which the map image itself refers, to the state. *It is North Carolina that is signified in the legend, not the elements of the map image,* though it *is* the selection of map elements and their disposition within the legend box that encourages the transformation of the legend into a sign. It is a sign only a cartographer (or graphic designer) could fail to understand. [...]

[...]

[...] *It is sincere.* We don't believe for a minute anyone sat down and cynically worked this thing out, carefully offsetting the presumptuousness of the overheated highway symbolism with the self-effacing quality of the children's encyclopaedia colours. But this is not to say that with this legend we are not in the presence of what Roland Barthes (1972) has called 'myth' – a kind of 'speech' better defined by its intention than its literal sense. Barthean myth is invariably constructed from signs which have been already constructed out of a previous alliance of a signifier and a signified. An example, an especially innocuous one, is given by the reading of a Latin sentence, '*quia ego nominor leo*', in a Latin grammar: There is something ambiguous about this statement: on the one hand, the words in it do have a simple meaning: *because my name is lion.* And on the other, the sentence is evidently there in order to signify something else to me. In as much as it is addressed to me, a pupil in the second form, it tells me clearly: I am a grammatical example meant to illustrate the rule about the agreement of the predicate. I am even forced to realise that the sentence in no way *signifies* its meaning to me, that it tries very little to tell me something about the lion and what sort of name he has; its true and fundamental signification is to impose itself on me as the presence of a certain agreement of the predicate. I conclude that I am faced with a particular, greater, semiological system, since it is co-extensive with the language: there is, indeed, a signifier, but this signifier is itself formed by a sum of signs, it is in itself a first semiological system (*my name is lion*). Thereafter, the formal pattern is correctly unfolded: there is a signified (*I am a grammatical example*) and there is a global signification, which is none other than the correlation of the signifier and the signified; for neither the naming of the lion nor the grammatical example is given separately.

The parallels with our legend are pronounced. On the one hand, it too is loaded with simple meanings: *where on the map you find a red square, on the ground you will find a point of interest.* But, as we have seen, the legend little commits itself to the unfurling of these meanings, even compared to the map image on which each is actually named – 'Singletary Lake Group Camp' or 'World Golf Hall of Fame'. The appearance of the red square on the legend thus adds nothing to our ability to understand the map. Instead it imposes itself on us as an assertion that *North Carolina has points of interest,* speaks *through* the map *about* the state. Yet as in Barthes' example, this assertion about North Carolina is constructed out of, stacked on top of, the simpler significance of the red square on the legend, namely, to be identified with the words, 'Points of Interest'. We thus have a two-tiered semiological system in which the simpler is appropriated by the more complex. [...]

In our case, at the level of language we have as signifier the various marks that appear on the legend: the red square, the black dashed line, the bright blue asterisk [...]. Taken together, the marks and phrases are *signs,* things which *in their sign function* are no longer usefully taken for themselves (there is no red square 350 yards on a side at Singletary Lake) but as indicative of or as pointing toward something else (a point of interest called Singletary Lake Group Camp). Collectively, these signs comprise the legend, *but this in turn is a signifier in another semiological system cantilevered out from the first.* At this level of myth we have as signified some version of what it might mean to be in North Carolina, some idea of its attractiveness [...]. The signifier is, of course, the legend appropriated from the level of language by this myth to be its sign. Insidiously, this myth is not required to declare itself in language: this is its power. At the moment of reception, it evaporates: the legend is only a legend after all. One sees only its neutrality, its innocence. [...] It is precisely this ambiguity that enables myth to work without being seen. Hidden on top of a primary semiological system, it resists transformation into symbols. [...]

This is why myth is experienced as innocent speech: not because its intentions are hidden – if they were hidden they could not be efficacious – but because they are naturalised. In fact, what allows the reader to consume myth innocently is that he does not see it as a semiological system but as an inductive one. Where there is only an equivalence, he sees a kind of causal process: the signifier and the signified have, in his eyes, a natural relationship.

[...] Of all the systems so not seen, is there one more invisible than the cartographic? The most fundamental cartographic claim is *to be a system of facts,* and its history has most often been written as the story of its ability to present those facts with ever increasing accuracy. That this system can be corrupted everyone acknowledges: none are more vehement in their exposure of the 'propaganda map' than cartographers, but having denounced this usage they feel but the freer in passing off their own products as untainted by the very values which alone constitute the structure of a semiological system. [...]

Nor does the map image escape the grasp of myth. On the contrary, it is more mythic precisely to the degree that it succeeds in persuading us that it is a natural consequence of perceiving the world.

[. . .] The map is always there to deny that the significations piled on top of it are there at all. It is only a map after all, and the pretence is that it is innocent, a servant of the eye that sees things as they really are.

[. . .]

Codes

It is, of course, an illusion: *there is nothing natural about a map.* It is a cultural artefact, a cumulation of choices made among choices every one of which reveals a value: not the world, but a slice of a piece of the world; not nature but a slant on it; not innocent, but loaded with intentions and purposes; not directly, but through a glass; not straight, but mediated by words and other signs; not, in a word, as it is, but in *code.* And of course it's in code: *all* meaning, *all* significance derives from codes, *all* intelligibility depends on them. [. . .]

More technically, a code can be said to be an assignment scheme (or rule) coupling or apportioning items or elements from a conveyed system (the signified) to a conveying system (the signifier).

[. . .]

[As] Umberto Eco (1976: 48–49) puts it: 'A sign is always an element of an *expression plane* conventionally correlated to one (or several) elements of a *content plane.* Every time there is a correlation of this kind, recognised by a human society, there is a sign. Only in this sense is it possible to accept Saussure's definition according to which a sign is the correspondence between a signifier and a signified. This assumption entails some consequences: *a sign is not a physical entity,* the physical entity being at most the concrete occurrence of the expressive pertinent element; *a sign is not a fixed semiotic entity* but rather the meeting ground for independent elements (coming from two different systems of two different planes and meeting on the basis of a coding correlation).'

Because signs neither have physical existence (unlike the signifier) nor permanence, they are frequently referred to as *sign functions,* or in Eco's (1976: 49) words: [. . .] 'A sign function is realised when two *functives* (expression and content) enter into a mutual correlation; the same functive can also enter into another correlation, thus becoming a different functive and therefore giving rise to a new sign function. Thus signs are the provisional result of coding rules which establish *transitory* correlations of elements, each of these elements being entitled to enter – under given coded circumstances – into another correlation and thus form a new sign.'

This is not a game of words. The vocabulary is not important (not to us). What *is* important is the notion that signs – or sign functions or symbols: what they are called *does not matter* – that signs, to repeat, are realised *only* when coding rules bring into correlation two elements or items (or functives) from two domains or systems (the one signifying, of expression; the other signified, of content); and that *whenever* there is such a correlation, there is a sign. You may call this resulting sign an icon. You may call it a pictogram. You may call it a word. You may call it an index. You may call it a symbol. You may call it a piece of sculpture. You may call it a sentence. You may call it a map. You may call it New York City. In every case, whatever else it is, it is, *in its sign function,* also a sign, that is, a creature of a code. *No signs without codes.* It must be insisted upon. That is, no self-explanatory signs. No signs which so resemble their referents as to self-evidently refer to them.

[. . .]

So it is the *codes* upon which one must fasten if the map is to be decoded (or if a map is to be encoded). We think it possible to distinguish ten of these (there are doubtless others), which either the map exploits, or by virtue of which the map is exploited. Neither class is independent of the other, and no map fails to be inscribed in (at least) these ten codes. Those which the map exploits we term *codes of intrasignification.* These operate, so to speak, within the map: at the level of language. Those by virtue of which the map is exploited we term *codes of extrasignification.* These operate, so to speak, outside the map: at the level of myth. Among the codes of intrasignification five at least are inescapable: the *iconic,* the *linguistic,* the *tectonic,* the *temporal* and the *presentational.* Under the heading *iconic* we subsume the code of 'things' ('events'), with whose relative location the map is enrapt: the streets of Genoa, rates of death by cancer, exports of French wine, the losses suffered in Napoleon's Russian campaign, airways, subways, the buildings of Manhattan, levels of air pollutants over six counties in Southern California, the rivers, roads, counties, airports, cities and towns of North Carolina. The iconic is the code of the inventory, of the world's fragmentation: into urban hierarchies, into hypsometric layers, into wet and dry. The *linguistic* is the code of the names: the *Via Corsica,* the *Corso Aurelio Saffi*; trachea, bronchus and lung cancer, white males, age-adjusted rate by county [. . .] The linguistic is the code of classification, of ownership: identifying, naming, assigning. The relationships of these things in space is given in the *tectonic* codes: in the *scalar* – in the number of miles (or feet) encoded in every inch – and in the *topological* – in the planimetry of cities, the stereometry of mountain ranges, the projective geometry of continents, the topographometry of the field traverse, the simple topology of the sketch map giving directions to the cocktail party. The tectonic is the code of finding, it is the code of getting there: it is the code of getting. Because there is no connection, no communication, except in time, the codes of filiation are *temporal*, codes of duration, codes of tense. The *durative* establishes the scale, the map's *durée*, its 'thickness': as the map of rates of death from cancer, 1950–1969, is 'thicker' than the 1978–79 North Carolina highway map, which is 'thicker' than the map of reactive hydrocarbons, 6 a.m. to 9 a.m., July 22, 1979. The durative reveals (or hides or is mute about) lapses in cosynchronicity.

The *tense* says when: some maps are in the past tense ('The World of Alexander the Great'), others in the future tense ('Tomorrow's Highways'), but most maps plump for the present ('State of the World Today'), or, if they can possibly get away with it, the aorist: no duration at all (no thickness), out of chronology (not lost – just out of it): free of time. These attain to myth at the very level of language. Each of these codes – iconic, linguistic, tectonic and temporal – is embodied in signs with all the physicality of the concrete instantiation of the expressive pertinent element. On the page, on the sheet of paper, on the illuminated display with its flashing lights, these concrete instantiations are ordered, arranged, organised by the *presentational* code: they are *presented.* [...]

[...] Among the codes of extrasignification, five again are inescapable: the *thematic*, the *topic*, the *historical*, the *rhetorical* and the *utilitarian.* All operate at the level of myth, all make off with the map for their own purposes (as they made the map), all distort its meaning (its meaning at the level of language), subvert it to their own. If the presentational code permits the map to achieve a level of discourse, the *thematic* code establishes its subject: *on what shall the map discourse? What shall it argue?* [...] And precisely as the thematic code runs off with the icons, so the *topic* code (with a long *o* from *topos*, place, as in *topography*, not *topicality*) runs off with the space established by the tectonic code, turns it from space to place, bounds it (binds it), gives it a name, sets it off from other space, asserts its existence: *this place is.* Ditto the *historical* code. Only it works on the time established in the map by the temporal code. Are there bounding dates to the map's *durée*? Then the historical code appropriates them to an era, assigns it a name, incorporates it in a vision of history.

[...] If the thematic code sets the subject for the discourse, if the topic and historical codes secure the place and time, it is the *rhetorical* code that sets the tone, that, having consumed the presentational code, most completely orients the map in its culture (in its set of values), pointing in the very act of pointing somewhere else (to the globe), to itself, to its maker, to the culture that produced it, to the place and time and omphalos of that culture [...]

[...]

As the map itself is finally used, picked up bodily by the *utilitarian* code to be carted off for any purpose myth might serve. [...] It is here that the academic model of the map with its scanning eyes and graduated circle-comparing minds breaks down most completely. It has no room for the real uses of most maps, which are to possess and to claim, to legitimate and to name.

These are the uses of maps as certainly as it is the most important function of maps in geographic journals to certify the geographic legitimacy of the articles they decorate. The anthropology of cartography is an urgent project: what *are* all those maps actually used for? [...]

Intrasignification

Clearly, the map is comprehended in two ways. As a medium of language (in the broadest sense) it serves as a visual analogue of phenomena, attributes and spatial relations: a model on which we may act, in lieu or anticipation of experience, to compare or contrast, mensurate or appraise, analyse or predict. It seems to inform, with unimpeachable dispassion, of the objects and events of the world. As myth, however, it refers to itself and to its makers, and to a world seen quite subjectively through their eyes. It trades in values and ambitions; it is politicised. Signing functions that serve the former set of purposes we have termed intrasignificant; those which serve the latter, extrasignificant. Whereas intrasignification consists of an array of sign functions indigenous to the map and which, taken jointly, constitute the map *assign*, extrasignification appropriates the complete map and deploys it *as expression* in a broader semiotic context. The map acts as a focusing device between these two planes of signification, as a lens that gathers up its internal or constituent signs and offers them up collectively *as a map.* But what offers from the map is not substantially different from what is afferent upon it – these have simply been repositioned in the semiological function – and, while extrasignification exploits the map in its entirety, we have seen how the initiatives of myth extend to even the most fundamental and apparently sovereign aspects of intrasignification, and are ultimately rooted in them. [...]

The map is the product of a spectrum of codes that materialise its visual representations, orient it in space and in time, and bind it together in some acceptable form. The actions of these codes are, if not entirely independent, reasonably distinct. *Iconic codes* govern the manner in which graphic expressions correspond with geographic items, concrete or abstract, and their attendant attributes. A *linguistic code* (occasionally two or several) is extended to the map to regulate the equivalence of typographic expressions and, via the norms of written language, a universe of terminology and nomenclature. As the space of the map is configured by *tectonic codes* – transformational procedures prescribing its topological and scalar relations to the space of the globe – *temporal codes* configure the time of the map in relation to the stream of events and observations from which it derives. The diversity of expressions that constitute the map are organised and orchestrated through a *presentational code* that fuses them into a coherent cartographic discourse.

[...]

Sign functions

Maps are about relationships. In even the least ambitious maps, simple presences are absorbed in multilayered relationships integrating and disintegrating sign functions, packaging and repackaging meanings. The map is a highly complex supersign (Ockerse and Van Dijk 1979; Ockerse 1984), a sign composed of lesser signs, or, more accurately, a synthesis of signs; and these are supersigns in their own right, systems of signs of more specific or individual function. It's not that the map conveys meanings so much as *unfolds* them *through a cycle of interpretation* in which it is continually torn down and rebuilt; and, to be truthful, this is not really the map's work but that of its user, who creates a wealth of meaning by selecting and subdividing, combining and recombining its terms in an effort to comprehend and understand. But, however elaborate, this is not an unbounded process. Inevitably, it has a lower bound, the most particular sign function that resists decomposition into constituent signs, and an upper bound, the integral supersign of the entire map that accesses the realm of extrasignification; and between these extremes it is stratified. Twofold stratifications have been repeatedly proposed (Head 1984; Petchenik 1979; Olson 1976; Bertin 1979), and widely accepted, but these don't go far enough. If we intend to explain how the map generates and structures the signing processes by virtue of which it is a map, then we need at least four strata or levels of signification, which we'll call *elemental, systemic, synthetic* and *presentational.*

[...]

Elemental signs

Elemental map signs, by definition, cannot be decomposed to yield lesser signs referring to *distinct geographic entities.* They are the least significant units which have specific reference to features, concrete (Omaha) or abstract (1000 pigs), within the map image. Appraised in terms of the map's graphic signifiers, this criterion is easily confused; and we must keep in mind that a sign is not its expression, but the marriage of expression and content. The elemental map sign operates at the lower bound of the map's content taxonomy, and below this bound reside connotation and characteristic but nothing which can be construed as feature. [...] As we have seen, the map is an iconic medium that imposes its behaviour on language, not the other way around; and there is no reason to expect graphic signs to observe the rigidly contrived, and separately evolved, protocol of phonetic representation.

[...]

We have tried to demonstrate why we must insist that map signs be considered in terms of both expression *and content*, and to point out the inadequacy of a Formalist perspective that regards only signifiers and not signs; as well as to suggest the degree to which our conceptualisation of phenomena structures, even dictates, the manner in which we represent them. Thus an elemental sign is a *sign of elemental meaning*, one which refers to an element of the landscape that, however artificial, we are not inclined to tear into constituent bits. With this premise it is possible to build systems of signs, and systemic meaning, from elements.

Sign systems

By sign system we mean a set or family of similar elemental signs *extensive in the space of the map image:* a distribution of statistical units, a network of channels, a matrix of areal entities, a nesting of isolines. In this respect, we identify a road system, a river system, or a system of cities. It requires that we interpret many like signs as one sign, again a syntactical product but now one of geographic syntax. The systemic signifier is shaped by the disposition of its corresponding set of phenomena in geodesic space and by the topological transformation that brings this space to the surface of the page. [...]

An arrangement of signifiers on the map constitutes a system only, of course, by virtue of our ability to perceptually organise its elements into something whole. At the systemic level, the bases of affinity among elements are those of implantation (yielding point, line, or area systems) and those formal and chromatic attributes variously termed qualitative, nominal, distinguishing or differential. [...] [C]artographic sign systems are typified by connectivity. Their elements link up, abut, cradle or nest within one another. They have anatomies. We recognise primarily their structure and utilise the characteristics of their elements mainly to highlight subsystems which would be otherwise undifferentiated, or to unstick systems of similar structure. That is to say, we attend more to the syntax of the system than the semantic import of its components.

[...]

Synthesis

There is no such thing as a monothematic map. Consider that emblem of thematic cartography: an array of graduated circles against the barest outline of subject area. This map image signifies at least the shoreline (usually elaborated beyond any conceivable utility), the water surface, the land surface, and probably one or more proprietary boundaries (in which case we've differentiated several political states as well), and – almost forgot – whatever it is the graduated circles represent. Stripping off the circles leaves us with an absolute minimum of three sign systems, and typically twice that many, lurking behind the ostensibly servile trace of the pen. Sure, cartographers design maps for cartographers – as architects design buildings for architects and politicians make laws for politicians – but to call this monothematic is going too far. Can we really take that much for granted? Are we so thoroughly hypnotised that we can't even *see* the map?

Maps are about relationships. In other words, they are about how one landscape – a landscape of roads, of rivers, of cities, government, sustenance, poison, the good life, of whatever – is positioned in relation to another. The map synthesises these diverse landscapes, projecting them onto and into one another, with less than subtle hints that one is correlative to another or that *this* is an agent or effect of *that.* The map can't simply say that something is present (present in what?) or that it is distributed in a certain way (distributed in relation to what?); it's after the *big picture,* the kind of insight that only comes with an omnipresent viewpoint and the power to choose what inhabits the world. At this level the map image as a whole (whole in content if not necessarily in scope) is the supersign, and the various systems it resolves to are its constituent signs. And signs can only have meaning in relation to other signs. [. . .]

[. . .] What the map does (and this is its most important internal sign function) is permit systems to open and maintain a dialogue with one another. It is obvious why a road folds back on itself when we can see the slope it ascends, or why two roads parallel one another a stone's throw apart when we can see them on opposite banks of a river, or why an interstate cramps into a tense circle when we can see the city and its rush-hour torment. We know the behaviour of this system so well, in fact, that we can take it as an index of other systems in the total absence of their direct representation. On the face of it, the map confirms these understandings; but they are understandings *that have already been created by maps.*

The *gestalt* of each sign system is positioned against the semiotic ground of another sign system, or a subsynthesis of systems. The roads in the state highway map aren't grounded against an insignificant white surface; they're grounded against North Carolina or Illinois or Texas. [. . .] *There is nothing in the map that fails to signify.* Not even in a map of the Moon. So the flow of water is interpreted against the ground of land form, and vice versa; and the pattern of forestation is interpreted against the ground of both, as both and each are interpreted against it. In the synthesised map image, every sign system is potentially figure and every sign system is potentially ground. [. . .]

The map image is a synthesis of spatially and temporally registered *gestelten,* each a synthesis in its own right; and to pretend that this whole is no more than the sum of its parts, or that we can do more than recommend a certain alignment of their priorities, is to reduce our concept of the map to that of a diagram. No degree of thematic constriction can silence the conversation among map signs. [. . .]

Presentation

In presentation the map attains the level of discourse. Its discursive form may be as simple as a single map image rendered comprehensible by the presence of title, legend and scale, or as complex as those in *The New State of the World Atlas* (Kidron and Segal 1984). [. . .] Presentation is more than placing the map image in the context of other signs, it's placing the map in the context of its audience. [. . .]

In bringing the map to this point we make it entirely accessible to the processes of extrasignification, and subject to their appropriation. It can be seized and carried off whole (necessarily whole) to serve the motives of mythic representation. [. . .] The map is simultaneously an instrument of communication – intrasignification, given the benefit of doubt – and an instrument of persuasion – extrasignification and its propensity toward myth.

Presentation locates the map front and centre in all this action, at the vertex of both planes of signification. It's not a quirk of house style that populates the *National Geographic* map with maize-laden Cherokee or the state highway map with trees, bees, civil war artefacts and cavorting tourists. It's the deliberate activation of popular visual discourse. It's not just pragmatism or objectivity that dresses the topographic map with reliability diagrams and magnetic error diagrams and multiple referencing grids, or the thematic map with the trappings of F-scaled symbols and psychometrically divided greys. It's the urge to claim the map as a scientific instrument and accrue to it all the mute credibility and faith that this demands. Presentation, as the end and the beginning of the map, closes the loop of its design. It makes the map whole and, in doing so, prepares it for a role that begins where its avowed attention to symbolism, geodesic accuracy, visual priority, and graphic organisation leaves off. It injects the map into its culture.

References

Barthes, R. (1972) *Mythologies,* Hill and Wang, New York.
Bertin, J. (1979) La test de base de la graphique. *Bulletin du Comité Français de Cartographie,* **79**, 3–18.

Eco, U. (1976) *A Theory of Semiotics*, Indiana University Press, Bloomington, IN.
Head, C.G. (1984) The map as natural language: a paradigm for understanding. *Cartographica*, **21** (1), 1–32.
Kidron, M. and Segal, R. (1984) *The New State of the World Atlas*, 2nd edn, Simon and Schuster, New York.
Ockerse, T. (1984) De-sign/super-sign. *Semiotica*, **52** (3/4), 251–252.
Ockerse, T. and Van Dijk, H. (1979) Semiotics and graphic design education. *Visible Language*, **13** (4), 358–378.
Olson, J.M. (1976) A co-ordinated approach to map communication improvement. *The American Cartographer*, **3**, 151–159.
Petchenik, B.B. (1979) From place to space: the psychological achievement of thematic mapping. *The American Cartographer*, **6**, 5–12.
Robinson, A. and Sale, R. (1969) *Elements of Cartography*, 3rd edn, John Wiley & Sons, Inc., New York.

Further reading

Crampton, J. (2001) Maps as social constructions: power, communication and visualization. *Progress in Human Geography*, **25** (2), 235–252. [This paper charts social constructivist challenges (including Wood's work) to scientific orthodoxy in approaches to mapping.]
MacEachren, A.M. (1995) *How Maps Work*, Guilford, New York. [A thorough scientific reworking of cognitive approaches to mapping incorporating semiotic ideas into a representational framework.]
Wood, D. (2010) *Rethinking the Power of Maps*, Guilford, New York. [Reprints the original *Designs on Signs* article along with interesting work on the history of cartography and a range of newer material around contextual studies of oppositional forms such as art, counter-mapping and indigenous cartography.]
Wood, D. and Fels, J. (2008) *The Natures of Maps: Cartographic Constructions of the Natural World*, University of Chicago Press, Chicago. [This monograph significantly extends their semiotic analysis, developed over twenty years earlier, and applies it to offer a series of readings of different natures constructed in a various printed and predominantly mass-market maps.]

See also

- Chapter 1.2: Semiology of Graphics
- Chapter 1.8: Deconstructing the Map
- Chapter 1.10: Cartography Without 'Progress': Reinterpreting the Nature and Historical Development of Mapmaking
- Chapter 1.13: Beyond the 'Binaries': A Methodological Intervention for Interrogating Maps as Representational Practices
- Chapter 3.6: Some Truth with Maps: A Primer on Symbolisation and Design
- Chapter 3.9: Mapping, Modernity: Art and Cartography in the Twentieth Century
- Chapter 3.10: Affective Geovisualisations
- Chapter 4.6: Reading Maps
- Chapter 5.3: Texts, Hermeneutics and Propaganda Maps

Chapter 1.8

Deconstructing the Map

J. B. Harley

Editors' overview

Towards the end of the 1980s scholarship in the history of cartography was an empirical pursuit without a social context and contemporary cartographic theory was locked into the application of cognitive scientific approaches grounded in communication theory. Neither was integrated into the intellectual mainstream of the critical social sciences and humanities. Meanwhile GIS was threatening to dispense with the paper map. Harley's article caught the time, with its passionate calls for a more engaged, critical and social constructivist approach to mapping. He seeks to achieve three things. Firstly, he examines cartography as a discourse in the light of Foucauldian ideas about the play of rules within discursive formations, critiquing the epistemic orthodoxy of the times. Secondly, he draws on Derridean ideas to examine the textuality of maps and how they might be deconstructed as rhetorical devices, before focusing, thirdly, on the wider social context of mapping, in a Foucauldian reading of cartography as a form of power-knowledge. The article proved influential and helped fostered new critical cartographies in the 1990s, better able to understand technological, epistemic and social changes to mapping.

Original published in 1989: *Cartographica*, **26** (2), 1–20.

[...]

The pace of conceptual exploration in the history of cartography – searching for alternative ways of understanding maps – is slow. Some would say that its achievements are largely cosmetic. Applying conceptions of literary history to the history of cartography, it would appear that we are still working largely in either a 'premodern' or a 'modern' rather than in a 'postmodern' climate of thought (Eagleton 1983; Ferraris 1988). A list of individual explorations would, it is true, contain some that sound impressive. Our students can now be directed to writings that draw on the ideas of information theory, linguistics, semiotics, structuralism, phenomenology, developmental theory, hermeneutics, iconology, Marxism and ideology. [...] Yet despite these symptoms of change, we are still, willingly or unwillingly, the prisoners of our own past. My basic argument in this essay is that we should encourage an epistemological shift in the way we interpret the nature of cartography. For historians of cartography, I believe a major roadblock to understanding is that we still accept uncritically the broad consensus, with relatively few dissenting voices, of what *cartographers* tell us maps are supposed to be. In particular, we often tend to work from the premise that mappers engage in an unquestionably 'scientific' or 'objective' form of knowledge creation. Of course, cartographers believe they have to say this to remain credible but historians do not have that obligation. It is better for us to begin from the premise that cartography is seldom what cartographers say it is.

[...]

The question becomes how do we as historians of cartography escape from the normative models of cartography? How do we allow new ideas to come in? How do we begin to write a cartographic history as genuinely revisionist as Louis Marin's 'The King and his Geometer' (in the context of a seventeenth century map of Paris) or William Boelhower's 'The Culture of the Map' (in the context of sixteenth century world maps showing America for the first time) (Marin 1988; Boelhower 1984, 1988) These are two studies informed by postmodernism. In this essay I also adopt a strategy aimed at the deconstruction of the map.

The Map Reader: Theories of Mapping Practice and Cartographic Representation, First Edition. Edited by Martin Dodge, Rob Kitchin and Chris Perkins.

The notion of deconstruction (Derrida 1976; Norris 1982) is also a password for the postmodern enterprise. Deconstructionist strategies can now be found not only in philosophy but also in localised disciplines, especially in literature, and in other subjects such as architecture, planning and, more recently, geography (Knox 1988; Gregory 1987; Dear 1988). I shall specifically use a deconstructionist tactic to break the assumed link between reality and representation which has dominated cartographic thinking, has led it in the pathway of 'normal science' since the Enlightenment, and has also provided a ready-made and 'taken for granted' epistemology for the history of cartography. The objective is to suggest that an alternative epistemology, rooted in social theory rather than in scientific positivism, is more appropriate to the history of cartography. It will be shown that even 'scientific' maps are a product not only of 'the rules of the order of geometry and reason' but also of the 'norms and values of the order of social... tradition.' (Marin 1988: 173). Our task is to search for the social forces that have structured cartography and to locate the presence of power – and its effects – in all map knowledge.

The ideas in this particular essay owe most to writings by Foucault and Derrida. My approach is deliberately eclectic because in some respects the theoretical positions of these two authors are incompatible. Foucault anchors texts in socio-political realities and constructs systems for organising knowledge of the kind that Derrida loves to dismantle (Said 1978; Hoy 1985). But even so, by combining different ideas on a new terrain, it may be possible to devise a scheme of social theory with which we can begin to interrogate the hidden agendas of cartography. Such a scheme offers no 'solution' to an historical interpretation of the cartographic record, nor a precise method or set of techniques, but as a broad strategy it may help to locate some of the fundamental forces that have driven map-making in both European and non-European societies. From Foucault's writings, the key revelation has been the omnipresence of power in all knowledge, even though that power is invisible or implied, including the particular knowledge encoded in maps and atlases. Derrida's notion of the rhetoricity of all texts has been no less a challenge (Derrida 1976). It demands a search for metaphor and rhetoric in maps where previously scholars had found only measurement and topography. Its central question is reminiscent of Korzybski's much older dictum, 'The map is not the territory' (Korzybski 1948), but deconstruction goes further to bring the issue of how the map represents place into much sharper focus. Deconstruction urges us to read between the lines of the map – 'in the margins of the text' – and through its tropes to discover the silences and contradictions that challenge the apparent honesty of the image. We begin to learn that cartographic facts are only facts within a specific cultural perspective. We start to understand how maps, like art, far from being 'a transparent opening to the world', are but 'a particular human way ... of looking at the world' (Blocker 1979: 43).

In pursuing this strategy I shall develop three threads of argument. Firstly, I shall examine the discourse of cartography in the light of some of Foucault's ideas about the play of rules within discursive formations. Secondly, drawing on one of Derrida's central positions I will examine the textuality of maps and, in particular, their rhetorical dimension. Thirdly, returning to Foucault, I will consider how maps work in society as a form of power-knowledge.

The rules of cartography

One of Foucault's primary units of analysis is the discourse. A discourse has been defined as 'a system of possibility for knowledge' (Philp 1985: 69). Foucault's method was to ask, it has been said, 'what rules permit certain statements to be made; what rules order these statements; what rules permit us to identify some statements as true and others as false; what rules allow the construction of a map, model or classificatory system ... what rules are revealed when an object of discourse is modified or transformed ... Whenever sets of rules of these kinds can be identified, we are dealing with a discursive formation or discourse' (Philp 1985). The key question for us then becomes, 'What type of rules have governed the development of cartography?' Cartography I define as a body of theoretical and practical knowledge that mapmakers employ to construct maps as a distinct mode of visual representation. The question is, of course, historically specific: the rules of cartography vary in different societies.

Here I refer particularly to two distinctive sets of rules that underlie and dominate the history of Western cartography since the seventeenth century. One set may be defined as governing the technical production of maps and is made explicit in the cartographic treatises and writings of the period (Crone 1953). The other set relates to the cultural production of maps. These must be understood in a broader historical context than either scientific procedure or technique. They are, moreover, rules that are usually ignored by cartographers so that they form a hidden aspect of their discourse.

The first set of cartographic rules can thus be defined in terms of a scientific epistemology. From at least the seventeenth century onward, European mapmakers and map users have increasingly promoted a standard scientific model of knowledge and cognition. The object of mapping is to produce a 'correct' relational model of the terrain. Its assumptions are that the objects in the world to be mapped are real and objective, and that they enjoy an existence independent of the cartographer; that their reality can be expressed in mathematical terms; that systematic observation and measurement offer the only route to cartographic truth; and that this truth can be independently verified (Campbell 1973; Hooykaas 1987; Woolgar 1988). The procedures of both surveying and map construction came to share strategies similar to those in science in general: cartography also documents a history of more precise instrumentation and measurement; increasingly complex classifications of its knowledge and a proliferation of signs for its representation; and, especially from the nineteenth century onward, the growth of institutions and a 'professional' literature designed to monitor the application and propagation of the rules (Wolter 1975a, 1975b). [...]

The acceptance of the map as 'a mirror of nature' (to employ Richard Rorty's (1979) phrase) also results in a number of other characteristics of cartographic discourse, even where these are not made explicit. Most striking is the belief in progress: that, by the application of science, ever more precise representations of reality can be produced. The methods of cartography have delivered a 'true, probable, progressive or highly confirmed knowledge' (Laudan 1977: 2). This mimetic bondage has led to a tendency not only to look down on the maps of the past (with a dismissive scientific chauvinism) but also to regard the maps of other non-Western or early cultures (where the rules of mapmaking were different) as inferior to European maps (Harley 1988a, 1988b). Similarly, the primary effect of the scientific rules was to create a 'standard' – a successful version of 'normal science' (Kuhn 1962; Lakatos and Musgrave 1970; Gauthier 1988) – that enabled cartographers to build a wall around their citadel of the 'true' map. Its central bastions were measurement and standardisation and beyond there was a 'not cartography' land where lurked an army of inaccurate, heretical, subjective, valuative and ideologically distorted images. Cartographers developed a 'sense of the other' in relation to nonconforming maps. [...]

In cases where the scientific rules are invisible in the map we can still trace their play in attempting to normalise the discourse. The cartographer's 'black box' has to be defended and its social origins suppressed. The hysteria among leading cartographers at the popularity of the Peters' projection (Cartographic Journal 1985; Loxton 1985; Peters 1983; Porter and Voxland 1986; Robinson 1985; Snyder 1988) or the recent expressions of piety among Western European and North American mapmakers following the Russian admission that they had falsified their topographic maps to confuse the enemy give us a glimpse of how the game is played according to these rules. [...]

This timely example also serves to introduce my second contention that the scientific rules of mapping are, in any case, influenced by a quite different set of rules: those governing the cultural production of the map. To discover these rules, we have to read between the lines of technical procedures or of the map's topographic content. They are related to values, such as those of ethnicity, politics, religion or social class, and they are also embedded in the map producing society at large. Cartographic discourse operates a double silence toward this aspect of the possibilities for map knowledge. In the map itself, social structures are often disguised beneath an abstract, instrumental space, or incarcerated in the coordinates of computer mapping. And in the technical literature of cartography they are also ignored, notwithstanding the fact that they may be as important as surveying, compilation or design in producing the statements that cartography makes about the world and its landscapes. Such an interplay of social and technical rules is a universal feature of cartographic knowledge. In maps it produces the 'order' of its features and the 'hierarchies of its practices' (Foucault 1973: xx). In Foucault's sense the rules may enable us to define an *episteme* and to trace an archaeology of that knowledge through time (Foucault 1973: xxii).

[...]

Why maps can be so convincing in this respect is that the rules of society and the rules of measurement are mutually reinforcing in the same image. Writing of the map of Paris, surveyed in 1652 by Jacques Gomboust, the King's engineer, Louis Marin (1988: 173) points to 'this sly strategy of simulation–dissimulation'. The knowledge and science of representation, to demonstrate the truth that its subject declares plainly, flow nonetheless in a social and political hierarchy. The proofs of its 'theoretical' truth had to be given, they are the recognisable signs; but the economy of these signs in their disposition on the cartographic plane no longer obeys the rules of the order of geometry and reason but, rather, the norms and values of the order of social and religious tradition.

[...]

The distinctions of class and power are engineered, reified and legitimated in the map by means of cartographic signs. The rule seems to be 'the more powerful, the more prominent'. To those who have strength in the world shall be added strength in the map. Using all the tricks of the cartographic trade – size of symbol, thickness of line, height of lettering, hatching and shading, the addition of colour – we can trace this reinforcing tendency in innumerable European maps. We can begin to see how maps, like art, become a mechanism 'for defining social relationships, sustaining social rules and strengthening social values' (Geertz 1983: 99).

[...] Much of the power of the map, as a representation of social geography, is that it operates behind a mask of a seemingly neutral science. It hides and denies its social dimensions at the same time as it legitimates. Yet whichever way we look at it the rules of society will surface. They have ensured that maps are at least as much an image of the social order as they are a measurement of the phenomenal world of objects.

Deconstruction and the cartographic text

To move inward from the question of cartographic rules – the social context within which map knowledge is fashioned – we have to turn to the cartographic text itself. The word 'text' is deliberately chosen. It is now generally accepted that the model of text can have a much wider application than to literary texts alone. [. . .] It has been said that 'what constitutes a text is not the presence of linguistic elements but the act of construction' so that maps, as 'constructions employing a conventional sign system' (McKenzie 1986: 34–39), become texts. With Barthes we could say they 'presuppose a signifying consciousness' that it is our business to uncover (Barthes 1973: 110). 'Text' is certainly a better metaphor for maps than the mirror of nature. Maps are a cultural text. By accepting their textuality we are able to embrace a number of different interpretative possibilities. Instead of just the transparency of clarity we can discover the pregnancy of the opaque. To fact we can add myth, and instead of innocence we may expect duplicity. Rather than working with a formal science of communication, or even a sequence of loosely related technical processes, our concern is redirected to a history and anthropology of the image, and we learn to recognise the narrative qualities of cartographic representation (Wood 1987) as well as its claim to provide a synchronous picture of the world. All this, moreover, is likely to lead to a rejection of the neutrality of maps, as we come to define their intentions rather than the literal face of representation, and as we begin to accept the social consequences of cartographic practices. I am not suggesting that the direction of textual enquiry offers a simple set of techniques for reading either contemporary or historical maps. In some cases we will have to conclude that there are many aspects of their meaning that are undecidable (Hoy 1985).

Deconstruction, as discourse analysis in general, demands a closer and deeper reading of the cartographic text than has been the general practice in either cartography or the history of cartography. It may be regarded as a search for alternative meanings. 'To deconstruct', it is argued, is to reinscribe and resituate meanings, events and objects within broader movements and structures; it is, so to speak, to reverse the imposing tapestry in order to expose in all its unglamorously dishevelled tangle the threads constituting the well-heeled image it presents to the world (Eagleton 1986).

The published map also has a 'well-heeled image' and our reading has to go beyond the assessment of geometric accuracy, beyond the fixing of location, and beyond the recognition of topographical patterns and geographies. Such interpretation begins from the premise that the map text may contain 'unperceived contradictions or duplicitous tensions' (Hoy 1985: 540) that undermine the surface layer of standard objectivity. Maps are slippery customers. In the words of W.J.T. Mitchell, writing of languages and images in general, we may need to regard them more as 'enigmas, problems to be explained, prison houses which lock the understanding away from the world.' We should regard them 'as the sort of sign that presents a deceptive appearance of naturalness and transparency concealing an opaque, distorting, arbitrary mechanism of representation' (Mitchell 1986: 8). Throughout the history of modern cartography in the West, for example, there have been numerous instances of where maps have been falsified, of where they have been censored or kept secret, or of where they have surreptitiously contradicted the rules of their proclaimed scientific status (Harley, 1988c).

As in the case of these practices, map deconstruction would focus on aspects of maps that many interpreters have glossed over. Writing of 'Derrida's most typical deconstructive moves', Christopher Norris (1987: 19) notes that

> 'deconstruction is the vigilant seeking-out of those 'aporias', blindspots or moments of self-contradiction where a text involuntarily betrays the tension between rhetoric and logic, between what it manifestly *means to say* and what it is nonetheless *constrained to mean.* To 'deconstruct' a piece of writing is therefore to operate a kind of strategic reversal, seizing on precisely those unregarded details (casual metaphors, footnotes, incidental turns of argument) which are always, and necessarily, passed over by interpreters of a more orthodox persuasion. For it is here, in the margins of the text - the 'margins', that is, as defined by a powerful normative consensus — that deconstruction discovers those same unsettling forces at work.'

A good example of how we could deconstruct an early map – by beginning with what have hitherto been regarded as its 'casual metaphors' and 'footnotes' – is provided by recent studies reinterpreting the status of decorative art on the European maps of the seventeenth and eighteenth centuries. Rather than being inconsequential marginalia, the emblems in cartouches and decorative title pages can be regarded as *basic to the way* they *convey their* cultural meaning (Clarke 1988; Harley 1984), and they help to demolish the claim of cartography to produce an impartial graphic science. But the possibility of such a revision is not limited to historic 'decorative' maps. A recent essay by Wood and Fels (1986) on the Official State Highway Map of North Carolina indicates a much wider applicability for a deconstructive strategy by beginning in the 'margins' of the contemporary map. They also treat the map as a text and, drawing on the ideas of Roland Barthes (1973) of myth as a semiological system, develop a forceful social critique of cartography which though structuralist in its approach is deconstructionist in its outcome.

[...]

A cartographer's stock response to this deconstructionist argument might well be to cry 'foul'. The argument would run like this: 'Well after all it's a state highway map. It's designed to be at once popular and useful. We expect it to exaggerate the road network and to show points of interest to motorists. It is a derived rather than a basic map'. It is not a scientific map. The appeal to the ultimate scientific map is always the cartographers' last line of defence when seeking to deny the social relations that permeate their technology. It is at this point that Derrida's strategy can help us to extend such an interpretation to all maps, scientific or non-scientific, basic or derived. Just as in the deconstruction of philosophy Derrida was able to show 'how the supposedly literal level is intensively metaphorical' (Hoy 1985: 44) so, too, we can show how cartographic 'fact' is also symbol. In 'plain' scientific maps, science itself becomes the metaphor. Such maps contain a dimension of 'symbolic realism' which is no less a statement of political authority and control than a coat-of-arms or a portrait of a queen placed at the head of an earlier decorative map. The metaphor has changed. The map has attempted to purge itself of ambiguity and alternative possibility (Eagleton 1983: 135). Accuracy and austerity of design are now the new talismans of authority culminating in our own age with computer mapping. We can trace this process very clearly in the history of Enlightenment mapping in Europe. The topography as shown in maps, increasingly detailed and planimetrically accurate, has become a metaphor for a utilitarian philosophy and its will to power. Cartography inscribes this cultural model upon the paper and we can examine it in many scales and types of maps.

Precision of instrument and technique merely serves to reinforce the image, with its encrustation of myth, as a selective perspective on the world. Thus maps of local estates in the European *ancien regime*, though derived from instrumental survey, were a metaphor for a social structure based on landed property. County and regional maps, though founded on scientific triangulation, were an articulation of local values and rights. Maps of the European states, though constructed along arcs of the meridian, served still as a symbolic shorthand for a complex of nationalist ideas. And world maps, though increasingly drawn on mathematically defined projections, nevertheless gave a spiralling twist to the manifest destiny of European overseas conquest and colonisation (Harley 1988b: 300). In each of these examples we can trace the contours of metaphor in a scientific map. This in turn enhances our understanding of how the text works as an instrument operating on social reality. In deconstructionist theory the play of rhetoric is closely linked to that of metaphor.

In concluding this section of the essay I will argue that notwithstanding 'scientific' cartography's efforts to convert culture into nature, and to 'naturalise' social reality (Eagleton 1983: 135–136), it has remained an inherently rhetorical discourse. [...] Even cartographers – as well as their critics – are beginning to allude to the notion of a rhetorical cartography, but what is still lacking is a rhetorical close-reading of maps (Wood and Fels 1986; Goffart 1988).

The issue in contention is not whether some maps are rhetorical, or whether other maps are partly rhetorical, but the extent to which rhetoric is a universal aspect of all cartographic texts. Thus for some cartographers the notion of 'rhetoric' would remain a pejorative term. It would be an 'empty rhetoric' which was unsubstantiated in the scientific content of a map. 'Rhetoric' would be used to refer to the 'excesses' of propaganda mapping or advertising cartography or an attempt would be made to confine it to an 'artistic' or aesthetic element in maps as opposed to their scientific core. My position is to accept that rhetoric is part of the way all texts work and that all maps are rhetorical texts. Again we ought to dismantle the arbitrary dualism between 'propaganda' and 'true', and between modes of 'artistic' and 'scientific' representation as they are found in maps. All maps strive to frame their message in the context of an audience. All maps state an argument about the world and they are propositional in nature. All maps employ the common devices of rhetoric such as invocations of authority (*especially* in 'scientific' maps) (Wood and Fels 1986) and appeal to a potential readership through the use of colours, decoration, typography, dedications,or written justifications of their method (Marin 1988). Rhetoric may be concealed but it is always present, for there is no description without performance.

The steps in making a map – selection, omission, simplification, classification, the creation of hierarchies and 'symbolisation' – are all inherently rhetorical. [...] Indeed, the freedom of rhetorical manoeuvre in cartography is considerable: the mapmaker merely omits those features of the world that lie outside the purpose of the immediate discourse. [...] I am not concerned to privilege rhetoric over science, but to dissolve the illusory distinction between the two in reading the social purposes as well as the content of maps.

Maps and the exercise of power

For the final stage in the argument I return to Foucault. In doing so I am mindful of Foucault's criticism of Derrida that he attempted 'to restrict interpretation to a purely syntactic and textual level' (Hoy 1985: 60; Norris 1987: 213–220), a world where political realities no longer exist. Foucault, on the other hand, sought to uncover 'the social practices that the text itself both reflects and employs' and to 'reconstruct the technical and material framework in which it arose' (Hoy 1985: 60). Though deconstruction is useful in helping to change the epistemological climate, and in encouraging a rhetorical reading of cartography, my final concern is with its social and political dimensions, and with understanding how the map works in society as a form of power-knowledge. This closes the circle to a context-dependent form of cartographic history. We have already seen how it is possible to view cartography as a discourse – a system which provides a set of rules for the representation of knowledge embodied in the images we define as maps and atlases. It is not difficult to find for maps – especially those produced and manipulated by the state – a niche in the 'power/knowledge matrix of the modern order' (Philp 1985: 76). Especially where maps are ordered by government (or are derived from such maps) it can be seen how they extend and reinforce the legal statutes, territorial imperatives and values stemming from the exercise of political power. Yet to understand how power works through cartographic discourse and the effects of that power in society further dissection is needed. A simple model of domination and subversion is inadequate and I propose to draw a distinction between *external* and *internal* power in cartography. This ultimately derives from Foucault's ideas about power-knowledge, but this particular formulation is owed to Joseph Rouse's (1987) book on *Knowledge and Power*, where a theory of the internal power of science is in turn based on his reading of Foucault.

The most familiar sense of power in cartography is that of power *external* to maps and mapping. This serves to link maps to the centres of political power. Power is exerted *on* cartography. Behind most cartographers there is a patron; in innumerable instances the makers of cartographic texts were responding to external needs. Power is also exercised *with* cartography. Monarchs, ministers, state institutions, the Church, have all initiated programmes of mapping for their own ends. In modern Western society maps quickly became crucial to the maintenance of state power – to its boundaries, to its commerce, to its internal administration, to control of populations and to its military strength. Mapping soon became the business of the state: cartography is early nationalised. The state guards its knowledge carefully: maps have been universally censored, kept secret and falsified. In all these cases maps are linked to what Foucault (1980: 88) called the exercise of 'juridical power'. The map becomes a 'juridical territory': it facilitates surveillance and control. Maps are still used to control our lives in innumerable ways. A map-less society, though we may take the map for granted, would now be politically unimaginable. All this is power *with* the help of maps. It is an external power, often centralised and exercised bureaucratically, imposed from above, and manifest in particular acts or phases of deliberate policy.

I come now to the important distinction. What is also central to the effects of maps in society is what may be defined as the power *internal* to cartography. The focus of inquiry therefore shifts from the place of cartography in a juridical system of power to the political effects of what cartographers do when they make maps. Cartographers manufacture power: they create a spatial panopticon. It is a power embedded in the map text. We can talk about the power of the map just as we already talk about the power of the word or about the book as a force for change. In this sense maps have politics (Winner 1980). It is a power that intersects and is embedded in knowledge. It is universal. Foucault writes of the omnipresence of power: not because it has the privilege of consolidating everything under its invincible unity, but because it is produced from one moment to the next, at every point, or rather in every relation from one point to another. Power is everywhere; not because it embraces everything, but because it comes from everywhere (Foucault 1978: 93).

Power comes from the map and it traverses the way maps are made. The key to this internal power is thus cartographic process. By this I mean the way maps are compiled and the categories of information selected; the way they are generalised, a set of rules for the abstraction of the landscape; the way the elements in the landscape are formed into hierarchies; and the way various rhetorical styles that also reproduce power are employed to represent the landscape. To catalogue the world is to appropriate it (Barthes 1980: 27), so that all these technical processes represent acts of control over its image which extend beyond the professed uses of cartography. The world is disciplined. The world is normalised. We are prisoners in its spatial matrix. For cartography as much as other forms of knowledge, 'All social action flows through boundaries determined by classification schemes' (Darnton 1984: 192–193). An analogy is to what happens to data in the cartographer's workshop and what happens to people in the disciplinary institutions – prisons, schools, armies, factories – described by Foucault (Rous, 1987): in both cases a process of normalistion occurs. Or similarly, just as in factories we standardise our manufactured goods, so in our cartographic workshops we standardise our images of the world. Just as in the laboratory we create formulaic understandings of the processes of the physical world, so, too, in the map, nature is reduced to a graphic formula (Monmonier and Schnell 1988: 15). The power of the mapmaker was not generally exercised over individuals but over the knowledge of the world made available to people in general. Yet this is not consciously done and it transcends the simple categories of 'intended' and 'unintended' altogether. I am not suggesting that power is deliberately or centrally exercised. It is a local knowledge which at the same time is universal. It usually passes unnoticed. The map is a silent arbiter of power.

What have been the effects of this 'logic of the map' upon human consciousness, if I may adapt Marshall McLuhan's (1962) phrase ('logic of print'). Like him I believe we have to consider for maps the effects of abstraction, uniformity, repeatability and visuality in shaping mental structures, and in imparting a sense of the places of the world. It is the disjunction between those senses of place, and many alternative visions of what the world is, or what it might be, that has raised questions about the effect of cartography in society. Thus, Theodore Roszac (1972: 410) writes, 'The cartographers are talking about their maps and not landscapes. That is why what they say frequently becomes so paradoxical when translated into ordinary language. When they forget the difference between map and landscape – and when they permit or persuade us to forget that difference – all sorts of liabilities ensue'.

One of these 'liabilities' is that maps, by articulating the world in mass-produced and stereotyped images, express an embedded social vision. Consider, for example, the fact that the ordinary road atlas is among the best selling paperback books in the United States (McNally 1987), and then try to gauge how this may have affected ordinary Americans' perception of their country. What sort of an image of America do these atlases promote? On the one hand, there is a patina of gross simplicity. Once off the interstate highways the landscape dissolves into a generic world of bare essentials that invites no exploration. Context is stripped away and place is no longer important. On the other hand, the maps reveal the ambivalence of all stereotypes. Their silences are also inscribed on the page: where, on the page, is the variety of nature, where is the history of the landscape, and where is the space–time of human experience in such anonymised maps (Barthes 1980; Szegö 1987; Roszak 1972).

The question has now become: do such empty images have their consequences in the way we think about the world? Because all the world is designed to look the same, is it easier to act upon it without realising the social effects? It is in the posing of such questions that the strategies of Derrida and Foucault appear to clash. For Derrida, if meaning is undecidable so must be *pari passu*, the measurement of the force of the map as a discourse of symbolic action. In ending, I prefer to align myself with Foucault in seeing all knowledge (Rabinow 1984: 6–7) – and hence cartography – as thoroughly enmeshed with the larger battles which constitute our world. Maps are not external to these struggles to alter power relations. The history of map use suggests that this may be so and that maps embody specific forms of power and authority. Since the Renaissance they have changed the way in which power was exercised. In colonial North America, for example, it was easy for Europeans to draw lines across the territories of Indian nations without sensing the reality of their political identity (Harley 1988d). The map allowed them to say, 'This is mine; these are the boundaries' (Boelhower 1984: 47). [. . .] While the map is never the reality, in such ways it helps to create a different reality. Once embedded in the published text the lines on the map acquire an authority that may be hard to dislodge. Maps are authoritarian images. Without our being aware of it, maps can reinforce and legitimate the status quo. Sometimes agents of change, they can equally become conservative documents. But in either case the map is never neutral. Where it seems to be neutral it is the sly 'rhetoric of neutrality' (Kinross 1985) that is trying to persuade us.

Conclusion

The interpretive act of deconstructing the map can serve three functions in a broad enquiry into the history of cartography. Firstly, it allows us to challenge the epistemological myth (created by cartographers) of the cumulative progress of an objective science always producing better delineations of reality. Secondly, deconstructionist argument allows us to redefine the historical importance of maps. Rather than invalidating their study, it is enhanced by adding different nuances to our understanding of the power of cartographic representation as a way of building order into our world. If we can accept intertextuality then we can start to read our maps for alternative and sometimes competing discourses. Thirdly, a deconstructive turn of mind may allow map history to take a fuller place in the interdisciplinary study of text and knowledge. Intellectual strategies, such as those of discourse in the Foucauldian sense, the Derridian notion of metaphor and rhetoric as inherent to scientific discourse, and the pervading concept of power/knowledge, are shared by many subjects. As ways of looking at maps they are equally enriching. They are neither inimical to hermeneutic enquiry nor antihistorical in their thrust. By dismantling we build. The possibilities of discovering meaning in maps and of tracing the social mechanisms of cartographic change are enlarged. Postmodernism offers a challenge to read maps in ways that could reciprocally enrich the reading of other texts.

References

Barthes, R. (1973) *Mythologies: Selected and Translated from the French by Annette Lavers*, Paladin, London.

Barthes, R. (1980) The plates of the encyclopedia, in *New Critical Essays*, Hill and Wang, New York.

Blocker, H.G. (1979) *Philosophy and Art*, Charles Scribner's Sons, New York.

Boelhower, W. (1984) *Through a Glass Darkly: Ethnic Semiosis in American Literature*, Edizioni Helvetia, Venezia.

Boelhower, W. (1988) Inventing America: a model of cartographic semiosis. *Word and Image*, **4** (2), 475–497.

Campbell, P.N. (1973) Scientific discourse. *Philosophy and Rhetoric*, **6** (1), 1–29.

Cartographic Journal (1985) The so-called Peters projection. *The Cartographic Journal*, **22** (2), 108–110.

Clarke, C.N.G. (1988) Taking possession: the cartouche as cultural text in eighteenth-century American maps. *Word and Image*, **4** (2), 455–474.

Crone, G.R. (1953) *Maps and Their Makers: An Introduction to the History of Cartography*, 1st edn, Dawson, Folkestone, UK.

Darnton, R. (1984) *The Great Cat Massacre and Other Episodes in French Cultural History*, Basic Books, New York.

Dear, M. (1988) The postmodern challenge: reconstructing human geography. *Transactions of the Institute of British Geographers NS*, **13**, 262–274.

Derrida, J. (1976) *Of Grammatology* (trans. G.C. Spivak), The John Hopkins University Press, Baltimore, MD.

Eagleton, T. (1983) *Literary Theory: An Introduction*, University of Minnesota Press, Minneapolis, MN.

Eagleton, T. (1986) *Against the Grain*, Verso, London.

Ferraris, M. (1988) Postmodernism and the deconstruction of modernism. *Design Issues*, **4** (1/2), 12–24.

Foucault, M. (1973) *The Order of Things: An Archaeology of the Human Sciences*, Vintage Books, New York.

Foucault, M. (1978) *The History of Sexuality: Volume I. An Introduction* (trans. R. Hurley), Random House, New York.

Foucault, M. (1980) *Power/Knowledge: Selected Interviews and Other Writings, 1972–1977* (ed. C. Gordon; trans. C. Gordon, L. Marshall, J. Mepham and K. Sopher), Pantheon Books, New York.

Gauthier, M. (1988) *Cartographie dans les Medias*, Presses de l'Université du Québec, Québec.

Geertz, G. (1983) Art as a cultural system, in *Local Knowledge: Further Essays in Interpretive Anthropology*, Basic Books, New York.

Goffart, W. (1988) The map of the barbarian invasions: a preliminary report. *Nottingham Medieval Studies*, **32**, 49–64.

Gregory, D. (1987) Postmodernism and the politics of social theory. *Environment and Planning D: Society and Space*, **5**, 245–248.

Harley, J.B. (1984) Meaning and ambiguity in Tudor cartography, in *English Map-Making, 1500–1650: Historical Essays* (ed. S. Tyacke), The British Library Reference Division Publications, London, pp. 22–45.

Harley, J.B. (1988a) L'histoire de la cartographie comme discourse. *Préfaces*, 5 December, 70–75.

Harley, J.B. (1988b) Maps, knowledge, and power, in *The Iconography of Landscape* (eds D. Cosgrove and S. Daniels), Cambridge University Press, Cambridge.

Harley, J.B. (1988c) Silences and secrecy: the hidden agenda of cartography in early modern Europe. *Imago Mundi*, **40**, 57–76.

Harley, J.B. (1988d) Victims of a map: New England cartography and the native Americans. Paper read at the Land of Norumbega Conference, Portland, Maine.

Hoy, D. (1985) Jacques Derrida, in *The Return of Grand Theory in the Human Sciences* (ed. Q. Skinner), Cambridge University Press, Cambridge, pp. 65–82.

Hooykaas, R. (1987) The rise of modern science: when and why? *The British Journal for the History of Science*, **20** (4), 453–473.

Kinross, R. (1985) The rhetoric of neutrality. *Design Issues*, **2** (2), 18–30.

Knox, P.L. (1988) *The Design Professions and the Built Environment*, Croom Helm, London.

Korzybski, A. (1948) *Science and Sanity: An Introduction to Non-Aristotelian Systems and General Semantics*, 3rd edn, The International Non-Aristotelian Library, Lakeville, CT.

Kuhn, T.S. (1962) *The Structure of Scientific Revolutions*, University of Chicago Press, Chicago.

Lakatos, I. and Musgrave, A. (1970) *Criticism and the Growth of Knowledge*, Cambridge University Press, Cambridge.

Laudan, L. (1977) *Progress and Its Problems: Toward a Theory of Scientific Growth*, University of California Press, Berkeley, CA.

Loxton, J. (1985) The Peters phenomenon. *The Cartographic Journal*, **22** (2), 106–108.

Marin, L. (1988) *Portrait of the King, Theory and History of Literature*, vol. 57 (trans. M.M. Houle), University of Minnesota Press, Minneapolis, MN, pp. 169–179.

McKenzie, D.F. (1986) *Bibliography and the Sociology of Texts*, The British Library, London.

McLuhan, M. (1962) *The Gutenberg Galaxy: The Making of Typographic Man*, University of Toronto Press, Toronto.

McNally, A. (1987) You can't get there from here, with today's approach to geography. *The Professional Geographer*, **39**, 389–392.

Mitchell, W.J.T. (1986) *Iconology: Image, Text, Ideology*, University of Chicago Press, Chicago.

Monmonier, M. and Schnell, G.A. (1988) *Map Appreciation*, Prentice Hall, Englewood Cliffs, NJ.

Norris, C. (1982) *Deconstruction: Theory and Practice*, Methuen, London.

Norris, C. (1987) *Derrida*, Harvard University Press, Cambridge, MA.

Peters, A. (1983) *The New Cartography*, Friendship Press, New York.

Philp, M. (1985) Michel Foucault, in *The Return of Grand Theory in the Human Sciences* (ed. Q. Skinner), Cambridge University Press, Cambridge, pp. 41–64.

Porter, P. and Voxland, P. (1986) Distortion in maps: the Peters' projection and other devilments. *Focus*, **36**, 22–30.

Rabinow, P. (1984) *The Foucault Reader*, Pantheon Books, New York.
Robinson, A.H. (1985) Arno Peters and his new cartography. *The American Cartographer*, **12**, 103–111.
Robinson, A.H., Sale, R.D., Morrison, J.L. and Muehrcke, P.C. (1984) *Elements of Cartography*, 5th edn, John Wiley & Sons, Inc., New York.
Rorty, R. (1979) *Philosophy and the Mirror of Nature*, Princeton University Press, Princeton, NJ.
Roszak, T. (1972) *Where the Wasteland Ends: Politics and Transcendence in Postindustrial Society*, Doubleday, New York.
Rouse, J. (1987) *Knowledge and Power: Toward a Political Philosophy of Science*, Cornell University Press, Ithaca, NY.
Said, E.W. (1978) The problem of textuality: two exemplary positions. *Critical Inquiry*, **4** (4), 673–714.
Snyder, J.P. (1988) Social consciousness and world maps. *The Christian Century, 24 February*, 190–192.
Soja, E.W. (1989) *Postmodern Geographies*, Verso, London.
Szegö, J. (1987) *Human Cartography: Mapping the World of Man* (trans. T. Miller), Swedish Council for Building Research, Stockholm.
Winner, L. (1980) Do artifacts have politics? *Daedalus*, **109** (1), 121–136.
Wolter, J.A. (1975a) The emerging discipline of cartography. Unpublished PhD dissertation. University of Minnesota.
Wolter, J.A. (1975b) Cartography – an emerging discipline. *The Canadian Cartographer*, **12** (2), 210–216.
Wood, D. (1987) Pleasure in the idea: the atlas as narrative form, in *Atlases for Schools: Design Principles and Curriculum Perspectives* (eds G.J.A. Carswell, N.M. de Leeuw and R.J.B. Waters), *Cartographica* **36**, 24–45.
Wood, D. and Fels, J. (1986) Designs on signs/myth and meaning in maps. *Cartographica*, **23** (3), 54–103.
Woolgar, S. (1988) *Science: The Very Idea*, Ellis Horwood, Chichester, UK.

Further reading

Belyea, B. (1992) Images of power: Derrida/Foucault/Harley. *Cartographica*, **29** (2), 1–9. [A critique of Harley that argues he did not really follow through on the deeper implications of Foucauldian and Derridian approaches to deconstruction.]
Dahl, E.H. (1991) Responses to J.B. Harley's article 'Deconstructing the map'. *Cartographica*, **26** (3/4), 89–127. [A series of immediate responses to Harley's article from a number of different intellectual positions.]
Edney, M.H. (2005) The origins and development of J.B. Harley's cartographic theories. *Cartographica*, **40** (1/2). [A systematic and rigorous assessment of the intellectual background to Harley's ideas.]
Harley, J.B. (2001) *The New Nature of Maps: Essays in the History of Cartography*, The Johns Hopkins University Press, Baltimore, MD. [A posthumous collection of Harley's writings including a useful introductory chapter from Andrews that critiques Harley's social constructivist overtones as unnecessary whilst celebrating Harley's empiricist, humanist and historical scholarship.]
Pickles, J. (2004) *A History of Spaces: Cartographic Reason, Mapping, and the Geo-coded World*, Routledge, London. [A Foucauldian history of spaces and the constitutive power of mapping to bring them to social life, strongly influenced by Harley's work.]
Wood, D. (2002) The map as a kind of talk: Brian Harley and the confabulation of the inner and outer voice. *Visual Communication*, **1** (2), 139–161. [An insightful response to Harley's challenge from his contemporary and the other major influence on the establishment of a more critical cartography in the 1990s.]

See also

- Chapter 1.7: Design on Signs/Myth and Meaning in Maps
- Chapter 1.10: Cartography Without 'Progress': Reinterpreting the Nature and Historical Development of Mapmaking
- Chapter 1.12: The Agency of Mapping
- Chapter 1.13: Beyond the 'Binaries': A Methodological Intervention for Interrogating Maps as Representational Practices
- Chapter 1.14: Rethinking Maps
- Chapter 2.3: Manufacturing Metaphors: Public Cartography, the Market, and Democracy
- Chapter 3.6: Some Truth with Maps: A Primer on Symbolisation and Design
- Chapter 4.5: The Map as Biography
- Chapter 4.6: Reading Maps
- Chapter 4.8: Refiguring Geography: Parish Maps of Common Ground

Chapter 1.9

Drawing Things Together

Bruno Latour

Editors' overview

Work from the tradition of Science Studies has become influential for scholars investigating mapping. As a technology and set of practices the social work of mapping depends upon the power of its transcriptions: mapping *draws upon* the world, but also *draws things together*. This excerpted chapter from Latour builds on his empirical work exploring the everyday practices of scientists and applies these ideas to the creative acts of inscription and their contexts. He argues for a consideration of the practices of knowledge construction and against an institutional or purely cognitive view of science. He establishes the critical role that inscriptions might play in the formation of scientific knowledge and practices, as devices to facilitate thinking and the communication and fixing of ideas. Using the example of the map deployed by explorer La Pérouse he establishes that we might usefully focus on the role of what he terms 'immutable mobiles', and traces historical and contextual processes through which these objects and practices emerge and are deployed. He develops and exemplifies nine key characteristics of 'paperwork' that allow interests to emerge around shared inscriptions, such as maps: mobile, immutable, flat, modifiable in scale, reproducible, capable of being recombined and layered, but also optically consistent and amenable to insertion into other texts. He argues that research might usefully focus on the centres of calculation and devices that allow the social power of mapping to be exercised.

Originally published in 1990: Chapter 2 in *Representation in Scientific Practice* (eds Michael Lynch and Steve Woolgar), MIT Press, Cambridge, MA, pp. 19–68.

Thinking with eyes and hands

It would be nice to be able to define what is specific to our modem scientific culture. It would be still nicer to find the most economical explanation (which might not be the most economic one) of its origins and special characteristics. [. . .]

[. . .] I was struck, in a study of a biology laboratory, by the way in which many aspects of laboratory practice could be ordered by looking not at the scientists' brains (I was forbidden access), at the cognitive structures (nothing special), nor at the paradigms (the same for thirty years), but at the transformation of rats and chemicals into paper (Latour and Woolgar 1979). Focusing on the literature, and the way in which anything and everything was transformed into inscriptions was [. . .] what the laboratory was made for. Instruments, for instance, were of various types, ages and degrees of sophistication. Some were pieces of furniture, others filled large rooms, employed many technicians and took many weeks to run. But their end result, no matter the field, was always a small window through which one could read a very few signs from a rather poor repertoire (diagrams, blots, bands, columns). All these inscriptions, as I called them, were combinable, superimposable and could, with only a minimum of cleaning up, be integrated as figures in the text of the articles people were writing.

My contention is that writing and imaging cannot by themselves explain the changes in our scientific societies. [. . .] Thus it is not all the anthropology of writing, nor all the history of visualisation that interests us in this context. Rather, we should concentrate on those aspects that help in the mustering, the presentation, the increase, the effective alignment or ensuring the fidelity of new allies. We need, in other words, to look at the way in which someone convinces someone else to take up a statement, to pass it along, to make it more of a fact, and to recognise the first author's ownership and originality. [. . .] If we remain at the level of

The Map Reader: Theories of Mapping Practice and Cartographic Representation, First Edition. Edited by Martin Dodge, Rob Kitchin and Chris Perkins.

the visual aspects only, we fall back into a series of weak clichés or are led into all sorts of fascinating problems of scholarship far away from our problem; but, on the other hand, if we concentrate on the agonistic situation alone, the principle of any victory, any solidity in science and technology escapes us forever. We have to hold the two eye pieces together so that we turn it into a real *bi*nocular; it takes time to focus, but the spectacle, I hope, is worth the waiting.

One example will illustrate what I mean. La Pérouse travels through the Pacific for Louis XVI with the explicit mission of bringing *back* a better map. One day, landing on what he calls Sakhalin he meets with Chinese and tries to learn from them whether Sakhalin is an island or a peninsula. To his great surprise the Chinese understand geography quite well. An older man stands up and draws a map of his island on the sand with the scale and the details needed by La Pérouse. Another, who is younger, sees that the rising tide will soon erase the map and picks up one of La Pérouse's notebooks to draw the map again with a pencil . . .

What are the differences between the savage geography and the civilised one? [. . .] The Chinese are quite able to think in terms of a map but also to talk about navigation on an equal footing with La Pérouse. Strictly speaking, the ability to draw and to visualise does not really make a difference either, since they all draw maps more or less based on the same principle of projection, first on sand, then on paper. So perhaps there is no difference after all and, geographies being equal, relativism is right? This, however, cannot be, because La Pérouse does something that is going to create an enormous difference between the Chinese and the European. What is, for the former, a drawing of no importance that the tide may erase, is, for the latter, the *single object* of his mission. What should be brought into the picture is how the picture is brought back. The Chinese does not have to keep track, since he can generate many maps at will, being born on this island and fated to die on it. La Pérouse is not going to stay for more than a night; he is not born here and will die far away. What is he doing, then? He is passing through all these places in order to take something *back* to Versailles, where many people expect his map to determine who was right and wrong about whether Sakhalin was an island, who will own this and that part of the world, and along which routes the next ships should sail. Without this peculiar trajectory, La Pérouse's exclusive interest in traces and inscriptions will be impossible to understand – this is the first aspect, but without dozens of innovations in inscription, in projection, in writing, archiving and computing, his displacement through the Pacific would be totally wasted – and this is the second aspect, as crucial as the first. We have to hold the two together. Commercial interests, capitalist spirit, imperialism, thirst for knowledge, are empty terms as long as one does not take into account Mercator's projection, marine clocks and their markers, copper engraving of maps, rutters, the keeping of 'log books', and the many printed editions of Cook's voyages that La Pérouse carries with him. [. . .] But, on the other hand, no innovation in the way longitude and latitudes are calculated, clocks are built, log books are compiled, copperplates are printed, would make any difference whatsoever if they did not help to muster, align and win over new and unexpected allies, far away, in Versailles. The practices I am interested in would be pointless if they did not bear on certain controversies and force dissenters into believing new facts and behaving in new ways. This is where an exclusive interest in visualisation and writing falls short, and can even be counterproductive. To maintain only the second line of argument would offer a mystical view of the powers provided by semiotic material [. . .]; to maintain only the first would be to offer an idealist explanation (even if clad in materialist clothes).

The aim of this paper is to pursue the two lines of argument at once. To say it in yet other words, we do not find all explanations in terms of inscription equally convincing, but only those that help us to understand how the mobilisation and mustering of new resources is achieved. We do not find all explanations in terms of social groups, interests or economic trends, equally convincing but only those that offer a specific mechanism to sum up 'groups', 'interests', 'money' and 'trends': mechanisms which, we believe, depend upon the manipulation of paper, print, images and so on. La Pérouse shows us the way, since without new types of inscriptions nothing usable would have come back to Versailles from his long, costly and fateful voyage; but without this strange mission that required him to go away and to come back so that others in France might be convinced, no modification in inscription would have made a bit of difference. The essential characteristics of inscriptions cannot be defined in terms of visualisation, print and writing. In other words, it is not perception which is at stake in this problem of visualisation and cognition. New inscriptions, and new ways of perceiving them, are the results of something deeper. If you wish to go out of *your* way and come back heavily equipped so as to force others to go out of *their* ways, the main problem to solve is that of *mobilisation*. You have to go and to come back *with* the 'things' if your moves are not to be wasted. But the 'things' have to be able to withstand the return trip without withering away. [. . .] The 'things' you gathered and displaced have to be presentable all at once to those you want to convince and who did not go there. In sum, you have to invent objects which have the properties of being *mobile* but also *immutable*, *presentable*, *readable* and *combinable* with one another.

On immutable mobiles

It seems to me that most scholars who have worked on the relations between inscription procedures and cognition, have, in fact, in their various ways, been writing about the history of these immutable mobiles.

A. *Optical consistency*

[. . .] The rationalisation that took place during the so-called 'scientific revolution' is not of the mind, of the eye, of philosophy, but of the *sight*. Why is perspective such an important invention? 'Because of its logical recognition of internal invariances through all the transformations produced by changes in spatial location' (Ivins 1973: 9). In a linear perspective, no matter from what distance and angle an object is seen, it is always possible to transfer it – to translate it – and to obtain the same object at a different size as seen from another position. In the course of this translation, its internal properties have not been modified. This immutability of the displaced figure allows Ivins to make a second crucial point: since the picture moves without distortion it is possible to establish, in the linear perspective framework, what he calls a 'two-way' relationship between object and figure. Ivins shows us how perspective allows movement through space with, so to speak, a return ticket. [. . .] With perspective, exactly as with La Pérouse's map – and for the same reasons – a new set of movements is made possible: you can go out of your way and come back with all the places you passed; these are all written in the same homogeneous language (longitude and latitude, geometry) that allows you to change scale, to make them presentable and to combine them at will.

Perspective, for Ivins, is an essential determinant of science and technology because it creates 'optical consistency'. [. . .] The shift from the other senses to vision is a consequence of the agonistic situation. You present absent things. No one can smell or hear or touch Sakhalin Island, but you can look at the map and determine at which bearing you will see the land when you send the next fleet. The speakers are talking to one another, feeling, hearing and touching each other, *but* they are now talking *with* many absent things presented all at once. This presence/absence is possible through the two-way connection established by these many contrivances – perspective, projection, map, log book and so on – that allows translation without corruption.

[. . .]

B. *Visual culture*

Still more striking than the Italian perspective described by Ivins and Edgerton, is the Dutch 'distance point' method for drawing pictures. [. . .] A new visual culture redefines both what it is to see, and what there is to see. [. . .] The new precise scenography that results in a world view defines at once what is science, what is art and what it is to have a world economy. [. . .] A little lowland country becomes powerful by making a few crucial inventions which allow people to accelerate the mobility and to enhance the immutability of inscriptions: the world is thus gathered up in this tiny country.

Alpers' description of Dutch visual culture reaches the same result as Edgerton's study of technical drawings: a new meeting place is designed for fact and fiction, words and images. The map itself is such a result. [. . .] The main quality of the new space is not to be 'objective' as a naïve definition of realism often claims, but rather to have optical consistency. This consistency entails the '*art of describing*' everything and the possibility of going from one type of visual trace to another. Thus, we are not surprised that letters, mirrors, lenses, painted words, perspectives, inventories, illustrated child books, microscope and telescope come together in this visual culture. All innovations are selected 'to secretly see and without suspicion what is done far off in other places' (cited in Alpers 1983: 201).

C. *A new way of accumulating time and space*

Another example will demonstrate that inscriptions are not interesting *per se* but only because they increase either the mobility or the immutability of traces. The invention of print and its effects on science and technology is a cliché of historians. But no one has renewed this Renaissance argument as completely as Elizabeth Eisenstein (1979). Why? Because she considers the printing press to be [. . .] a device that makes both mobilisation and immutability possible at the same time. Eisenstein does not look for one cause of the scientific revolution, but for a secondary cause that would put all the efficient causes in relation with one another. [. . .] Immutability is ensured by the process of printing many identical copies; mobility by the number of copies, the paper and the movable type. The links between different places in time and space are completely modified by this fantastic acceleration of immutable mobiles which circulate everywhere and in all directions in Europe. As Ivins has shown, perspective *plus* the printing press *plus* aqua forte is the really important combination, since books can now carry with them the realistic images of what they talk about. For the first time, a location can accumulate other places far away in space and time, and present them: [. . .] this synoptic presentation, once reworked, amended or disrupted, can be spread with no modification to other places and made available at other times.

[. . .]

On inscriptions

What is so important in the images and in the inscriptions scientists and engineers are busy obtaining, drawing, inspecting, calculating and discussing? It is, first of all, the unique advantage they give in the rhetorical or polemical situation. 'You doubt of what I say? I'll show you.' And, without moving more than a few inches, I unfold in front of your eyes figures, diagrams, plates, texts, silhouettes, and then and there present things that are far away and with which some sort of two-way connection has now been established. I do not think the importance of this simple mechanism can be overestimated. [. . .] We are so used to this world of print and images, that we can hardly think of what it is to know something without indexes, bibliographies, dictionaries, papers with references, tables, columns, photographs, peaks, spots, bands.

One simple way to make the importance of inscriptions clearer is to consider how little we are able to convince when deprived of these graphisms through which mobility and immutability are increased. As Dagognet (1969, 1973) has shown in two excellent books, no scientific discipline exists without first inventing a visual and written language which allows it to break with its confusing past. [. . .]

In another suggestive book, Fabian (1983) tries to account for anthropology by looking at its craftsmanship of visualisation. The main difference between us and the savages, he argues, is not in the culture, in the mind, or in the brain, but in the way *we* visualise *them*. An asymmetry is created because we create a space and a time in which we place the other cultures, but they do not do the same. For instance, we map their land, but they have no maps either of their land or of ours. [. . .] Fabian's argument [. . .] is that once this first violence has been committed, no matter what we do, we will not understand the savages any more. Fabian [. . .] wishes to find another way to 'know' the savages. [. . .] The constraints imposed by convincing people, going out and coming back, are such that this can be achieved only if everything about the savage life is transformed into immutable mobiles that are easily readable and presentable. [. . .]

There is no detectable difference between natural and social science, as far as the obsession for graphism is concerned. [. . .] Scientists start seeing something once they stop looking at nature and look exclusively and obsessively at prints and flat inscriptions. In the debates around perception, what is always forgotten is this simple drift from watching confusing three-dimensional objects to inspecting two-dimensional images which have been *made less confusing*. Lynch, like all laboratory observers, has been struck by the extraordinary obsession of scientists with papers, prints, diagrams, archives, abstracts and curves on graph paper. No matter what they talk about, they start talking with some degree of confidence and being believed by colleagues only once they point at simple geometrised two-dimensional shapes. [. . .] What is extracted from them is a tiny set of figures. This extraction, like the few longitudes and latitudes extracted from the Chinese by La Pérouse, is *all that counts*. [. . .] Every time there is a dispute, great pains are taken to find, or sometimes to invent, a new instrument of visualisation, which will enhance the image, accelerate the readings, and, as Lynch has shown, conspire with the visual characteristics of the things that lend themselves to diagrams on paper (coast lines, stars which are like points, well aligned cells, etc.).

[. . .]

So, the phenomenon we are tackling is *not* inscription *per se*, but the *cascade* of ever simplified inscriptions that allow harder facts to be produced at greater cost. [. . .] Although more empirical studies should be made in many different fields, there seems to be a trend in these cascades. They always move on the direction of the greater merging of figures, numbers and letters, merging greatly facilitated by their homogeneous treatment as binary units in and by computers.

This trend toward simpler and simpler inscriptions that mobilise larger and larger numbers of events in one spot cannot be understood if separated from the agonistic model that we use as our point of reference. [. . .] Knorr (1981) has criticised this argument by taking an ethno-methodological standpoint. She argues, and rightly so, that an image, a diagram, cannot convince anyone, both because there are always many interpretations possible, and, above all, because the diagram does not force the dissenter to look at it. She sees the interest in inscription devices as an exaggeration of the power of semiotics (and a French one at that). But such a position misses the point of my argument. It is precisely because the dissenter can always escape and try out another interpretation, that so much energy and time is devoted by scientists to *corner* him and surround him with ever more dramatic visual effects. Although *in principle* any interpretation can be opposed to any text and image, *in practice* this is far from being the case; the cost of dissenting increases with each new collection, each new labelling, each new redrawing. This is especially true if the phenomena we are asked to believe are invisible to the naked eye; quasars, chromosomes, brain peptides, leptons, gross national products, classes, coast lines are never seen but through the 'clothed' eye of inscription devices. Thus, one *more* inscription, one more trick to enhance contrast, one simple device to decrease background, one colouring procedure, might be *enough*, all things being equal, *to swing the balance* of power and turn an incredible statement into a credible one. [. . .] It is possible to overestimate the inscription, but not the setting in which the cascade of ever more written and numbered inscriptions is produced. What we are really dealing with is the *staging* of a scenography in which attention is focused on one set of dramatised inscriptions. The setting works like a giant 'optical device' that creates a new laboratory, a new type of vision and a new phenomenon to look at. [. . .]

Capitalising inscriptions to mobilise allies

Can we summarise why it is so important [. . .] to work on two-dimensional inscriptions instead of the sky, the air, health or the brain? [. . .] Let me list a few of the advantages of the 'paperwork'.

(1) Inscriptions are *mobile*, as I indicated for La Pérouse's case. Chinese, planets, microbes – none of these can move; however, maps, photographic plates, and Petri dishes can.

(2) They are *immutable* when they move, or at least everything is done to obtain this result: specimens are chloroformed, microbial colonies are stuck into gelatine, even exploding stars are kept on graph papers in each phase of their explosion.

(3) They are made *flat*. There is nothing you can dominate as easily as a flat surface of a few square metres; there is nothing hidden or convoluted, no shadows, no 'double entendre'. In politics, as in science, when someone is said to 'master' a question or to 'dominate' a subject, you should normally look for the flat surface that enables mastery (a map, a list, a file, a census, the wall of a gallery, a card index, a repertory); and you will find it.

(4) The *scale* of the inscriptions may be *modified* at will, without any change in their internal proportions. Observers never insist on this simple fact: no matter what the (reconstructed) size of the phenomena, they all end up being studied only when they reach the same average size. [. . .] This trivial change of scale seems innocuous enough, but it is the cause of most of the 'superiority' of scientists and engineers. [. . .]

(5) They can be *reproduced* and spread at little cost, so that all the instants of time and all the places in space can be gathered in another time and place. This is 'Eisenstein's effect'.

(6) Since these inscriptions are mobile, flat, reproducible, still and of varying scales, they can be reshuffled and *recombined*. Most of what we impute to connections in the mind may be explained by this reshuffling of inscriptions that all have the same 'optical consistency'. [. . .]

(7) [. . .] It is possible to *superimpose* several images of totally different origins and scales. To link geology and economics seems an impossible task, but to superimpose a geological map with the printout of the commodity market at the New York Stock Exchange requires good documentation and takes a few inches. Most of what we call 'structure', 'pattern', 'theory' and 'abstraction' are consequences of these superimpositions (Bertin 1973). [. . .]

(8) But one of the most important advantages is that the inscription can, after only little cleaning up, be *made part of a written text*. [. . .] This characteristic of scientific texts has been shown by Ivins and Eisenstein for the past. A present day laboratory may still be defined as the unique place where a text is made to comment on things which are all present in it. Because the commentary, earlier texts (through citations and references) and 'things' have the same optical consistency, and the same semiotic homogeneity, an extraordinary degree of certainty is achieved by writing and reading these articles (Latour and Bastide 1985; Lynch 1985; Law 1983). The text is not simply 'illustrated', it carries all there is to see in what it writes about. Through the laboratory, the text and the spectacle of the world end up having the same character.

(9) But the last advantage is the greatest. The two-dimensional character of inscriptions allows them to merge *with geometry*. As we saw for perspective, space on paper can be made continuous with three-dimensional space. The result is that we can work on paper with rulers and numbers, but still manipulate three-dimensional objects 'out there' (Ivins 1973). Better still, because of this optical consistency, everything, no matter where it comes from, can be converted into diagrams and numbers, and combination of numbers and tables can be used which are still easier to handle than words or silhouettes (Dagognet 1973). You cannot measure the sun, but you can measure a photograph of the sun with a ruler. Then the number of centimetres read can easily migrate through different scales, and provide solar masses for completely different objects. [. . .]

These nine advantages should not be isolated from one another and should always be seen in conjunction with the mobilisation process they accelerate and summarise. [. . .]

It is especially interesting to focus on the ninth advantage. [. . .] To go from 'empirical' to 'theoretical' sciences is to go from slower to faster mobiles, from more mutable to less mutable inscriptions. The trends we studied above do not break down when we look at formalism but, on the contrary, increase fantastically. [. . .] The mobilisation of many resources through space and time is essential for domination on a grand scale. [. . .] [B]y themselves the inscriptions do *not* help a location to become a centre that dominates the rest of the world. Something has to be done to the inscriptions which is similar to what the inscriptions do to the 'things', so that at the end a few elements can manipulate all the others on a vast scale. [. . .]

Let us take as example 'the effectiveness of Galileo's work', as it is seen by Drake (1970). Drake does indeed use the word formalism to designate what Galileo is able to do that his predecessors were not. But what is described is more interesting than that. Drake compares the diagrams and commentaries of Galileo with those two older scholars, Jordan and Stevin. [. . .] What seems to happen is that Galileo's two predecessors could not visually *accommodate* the problem on a paper surface and see the result simultaneously as both geometry and physics. A simple change in the geometry used by Galileo allows him to connect many different problems, whereas his two predecessors worked on disconnected shapes over which they had no control. [...] This ability to connect might be located in Galileo's mind. In fact, what gets connected are three different visual horizons held synoptically because the surface of paper is considered as geometrical space. [...]

[...]

We now come closer to an understanding of the matter that constitutes formalism. The point of departure is that we are constantly hesitating between several often contradictory indications from our senses. Most of what we call 'abstraction' is in practice the belief that a written inscription must be believed more than any contrary indications from the senses. Koyré, for instance, has shown that Galileo believed in the inertia principle on mathematical grounds even against the contrary evidences offered to him not only by the Scriptures, but also by the senses. Koyré claims that this rejection of the senses was due to Galileo's Platonist philosophy. This might be so. But what does it mean practically? It means that faced with many contrary indications, Galileo, in the last instance, believed *more* in the triangular diagram for calculating the law of falling bodies then in any *other* vision of falling bodies (Koyré 1966: 147). When in doubt, believe the inscriptions, written in mathematical terms, *no matter* to what absurdities this might lead you.

[...]: What is this society in which a written, printed, mathematical form has greater credence, in case of doubt, than anything else: common sense, the senses other than vision, political authority, tradition and even the Scriptures? [...] Without this peculiar tendency to privilege what is written, the power of inscription would be entirely lost, as Edgerton hints in his discussion of Chinese diagrams. No matter how beautiful, rich, precise or realistic inscriptions may be, no one would believe what they showed if they could be contradicted by other evidence of local, sensory origin or pronouncements of the local authorities. I feel that we would make a giant step forward if we could relate this peculiar feature of our culture with the requirement of mobilisation I have outlined several times. Most of the 'domain' of cognitive psychology and epistemology does not exist but is related to this strange anthropological puzzle: a training (often in schools) to manipulate written inscriptions, to array them in cascades and to believe the last one on the series more than any evidence to the contrary. It is in the description of this training that the anthropology of geometry and mathematics should be decisive (Livingston 1983; Lave 1985, 1986; Serres 1982).

Paperwork

[...]

Th[e] role of the bureaucrat *qua* scientist *qua* writer and reader, is always misunderstood because we take for granted that there exist, somewhere in society, macro-actors that naturally dominate the scene: Corporation, State, Productive Forces, Cultures, Imperialisms, 'Mentalités' and so on. Once accepted, these large entities are then used to explain (or to not explain) 'cognitive' aspects of science and technology. The problem is that these entities could not exist at all without the construction of long networks in which numerous faithful records circulate in both directions, records which are, in turn, summarised and displayed to convince. A 'state', a 'corporation', a 'culture', an 'economy' are the result of a punctualisation process that obtains a few indicators out of many traces. In order to exist these entities have to be *summed up* somewhere. Far from being the key to the understanding of science and technology, these entities are the very things a new understanding of science and technology should explain. The large scale actors to which sociologists of science are keen to attach 'interests' are immaterial in practice as long as precise mechanisms to explain their origin or extraction and their changes of scale have not been proposed.

A man is never much more powerful than any other – even from a throne; but a man whose eye dominates records through which some sort of connections are established with millions of others may be said to *dominate*. This domination, however, is not a given but a slow construction and it can be corroded, interrupted or destroyed if the records, files and figures are immobilised, made more mutable, less readable, less combinable or unclear when displayed. [...] Even the very notion of scale is impossible to understand without an inscription or a map in mind. The 'great man' is a little man looking at a good map. In Mercator's frontispiece Atlas is transformed from a god who carries the world into a scientist who holds it in his hand (Mukerji 1995).

Since the beginning of this presentation on visualisation and cognition, I have been recasting the simple question of power: how the few may dominate the many. After McNeill's (1992) major reconceptualisation of the history of power in terms of mobilisation, this age-old question of political philosophy and sociology can be rephrased in another way: how can distant or foreign places and times be gathered in one place in a form that allows all the places and times to be presented at once, and which allows orders to move back to where they came from? Talking of power is an endless and mystical task; talking of distance, gathering, fidelity, summing up, transmission and so on is an empirical one, as has been illustrated in a recent study by John Law of the Portuguese spice road to India (1986). Instead of using large scale entities to explain science and technology as most sociologists of science do, we should start from the inscriptions and their mobilisation and see how they help small entities to become large ones. In this shift from one research programme to another, 'science and technology' will cease to be the mysterious cognitive object to be explained by the social world. [...]

If this little shift from a social/cognitive divide to the study of inscriptions is accepted, then the importance of *metrology* appears in proper light. Metrology is the scientific organisation of stable measurement and standards. Without it no measurement is stable enough to allow either the homogeneity of the inscriptions or their return. It is not surprising then to learn that metrology costs up to three times the budget of all Research and Development, and that this figure is for only the first elements of the metrological chain (Hunter 1980). Thanks to metrological organisation the basic physical constants (time, space, weight, wavelength) and many biological and chemical standards may be extended 'everywhere' (Zerubavel 1982; Landes 1983). The universality of science and technology is a cliché of epistemology but metrology is the practical achievement of this mystical universality. In practice it is costly and full of holes. [...] Metrology is only the official and primary component of an ever increasing number of measuring activities we all have to undertake in daily life. [...] 'Rationalisation' has very little to do with the reason of bureau- and techno-crats, but has a lot to do with the maintenance of metrological chains (Uselding 1981). This building of long networks provides the stability of the main physical constants, but there are many other metrological activities for less 'universal' measures (polls, questionnaires, forms to fill in, accounts, tallies).

[...]

[But] capitalism is not to be used to explain the evolution of science and technology. It seems to me that it should be quite the contrary. Once science and technology are rephrased in terms of immutable mobiles it might be possible to explain economic capitalism, as another process of mobilisation. What indicates this are the many weaknesses of money; money [...] circulates from one point to another but it carries very little with it. If the name of the game is to accumulate enough allies in one place to modify the belief and behaviour of all the others, money is a poor resource as long as it is isolated. It becomes useful when it is combined with all the other inscription devices; then, the different points of the world become really transported in a manageable form to a single place which then becomes a *centre.* [...] What counts is not the capitalisation of money, but the capitalisation of all compatible inscriptions. Instead of talking of merchants, princes, scientists, astronomers and engineers as having some sort of relation with one another, it seems to me it would be more productive to talk about '*centres of calculation*'. The currency in which they calculate is less important than the fact that they calculate only with inscriptions and mix together in these calculations inscriptions coming from the most diverse disciplines. The calculations themselves are less important than the way they are arrayed in cascades [...]. This qualification should be granted to centres of calculation and to the peculiarity of written traces, which makes rapid translation between one medium and another possible.

Many efforts have been made to link the history of science with the history of capitalism, and many efforts have been made to describe the scientist as a capitalist. All these efforts (including mine – Latour and Woolgar 1979, Chapter 5; Latour 1984a) were doomed from the start, since they took for granted a division between mental and material factors, an artefact of our ignorance of inscriptions. There is not a history of engineers, then a history of capitalists, then one of scientists, then one of mathematicians, then one of economists. Rather, there is a single history of these centres of calculation. It is not only because they look exclusively at maps, account books, drawings, legal texts and files, that cartographers, merchants, engineers, jurists and civil servants get the edge on all the others. It is because all these inscriptions can be superimposed, reshuffled, recombined and summarised, and that totally new phenomena emerge, hidden from the other people from whom all these inscriptions have been exacted.

More precisely we should be able to explain, with the concept and empirical knowledge of these centres of calculation, how insignificant people working only with papers and signs become the most powerful of all. Papers and signs are incredibly weak and fragile. [...] La Pérouse's map is not the Pacific, anymore than Watt's drawings and patents are the engines [...]. This is precisely the paradox. By working on papers alone, on fragile inscriptions which are immensely less than the things from which they are extracted, it is still possible to dominate all things, and all people. What is insignificant for all other cultures becomes the most significant, the only significant aspect of reality. The weakest, by manipulating inscriptions of all sorts obsessively and exclusively, become the strongest. This is the view of power we get at by following this theme of visualisation and cognition in all its consequences.

References

Alpers, S. (1983) *The Art of Describing: Dutch Art in the 17th Century*, University of Chicago Press, Chicago.

Bertin, P. (1973) *Sémiologie Graphique*, Mouton, Paris.

Booker, P.J. (1982) *A History of Engineering Drawing*, Northgate Publishing, London.

Clanchy, M.T. (1979) *From Memory to Written Records 1066–1300*, Harvard University Press, Cambridge, MA.

Dagognet, F. (1969) *Tableaux et Langages de la Chimie*, Le Seuil, Paris.

Dagognet, F. (1973) *Ecriture et Iconographie*, Vrin, Paris.

de Roover, R. (1963) *The Rise and Decline of the Medici Bank*, Harvard University Press, Cambridge, MA.

Drake, S. (1970) *Galileo Studies*, University of Michigan Press, Ann Arbor, MI.

Edgerton, S. (1976) *The Renaissance Discovery of Linear Perspective*, Harper and Row, New York.

Edgerton, S. (1980) The Renaissance artist as a quantifier, in *The Perception of Pictures*, vol. 1 (ed. M.A. Hagen), Academic Press, New York.

Eisenstein, E. (1979) *The Printing Press as an Agent of Change*, Cambridge University Press, Cambridge.

Fabian, J. (1983) *Time and the Other: How Anthropology Makes Its Object*, Columbia University Press, New York.

Foucault, M. (1963) *Naissance de la Clinique: Une Archéologie du Regard Médical*, PUF, Paris.

Foucault, M. (1966) *Les Mots et Les Choses*, Gallimard, Paris.

Foucault, M. (1975) *Surveiller et Punir*, Gallimard, Paris.

Fourquet, M. (1980) *Le Comptes de la Puissance*, Encres, Paris.

Hunter, P. (1980) The national system of scientific measurement. *Science*, **210**, 869–874.

Ivins, W.M. (1953) *Prints and Visual Communications*, Harvard University Press, Cambridge, MA.

Ivins, W.M. (1973) *On the Rationalization of Sight*, Plenum Press, New York.

Knorr, K. (1981) *The Manufacture of Knowledge*, Pergamon Press, Oxford.

Koyré, A. (1966) *Etudes Galiléennes*, Hermann, Paris.

Landes, D. (1983) *Revolution in Time: Clock and the Making of the Modern World*, Harvard University Press, Cambridge, MA.

La Pérouse, J.F. (no date) *Voyages Autour du Monde*, Michel de l'Ormeraie, Paris.

Latour, B. (1984a) Le demier des capitalistes sauvages, interview d'un biochimiste. *Fundamenta Scientia*, **4** (3/4) 301–327.

Latour, B. (1984b) *Le Microbes: Guerre et Paix suivi de Irréductions*, A.M. Métallié, Paris.

Latour, B. and Bastide, F. (1985) Science-fabrication, in *Qualitative Scientometrics: Studies in the Dynamic of Science* (eds M. Callon, J. Law and A. Rip), Macmillan, London.

Latour, B. and Woolgar, S. (1979) *Laboratory Life: The Social Construction of Scientific Facts*, Sage, London.

Lave, J. (1985) Arithmetic practice and cognitive theory: an ethnographic enquiry. Unpublished manuscript. University of California, Irvine, CA.

Lave, J. (1986) The values of quantification, in *Power, Action and Belief* (ed. J. Law), Routledge, London, pp. 88–111.

Law, J. (1983) Enrôlement et contre-enrôlement: les luttes pour la publication d'un article scientifique. *Social Science Information*, **22** (2), 237–251.

Law, J. (1986) On the methods of long-distance control: vessels, navigations and the Portuguese route to India, in *Power, Action and Belief* (ed. J. Law), Routledge, London, pp. 236–263.

Levi-Strauss, C. (1962) *La Pensée Sauvage*, Plon, Paris.

Livingston, E. (1993) An ethnomethodological investigation of the foundations of mathematics. Unpublished PhD thesis. University of California, Los Angeles.

Lynch, M. (1985) Discipline and the material form of images: an analysis of scientific visibility. *Social Studies of Science*, **15**, 37–66.

McNeill, W. (1992) *The Pursuit of Power, Technology, Armed Forces and Society Since A.D. 1000*, University of Chicago Press, Chicago.

Mukerji, S. (1995) Voir le pouvoir, in *Les 'Vues' de l'Esprit* (ed. B. Latour), pp. 208–223

Pinch, T. (1985) Toward an analysis of scientific observations: the externality of evidential significance of observational reports in physics. *Social Studies of Science*, **15**, 3–37.

Serres, M. (1982) *Hermes*, The Johns Hopkins University Press, Baltimore, MD.

Simon, H. (1982) Cognitive processes of experts and novices. *Cahiers de la Fondation Archives Jean Piaget*, **2** (3), 154–178.

Uselding, P. (1981) Measuring techniques and manufacturing practice, in *Yankee Enterprise* (eds O. Mayr and R. Post), Smithsonian Institute Press, Washington, DC, pp. 13–26.

Zerubavel, E. (1982) The standardization of time: a sociohistorical perspective. *American Sociological Review*, **88** (1), 1–29.

Further reading

Edney, M. (1997) *Mapping an Empire: The Geographical Construction of British India, 1765–1843*, University of Chicago Press, Chicago. [An in-depth consideration of the assemblage of paperwork, and specifically mapping, involved in the management and development of imperial power in India.]

Harvey, F. (2001) Constructing GIS: actor networks of collaboration. *Journal of the Urban and Regional Information Systems Association*, **13** (1), 29–37. [An analysis of the nature of GIS informed by Latour's conception of the inscription.]

Latour, B. (2005) *Reassembling the Social: An Introduction to Actor-Network-Theory*, Oxford University Press, Oxford. [A re-appraisal of Actor-Network approaches and the role that inscriptions play in this kind of work.]

Law, J. (2003) *After Method: Mess in Social Science Research*, Routledge, London. [An exploration of how to research the everyday practices associated with technologies and inscriptions and their deployment as actants in different fields.]

See also

- Chapter 1.8: Deconstructing the Map
- Chapter 1.10: Cartography Without 'Progress': Reinterpreting the Nature and Historical Development of Mapmaking
- Chapter 1.12: The Agency of Mapping
- Chapter 1.14: Rethinking Maps
- Chapter 4.8: Refiguring Geography: Parish Maps of Common Ground
- Chapter 5.2: The Time and Space of the Enlightenment Project
- Chapter 5.8: Cartographic Rationality and the Politics of Geosurveillance and Security

Chapter 1.10

Cartography Without 'Progress': Reinterpreting the Nature and Historical Development of Mapmaking

Matthew H. Edney

Editors' overview

Edney's article is a sustained criticism of a narrowly defined way of writing about the history of mapping. It comes in the aftermath of the social constructivist challenge to positivist framework in the history of cartography and can be seen to represent a new methodological and philosophical framework. He explores the limitations of a monolithic, linear and progressive view of mapping history, arguing that it is more productive to view histories of mapping as assemblages of different 'modes', situated in different cultural, social and technological relationships. Edney suggests these 'modes' are best understood as responses to particular, and often local, contexts and illustrates the rich potential of an anti-progressive approach to cartographic history to generate new insights.

Originally published in 1993: *Cartographica*, **30** (2/3), 54–68.

The recent literature of cartographic theory constitutes an extended critique of the supposition that modern cartography is an empiricist practice. Modern western culture has established a direct association between real-world phenomena and their cartographic representations and has then privileged those representations with a correctness derived from the act of observation rather than from the social and cultural conditions within which the representations are grounded. Harley (1988a [...]), Wood (1992a [...]) and others (Belyea 1992; Pickles 1992; Rundstrom 1991; Woodward 1992) have argued from a variety of philosophical positions that mapmaking is inherently ideological and its 'facts' are not as value free as our culture has supposed. All maps – whatever their claims to the contrary – serve a larger purpose; mapmaking is not a neutral activity divorced from the power relations of any human society, past or present; there is no single nor necessarily best way in which to represent either the social or physical worlds.

One issue that has yet to be addressed in detail by this critique is the constitution of cartography as a practice. We generally conceive of cartography as a singular and monolithic enterprise and we derive that conception from the history of cartography in a recursive manner. That is, the modern discipline of cartography justifies and legitimates its empiricist claims to objectivity and neutrality by pointing to its past progress (which it also extrapolates into the future); conversely, historians of cartography have defined their subject in terms of their *a priori* assumptions of mapmaking's objectivity, neutrality and progressiveness (Harley 1989). This paper extends the ideological construction of cartography to redefine the nature of cartography as a practice without the empiricist emphasis upon observation and the sense of progress that it entails. In doing so, it also extends Rundstrom's (1991) idea of 'process cartography'. [...]

[...]

The Map Reader: Theories of Mapping Practice and Cartographic Representation, First Edition. Edited by Martin Dodge, Rob Kitchin and Chris Perkins.

Cartography's information emphasis

The philosophical concept of most significance for empiricist cartography is the ontological assumption that the world possesses a quite unambiguous existence and can therefore be objectively known. There is a world of geographic facts 'out there' – separate and distant from the observer – which are to be 'discovered' by the explorer and surveyor just as scientists 'discover' new facts about the way in which the world functions. Indeed, exploration and mapmaking were important factors in the development of the modern western scientific tradition, in which respect maps serve as a paradigmatic document of empiricist science (Grafton 1992; Livingstone 1992: 32–101).

Spatial location – the geographic fact – is deemed an independent attribute of discrete entities in the real world. The geometric expression of location might be inaccurate or imprecise, but it is never ambiguous; each place exists in only one location. The map, as the repository of unambiguous geographic facts, becomes itself an unambiguous, objective and factual document which is subject to variation only with respect to the geometrical accuracy and comprehensiveness of those facts. Each map is thus defined by its content and is evaluated by comparing that content to an idealised, distortion-free replication of the world. 'We remain caught', Pickles (1992: 199) has observed, 'within the metaphysics of presence, which presupposes some foundational object against which the distortions and interpretations can be measured: that some interpretation-free image could be produced that does not distort the world.'

The mediation between the cartographic image and real world is through discrete geographic facts. Conceptually, the 'geographic quanta' constitute a single, monolithic corpus of data which is directly related to the world according to the world's spatial structure (quantified by their latitude and longitude). Each map is simply a selection of these autonomous data in graphic form. [. . .] In this system, questions of scale influence the selection of the detailed information to be mapped, but they do not alter the sense that there exists a pristine, scale-less database that is equivalent to the world itself.

[. . .]

The monolithic corpus of geographic data is inherently accretive: it grows and develops in ways that maps themselves do not (Wood 1992b). Each survey adds new data and it corrects existing data. The database becomes progressively more comprehensive, more precise and more accurate. Given that a map is defined by its content, no distinction is made between maps made of, and maps made of *and made in*, an area (criticised by Stone 1988; Woodward 1974). Any two maps of the same area can accordingly be compared with and judged against each other. Consider the example of a cartometric analysis of the South Carolina coastline on five maps from 1757 to 1865. This analysis started with the hypothesis that 'the accuracy with which the South Carolina coastline was represented on maps would improve through time, with each new map depicting it more accurately than earlier ones. *This may sound like a self-evident notion,* but other authors have noted that the eventual improvement in map accuracy over time is not a steady transition. Rather it is marked by examples of more recent, but less accurate, maps' (Lloyd and Gilmartin 1987: 20, emphasis added).

[. . .]

By confusing maps of potentially wildly variant scale and source materials, historians and cartographers have created a linear trend defined by a canon of Great Maps which represent the great advances in geographic knowledge. Each map can be judged and ranked against other maps according to how well its content matches the overall corpus of data. Thus, Crone (1978: xi) stated that 'the history of cartography is largely that of the increase in the accuracy with which . . . elements of distance and direction are determined and . . . [in] the comprehensiveness of the map content', and Buczek (1982: 7) hypothesised a 'normal' sequence of cartographic coverage for any European country, from small-scale, general maps in the sixteenth century to medium-scale topographic mapping of particular provinces during the seventeenth and eighteenth centuries to the advent of large-scale national surveys after 1800. Most historians, however, pay closer attention to the historical record and agree with Skelton (1972: 5) that simplistic assumptions of *continuous* progress are perhaps unjustified and prefer a punctuated progress: 'the development of mapmaking . . . is marked by periods of rapidly accelerated advance, followed by periods of standstill or even retrogression.'

Historical vignettes are important features of the introductions to cartographic texts and atlases, where they serve to define the field as a progressive science for the benefit of the novitiates (Harley and Woodward 1989). For example, Tyner (1992: 4–7) includes a graph of ever increasing 'cartographic activity' over time; Hammond's new world atlas is wrapped in a rhetoric of progress which culminates in their own 'five-year effort to revolutionise how maps are made' (Hammond 1993: 6–9 and back cover). Significantly, the key to these historical vignettes is the juxtaposition of maps from widely different eras. Hammond (1993: 6–7) set a 1570 map of the Netherlands next to a Landsat image of the same region; foregoing a lengthy sketch, Campbell (1991: 1–3) nonetheless established a sense of historical progress by contrasting a 2500 BC Mesopotamian map against modern maps. As long as cartography defines its intellectual arena around issues of geometrically defined geographic data, cartography's progressive past will always be evident when we place, side by side, two maps of the same region but from such different eras.

Such contrasts rely for their effect upon the differences in *form* between the maps: clay tablet against modern drafting films and scribe-coat; decorative compass roses and ship-filled seas against austere and information-heavy pixels. These comparisons rely for their effect upon the idea that cartography has progressed from being an 'art' to being a 'science'. The occasion of this change is often called the 'cartographic reformation' and is broadly located in the century between 1670 and 1770 (Brannon 1989; Rees (1980)). The decline in florid decoration and the rise of the factual neutrality of white space are used as surrogates for the decline in cartography's artistry (as the product of individual workmen) and the rise in cartography's science (as the product of large-scale, institutional surveys). Thus, Bagrow's (1985: 22) oft-quoted statement – that he ended his general history in the late eighteenth century when 'maps ceased to be works of art, the products of individual minds, and [when] craftsmanship was finally superseded by specialised science and the machine' – has been reprised by Pelletier (1986: 26): 'from the end of the seventeenth century and in the eighteenth century the practice of cartography has moved out from the cabinet to the field.' [. . .]

Once we decide, however, that there is more to geographic data than their geometrical definition, and that the practice of cartography is more than the collection and replication of these data, then the empiricist conception of cartography rapidly disintegrates. Hammond (1993: 7) might claim that his new atlas contains data for 'nearly every important geographic feature on earth', yet the layout reveals the usual biases so that 'important' continues to mean 'important for the Euro-American economies'. Geographical information is not scale-less (or 'scale independent') as has been supposed but is scale-defined; most fundamentally, a clear distinction can be drawn between the nature of small-scale and large-scale data (Carstensen 1989; Robinson and Petchenik 1976: 108–123; Woodward 1992). There is no hard and fast distinction between the 'art' and the 'science' of cartography; nor is it that 'cartography is *both* an art and a science', as the usual vapid compromise would have it, but that each cartographic mode is 'socially constructed' in the same manner as art, science, technology and all their respective subdisciplines; as such, 'the boundary between' science, technology, art, cartography and so on 'is a matter for social negotiation and represents no underlying distinction' (Hughes 1986: 284).

The linear view of cartographic history must also be rejected. Blakemore and Harley (1980: 14–32) present the most succinct critique of the manner in which presumptions of cartography's past progress have forcibly distorted the historical record to fit a simple and chronologically linear sequence. Perceived precocity is lauded whereas hysteresis is denigrated; spectral complexity is shunned in favour of monochromatic hindsight. More particularly, Stone (1988) argued that cartography's reformation was not reflected in the European mapping of Africa: neither map production techniques nor map use appreciably altered between 1650 and 1850; what did change in the eighteenth century was the *look* as opposed to the content of the maps. We might draw an analogy to the nineteenth century 'Whig' interpretation of history, an interpretation 'which celebrated [the English past] as revealing a continuous, on the whole uninterrupted, and generally glorious story of constitutional progress, all leading up to the triumph of liberty and representative institutions' (Clive 1989: 129). Yet when Victorian historians turned from constitutional to social and economic history, the Whig interpretation collapsed under the weight of contrary evidence. Similar re-evaluations are currently occurring in the histories of science and technology. The 'quasilinear model' of technological history – in which successful innovations form a single sequence from which stem failed offshoots – is being replaced by much more complex models which stress the interaction of society with technology (Bijker *et al.* 1987). Similarly, historians of science are now looking beyond the absolutes of scientific knowledge to the processes by which that knowledge was created. In doing so, they have realised that the once unambiguous 'scientific revolution', the defining event for modem science and the inspiration for cartography's own reformation, actually presents a mass of contradictions and frequently 'unscientific' behaviour (Lindberg and Westman 1990; Toulmin 1990). [. . .]

Cartographic modes

The view that cartography is a singular enterprise is true only in its most general sense as the construction of map artefacts. We habitually distinguish between specific types of mapmaking. We identify conceptual differences in the information content and design of 'general' and 'thematic' maps (Castner 1980; Petchenik 1979). We use 'atlas' to define a very specific intermingling of written and cartographic texts whose whole is more man just the sum of its maps: the atlas is a symbol of both its maker's professional status and the social worth of its owner (because of the greater financial capital involved in atlas production) and is also a metaphor for the encyclopaedic sum of geographic knowledge (Akerman 1991). We contrast the topographic map with marine and aeronautical charts, or terrestrial with celestial maps: even our normal, everyday words convey the differences. We also draw a boundary between the topographic map and other 'scientific illustration', such as engineering plans or blueprints, botanical or anatomical illustrations, and multivariate graphs. The list of different types of cartography is long, and is getting longer (Hall 1992). These distinctions do not simply reflect variations in cartographic form which have been made to fulfil different functions. They also make manifest the requirements of different social organisations for graphic representations to aid their understanding of the human world. That is, they are the artefactual manifestations of different cartographic *modes*.

Each cartographic mode is a set of specific relations which determine a particular cartographic practice. There are three sorts of relations: cultural, social and technological. Each mode is thoroughly enmeshed in and is an integral part of these relations. In this respect, my classification of relations is a didactic device. On the other hand, when the interconnections between modes are considered, these relations form a hierarchy of definition. Cultural relations are the most subtle, as they bind maps to spatial conceptions. Each human culture perceives space in its own way and those perceptions determine the culture's maps . . . and those maps determine the culture's spatial perceptions: 'we see the earth's surface in terms of the cartographic convention we are familiar with: "[the discourse] constitutes its own object"' (Belyea, 1992: 6).

Further relations exist between maps and other spatial representation (landscapes, poetry, dance, architecture, cosmography, travel diary etc.). These relations govern the production of space: do we objectify and commodity space in order to exploit it and its contents, or do we personify and empathise with it? The cultural relations of cartography form the arena of the map's subliminal geometry (Harley 1988b), of the map's potential as an iconic device and of the map's very existence as a complex sign system (Wood and Fels 1986) and as a discourse. They confer both cognitive and symbolic meaning on spatial configurations; they define the most fundamental constitution of 'geographic information'; they govern cartographic conventions within the scope of a given culture, there are numerous social institutions which have a potential need for map use and so for mapmaking, if only in a limited way. The relations between the institutions and cartographic practices are either enabling or constraining, but their precise nature will, of course, vary. The following statements concern he social formations only of modern Western culture.

[. . .]

The mode is thus the combination of cartographic form and cartographic function, of the internal construction of the data and their representation on the one hand and the external *raison d'être* of the map on the other. No sense of priority is intended, lest we indulge in futile chicken-and-egg arguments. Rather, in cartography, as in any other process-oriented discipline, form follows function as much as function follows form.

Nor are cartographic modes independent. Within any given culture, there will probably be some overlap between the modes' cultural relations. Cartographic modes, the practices they determine, and the geographic information they encompass, do interact with each other. For example, when present day cartographers set out to make small-scale maps of the United States, they do not start with the original USGS topographic quadrangles at 1:24 000 and generalise down to 1:5 000 000; instead, they use existing maps or databases already established for the small scale. Yet the present day, small-scale database is the result of one or two centuries of refinement informed by large-scale surveys.

The overall history of cartography is recast as the history of the creation, internal change, interaction, merging, bifurcation and overlapping of cartographic modes. At certain periods in the history of cartography the various modes are easily distinguished; at other times they are so confused and interwoven as to *appear* to be one, as has been the case under the dominance of empiricist cartography during the past 200 years.

It is with respect to the issue of the interaction of concurrent modes that I find myself to be most in sympathy with Rundstrom's recent plea for a processual cartography. He seeks to broaden the study of maps from the current narrow focus on the 'end product' of a closed process to take into account the position that maps are contingent for their power upon continual, open ended 'cultural, social, political and technical processes'. Following on, Rundstrom states that process cartography 'situates the map artefact within the mapmaking process, *and* it places the entire mapmaking process within the context of intracultural and intercultural dialogues . . .' (Rundstrom 1991: 6). I would change these statements by subsuming the political under the social (a minor point) and by replacing 'the entire mapmaking process' with 'all mapmaking processes'. The difference is subtle yet significant. Also of interest is Rundstrom's argument that cultures cannot be strictly classified as either 'incorporating' or 'inscribing'. Even predominantly inscribing cultures (those, such as our own, which hold and fix meaningful information through material artefacts) possess incorporating traits (in which meaning is borne by oral communication and ephemeral human actions, e.g. dance and ritual). Much of the cultural relevance of the British mapping of India, for example, resided in the act (ceremony?) of surveying (Edney 1993). That is, 'process cartography allows us to see that *acts* empower artefacts' (Rundstrom 1991 7). Cartographic practices not only lead to an end product which is wrapped in the various cultural, social and technological relations of its mode, but the practice itself (the process) is wrapped up in those relations.

[. . .]

The historiographic identification of cartographic modes is a problematic task. As noted, the modes themselves are interrelated, sometimes very closely. The autonomy and distinctiveness of modes within a given society will also depend upon its degree of economic articulation and social complexity (Wood 1993). It is accordingly impossible (and certainly undesirable) to create a formal list of attributes and properties whose permutations define any mode of mapmaking. We cannot treat cartographic modes as being as distinct and as non-overlapping – as rigorous and as unambiguous – as we treat the categories of a choropleth map. Instead, modes must be defined radially and with the recognition that their interfaces are fuzzy and permeable and are not hard and discrete. There are prototypical instances of each mode about which we can place other instances in an abstract, conceptual space; we can then draw bounds about each set of instances, so defining the modes and their interrelations (Lakoff 1987). [. . .]

The modes of formal European cartography, 1500–1850

To illustrate the workings and interactions of cartographic modes, I turn to the broad sweep of cartographic history in Europe, between the sixteenth and the mid-nineteenth centuries, and in particular across the period of cartography's scientific reformation. As noted, my particular interest is in the formal cartographic modes. I do not wish to imply that the following sequence of cartographic history was teleologically determined; my intention is simply to summarise how the cartographic modes developed. [. . .]

With the Renaissance, the introduction of the printing press, the rise of the modern territorial state, the establishment of mercantile classes, and the creation of a commercial land market, there evolved three formal cartographic modes: chorography, charting and topography. Europe's social and economic structures became increasingly complex and articulated. The burgeoning urban populations supported an increasingly specialised work force; the medieval artists' and scribes' guilds of the Italian cities spawned commercial print makers and cartographers.

[. . .]

[T]he three main modes are quite distinct in terms of their scales of inquiry, spatial conceptions, social institutions, technologies and corpora of information. Chorography is the mode of small-scale mapping both of regions and the world (1: 1 000 000 or less). [. . .] The information of chorography is concerned with places as points, with roads and rivers as lines, and with regions. Chorography itself is concerned as much with geographic generalisation as with cartographic representation; its essence is the evaluation and reconciliation of geographic information of different types (textual, itinerary, other cartographic, pictorial) and of different dates (both ancient and contemporary) to produce a single map. A map projection would first be constructed in a form that was aesthetically pleasing or easy to draw (whose mathematical rigour, if any, was a bonus), the key points whose coordinates were well known would be plotted, and then the rest of the map's data would be interpolated between. The fact that raw geographic data were now quantified (being ascribed latitudes and longitudes) does not mean that small-scale mapping itself became an objective practice; the imprecision of astronomical observations was well known, and the number of properly fixed places was very small, even into the middle of the eighteenth century. For example, Tobias Mayer had only 33 control points for his *Germaniae atque in ea locorum principaliorum Mappa Critica* of 1753, and of these none had an acceptable longitude value! (Forbes 1980a: 65) Small-scale mapping was thus a consciously intellectual process – the 'Science of Princes' – which has been the preserve of geographers who have never visited the regions that they map yet who have nonetheless debated the relative merits and consistencies of different reports about those regions. [. . .]

The early modern mode of charting was the lineal descendent of the medieval mode of constructing charts from *portolani*, or intricate lists of navigational instructions. The information base of each chart comprised sailing distances and directions, with very few positions located by their latitude and longitude. The social institutions of charting were, first and foremost, the maritime governments and the bureaucratic entities charged with regulating navigation and with preserving the secrecy of the new routes to Asia and the Americas. [. . .]

The third mode of early modern cartography – topography – is that of representing limited portions of land at a large scale (larger than 1: 100 000). It is dominated by surveying, the process of directed inquiry for the purposes of the management and supervision of the place concerned. [. . .] The information base of topography is that of the boundaries (walls, edges and fences) which enclose and define fields, lakes, buildings, woodlands, estates and so on, and that of the lie of the land for drainage and engineering purposes. It is the concrete and rarely abstract information of property ownership and of those human activities which are literally built into the landscape. The actual techniques employed depend upon the individual's level of education and experience, from simple pacing and sketching, through direct measurement with rod and chain, to complex techniques, such as triangulation, which require sophisticated trigonometrical and geometrical knowledge to be converted to a paper image. The representational strategies may ideally involve planimetric space but they merge into landscape images. Nor does the planimetric large-scale map necessarily share the projective space of chorography and charting. Topography's techniques of representation are also direct and involve little cartographic generalisation: the decision of which features are important is made in the field so that they are measured and recorded; the map itself and the landscape painting can be sketched directly in the field before the final design and polish is added in the workroom. [. . .]

The projective space of chorography and the techniques of topography merged after 1500 in a new mode, that of geodesy. In terms of the amount of effort expended before the later seventeenth century, geodesy was quite a minor mode, yet it is intellectually significant. Geodetic surveys involve the comparison of the terrestrial and astronomical lengths of the same meridional arc. [. . .] Geodesy was intimately tied to astronomy, mathematics and natural philosophy; its practitioners were scientists of international renown. While the geodetic surveys were concerned with the fundamental cartographic issue of the earth's figure, they rarely produced maps other than abstract triangulation diagrams. Geodesy as a cartographic practice was situated midway between the two modes of chorography and topography: it relied upon surveying techniques for the actual measurement, but its results were used in chorography, especially for the provision of scale to map projections.

The Enlightenment convergence and mathematical cosmography

By the early eighteenth century, these four modes had all gradually converged into the single, if ambiguous, mode of 'mathematical cosmography' (Forbes 1980b). The new mode's earliest manifestations are in the later seventeenth century (with for example the first Cassini survey) and reach a peak, as it were, in the period 1750–1800, but instances of the mode continue into the present. Again, while it is possible to identify 'pure' instances of cosmography, topography and charting throughout the 1700s, it is nonetheless impossible to draw hard and fast distinctions between each of the existing modes and the new, super-encompassing mathematical cosmography. The ambiguity derives from the varying degree of merger of each of the three components of the established modes. The culture of the Enlightenment promoted a unified and geometrical philosophy of mapmaking. Socially, the eighteenth century expansion of *the* role of the state provided a more coherent social basis for cartography, although this was by no means total. Least homogeneously, the techniques and informational corpora of the existing modes did not fuse but instead each blurred and merged to create continua of information sets and of practices in parallel to the continuum of cartographic scale.

[. . .]

[T]he cartographic modes of early modern Europe fused most completely in terms of the philosophical propositions which gave them a theoretical unity to match their practical convergence. [. . .]

The multiple perspectives of the 1500s gave way to a single perspective of rationality, which was explicitly sanctioned by the European states when they established scientific academies after 1660 (McClellan 1985: 41–66). Ortelius could put two quite different maps of Denmark on one plate in his *Theatrum Orbis Terrarum* (1570) without fear of confusing a reader, but the new intellectual regime transformed this epistemological multiplicity into self contradiction (reflected in Hodgkiss 1981: 14); the new epistemological singularity required that at any *one* time, there should be only *one* map of *one* territory. From this perspective, maps and mapmaking epitomised the Enlightenment practice of science. While the Enlightenment was a multifaceted and often contradictory intellectual period, we can still come to some broad understanding of this period as one which championed the application of rational thought in the form of experience and experiment, of observation and measurement (Porter 1990: 1–11). Implicit within this mentality was the perspective of 'encyclopaedism', that properly conducted rational debate can reconcile conflicting points of view, so that all knowledge could ultimately be brought into a single whole and there be systematically described (Macintyre 1990: 170–172). Combined with a mechanistic conception of the cosmos, this perspective led naturally to a formalisation of the sciences, and indeed human society generally, as being inherently progressive. [. . .]

The cosmographer's habitual reconciliation of information from varying sources and of varying quality to produce a single map of undeniable worth, comprehensiveness and utility has been recast as a metaphor for the fundamental belief of Enlightenment philosophy and the encyclopaedic mentality that rational debate and enquiry will lead ultimately to unified knowledge. Belyea's (1992) identification of contrasting archives of imaginative geographers and empirical explorers reflects a duality inherent to mathematical cosmography. Encyclopaedism encouraged the reconciliation of conflicting data through 'rational thought,' which sometimes involved large conclusions being logically deduced from a small database, as when Guillaume Delisle 'discovered' the Mer de l'Ouest in the interior of North America. Ordered knowledge was achieved by reconstituting already existing knowledge within a single framework provided by the map. Cartographic truth is created by the cartographer. But, at the same time, mapmaking had incorporated astronomical and geodetic measurements, careful and meticulous observation, the newest instruments (Frangsmyr *et al.* 1990), mathematical principles and the Enlightenment's *ésprit géometrique,* all of which offered a second avenue to ordered knowledge via its systematic creation right from the start. This is the archive of Cook and Vancouver.

By the end of the eighteenth century, this second approach to mapmaking was widely accepted by European states as the proper way to create entirely *new* corpora of structured and ordered geographic information. The result was the birth of a new cartographic mode – systematic mapmaking – which involves mapping an entire country at the same scale and with consistent survey techniques. More particularly, systematic surveys feature new detailed surveys which depend for their accuracy upon a prior-established control network of triangulation.

Modern systematic mapmaking was not the natural and inevitable development after cartography's supposed scientific reformation of the seventeenth and eighteenth centuries. It was instead the creation of Europe's burgeoning militaristic states (Brewer 1988). The first systematic surveys were established by the French state to create a uniform and systematic corpus of knowledge about French territory and so allow an increasingly centralised government to support a huge military machine.

[...]

When states lacked the bureaucratic and institutional ability to support complex systematic surveys, those surveys simply did not succeed. Giuseppe Piazzi, for example, attempted to establish a systematic survey of Sicily in 1808, but the state's bureaucracy was too undeveloped to support the survey after the political demise of its patron in 1810 (Fodera'Serio and Nastasi 1985). Those states perpetuated the chorographic aspects of mathematical cartography and sought to reconstitute existing data according to a new framework, thereby creating a single image.

[...]

The needs of the centralised, militaristic state also spawned another new mode from mathematical cosmography: thematic mapmaking. While examples can be found from before 1700 of several of the basic symbolisation strategies commonly used on thematic maps (e.g. the isoline, the flow line, or the shading of socially defined regions) the necessary institutional elements were not established until the later 1700s. There were two salient factors. Firstly, the seventeenth century realisation that the physical world can best be studied through mathematics had merged early in the eighteenth century with the desire to study society and its aspects in a rational, systematic and encyclopaedic manner, a methodology which required the quantification and measurement of the object of study (Frangsmyr *et al.* 1990; Toulmin 1990). Thus there had developed the intellectual assumption that it is meaningful to map data about the social world in the same manner as data about the physical world. Secondly, the later Enlightenment featured a dramatic increase in the collection of economic, demographic and social data by European states – or by members of the educated elites working for the state's benefit – in order to better understand and so control their societies and economies (Nadal and Urteaga 1990; Porter 1986: 18–39). [...] The connection between the development of systematic mapping and of thematic mapping is clear: while the first allowed the state to understand and control the physical territory of the European state or of European colonies, the second allowed the state to understand and control the social contents of those territories.

The development of the two new modes and the continuation of mathematical cosmography into the present does not mean that the older modes of chorography, charting and topography have ceased to exist. However, they underwent quite substantial changes. Topography was affected by increasing professional specialisation during the nineteenth century, so that cartographic surveyors slowly surrendered their non-mapping functions to other groups. [...]

For its part, chorography is still tied closely to the state and to the established, educational elites. Increasing literacy in European and North American societies plus decreasing production costs (with lithography) have produced a sub-mode of truly commercial cartographers unsubsidised by the state. Nonetheless, most commercial cartographers have remained close to the state, using government contracts to subsidise commercial ventures or repackaging and reselling government maps (today, digital geographic information) at a profit. Charting is very much a maritime version of the systematic surveys, although it too has a commercial component which repackages government information for the civilian mariner.

It should also be noted that none of the post-1800 formal cartographic modes enjoyed the same cultural relevancy and intellectual glitter that mathematical cosmography had at its peak. The Romantic movement and the early nineteenth century religious revival challenged most of the Enlightenment's assumptions and social philosophies. Certainty about being able to find the precise size and shape of the earth gave way to statistical representations under the weight of irreconcilable results (Stigler 1986: 11–158), while the dramatic progress made in instrument manufacture during the eighteenth century made high accuracy and high precision available to many surveyors rather than the select few. Geographers and geodesists turned to new theoretical issues and left cartography behind as mere technique (Bialas 1982; Godlewska 1989).

Explaining the rhetoric of empiricist cartography

The fragmentation of mathematical cosmography has left mapmaking in a schizophrenic state. On the one hand, the systematic mapping activities of most of the world's wealthier countries have continued to advertise and promote the philosophical assumptions about maps and nature as established under mathematical cosmography [...]. The official, systematic survey proclaims that there is one world which can be progressively described and known, and that the survey map is that description. It perpetuates the eighteenth century blurring of the boundary between the chorographic and topographic corpora of data by subsuming the generalised and abstracted data of the one beneath the concrete and hard edged data of the other. Geographic data are awarded an epistemological coherence to match the ontological coherence of the world itself. The other modes have adopted the same rhetoric of certainty, both in their representational strategies and in their social acceptance. [...] All the formal modes of modern western cartography share the same cultural expectations that the map replicates the territory's structure precisely and accurately: *this* is the empiricist conception of cartography. On the other hand, none of those modes can entirely or properly fulfil those expectations.

The legacy of the Enlightenment's mathematical cosmography is cartography's empiricism. The key to the discipline's self-image and to its rhetoric is accordingly its perception of eighteenth century cartography. That is, cartography's supposed reformation is actually the historiographic misunderstanding of empiricism's ascendancy, a misunderstanding justified by the scientific revolution which serves both as the necessary pre-cursor to the cartographic reformation and as a demonstration that such intellectual shifts can occur and that they have indeed occurred. Empiricist cartography presupposes that maps are graphic selections from a single corpus of geographic data which relate directly to the world via the global coordinate system. The history of such a cartography therefore stresses those cartographic modes which feature corpora of information which are organised by latitude and longitude. Before 1700, these were only chorography and charting; after 1800 the empiricist self-image postulates a single mode based on systematic mapping. Cartography's supposed reformation comprises the switch from one to the other, from small-scale ('artistic') to apparently scaleless (and 'scientific') geography.

That these otherwise quite distinct cartographic modes are historiographically comparable is the result of mathematical cosmography's internal duality. Chorography, incorporated within mathematical cosmography via encyclopaedism, is made to be directly equivalent to the instrumentalist-driven systematic survey. The uncertain art of the regional mapmaker is held to be comparable to the exact science of the surveyor. Yet, in the formulation presented here, cartography comprises a number of modes which should be compared only with caution. Comparison between modes requires the consideration of all their constituent relations – cultural, social and technological – and of their respective corpora of data. It is insufficient to assume that commonality in one component indicates commonality in the rest. Although today's formal modes share a common cultural grounding, they do not constitute the monolithic enterprise that empiricism suggests. Because two temporally separated modes organise data in a similar manner, one is not the antecedent of the other.

The approach suggested here, of considering cartography to be constituted from many modes, is more satisfactory than the traditional means for envisioning the broad sweep of cartographic history. It is more consistent with the complexity of both the historical record and the intellectual character of mapmaking as an intellectual, technological, social and cultural process. This approach is neither prescriptive nor proscriptive and seeks only to broaden our discussions of the nature and history of cartography to encompass the myriad forms in which maps have been – and in which they continue to be – constructed and used.

References

Akerman, J.R. (1991) On the shoulders of a Titan: viewing the world of the past in atlas structure. Unpublished PhD dissertation. Pennsylvania State University.

Bagrow, L. (1985) *The History of Cartography*, 2nd edn, Precedent Publishing, Chicago.

Belyea, B. (1992) Images of power: Derrida/Foucault/Harley. *Cartographica*, **29** (2), 1–9.

Bialas, V. (1982) *Erdgestalt, Kosmologie und Weltanschauung: the Geschichte der Geodäsie als Teil der Kulturgeschicht der Menschheit*, Konrad Winter, Stuttgart.

Bijker, W.E., Hughes, T.P. and Pinch, T.P. (1987) *The Social Construction of Technological Systems: New Directions in the Sociology and History of Technology*, MIT Press, Cambridge, MA.

Blakemore, M. and Harley, J.B. (1980) Concepts in the history of cartography: a review and perspective. *Cartographica*, **17** (4), 1–120.

Brannon, G. (1989) The artistry and science of map-making. *Geographical Magazine*, **61** (9), 37–40.

Brewer, J. (1988) *The Sinews of Power: War, Money and the English State*, Century Hutchinson, London.

Buczek, K. (1982) *The History of Polish Cartography from the 15th to the 18th Century* (trans. A. Polocki), Meridian Publishing, Amsterdam.

Campbell, J. (1991) *Introductory Cartography*, Prentice Hall, Englewood Cliffs, NJ.

Carstensen, L.W. (1989) A fractal analysis of cartographic generalization. *The American Cartographer*, **16**, 181–89.

Castner, H.W. (1980) Special purpose mapping in 18th century Russia: a search for the beginnings of thematic mapping. *The American Cartographer*, **7**, 163–75.

Clive, J. (1989) *Not by Fact Alone: Essays on the Writing and Reading of History*, Houghton Mifflin, Boston.

Crone, G.R. (1978) *Maps and Their Makers: An Introduction to the History of Cartography*, Dawson, Folkestone, UK.

Edney, M.H. (1993) The patronage of science and the creation of imperial space: the British mapping of India, 1799–1843. *Cartographica*, **30** (1), 61–67.

Fodera'Serio, G. and Nastasi, P. (1985) Giuseppe Piazzi's survey of Sicily: the chronicle of a dream. *Vistas in Astronomy*, **28**, 269–76.

Forbes, E.G. (1980a) *Tobias Mayer (1723–62): Pioneer of Enlightened Science in Germany*, Vandenhoeck and Ruprecht, Gottingen.

Forbes, E.G. (1980b) Mathematical cosmography, in *The Ferment of Knowledge: Studies in the Historiography of Eighteenth-Century Science* (eds G.S. Rousseau and R. Porter), Cambridge University Press, Cambridge, pp. 417–448.

Frangsmyr, T., Heilbron, J.L. and Rider, A. (1990) *The Quantifying Spirit in the Eighteenth Century*, University of California Press, Berkeley, CA.

Godlewska, A. (1989) Traditions, crisis, and new paradigms in the rise of the modern French discipline of geography, 1760–1850. *Annals of the Association of American Geographers*, **79**, 192–213.

Grafton, A. (1992) *New Worlds, Ancient Texts: The Power of Tradition and the Shock of Discovery*, Harvard University Press, Cambridge, MA.

Hall, S.S. (1992) *Mapping the Next Millennium: The Discovery of New Geographies*, Random House, New York.

Hammond (1993) *Atlas of the World*, concise edn, Hammond, Maplewood, NJ.

Harley, J.B. (1988a) Silences and secrecy: the hidden agenda of cartography in early modern Europe. *Imago Mundi*, **40**, 57–76.

Harley, J.B. (1988b) Maps, knowledge and power, in *The Iconography of Landscape* (eds D. Cosgrove and S.J. Daniels), Cambridge University Press, Cambridge, pp. 277–312.

Harley, J.B. (1989) The myth of the great divide: art, science, and text in the history of cartography. Paper presented at 13th International Conference on the History of Cartography, Amsterdam.

Harley, J.B. and Woodward, D. (1989) Why cartography needs its history. *The American Cartographer*, **16**, 5–15.

Hodgkiss, A.G. (1981) *Understanding Maps: A Systematic History of their Use and Development*, Dawson, Folkestone, UK.

Hughes, T.P. (1986) The seamless web: technology, science, etcetera, etcetera. *Social Studies of Science*, **16**, 281–92.

Lakoff, G. (1987) *Women, Fire, and Dangerous Things: What Categories Reveal About the Mind*, University of Chicago Press, Chicago.

Lindberg, D.C. and Westman, R.S. (1990) *Reappraisals of the Scientific Revolution*, Cambridge University Press. Cambridge.

Livingstone, D.N. (1992) *The Geographical Tradition: Episodes in the History of a Contested Enterprise*, Basil Blackwell, Oxford.

Lloyd, R. and Gilmartin, P. (1987) The South Carolina coastline on historical maps: a cartometric analysis. *The Cartographic Journal*, **24** (1), 19–26.

Macintyre, A. (1990) *Three Rival Versions of Moral Enquiry: Encyclopaedia, Genealogy, and Tradition*, University of Notre Dame Press, Notre Dame, IN.

McClellan, J.E. (1985) *Science Reorganized: Scientific Societies in the Eighteenth Century*, Columbia University Press, New York.

Nadal, F. and Urteaga, L. (1990) *Cartography and State: National Topographic Maps and Territorial Statistics in the Nineteenth Century* (Catedra de Geografia Humana, Facultad de Geografia e Historia, Universitat de Barcelona, *Cuadernos Criticos de Geografia Humana 88*, English Parallel Series 2: 9–67).

Pelletier, M. (1986) La France mesuree. *Mappemonde*, **3**, 26–32.

Petchenik, B.B. (1979) From place to space: the psychological achievement of thematic mapping. *The American Cartographer*, **6**, 5–12.

Pickles, J. (1992) Texts, hermeneutics and propaganda maps, in *Writing Worlds: Discourse, Text & Metaphor in the Representation of Landscape* (eds T.J. Barnes and J.S. Duncan), Routledge, London, pp. 193–230.

Porter, T.M. (1986) *The Rise of Statistical Thinking, 1820–1900*, Princeton University Press, Princeton, NJ.

Porter, R. (1990) *The Enlightenment. Studies in European History*, Humanities Press International, Atlantic Highlands, NJ.

Rees, R. (1980) Historical links between cartography and art. *Geographical Review*, **70**, 60–78.

Robinson, A.H. and Petchenik, B.B. (1976) *The Nature of Maps: Essays Toward Understanding Maps and Mapping*, University of Chicago Press, Chicago.

Rundstrom, R.A. (1991) Mapping, postmodernism, indigenous people and the changing direction of North American cartography. *Cartographica*, **28** (2), 1–12.

Skelton, R.A. (1972) *Maps: A Historical Survey of Their Study and Collecting*, University of Chicago Press, Chicago.

Stigler, S.M. (1986) *The History of Statistics: The Measurement of Uncertainty Before 1900*, Harvard University Press, Cambridge, MA.

Stone, J.C. (1988) Imperialism, colonialism and cartography. *Transactions of the Institute of British Geographers NS*, **13**, 57–64.

Toumlin, S. (1990) *Cosmopolis: The Hidden Agenda of Modernity*, The Free Press, New York.

Tyner, J. (1992) *Introduction to Thematic Cartography*, Prentice-Hall, Englewood Cliffs, NJ.

Wood, D. (1992a) *The Power of Maps*, Guilford, New York.

Wood, D. (1992b) How maps work. *Cartographica*, **29** (3/4), 66–74.

Wood, D. (1993) Maps and mapmaking. *Cartographica*, **30** (1), 1–9.

Wood, D. and Fels, J. (1986) Designs on signs / myth and meaning in maps. *Cartographica*, **23** (3), 54–103.

Woodward, D. (1974) The study of the history of cartography: a suggested framework. *The American Cartographer*, **1**, 101–115.

Woodward, D. (1992) Representations of the world, in *Geography's Inner Worlds: Pervasive Themes in Contemporary American Geography* (eds R. Abler, M.G. Marcus and J.M. Olson), Rutgers University Press, New Brunswick, NJ, pp. 50–73.

Further reading

Edney, M.H. (1997) *Mapping an Empire: The Geographical Construction of British India*, University of Chicago Press, Chicago. [A detailed illustration of the application of these ideas outlined in the above excerpt to the emergent and mutable forms of an imperial mapping enterprise.]

Harley, J.B. and Woodward, D. (1997) *The History of Cartography*, University of Chicago Press, Chicago. [The definitive multi-volume history, as yet incomplete, now led by Edney and offering a richly detailed empirical overviews of cartography for different time periods.]

Jacob, C. (2006) *The Sovereign Map: Theoretical Approaches in Cartography Through History*, University of Chicago Press, Chicago. [A challenging and sophisticated exploration of different approaches to the history of mapping.]

Wood, D. (2010) *Rethinking the Power of Maps*, Guilford, New York. [The first chapter of Wood's book is a very effective illustration of Edney's arguments for a contextual approach to the history of mapping.]

See also

- Chapter 1.8: Deconstructing the Map
- Chapter 1.12: The Agency of Mapping: Speculation, Critique and Invention
- Chapter 1.14: Rethinking Maps
- Chapter 2.2: A Century of Cartographic Change
- Chapter 2.3: Manufacturing Metaphors: Public Cartography, the Market, and Democracy
- Chapter 5.2: The Time and Space of the Enlightenment Project
- Chapter 5.4: Mapping: A New Technology of Space; Geo-Body

Chapter 1.11

Exploratory Cartographic Visualisation: Advancing the Agenda

Alan M. MacEachren and Menno-Jan Kraak

Editors' overview

By the early 1990s theoretical groundings for cartographic research based around notions of cartographic communication were becoming increasingly problematic. On the one hand, social constructivist critique was questioning the neutrality of mapping. But the new potential of computer display was also rendering notions of optimal map design redundant – mapping was increasingly exploratory and interactive. Building upon a focus upon scientific representation, and upon analytical work from statistical researchers, new approaches to mapping were framed by MacEachren and co-workers at Penn State University, and picked up by many academics, offering a revived potential for applied mapping research. In this influential article key concerns of this new research field are charted. MacEachren and Kraak explore the history of visualisation research, explain its central visualising device – the (cartography cubed) – and suggest priorities for future research directions.

Originally published in 1997: *Computers & Geosciences*, **23** (4), 335–343.

Introduction

The nature of maps and of their use in science and society is in the midst of remarkable change – change that is stimulated by a combination of new scientific and societal needs for geo-referenced information and rapidly evolving technologies that can provide that information in innovative ways. A key issue at the heart of this change is the concept of 'visualisation'.

At one level, all mapping can be considered a kind of visualisation – in the sense of 'making visible'. From this perspective, cartography has always been about visualisation. Over the past several decades, cartographers have devoted considerable attention to understanding how *presentation* maps work to make aspects of the world visible – what Freitag (1993) terms the 'communication function' of maps. We know relatively little, however, about how maps that facilitate thinking, problem solving and decision making work (uses that, according to Freitag's typology, are representative of the 'cognitive function' and the 'decision support function' for maps), nor what the implications of 'working' are in these contexts (implications associated with Freitag's 'social function' and Török's (1993) 'social context' – the latter derived largely from Harley's (1988, 1989) approach to the ideology of mapping). It is to the cognitive and decision support functions that much of the new geo-information technologies are directed – particularly those technologies which include maps having dynamic and interactive components. It is also in these functions that scientific visualisation and cartography share the greatest overlap (MacEachren and Monmonier 1992).

Whereas cartographers can argue that we have always been involved with visualisation (Freitag 1993; Rimbert 1993), treating current conceptions of visualisation as 'nothing new' ignores the most important implications of the alternative definitions of visualisation used beyond the discipline. These new definitions of visualisation are linked to specific ways in which modem computer technology facilitates the process of 'making visible' in real time (Taylor 1991, 1994). The ability

The Map Reader: Theories of Mapping Practice and Cartographic Representation, First Edition. Edited by Martin Dodge, Rob Kitchin and Chris Perkins.

to prompt instantaneous changes in maps results not only in a quantitative difference in the number of things a user can make visible, but a qualitative difference in the way users think – and in turn in the way maps function as prompts to thinking and decision making (Wood 1994).

Cartography as a discipline has a significant stake in the evolving role of maps within systems for scientific visualisation, within spatial decision support systems, within hypermedia information access systems, and within virtual reality environments. Cartography has much to offer the scientific community through its long history of design and production of visual representations of the Earth, its knowledge of geographical (and cartographical) information systems, and its experience with linking digital and visual geographic representations. On the other hand, cartography has much to gain from collaborations with the wider scientific visualisation community where approaches to interactive computer tool development, interface design, three-dimensional computer modelling, and related methods and technologies are more fully developed.

[...]

Visualisation in cartography: evolution or revolution?

Use of the term visualisation in the cartographic literature can be traced back at least four decades (Philbrick 1953). It was the 1987 publication of a report by the US National Science Foundation, however, that established a new meaning for this term in the context of scientific research (McCormick *et al.* 1987). The report, produced by a committee containing no cartographers, emphasised the role of computer display technology in prompting mental visualisation – and subsequent insight. Scientific visualisation has, thus, been defined as the use of sophisticated computing technology to create visual displays, the goal of which is to facilitate thinking and problem solving. Emphasis is not on storing knowledge but on knowledge construction. In relation to the spatial information processing goals for maps delineated by Rimbert (1993), 'spatial analysis' and 'spatial simulation' could be considered prototypical components of scientific visualisation. Rimbert's 'travel guide', 'spatial inventory' and 'secondary information source' goals, in contrast, would be considered (at least by some researchers in scientific visualisation) to be ancillary to the visualisation process – although they may benefit from information processing and display methods developed to support the core visualisation goals.

Following publication of the McCormick report on visualisation in scientific computing, several cartographers took up the challenge of trying to grapple with the cartographic implications of this new (or renewed) reliance on visual representation in science. DiBiase (1990) borrowed from the exploratory data analysis (EDA) literature of statistics to propose a graphic model of stages in map-based scientific visualisation applied to the earth sciences. The model presented visualisation as a four-stage process consisting of two private visual thinking stages (exploration and confirmation) and two public visual communication stages (synthesis and presentation). An intent of the model was to encourage cartographers to direct attention to the role of maps at the early (private) stages of scientific research where maps and map-based tools are used to facilitate data sifting and exploration of extremely large data sets.

MacEachren and Ganter (1990), in a parallel effort, developed a simple cognitive model to identify key parts of the user-display interaction that occurs during exploratory map-based visual analysis. Their emphasis was on developing cartographic tools that prompt pattern identification and on the potential for visualisation errors (errors that are similar in nature to the Type I and Type II errors associated with traditional statistically-based hypothesis testing). The related topic of visualising data quality/reliability has proved to be a particularly active research direction within cartographic visualisation (MacEachren 1992; Beard and Buttenfield 1991; Buttenfield and Beard 1994; Fisher 1994; van der Wel *et al.* 1994). MacEachren (1995) has subsequently grounded the pattern identification approach to visualisation in an integrated cognitive-semiotic conceptualisation of how maps (and other geo-referenced displays) work as visualisation tools and has applied that approach to the design of visual interfaces directed to reliability representation.

Although these conceptions of visualisation in cartography put emphasis on the private cognitive processes of visual thinking (particularly those associated with scientific hypothesis formulation and confirmation), Taylor (1991) directed attention to the place of visualisation in the structure of cartography as a discipline. His approach presented visualisation as the intersection of research on cognition, communication and formalism (with the latter implying strict adherence to rule structures dictated by digital computer systems). In a recent revision of this approach, Taylor (1994) made it clear that he does not equate 'visualisation' with 'cartography'. Instead, he argues for a view of visualisation as a distinct development in cartography, and in science in general, that will have an impact on three major aspects of cartography – aspects that he defines as the sides of a 'conceptual basis' triangle (cognition and analysis, communication and formalism).

Exploratory cartographic visualisation: emphasising map use

Building from the perspectives on georeferenced/cartographic visualisation outlined above, a conception of visualisation emphasising use of visual displays was developed (MacEachren 1994). [...] The approach treats map use as a 'space' referred to by MacEachren as (CARTOGRAPHY)^3 – a reference to the three axes along which map use was characterised (Figure 1.11.1). In this space, visualisation is considered to be the complement of communication. All map use involves both visualisation (defined loosely as the prompting of visual thinking and knowledge construction) and communication (defined loosely as the transfer of information), but map use can differ considerably in which of these activities is emphasised. The axes of the use-space are delineated as private versus public, high interaction versus low interaction, presenting knowns (i.e. simple information retrieval) versus revealing unknowns.

Past communication-oriented cartographic research has emphasised the use of static maps designed for public consumption with the emphasis on extracting specific pieces of information (e.g. research on communication effectiveness of textbook or topographic maps). [...]

As Kraak (1999) has discussed, strategies must differ for design of visual displays and display systems to support map use at various positions in the use-space identified. Building from the previous use-oriented discussion of visualisation, combined with both Kraak's delineation of visualisation strategies and DiBiase's (1990) conceptions of the stages that underpin EDA adapted to geo-referenced data, four use *goals* can be recognised. Each requires its own approach, characteristics of which are implied by position of the goal within the use cube (Figure 1.11.2). Since a particular map, whether static or dynamic, might be used to meet all of the goals, the distinction between the goals is based, not on types of map, but on the audience, data relations and interaction level that is typical for pursuit of the goal. The location of the spheres in the diagram represents current typical applications of visualisation methods to each goal.

Over the past decade interactivity has become increasingly important in strategies to achieve all goals, and we envision a future in which the dominant strategies for pursuing all four goals are arrayed along the left wall of the use cube – a future in which high interaction is as typical of presentation uses as of exploration uses.

Visualisation to explore, in order to examine unknown and often raw data creatively, is the dominant strategy at the *private–high instruction–exploration of unknown* corner of the cube (Figure 1.11.2). Several applications, such as those dealing with remotely sensed data, there are abundant (often temporal) data available. Questions such as 'what is the nature of the data set?', 'which of those data sets reveal patterns related to the current problem studied?' and 'what if...?' have to be answered before the data can actually be used in a spatial analysis operation. Required to support mapping strategies here are functions to allow the user to explore the spatial data visually (e.g. by

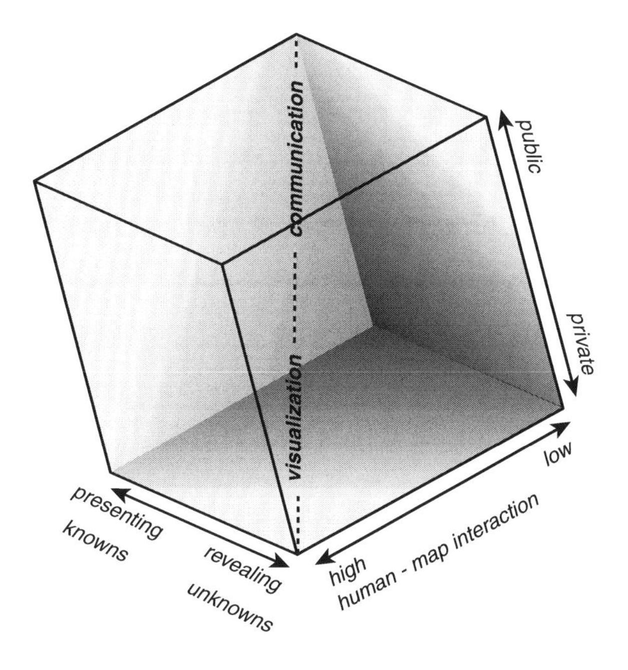

Figure 1.11.1 (CARTOGRAPHY)^3 – representation of three-dimensional 'space' of map use. Relative emphasis on visualisation and communications activities is depicted. Axes relate to audience or user of map (ranging from private to public), objectives of use (from revealing unknowns to presenting knowns) and degree of interactivity with map or mapping environment (from high to low).

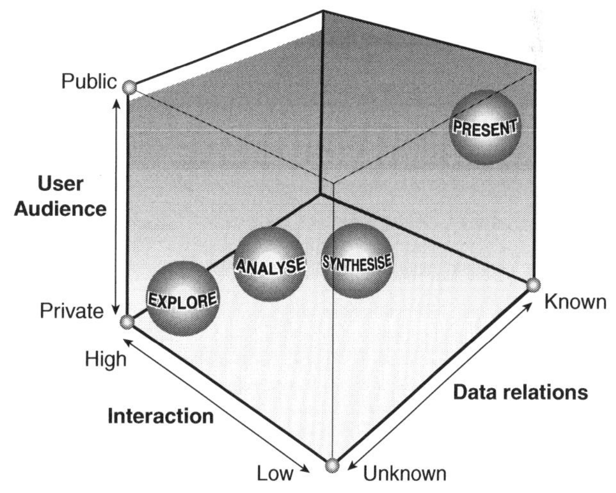

Figure 1.11.2 Goals of map use. Sequence of goals that can be facilitated by visualisation methods are arrayed in a map-use cube. Here, emphasis is on distinguishing among use goals that may require different visualisation strategies.

animation), to identify relationships among variables and to 'see' data from multiple perspectives (both spatially and conceptually).

Therefore, specific visual exploration functions and tools are needed that facilitate use of these 'new' dynamic mapping methods. It also should be noted that, in a data exploration environment, it is likely that the user, at least at the start of a visualisation session, is unfamiliar with the exact nature of the data, a situation that calls for flexible visualisation that can be adapted by the user.

Visualisation applied to analysis generally involves manipulation of known data in a search for unknown relationships and answers to questions. The use strategies here also emphasise relatively private use that is facilitated by interactive systems. In a planning environment, for example, the nature of two separate data sets might initially be fully understood, but not their relationship. A spatial analysis operation, such as (visual) overlay, combines both data sets to determine their possible spatial relationship. Questions such as 'is there a spatial correlation?', 'what is the best site?' or 'what is the shortest route?' are typical of analytical uses of visualisation tools and methods. Required to support these uses are functions to target individual map components to extract information from them (e.g. the EDA method of 'focusing'), and functions to process and manipulate that information.

Visualisation applied to synthesis of information moves the emphasis away from single investigators toward groups (perhaps of specialists) and from revealing unknowns toward presenting knowns. With synthesis, however, there is still considerable chance for new insights. Questions to be answered here include: 'what general result(s) does the analysis suggest?', 'how different are a pair of problem solutions?' or 'how can we best summarise many, perhaps conflicting, results?'. Functions required are ones that extract the most salient patterns and relationships from a data set and display them as a coherent abstraction of the exploration and analysis process. Tools to filter out local details, random noise and so on are required. Monmonier's (1992) concept of 'summary graphics' provides a model strategy for the generation of displays that support synthesis. Both he and DiBiase (1990) emphasise giving up detail in exchange for a useful abstraction as a key to effective synthesis. Geoinformation use that emphasises synthesis, just as exploration and analysis, can benefit from interactive tools that allow a user to experience more than one possible synthesis, perhaps ones that emphasise different parameters of a problem or different philosophical perspectives on that problem.

Presentation is often equated with cartographic communication (in the information theory sense that underlies the communication model approach to cartography). Presentation can, however, include both transfer of some predetermined 'message' and the prompting of new insight on the part of the person who experiences or accesses the presentation.

Visualisation applied to presentation emphasises public use and the 'presentation' of information that is largely known to the information designer, but not to the user of the presentation. Similar to the goals detailed previously, presentation can derive considerable benefit from use of interactivity. Thus the sphere used in our diagram to characterise presentation uses of visualisation methods is not at the extreme opposite corner of the use-space from the one representing exploration. As Cartwright's (1997) analysis of the Web's impact on cartography makes clear, presentation uses are rapidly incorporating interactivity.

Whether interactive or not, presentation strategies should emphasise the transfer of spatial knowledge (rather than creation of new knowledge). If cartographic displays are well designed to support presentation uses, the results of spatial analysis operations can be easily understood by a wide audience. Design strategies must support queries such as 'what is?' or 'where is?', and 'what belongs together?'. The cartographic discipline offers a comprehensive cartographic 'language' with associated rules, strategies and conventions with which to generate maps supporting these uses (Kraak and Ormeling 1996). Depending on the nature of a spatial distribution, this language will suggest particular presentation mapping solutions. A number of authors have proposed methods to formalise these rules into expert systems that support dynamic visualisation (Buttenfield and Mark 1991; Jung 1995). Caution must be exercised, however, when applying cartographic rules developed for presentation to the design of visualisation environments intended to support the full range of use goals identified.

Research directions for exploratory cartographic visualisation

The evolving conceptualisation of visualisation, as it relates to cartography and to science in general, raises a wide variety of new research questions for cartography, as well as for other sciences with an interest in georeferenced visualisation.

Investigate the implications of a change in cartographic emphasis from a focus on designing optimal maps to one that favours mapping systems that facilitate multiple perspectives [...]

Develop a conceptual model and associated tools for the visualisation of spatial–temporal process information [...]

Develop a conceptual model and associated tools for the visualisation of data quality/reliability information [...]

Study of methods for, and implications of, linking cartographic visualisation tools to GIS [. . .]

Explore the impact of map-based spatial decision support tools on decision making strategies and on the outcome of decision making [. . .]

Study the potential of three-dimensional representation tools and the corresponding implications of both three-dimensional display and the associated general trend toward realism (versus abstraction) in scientific representation [. . .]

Address implications, for our approaches to map design, of the ability to link many representation forms together in hypermedia documents (e.g. maps, graphs, text, audio narratives, sonic data representation, etc.) [. . .]

Investigate alternative computer interface design strategies as they relate to use of visualisation tools for hypothesis formulation and decision support [. . .]

[. . .]

References

Beard, M.K. and Buttenfield, B.P. (1991) NCGIA Research Initiative 7: Visualization of Spatial Data Quality, Technical Paper 91–26, NCGIA.

Buttenfield, B.P. and Beard, M.K. (1994) Graphical and geographical components of data quality, in *Visualization in Geographic Information Systems* (eds D. Unwin and H. Hearnshaw), John Wiley & Sons, Ltd, Chichester, UK, pp. 150–157.

Buttenfield, B.P. and Mark, D.M. (1991) Expert systems in cartographic design, in *Geographic Information Systems: The Microcomputer and Modern Cartography* (ed. D.R.F. Taylor), Pergamon Press, Oxford, pp. 129–150.

Cartwright, W. (1997) New media and their application to the production of map products. *Computers & Geosciences*, **23** (4), 447–456.

DiBiase, D. (1990) Visualization in the earth sciences. *Bulletin of the College of Earth and Mineral Sciences*, **59** (2), 13–18.

Fisher, P. (1994) Randomization and sound for the visualization of uncertain spatial information, in *Visualization in Geographical Information Systems* (eds D. Unwin and H. Hearnshaw), John Wiley & Sons, Ltd, Chichester, UK, pp. 181–185.

Fisher, P. Dykes, J. and Wood, J. (1993) Map design and visualization. *The Cartographic Journal*, **30** (20), 136–142.

Freitag, U. (1993) Map functions, in *The Selected Main Theoretical Issues Facing Cartography: Report of the ICA Working Group to Define the Main Theoretical Issues on Cartography* (ed. T. Kanahubo), International Cartographic Association, Cologne, Germany, pp. 9–19.

Harley, J.B. (1988) Maps, knowledge, and power, in *The Iconography of Landscape: Essays on the Symbolic Representation, Design and Use of Past Environments* (eds D. Cosgrove and S. Daniels), Cambridge University Press, Cambridge, pp. 277–311.

Harley, J.B. (1989) Deconstructing the map. *Cartographica*, **26** (2), 1–20.

Jung, V. (1995) Knowledge-based visualization design for geographic information systems. *Proceedings of the ACM-GIS'95 3rd ACM International Workshop on Advances in Geographic Information Systems*, ACM Press, New York, pp. 101–108.

Kraak, M.J. (1999) Visualizing spatial distributions, in *Geographical Information Systems: Principles, Techniques, Management, and Applications*, 2nd edn (eds P. Longley, M. Goodchild, D.M. Maguire and D. Rhind), Geoinformation International, Cambridge.

Kraak, M.J. and MacEachren, A.M. (1994) Visualization of spatial data's temporal component. *SDH 94 Sixth International Symposium on Spatial Data Handling*, 5–9 September, Edinburgh, pp. 391–409.

Kraak, M.J. and Ormeling, F.J. (1996) *Cartography: Visualization of Spatial Data*, Addison-Wesley-Longman, Harrow, UK.

MacEachren, A.M. (1992) Visualizing uncertain information. *Cartographic Perspectives*, **13**, 10–19.

MacEachren, A.M. (1994) Visualization in modern cartography: setting the agenda, in *Visualization in Modem Cartography* (eds A.M. MacEachren and D.R.F. Taylor), Pergamon, Oxford, pp. 1–12.

MacEachren, A.M. (1995) *How Maps Work: Representation, Visualization and Design*, Guilford, New York.

MacEachren, A.M. and Ganter, J.H. (1990) A pattern identification approach to cartographic visualization. *Cartographica*, **27** (2), 64–81.

MacEachren, A.M. and Monmonier, M. (1992) Geographic visualization: introduction. *Cartography and Geographic Information Systems*, **19** (4), 197–200.

McCormick, B.H., DeFanti, T.A. and Brown, M.D. (eds) (1987) Visualization in scientific computing. *Computer Graphics*, **21** (6).

Monmonier, M. (1992) Summary graphics for integrated visualization in dynamic cartography. *Cartography and Geographic Information Systems*, **19** (1), 23–36.

Philbrick, A.K. (1953) Toward a unity of cartographical forms and geographical content. *Professional Geographer*, **5** (5), 11–15.

Rimbert, S. (1993) Social context, in *The Selected Main Theoretical Issues Facing Cartography: Report of the ICA Working Group to Define the Main Theoretical Issues on Cartography* (ed. T. Kanahubo), International Cartographic Association, Cologne, Germany, pp. 29–28.

Taylor, D.R.F. (1991) Geographic information systems: the microcomputer and modem cartography, in *Geographic Information Systems: The Microcomputer and Modern Cartography* (ed. D.R.F. Taylor), Pergamon, Oxford, pp. 1–20.

Taylor, D.R.F. (1994) Perspectives on visualization and modern cartography, in *Visualization in Modem Cartography* (eds A.M. MacEachren and D.R.F. Taylor), Pergamon, Oxford, pp. 333–342.

van der Wel, F.J.M., Hootsman, M. and Ormeling, F. (1994) Visualization of data quality, in *Visualization in Modern Cartography* (eds A.M. MacEachren and D.R.F. Taylor), Pergamon, Oxford, pp. 313–331.

Wood, M. (1994) Visualization in historical context, in *Visualization in Modem Cartography* (eds A.M. MacEachren and D.R.F. Taylor), Pergamon, Oxford, pp. 13–26.

Further reading

Andrienko, G.L. and Andrienko, N.V. (1999) Interactive maps for visual data exploration. *International Journal of Geographical Information Science*, **13** (4), 355–374. [An influential and practical example of a software system designed to tackle many of the research challenges identified in MacEachren and Kraak's paper.]

Crampton, J. (2001) Maps as social constructions: power, communication and visualization. *Progress in Human Geography*, **25** (2), 235–252. [Situates the development of the visualisation in relation to the communications paradigm in cartography and subsequent social constructivist challenges.]

Dykes, J., MacEachren, A. and Kraak, M.J. (2005) *Exploring Geovisualization*, Elsevier, London. [The wide ranging edited volume presents examples of interactive mapping, cartographic exploration and data visualisation research that build upon the agenda.]

MacEachren, A.M. and Kraak, M.J. (2001) Research challenges in geovisualization. *Cartography and Geographic Information Science*, **28** (1). [A themed issue of the journal developing the 1997 agenda and presenting the results of international collaboration over research priorities.

See also

- Chapter 1.2: Semiology of Graphics
- Chapter 1.3: On Maps and Mapping
- Chapter 1.4: The Science of Cartography and its Essential Processes
- Chapter 1.5: Analytical Cartography
- Chapter 1.6: Cartographic Communication
- Chapter 2.6: Cartographic Futures on a Digital Earth
- Chapter 3.5: Strategies for the Visualisation of Geographic Time-Series Data
- Chapter 3.6: Some Truth with Maps: A Primer on Symbolisation and Design
- Chapter 3.10: Affective Geovisualisations
- Chapter 4.3: Cognitive Maps and Spatial Behaviour: Process and Products
- Chapter 4.9: Cognitive Maps: Encoding and Decoding Information
- Chapter 4.11: Usability Evaluation of Web Mapping Sites

Chapter 1.12

The Agency of Mapping: Speculation, Critique and Invention

James Corner

Editors' overview

Corner's writing evokes the emancipatory potential of mapping, at a time when it was much more usual to demonise it as a form of elite discourse, facilitating governance by the powerful. Corner draws instead on the creative potential of the medium, deploying the figures of Gille Deleuze and Félix Guattari to demonstrate the constructive agency that can be enacted through cartographic practice in the fields of architecture, landscape and urban planning. He explores four ways in which new practices of mapping are emerging in contemporary design and planning, which he terms as: 'drift', 'layering', 'game-board' and 'rhizome'. Corner concludes that mapping is not endless data accumulation but is rather better seen as a practice of relational reasoning that intelligently unfolds new realities out of existing constraints.

Originally published in 1999: Chapter 10 in *Mappings* (ed. Denis Cosgrove), Reaktion, London, pp. 213–252.

Introduction

Mapping is a fantastic cultural project, creating and building the world as much as measuring and describing it. Long affiliated with the planning and design of cities, landscapes and buildings, mapping is particularly instrumental in the construing and constructing of lived space. In this active sense, the function of mapping is less to mirror reality than to engender the re-shaping of the worlds in which people live. While there are countless examples of authoritarian, simplistic, erroneous and coercive acts of mapping, with reductive effects upon both individuals and environments, I focus in this essay upon more optimistic revisions of mapping practices (Wood 1992; Monmonier 1991; Pickles 1992; Scott 1998; Hall 1992). These revisions situate mapping as a collective enabling enterprise, a project that both reveals and realises hidden potential. Hence, in describing the 'agency' of mapping, I do not mean to invoke agendas of imperialist technocracy and control but rather to suggest ways in which mapping acts may emancipate potentials, enrich experiences and diversify worlds. We have been adequately cautioned about mapping as a means of projecting power knowledge, but what about mapping as a productive and liberating instrument, a world-enriching agent, especially in the design and planning arts?

As a creative practice, mapping precipitates its most productive effects through a finding that is also a founding; its agency lies in neither reproduction nor imposition but rather in uncovering realities previously unseen or unimagined, even across seemingly exhausted grounds. Thus, mapping unfolds potential; it re-makes territory over and over again, each time with new and diverse consequences. Not all maps accomplish this, however; some simply reproduce what is already known. These are more 'tracings' than maps, delineating patterns but revealing nothing new. In describing and advocating more open-ended forms of creativity, philosophers Gilles Deleuze and Félix Guattari (1987: 12) declare: 'Make a map not a tracing!' They continue:

> What distinguishes the map from the tracing is that it is entirely oriented toward an experimentation in contact with the real. The map does not reproduce an unconscious closed in upon itself; it constructs the unconscious. It fosters connections between fields, the removal of blockages on

The Map Reader: Theories of Mapping Practice and Cartographic Representation, First Edition. Edited by Martin Dodge, Rob Kitchin and Chris Perkins.

> bodies without organs, the maximum opening of bodies without organs onto a plane of consistency . . . The map has to do with *performance*, whereas the tracing always involves an 'alleged competence.'

The distinction here is between mapping as equal to what is ('tracing') and mapping as equal to what is and to what is not yet. In other words, the unfolding agency of mapping is most effective when its capacity for description also sets the conditions for new eidetic and physical worlds to emerge. Unlike tracings, which propagate redundancies, mappings discover new worlds within past and present ones; they inaugurate new grounds upon the hidden traces of a living context. The capacity to reformulate what already exists is the important step. And what already exists is more than just the physical attributes of terrain (topography, rivers, roads, buildings) but includes also the various hidden forces that underlie the workings of a given place. These include natural processes, such as wind and sun; historical events and local stories; economic and legislative conditions; even political interests, regulatory mechanisms and programmatic structures. Through rendering visible multiple and sometimes disparate field conditions, mapping allows for an understanding of terrain as only the surface expression of a complex and dynamic imbroglio of social and natural processes. In visualising these interrelationships and interactions, mapping itself participates in any future unfoldings. Thus, given the increased complexity and contentiousness that surrounds landscape and urbanism today, creative advances in mapping promise designers and planners greater efficacy in intervening in spatial and social processes. Avoiding the failure of universalist approaches toward master planning and the imposition of state controlled schemes, the unfolding agency of mapping may allow designers and planners not only to see certain possibilities in the complexity and contradiction of what already exists but also to *actualise* that potential. This instrumental function is particularly important in a world where it is becoming increasingly difficult to both *imagine* and actually to *create* anything outside of the normative.

The agency of mapping

Mappings have agency because of the double-sided characteristic of all maps. Firstly, their surfaces are directly *analogous* to actual ground conditions, as horizontal planes, they record the surface of the earth as direct impressions. As in the casting of shadows, walks and sightings across land may be literally *projected* onto paper through a geometrical graticule of points and lines drawn by ruler and pen. Conversely, one can put one's finger on a map and trace out a particular route or itinerary, the map projecting a mental image into the spatial imagination. Because of this directness, maps are taken to be 'true' and 'objective' measures of the world, and are accorded a kind of benign neutrality. By contrast, the other side of this analogous characteristic is the inevitable *abstractness* of maps, the result of selection, omission, isolation, distance and codification. Map devices, such as frame, scale, orientation, projection, indexing and naming, reveal artificial geographies that remain unavailable to human eyes. Maps present only one version of the earth's surface, an eidetic fiction constructed from factual observation. As both analogue and abstraction, then, the surface of the map functions like an operating table, a staging ground or a theatre of operations upon which the mapper collects, combines, connects, marks, masks, relates and generally explores. These surfaces are massive collection, sorting and transfer sites, great fields upon which real material conditions are isolated, indexed and placed within an assortment of relational structures.

The analogous-abstract character of the map surface means that it is doubly projective: it both captures the projected elements off the ground and projects back a variety of effects through use. The strategic use of this double function has, of course, a long alliance with the history of mapping, and not only militaristically (*reconnaissances militaires*) but also ideologically (Harley 1988). Surprisingly, however, the strategic, constitutive and inventive capacities of mapping are not widely recognised in the urban design and planning arts, even though cartography and planning have enjoyed a long and mutually influential relationship since the fifteenth century (Buisseret 1998; Söderström 1996). Throughout the twentieth century, mapping in design and planning has been undertaken conventionally as a quantitative and analytical survey of existing conditions made prior to the making of a new project. These survey maps are both spatial and statistical, inventorying a range of social, economic, ecological and aesthetic conditions. As expertly produced, measured representations, such maps are conventionally taken to be stable, accurate, indisputable mirrors of reality providing the logical basis for future decision making, as well as the means for later projecting a designed plan back onto the ground. It is generally assumed that if the survey is quantitative, objective and rational, it is also true and neutral, thereby helping to legitimise and enact future plans and decisions (Giddens 1994; Porter 1995). Thus, mapping typically precedes planning because it is assumed that the map will objectively identify and make visible the terms around which a planning project may then be rationally developed, evaluated and built (Scott 1988; Söderström 1996).

What remains overlooked in this sequence, however, is the fact that maps are highly artificial and fallible constructions, virtual abstractions that possess great force in terms of how people see and act. One of the reasons for this oversight derives from a prevalent tendency to view maps in terms of what they represent rather than what they do. [. . .] [M]ost designers and planners consider mapping a rather unimaginative, analytical practice, at least compared to the presumed 'inventiveness' of the designing activities that occur *after* all the relevant maps have been made (often with the contents of the maps ignored or forgotten).

This indifference towards mapping is particularly puzzling when one considers that the very basis upon which projects are imagined and realised derives precisely from how maps are made. The conditions around which a project develops originate with what is selected and prioritised in the map, what is subsequently left aside or ignored, how the chosen material is schematised, indexed and framed, and how the synthesis of the graphic field invokes semantic, symbolic and instrumental content. Thus, the various cartographic procedures of selection, schematisation and synthesis make the map *already* a project in the making (Arnheim 1970; Robinson and Petchenik 1976). This is why mapping is never neutral, passive or without consequence; on the contrary, mapping is perhaps the most formative and creative act of any design process, firstly disclosing and then staging the conditions for the emergence of new realities.

In what follows, I discuss mapping as an active agent of cultural intervention. Because my interests lie in the various processes and effects of mapping, I am less concerned with what mapping *means* than with what it actually *does*. Thus, I am less interested in maps as finished artifacts than I am in mapping as a creative *activity*. It is in this participatory sense that I believe new and speculative techniques of mapping may generate new practices of creativity practices that are expressed not in the invention of novel form but in the productive reformulation of what is already given. By showing the world in new ways, unexpected solutions and effects may emerge. However, given the importance of representational technique in the creative process, it is surprising that whilst there has been no shortage of new ideas and theories in design and planning there has been so little advancement and invention of those specific tools and techniques – including mapping – that are so crucial for the effective construal and construction of new worlds (Corner 1999a, 1992).

The efficacy of technique

A comparison between Mercator's projection of the earth's surface and Buckminster Fuller's Dymaxion projection reveals radically different spatial and socio-political structures. The same planet, the same places, and yet significantly dissimilar relationships are revealed or, more precisely, *constructed*. The Mercator map stretches the surface of the globe without excision onto a flat surface, oriented 'upwards' to the north. The compass directions are made parallel, leading to gross distortions of land area and shape, especially as one moves towards the poles. The northern hemisphere dominates, with Greenland more than twice the size of Australia, even though the southern island is in fact greater than three times the land area of the northern. Needless to say, this view has well suited the self-image of Europeans and North Americans in an era of Western political hegemony. By contrast, Fuller's Dymaxion Airocean World Map of 1943 cuts the earth into triangular facets that are then unfolded as a flat polyhedron (Figure 1.12.1). Both the north and south poles are presented frontally and equally with little distortion, although the typical viewer is at first likely to be disoriented by this unusual, polydirectional arrangement of countries. Only the graphic graticule of latitude and longitude allows the reader to comprehend the relative orientation of any one location (Marks and Buckminster Fuller 1973).

[...]

Unlike the scientific objectivism that guides most modern cartographers, artists have been more conscious of the essentially fictional status of maps and the power they possess for construing and constructing worlds (Storr 1994). In the same year as Fuller's projection, the Uruguayan artist Joaquin Torres-Garcia drew the *Inverted Map of South America* with a very distinct 'S' at the top of the drawing (Figure 1.12.2). This remarkable image reminds us of the ways in which habitual conventions (in this case the unquestioned domination of north on top) condition spatial hierarchies and power relations. The convention of orienting the map to the north first arose early in the global and economic expansion of Northern Europe and in response to practices of navigation. But there are many instances of other societies at different times orienting their maps towards one of the other cardinal points, or making them circular without top and bottom (the Dymaxion map is perhaps one of the few modern instances where singular orientation is not a prerequisite). Maps of this sort are still legible and 'correct' in their depiction of spatial relationship, but the reader must first learn the relevant mapping codes and conventions.

Another instance of critique and invention of the modern map is Waltercio Caldas's *Japão*, of 1972 (Storr 1994). Here, the artist is mapping a territory that is foreign, or 'unimaginable' for many in the West. Rather than colonising this territory through survey and inventory, typically Western techniques of power knowledge, Caldas simply marks an otherwise empty map surface with very small inscriptions and numbers. These are contained by a very prominent, classical cartographic frame. There are no other outlines, shapes or forms, just small type and a few scribbles. There is no scale, no identifiable marks, no graticule of orientation, just a square ink frame. In this stark, minimal cartographic field, Caldas presents an elusive geography, an open and indeterminate field of figures that returns *terra incognita* to an otherwise excessively mapped planet. The image is also a commentary on the cage-like power of the imperialising frame: the graphic square surrounds, captures and holds its quarry, but at the same time its contents remain foreign, evasive and

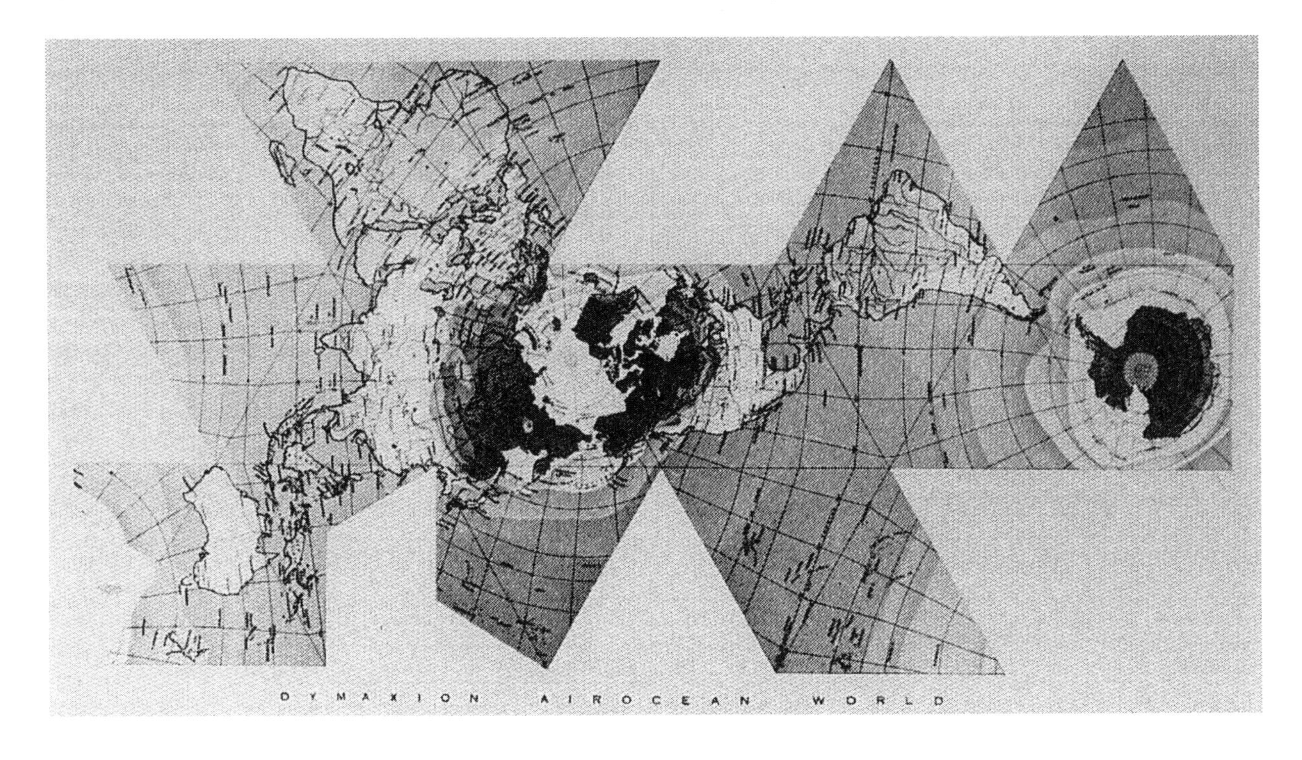

Figure 1.12.1 R. Buckminster Fuller and Shoji Sadao. (Source: Dymaxion Airocean World Map 1954.)

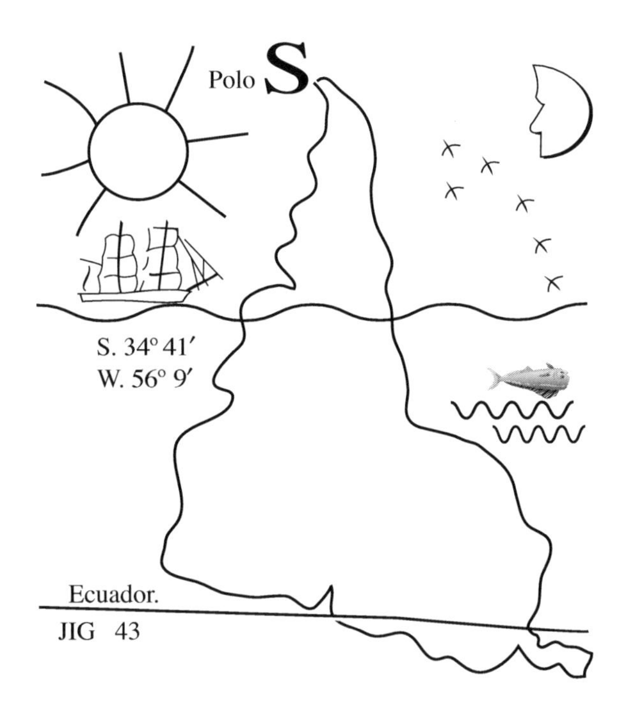

Figure 1.12.2 Joaquín Torres-García. (Source: Inverted Map of South America 1943.)

autonomous. This blank, non-figured space raises both anxiety and a certain promise – promise because its potential efficacy lies in the emancipation of its contents. The autonomous, abstract structure suggests how mystery and desire might be returned to a world of places and things that have been otherwise excessively classified and structured. In Caldas's image, such places are liberated through precisely the same measures that first captured them.

Whereas certain artists have engaged creatively with cartographic techniques, planners and designers have been less ambitious (Harrison and Turnbull 1996). Techniques of aerial-oblique and zenithal views – planimetry, ichnography and triangulation – were most developed during the early sixteenth century, and have since become the primary tools with which cities and landscapes are analysed, planned and constructed. Quantitative and thematic mapping techniques originated with the Enlightenment enthusiasm for rational progress and social reform, and these were later complemented by various statistical, comparative and 'zoning' techniques during the late nineteenth and early twentieth centuries (Hall 1988). Some advances in these techniques have occurred over the past 30 years with the rise of satellite and remote sensing capabilities, together with new computer technologies such as Geographic Information Systems, but in principle they remain unchanged. [. . .] With only a handful of exceptions, the relationship of maps to world-making is surprisingly under-thought.

[...]

Maps and reality

Jorge Luis Borges's tale of a fully detailed and life-sized map that eventually tore and weathered to shreds across the actual territory it covered is frequently quoted in essays on mapping (Borges 1933). Not only does the tale beautifully capture the cartographic imagination, it goes to the heart of a tension between reality and representation, between the territory and the map. Equally referenced is Lewis Carroll's tale in *Sylvie and Bruno*, also of a life-sized map, in this case folded, thus preventing it being unfolded for practical application. The map was useless, allowing Carroll's character Mein Herr to conclude, 'so now we use the country itself, as its own map, and I assure you it does nearly as well'. In these two fables, not only is the map an inferior, secondary representation of territory but the more detailed and life-like the map strives to be, the more redundant or unnecessary it becomes. Unlike paintings or photographs, which have the capacity to bear a direct resemblance to the things they depict, maps must by necessity be abstract if they are to sustain meaning and utility. And such abstraction, the bane of untrained map readers, is not at all a failing of maps but rather their virtue.

Jean Baudrillard (1983: 2) reverses Borges's tale to make another point:

> Simulation is no longer that of a territory, a referential being or substance. It is the generation by models of a real without origin or reality: a hyper-real. The territory no longer precedes the map, nor survives it. Henceforth, it is the map that precedes the territory.

Arguably, of course, the map *always* precedes the territory, in that space only becomes territory through acts of bounding and making visible, which are primary functions of mapping. But Baudrillard is going one step further here, claiming that late twentieth century communication and information technologies have produced such a blurring of what is real and what is a representation that the two can no longer be distinguished. He inverts Borges's fable to proclaim that 'it is the real and not the map whose vestiges subsist here and there' (Baudrillard 1983). Here, Baudrillard is careful to explain that this reversal does not mean that the world is scarcely more than a vast simulacrum, but rather that the act of differentiating between the real and the representation is no longer meaningful.

[...]

Reality, then, as in concepts such as 'landscape' or 'space', is not something external and 'given' for our apprehension; rather it is constituted, or 'formed', through our participation with things: material objects, images, values, cultural codes, places, cognitive schemata, events and maps. As the philosopher of science Jacob Bronowski pointedly observes, 'there are no appearances to be photographed, no experiences to be copied, in which we do not take part. Science, like art, is not a copy of nature but a re-creation of her' (1965). [...] The application of judgement, subjectively constituted, is precisely what makes a map more a project than a 'mere' empirical description.

[...]

For the landscape architect and urban planner, maps are sites for the imaging and projecting of alternative worlds. [...] The map 'gathers' and 'shows' things presently (and always) invisible, things which may appear incongruous or untimely but which may also harbour enormous potential for the unfolding of alternative events. In this regard, maps have very little to do with representation as depiction. After all, maps look nothing like their subject, not only because of their vantage point but also because they present all parts at once, with an immediacy unavailable to the grounded individual. But more than this, the function of maps is not to depict but to enable, to precipitate a set of effects in time. Thus, mappings do not *represent geographies* or ideas; rather they *effect* their actualisation.

Mapping is neither secondary nor representational but doubly operative: digging, finding and exposing on the one hand, and relating, connecting and structuring on the other. Through visual disclosure, mapping both sets up and puts into effect complex sets of relationship that remain to be more fully actualised.

[...]

Space and time today

A creative view of mapping in the context of architectural, landscape and urban production is rendered all the more relevant by the changing nature of spatial and temporal structures in today's world. Events occur with such speed and complexity that nothing remains certain. Large numbers live in a world where local economies and cultures are tightly bound into global ones, through which effects ripple with enormous velocity and consequence. Surrounded by media images and an excess of communication that makes the far seem near and the shocking merely normal, local cultures have become fully networked around the world. Air travel and other modes of rapid transportation have become so accessible that localities can be more closely connected to sites thousands of miles away than to their immediate surroundings. Today, structures of community life are shifting from spatial stability towards shifting, temporal coordination. Public life is now scheduled and allocated more by time than centred according to place, while the circulation of capital demands an ever-more mobile and migratory workforce. [...]

Mapping and contemporary spatial design techniques more generally have yet to find adequate ways to engage creatively with the dynamic and promiscuous character of time and space today. Most design and planning operations appear somewhat outmoded, overwhelmed or incongruent in comparison to the rapidly metabolising processes of urbanisation and communication.

[...]

Through such urbanists as Reyner Banham, Edward Soja, David Harvey, Rem Koolhaas and Bernard Tschumi, anthropologists such as Marc Auge, or philosophers such as Henri Lefebvre or Gilles Deleuze, it is becoming clearer to architects and planners that 'space' is more complex and dynamic than previous formal models allowed. Ideas about spatiality are moving away from physical objects and forms towards the variety of territorial, political and psychological social processes that flow through space. The interrelationships amongst things in space, as well as the effects that are produced through such dynamic interactions, are becoming of greater significance for intervening in urban landscapes than the solely compositional arrangement of objects and surfaces.

The experiences of space cannot be separated from the events that happen in it; space is situated, contingent and differentiated. It is remade continuously every time it is encountered by different people, every time it is represented through another medium, every time its surroundings change, every time new affiliations are forged. [...] Thus, the emphasis shifts from static object–space to the space–time of relational systems. And, it is here, in this complex and shifty *milieu*, that *maps*, not *plans*, achieve a new instrumental significance.

Mapping

'To plan a city is both to think the very plurality of the real and to make that way of thinking effective,' writes the philosopher of the everyday Michel de Certeau (1984: 94) 'it is to know how to articulate it and be able to do it.' Mapping is key here for it entails processes of gathering, working, reworking, assembling, relating, revealing, sifting and speculating. In turn, these activities enable the inclusion of massive amounts of information that, when articulated, allow certain sets of possibility to become actual. In containing multiple modes of spatio-temporal description, mapping precipitates fresh insights and enables effective actions to be taken. Thus mapping differs from 'planning' in that it entails searching, finding and unfolding complex and latent forces in the existing *milieu* rather than imposing a more-or-less idealised project from on high. Moreover, the synoptic imposition of the 'plan' implies a consumption (or extinguishing) of contextual potential, wherein all that is available is subsumed into the making of the project. Mapping, by contrast, discloses, stages and even adds potential for later acts and events to unfold. Whereas the plan leads to an end, the map provides a generative means, a suggestive vehicle that 'points' but does not overly determine.

A particularly important aspect of mapping in this regard is the acknowledgement of the maker's own participation and engagement with the cartographic process.

[...]

[M]apping precedes the map, to the degree that it cannot properly anticipate its final form. Robinson and Petchenik (1976: 74) claim that 'in mapping, one objective is to discover (by seeing) meaningful physical and intellectual shape organisations in the *milieu*, structures that are likely to remain hidden until they have been mapped ... plotting out or mapping is a method for searching for such meaningful designs'. In other words, there are some phenomena that can only achieve visibility through representation rather than through direct experience. Furthermore, mapping engenders new and meaningful relationships amongst otherwise disparate parts. The resultant relational structure is not something already 'out there', but rather something constructed, bodied forth through the act of mapping. As the philosopher Brand Blanshard (1948: 525) observes, 'space is simply a relation of systematised outsideness, by itself neither sensible nor imaginable'; it is *created* in the process of mapping.

Mapping operations

The operational structure of mapping might be schematised as consisting of 'fields', 'extracts' and 'plottings'. The field is the continuous surface, the flat bed, the paper or the table itself, schematically the analogical equivalent to the actual ground, albeit flat and scaled. The field is also the graphic system within which the extracts will later be organised. The system includes the frame, orientation, coordinates, scale, units of measure and the graphic projection (oblique, zenithal, isometric, anamorphic, folded, etc). The design and set-up of the field is perhaps one of the most creative acts in mapping, for as a prior system of organisation it will inevitably condition how and what observations are made and presented. Enlarging the frame, reducing the scale, shifting the projection or combining one system with another are all actions that significantly affect what is seen and how these findings are organised. Obviously, a field that has multiple frameworks and entryways is likely to be more inclusive than a singular, closed system. Also, a field that breaks with convention is more likely to precipitate new findings than one that is more habitual and routine. And thirdly, a field that is designed to be as non-hierarchical and inclusive as possible – more 'neutral' – is likely to bring a greater range of conditions into play than a field of restrictive scope.

Extracts are the things that are then observed within a given *milieu* and drawn onto the graphic field. We call them extracts because they are always selected, isolated and pulled out from their original seamlessness with other things; they are effectively 'de-territorialised'. They include objects but also other informational data: quantities, velocities, forces, trajectories. Once detached they may be studied, manipulated and networked with other figures in the field. As described above, different field systems will lead to different arrangements of the extracts, revealing alternative patterns and possibilities.

Plotting entails the 'drawing out' of new and latent relationships that can be seen amongst the various extracts within the field. There are, of course, an infinite number of relationships that can be drawn depending upon one's criteria or agenda. Richard Long, for example, who has made an art form of walking, may plot a line upon a map to connect the highest to the lowest summit in sequential order, for example, revealing a latent structural line across a given terrain. Upon the same map, however, it is possible to plot a line that connects all south-facing aspects in sequential order from large to small areas, or to find a range of wet conditions that can then be set into relationship by plotting a comparative index of water characteristics. In addition to geometrical and spatial plotting, taxonomic and genealogical procedures of relating, indexing and naming can often be extremely productive in revealing latent structures. Such techniques may produce insights that have both utility and metaphoricity. In either case, plotting entails an active and creative interpretation of the map to reveal, construct and engender latent sets of possibility. Plotting is not simply the indiscriminate listing and inventorying of conditions, as in a tracing, a table or a chart, but rather a strategic and imaginative drawing out of relational structures. To plot is to track, to trace, to set in relation, to find and to found. In this sense, plotting produces a 're-territorialisation' of sites.

Thus we can identify three essential operations in mapping; firstly, the creation of a field, the setting of rules and the establishment of a system; secondly, the extraction, isolation or 'de-territorialisation' of parts and data; and, thirdly, the plotting, the drawing out, the setting up of relationships, or the 're-territorialisation' of the parts. At each stage, choices and judgements are made, with the construing and constructing of the map alternating between processes of accumulation, disassembly and reassembly. By virtue of the mapmaker's awareness of the innately rhetorical nature of the map's construction as well as of personal authorship and intent, these operations differ from the mute, empirical documentation of terrain so often assumed by cartographers.

We may now identify four thematic ways in which new practices of mapping are emerging in contemporary design and planning, each producing certain effects upon perceptions and practices of space. I label these techniques 'drift', 'layering', 'game-board' and 'rhizome'.

Drift

The Situationists were a European group of artists and activists in the 1950s and 1960s. [...] Guy Debord, a key Situationist theorist, made a series of maps, or 'psychogeographic guides', of Paris. These were made after Debord had walked aimlessly around the streets and alleys of the city, turning here and there wherever the fancy took him. [...] More a form of cognitive mapping than mimetic description of the cityscape, Debord's maps located his own play and representation within the recessive nooks and crannies of everyday life. Such activity became known as the *dérive*, or the dream-like drift through the city, mapping alternative itineraries and subverting dominant readings and authoritarian regimes (Figure 1.12.3).

What is interesting about the *dérive* is the way in which the contingent, the ephemeral, the vague, fugitive eventfulness of spatial experience becomes foregrounded in place of the dominant, ocular gaze. [...]

It is important to understand that the primacy of [...] the Situationist's use of maps belongs to the their *performative* aspects, that is to the way in which mapping directs and enacts a particular set of events, events that derive from a given *milieu*. But, of course, there are the recordings that come after the proceedings, and these are neither passive nor neutral in their effects either. [...]

These various practices of 'drift' use maps as instruments for establishing and aligning otherwise disparate, repressed or unavailable topographies; they are 'set-ups' that both derive from and precipitate a series of interpretative and participatory acts. Their highly personal and constructive agency make them quite unlike the detached work of conventional mapmakers. They are openly cognitive, mental maps, rendering new images of space and relationship. Moreover, the drift permits a critique of contemporary circumstances, not from outside and above (as a master plan) but from participation within the very contours and fabric of political and institutional reality. [...]

Layering

A relatively new development in the design of large-scale urban and landscape fabrics has been 'layering'. This involves the superimposition of various independent layers one upon the other to produce a heterogeneous and 'thickened' surface. Architects Bernard Tschumi and Rem Koolhaas were amongst the first to develop layering strategies in design and planning in their respective proposals for the Parc de la Villette in Paris, 1983 (Tschumi 1987; Koolhaas and Man 1995). Generally, these projects dismantle the programmatic and logistical

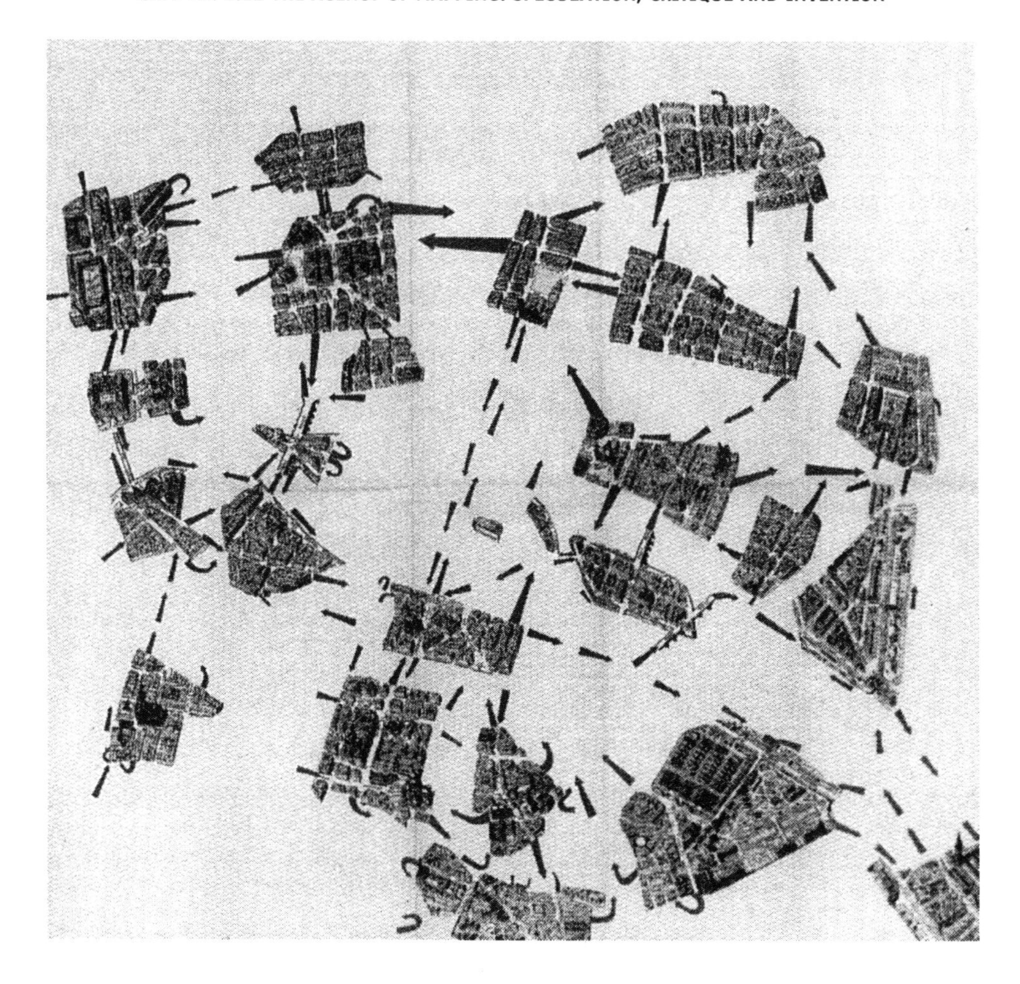

Figure 1.12.3 Guy Debord, *Discours sur les Passions de l'amour* 1957.

aspects of the park into a series of layers, each of which is then considered independently from the other layers. There is an internal logic, content and system of organisation to each layer, depending on its function or intended purpose. The layers are not mappings of an existing site or context, but of the complexity of the intended programme for the site. [...] When these separate layers are overlaid together, a stratified amalgam of relationships amongst parts appears. The resulting structure is a complex fabric, without centre, hierarchy or single organising principle (Figure 1.12.4). The composite field is instead one of multiple parts and elements, cohesive at one layer but disjunct in relation to others. Such richness and complexity cannot be gained by the limited scope of the single master plan or the zoning plan, both of which group, hierarchicalise and isolate their component parts.

[...]

Game-board

A third thematic development of mapping in contemporary design practice, and one related to the notions of performance mentioned above, has been the projection of 'game-board' map structures. These are conceived as shared working surfaces upon which various competing constituencies are invited to meet to work out their differences. As a representation of contested territory the map assumes an enabling or facilitating status for otherwise adversarial groups to try and find common ground while 'playing out' various scenarios. [...]

Raoul Bunschoten is a London-based architect who has engaged with a number of complex and contentious urban regions in Europe, and has developed a number of innovative mapping techniques for working with such sites. For Bunschoten (1996, 1997, 1998), cities are dynamic and multiple; they comprise a vast range of 'players' and 'agents'

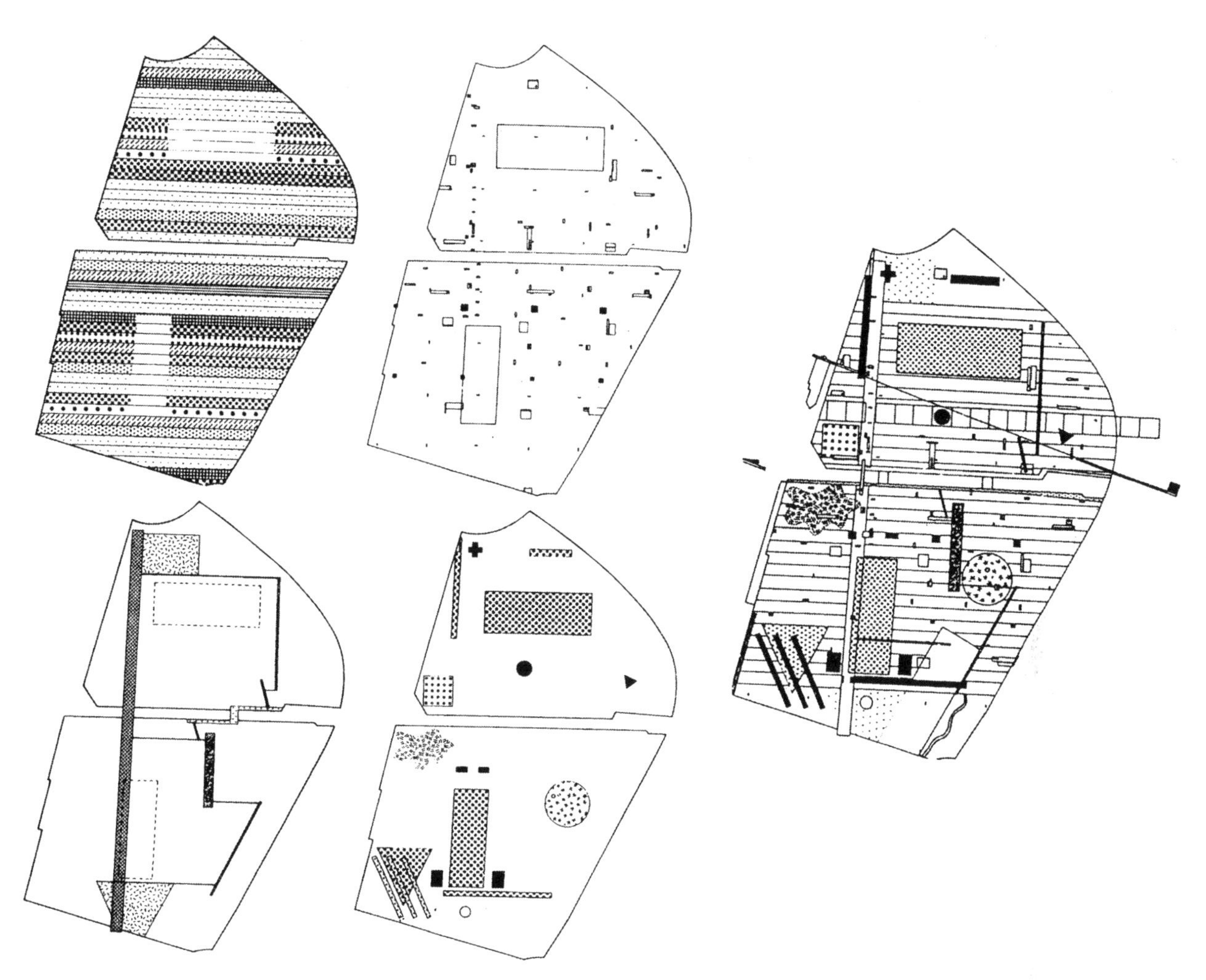

Figure 1.12.4 Rem Koolhaas layer diagrams for the Parc de la Villette (Office for Metropolitan Architecture 1983).

whose 'effects' flow through the system, continually reworking the variety of urban spaces in any given field. His approach is aimed first towards identifying and then redirecting the temporal play of these various forces. Consequently, urban design is practised less as spatial composition and more as orchestrating the conditions around which processes in the city may be brought into relationship and 'put into effect'. Bunschoten calls this 'stirring'.

[...]

In order to employ and operationalise these various conditions, they must first be made visible. Bunschoten accomplishes this by setting up a number of map frames, within which certain processes or conditions are graphically identified (Figure 1.12.5). He is careful to link the various cultural aspirations of each group to a physical space or territory distinguishing amongst 'local authorities' who anchor conditions into specific institutions or places, 'actors' who participate with stated desires and 'agents' who have the power and capacity to make things happen. Each frame permits the play of certain thematic conditions (preservation, ecology economic development or cultural memory for instance), whilst the composite overlay of all of the frames more accurately conveys the plural and interacting nature of the urban theatre.

[...]

The graphic map provides the game-board for playing out a range of urban futures. Identified players and actors are brought together to try to work out complex urban issues within an open-ended generative structure. Diverse forms of negotiation are promoted as the survival strategies of each player unfold and become interwoven with others in reaction to changing interests and situations. Thus the

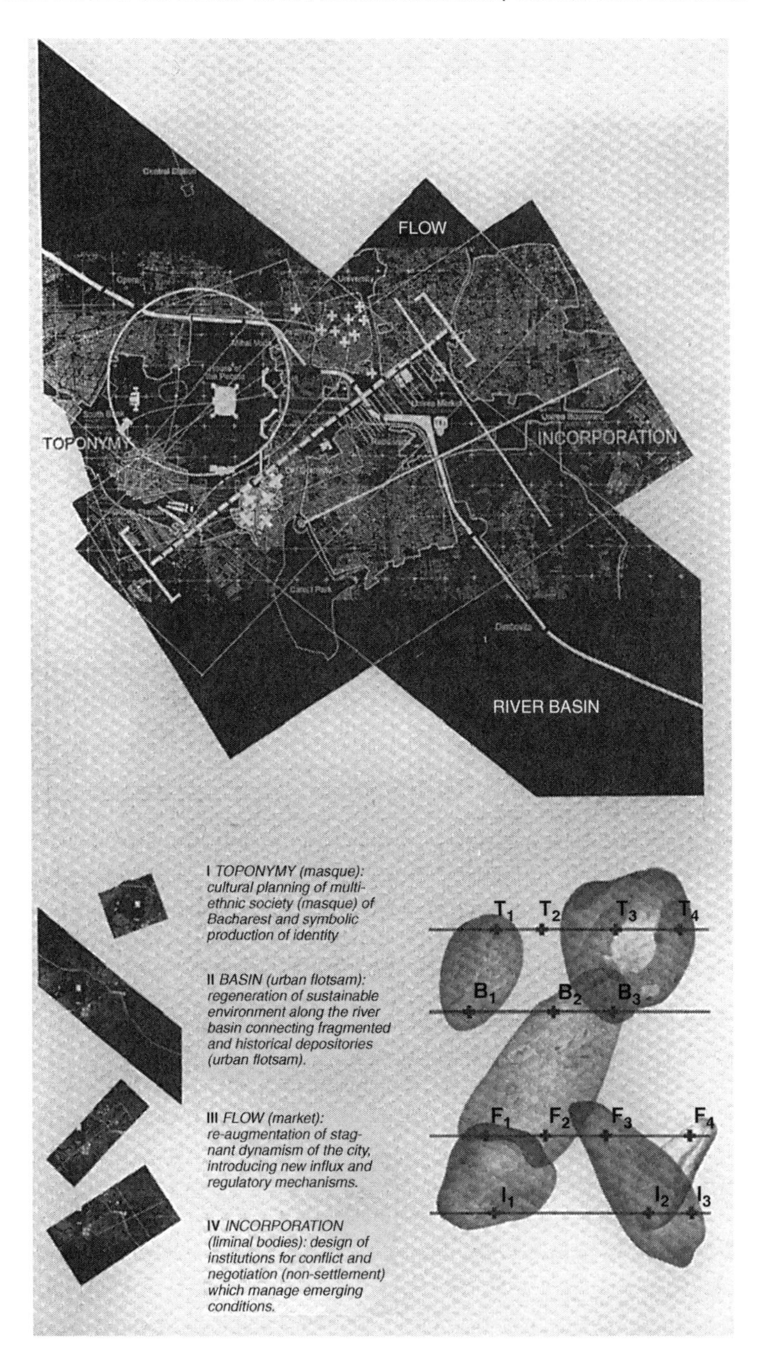

Figure 1.12.5 Raoul Bunschoten/CHORA, Four Planning Fields for Bucharest, 1996.

maps themselves are evolving structures, drawn and redrawn by the urban planner so as to permit the game to continue while also generating the necessary conditions for the emergence of an enterprising urbanity.

[. . .]

Rhizome

Open-ended and indeterminate characteristics can be likened to the process-form of the rhizome. 'Unlike trees or their roots,' write Deleuze and Guattari (1987: 6), 'the rhizome connects any point to any other point ... It has neither beginning nor end, but always a middle (*milieu*) from which it grows and overspills, [constituting] linear mulitiplicities.' In contrast to centric or tree-like, hierarchical systems, the rhizome is acentred, non-hierarchical and continually expanding across multiplicitous terrains. [...]

Deleuze and Guattari (1987: 6) draw an important distinction between 'maps' and 'tracings', describing the former as open, connectable, 'experimentations with the real', and the latter as repetitive redundancies that 'always come back to "the same"'. Hence, tracings belong to hierarchical systems of order that ultimately limit any hope of innovation – 'all of tree logic is a logic of tracing and reproduction' (Deleuze 1987: 12). By contrast, the infinitely open, rhizomatic nature of mapping affords many diverse entryways, exits and 'lines of flight', each of which allows for a plurality of readings, uses and effects.

The significance of the rhizome for mapping is encapsulated in Deleuze and Guattari's belief that 'the book' (and we might equally say the map, the city or the landscape) 'has no object. As an assemblage [it] has only itself, in connection with other assemblages and in relation to other bodies without organs.' Thus, Deleuze and Guattari (1987: 4) conclude:

> We will never ask what a book means, as signifier or signified; we will not look for anything to understand in it. We will ask what it functions with, in connection with what other things it does or does not transmit intensities, in which other multiplicities its own are inserted and metamorphosed, and with what other bodies it makes its own converge.

This viewpoint privileges actions and effects over representation and meaning; the concern is for how things work and what they do. Moreover, there is an explicit interest here for new kinds of affiliative relationship and interconnection. The argument emphasises probing practices of interpretation that extend previous products of culture (maps and landscapes, for instance) towards more diverse and interconnected fields of possibility their 'becoming' bodied-forth through various acts of mapping and relating.

One especially important principle with regard to mapping as a rhizomatic (burrowing and extending) activity is what Deleuze and Guattari refer to as the 'plane of consistency'. While this assumes a rich and complex array of meanings for the authors, I shall summarise plane of consistency here as a surface that is both inclusive (even of things that may not normally fit or 'belong' to any given scheme, including arbitrary 'debris') and *structuring* of new and open-ended series of relationships. Obviously if such a surface is both inclusive and structuring, the techniques and modes of representation must be both multiple and flexible.

[...]

[M]appings construct 'planes of consistency' that present analytical information while also allowing for suggestive readings/projections. They 'draw out' of common maps and landscapes certain figural and processual relationships that might occasion new landscapes. Admittedly, these mappings are not as open or rhizomatic as they might be, owing to their thematic focus, but their inclusion and incorporation (synthesis) of diverse kinds of information and possibility as well as their utilisation and subversion of dominant conventions, illustrates two important ways in which mapping might move towards more polymorphous and creative ends. They are also suggestive of how temporal, systemic, performance networks can be rendered distinct from traditional cartographic concerns with static space.

[T]he experience of spatial life today is as much immaterial as it is physical, as much bound into time and relational connections as it is to traditional notions of enclosure and 'place.' By extension, the principle of rhizomatic planes of consistency – together with the above-mentioned and closely allied themes of drift, *dérive*, layering, scaling, *milieu* and game-board structures – provides a useful model for mapping as a creative form of spatio-temporal practice in urban planning and design. In this way, we move away from urbanistic projects as authoritative master plans, concerned solely with the composition and order of static parts, toward practices of self-reflexive organisation. [...] Instead of designing relatively closed systems of order, rhizomatic mappings provide an infinite series of connections, switches, relays and circuits for activating matter and information. Hence *mapping*, as an open and inclusive process of disclosure and enablement, comes to replace the reduction of *planning*.

Conclusion

[...]

If maps are essentially subjective, interpretative and fictional constructs of facts, constructs that influence decisions, actions and cultural values generally, then why not embrace the profound efficacy of mapping in exploring and shaping new realities? Why not embrace the fact that the potentially infinite capacity of mapping to find and found new conditions might enable more socially engaging modes of exchange within larger *milieux*? The notion that mapping should be restricted to empirical data sorting and array diminishes the profound social and orienting sway of the cartographic enterprise. And yet the power of 'objective analysis' in building consensus and representing collective responsibility is not something to be abandoned for a free-form 'subjectivity'; this would be both naive and ineffective. The power of maps resides in their facticity. The analytical measure of factual objectivity (and the credibility that it brings to collective discourse) is a characteristic of mapping that ought to be embraced, co-opted and *used* as the means by which critical projects can be realised (Corner and MacLean 1996). After all, it is the apparent rigour of objective analysis and logical argument that possesses the greatest efficacy in a pluralistic, democratic society Analytical research through mapping enables the designer to *construct* an argument, to embed it within the dominant practices of a rational culture, and ultimately to turn those practices towards more productive and collective ends. In this sense, mapping is not the indiscriminate, blinkered accumulation and endless array of data, but rather an extremely shrewd and tactical enterprise, a practice of relational reasoning that intelligently unfolds new realities out of existing constraints, quantities, facts and conditions (Allen 1998; Beck 1994; Corner 1999a; Koolhaas and Man 1994).[...]

[...]

Instances of drift, strata, game-board and rhizome represent only a handful of techniques that mapping practices might assume if they are to play more creative roles in design and planning, and in culture more generally. These techniques presuppose any number of variations and enhancements as issues of framing, scaling, orientation, projection, indexing and coding become more flexible and open-ended, especially in the context of powerful new digital and animation media. As we are freed from the old limits of frame and boundary – preconditions for the survey and 'colonisation' of wilderness areas – the role of mapping will become less one of tracing and re-tracing already known worlds, and more one of inaugurating new worlds out of old. Instead of mapping as a means of appropriation, we might begin to see it as a means of emancipation and enablement, liberating phenomena and potential from the encasements of convention and habit. What remains unseen and unrealised across seemingly exhausted grounds becomes actualised anew with the liberating efficacy of creatively aligned cartographic procedures. Mapping may thus retain its original entrepreneurial and exploratory character, actualising within its virtual spaces new territories and prospects out of pervasive yet dormant conditions.

References

Allen, S. (1998) Artificial ecologies. *El Croquis*, **86**, 26–33.

Arnheim, R. (1970) *Visual Thinking*, University of California Press, Berkeley, CA.

Baudrillard, J. (1983) *Simulations*, Semiotext(e), New York.

Beck, U. (1994) *Reflexive Modernization: Politics, Tradition and Aesthetics in the Modern Social Order*, Polity Press, Cambridge.

Blanshard, B. (1948) *The Nature of Thought*, Allen & Unwin, London.

Borges, J.L. (1933) Of exactitude in science, reprinted in *A Universal History of Infamy*, Penguin, London.

Bronowski, J. (1965) *Science and Human Values*, Harper Torch, New York.

Buisseret, D. (1998) *Envisioning the City: Six Studies in Urban Cartography*, University of Chicago Press, Chicago.

Bunschoten, R. (1996) Proto-urban conditions and urban change, in *Beyond the Revolution: The Architecture of Eastern Europe: Architectural Design Profile 119* (ed. T. Toy), London Architectural Design, London, pp. 17–21.

Bunschoten, R. (1997) Black Sea: Bucharest stepping stone, in *Architecture After Geometry: Architectural Design Profile 127* (eds P. Davidson and D. Bates), Academy Additions, London, pp. 82–91.

Bunschoten, R. (1998) *Urban Flotsam*, Chora Publishers, Rotterdam.

Corner, J. (1992) Representation and landscape. *Word & Image*, **8** (3), 243–275.

Corner, J. (1999a) Landscape and ecology as agents of creativity, in *Ecological Design and Planning* (eds G.F. Thompson and R.F. Steiner), John Wiley & Sons, Inc., New York, pp. 80–108.

Corner, J. (1999b) Operational eidetics in forging new landscapes, in *Recovering Landscape: Essays in Contemporary Landscape Architecture*, Princeton Architectural Press, Princeton, NJ.

Corner, J. and MacLean, A. (1996) *Taking Measures Across the American Landscape*, Yale University Press, New Haven, CT.

de Certeau, M. (1984) *The Practice of Everyday Life*, University of California Press, Berkeley, CA.

Deleuze, G. and Guattari, F. (1987) *A Thousand Plateaus: Capitalism and Schizophrenia*, University of Minnesota Press, Minneapolis, MN.

Giddens, A. (1994) Living in a post-traditional society, in *Reflexive Modernization: Politics, Tradition and Aesthetics in the Modern Social Order* (eds U. Beck, A. Giddens and S. Lasch), Polity Press, Cambridge.

Hall, P. (1988) *Cities of Tomorrow: An Intellectual History of Urban Planning and Design in the Twentieth Century*, Blackwell, Oxford.

Hall, S. (1992) *Mapping the Next Millennium*, Random House, New York.

Harley, J.B. (1988) Maps, knowledge, and power, in *The Iconography of Landscape* (eds D. Cosgrove and S. Daniels), Cambridge University Press, Cambridge, pp. 277–312.

Harrison, J. and Turnbull, D. (1996) *Games of Architecture: Architectural Design Profile 121*, Academy Editions, London.

Koolhaas, R. and Man, B. (1995) *S,M,L,XL*, Monacelli Press, New York.

Marks, R. and Buckminster Fuller, R. (1973) *The Dymaxion World of Buckminster Fuller*, Doubleday & Co, New York.

Monmonier, M. (1991) *How to Lie with Maps*, University of Chicago Press, Chicago.

Pickles, J. (1992) Texts, hermeneutics and propaganda maps, in *Writing Worlds* (eds T.J. Barnes and J.S. Duncan), Routledge, London, pp. 193–230.

Porter, T.M. (1995) *Trust in Numbers: The Pursuit of Objectivity in Science and Public Life*, Princeton University Press, Princeton, NJ.

Robinson, A.H. and Petchenik, B.B. (1976) *The Nature of Maps*, University of Chicago Press, Chicago.

Scott, J. (1998) *Seeing Like a State: Why Certain Schemes to Improve the Human Condition Have Failed*, Yale University Press, New Haven CT.

Söderström, O. (1996) Paper cities: visual thinking in urban planning. *Ecumene*, **11** (3), 249–281.

Storr, R. (1994) *Mapping*, Harry N Abrams, New York.

Tschumi, B. (1987) *Cinegramme Folie: le Parc de la Villette*, Princeton University Press, Princeton, NJ.

Wood, D. (1992) *The Power of Maps*, Guildford, New York.

Further reading

Abrams, J. and Hall, P. (2006) *Else/Where: Mapping New Cartographies of Networks and Territories*, University of Minnesota Design Institute, Minneapolis, MN. [An inventive edited collection that demonstrates the creative and artistic potential for mapping.]

Anderson, B. and Harrison, P. (2010) *Non-Representational Theories and Geography*, Ashgate, London. [The useful introduction to non-representational ways of approaching the world, including many examples of work inspired by Deleuzian ideas.]

Grasseni, C. (2004) Skilled landscapes: mapping practices on locality. *Environment and Planning, D* **22**, 699–717. [An interesting example of the practical application of Corner's ideas in an Italian community mapping initiative.]

Pickles, J. (2004) *A History of Spaces: Cartographic Reason, Mapping, and the Geo-coded World*, Routledge, London. [A largely Foucauldian exploration of the constitutive power of mapping: social chapters informed by Corner's work.]

See also

- Chapter 1.7: Design on Signs / Myth and Meaning in Maps
- Chapter 1.8: Deconstructing the Map
- Chapter 1.9: Drawing Things Together
- Chapter 1.10: Cartography Without 'Progress': Reinterpreting the Nature and Historical Development of Mapmaking
- Chapter 1.13: Beyond the 'Binaries': A Methodological Intervention for Interrogating Maps as Representational Practices
- Chapter 1.14: Rethinking Maps
- Chapter 3.9: Mapping, Modernity: Art and Cartography in the Twentieth Century
- Chapter 3.10: Affective Geovisualisations
- Chapter 4.5: The Map as Biography
- Chapter 4.6: Reading Maps
- Chapter 4.8: Refiguring Geography: Parish Maps of Common Ground

Chapter 1.13

Beyond the 'Binaries': A Methodological Intervention for Interrogating Maps as Representational Practices

Vincent J. Del Casino Jr. and Stephen P. Hanna

Editors' overview

Del Casino and Hanna offer one of the first rejections of social constructivist orthodoxy regarding approaches to mapping and power. They suggest it may be more useful to move beyond representation and discourse, and instead adopt a more performative approach to mapping, which rejects binaries such as representation/practice, production/consumption, conceptualisation/interpretation, and corporeality/sociality. Their article provides a methodology informed by feminist critique and, in particular, the work of Judith Butler and Sandra Harding. Del Casino and Hanna urge scholars to consider practice and the ways in which cartography co-constructs the places that it represents through the everyday enrolment of mapping activities into other social tasks. Using a case study of tourist mapping practices in Fredricksburg, Virginia, USA, they exemplify the rich potential of moving beyond the binaries for map scholarship.

Originally published in 2006: *ACME: An International E-Journal for Critical Geographies*, **4** (1), 34–56.

Introduction

[...]

When does the moment of map production end? At the time when the printing press stops rolling, or the crayon leaves the page, or when a *yahoo map* stops loading, or perhaps when the finished map is found embedded between columns two and three of the *New York Times*? Or, maybe, production never stops. Maybe maps are constantly produced and (re)produced as is suggested through the democratisation of production in participatory GIS. Such questioning of production, however, should be accompanied by similar questions about consumption. When is a map first consumed? After it leaves the hands of its authors? Do authors not consume their own representation, see themselves in its images, reconstruct their own desires through this object, or dare we say subject? Still, many critical cartographers (Black 1997; Harley 1988, 1989, 1990; Wood 1992) maintain an implicit duality between production and consumption, author and reader, object and subject, design and use, representation and practice. They still focus on how maps are produced in particular social, political and economic contexts; or they concentrate on the consumption and use of these particular objects in their post-production phase. Yet, maps, to borrow from Gibson (2001), are in state of 'becoming'. As such, maps stretch beyond their physical boundaries; they are not limited by the paper on which they are printed or the wall upon which they might be scrawled. Each crease, fold and tear produces a new rendering, a new possibility, a new (re)presentation, a new moment of production *and* consumption, authoring *and* reading, objectification *and* subjectification, representation *and* practice.

Maps are thus not simply representations of particular contexts, places and times. They are mobile subjects, infused with meaning through contested, complex, intertextual and interrelated sets of socio-spatial practices. As Deleuze and Guarttari (1987: 12) suggest, 'the map has multiple entryways' and a myriad number of

The Map Reader: Theories of Mapping Practice and Cartographic Representation, First Edition. Edited by Martin Dodge, Rob Kitchin and Chris Perkins.

possibilities because it operates at the margin and centre simultaneously. Maps are also not, as some may argue (Harley 1989), fixed at the moment of production, a result of the hegemonic authority embedded by the mapmaker in/on the representation. Thus, while maps may be infused with power, and thus ripe for deconstruction, it is not enough to demythologise the map (Sparke 1995). Instead, maps ought to be theorised as processes, 'detachable, reversible, susceptible to constant modification.' It is therefore appropriate to say that maps, as representations, 'work' (Wood 1992). As we contend, representations, such as maps, work because 'they help make connections to other representations and to other experienced spaces' (Hanna *et al.* 2004: 464), suggesting that maps do, indeed, provide multiple entryways into how they are produced and consumed as well as how they are used, interpreted and constituted.

[...][W]e are particularly concerned with interrogating the binaries – such as representation/practice, production/consumption, map/space – upon which critical cartography and critical GIS are partially based. This is a methodological intervention because, like Harding (1987: 3), we view methodology as 'a theory and analysis of how research does or should proceed.' This suggests that geographers must first ask how they conceptualise and frame their questions and data. In offering this critique, however, we are not simply moving toward a non-representational theory of maps and mappings as practices (Perkins 2004), at least as Thrift (1996) has conceptualised the non-representational [...]. Instead, we are interested in pushing our methodological (and by default our epistemological and ontological) assumptions and processes toward thinking of maps as spaces. As such, maps and mappings are both representations and practices (read: performances) simultaneously. Neither is fully inscribed with meaning as representations nor fully acted out as practices. As Nash (2000: 662) avers, we must 'allow room for considering visual and textual forms of representations as practices themselves.' [...] 'Any critical perspective of mapping must...investigate the multiple historical and spatial referents that are part and parcel of any...map. This compilation of referents comes together and constitutes what we call a map space' (Del Casino and Hanna 2003). [...]

[...]

On the 'critical' in critical cartography (and GIS)

[...]

The production of 'multiple maps' offers the opportunity to democratise cartographic production, at least theoretically (Ghose and Elwood 2003). It also exposes the 'hidden' *a priori* decisions that are made before and during map production and thus challenges the hegemony of cartographers over the mapmaking process (Crampton 2001). Crampton's suggestion that critical cartographers use the approaches of geographic visualisation thus partially upends the binary logic of map producer and map user. This echoes the earlier calls of Wood (1992), who ends *The Power of Maps* by criticising the arrogance of professional cartographers who worry about 'cartographic problems' and praising people who make their own maps to serve their own interests. In fact, the push toward more collaborative cartographies and geographic information systems (GIS) has become an important part of the growing critical literature on maps and GI/Science (Ghose and Elwood 2003; MacEachren 2000 [...]). For the most part, this work is focused on how the participatory context functions between various actors. But, like Crampton's and MacEachren's discussion of geovisualisation, there are several potential limitations to this analysis. Most importantly, there remains in this work a lingering emphasis on authoring, production and writing. The only way a map becomes something more than what Harley envisioned is if users become involved in the production process.

[...]

The turn to poststructuralism suggests that a focus on the democratisation of mapping, albeit unintentionally, may marginalise the consumptive processes of mapmaking, maps and mappings. It may also fail to take into account the constitutive nature of consumption *as* production and reading *as* authoring, which is critical to poststructuralist theories of representational practices (Barthes 1978). Further, it delinks maps from spaces by pushing aside the intertextuality and relationality that are part and parcel of the relationships between spaces and representations. This is not to suggest that there is some 'real' space out there. Rather, it is to argue that all spaces are always already representations that are produced by and productive of a myriad number of bodily practices and performances as social actors draw on their intertextual knowledges while reauthoring both space and representation simultaneously.

[...]

[W]e suggest that the reiterative processes of consuming maps may temporarily 'naturalise' maps as somehow fixed in their meaning. And yet, the very act of consumption through the performance of maps offers productive possibilities, particularly through the emergence of constitutive spaces at/in the 'gaps and fissures' of the mapping processes. And, yet, there is no deep underlying structure that somehow organises this overall process.

Thus, investigating cartographic collaboration may involve, as Harvey (2001: 30) notes, studying 'the traces of relationships between people, institutions, and artefacts [as they are] connected by agreements and exchanges.' Working through actor-network theory, Harvey emphasises that there is nothing 'natural' (2001: 31) about the separations between map producers and users, representations and practices, or the technologies and artefacts that mediate the relationships between the various 'nodes' in any given network of social actors. As Harvey (2001: 30) further explains, actor-network theory has real methodological potential for breaking down the boundaries between production and consumption, author and reader, map and space because '[t]echnical artefacts [such as maps] are much more than surrogates for certain humans; they are actors who bundle multiple intentions and act in ways that complement and extend humans.' In actor-network theory, we are interested not in some deep relationship between structures and agency; rather, attention is paid to the 'traces' between networks and actors (Mol and Law 1994, as cited in Harvey 2001: 30). These traces, discursive and material residuals left by the reiterative process, are part of actor-network processes that produce temporarily perhaps, a 'common ground' (Harvey 2001: 31) among actors in the network. They do not, however, point to any original moment, but rather constitute a complex and spiralling web of socio-spatial relations informed by a myriad number of discourses and practices. Following Harvey's theorisation, we are closer, perhaps, to critiquing the binary logic of production and consumption in critical cartography that typically privileges the production context over the mundane uses of maps and mappings.

Poststructuralist actor-network theories parallel, in some regard, the work of Matthew Sparke (1995: 4) who suggests that 'Following Derrida...enables us to demythologise while being more wary, and less romantic, about uncovering 'true' places and constructing 'true' maps'. Ultimately, he hopes that deconstruction 'urges us to go back and look for other ways in which the map, and what is supposed to lie outside of it – power relations, interpretations of the 'real' world, and so on – might actually be still more complexly interrelated' (Sparke 1995). In both cases, maps are not theorised as closed off objects nor are their meanings and uses fixed by either production or by academic critique. Instead, maps are conceptualised as working because they are practiced – reading, like authoring, is a practice – and performed in various socio-cultural and political-economic contexts.

These poststructuralist turns in critical cartographic and GIS analysis pose interesting challenges to the traditional binary logic of re-presentational – as opposed to representational – theories of map communication and production. As Dixon and Jones (2004: 88) argue '[p]oststructuralists take note of and critique forms of thought that distinguish between the 'real' world and its 'mere' re-presentation in communication, whether conceived in terms of language, sensory perceptions, or electronic media.' Poststructuralists are interested in not only challenging the binary logic of 'real' and 're-presented,' they are also interested in deconstructing the tensions that are part and parcel of all binaries. As in actor-network theory, poststructuralist representational theories are not interested in how we communicate truth or reality. Rather, there is 'no signifier [that] can be presumed to stand in a one-to-one relationship with a real-world referent' (Dixon and Jones 2004: 88). As such, representations, such as maps, 'work' because social actors use them in ways that may reify dominant ideologies. At the same time, maps 'work' as social actors perform the very mundane act of using maps and mappings in everyday interactions in ways that open up productive moments of resistance (Foucault 1997). Such a theorisation suggests that we might turn our attention toward how maps and mappings are practiced beyond the productive moments of making a map or constructing various mappings. Some have suggested that non-representational theories might open up the possibility for rethinking how maps as objects are tied to various practiced contexts (Perkins 2004). We, therefore, want to briefly turn to non-representational theories of practice and ask if they help us move from maps as products to maps as practices.

On a non-non-representational geography of maps and mappings

[...] It seems to us rather odd [...] that the uses and practices of mapping might be framed by theories that purport to be non-representational. What we want to suggest is that it is not necessary to conflate all theories of practice and performance with non-representational theory. And, in the case of maps and mappings, it may be somewhat counterproductive to do so because it may re-inscribe the binary logic of representation and practice that limits our theoretical possibilities.

The appropriation of Thrift's (1996) non-representational theory by cultural geographers tends to presuppose an ontology of real emotions, experiences, and senses that somehow make representations less real. As Nash (2000: 655) argues, 'non-representational theory moves away from a concern with representation and especially text since, Thrift argues, text only inadequately commemorates ordinary lives since it values what is written or spoken over multisensual practices and experiences.' In this way, Nash suggests that Thrift is suturing representations to texts theoretically. And, indeed, Thrift does critique the textual turn in cultural geography for its focus on the textual and visual at the detriment of the other senses. Deconstruction is thus a limited methodological framework for Thrift because of its focus on textuality (Nash 2000: 656). Nash argues that Thrift's appropriation of performance theory is potentially limiting, however, because he positions one end of the binary over the other, practices are more experiential than representations. We tend to agree: 'a dichotomy between representation and non-representation, while perhaps a valuable heuristic, does little to forward our thinking about how place is performed. Rather, we argue that representations are material practices and are intertextually constructed through the daily actions' of various social actors (Hanna *et al.* 2004: 462). Moreover, Thrift's non-representational theory detaches representations from the spaces they claim to represent. Following Smith (2003: 76), however, 'it is no longer the case of having either maps [e.g. representations] or territories [e.g. spaces] because the distinction between the two has vanished'. Instead, it is better to theoretically consider maps and spaces are co-constitutive. Methodologically, Thrift's approach thus marginalises the value of deconstruction, which we want to argue usefully teases out the constitutive and relational nature of discourses and practices, social and material experiences, and representations and spaces.

If we follow Thrift's logic, maps are merely textual and not spatial objects. They are constituted through the production of visual representations of space and not through mundane spatial practices. It is our contention, however, that if we turn to the performance theory of Butler, as Nash does, and work through identity theory (Natter and Jones 1997), we come to a different rendering of the map-space relationship – one which collapses the binary logic of text and space and experience and representation around which Thrift's non-representational theory is centred. These theories of performance and identity are also important to understanding the emergence of what we call map spaces. We use this term to denote the theoretical impossibility of disentangling representations from performances. Map spaces are always partial and incomplete, contested sites where the collisions of various identity and subject positions blur the boundaries of centre and margin. 'While maps draw meaning from and help to define the spaces and identities...they contain uncertainties and traces of excluded others that introduce potential ambiguities in their relationships with space and identity' (Del Casino and Hanna 2000: 24). Those ambiguities are performed through various bodily actions that often push the meaning of the map beyond its intended use even in a participatory context (Curran 2003; DeLyser 2003; Shields 2003). We can, however, deconstruct the ways in which these ambiguities become temporarily sutured to various spaces and identities through the performative and repetitive practices of social actors.

Thinking about map spaces in this way means neither the production nor the consumption of maps is separable from space in the most mundane of settings. Maps that people simultaneously make and use mediate their experiences of space. People's bodily practices of walking, driving, touching, smelling and gazing, as well as their understandings of landscapes and places can be guided and informed by maps and by the innumerable intertextual and experiential references always present in any map. At the same time, spaces mediate people's experiences of maps. For example, the physicality of a monument and the emotions it brings forth lends meaning to its symbol on a map (Dwyer 2000, 2003). The everyday movements and distractions always present in spaces interrupt and interfere with both the reading and making of maps. When driving we must pull over and stop to look at a map or risk a traffic accident. Our theorisation, therefore, does not prioritise writing over reading or production over consumption in the constant recreation of the map space. Nor do we wish to argue that map spaces as representation are separable from map spaces as practiced, worked or performed.

Possible openings and methodological interventions

In this section we want to suggest that there might be an alternative to thinking through the binaries of design and use, representation and practice, objectivity and subjectivity that undergirds much of the critical cartography literature by suggesting how map spaces as performed 'work.' We do this through a brief examination of the use of a tourism map of Fredericksburg, Virginia, USA. [...] Here, we want to centre our discussion on the map space, as written, read, performed and consumed, and use it as a site through which to offer our methodological intervention.

[...]

Consider the tourist who begins her exploration of Fredericksburg at the Visitors Center. She will be greeted by one of Visitor Councillors who are trained to help tourists match their desires and interests to the right numbers on the map. As Bill, a Fredericksburg Visitor's Center Volunteer explains, '... I would suggest that they see our fourteen minute video that we show on Fredericksburg. It is much easier to draw a map and tell people about the map if they have some knowledge of what's it all about.'

[...]

[B]oth tourism worker and tourist are consumers *and* producers. The Councillor needs the map to perform her job; she helps tourists explore Fredericksburg. A tourist uses the map to find attractions, but also to begin organising her or his knowledge and imagination about the town's history. In this way, the bodily interactions between representation, reader, and the various spaces the representation claims to mark reproduce both Fredericksburg and the map as both a city and representation simultaneously.

[...]

This is a productive process. Tourists engage in the (re) production and consumption of this space simultaneously through the asking of questions, reading along on a paper mapped representation of the city, marking the map and making other notes. Alternatively, tourists may focus on creating their own map space of historic places and sites of interest as they listen to, or ignore, the directions of the trolley driver.

[...]

These bodily interactions do not fix this map space for tourists. When they leave the Visitors Center, map in hand, they are free to explore the map space as they desire. To some, this means boarding the tourist trolley and experiencing the sites on the map, and many others, while gazing out the window and listening to the driver's narration. Others may choose to walk the route – a physical act which undoubtedly changes their experience. Still others may limit themselves to sites, homes and monuments related to George Washington's life history, perhaps suturing this representation temporarily to a series of emotions including nationalism and patriotism.

[...]

Through performances [...] tourists reproduce and alter the map space with their bodies and minds. They also employ other senses as they absorb the smells of foods from the restaurants or touch the antiques that are ubiquitously placed throughout the city's downtown shops. In some cases, the mapped representation in their hand becomes part and parcel of those experiences, a keepsake used to remember Fredericksburg as a place.

[...]

Thus, the mapped representation can always be exceeded and used in different ways as individual social actors mark the map with restaurants, antique shops, or new objects of their own personal historical interest. As such, maps are never fully complete nor are they ever completely inscribed with meaning through production. Rather, consumption is production. Map spaces are processes, fluid and contested, although they find themselves temporarily fixed through certain practices of consumption that (re)produce these objects in new and unique ways.

Conclusion

If cartography has always already been a partial and contested set of practices and performances, then it makes sense that maps, such as the one discussed above, are also always in the process of production and consumption, authoring and reading. Methodologically, what this suggests is that ethnographic and other qualitative methodologies may be deployed to 'capture' the performed identities of map spaces. As Perkins (2004: 386) argues, 'recent ethnographic approaches have investigated everyday social experiences of places and the role that mapping practices play in identity and knowledge construction...An ethnographic approach [thus] reorientates theory so that mapping becomes a social activity, rather than an individual response.' But, what we want to further suggest is that ethnographic studies of maps and mapping need not work on the side of either production or consumption, representation or practice. Rather, the two sides of these binaries are constitutive. While we have suggested how maps 'work' after a cartographer has designed a tourism map, we could examine how cartographers also consume objects, spaces and representations in their reworkings of their own subjects. A critical cartography and/or GIS thus needs to further blur the boundaries of production and consumption, author and reader, subject and object if it is to truly become participatory and democratic.

This is a methodological concern because it is about that meso-level moment when we temporarily suture our own epistemological assumptions to the question of what are data and how we might go about collecting those data (Harding 1987). In our own epistemological view, to parody Barthes (1978), the author of the map is dead; reading produces and reproduces map spaces through the multiplicity of performances that social actors deploy in their mundane interactions with maps. This suggests that we might ask the question: to what extent do cartographers actually author space? In turning even further toward theories of performance in critical cartography and GIS, however, we want to caution that non-representational theory as it is currently operationalised in geography might limit rather than expand the possibilities for breaking down the binaries that currently operate in the field. In so doing, we suggest deploying the theoretical concept of map space as opposed to studying maps and spaces as somehow ontologically separate sites of inquiry. Methodologically, this theoretical shift suggests that our objects of analyses are not simply maps but are instead the myriad interconnections that make the production and consumption of map spaces a process of both authoring and reading simultaneously. In offering this intervention, therefore, we suggest that that it is necessary to consider the epistemological and ontological assumptions that undergird notions of production and consumption in cartographic and GIS analysis as we think through our own methodological frameworks and, eventually, methods of analysis.

References

Barthes, R. (1978) The death of the author, in *Image, Music, Text* (ed. and trans. S. Heath), Farrar, Straus, Giroux, New York, pp. 142–148.

Black, J. (1997) *Maps and Politics*, University of Chicago Press, Chicago.

Crampton, J. (2001) Maps as social constructions: power, communication, and visualization. *Progress in Human Geography*, **25** (2), 235–252.

Curran, M. (2003) Dialogues of difference: contested mappings of tourism and environmental protection, in *Mapping Tourism* (eds S. Hanna and V.J. Del Casino), University of Minnesota Press, Minneapolis, MN, pp. 132–160.

Del Casino, V.J. and Hanna, S.P. (2000) Representations and identities in tourism map spaces. *Progress in Human Geography*, **24** (1), 23–46.

Del Casino, V.J. and Hanna, S.P. (2003) Mapping identities, reading maps: the politics of representation in Bangkok's sex tourism industry, in *Mapping Tourism* (eds S. Hanna and V.J. Del Casino), University of Minnesota Press, Minneapolis, MN, pp. 161–186.

Deleuze, G. and Guattari, F. (1987) *A Thousand Plateaus: Capitalism and Schizophrenia*, University of Minnesota Press, Minneapolis, MN.

DeLyser, D. (2003) A walk through Old Bodie: presenting a ghost town in a tourism map, in *Mapping Tourism* (eds S. Hanna and V.J. Del Casino), University of Minnesota Press, Minneapolis, MN, pp. 79–107.
Dixon, D. and Jones, J.P. (2004) Poststructuralism, in *A Companion to Cultural Geography* (eds J. Duncan, N. Duncan and R. Schein), Blackwell, London, pp. 79–107.
Dwyer, O. (2000) Interpreting the civil rights movement: place, memory, and conflict. *The Professional Geographer*, **52** (4), 660–71.
Dwyer, O. (2003) Memory on the margins: Alabama's civil rights journey as a memorial text, in *Mapping Tourism* (eds S. Hanna and V.J. Del Casino), University of Minnesota Press, Minneapolis, MN, pp. 28–50.
Foucault, M. (1997) Governmentality, in *The Foucault Effect: Studies in Governmentality* (eds C. Gordon and P. Miller), University of Chicago Press, Chicago, pp. 87–104.
Ghose, R. and Elwood, S. (2003) Public participation GIS and local political context: propositions and research directions. *URISA Journal*, **15**, 17–24.
Gibson, K. (2001) Regional subjection and becoming. *Environment and Planning D: Society and Space*, **19** (6), 639–667.
Hanna, S., Del Casino, V.J., Selden, C. and Hite, B. (2004) Representation as work in 'America's most historic city'. *Social and Cultural Geography*, **5** (3), 459–482.
Harding, S. (1987) *Feminism and Methodology*, Indiana University Press, Bloomington, IN.
Harley, J.B. (1988) Maps, knowledge, and power, in *The Iconography of Landscape: Essays on Symbolic Representation, Design and Use of Past Environments* (eds D. Cosgrove and S. Daniels), Cambridge University Press, Cambridge, pp. 277–312.
Harley, J.B. (1989) Deconstructing the map. *Cartographica*, **26** (2), 1–20.
Harley, J.B. (1990) Cartography, ethics, and social theory. *Cartographica*, **27** (2), 1–23.
Harvey, F. (2001) Constructing GIS: actor networks of collaboration. *URISA Journal*, **13** (1), 29–37.
MacEachren, A.M. (2000) Cartography and GIS: facilitating collaboration. *Progress in Human Geography*, **24** (3), 445–456.
Mol, A. and Law, J. (1994) Regions, networks, and fluids: anemia and social topology. *Social Studies of Science*, **24**, 641–671.
Nash, C. (2000) Performativity in practice: some recent work in cultural geography. *Progress in Human Geography*, **24** (4), 653–664.
Natter, W. and Jones, J.P. (1997) Identity, space, and other uncertainties, in *Space and Social Theory: Interpreting Modernity and Postmoderity* (eds G. Benko and U. Strohmayer), Blackwell, Oxford, pp. 43–161.
Perkins, C. (2004) Cartography – cultures of mapping: power in practice. *Progress in Human Geography*, **28** (3), 381–391.
Shields, R. (2003) Political tourism: mapping memory and the future of Québec City, in *Mapping Tourism* (eds S. Hanna and V.J. Del Casino), University of Minnesota Press, Minneapolis, MN, pp. 1–27.
Smith, R.G. (2003) Baudrillard's nonrepresentational theory: burn the signs and journey without maps. *Environment and Planning D: Society and Space*, **21** (1), 67–84.
Sparke, M. (1995) Between demythologizing and deconstructing the map: Shawnadithit's New-Found-Land and the alienation of Canada. *Cartographica*, **32**, 1–21.
Thrift, N. (1996) *Spatial Formations*, Sage, London.
Wood, D. (1992) *The Power of Maps*, Guilford, New York.

Further reading

Anderson, B. and Harrison, P. (2010) *Non-Representational Theories and Geography*, Ashgate, London. [The useful introduction to the performative potential of non-representational ways of approaches the world.]
Brown, B. and Laurier, E. (2005) Maps and journeys: an ethnomethodological investigation. *Cartographica*, **40** (3), 17–33. [The article focuses on the practices around maps, through an ethnographic and participative study of mapping in the conduct of everyday journeys.]
Perkins, C. (2009) Performative and embodied mapping, in *International Encyclopedia of Human Geography* (eds R. Kitchin and N. Thrift), Elsevier, Oxford. [A useful overview of the potential of more performative approaches to mapping.]
Whatmore, S. (2002) *Hybrid Geographies: Natures, Cultures and Space*, Sage, London. [Argues for a more complex and hybrid approach to notions of nature and cultures that transcends binary logic and parallels Hanna and Del Casino's arguments in relation to mapping.]

See also

- Chapter 1.7: Design on Signs/Myth and Meaning in Maps
- Chapter 1.8: Deconstructing the Map
- Chapter 1.9: Drawing Things Together
- Chapter 1.10: Cartography Without 'Progress': Reinterpreting the Nature and Historical Development of Mapmaking
- Chapter 1.12: The Agency of Mapping
- Chapter 1.14: Rethinking Maps
- Chapter 3.9: Maps, Mapping, Modernity: Art and Cartography in the Twentieth Century
- Chapter 4.2: Map Makers are Human: Comments on the Subjective in Maps
- Chapter 4.5: The Map as Biography
- Chapter 4.6: Reading Maps
- Chapter 4.8: Refiguring Geography: Parish Maps of Common Ground
- Chapter 5.9: Affecting Geospatial Technologies: Toward a Feminist Politics of Emotion

Chapter 1.14

Rethinking Maps

Rob Kitchin and Martin Dodge

Editors' overview

A profusion of new ways of thinking about mapping has emerged in the last two decades, including most recently a focus upon mapping practices and performance. Kitchin and Dodge's article suggest that this 'turn' might usefully be developed if researchers were to focus upon the status of the map as artefact and mapping as a practice. They argue that maps might best be thought of as 'emergent', called into being to allow particular social tasks to be enacted, but changeable according to the context and problems encountered. Invoking Mackenzie's ideas of 'transduction' and the technicity of mapping, they demonstrate the potential of this rethinking against two mapping vignettes. They characterise the map as ontogenic, rather than ontic, and argue this might allow a reconciliation between social theoretical approaches to cartographic scholarship and practical, pragmatic mapmaking.

Originally published in 2007: *Progress in Human Geography*, **31** (3), 331–344.

Cartography's ontological crisis

Maps have long been seen as objective, neutral products of science. Cartography is the means by which the surface of the earth is represented as faithfully as possible. The skill of the cartographer is to accurately capture and portray relevant features. Cartography as an academic and scientific pursuit then largely consists of theorising how best to represent spatial data (through new devises, e.g. choropleth maps, contour lines; through the use of colour; through ways that match how people think, e.g. drawing on cognitive science; and so on). In the latter part of the twentieth century, the science of cartography was influenced deeply by Arthur Robinson. He re-cast cartography, focusing in particular on systematically detailing map design principles with the map user in mind. His aim was to produce what he termed 'map effectiveness' – that is, maps that capture and portray relevant information in a way that the map reader can analyse and interpret (Robinson *et al.* 1995). Since the mid-1980s this particular view of cartography has been under challenge. On the one side has been other 'scientific' cartographers seeking to replace Robinson's model with one more rooted in cognitive science (MacEachren 1995) or visualisation principles (Antle and Klickenberg 1999); on the other side have been critical cartographers who, drawing on critical social theory, have questioned the rationale and principles of cartography, but often have little say about the technical aspects of how to create a map or how maps work (Crampton 2003).

[...]

As Denis Wood (1993) and Jeremy Crampton (2003) outline [...] Harley's (1989) application of Foucault to cartography, and therefore nearly all critical cartography that follows, is limited. Harley's observations, whilst opening a new view onto cartography, stopped short of following Foucault's line of inquiry to its logical conclusion. Instead, Crampton (2003: 7) argues that Harley's writings 'remained mired in the modernist conception of maps as documents charged with "confessing" the truth of the landscape'. In other words, Harley believed that the truth of the landscape could still be revealed if we took account of the ideology inherent in the representation. The problem was not the map *per se*, but 'the bad things people *did* with maps' (Wood 1993: 50, original emphasis). Harley's strategy was to identify the politics of representation in order to circumnavigate them (to reveal the truth lurking underneath), not fully appreciating, as with Foucault's observations, that there

The Map Reader: Theories of Mapping Practice and Cartographic Representation, First Edition. Edited by Martin Dodge, Rob Kitchin and Chris Perkins.

is no escaping the entangling of power/knowledge. Another strategy to address the crisis of representation has been the production and valuing of counter-mappings – maps made by diverse interests that provide alternative viewpoints to state-sanctioned and commercial cartography (Wood 1992). Again, this strategy does not challenge the ontological status of the map, rather it simply reveals the politics of mapping.

Crampton (2003) [...] argues [for a] move from understanding cartography as a set of ontic knowledges to examining its ontological terms. Ontic knowledge consists of the examination of how a topic should proceed from within its own framework where the ontological assumptions about how the world can be known and measured is implicitly secure and beyond doubt (Crampton, 2003). In other words, there is a core foundational knowledge – a taken for granted ontology – that unquestioningly underpins ontic knowledge. With respect to cartography this foundational ontology is that the world can be objectively and truthfully mapped using scientific techniques that capture and display spatial information. Cartography in these terms is purely technical and develops by asking self-referential, methodological questions of itself that aim to refine and improve how maps are designed and communicate (Crampton (2003) gives the examples of what colour scheme to use, the effects of scale, how maps are used historically and politically). In these terms a book like Robinson *et al.* (1995) is a technical manual that does not question the ontological assumptions of the form of mapping advocated, rather it is a 'how to do "proper" cartography' book that in itself perpetuates the security of cartography's ontic knowledge. In this sense, Harley's questioning of maps is also ontical (Harley 1992), as his project sought to highlight the ideology inherent in maps (and thus expose the truth hidden underneath) rather than to question the project of mapping *per se*; 'it provided an epistemological avenue into the map, but still left open the question of the ontology of the map' (Crampton 2003: 90). In contrast, Crampton details that examining cartography ontologically consists of questioning the project of cartography itself.

[...]

For Crampton (2003) this means that politics of mapping should move beyond a 'critique of existing maps' to consist of 'a more sweeping project of examining and breaking through the boundaries on how maps are, and our projects and practices with them' (p. 51); it is about exploring the 'being of maps'; how maps are conceptually framed in order to make sense of the world.

Similarly, John Pickles (2004: 12) seeks to extend Harley's observations beyond ontic status, focusing on 'the work that maps do, how they act to shape our understanding of the world, and how they code that world'. Pickles' project is to chart the 'practices, institutions and discourses' of maps and their social roles within historical, social and political contexts from within a poststructural framework that understands maps as complex, multivocal and contested, and which rejects the notion of some 'truth' that can be uncovered by exposing ideological intent. Pickles' detailed argument unpicks the science of representation, calling for a post-representational cartography that understands maps not as mirrors of nature, but as producers of nature. [...]

While we think Crampton's and Pickle's ideas are very useful, and we are sympathetic to their projects, we are troubled by the ontological security the map still enjoys within their analysis. Despite the call for seeing maps as 'beings in the world', as non-confessional spatial representations, post-representational or de-ontologised cartography, and non-progressivist or denaturalised histories of cartography, maps within Crampton and Pickles view remains secure as spatial representations that say something about spatial relations in the world (or elsewhere). The map might be seen as diverse, rhetorical, relational, multivocal and having effects in the world, but is nonetheless a coherent, stable product – *a* map. While in some respects Crampton and Pickles demonstrate that maps are not, in Latour's (1989) terms, immutable mobiles (that is, stable and transferable forms of knowledge that allow them to be portable across space and time), they nonetheless slip back into that positioning, albeit with maps understood as complex, rhetorical devices not simply representations. In this sense, Figure 1.14.1 is unquestioningly a map.

We think it productive to take a different tack to think ontologically about cartography. For us, maps [...] have no ontological security, they are *ontogenetic* in nature. Maps are of-the-moment, brought into being through practices (embodied, social, technical), *always* re-made every time they are engaged with; mapping is a process of constant re-territorialisation. As such, maps are transitory and fleeting, being contingent, relational and context-dependent. *Maps are practices* – they are always *mappings*; spatial practices enacted to solve relational problems (e.g. how best to create a spatial representation, how to understand a spatial distribution, how to get between A and B, and so on). From this position, Figure 1.14.1 is not unquestioningly a map (an objective, scientific representation (Robinson) or an ideologically laden representation (Harley), or an inscription that does work in the world (Pickles)), it is rather a set of points, lines and colours that takes form as, and is understood as, a map through mapping practices (an inscription in a constant state of re-inscription). Without these practices a spatial representation is simply coloured ink on a page (this is not a facetious statement – without the knowledge of what constitutes a map is or how a map works how can it be otherwise?). Practices based on learned knowledge and skills (re)make the ink into a map and this occurs *every time* they are engaged with – the set of points, lines and

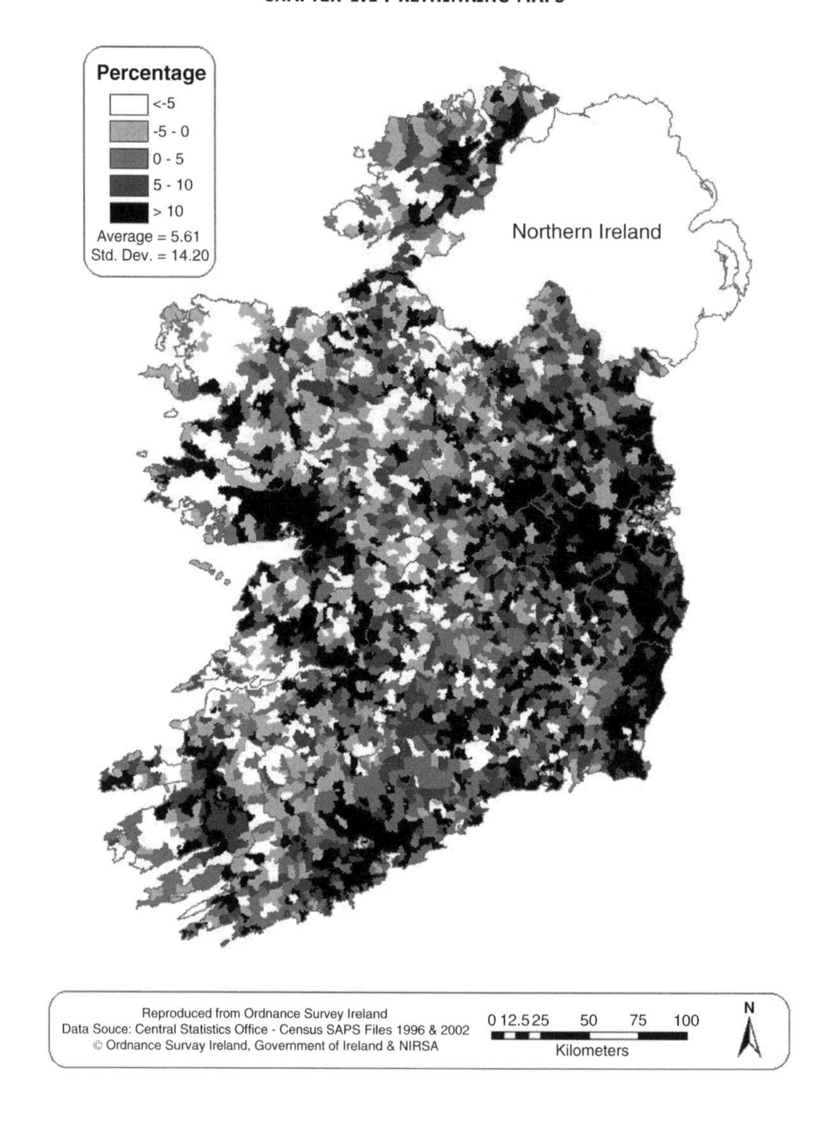

Figure 1.14.1 Is this image a map? Population change in Ireland, 1996–2002.

areas are recognised as a map; they are interpreted, translated and made to do work in the world. As such, maps are constantly in a state of becoming; constantly being remade.

At the heart of our analysis are two fundamental questions. One, how do individuals know that an arrangement of points, lines and colours constitute a map (rather than a landscape painting or an advertising poster)? How does the idea of a map and what is understood as a map gain ontological security and gain the semblance of an immutable mobile? Our thesis is that ontological security is maintained because the knowledge underpinning cartography and map use is learned and constantly reaffirmed. A map is never a map with ontological security assumed, it is bought into the world and made to do work through practices such as recognising, interpreting, translating, communicating, and so on. It does not re-present the world or make the world (by shaping how we think about the world), it is a co-constitutive production between inscription, individual and world; a production that is constantly in motion, always seeking to appear ontologically secure. Two, how do maps become? How does the constant, co-constitutive production of a map occur? We seek to answer this question by examining two vignettes outlining the unfolding nature of mapping and by the drawing on the concepts of transduction (that understands the unfolding of everyday life as sets of practices that seek to solve on-going relational problems) and technicity (the power of technologies to help solve those problems) (Dodge and Kitchin 2005).

The argument we forward is not being made to demonstrate clever word play or to partake in aimless philosophising. In contrast, we are outlining what we believe is a significant conceptual shift in how to think about maps and cartography (and by implication what are commonly understood as other representational outputs and endeavours); that is a shift from ontology (how things are) to ontogenesis (how things become) – from (secure) representation to (unfolding) practice. This is not minor argument with little theoretical or practical implications. Rather it involves adopting a radically different view of maps and cartography. In particular, we feel that the ontological move we detail has value for five reasons. Firstly, we think it is a productive way to think about the world, including cartography. It acknowledges how life unfolds in multifarious, contingent and relational ways. Secondly, we believe that it allows us a fresh perspective on the epistemological bases of cartography – how mapping and cartographic research is undertaken. Thirdly, it 'denaturalises and deprofessionalises cartography' (Pickles 2004: 17) by re-casting cartography as a broad set of spatial practices, including gestural and performative mappings such as Aboriginal song lines, along with sketch maps, counter-maps, and participatory mapping, moving it beyond a narrowly defined conception of mapmaking (this is not to denigrate the work of professional cartographers, but to recognise that they work with a narrowly defined set of practices that are simply a subset of all potential mappings). As such, it provides a way to think critically about the practices of cartography and not simply the end product (the so-called map). Fourthly, it provides a means to examine the effects of mapping without reducing such analysis to theories of power, instead positioning maps as practices that have diverse effects within multiple and shifting contexts. Fifthly, it provides a theoretical space in which 'those who research mapping as a practical form of applied knowledge, and those that seek to critique the map and mapping process', can meet, something that Perkins (2003: 341) feels is unlikely to happen as things stand. Perkins (2003: 342) makes this claim because he feels 'addressing *how* maps work ... involves asking different questions to those that relate to *power* of the medium' – one set of questions being technical the other ideological. We do not think that this is the case – both are questions concerning practice.

Maps as practice – always mapping

[...]

Brown and Laurier (2005) note that people are never simply mappers, but rather mapping is part of finding a solution to a wider problem. We think conceptualising mapping as a set of practices aimed at solving spatial problems is highly productive.

Vignette 1

John Doe has been given the task by a government department of reporting on the distribution of population change in Ireland between the 1996 and 2002 census. There are several potential solutions to this problem, such as producing statistical tables, figures or narrative description, each of which consists of a set of technical practices which can be used to complete the task. Given the spatial nature of the problem, producing what is commonly understood to be a map provides one viable solution. John's task as a cartographer is to *construct* a spatial representation using available data that conforms to agreed standards and conventions and effectively communicates the pattern of population change.

Starting from a position of having specialised tools (scientific instruments or software) and resources (boundary and attribute data, previously mapped information), and a degree of knowledge, experience and skills, John works to create a map. The map thus *emerges* through a set of iterative and citational practices – of employing certain techniques that build on and cite previous plottings or previous work (other spatial representations) or cartographic ur-forms (standardised forms of representation). This process is choreographed to a certain degree, shaped by the scientific culture of conventions, standards, rules, techniques, philosophy (its ontic knowledge), and so on, but is *not* determined and essential. Rather, instead of there being a teleological inevitability in how the map is constructed and or how the final product will look, the map is contingent and relational in its production through the decisions made by John with respect to what attributes are mapped, their classification, the scale, the orientation, the colour scheme, labelling, intended message, and so on, *and* the fact that the construction is enacted through affective, reflexive, habitual practices that remain outside of cognitive reflection. Important here is the idea of play – of 'playing' with the possibilities of how the map will become, how it will be remade by its future makers – and of arbitrariness, of unconscious and affective design. John thus experiments with different colour schemes, different forms of classification, and differing scales to map the same data. Making maps then is inherently creative – it can be nothing else; and maps emerge in process.

[...]

Vignette 2

Jane Doe is travelling from Manchester Piccadilly train station to the town hall. Ten minutes after she leaves the station she realises she has taken a wrong turn somewhere and is now lost. Jane's problem is to determine where she is and then to compute a new route to her destination. One solution to this problem is consult the street map she is carrying in her bag. This consultation consists of more than reading an immutable mobile. Jane does not simply receive information from the map, rather she brings her own map into being in the moment through an engagement with the printed representation. In other words, the map is re-made anew, emerging from the intersections between the knowledge, skills and experience of Jane to understand the language of cartographic representation and the spatial data within the representation. Jane *makes* the coloured ink on the page into a map through praxis; she works with the spatial representation to try and make sense of the world.

In Jane's case, making sense of the world is undertaken by making correspondences between the map and the streetscape. She looks at the map, then at the road, then back at the map. She tries to find objects such as street names or landmarks in the landscape that she can match to the map and vice versa. She locates the train station on the map then traces her finger along the roads she thinks she might have taken, trying to locate herself. She then twists the map, changing its orientation and glances back at the street. She follows this by changing her own orientation turning to face a new direction, shifting her vision from map to street, gaining her bearings as she starts to make correspondence between her surroundings and the lines and symbols of the map. Jane is thus placing herself both in the material geography of the street and the map. In so doing, the map and the world gain legibility; they get re-made in new ways. The process of mapping then alters Jane's imaginative geography of Manchester city centre and also the spatiality of the street in which she resides. Map and landscape are folded into each other to solve the problem of determining where she is. In other words, the map Jane beckons into being does not represent a space, or simply re-present a space, it brings space into being (Dodge and Kitchin 2005). This beckoning is not determinate and teleological but is contingent and relational, embedded with the context of the moment (e.g. anxiety, frustration) and as an aspect of other tasks (e.g. attending a job interview, meeting friends, etc.).

[. . .]

An ontogenetic understanding of maps

From our examples we would argue that maps *emerge in process* through a diverse set of practices. Given that practices are an on-going series of events, it follows that maps are constantly in a state of becoming; they are ontogenetic (emergent) in nature. Maps have no ontological security, they are of-the-moment, transitory, fleeting, contingent, relational and context-dependent. They are never fully formed and their work is never complete. Maps are profitably theorised, not as mirrors of nature (as objective and essential truths) or as socially constructed representations, but as emergent. In this section we want to start to think through *how* maps emerge through practices, drawing on the concepts of transduction and technicity; to provide a starting point for conceptually framing the process by which John and Jane into being to solve their relational problems.

According to Adrian Mackenzie (2003: 10) 'transduction is a kind of operation, in which a particular domain undergoes a certain kind of ontogenetic modulation. Through this modulation *in-formation* or individuation occurs. That is, transduction involves a domain taking-on-form, sometimes repeatedly' (his emphasis). Simondon (1992: 313) explains '[t]he simplest image of the transductive process is furnished if one thinks of a crystal, beginning as a tiny seed, which grows and extends itself in all directions in its mother water. Each layer of molecules that has already been constituted serves as the structuring basis for the layer that is being formed next, and the result is amplifying reticular structure'. In other words, the crystal grows through individuations, that cite previous individuations, to transduce elements into a crystal. Using this idea, if we think of John creating a map of population change, we can say that the plotting of lines, colours and so on consists of a series of individuations that transduces the blank page into a map, with each individuation citing previous plottings.

Transduction occurs because we are endlessly confronted with sets of relational problems – practices in effect aim to solve these problems (Mackenzie 2002). In the case of mapping, those problems include meta-problems such as the production of maps or finding one's way, that in themselves are made up of hundreds of smaller problems such as where to place a label, what colour scheme to use, or how to orientate or make correspondence between map and territory. The solving of problems is always partial, opening up new problems (e.g. the plotting of one line, leads to the plotting of the next, and so on), and contextual (embedded within standards, conventions, received wisdom, personal preferences, direction by others, and so on). In this sense, transduction is the means by which a 'domain structures itself as a partial, incomplete solution to a relational problem' (Mackenzie 2003: 10). In the examples above, the meta-problem for John is one of providing information with respect to population change in Ireland in a meaningful form that can be used by the contracting party in a policy document. This document in itself has transductive effects, alternatively modulating how the world is understood, and this understanding can then be used to enact policy initiatives and to transduce material geographies. The meta-problems for Jane are to locate herself with respect to map and location, and then to make her way to the town hall. Similarly, both imagined and material geographies are alternatively modulated.

Without the map the problem getting from A to B might not be solved, or will be solved less efficiently or in a more costly manner.

As these relational problems make clear, maps are the product of transduction and they enable further transductions in other places and times; they are always in the process of mapping; of solving relational problems such as how best to present spatial information, how to understand a spatial distribution, how to find one's way. Here, we want to make it clear that we are *not* drawing a distinction between what traditionally has been divided into mapmaking and map use. Instead, to us all engagement with 'maps' are emergent – all maps are beckoned into being to solve relational problems; *all* are (re)mappings – the (re)deployment of spatial knowledges and practices. And all emergence is contextual and a mix of creative, reflexive, playful, affective and habitual practices; affected by the knowledge, experience and skill of the individual to perform mappings and apply them in the world. Conceiving of mapping in this way reveals the mutability of maps; that they are re-made as opposed to mis-made, mis-used or mis-read.

Mapping works because its set of practices have been learnt by people[1], and because maps are the product of technicity (made by tools) and they possess technicity (they are a tool themselves). Technicity refers to the extent to which technologies mediate, supplement, and augment collective life; the unfolding or evolutive power of technologies to make things happen *in conjunction* with people (Mackenzie 2002). For example, mapping practices used to produce a spatial representation understood as a map by its creator are the product of cartographic instruments (pens, paper, rulers, software packages, etc.) used in conjunction with people, where the outcome is co-dependent on both instruments and individual, and embedded within a particularised context. A spatial representation can be said to possess technicity when it is used by a person to solve relational problems; to alternatively modulate (transduce) activity and space. The solution arises from the conjunction of person and representation; they are produced through, or *folded* into, each other in complex ways. Maps thus should be understood 'processually . . . as events rather than objects, as contingent the whole way down'; 'as networks of social–material interactions rather than simply reflections of human capacities or innately alien objects' (Mackenzie 2003: 4, 8). Cartography as a profession is thus repositioned as a processual[2], as opposed to representational, science.

From this perspective, the important question is not is not what a map is (a spatial representation or performance), nor what a map does (communicates spatial information), but *how the map emerges* through contingent, relational, context embedded practices to solve relation problems (their ability to make a difference to the world); to move from essentialist and constructivist cartography to what we term emergent cartography.

Epistemologically, what this means is that the science of cartography (how maps are produced) and critical analysis of cartography (the history and politics of cartography) are both positioned as processual in nature. Rather than one asking technical questions and the other ideological, both come to focus on how maps *emerge* through practices; how they come to be in the world. With respect to both, as Brown and Laurier (2005: 23, original emphasis) note, this calls for a radical shift in approach from '*imagined* scenarios, *controlled* experiments or *retrospective* accounts' to examine how maps emerge as solutions to relational problems; to make sense of the 'unfolding action' of mapping. As such, cartographic research becomes re-focused as a science of practices, not representations; on how a map is produced, how mapping is contextually co-constituted (within individual, collective and institutional frameworks), how maps are made to do work in the world, how the craft of cartographers and the lexicon they develop and use influence how maps are (re)made, how this work varies between people and the relational problems being solved, how maps gain the status of immutable mobiles and how this varies, and has varied, over time and space. Within this conceptual view, technical questions (ontic knowledge) concerning such things as accuracy and standards, remain an important focus of study, but are appreciated to be contingent, relational and context-dependent; that addressing technical questions is in itself a process of seeking to solve a set of relational problems. In other words, the focus of attention shifts to the relationship between cartographer, individuals, and a potential solution, and how mapping is employed to *solve diverse and context-dependent problems* (e.g. how John produced a map of population change and Jane produced a map – using a published spatial representation – to get from one location to another), rather than a single map being viewed as a universal and essential solution to a range of questions (that there can be a 'best' or 'most accurate' map

[1] Pickles (2004: 60–61) explains,

'Maps work by naturalising themselves by reproducing a particular sign system and at the same time treating that sign system as natural and given. But, map knowledge is never naïvely given. It has to be learned and the mapping codes and skills have to be culturally reproduced . . . The map opens a world to us through systems and codes of sedimented, acculturated knowledge.'

[2] Our vision of processual cartography moves beyond Rundstrum's (1991) process cartography – which sees mapmaking a subjective process embedded within a wider social context wherein the resultant map is immutable mobile – to recognise mapping as contingent, relational and continually emergent process. That is, is more profoundly poststructural in nature.

that all people understand and use in the same way to address a range of problems). This is, we believe, a subtle but important distinction as it recognises a fundamental shift in conceptualising the foundational knowledge underpinning cartography, and reconfigures the epistemology appropriately, without necessarily fundamentally altering many of the key technical questions at a technical level (but clearly at a philosophical level) that professional cartographers are interested in, while also opening up a set of wider issues and concerns that we believe deserves wider attention.

[. . .]

References

Antle, A. and Klickenberg, B. (1999) Shifting paradigms: from cartographic communication to scientific visualization. *Geomatica*, **53** (2), 149–155.

Brown, B. and Laurier, E. (2005) Maps and journeys: an ethno-methodological investigation. *Cartographica*, **4** (3), 17–33.

Crampton, J. (2003) *The Political Mapping of Cyberspace*, Edinburgh University Press, Edinburgh.

Dodge, M. and Kitchin, R. (2005) Code and the transduction of space. *Annals of the Association of American Geographers*, **95** (1), 162–180.

Edney, M.H. (1993) Cartography without 'progress': reinterpreting the nature and historical development of mapmaking. *Cartographica*, **30** (2/3), 54–68.

Harley, J.B. (1989) Deconstructing the map. *Cartographica*, **26**, 1–20.

Harley, J.B. (1992) Rereading the maps of Columbian encounters. *Annals of the Association of American Geographers*, **82** (3), 522–536.

Latour, B. (1989) *Science in Action*, Harvard University Press, Cambridge, MA.

MacEachren, A.M. (1995) *How Maps Work: Representation, Visualization and Design*, Guildford, New York.

Mackenzie, A. (2002) *Transductions: Bodies and Machines at Speed*, Continuum Press, London.

Mackenzie, A. (2003) Transduction: invention, innovation and collective life. http://www.lancs.ac.uk/staff/mackenza/papers/transduction.pdf.

Perkins, C. (2003) Cartography: mapping theory. *Progress in Human Geography*, **27** (3), 341–351.

Pickles, J. (2004) *A History of Spaces: Cartographic Reason, Mapping and the Geo-Coded World*, Routledge, London.

Robinson, A.H., Morrison, J.L., Muehrcke, P.C. *et al.* (1995) *Elements of Cartography*, 6th edn, John Wiley & Sons, Inc., New York.

Rundstrom, R.A. (1991) Mapping, postmodernism, indigenous people and the changing direction of North American cartography. *Cartographica*, **28** (2), 1–12.

Simondon, G. (1992) The genesis of the individual, in *Incorporations 6* (eds J. Crary and S. Kwinter), Zone Books, New York, pp. 296–319.

Wood, D. (1992) *The Power of Maps*, Guilford, New York.

Wood, D. (1993) The fine line between mapping and mapmaking. *Cartographica*, **30** (4), 50–60.

Further reading

Dodge, M., Kitchin, R. and Perkins, C. (2009) *Rethinking Maps: New Frontiers in Cartographic Theory*, Routledge, London. [This edited volume offers a variety of different approaches to the status of the map and mapping.]

Laurier, E. and Brown, B. (2008) Rotating maps and readers: Praxiological aspects of alignment and orientation. *Transactions of the Institute of British Geographers*, **33**, 201–221. [Provides a gestural, social and performative alternative to cognitive ways of understanding map use.]

Olson, G. (2007) *Abysmal: A Critique of Cartographic Reason*, University of Chicago Press, Chicago. [Theorises the deep cultural associations between mapping and ways of thinking about the world.]

Perkins, C. (2009) Performative and embodied mapping, in *International Encyclopedia of Human Geography* (eds R. Kitchin and N. Thrift), Elsevier, Oxford. [A useful overview of the potential of more performative approaches to mapping.]

See also

- Chapter 1.7: Design on Signs / Myth and Meaning in Maps
- Chapter 1.8: Deconstructing the Map
- Chapter 1.9: Drawing Things Together
- Chapter 1.10: Cartography Without 'Progress': Reinterpreting the Nature and Historical Development of Mapmaking
- Chapter 1.12: The Agency of Mapping
- Chapter 1.13: Beyond the 'Binaries': A Methodological Intervention for Interrogating Maps as Representational Practices
- Chapter 3.9: Maps, Mapping, Modernity: Art and Cartography in the Twentieth Century
- Chapter 4.2: Map Makers are Human: Comments on the Subjective in Maps
- Chapter 4.5: The Map as Biography
- Chapter 4.6: Reading Maps
- Chapter 4.8: Refiguring Geography: Parish Maps of Common Ground
- Chapter 5.7: A Map that Roared and an Original Atlas: Canada, Cartography, and the Narration of Nation

SECTION TWO

Technologies of Mapping

Chapter 2.1

Introductory Essay: Technologies of Mapping

Martin Dodge, Rob Kitchin and Chris Perkins

Introduction

Technological foundations of cartography are crucial to understanding the contemporary nature of maps. Over hundreds of years there have been many new technical developments concerning the capture of data about the world, the processing of geographic information, and the production and design of representational media. Earlier shifts in the mode of production of mapping focused upon the emergence of printing technologies in Western Europe in the Renaissance period, which facilitated the mass production and dissemination of maps printed on paper. A progressive shift took place from manuscript production, to printing based on woodblocks, copper engraved plates, lithography and, by the twentieth century, to photo-mechanical technologies (Cook 2002; Mukerji 2006). Meanwhile changes in data collection were reflected in changing modes of surveying, such as the systematic development of triangulation associated with the rise of national and military mapping agencies (Biggs 1999; Seymour 1980), and the application of photogrammetry in the early twentieth century (Collier 2002). New technologies were also deployed in the projection of data (Snyder 1993, excerpted as Chapter 2.9).

These developments, and how they were exploited by individuals and institutions to their advantage (e.g. different sea charts aiding more successful navigation and the expansion of trading empires), have profoundly affected the mapping process at different times resulting in many distinct modes of cartography (Edney 1993, excerpted as Chapter 1.10). This introductory chapter focuses in detail upon just one of many technological transitions (Monmonier 1985, excerpted as Chapter 2.2), the latest in a series of shifts through which mapping has passed and explores how different technologies are enrolled into a working series of practices and mapped artefacts.

The dominant technology of contemporary mapping is computing, which has emerged over the last fifty years to underpin digital cartography. Various specialised hardware, sophisticated software applications, databases and video displays operate as powerful socio-technological agents because they provide means to automate and augment existing cartographic processes as well as opening new channels for mapping to be undertaken. As Tobler noted in his prescient article in the 1959 at the beginnings of the transition: 'It seems that some basic tasks, common to all cartography, may in the future be largely automated and that the volume of maps produced in a given time will be increased while the cost is reduced' (p. 534; excerpted as Chapter 2.5).

Digital cartography then exploits processes of automation and augmentation through technologies for data capture (e.g. satellite imagery, GPS, laser ranging tools), the handling and processing of data (e.g. CAD, GIS and desktop publishing applications), the efficient storage and rapid distribution of vast quantities of data (e.g. database software, hard drives, servers, data networks, the Internet) and the delivery, presentation and interactive uses of maps (e.g. widespread

The Map Reader: Theories of Mapping Practice and Cartographic Representation, First Edition. Edited by Martin Dodge, Rob Kitchin and Chris Perkins.

availability of high resolution display screens, affordable laser printing, embedded multimedia documents, streaming 'live' to location-aware mobile devices).

Computers, as so-called 'universal machines', appear to offer unprecedented advantages in the quest for more accuracy and efficacy in map production. In terms of technologies of data capture, for example, it can be argued that computers, and the assemblage of measurement/imaging/sensing technologies, have brought improved and more mimetic ways of knowing the world and appear to be the next step on the 'path to perfection' in mapping. For example, locational precision has become widely and easily available through GPS and the ever-increasing spatial resolution of satellite imaging. Spatial data can be logged automatically and continuously without human intervention. Indeed, cartography's ability to accurately capture the world has been transformed by digital photogrammetry, remote sensing, GPS-based surveying and mobile mapping (Jensen and Cowen 1999, excerpted as Chapter 2.8; Li 1997, excerpted as Chapter 2.10). Advances in digital data capture, processing and geovisualisation not only enable us to 'see' the world in greater depth (Pickles 2004), but also to 'see' new things (including virtual spaces), in new temporal registers.

Technologies of cartographic production have often been explained through narratives of scientific progress. As a consequence, the history of cartography tends to be written as a history of technique (Crone 1953), with an underlying assumption that rational decision making leads to the adoption of improved technologies and updating institutional practices when they become available. For example, in much of the writing – both applied and scholarly – the computerisation of cartography is bound-up in progressive discourses of scientific advancement and increasing accuracy and depth of knowledge (Goodchild 1999, excerpted as Chapter 2.6; Monmonier 1985, excerpted as Chapter 2.2). This fits within with a long running storyline of progress in cartography: art becomes science, florid designs become formal display, the named cartographer becomes an anonymous technician; see also discussion in Chapter 1.1.

However, whilst it is clear that digital cartography has some distinctive qualities with respect to previous modes of mapping, we argue it would be naive to assert that computers give rise to ostensibly *superior* mapping to other modes. The ideas and techniques underpinning cartographic practice has always been contested across time and space. As such, we should be careful not be read the present prevalence of digital cartography as a simple and progressive path of innovation and adoption, that inevitably leads to better mapping of the world, any more than earlier applications of technologies *inherently* led to progress. Rather we would argue that change is messy, contingent and partial. Developments unfold in fits and starts, proceeding with leaps and failures. Whilst undoubtedly digital data capture and new computerised mapping systems can supply more detail and more cartographic data to be displayed on-demand, it is questionable as to whether they deliver better or more objective representations of the world than previous methods and technologies of mapping. Maps tend to be judged on how well they communicate, not according to their level of detail. Further, many spaces of human culture remain unmapped and are perhaps unmappable, despite improved sensors and sophisticated GIS software (Muehrcke 1990, excerpted as Chapter 2.7). Moreover, as a new technology is adopted, the role and power of individuals and institutions is reconfigured: there are always winners and losers due to innovations and new practices and relations (see discussion in McHaffie 1995, excerpted as Chapter 2.3). For example, with the rise of internet-based mapping, the role of national mapping agencies is weakened with respect to commercial data providers, and software engineers and interface designers start to displace professionally-trained cartographers (Wood 2003).

Characteristics of digital cartographies

The development and rapid diffusion of digital technologies in the last three decades has affected all aspects of mapping, changing methods of data collection, cartographic production and the dissemination and use of maps. This has been termed the 'digital transition' in cartography (Goodchild 1999, excerpted as Chapter 2.6; Pickles 1999; Rhind 1999) and it is continuing apace (for example, developments in mass market satnav systems or innovative mobile mapping services; see later). As such the computer is a vital component in understanding the milieu in which new forms of mapping practice are emerging.

While the detailed social and technical histories of the digitisation of the cartographic industry are complex and largely unwritten, it would be fair to say that, in the last couple of decades, mapping practice has been almost wholly subsumed in a rapid convergence of spatial technologies, such that today professional cartography operates as a rather marginal 'end service' component of the multi-billion dollar GI industry. Nowadays, the majority of maps are digital and created only 'on demand' from geospatial databases for temporary display on screens. The heyday of published unwieldy folded map sheets and heavy paper atlases is past: they are being replaced by the rapid technological development of GIS, spatial databases and real time mapping systems; the potency of these developments is most evident perhaps in terms of web mapping.

Developments in networking technologies and computer-mediated communications, and the rise of the World-Wide Web from the early 1990s, have meant that digital maps are now very easy to distribute at marginal cost and can be accessed 'on demand' by many (Peterson 2003, 2008). One of the first examples was the Xerox PARC Map Viewer, launched online in June 1993 by Steve Putz. (The map is no longer online, however background details are available at <www2.parc.com/istl/projects/www94/iisuwwwh.html>). Commercial online mapping and driving instructions were pioneered by the internet portal MapQuest.com in the mid 1990s, which by the turn of the century had already generated more digital maps than any other publisher in the history of cartography (Peterson 2001). Since launching in 2005 the popularity of Google Maps with its open API (Application Programming Interface), has inspired an explosion of new online mapping tools and hacks (Geller 2007, excepted as Chapter 2.12; Gibson and Erle 2006). These web mapping services are seemingly 'free' at the point of use and are encouraging the casual use of cartography (the substantial capital costs of granting no-cost public access to detailed topographic maps and high resolution satellite imagery are being met, in part, by revenues from geographically-targeted advertising, but they are also being heavily subsidised at the moment by large corporations, like Google and Microsoft as they seek to entice users to their sites and to dominate the marketplace for online mapping). There is even the prospect that expensive, complex, standalone GIS will begin to adapt and evolve around a web services mapping model (Sui 2008).

Digital cartography has exploited the affordances offered by computer software and the flexibility of screen display to deliver maps in new media forms and offering novel modes of user interactivity. As the map itself became a fully digital text, many of its basic properties changed. It became almost infinitely malleable and responsive to the user, such that pre-digital, paper mapping seems stilted and somewhat lifeless. A multitude of maps can be generated from a single database in GIS, many design options can be explored at marginal additional cost. The map itself is an interface to the world that can be directly manipulated by users – zooming, panning, selecting layers, querying (Cartwright 1999, excerpted as Chapter 2.11). Rather than reading off the surface of a map, we become increasingly immersed within the mapping experience. Just as the word processor has reconfigured the practices of composing text, so the GIS has profoundly changed the making of maps. Of course this does not necessarily mean better maps (Muehrcke 1990, excerpted as Chapter 2.7) just as using Microsoft Word does not guarantee readable prose. Cheap, powerful computer graphics on PCs and increasingly mobile devices, however, do enable a much more expressive and interactive cartography, potentially available to a growing number of people.

The pervasive paradigm of hypertext as a way to structure and navigate digital information has also influenced digital cartography. Increasingly, maps are used as core components in larger multimedia information resources where locations and features on the map are hot-linked to pictures, text and sounds, to create distinctively new modes of map use (Cartwright 1999, excerpted as Chapter 2.11). In design terms, the conventional planar map form itself is, of course, only one possible representation of spatial data and new digital technologies have contributed to much greater diversity of cartographic-related forms including, pseudo three-dimensional landscape views, interactive panoramic image-maps, fully three-dimensional flythrough models (Dodge *et al.* 2008; Fisher and Unwin 2001; Geller 2007, excerpted as Chapter 2.12). It has also reinvigorated long standing but marginal forms of mapping, including cartograms and globes, and facilitated the construction of many new kinds of cartographic projection that could not have been calculated without computers (Snyder 1993, excerpted as Chapter 2.9).

Developments in computer graphics, computation and user interfaces have also begun to fundamentally transmute the role of the map from the finished product to a visual tool to be used interactively for exploratory data analysis (typically with the interlinking of multiple representations such as statistical charts, three-dimensional plots, tables and so on). This changing conceptualisation of the map is at the heart of the emerging field of geovisualisation, which in the last five years or so has been one of the leading areas of applied cartographic research (Dykes and Wood 2009, excerpted as Chapter 3.12; MacEachren and Kraak 2001, excerpted as Chapter 1.11).

Although not universally the case, it is evident that the emergence of digital cartography has also made mapping much more available, fostered a good deal of creativity and widened participatory options (Goodchild 2007, excerpted as Chapter 4.10; see also discussion in Chapters 4.1 and 5.1). More people have the option to become mapmakers themselves, without needing to master a wide range of technical and technological skills, be it via simple 'map charting' options in spreadsheets to produce basic thematic maps of their own data, through desktop GISs such as MapInfo and, of course, with a plethora of online tools (Geller 2007, excerpted as Chapter 2.12). As more and more people 'bypass' professional cartographers to make their own maps as and when required, it is possible that the diversity of map forms and usage will expand; although access to 'point-and-click' mapping software itself is no guarantee that the maps produced will be as effective as those hand-crafted by professionally-trained cartographers (see Chapter 3.1). More recent developments in so-called 'volunteered geographic information' are also dependent on a raft of digital technologies for collaboration (Goodchild 2007, excerpted as Chapter 4.10; Elwood 2008). The emergence of open-source cartography, exemplified by the OpenStreetMap project, also has the potential to challenge the commercial commodification of geospatial data by developing a 'bottom-up' capture infrastructure that is premised on a co-operative and non-profit philosophy (see also Colour Plate Five).

The widespread provision of GIS tools and online mapping services is significantly shifting access to mapping and spatial data, as well as altering user perception of what a map should be. There are clear signs that cartography will be seen as simply one of many available 'on demand' web services. As the digital map display becomes more flexible and accessible, it is also, in some respects, granted a less reified status than the analogue paper map of the past. Maps are increasingly treated as transitory information resources, created in the moment, and discarded immediately after use. In some senses, this devalues the map, as it becomes just another ephemeral medium, one of the multitude of screen images that people encounter everyday. Cartographic knowledge itself is just another informational commodity to be bought and sold, repackaged and endlessly circulated (McHaffie 1995, excerpted as Chapter 2.3; Pickles 1999).

However, technological innovation also seems to be pushing digital cartography towards more personal mapping. Here, web mapping tools generate maps tailored to answer specific queries with the point of interest lying at the centre of the display, whilst directional controls mean one can move about the map seemingly at will and without arbitrary constraints of sheet boundaries as with paper products. The mundane power of the so-called 'slippy' map is now so common as to be noticed only when it is not available on a digital mapping system. Mobile devices, locational awareness and ubiquitous mapping delivered to the palm of one's hand seem to put the user at the very heart of the map, and crucially this kind of 'me-map' can dynamically update in time with the moving user. The synchronisation of map and body makes for a new and highly compelling form of cartography (Meng 2005, excerpted as Chapter 3.11). The perceptual power of the digital 'me-map' to intimately connect people to place can be further enhanced by use of the first person perspective display: one is looking *into* the world, rather than down *onto* it. This can be seen, for example, in the scrolling isometric view pioneered by TomTom satnavs and the ground-level Google Street View mapping. Such views present the world in new ways and the sense of interactivity seems to change who controls the viewing. They are also, importantly, fun to use with game-like qualities of exploration and play (Churchill 2008). It is somewhat ironic that making maps more personally focused also serves the interests of corporations and states, as they can operate as surveillant technologies – typing a postcode into a search boxes generates a unique map for the individual but also reveals to the mapping site what that individual is interested in at that moment in time. In contrast, looking up an address in a paper street atlas leaves behind no trace of mapping intent.

Interestingly, in the future, much of the growth in personal mapping will come from people gathering geospatial data as they go about their daily activity, automatically captured by location-aware devices that they will carry and use (Ratti *et al.* 2006; for overview discussion see Thielmann 2010). From this kind of emergent mobile spatial data capture it will be possible to 'hack' together new types of maps, rather than be dependent on the map products formally published by governments or commercial firms. Such individually made, 'amateur' mapping may be imperfect in many respects (not meeting the positional accuracy standards or adhering to TOPO-96 surveying specifications for example), but could well be more fit-for-purpose than professionally produced, general cartographic products. There is also exciting scope for using locative media to annotate personal maps with ephemeral, micro-local details, personal memories, messages for friends and so on, that are beyond the remit of governmental cartography or the profitability criteria for commercial cartographic industry. An example would be the work of artist Christian Nold's ongoing emotion mapping project (www.emotionmap.net), as well innovative work in affective mapping (Aitken and Craine 2006, excerpted as Chapter 3.10).

Cautions and caveats in digital cartographic developments

In some respects, then, the outcome of the digital transition can be read as a democratisation of cartography (Rød *et al.* 2001), widening access to mapping and breaking the rigid control of authorship by an anonymised professional elite. However, if one looks more closely (and sceptically), the freedom for people to make their own maps with these types of software tools is strongly inscribed in the design and functionality of the software itself. The maps one can make online are only the maps the services allows one to make. Many people make their own maps with Google's service but these all ultimately still have the look and feel of a Google Map and are constrained by the tools that the corporation provides. Indeed, the majority of people still do not have the time or skills to break free from the functional constraints that the software imposes (also see Fuller's (2003) analysis of the framing power of Microsoft Word on writing and Tufte's (2003) critique of Microsoft PowerPoint on how people give presentations). Google may currently make a vast amount of spatial data freely available online (supported by advertising) but it is subject to their terms and conditions of use and raises the risk of monopolistic provision (Farman 2010, excerpted as Chapter 5.11; Zook and Graham 2007).

Further, interpreting the digital transition should not merely be about plotting technical 'impacts', but should also involve assessing the political implications of changing social practices in data capture and map authorship. Being wary of linear narratives of progress, one should not read the digitisation of the map as seamless, unproblematic or inevitable (Pickles 1999). Technological change is always contested, driven by competing interests and received in different ways and at different speeds in particular institutional settings (McHaffie 1995, excerpted as Chapter 2.3; Harvey 2001). Technology is never a neutral actor. It is shaped by social forces and bound up in networks of power, capital and control of new institutional practices in the processes of cartographic digitisation. The benefits and costs of change are always uneven. Government agencies and large commercial mapping firms have invested heavily in digitisation not from enlightened ideals to improve cartography, but because it serves their interests by maximising efficiency, reducing costs by deskilling production and by boosting revenues. The popular discourses of digitisation in cartography and elsewhere are often uncritical, driven in large part by the hype of the vendors of hardware and software, and IT consultants offering 'solutions'. The reality of the 'messy' social aspects of digitisation are glossed over in techno-utopian fantasies. There are risks, uncertainties and resistance to technological change that rarely get reported or recorded (e.g. the loss of craft skills; the risks of investing in technology instead of labour; the industrial disputes that often follow from technological innovation etc.).

The digital transition in cartography has made it more urgent to understand the wider social milieu in which maps are produced and disseminated. One needs to realise that the path of digitisation in cartography has been driven in large part by militaristic interests in various guises (Clarke 1992, excerpted as Chapter 2.4; McHaffie 1995, excerpted as Chapter 2.3; Cloud 2002). The underlying geospatial technologies and capture infrastructures (such as satellite imaging and GPS) are still dependent on state funding and imperatives of territorial security. Rather than becoming more democratic, one could argue that the surveillant power of the cartographic gaze is deepening, particularly after 9/11 (Monmonier 2002), accompanied by a fetishisation of the capability of geospatial technologies to 'target terrorism'. The mundane disciplining role of digital maps in systems of computerised governmentality continues to grow, for example in consumer marketing and crime mapping (Crampton 2003, excerpted as Chapter 5.8; Farman 2010, excerpted as Chapter 5.11). Such surveillance requirements are also a major hidden driver in the development of new mapping techniques for internet and mobile services. In conclusion, Pickles (2004: 146) notes cautiously: 'As the new digital mappings wash across our world, perhaps we should ask about the worlds that are being produced in the digital transition of the third industrial revolution, the conceptions of history with which they work, and the forms of socio-political life to which they contribute.'

References

Aitken, S. and Craine, J. (2006) Guest editorial: Affective geovisualizations. *Directions Magazine*, 7 February, www.directionsmag.com. (Excerpted as Chapter 3.10.)

Biggs, M. (1999) Putting the state on the map: cartography, territory, and European state formation. *Comparative Studies in Society and History*, **41**, 374–405.

Cartwright, W. (1999) Extending the map metaphor using web delivered multimedia. *International Journal of Geographical Information Science*, **13** (4), 335–353. (Excerpted as Chapter 2.11.)

Churchill, E.F. (2008) Maps and moralities, blanks and beasties. *ACM Interactions*, July/August, 40–43.

Clarke, K.C. (1992) Maps and mapping technologies of the Persian Gulf war. *Cartography and Geographic Information Systems*, **19** (2), 80–87. (Excerpted as Chapter 2.4.)

Cloud, J. (2002) American cartographic transformations during the Cold War. *Cartography and Geographic Information Science*, **29** (3), 261–282.

Collier, P. (2002) The impact on topographic mapping of developments in land and air survey: 1900–1939. *Cartography and Geographic Information Science*, **29** (3), 155–174.

Cook, K.S. (2002) The historical role of photo-mechanical techniques in map production. *Cartography and Geographic Information Science*, **29** (3), 137–154.

Crampton, J. (2003) Cartographic rationality and the politics of geosurveillance and security. *Cartography and Geographic Information Science*, **30** (2), 135–148. (Excerpted as Chapter 5.8.)

Crone, G.R. (1953) *Maps and Their Makers: An Introduction to the History of Cartography*, Hutchinson's University Library, London.

Dodge, M., McDerby, M. and Turner, M. (2008) *Geographic Visualization: Concepts, Tools and Applications*, John Wiley & Sons, Ltd, Chichester, UK.

Dykes, J. and Wood, J. (2009) The geographic beauty of a photographic archive, in *Beautiful Data* (eds T. Segaran and J. Hammerbacher), O'Reilly, Sebastapol, CA, pp. 85–102. (Excerpted as Chapter 3.12.)

Edney, M.H. (1993) Cartography without 'progress': reinterpreting the nature and historical development of mapmaking. *Cartographica*, **30** (2/3), 54–68. (Excerpted as Chapter 1.10.)

Elwood, S. (2008) Volunteered geographic information: Future research directions motivated by critical, participatory, and feminist GIS. *GeoJournal*, **72** (3/4), 173–183.

Farman, J. (2010) Mapping the digital empire: Google Earth and the process of postmodern cartography. *New Media & Society*, **12**, doi: 10.1177/1461444809350900. (Excerpted as Chapter 5.11.)

Fisher, P. and Unwin, D. (2001) *Virtual Reality in Geography*, Taylor & Francis, London.

Fuller, M. (2003) It looks like you're writing a letter: Microsoft Word, in *Behind the Blip: Essays on the Culture of Software* (ed. M. Fuller), Autonomedia, Brooklyn, NY, pp. 137–165.

Geller, T. (2007) Imaging the world: The state of online mapping. *IEEE Computer Graphics and Applications*, March/April, 8–13. (Excerpted as Chapter 2.12.)

Gibson, R. and Erle, S. (2006) *Google Mapping Hacks*, O'Reilly, Sebastopol, CA.

Goodchild, M.F. (1999) Cartographic futures on a digital Earth. *Proceedings of the 19th International Cartographic Association*, 14–21 August 1999, Ottawa, Canada. (Excerpted as Chapter 2.6.)

Goodchild, M.F. (2007) Citizens as sensors: The world of volunteered geography. *GeoJournal*, **69** (4), 211–221. (Excerpted as Chapter 4.10.)

Harvey, F. (2001) Constructing GIS: actor networks of collaboration. *URISA Journal*, **13** (1), 29–37.

Jensen, J.R. and Cowen, D.C. (1999), Remote sensing of urban/suburban infrastructure and socio-economic attributes. *Photogrammetric Engineering & Remote Sensing*, **65** (5), 611–622. (Excerpted as Chapter 2.8.)

Li, R. (1997) Mobile mapping: An emerging technology for spatial data acquisition. *Photogrammetric Engineering & Remote Sensing*, **63** (9), 1085–1092. (Excerpted as Chapter 2.10.)

MacEachren, M.A. and Kraak, M.J. (1997) Exploratory cartographic visualization: advancing the agenda, *Computers & Geosciences*, **23** (4), 335–343. (Excerpted as Chapter 1.11.)

McHaffie, P.H. (1995) Manufacturing metaphors: Public cartography, the market, and democracy, in *Ground Truth: The Social Implications of Geographical Information Systems* (ed. J. Pickles), Guilford, New York, pp. 113–129. (Excerpted as Chapter 2.3.)

Meng, L. (2005) Egocentric design of map-based mobile services. *The Cartographic Journal*, **42** (1), 5–13. (Excerpted as Chapter 3.11.)

Monmonier, M.S. (1985) *Technological Transition in Cartography*, University of Wisconsin Press, Madison, WI. (Excerpted as Chapter 2.2.)

Monmonier, M.S. (2002) *Spying with Maps: Surveillance Technologies and the Future of Privacy*, University of Chicago Press, Chicago.

Muehrcke, P.C. (1990) Cartography and geographic information systems. *Cartography and Geographic Information Systems*, **17** (1), 7–15. (Excerpted as Chapter 2.7.)

Mukerji, C. (2006) Printing, cartography and conceptions of place in Renaissance Europe. *Media, Culture & Society*, **28** (5), 651–669.

Peterson, M.P. (2001), The development of map distribution through the Internet. *Proceedings of the 20th International Cartographic Conference*, vol. 4, pp. 2306–2312.

Peterson, M.P. (2003) *Maps and the Internet*, Elsevier, Amsterdam.

Peterson, M.P. (2008) *International Perspectives on Maps and the Internet*, Springer, New York.

Pickles, J. (1999) Cartography, digital transitions, and questions of history. *Proceedings of the 19th International Cartographic Association Conference*, 14–21 August 1999, Ottawa, Canada.

Pickles, J. (2004) Computing geographical futures, in *Envisioning Human Geographies* (eds P. Cloke, P. Crang and M. Goodwin), Hodder Arnold, London, pp. 172–194.

Ratti, C., Williams, S., Frenchman, D. and Pulselli, R.M. (2006) Mobile landscapes: using location data from cell phones for urban analysis. *Environment and Planning B: Planning And Design*, **33** (5), 727–748.

Rhind, D. (1999) Business, governments and technology: inter-linked causal factors of change in cartography. *Proceedings of the 19th International Cartographic Association Conference*, 14–21 August 1999, Ottawa, Canada.

Rød, J.K., Ormeling, F. and van Elzakker, C. (2001) An agenda for democratising cartographic visualisation. *Norsk Geografisk Tidsskrift (Norwegian Journal of Geography)*, **55** (1), 38–41.

Seymour, W.A. (1980) *A History of the Ordnance Survey*, Dawson, Folkestone, UK.

Snyder, J.P. (1993) *Flattening the Earth: Two Thousand Years of Map Projections*, University of Chicago Press, Chicago. (Excerpted as Chapter 2.9).

Sui, D.Z. (2008) The Wikification of GIS and its consequences: Or Angelina Jolie's new tattoo and the future of GIS. *Computers, Environment and Urban Systems*, **32** (1), 1–5.

Thielmann, T. (2010) Locative media and mediated localities. *Aether: The Journal of Media Geography*, **5a**, 1–18.

Tobler, W.R. (1959) Automation and cartography. *Geographical Review*, **49** (4), 526–534. (Excerpted as Chapter 2.5.)

Tufte, E.R. (2003) *The Cognitive Style of Powerpoint*, Graphics Press, Cheshire, CT.

Wood, D. (2003) Cartography is dead (thank god!). *Cartographic Perspectives*, **45**, 4–7.

Zook, M.A. and Graham, M. (2007) The creative reconstruction of the Internet: Google and the privatization of cyberspace and DigiPlace. *Geoforum*, **38** (6), 1322–1343.

Chapter 2.2

A Century of Cartographic Change, from *Technological Transition in Cartography*

Mark S. Monmonier

Editors' overview

This excerpt, from the introduction of Monmonier's 1985 book, makes the case for significant improvement in the nature of cartographic representation through the application of increasingly sophisticated technologies. Cartographic progress can be assessed in terms of geographic coverage, increasing map scale, spatial accuracy, thematic diversity, and widening availability and use. His book reviews developments in a range of technologies underpinning cartography, looking at changing means for determining location, for surveying, capturing imagery, data storage and publishing. The combination of these technologies has meant a marked acceleration in progress of mapping through the twentieth century. Monmonier sees the 'electronic transition' as likely to continue to drive forward cartography at a rapid pace.

Originally published in 1985: Part of the Introduction in Mark S. Monmonier, *Technological Transition in Cartography*, University of Wisconsin Press, Madison, WI, 4–14.

How much progress is likely in how short a time? Is a prediction of the dominance of the digital map early in the twentyfirst century not perhaps looking ahead too far too soon? An examination of past cartographic progress might relieve some of the sceptical reader's doubts about the rate of improvement possible in maps and mapping technology. The following two examples, each covering approximately a century, illustrate the significant recent change in both geographic detail and the richness of map information.

The first example focuses on the village of Manchester, Vermont, a popular resort in the Taconic Mountains. One of the most detailed early maps of the area is the portion shown in Figure 2.2.1 of a sheet in the Atlas of Bennington County, Vermont (Beers *et al.* 1869). The original scale of this map is approximately 1:48 000. Its symbols and type emphasise political units, the transport system and individual residences. Surveying methods were crude, and many atlases of this cartographic genre were hastily produced advertising vehicles replete with self-flattering engravings of the residences of those landed gentry who were persuaded to buy subscriptions from the atlas publisher's field corps of salesman surveyors (Thrower 1972: 102–105).

The publication in 1894 of the Equinox, Vermont 15-minute topographic quadrangle map marked a significant advance. Figure 2.2.2 shows the Manchester portion of this map sheet. With a scale of 1:62 500 and a contour interval of 20 feet [6 m], the Equinox sheet was part of the US Geological Survey's attempt to provide an accurate, systematic, standardised 'mother map' of the nation's terrain, hydrography, political boundaries, transport network, landmarks and place names. Progress was slow, and coverage was not complete for Vermont until the 1950s. But by that time the Geological Survey had undertaken an even more ambitious project, its 7.5-minute topographic series with four, more detailed 1:24 000-scale maps

The Map Reader: Theories of Mapping Practice and Cartographic Representation, First Edition. Edited by Martin Dodge, Rob Kitchin and Chris Perkins.

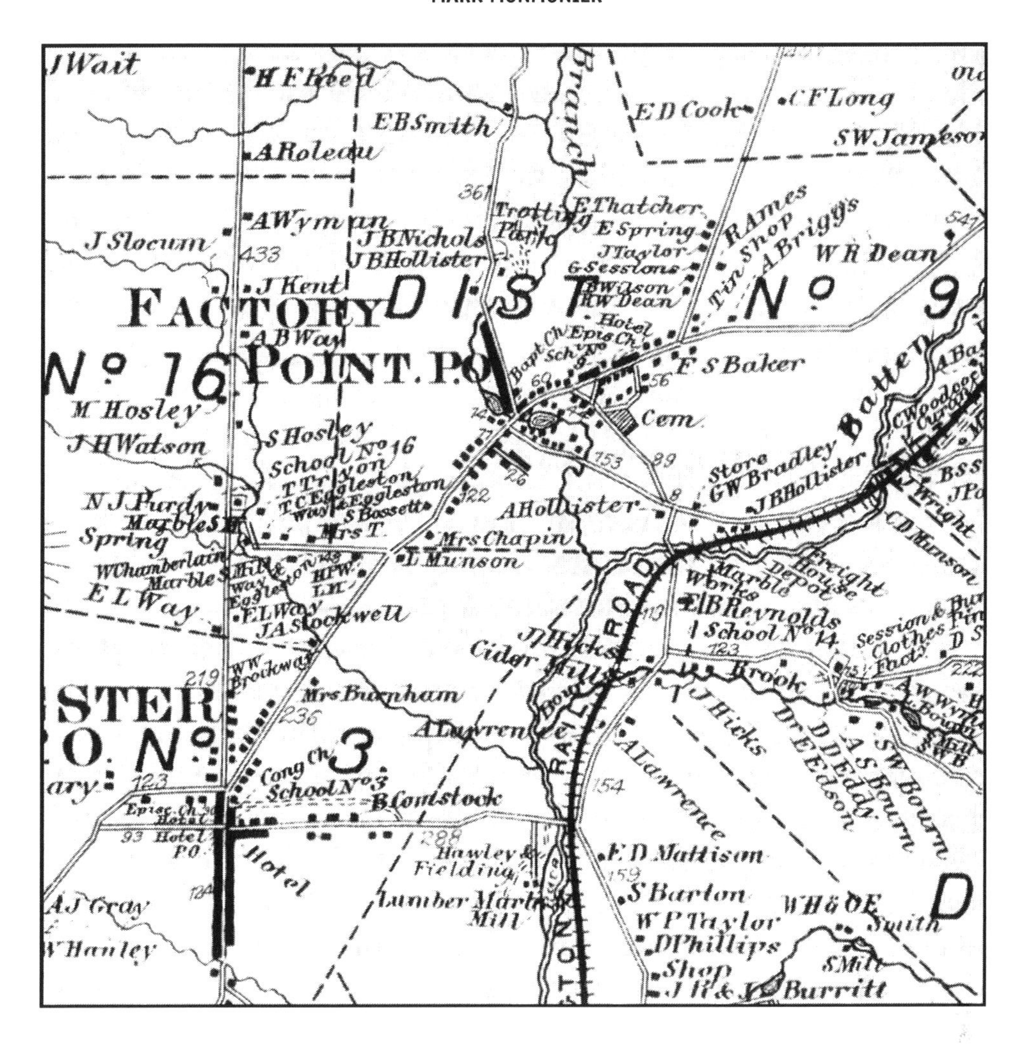

Figure 2.2.1 The vicinity of Factory Point, Vermont, later called Manchester, as shown in the Atlas of Bennington County, Vermont, published in 1869. This portion of a larger map has been reduced photographically from a scale of approximately 1:48 000 and formatted to match Figures 2.2.2 and 2.2.3 in scale and areal extent. (Source: Beers *et al.* 1869: Plate 13.)

covering the same area represented on a single sheet at 1:62 500 in the old 15-minute series. The Manchester 7.5-minute sheet was not published until 1968. Aside from a green area pattern showing the distribution of woodland, the most noteworthy changes were the increased accuracy of the contours and horizontal positions and the revision of the expanded network of village streets and town roads. Figure 2.2.3 shows transportation routes and other cultural features from the 7.5-minute sheet.

Improved accuracy, increased detail and more current cultural information are not the only hallmarks of cartographic progress represented in Figure 2.2.3. More significant than the updated landmarks and revised road network are the boundaries and numeric codes representing categories of land use and land cover. These symbols are bold in this example because they have been enlarged and transferred photographically from the Geological Survey's 1:250 000-scale Glens Falls, New York, Land Use and Land Cover map sheet, which covers a quadrangle encompassing one degree of latitude by two degrees of longitude. Although larger and more generalised, this sheet is far richer in thematic detail than a standard topographic map. It is also significant as part of a systematic attempt to map land use and land cover according to 37 categories for the entire lower 48 states and Hawaii (Place 1977). Furthermore, this land classification coverage is only one part of a digital geographic data base that includes representations of drainage basins, political

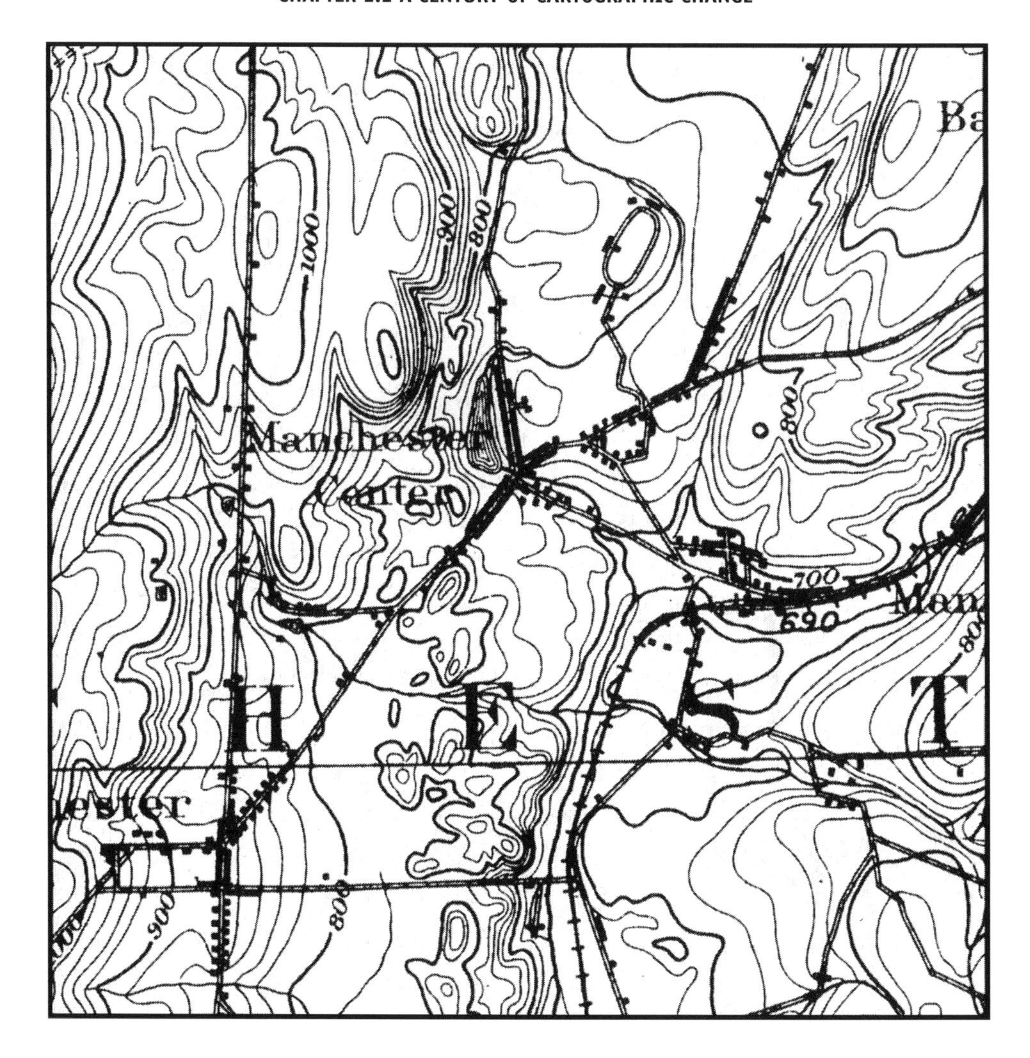

Figure 2.2.2 Topography in the vicinity of Manchester, Vermont, as shown on a portion of the Equinox, Vermont US Geological Survey 15-minute quadrangle map, 1894. Map has been enlarged from original scale of 1:62 500. Area shown is a square approximately 3.9 km [2.4 miles] on a side.

units, census reporting areas and federal land ownership (Mitchell *et al.* 1977). Begun in the early 1970s, national coverage should be complete – and ready for revision – in the late 1980s.

Were this book printed in colour, more visually dramatic examples of cartographic progress could be presented. For the reader with a personal graphic display system, it would be possible also to demonstrate that no longer need cartographers think of maps largely as 'sheets', bounded by fixed rectangles and joined to cover larger areas only with difficulty. Digital cartography can readily provide displays for irregular areas and regions heretofore divided among several adjoining quadrangles.

The Commonwealth of Pennsylvania is the focus of the second case study, an examination of copyrighted maps, produced principally by private firms and individuals. Table 2.2.1 shows percentage frequencies of copyrights by map type for three five-year periods spanning a century of progress in commercial, private sector cartography (Monmonier 1981). The differences between periods are noteworthy. The early 1870s were marked by the county atlas, the state outline map and the mineral survey; the early 1920s were dominated by highly detailed fire underwriters maps for urban areas; and the early 1970s were marked by city street guides and various road maps. Particularly significant, though, is the absence for 1920–1924 and 1970–1974 of the county atlas, and the absence in 1970–1974 of fire insurance maps. Needs and capabilities change, and a half century can alter radically both the mix of maps and the dominant type.

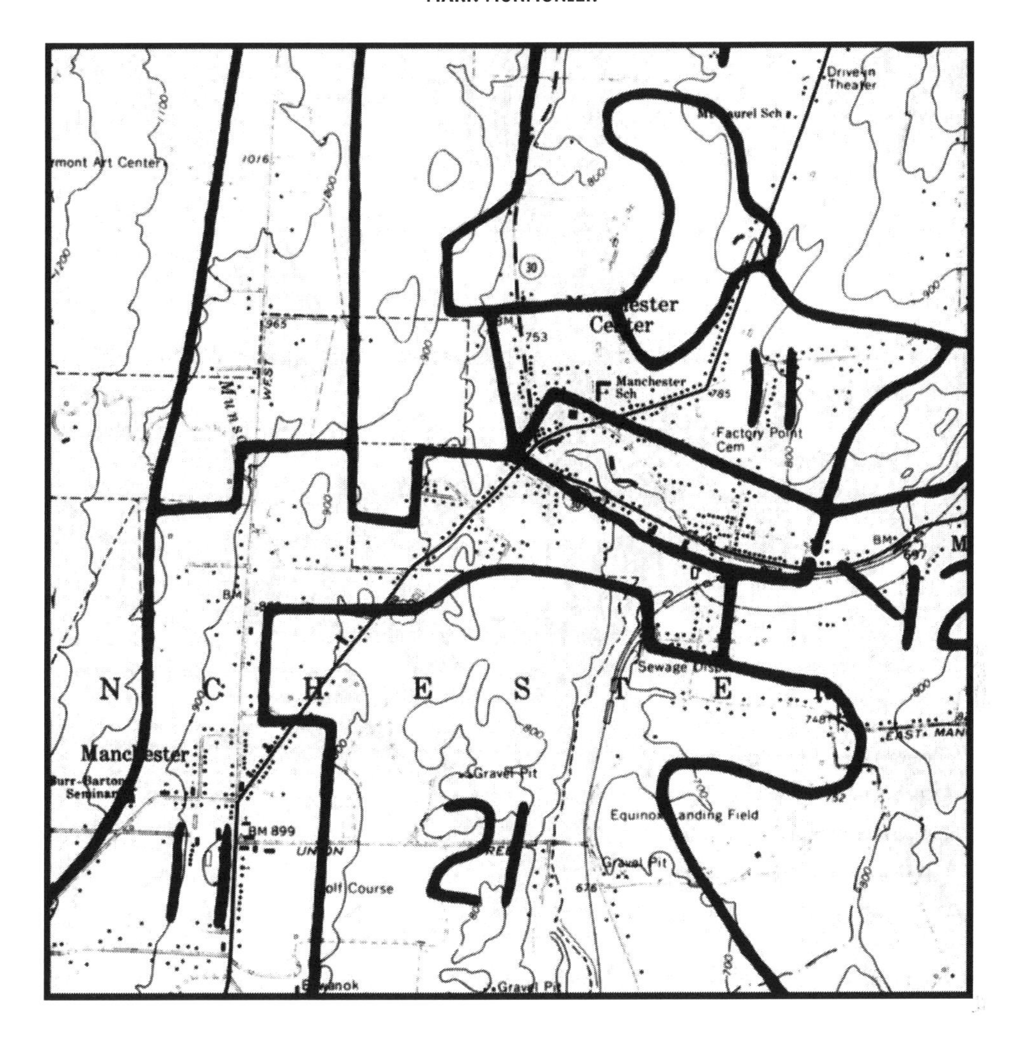

Figure 2.2.3 Land use and land cover in the vicinity of Manchester, Vermont, as shown on a portion of the Glens Falls, New York, US Geological Survey Land Use and Land Cover Map, no. L-185, 1972. Map has been enlarged from original scale of 1:250 000 and superimposed upon features from the Manchester, Vermont 7.5-minute US Geological Survey topographic quadrangle map, 1968. Topographic map has been reduced from original scale of 1:24 000, and hydrography and some contours have been suppressed. Land categories labelled include Residential land (1 l), Commercial and Services (12, to the right of the centre of the figure), and Cropland and Pasture (21). Closed polygon in the upper right is Evergreen Forest land. Land categories around the perimeter of the figure, starting at the lower left and proceeding clockwise, are Residential (11), Mixed Forest land, Cropland and Pasture, Evergreen Forest land, Cropland and Pasture, Evergreen Forest, Mixed Forest, and Cropland and Pasture (21).

The apparent increase in the variety of maps after the 1870s had its roots in the first half of the nineteenth century, and a bit before, when numerous technical and conceptual innovations spawned many new symbols, especially for thematic maps. As cartographic historian Arthur Robinson has carefully documented, the two hundred years between the mid-seventeenth and mid-nineteenth centuries witnessed a major revolution in cartography. Indeed, this period produced most of the symbols now employed widely on statistical maps and in atlases (Robinson 1982). Cartography has seen other revolutions as well, most notably that following the development of printing in the fifteenth century. The printing revolution increased the number of maps in circulation, and the thematic revolution increased their information content. The 'electronic transition', which mapping has recently entered, should bring changes as deep and broad as these previous upheavals.

Table 2.2.1 Variety of maps for the Commonwealth of Pennsylvania and places therein copyrighted 1870–1874, 1920–1924 and 1970–1974

	Percentage Frequency		
Type of Map	1870–1874	1920–1924	1970–1974
County atlas	21		
State atlas	18	2	
Mineral	13	2	
City atlas	8		
City map	8	8	56
County topographic	5		
Facsimile or historic	5		1
Fire insurance		58	
State road		6	12
County road		10	8
State wall or folded		4	
Tourism		3	5
Utility and railroad		3	4
Local engineering/planning		2	1
County outline		1	
Advertising market area			3
Recreational			4
Urban transit			1
Other	22	1	5

(*Source:* Generalised from Tables 1, 2, and 3 in Monmonier 1981.)

Mapping and the rate of technological progress

Astute cartographers have long been aware of the accelerating development of mapping technology. In 1937, for example, Erwin Raisz published an illuminating, compact graphic summary of cartographic innovation. His paper includes six full quarto-size pages with the names of innovations and innovators plotted against a vertical axis representing time. His first chart, Antiquity, covers the nine and a half centuries from 600 BC to 350 AD and mentions 31 separate persons or concepts. His last two pages, labelled Modern, cover the 230 years between 1700 and 1930, and mention over 250 separate persons, concepts or events. Equal amounts of space accommodate four decades of Antiquity and a single decade of the Modern period. Were Raisz alive today to extend his 'Chart of Historical Cartography' into the 1980s, a comparable level of detail might well require an extension four pages wide to cover but a single year in each vertical centimetre.

A full appreciation of the Electronic Transition and its probable effects must recognise the increased integration of mapping with other, larger technologies such as computer hardware, telecommunications and solid state physics. The accelerated pace of development in these areas reflects the society's resources and its attitudes, objectives, resources and technical abilities. Cultures grow and become more complex, thereby establishing new needs that in turn lead to new priorities. These new priorities also create new opportunities, with high profits for successful innovators (Starr and Rudman 1973). The resources made available for research and development also are important. Because of its value to both military defence and economic development, mapping has received an especially abundant share of research and development resources.

John Lienhard's (1979) concise graphic summary of technological progress illustrates the explosive growth in science and engineering underlying the accelerated development of mapping technology. Lienhard calculated a time constant of change for 14 innovations, from the mechanical clock to the printed circuit. His time constant is roughly the number of years needed for a technology to improve about 2.71828 times – he called this one '*e*-folding'. It is based on the theory of exponential growth and the mathematical constant *e* (2.71828…), the base of the natural logarithms. Lienhard plotted this comparative measure of

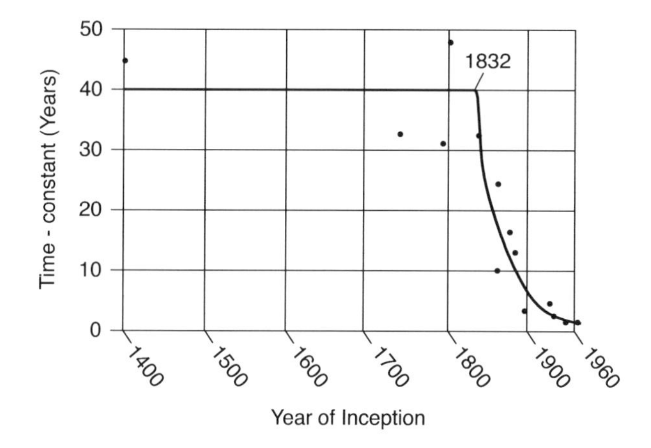

Figure 2.2.4 Increased rate of technological improvement as illustrated by generalised relationship between year of inception of a technology and Lienhard's time constant of change – roughly the number of years required for the technology to improve about 2.7 times. Dots represent various technologies upon which Lienhard's study is based. (Source: Plotted from formula and data reported in Lienhard (1979); his original graph portrayed a log-linear relationship.)

technological progress against the date of each innovation's inception. A significant shift was noted after 1832, as shown in Figure 2.2.4. From 1400 until well into the nineteenth century, about forty years was required for one *e*-folding. As shown by the generalised curve in Figure 2.2.4, the time constant has declined steadily since 1832, levelling off somewhat in the mid-twentieth century. This curve illustrates continued progress in the overall rate of technological development, with more rapid improvement for such newer technologies as digital computers, particle accelerators, liquid-fuel rockets and printed circuits.

Technological historian Donald Cardwell (1972: 215–217) attributes this accelerated progress to the evolution of a class of professional scientists and the establishment of research institutes. To be sure, today's scientists and engineers are quite different from the largely self-taught amateurs and entrepreneurs of the nineteenth century. He identified four stages in the development of technology, beginning with the Middle Ages. Cardwell's last phase began about 1860 with the advent of the industrial research laboratory. He aptly quotes philosopher Alfred North Whitehead's observation that 'the greatest invention of the nineteenth century was the invention of the method of inventions'. Government and university research programmes also marshalled the personnel, information base and management structure required for rapid technological advance. Mapping, a field long allied with the military as well as with civilian public works agencies, had an institutional foundation that fostered its participation in technological progress.

Adapting and institutions

In the late twentieth century, as indeed throughout much of its earlier history, cartography has been less a field of original innovation than one of clever adaptation and application. This view in no way degrades the originality of recognising an application to geographic measurement or representation of some new or long overlooked principle of mathematics, engineering, art or polymer chemistry. Progress through secondary or transfer innovation is perhaps inevitable given the small number of cartographers in comparison to the much larger number of engineers and physical scientists with at least a peripheral interest in things that might be employed in mapping. Thus, invention of the method of technological invention, as observed by Whitehead, is not sufficient for progress in cartography: innovation is needed as well in the organisation and management of programmes for the collection and dissemination of geographic information. If cartography is to continue to advance, its institutions as well as its researchers, operatives and theories must also advance.

In a recent monograph examining diverse approaches to the history of cartography, Blakemore and Harley (1980) suggest that maps might best be studied within the conceptual framework of a 'living language'. Modem technology suggests with equal vigour that mapping – the process more than the image – should also be a focus, to be examined through its living institutions, and their economics, geography, politics and sociology. Cartographic scholars must develop for public policy, government agencies, the private sector, the professional press and the professional organisation the same scrutiny they have lavished upon graphic symbology, map design, data collection and other superficially more enigmatic problems. If the fifteenth century saw a revolution in publishing and the nineteenth century a revolution in design, the late twentieth and early twentyfirst centuries must see a revolution in organisation and management.

[. . .] Mapping's future will precipitate new problems and societal expectations, and will demand new skills and management strategies. Exactly when that future will arrive is, of course, moot, yet there can be little doubt that mapping and geography are being propelled forward by the technological currents in which they have become engulfed (Dobson 1983). What is not at all certain is whether maps will be better designed and more accessible, and whether society will obtain the best return for its investment in mapping infrastructure. Successful, economical adaptation is cartography's foremost challenge.

References

Beers, F.W. *et al.* (1869) *Atlas of Bennington County, Vermont*, F. W. Beers, New York.

Blakemore, M.J. and Harley, J.B. (1980) *Concepts in the History of Cartography: A Review and Perspective*, University of Toronto Press, Toronto.

Cardwell, D.S.L. (1972) *Turning Points in Western Technology*, Science History Publications, New York.

Dobson, J. (1983) Automated geography. *Professional Geographer*, **35**, 135–143.

Lienhard, J.H. (1979) The rate of technological improvement before and after the 1930s. *Technology and Culture*, **20**, 515–530.

Mitchell, W.B. *et al.* (1977) GIRAS: a geographic information retrieval and analysis system for handling land use and land cover data. US Geological Survey Professional Paper No. 1059, US Government Printing Office, Washington, DC.

Monmonier, M.S. (1981) Private sector mapping of Pennsylvania: a selective cartographic history for 1870 to 1974. *Proceedings of the Pennsylvania Academy of Science*, **55**, 69–74.

Place, J.L. (1977) The land use and land cover map and data program of the US geological survey: an overview. *Remote Sensing of the Electromagnetic Spectrum*, **4** (4), 1–9.

Raisz, E. (1937) Charts of historical cartography. *Imago Mundi*, **2**, 9–16.

Robinson, A.H. (1982) *Early Thematic Mapping in the History of Cartography*, University of Chicago Press, Chicago.

Starr, C. and Rudman, R. (1973) Parameters of technological growth. *Science*, **182**, 358–364.

Thrower, J.W. (1972) *Maps and Man: An Examination of Cartography in Relation to Culture and Civilization*, Prentice-Hall, Englewood Cliffs, NJ.

Further reading

Brunn, S.D., Cutter, S.L. and Harrington, J.W. (2004) *Geography and Technology*, Springer, London. [A wide ranging edited collection looking at many different technologies and how they have impacted geographic practices and ways of knowing the world, including mapping.]

Calvert, C., Murray, K. and Smith, N. (1997) New technology and its impact on the framework for the world, in *Framework for the World* (ed. D. Rhind), GeoInformation International, Cambridge, pp. 133–159. [A systematic analysis of different technological changes in the context of late twentieth century cartographic production by national mapping agencies.]

Goodchild, M.F. (1988) Stepping over the line: technological constraints and the new cartography. *Cartography and Geographic Information Science*, **15** (3), 311–319. [Discusses a range of ways the development in digital mapping can move beyond conventional constraints of pen and paper cartography.]

Pickles, J. (1999) Cartography, digital transitions, and questions of history. *Proceedings of the 19th International Cartographic Association Conference*, 14–21 August 1999, Ottawa, Canada. [This paper applies a more political focus to notions of computerisation of cartographic production.]

See also

- Chapter 1.9: Drawing Things Together
- Chapter 1.10: Cartography Without 'Progress': Reinterpreting the Nature and Historical Development of Mapmaking
- Chapter 2.3: Manufacturing Metaphors: Public Cartography, the Market, and Democracy
- Chapter 2.6: Cartographic Futures on a Digital Earth
- Chapter 2.12: Imaging the World: The State of Online Mapping
- Chapter 5.2: The Time and Space of the Enlightenment Project

Chapter 2.3

Manufacturing Metaphors: Public Cartography, the Market, and Democracy

Patrick H. McHaffie

Editors' overview

McHaffie's chapter provides a useful discussion of the connections between technological change and cartographic labour processes. How maps come into being has always been influenced by a range of technologies for data capture, production and publishing, and it could be argued that the pace of technological change in map production has increased significantly in the twentieth century. McHaffie's work is historically rooted in the analysis of topographic mapping in the United States, detailing how new technologies of aerial photography for large-scale data capture and mechanised desktop photogrammetry contributed to reconfiguring mapmaking into a production line manufacturing process. This had significant implications for the practice of cartography, as expert knowledge was broken down into narrowly defined tasks that can be undertaken by technicians, and this kind of industrialisation also led to an increasing commodification of products and changed democratic control and access to mapping.

Originally published in 1995: Chapter 6 in *Ground Truth: The Social Implications of Geographical Information Systems* (ed. John Pickles), Guilford, New York, 113–129.

[...] In this chapter, I extend earlier work on cartographic information as a commodity, an economic good and the product of a specific labour process (McHaffie 1993) by focusing on the map as a commodity in, and a product of, a mixed public–private institutional matrix. In this setting the chapter seeks to problematise the current focus on the text and opens a dialogue concerned with the 'author' of the text and the nature of the cartographic labour process.

My purposes here are twofold. The emergence of new communications technologies has created a climate where a conflation of access to cartographic information and democracy has occurred in the disciplinary mind of many geographers and geographic information systems (GIS) specialists. The potential of this developing technology has created a false sense of egalitarianism among champions of the 'new' cartography and GIS. I hope to recast this issue into a more realistic mould, exposing the commodified nature of digital cartographic information. In addition, I hope to open a discourse that contextualises the actual human activities surrounding the production of public cartographic information, reconnecting the 'cartographer/author' to the 'map/text' at the point of production and portraying cartographic information as the product of an industrial process, while at the same time offering a sympathetic alternative to the 'map as text' metaphor.

The cartographic labour process as a state entity

[...]

During the late nineteenth and early twentieth centuries industrial capitalism was able, to a degree, to resolve crises in Western Europe and North America through the implementation of various configurations of Taylorist and Fordist 'scientific management' schemes (Gartman 1979; Littler 1982). These revolutionised the industrial production process and restructured industrial production into fragmented but linked serial processes. A key element

The Map Reader: Theories of Mapping Practice and Cartographic Representation, First Edition. Edited by Martin Dodge, Rob Kitchin and Chris Perkins.

of Taylorist formulations is the physical and social separation of the conception of the task and its execution, as well as the reduction of the production process to a number of more simple connected tasks (Taylor 1911; Braverman 1974). This type of production is commonly known as 'production line' or 'assembly line' production.

Not only is the production process divided, however, but the social relations of production are fragmented into horizontal and vertically integrated 'cells'. It becomes more difficult for a production worker to understand just who he or she is working for and where and how his or her contribution to the process fits into 'the big picture'. The separation of conception and execution, hand and brain, opens the door to the development of antagonistic social relations between managers and workers.

[...]

The first major differentiation of cartographic labourers occurred with the invention and adoption of the printing press, when compilation and reproduction of the map were separated (Monmonier 1985: 145–146). Mass reproduction commodified the map image and created a distinctive cartographic labour process. The map became one of a growing list of 'objects' deemed to be necessary to rational society and the expanding space economies of Western Europe. The map's role in consolidating state power and the need for military secrecy ensured that maps would be produced within the state apparatus. Private sector mappers developed expertise in small-scale, highly generalised representations or local area maps with limited local markets, while the production of expensive large-scale 'general purpose' (topographic and geodetic) cartographic systems was left to the state. [...]

In the twentieth century the US federal government developed large-scale public mapping systems as the necessary topographic precursor to geological, hydrological and botanical investigations west of the Mississippi River. The federal commitment to this policy was created within the context of bureaucratic/institutional tension between the federal and state governments regarding responsibility for the funding of scientific endeavours within their respective borders (Edney 1986). The development of a national mapping policy and the establishment and expansion of the two major twentieth century civilian mapping agencies, the US Coast and Geodetic Survey and the US Geological Survey (USGS), came as the result of *ad hoc* legislation designed to delineate responsibilities between agencies at the federal and state level.

The transformation of the public cartographic labour process in the United States displays elements of a corporatist logic that borrowed the most efficient components of developing. industrial labour (Taylorist) processes in the late nineteenth and early twentieth centuries. The resulting system would more closely resemble a modern factory than the craft/apprentice organisation that had predominated prior to World War I (McHaffie 1993). Before World War I, most of the labour involved in topographic mapping was accomplished in the field. The tradition and myth of the mapper as a rugged surveyor single-handedly mapping and taming the vast Western wilderness was created during this period. However, after World War I, many USGS topographers returned from military service with experience gained in the use of aerial photography in reconnaissance and mapping. They 'applied their interest in aerial photography to its potential use in the civilian topographic mapping programme. Throughout the 1920s they experimented with applications of the relatively new science of photogrammetry and succeeded in making a few maps from aerial photographs' (Thompson 1981: 7).

The incorporation of aerial photography into the cartographic labour process allowed the restructuring of the relations of production within the USGS to a more efficient, more easily controlled Taylorist formulation. The aerial photographer could, in a single day, photograph hundreds of square miles and supply technicians with the necessary materials for map compilation. The mappers were no longer required to 'slog' into the messy reality of the field in order to produce the 'map'; they no longer were required to compile their manuscripts in the field. In fact a 'new' cartographer was created, one labouring in dark rooms using complicated optical–mechanical instruments, embedded within a process that was more easily watched, more easily controlled. Within agencies such as the USGS and the Defense Mapping Agency (DMA) this process has since been refined and the technology has since been modernised through improvements in aerial platforms, cameras, film, optics and electronic digital storage and processing, but the conditions of work for labourers in this stage of cartographic production have remained fundamentally the same. The need for field survey and the actual contact of the cartographer with the object of his or her work was, as a consequence, greatly reduced. A new technique of aerial photograph interpretation was born, elaborated, codified and institutionalised within the state.

The labour process is the instrument that shapes the map as an embodiment of larger societal power relations. Cartographers serve not merely as channels of existing power relations, but also act independently within the constraints of a particular production process to constantly redefine and reshape the cartographic messages that pass from the state to the public. The specificity of the labour process, as constituted within the state apparatus, determines the conditions under which public cartographic information is produced. But the public cartographic labour process is peculiar as a state formation in that it assumes a Taylorist configuration within the state *in the absence of the logic of accumulation that is assumed to drive similar schemes in the private sector.* This situation is comparable to other industrial production processes that have been promulgated under the direction of the state, notably the production of actual specie within government mints and the production of documents by government printing offices.

However, both of these processes are merely concerned with the reproduction of a preformed message. In the production of cartographic information the cartographer is, to a greater or lesser degree, the author of the created information.

The history of the cartographic labour process has been characterised by the division of labour into a series of disconnected operations, with each operation contributing to the final message embodied within the map. This history has been one of fragmentation, rationalisation of production procedures, and progressive de-skilling of the cartographic labourer. If 'cartographers manufacture power' as Harley (1989) states, then public sector cartographers do so within circumstances that are shaped by the particular institutional configuration of the state cartographic production process. This configuration can take many forms, from the one-person shop in the smallest local government office producing representations using pen-and-ink techniques, to huge agencies such as the DMA responsible for producing global mapping systems in a significantly different technological environment.

[. . .]

Figure 2.3.1 shows a schematic diagram of the cartographic production process as conceived and represented by the USGS (Thompson 1981: 24). Production is organised as a linear, one-way process consisting of a number of tasks along the way. These tasks are particular moments in the production process, sites of transformation of the incipient map into something closer to the 'final' product. They are sites of the creation of public cartographic information. Each station along the line is only a part of the labour process, connected through the production sequence to two other parts of the production process; that which immediately transformed the map before it arrived in your shop, and that which requires the transformed product you will produce. It is significant that the process is schematically represented as a conveyor belt, a machine used for moving product from fixed site to fixed site inexorably closer to the point of contact with the user, 'Distribution'. The seven sites are alternately peopled (manned?) by workers or by machines. Integration of technology and human labour progressively devalues the human contribution to the map. Workers are increasingly

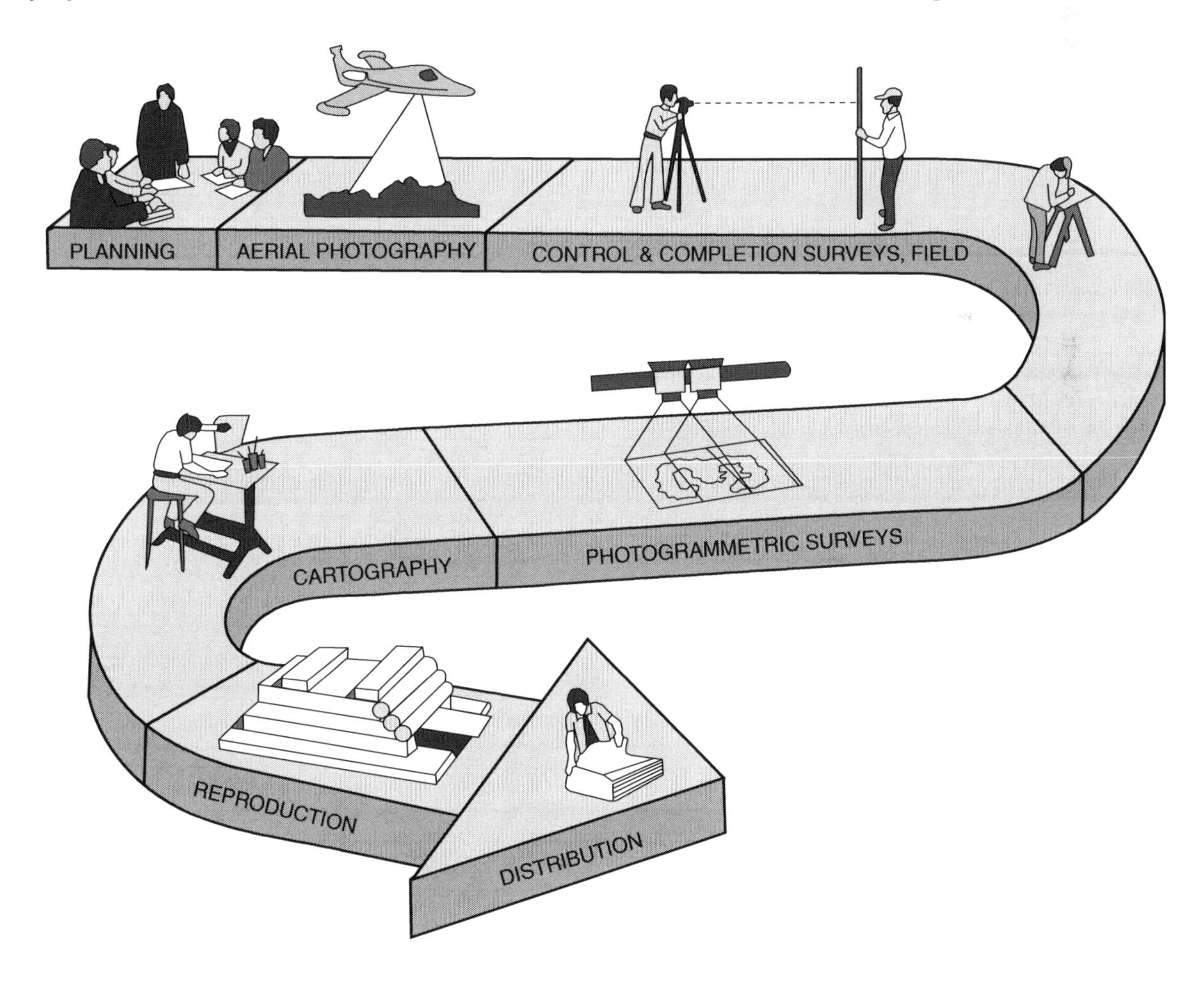

Figure 2.3.1 The cartographic product process (Thompson 1981: 24).

viewed as 'technicians', 'operators' or 'warm bodies'. At this point, the job category of 'topographer', the maker of the topographic map, has disappeared and has been replaced by specialists. This view of the cartographic labourer reflects a historical de-skilling of production workers that has proceeded hand in hand with the fragmentation of the labour process achieved under 'scientific management' schemes.

The incorporation of aerial photography and photogrammetric techniques into public sector mapping has been the most transformative of all technological developments in the cartographic labour process. It has, to a large extent, created and furthered the aura of scientific objectivity that enshrouds topographic mapping today. If the cartographer is physically separated through the abstraction of the photograph from the object of his or her attention, then the resulting product, the topographic map, will somehow be objective, value free and scientific. [...]

The pioneers of photogrammetry – Fairchild, Abrams, Goddard, Kauffman and Brock – sold the new technology of aerial mapping to the USGS based on the improvements in accuracy that it produced (Heiman 1972). This, and other new technologies in the area of photographic processes and reproduction, further increased the need for linked systems of specialists. They took on new names (e.g. operators, scribers, photographers, photogrammetrists), assumed newly de-skilled roles and conformed to the Taylorist fragmentation of the labour process in a corporatist spirit of increased productivity, loyalty and efficiency. The 'scientific management' project of control, projected into the cartographic labour process, was reflected back onto society in the form of rationalised and increasingly depersonalised representations of the modern world. To witness the transformation of cartographic representation as technology is introduced one needs only study, for a particular area, the 30-minute maps produced by the USGS in the 1880s, the 15-minute maps produced by the USGS in the 1920s, the 7.5-minute maps produced by the USGS in the 1950s and the DLGs (Digital Line Graphs) produced by the USGS in the 1980s. Each successive map series presents a more standardised, more accurate, more codified product. The 'mistakes' or artefacts produced by a closer human involvement disappear.

[...]

Cartographic information + the market = democracy?

[...]

The Landsat programme can serve as an illustration of the inherent problems of mixing a 'free market' ideology with subsidised information production. The programme was established as an experimental system for gathering repetitive worldwide multispectral data for civilian uses. Through the 1970s and into the 1980s the Landsat programme was managed by the USGS through the EROS Data Center and provided low cost photographic products and digital data to corporations, government agencies, educators, students, researchers and foreign governments. In the 1970s satellite photographic images were priced from $8 to $50, while digital scenes were sold for $200. Pricing under this regime was limited to the cost of reproduction, rather than the high cost of acquiring the data. The cost of system development and maintenance was subsidised through annual appropriations from Congress. During the 1980s, however, the programme was first transferred to the National Oceanic and Atmospheric Administration (NOAA) and then it was finally privatised under the Earth Observation Satellite (EOSAT) Company. Price increases during the NOAA transition and after commercialisation were intended eventually to pass on the full cost of acquiring the data and maintaining the system to the consumer. During the period between 1980 and 1990 the cost of photographic products, arguably the most 'democratic' product (requiring the least capital investment to use the image), has increased as much as 50 times for some comparable items. During the same period the prices of digital products were increased up to five times their pre-commercial cost. In addition, copyright restrictions were placed on the use of both photographic and digital products, effectively removing all imagery produced after 1985 from the public domain and severely limiting their use. During the same period several other nations have launched, successful private venture and state subsidised remote sensing satellites. One result of these developments has been a precipitous drop in the number of photographic products sold. In 1980, at an average price of $15, there were over 128 000 Landsat film products sold. This compares to just over 4000 sold in 1989 at an average price of $150. Digital products have seen a slow but steady increase, still amounting to only 7374 computer-compatible tapes sold in 1989. Who are the consumers?

Landsat data are now being purchased primarily by only a few government agencies and a number of aggressive corporations. Research facilities, academic institutions, educators, students, State and local governments, and the governments of less developed nations are now purchasing considerably less data than they did a few years ago. (Lauer *et al.* 1991: 649.)

[...]

The economist J. M. Keynes said, 'Capitalism is the extraordinary belief that the nastiest of men for the nastiest of motives will somehow work for the benefit of us all' (quoted in Sherrill 1984: 355). Whether one agrees with this rather dour view of the market economy and its workings or not, even the most ardent supporters of free enterprise must admit that the benefits of *laissez-faire* capitalism are unevenly distributed. The 'belief' cited above is a good example of an *ideology*: 'an inverted, truncated, distorted reflection of reality' (a paraphrase of Jorge Larrain, quoted by Smith 1984: 15). If the production of cartographic information is organised and controlled with reference to this ideology, albeit in a mixed public/private institutional matrix, and within a scientific/technological milieu that is in a state of constant transformation, it remains to be seen how this economy will interact or draw sustenance from another enduring ideology, that of democracy.

Many have claimed that we are on the verge of a new era of *electronic democracy.* Jim Warren (1993), writing in *Government Technology* under the heading 'Electronic Democracy: The Year of the Internet (Networking Hits Critical Mass)', refers to electronic mail, international forums or discussion lists of 'experts,' computer conferencing and the proliferation of on-line data as a signal that new forms of access to information will result in a more democratic society. Indeed, the networking revolution and the growth of electronic networks such as Internet, Bitnet and Fidonet has reconfigured the technology of information distribution and created new avenues for 'grassroots' initiatives to become more generally distributed. [...] Perhaps those of us with 'access' have become jaded and forgotten that the 14.2 million individuals claimed to be connected in some way to one of the international computer networks represent only a small fraction of the five billion or so individuals on the planet (substantially less than 1%). [...]

For billions the possibility of accessing the best technology and information made available through digital communications networks will always be a luxury. Cartographic information, digital or otherwise, becomes a commodity in its mass production and marketing. If market forces are guiding the production of this commodity, even though it is portrayed as a 'general purpose' product, then it will be produced to serve a particular constituency, in particular those who serve to gain from the 'digging', 'gathering' or 'harvesting' the wealth of America and of the world.

References

Braverman, H. (1974) *Labor and Monopoly Capital: The Degradation of Work in the 20th Century*, Monthly Review Press, New York.

Edney, M. (1986) Politics, science and government mapping policy in the United States, 1800–1925. *American Cartographer*, **13** (4), 295–306.

Gartman, D. (1979) Origins of the assembly line and capitalist control of work at Ford, in *Case Studies on the Labor Process* (ed. A.S. Zimbalist), Monthly Review Press, New York, pp. 193–205.

Harley, J.B. (1989) Deconstructing the map. *Cartographica*, **26** (2), 1–20.

Heiman, G. (1972) *Aerial Photography: The Story of Aerial Mapping and Reconnaissance*, Macmillan, New York.

Lauer, D.T., Estes, J.E., Jensen, J.R. and Greenlee, D.D. (1991) Institutional issues affecting the integration and use of remotely sensed data and geographic information systems. *Photogrammetric Engineering & Remote Sensing*, **57** (6), 647–654.

Littler, C.R. (1982) *The Development of the Labour Process in Capitalist Societies*, Heinemann, London.

McHaffie, P. (1993) The public cartographic labor process in the United States: rationalization then and now. *Cartographica*, **30** (1), 55–60.

Monmonier, M. (1985) *Technological Transition in Cartography*, University of Wisconsin Press, Madison, WI.

Sherrill, R. (1984) *Why They Call It Politics: A Guide to America's Government*, Harcourt Brace Jovanovich, New York.

Smith, N. (1984) *Uneven Development: Nature, Capital, and the Production of Space*, Blackwell, Oxford.

Taylor, F.W. (1911) *The Principles of Scientific Management*, W.W. Norton, New York.

Thompson, M.M. (1981) *Maps for America*, 2nd edn, US Geographical Survey, Reston, VA.

Warren, J. (1993) Electronic democracy: the year of the Internet. *Government Technology*, **6** (7), 22.

Further reading

McHaffie, P.H. (2002) Towards the automated map factory: early automation at the US Geological Survey. *Cartography and Geographic Information Science*, **29** (3), 193–206. [Additional analysis by McHaffie along similar lines to this excerpt.]

Pedley, M.S. (2005) *The Commerce of Cartography*, University of Chicago Press, Chicago. [An introduction to commercial concerns underpinning the pre-industrial eighteenth century development of cartographic production.]

Pickles, J. (1999) Cartography, digital transitions, and questions of history. *Proceedings of the 19th International Cartographic Association Conference*, 14–21 August 1999, Ottawa, Canada. [A social theoretical critique of technological explanations for transitions in the mapping industries.]

Rhind, D. (1999) Business, governments and technology. *Proceedings of the 19th International Cartographic Association Conference*, 14–21 August 1999, Ottawa, Canada. [An organisational explanations for transitions in cartography arguing that institutional power mediated by government policy is the most useful explanation for change.]

See also

- Chapter 1.8: Deconstructing the Map
- Chapter 1.10: 'Cartography Without 'Progress': Reinterpreting the Nature and Historical Development of Mapmaking
- Chapter 2.2: A Century of Cartographic Change
- Chapter 2.6: Cartographic Futures on a Digital Earth
- Chapter 2.8: Remote Sensing of Urban/Suburban Infrastructure and Socio-Economic Attributes
- Chapter 4.10: Citizens as Sensors: The World of Volunteered Geography

Chapter 2.4

Maps and Mapping Technologies of the Persian Gulf War

Keith C. Clarke

Editors' overview

Clarke's paper suggests how the technological underpinnings of cartographic production and use advanced at the end of the twentieth century under the intense pressure of war fighting. Focusing on the United States military campaign in the Persian Gulf in 1990–1991, he describes the impact of wholesale computerisation in production, updating and dissemination of a range of geographic data as well as the integration of GPS and overhead imagery that has subsequently become central to both military and, latterly, civilian mapping.

Originally published in 1992: *Cartography and Geographic Information Systems*, **19** (2), 80–87.

Introduction

On 2 August 1990, Iraq invaded Kuwait, drawing immediate United Nations Security Council condemnation and prompting the United States to seize Iraqi assets and ban trade with Iraq. The following day, President George Bush [Senior] announced that naval forces would be sent to the Persian Gulf as a deterrent and, three days later, Operation Desert Shield began, with the United Nations' approval of worldwide sanctions against Iraq. On 8 November 1990, Bush ordered an additional 200 000 troops to the Gulf, making the build-up of equipment and troops the largest military operation in history. The air and ground combat that followed proved to be full of surprises, both for the outcome of the war and for the multitude of new military technologies that were tested for the first time. As no surprise to cartographers, the mapping sciences technology emerged as central to the conduct of the war. Compared with previous wars, the effectiveness of mapping technology was unprecedented and its performance during the war was close to perfect, giving the military and the public a new respect for the tools and methods of our profession.

[. . .] Historically, war has 'forced' new technologies into operation. This time, due to the sheer magnitude of the operation and the speed with which events unfolded, the effect was, instead, to prove the effectiveness of existing cartographic and mapping technology. Some of the lessons learned from the war, therefore, have implications for how mapping should be done both now and in the years to follow.

[. . .]

Maps for the military

The primary agency for the production of military maps is the Defense Mapping Agency (DMA), which dates from 1972, and is the centralised source of all air, sea and land maps and images for military use. With the start of Operation Desert Shield, the DMA moved into 24-hour production. The agency, primarily at its two plants in St. Louis and in the Brookmont area of suburban Maryland, had employees work 10- to 12-hour shifts, seven days a week, with even the DMA school's map printing press drafted into 24-hour use. Two hundred years of overtime were expended in producing 12 000 new map products, 600 of them all-digital, and in printing more than 100 million

The Map Reader: Theories of Mapping Practice and Cartographic Representation, First Edition. Edited by Martin Dodge, Rob Kitchin and Chris Perkins.

sheet maps for the field. Why so many? According to A. Clay Ancell, deputy director of the St. Louis DMA centre, 'You've got these things folded up and stuffed in your pocket, so after a sweaty day in the desert, they literally come apart.' In addition, many maps intended for air-sortie support were used and destroyed after the completion of each mission (Associated Press 1991).

[...]

Such a massive mapping effort raised problems, of course. Among them were the lack of trained digital cartographers in the field, the inflexibility of propriety software when systems integration was necessary, the need to sacrifice precision and accuracy for speed, the need to buy imagery from commercial satellite companies to produce image maps, the backlog in cataloguing and storing the flood of map products, and the need to rectify generalised and inaccurate maps on different projections and datums with extremely high precision GPS data.

[...]

In general, the military map component of the war effort was a remarkable example of organised, large-scale production and flexibility with very new technology, often against severe obstacles. Clearly, the importance of sheet maps and digital map products came to the forefront during the war. This, however, is only part of the story.

GPS and the Gulf War

Perhaps more than any other technology, GPS, also known by its military name, NAVSTAR, more than proved its worth during the Gulf War. GPS is a $10 billion Air Force system designed to provide almost instantaneous, precise locational information in the field in all weather. It eventually will consist of 21 operational orbiting satellites and three orbiting spares. The satellites orbit at an altitude of 10 900 nautical miles, once every 12 hours, with three satellites on a single track. Atomic clocks on the satellites transmit a signal continuously. A GPS decoder/receiver interprets signals from three or four satellites, and in conjunction with ephemeris information about the satellite's orbital track, trilateration from distances allows precise two- and three-dimensional positioning. Small, portable, hand-held units can provide accuracy of about 18 metres horizontally, and 28 metres vertically, although higher precision is possible using two units together in differential mode. Larger backpack-size units are capable of precision at the sub-metre level. Hand-held units have been broadly available and relatively inexpensive – less than $5000 – for some time.

GPS was used to some extent in the mine sweeping operation in the Persian Gulf during the Iran–Iraq War, and in the invasion of Panama. However, GPS was used extensively by all branches of the forces and by the Iraqis in the Gulf War.

[...]

More than 60 different air, sea and land systems and vehicles used GPS, including planes, AWACS, helicopters, tanks, missile launchers and even missiles themselves. Four thousand units had been delivered to the Air Force by 3 August 1990, and an additional 10 000 were supplied. The Army rounded up and shipped to the Gulf 5000 hand-held units, which were being evaluated and used in training. [...] The use of GPS systems on the French Puma helicopter allowed some effective search-and-rescue operations, including the rescue of a US Air Force pilot on 13 January 1991, in stormy weather (Aviation Week and Space Technology 1991a).

Kiernan (1991) quotes Vice Admiral Jerry Tuttle, the US Navy's director of space, command and control, as saying that GPS will 'revolutionise tactics in every warfare area', a fact that indeed seems true of the Gulf War. The broad availability of hand-held units will place an especially rigid demand on digital and printed cartographic products, especially at large scales such as 1:50 000. Pratt of the DMA noted that GPS units caused some confusion in the field over the use of datums and map projections, and forced the revision of many older map products to meet the increased need for geodetic accuracy.

GIS, remote sensing, and the war

The Persian Gulf War also led to different uses of remotely sensed imagery and geographic information systems (GIS) than those in peacetime.

[...]

In terms of space-based remote sensing, the primary tool was a series of KH-11 satellites in elliptical low orbits, which provided overlapping coverage of the Gulf area (Denton 1991). These satellites provided high resolution infrared imagery, primarily for bomb damage and other assessments. Denton noted that other satellites, such as the Lacrosse radar imaging satellite and other defence support programme satellites in geosynchronous orbits, were used, but there were gaps in coverage, caused in large part by poor weather.

More traditional commercial satellite imagery also was used. SPOT and Landsat coverage was used to assemble several image maps of areas of interest, even by commercial companies such as Intergraph. This imagery was processed into annotated and printed maps by the Army's topographic engineers, with limited early success. Where no map coverage existed, these image maps, of which the DMA produced 125, proved helpful.

By far the greatest success story, however, was an imaging radar system developed by the Air Force and known as JointSTARS (JSTARS), an aircraft-based synthetic aperture radar (SAR) system. The two JSTARS aircraft were ordered to the Gulf area on 18 December 1990, and required extensive crash training in the field. The two planes flew one 10- to 12-hour mission per day during the war, providing large amounts of radar imagery (Sweetman 1991). The radars operate in two modes – a moving vehicle detection mode, capable of covering a million square kilometres per mission, and the regular SAR imaging mode for fixed targets. A particular advantage of the system over other satellite and aircraft imaging systems was its ability to work at night and in bad weather. While detecting whether moving vehicles were wheeled or tracked proved impossible (Boatman 1991), the system was judged to have performed above expectations. Aviation Week and Space Technology (1991b) reported that the system worked despite jamming by the Iraqis.

Conclusion

This paper began by noting that cartography traditionally receives a technological boost in wartime. Aerial photography, radar, digital photogrammetry, digital cartography in general, and even the DMA itself all have early origins in the major wars of this century. The Gulf War, however, was an exception to this rule. Perhaps because the war was over so quickly, and was prepared for over an extended period, it proved a testing ground for new technologies that already had been developed. Many of these technologies were rushed from testing to first field use in a matter of weeks, and most performed impeccably. As a result, the broad scale acknowledgment that advanced, high technology cartographic systems are essential was probably the single most important outcome of the war, from a cartographic standpoint.

Post-war analysis has pointed to some easily attainable goals for cartography, including: (1) more, and better trained, digital cartographers and photogrammetrists; (2) better, faster, more flexible and reliable digital cartographic systems; and (3) more large-scale digital map coverage of unmapped parts of the world. Pratt's call for better systems integration is itself an endorsement for current trends within cartography, where common data formats and spatial data transfer standards have been years in development, and are now close to reality.

Just as interesting, the mapping at 1:50 000 of extensive land areas was achieved quickly with a high technology, concentrated effort. With an even larger global effort, using satellite imagery and GPS ground truth, it now seems possible that the entire land surface area of the earth could be mapped at 1:50 000 with a standard symbol set and identical accuracy standards. Such a map would be extremely useful, the cartographic equivalent of the current biological project to complete the DNA mapping of the human genome.

[...]

The repercussions of the Persian Gulf War for the Middle East will last a lifetime at least, and are at best indistinct. For cartography, however, the war's effects have been immediate – accelerating technological change; proving the value of new mapping-science technologies; and moving cartography ever further from the pen-and-ink, scribe coat and stereo viewers of only a generation ago.

References

Associated Press (1991) Agency shows way for Gulf Forces. *The Washington Post* (Federal Page), 2 January, A13.

Aviation Week and Space Technology (1991a) French search rescue units aided by GPS receivers, 25 February, 44.

Aviation Week and Space Technology (1991b) Washington roundup: who gets what, 22 April, 19.

Boatman, J. (1991) Joint STARS was 100% reliable. *Jane's Defence Weekly*, 30 March, 466.

Denton, R.V. (1991) Electronic imaging's role in the Gulf War. *Advanced Imaging*, **6** (4), 57–60.

Kiernan, V. (1991) Guidance from above in the Gulf War. *Science*, **251** (1 March), 1012–1014.

Sweetman, B. (1991) Flying start for USAF systems. *Jane's Defence Weekly*, 30 March, 499.

Further reading

Barnes, T.J. (2008) Geography's underworld: the military-industrial complex, mathematical modelling, and the quantitative revolution. *Geoforum*, **39**, 3–16. [A considered analysis of the connections between military institutions and their agendas, and the development of United States academic geography and cartography.]

Cloud, J. (2002) American cartographic transformations during the Cold War. *Cartography and Geographic Information Science*, **29**, 261–282. [This paper provides a useful discussion of the influence of the United States military in the development of cartographic production through the second half of the twentieth century.]

Monmonier, M. (2002) *Spying With Maps: Surveillance Technologies and the Future of Privacy*, University of Chicago Press, Chicago. [This accessible book reviews the many ways that maps and allied geospatial technologies are deployed to govern people.]

Pickles, J. (1995) Representations in an electronic age: geography, GIS, and democracy, in *Ground Truth: The Social Implications of Geographic Information Systems* (ed. J. Pickles), Guilford, New York, pp. 1–30. [A critical commentary on the impacts of digital cartography on democratic discourse and civil rights.]

Woodward, R. (2004) *Military Geographies*, John Wiley & Sons, Ltd, Chichester, UK. [A thorough discussion of the geographical nature of the military institution and its role in contemporary society.]

See also

- Chapter 2.3: Manufacturing Metaphors: Public Cartography, the Market, and Democracy
- Chapter 2.8: Remote Sensing of Urban/Suburban Infrastructure and Socio-Economic Attributes
- Chapter 2.10: Mobile Mapping: An Emerging Technology for Spatial Data Acquisition
- Chapter 5.8: Cartographic Rationality and the Politics of Geosurveillance and Security

Chapter 2.5
Automation and Cartography

Waldo R. Tobler

Editors' overview

Tobler's paper, published over fifty years ago, was a pioneering consideration of the potentially revolutionary role of computing in various cartographic practices (collection of map data, storage, analysis, output display). While digital computation technologies have improved dramatically in succeeding decades, making some of Tobler's discussion seem naïve, his analysis on the fundamentals of map automation is still very valid, in particular his consideration of the aspects of mapping that cannot be appropriately conducted by software algorithms remain applicable today.

Originally published in 1959: *The Geographical Review,* **49** (4), 526–534.

Automation, it would seem, is here to stay. Advantages in speed and accuracy seem likely to make the use of computing machinery more common, despite the relatively high initial cost. In view of recent developments in automation and high speed data processing, it is appropriate to ask, 'Do possibilities for automation exist in cartography? And, if so, where can these possibilities be found?' In order to answer these questions, the preparation of maps should be viewed as a complex data processing system. Certain similarities then become apparent between data processing in general and cartographic processing in particular.

The map as a data storage element

A data processing system can be said to consist of four major steps: gathering of the data, manipulation, storage, and utilisation. [...] [T]he equating of a map to a data storage element in a data processing system is the most comprehensive, and probably the most accurate. Here the function of the map is similar to that of a book. It contains information in symbolic form, and, in particular, it serves as a graphic storehouse of selected information. It can be compared with the magnetic tape in a computer system with insufficient internal memory: the tape is used for interim storage of data and is called up as required. Obviously, data concerning spatially distributed phenomena need not be stored on a map but can be coded into other symbologies. Whether advantages accrue is another question.

The map as a computer input

The conceptualisation of a map as a data storage medium leads directly to the concept of it as a computer input element. Here two methods of use seem possible. In the simpler, data are extracted from a map, translated into some symbology that available machinery will accept, and then operated upon by the data manipulation unit. Examples would include the recording of positional data in terms of some coordinate system, punching this information on Hollerith-type cards, and feeding the cards to a data manipulation system. A variety of equipment (often referred to as data reduction equipment) has been built that will do this very thing. It will extract information from such material as graphs or photographs of oscilloscope traces – which to the machine are the same as maps – and automatically prepare punched cards or tapes. Little use of such equipment has been made in cartography, with the possible exception of photogrammetry, perhaps because the data are often available in more readily convertible form – that is, the information from which the maps are prepared.

The Map Reader: Theories of Mapping Practice and Cartographic Representation, First Edition. Edited by Martin Dodge, Rob Kitchin and Chris Perkins.

The second method of using maps as a data manipulation input consists in transferring the map data directly into a computation system. This is much more complicated technically, and less well developed today. By this method a map might be fed into a 'slot', processed, and 'discarded', its entire informational content having been stored in the memory element of the machine. [. . .] Of course, a similar processing occurs in the transformation of colour separations into a multicoloured product by photo-offset lithography, but there is a great difference in the amount and kind of data manipulation that can take place. An offset camera can change the scale of a map and, to a limited extent, the projection. It cannot, however, give the distance between two points on the map, nor can it choose the optimum location, route or area for a given purpose. Certain stored-program computers can do all these logical operations and many more, if the map image is stored in a convenient manner. But herein lies the difficulty. Although many techniques are now in use for the wholesale transformation of maps into optical, electrical, magnetic or other images, the conceptual problems of machine pattern recognition have not been solved. The inputting of geographic information to the computer in analytic rather than graphic form avoids this problem but in many cases is quite difficult.

Figure 2.5.1 Map of the United States drawn directly by machine from a deck of 343 punched cards. Plotting time, approximately 15 minutes. The map has been reduced, but not retouched. Bipolar oblique conic conformal projection (outline of original map from the American Geographical Society's Map of the Americas, 1:5 000 000). (Plotter courtesy the Benson-Lehner Corporation, Los Angeles.)

The map as data processing output

The third possibility of automating cartography derives from the conceptualisation of a map as an output of a data processing system. This is perhaps of most practical interest. Most of the currently available data processing output units provide information for viewing or reading, and, as maps are visual displays, it follows that they can be prepared with this equipment. Radar mapping, for example, makes use of a cathode ray tube, which is much like the tube used in a television set. [. . .] High speed permanent recording of the display from such a device on film or offset plates is, of course, possible. Other available mechanical devices plot symbols and/or draw lines automatically. Figure 2.5.1 was drawn by just such a machine.

Two types of computer generated map output may now be distinguished. In the first, the information is recorded on a prepared base map. [. . .] Maps prepared in this way might be used as compilation copy, for preliminary analyses of data in a spatial context, or for the recording of census data. When used in conjunction with a computer, for example, the generation and printing directly on a map of such data as index numbers, sales figures and average rainfall could become almost one operation. It seems probable that maps printed on continuous form paper will become common in the future in this type of application. Other examples of plotting on a prepared base map can be cited; an interesting one has been recorded in the engineering literature (Locke 1955: 46). The impact point of a missile being tested was continuously predicted by a computer, and a trace of this point was automatically drawn on a map. If the trace crossed the boundary of the firing range as shown on the map, the missile was exploded in the air. Another current use is to plot the course of a moving vehicle on a map of the general vicinity, so that its position is known at all times.

In the second type of computer generated map output, the map is reproduced on a blank recording medium, probably film or paper, instead of on a prepared base. Figure 2.5.1 is an example of such a map. Its preparation embodied the three elements discussed in this paper: use of a map as a storehouse of data; translation and inputting of the map data to a machine; and automatic preparation of a map from these data. Firstly, an appropriate map of the United States was chosen, and a transparent overlay with an orthogonal grid was placed over the map. The coastline was then broken into straight line segments, on the principle that any curve can be approximated by a series of straight lines. The end points of the segments were sequentially recorded in terms of the grid overlay and these coordinates recorded as punched holes in IBM cards, which were subsequently fed to the plotting machine that drew the map. No computer was used, and, although the plotting machine contains controls for changing scale, all the machine did was to draw straight lines between the points. Computer manipulation of the input data in terms of a stored program could have changed the projection and reduced the scale, generalising the outline if necessary. The outline could also have been made to look 'better' if the coastline had been taken from a larger-scale map.

Cartographic establishments may someday stock standard decks of punched cards containing basic geographic information for such plotting. A separate deck for each category of geographic information, such as coastlines, state boundaries, cities, contours, railroads and population, is entirely feasible. Possibly some such method will be utilised to alleviate storage problems where maps are used as legal records (for example, platting). Many other applications should be apparent to the reader.

Other applications of data processing procedure

So far, this discussion has been concerned with the conceptualisation of a map as a storage medium, a data processing input, or a processed output. Other parts of the cartographic data processing procedure might also be automated. In topographic surveying, for example, partial automation is taking place in precise data gathering (geodimeter, etc.). Feedback control and inertial guidance are evolving toward complete navigation systems – an example of automated data utilisation. Many additional kinds of data manipulation are currently feasible with stored-program digital computers. These machines have been popularly referred to as 'giant brains'. They can compute intersections of planes and surfaces (on a flat map these intersections might be contours), change map scales and projections, and perform other logical operations easily, rapidly and accurately. Indeed, many much more complex transformations of data can be performed with currently available techniques.

There are, however, a large number of difficulties in any attempt to automate cartographic procedures. For example, the drawing of isolines from sampled data has been shown by Mackay (1951) to contain a logical ambiguity in certain cases. Still other problems are involved in the artistic aspect of cartography. The machines cannot at present do as nice a job of relief shading as an airbrush artist. However, new equipment and procedures are being developed so rapidly that this may soon be possible. As another example, the heuristics of generalisation, balance and contrast seem complex and difficult to define. The history of cartography also shows clearly that opinion as to what constitutes a 'good' map can, and does, change. [...]

The use of high speed computing devices will often demand a more extensive mathematical understanding of map projections than has been commonly required in the past. Considerable savings can frequently be effected by converting back and forth among several projections in an internal machine program to solve specific problems (thus taking advantage of the properties of each projection) before arriving at a projection for display purposes. It is well known that projections having desirable visual properties may require involved computations. The use of Mercator's projection to solve navigation problems graphically provides the classic example.

The possibilities for automation

Let us now return to our initial questions: Do possibilities for automation exist in cartography? If so, where can they be found? It has been demonstrated that possibilities for automation do exist, that machinery is currently available to perform many of the more tedious tasks in cartography, and that certain problems have to be solved, a few of which have been indicated. It seems that some basic tasks, common to all cartography, may in the future be largely automated, and that the volume of maps produced in a given time will be increased while the cost is reduced. The machinery is expensive, but it is becoming more common and less expensive. Many of the cartographic uses of computing machines take only a few minutes of machine time, and in these cases the practice of renting time on a machine will probably continue. In other cases, available machinery purchased for other purposes can be used to advantage in preparing maps. Tabulating equipment in particular is common in this country. Other types of plotting equipment will possibly find their greatest use in large cartographic establishments, initially in compilation and map plotting. Other applications appear likely in meteorologic, geologic, geographic or other research in which maps are used extensively.

References

Locke, A.S. (1955) *Guidance: Principles of Guided Missile Design*, vol. 2, Van Nostrand, Princeton, NJ.

Mackay, J.R. (1951) Some problems and techniques in isopleth mapping. *Economic Geography*, **27**, 1–9.

Further reading

Dobson, J.E. (1983) Automated geography. *The Professional Geographer*, **35** (2), 135–143. [This paper considered how geography would be radically remade by the integration of computer-based techniques and digital data.]

Foresman, T. (1998) *The History of Geographic Information Systems: Perspectives from the Pioneers*, Prentice Hall. [A useful edited volume of essays on the early development of GIS in different contexts.]

Hägerstrand, T. (1967) The computer and the geographer. *Transactions of the Institute of British Geographers*, **42**, 1–19. [An early call by a leading quantitative geographer for greater exploitation of computer processing to tackle challenging empirical research problems of the time.]

Klinkenberg, B. (2007) Geospatial technologies and the geographies of hope and fear. *Annals of the Association of American Geographers*, **97** (2), 350–360. [An insightful review paper that considers how digital mapping and allied GIS tools can be applied in progressive and humanitarian ways.]

Sui, D.Z. (2004) GIS, cartography, and the 'Third Culture': geographic imaginations in the computer age. *The Professional Geographer*, **56** (1), 62–72. [Identifies possible ways forward for GIScience research that involve a recognition of the creative potential of mapping practice and geospatial technology.]

See also

- Chapter 1.5: Analytical Cartography
- Chapter 2.2: A Century of Cartographic Change
- Chapter 2.6: Cartographic Futures on a Digital Earth
- Chapter 2.7: Cartography and Geographic Information Systems
- Chapter 4.10: Citizens as Sensors: The World of Volunteered Geography

Chapter 2.6

Cartographic Futures on a Digital Earth

Michael F. Goodchild

Editors' overview

Goodchild's paper delineates the so-called 'digital transition' in the creation, processing and use of geographic information, and plots some of the implications for the future production of maps that were evident at the end of the 1990s. Importantly, he highlights the contradictory impacts of digitisation on mapping in that technologies can, at one level, open up more opportunities for people to make their own maps, and yet also work to further marginalise professional cartographic practice from the burgeoning field of GIS analysis. The result, according to Goodchild, is that digital mapping technologies deliver many more maps to computer screens, but also increasingly means the loss of cartographic knowledge and skills in effective visual communication.

Originally published in 1999: *Proceedings of the 19th International Cartographic Association Conference*, 14–21 August 1999, Ottawa, Canada.

Introduction

This paper is written with some trepidation, since I am not a cartographer, and would certainly not want to be perceived as trying to prescribe cartography's future. [. . .] It is written from the perspective of someone who cares greatly about the cartographic aspects of what we do, who like many of us grew to love maps at an early age, and who sees cartography as an indispensable part of any future for my own discipline of geography, and the broader enterprise that we variously know as geomatics, geoinformatics or geographic information science.

The paper begins by introducing two broad trends that provide a context of vital importance for cartography: the digital transition, which began some decades ago but seems to dominate more and more of our vision of the future; and what appears to be an increasing interest in society generally in geography, the stuff of maps, and in all things geographic. These two themes come together in a discussion of how the digital transition will affect the production, dissemination and use of maps; the institutions that manage and regulate those activities; and, ultimately, the nature of maps themselves. This leads to the identification of a basic paradox, between the increasing marginalisation of cartography within the larger digital geographic enterprise and the increasing need for good cartographic practice in visual communication, as more and more people are empowered by new technology to make maps. [. . .]

The digital transition

The idea of communicating in code is as old as language itself, requiring only the establishment of standards within a community regarding the code's meaning. An even older code is the alphabet of four bases used to communicate genetic information between parent and offspring; incredible as it may seem, the entire architecture of the human body, and the instruction to a chick to begin pecking after 21 days of incubation, are somehow successfully coded in a permutation of A, C, G and T. But the explosive growth of digital communication that has occurred in the past thirty years relies on several other factors besides a universal code of zeroes and ones. The code is readily processed at great speed by digital computers; it can be stored virtually indestructibly (although practice often falls well short of theory); modem standards include automatic error

The Map Reader: Theories of Mapping Practice and Cartographic Representation, First Edition. Edited by Martin Dodge, Rob Kitchin and Chris Perkins.

checking; and it can be transmitted at close to the speed of light. Today, virtually all human communication-at-a-distance passes through a digital coding and decoding at some point. Telephones, fax, written text, photographs, music, all have associated and generally accepted standards of coding in digital form. Only the mail remains as a predominantly analogue method of communication, although most sorting of the packages themselves is now digital. [...]

Digital technology is already pervasive, but its impacts are only just beginning to be felt in the ways humans organise and conduct their activities. Take, for example, the case of geologic mapping. Figure 2.6.1 shows the stages of mapping, from the work of the field geologist through to eventual use, storage in libraries and archiving. Each person or group in the chain communicates with the next person or group: the field geologist gives notes and sketches to the cartographer, while the printer sends paper maps to the distributor and on to the library and user.

The first infection by the digital virus occurred among cartographers, who were persuaded as early as the late 1960s that the time and cost of preparing and editing maps could be greatly reduced by adopting digital technology, initially by fixing simple encoders to the arms of plotters to capture locations, and later by replacing drafting tables by digitisers. Today, it is hard to find a single drafting pen in many map production operations and cartography classrooms. Then users began to demand digital product, because of the obvious potential of digital analysis and the simultaneous growth of geographic information systems (GIS) as analysis engines for map data. But this second round of infection had a more significant impact, since it created a new path that bypassed the traditional printing and dissemination arrangements. More recently, the World Wide Web strain of the digital virus has further infected the distribution function, as digital spatial data libraries (such as the Alexandria Digital Library) and spatial data clearinghouses (such as the US National Geospatial Data Clearinghouse) provided an alternative to the traditional library as a source of archived information.

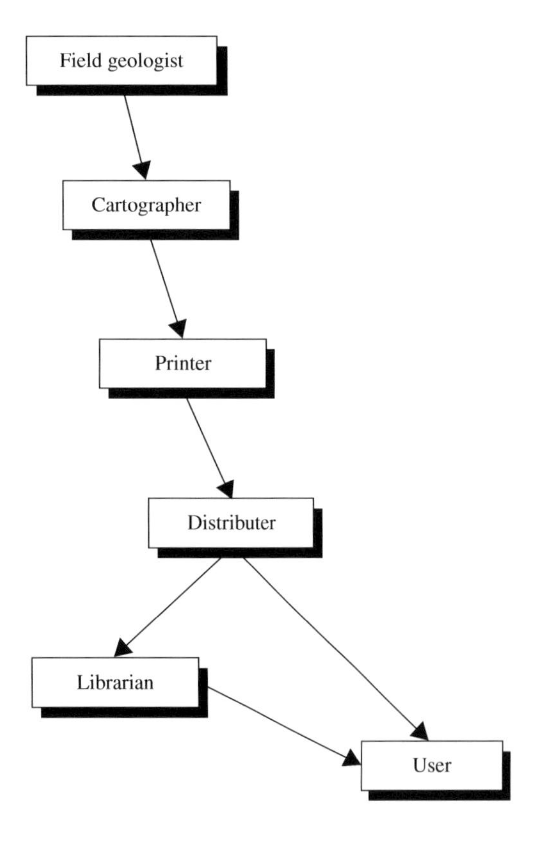

Figure 2.6.1 Schematic of the geological mapping process as communication.

Digital technology has yet to infect the work of the field geologist to a significant degree, although it is common today to find laptops at field sites. The sketches and field notes that a geologist passes to a cartographer are still largely in analogue form, and suitable software for capturing and processing such information is still primitive. But the technology already exists to allow the field scientist to download images of a project area from the Web, to annotate it digitally with sketches and notes, and to link digital photographs to field locations. Information technology in the field promises to improve greatly one of the most severe impediments to the various stages of communication shown in Figure 2.6.1, and one that underlies much of the subsequent discussion in this paper: the inability of the field geologist to communicate more than a small fraction of what he or she discovers in the field to the eventual end user, because of the highly restricted nature of the traditional communication channels. In the longer term, extensive application of information technology in the field promises to open up novel channels of communication. For example, it will be possible for field scientists to share information remotely as soon as it is collected or interpreted; to communicate directly with end users; and in the long run to bypass entirely the traditional stages of cartographic production.

In recent years massive investments have been made in digital libraries, metadata (data about data, the digital equivalent of the catalogue record), new search mechanisms and other developments aimed at making it possible to find geographic data in the massive, distributed archives of electronic networks. Moreover, it seems clear that investments to date are tiny compared to what is to come as the information economy heats up. Surfing the Web for data is providing an increasingly effective alternative to visiting one's local map librarian.

The digital transition is affecting the geography of map production as well, as the traditional arrangements break down or are modified by new technology and changing economics. Much cartographic software is now cheap and affordable, allowing anyone with a personal computer and access to the Web to make maps. Farmers with access to the technology of precision agriculture can build maps of their fields at much higher resolution than traditional soil maps, and can capture and compile detailed spatial information on inputs and yields using devices attached to harvesters and tractors. Local governments can rent vans equipped with GPS units, drive along every street and produce street maps at higher accuracy and much lower cost than the traditional production arrangements of central governments. In short, changing technology and economics are moving map production from a system of unified central production to a local patchwork, and the old radial system of dissemination is being replaced with a complex network.

In the early stages of the digital transition much use was made of the new technology to perform operations more quickly, at lower cost. But as the transition advances it is the operations themselves that come into question, along with the organisational structures and arrangements that evolved around them. The survivors in this world will be those who can think beyond past practices, and adapt quickly to new opportunities.

The stuff of maps

[...]

In the past few years many new services based on geographic information have appeared on the Web. Microsoft's Terraserver began as an effort to build and demonstrate a capability to serve information at a massive scale, with geographic data chosen as the content because it was cheap and comparatively unencumbered by issues of intellectual property. But Terraserver has been very successful as a pioneering effort to serve imagery to a vast population of users, many of whom had never had access to easy-to-use Earth imagery before. Microsoft's Home Advisor provides GIS-like services in the form of home listings and social data about surrounding neighbourhoods. MapQuest is one of many sites offering maps, georeferencing and optimal routing services.

One of the greatest impediments to effective use of geographic data has been the inability to integrate information about a place. Our traditional arrangements for production of geographic information emphasised horizontal uniformity; one government programme produced all topographic maps, another all soil maps. These arrangements have been largely inherited by the digital world, so that one goes to one site to obtain an image (e.g. the MIT server of digital orthophoto quadrangles), and another site to obtain a topographic map, and still a third to obtain a soil map. It has been virtually impossible to approach a library, or the Web, and ask for all information about one place, and equally difficult to integrate such data once obtained because of variations in formatting and terminology, and positional inaccuracies.

The US Geological Survey provides an interesting case in point. The traditional organisation of the Survey into four divisions (with responsibilities for mapping, geology, water resources and biological resources) has also determined the face it presents to the world as a source of information. Thus a user approaching the USGS Web site finds it much easier to obtain information about one theme for places, than information about many themes for the same place. The Survey's Gateway to the Earth project proposes to replace this external view by a more integrated one that will allow the user to find everything the Survey knows about a place. This idea of place-based search is already implemented to a degree in the US Environmental Protection Agency's site (see the ZIP code search feature of www.epa.gov).

These place-based search mechanisms resemble what I have called a *geolibrary* (Goodchild 1998), or a library that one can approach with the query 'what do you have about there?' Place-based search has been very difficult in the traditional library, for numerous reasons, but promises to be comparatively easy in a digital world, and provides yet another instance of how the digital transition is changing our arrangements for producing, disseminating and using geographic information.

Geographic information and maps

But maps are only one form of expression of geographic information. Very broadly, one could define geographic information as information about well-defined locations on the Earth's surface; in other words, information associated with a geographic *footprint*. But that definition fits guidebooks, photographs of landscapes, even pieces of music with geographic associations. Possession of a footprint is the minimal requirement for place-based retrieval. Maps, on the other hand, are:

- *Visual* forms of geographic information, rather than textual, verbal, acoustic, tactile or olfactory; though tactile maps have been developed to address the needs of the visually impaired.
- *Flat*, requiring the Earth's surface to be expressed in a distorted manner.
- *Exhaustive*, expressing a uniform level of knowledge about every part of the area covered by the map.
- *Uniform in level of detail*, although no flat map can ever have a perfectly uniform scale.

- *Static*, since a map once drafted and printed cannot be substantially changed.

- *Generic*. The strong economies of scale in map production ensure that maps will be produced only when demand reaches a sufficient level. Thus, maps tend to be produced to satisfy many uses and users simultaneously. They present a shared perspective that cannot be user-centred, and thus is almost always vertical.

- *Precise*, few methods being available for display of uncertainty.

- *Slow*, due to the lengthy time required for all of the different stages of production, as shown for example in Figure 2.6.1.

None of these constraints is inherent to geographic information, however, especially in a digital world that changes the economics of map production, provides the tools for interacting directly with information and allows information to be compiled and delivered at electronic speed. In the digital world all information is expressed in bits, whatever its origins; it is stored in the same places and transmitted using the same techniques. It matters not at all to the Internet whether a 'bag of bits' represents text, an image or a map. Thus we hear more and more about multimedia systems that handle data irrespective of the media on which they were traditionally stored in the analogue world. The term *multivalent document* (Phelps and Wilensky 1996) refers to the ability to link images, text, sketches, notes and so on, that refer to the same subject, and to handle them as if they were a single unit. In this multimedia world, old arrangements based on the problems of handling different media will be challenged, and may be abandoned completely.

The paradox of contemporary cartography

It follows from the previous discussion that the map[1] is only one form of expression of geographic information in a digital world, competing with other forms on a playing field that is increasingly level. For example, flat views of the world and the distortions they embody must compete with orthographic views, such as those provided by the current version of Microsoft's Encarta atlas. To the average citizen, working with and manipulating a digital globe may be far more straightforward and comprehensible than working with a digital Mercator projection; and children may be able to understand a globe more easily and at an earlier age than its projected version.

It is widely believed that geographic information systems are being absorbed into the information technology mainstream: that in the near future such standard applications as spreadsheets, e-mail and word processors will include support for geographic information. [. . .] GIS has made it possible for anyone who can afford a basic personal computer and cheap software to display geographic information in the form of a map; by offering these functions through such common applications as Microsoft's Excel, software developers are now putting these tools into the hands of everyone, irrespective of their credentials and sensitivity to cartographic principles. In a world in which everyone can make a map, who needs the cartographer?

Only a few institutions have had the wisdom to elevate cartography to the status of a department; instead, cartographers have historically found themselves rubbing shoulders with geographers or surveyors. Today, rapid growth in interest in geographic information systems and related fields has led to an emergence of new collaborations, under a variety of names: geomatics, geoinformatics or geographic information science. Cartography finds itself a small part of a larger academic enterprise, and at the same time increasingly marginalised by the rapid spread of mapmaking tools.

Much of the attraction of GIS lies in its visual focus: colourful maps appear on the screen, to be manipulated and explored by the user at the touch of a mouse. GIS communicates primarily through the visual channel, especially when used in efforts to influence public opinion and policy. But such communication is never simple and straightforward. Only one colour can be assigned to each location in the visual field, which is then mapped through the optics of the visual system to the human retina. While a database makes the link between an object and its name explicit, the eye–brain system does this through complex rules of visual association that must be understood by the map designer and made the basis of map design. Map designers must devise complex and sophisticated rules to make it possible for a map to communicate more than one attribute of an object; and yet none of these rules are needed when information is communicated in other forms, such as through tables.

Yet despite this complexity, or perhaps because of it, a large number of the maps produced using today's software are simply awful. As David Rhind has been known to remark, GIS technology lets us produce rubbish faster, more cheaply and in greater volume than ever before. Paradoxically, then, the world that is marginalising cartography is also the world that needs cartographic principles and skills more than ever. It is also a world of

[1] As used here, the term implies a paper product or its direct digital equivalent produced by scanning or digitising; both are therefore subject to the constraints identified earlier.

unprecedented opportunities for cartography as the digital transition removes many of the inherent barriers and impediments of the traditional map, and makes communication of geographic information between people richer and more efficient than was previously possible. To restate the previous list, communication of geographic information need no longer restrict itself to the visual field; flatten the Earth; cover every part of an arbitrarily defined area that happens to include the area of interest in uniform detail; maintain a uniform level of detail and a vertical perspective irrespective of the content's focus; remain static irrespective of the acquisition of new information or change in the landscape; serve the interests of a large number of users to overcome high fixed costs; fail to reveal anything of its own inherent uncertainties; or take substantially longer than any other form of communication.

[...]

Conclusion

A technology that began as a way of making large numbers of numerical calculations possible has turned into something that, if the pundits are to be believed, has the potential to reorganise much of what we do. Cartographers first felt that impact in the early 1970s when computers began to be used to reduce the production costs of paper maps. Since then, digital technology has affected almost all aspects of mapping, created an entirely new application in geographic information systems, facilitated many other new geographic information technologies, and spawned a new partnership of the mapping sciences known variously as geomatics or geographic information science.

For the traditional cartographer, what began as a useful aid has turned into a monster, empowering virtually everyone with access to the tools that used to be the exclusive preserve of specialists. In a world in which everyone can make a map, who needs cartography? Or as Judy Olson titled her Presidential Session at the Association of American Geographers annual meetings in Charlotte in 1996, 'GIS has killed Cartography'. But, paradoxically, the need for good cartographic design is now stronger than ever.

Faced with this situation, it seems to me that a suitable strategy for visualising the future while touching the past would have the following components:

(1) Establishment of clear principles underlying the communication of geographic information. Such principles should be independent of media and technology, and thus robust against a major technological transition such as the one we are currently experiencing. These would not be the principles of paper maps alone, or of digital displays alone, or of visual communication alone, but of the communication of geographic information from one person to another. Different communication channels and media would be represented through different parameters, constraints and rules within this general framework.

(2) Anticipation of the full impact of the digital transition. It is hard to see the wood for the trees in times of fundamental change, and to think beyond the immediate impact of computerisation on a single task. But in the long term the world will reorganise itself according to principles such as those suggested in (1) that are truly fundamental. These include function, economics and the basic forces that drive society, including the distribution of power and influence.

(3) Identification of a moonshot, an articulated vision of what communication of geographic information might mean at some point in the future. Without such a vision it is difficult to see how a prioritised agenda for cartographic research can emerge. Curiosity will always be around to drive research, as will immediate economic gain, but it is much more difficult to identify a clear sense of common purpose that can both drive research and appeal to potential sources of funding.

References

Goodchild, M.F. (1998) The geolibrary, in *Innovations in GIS 5* (ed. S. Carver), Taylor and Francis, London, pp. 59–68.

Phelps, T.A. and Wilensky, R. (1996) Multivalent documents: inducing structure and behaviors in online digital documents. *Proceedings of IEEE International Conference on System Sciences*, January 1996, University of Hawaii, Maui, HI.

Further reading

Pickles, J. (1999) Cartography, digital transitions, and questions of history. *Proceedings of the 19th International Cartographic Association Conference*, 14–21 August 1999, Ottawa, Canada. [A critical reaction to the technological arguments advanced by Goodchild, arguing instead that it is social factors, and not the inevitable deployment of technology, that influences change.]

Silver, M. and Balmori, D. (2003) *Mapping in the Age of Digital Media*, Wiley-Academy, London. [Reflects on the encounter between technology, mapping and the arts.]

Sui, D.Z. (2004) GIS, cartography, and the 'third culture': geographic imaginations in the computer age. *Professional Geographer*, **56**, 62–2. [Identifies possible ways forward for GIScience research that involve a recognition of the creative potential of mapping practice and geospatial technologies.]

Wood, D. (2003) Cartography is dead (thank God!). *Cartographic Perspectives*, **45**, 4–7. [A trenchant critique of the cartographic profession arguing its rationale is tied to modernism and the nation state, instead of some set of bureaucratised skills.]

See also

- Chapter 1.10: Cartography Without 'Progress': Reinterpreting the Nature and Historical Development of Mapmaking
- Chapter 2.2: A Century of Cartographic Change
- Chapter 2.3: Manufacturing Metaphors: Public Cartography, the Market, and Democracy
- Chapter 2.5: Automation and Cartography
- Chapter 2.7: Cartography and Geographic Information Systems
- Chapter 2.12: Imaging the World: The State of Online Mapping
- Chapter 4.10: Citizens as Sensors: The World of Volunteered Geography

Chapter 2.7

Cartography and Geographic Information Systems

Phillip C. Muehrcke

Editors' overview

Muehrcke's paper considers the mutually beneficial relationship between the practices of cartography and the rapidly growing field of GIS. Writing in 1990, he makes the case that the information systems thinking underlying GIS has the potential to lead to a major technological reinvigoration of cartography, likely to influence what gets mapped, how these maps appear and how they are used. While GIS is built upon databases and mathematical algorithms, Muehrcke discusses the how the map remains central to it, providing a powerful framework for analysis and the primary mode of user interaction.

Originally published in 1990: *Cartography and Geographic Information Systems*, **17** (1), 7–15.

Centuries ago, people devised maps to assist them in environmental thought, communication and mobility. Since that time, the maps of every age have been closely linked to the current state of environmental thinking and the nature of available representational tools. Since environmental thought and representation have both undergone evolutionary as well as revolutionary changes through time, the history of cartography is likewise characterised by often dramatic changes. Indeed, the history of cartography could be described as a see-saw adjustment between shifting environmental thought on one hand and developments in technology on the other. A change in one leads quickly to a change in the other, and enhanced cartographic activity soon follows. The synergy created in this thought–technology spiral produces an atmosphere receptive to creative, productive changes.

The most recent shift in thought focuses on the need to deal with our environment in an integrated, holistic way. This ecological, systems mentality has put a high premium on complex information handling (Pagels 1988). The electronic technology of computers and telecommunications makes the modern information age possible (Brand 1987). The mapping implications of information systems thinking and electronic technology are profound. We can see the emergence of automated geographic information systems (GIS) as dominating cartographic development in the foreseeable future. Conversely, it is apparent cartography also exerts considerable influence on development and use of GIS technology.

From map to mapping

In the decades following World War II, cartography emerged as a university discipline in the United States (Kimerling 1989). Robinson's (1952) monograph, *The Look of Maps,* set the dominant research agenda. In most simplistic terms, the aim of this research was to determine how map users responded to elements of map design. Research findings were to be used to redesign maps so their information would be as assessable as possible to the broadest user community. Although this was a worthy goal it did have a negative side, which was to encourage people to shift focus from the map as geographic thinking to the map as geographic illustration (Muehrcke 1981). Under these conditions, the map became an object.

One effect of this geographic illustration mentality was to reduce emphasis on the value of geographic cartography or,

The Map Reader: Theories of Mapping Practice and Cartographic Representation, First Edition. Edited by Martin Dodge, Rob Kitchin and Chris Perkins.

as it was later called, analytical cartography (Tobler 1976), where the process of mapping was recognised often to be more insightful than the map itself. A person attempting to structure the environment in map-like terms is forced to make judgments that can greatly enhance understanding of the material being mapped.

Analytical cartography emerged from experiences with digital mapping, including the recognition that extensive manipulation of information in geographic databases prior to actual mapping could produce useful dividends. Remote sensing image processing is based on this notion. The analytical component of geographic information systems further expands on this concept. Analysis with GIS technology is the latest embodiment of geographic cartography. Not all procedures supported in GIS analysis require a knowledge of map geometry, however. A variety of facility site and route selection problems can be handled with map topology rather than map geometry as input. In any case, GIS analysis is performed on a cartographic database (a digital record of mapped features) rather than on a map itself.

Computer-assisted mapping to GIS

Maps today are routinely produced through digital, electronic means. So where does GIS technology fit into the picture? To answer this question, we can say the same technology used to develop a geographic information system is needed to facilitate computer-assisted mapping. Maps are one of several possible GIS products. GIS technology goes beyond automated mapping, however, especially regarding the way data are structured in a database (Chrisman *et al.* 1989). Computer applications in mapping began with geometry-only operations supported by entity-oriented geocoding. It was sufficient to use the computer to draw map symbols for geographic features coded into the database. Early proof-of-concept maps were sometimes mere digital recreations of existing analogue maps. Cartographers soon realised that adding topological relations to the coding of geometric primitives (points, lines and areas) made it possible to use computer software to carry out a variety of data coding and manipulation chores, including accuracy checks. [. . .]

A desire to integrate and move freely about and between databases led to the next advance in data structures. Digital records have to be cross-referenced to each other if cartographers are to have ready access to various geographic feature files for different time frames and geographic regions. Spatial cross-referencing of data files vertically (hierarchical structures) or horizontally (network structures) produces relational databases. Databases optimised for flexible, integrated applications would exhibit both types of relational structures. [. . .]

It is at this point that we can see the distinction between automated mapping and geographic information systems. The ideal mapping system would make it possible to create any map form with ease and speed, using any mapping technique, for any part of the world, for any time frame, highlighting any combination of geographic features. An automated mapping system with these capabilities uses information age concepts and technology. Such a system requires the following: a means for entering new and updating existing geocoded data; a fully integrated global data base that provides direct access to all known non-geographic as well as spatial data sources; a wide array of procedures for merging and spatially manipulating database information; a full statistical and mathematical data analysis capability; a full graphical (including cartographical) design, analysis and display capability; and a 'friendly' system interface with prompting tutorials to make use of the system physically convenient and intellectually stimulating. A geographic information system is the current approximation to this automated mapping environment.

Information systems are multifaceted by nature. They are characterised by flexibility at the information input, storage, manipulation and output stages. GIS technology is no exception. It can utilise cartographic data and concepts at the input, manipulation and output stages, but as an information system it can also do much more. Cartography may be necessary but is not sufficient for a geographic information system. Maps will continue to provide dramatic graphic expressions of what GIS technology can offer, however. They will be much sought after by those taking advantage of this information age technology. To our good fortune, mapping is getting help from growing recognition of the value of graphics in general.

[. . .]

One map of many

An important consequence of electronic visualisation technology is that it draws attention to the rich variety of ways information can be expressed graphically. The more visualisation options we have, the better we understand the nature of individual visual expressions. After centuries of experience using maps, for example, it is surprising how few people grasp what it means to map. People commonly speak of 'the' map when they mean 'a' map. Even environmental professionals are prone to use true, perfect, real, or similar words in describing a map. These examples seem to be more than careless language usage. They appear to reflect a feeling that maps show the environment as accurately as technically possible given the available tools. The danger in this attitude is it encourages a person to attribute more status to a particular map than is warranted. A healthier attitude is to consider a given map as but one of several ways of looking at the situation. Usually an infinite variety of maps can be made using the same environmental data, and the impressions they give can differ markedly. In most, if not all, situations a more thorough understanding will be gotten by using several maps along with supporting textual and tabular materials. GIS technology should help make this point clear because it introduces map users often having no formal background in cartography to the mapping process. Once these people are faced with making the various decisions required to specify which map, they are sure to better understand the nature of cartography (Robinson and Petchenik 1976; Keates 1982).

Metacartography

Once we free ourselves of the single-map notion, we are prepared to take advantage of what cartographic abstraction really offers (Bunge 1962). To appreciate this, it may help to consider the art world. In the first decades of the twentieth century, Picasso and Braque introduced us to the concept of cubism through their art. A critical idea underlying cubism is the ability to transcend constraints imposed by a single space/time vantage point. This means the ability not only to get outside our head with its normal ground level perspective, but the ability to get outside the space/time domain of our life as well. This amounts to a tearing apart and fragmentation of our sensory world. Cubism adds the freedom to re-assemble individual images into a composite that lets us see the environment as it might appear if viewed from several viewpoints at once.

This synthetic juxtaposition of environmental images in time or space provides a fuller visualisation than would be possible by confronting reality directly with our unaided senses. We thereby enhance our chances of capturing the essence of what is being depicted. What we gain through this multiplicity of space/time vantage points can be significant.

From a cubist perspective, cartography takes on an improvisational 'What if... ?' character. We are free to make various cartographic assumptions, and then see what the environment looks like. We are seeking cartographic effect, not geographic truth. If the effect is uninteresting, the map can be discarded (or erased from the screen), since it has little inherent value. We can get at pressing environmental issues from different angles and dimensions merely by trying different combinations of cartographic assumptions.

The important but also demanding multiperspective mapping trend has been enhanced by GIS technology. Maps of this genre require prolonged scrutiny and contemplation, as opposed to simpler forms that give up their information 'at a glance'. Multiperspective maps require more effort to fathom because they say as much about how we grasp the environment as they do about the environment itself. These maps tell us something about our perceptions and conceptions, and about the way we represent environmental knowledge with our language vehicles (words, numbers, graphics). Multiperspective mapping gets at the heart of environmental thought and communication.

It can come as a shock when one first realises that the real power of maps lies not in their ability to give us a realistic picture of geographic reality, but rather in their power to free us from the constraints imposed by this physical reality. The transformative nature of abstraction has many benefits (Rucker 1987). It lets us visualise the world in insightful ways not otherwise possible because of the constraints of our physical concreteness. We need to keep this legacy of cubism in mind when we use powerful GIS tools to create artificial realities of a cartographic kind.

Mappability

Optimised GIS technology can make a lot of things possible in mapping that could not be done with previous technologies. But no matter how sophisticated our geographic information system, our visualisation tools, or our multiperspective representations, the fact remains that it is only our method, not the environment, that has changed. And, by their nature, some environmental phenomena are more mappable than others.

The relation between the nature of environmental phenomena and the nature of their cartographic representation is critical to the success of mapping. The correspondence between map and reality can vary significantly from one feature to another. The traditional symbol of geometric/line mapping is sharply defined. When such a symbol is used to represent a discrete feature, such as a road, the depiction appears realistic. This is also the case when using geometric/line symbols to depict a variety of other discrete point, linear and areal features.

Features that are inherently transitional or fuzzy toward their edge are considerably less mappable using sharply defined point, line and areal symbols (Goodchild 1988). Cartographers working by hand or with photomechanical vignetting techniques are able to create shading gradients between symbols, but the great labour involved prevents wide adoption of the practice. In contrast, electronic dithering and vignetting is simple, and should grow dramatically in popularity as people explore its potential effects. A GIS is not needed for this task, but any sophisticated GIS should incorporate the capability in its graphical methods base.

The class of discontinuous/dispersed environmental features presents an even greater challenge in mapping. Consider, for instance, the journalistic maps you have seen of the wars in Lebanon or Vietnam. In both cases, there is no broad brush division of territory, no convenient we/they dividing line. Instead, there exists a 'pox-like' distribution that is too finely textured to be represented in meaningful geographic detail at media mapping scales. When journalistic mapping is applied to this type of geographic reality, the cartographic product is usually so abstract that most geographic meaning is lost in the representation.

The reason for dwelling on the subject of mappability is that GIS technology simplifies and thereby encourages integration and manipulation of all variety of environmental features with little, if any, way of dealing with the mappability of each. Therefore, the cartographic quality of some aspects of the integrated representations will be superior to others. Problems may arise if unequals are weighted or otherwise treated equally, especially in the analysis phase of GIS activity. The quality of the overall result will be determined by the component with the least mappability. We can control the readability of a map, but not the nature of the reality mapped. Making maps simple does not make the geography simple.

[...]

Interactive maps

Great strides have been made in map effectiveness since the probable ice-age origin of the mapping idea (Davis 1986; White 1989). But for these many centuries maps have remained passive and dumb. Significant effort is required to retrieve information from a printed map. Procedures range from straightforward cartometric measures to more complex indices of spatial order and association. Cartometric analysis has received the most attention (Maling 1989). Although a number of labour-saving tools (distance measuring wheels, area measuring planimeters, height measuring parallax bars etc.) have been invented, cartometric analysis remains a laborious chore.

Pattern comparison procedures that try to capture the internal order (auto-correlation) of patterns and spatial association (cross-correlation) between patterns are at least equally demanding of the map user's energy. Pattern comparison normally is done mathematically, which usually means data first must be digitised from a map or other graphic product and then processed with a computer program.

One problem with the traditional approach to map analysis is the buffer role maps play between environment and map user. Geographic distortions occur because maps depict the environment abstractly. Furthermore, map analysis and geographic analysis are not the same thing. Another problem with map analysis is that the only data available for scrutiny are those reflected in the map symbology. The full scope and depth of geographic information available to the cartographer does not survive the mapping process. A map user has no direct way to gain access to original data filtered out by the cartographer.

Information technology finally gets us beyond the limitations of using passive, dumb maps. Since all the information needed to display a map on an electronic screen is held digitally in computer-readable files, the map user potentially has access to these data as well as to their graphic depiction. Given a proper interface design, map users can direct questions about mapped features to the digital data used to make a map. Additional data, linked only through geocoding procedures to data actually used to make the map, are also accessible. This gives rise to an exciting new map form, the interactive map. Given appropriate software, a map user will be able to identify a lake on the map and ask the database for such things as its surface area, perimeter, greatest depth or volume. All other data that might be associated with the lake would also be retrievable from an intelligent database.

An interactive capability also provides the mapmaker with a powerful design tool. By moving back and forth between initial design concepts and the database, cartographers will be able to refine their abstraction procedures as never before. In particularly difficult design situations, the ability to iterate easily through several design options may greatly enhance the final mapping effect. Electronic integration of graphics and data also makes it possible to consider different combinations of environmental variables. A robust GIS would be able to suggest additional non-conventional ways of representing the data as well.

The term intelligent graphics is commonly used in discussing electronic map displays linked interactively with multifaceted databases. This of course is the promise of GIS technology for cartography. 'Supermaps' of this type have the potential to vastly expand the topics, users and applications of mapping. The design of these interactive maps will have to reflect this role as graphic interface to integrated database.

Map as interface

Interactive maps do not have to bear the data storage burden of conventional printed maps. Maps serving as links to databases, not as ends in themselves, require design considerations that enhance this interface function. If we are to ask questions of the database through the help of a map, the cartographic interface should be as convenient to use and as intellectually stimulating as possible. It will pay to remember that maps play the valuable role of a trigger device in environmental thought. Much of what we think we learn from printed maps really comes from our mind being stimulated by the map, and not from the map itself (Dreyfus and Dreyfus 1986). We want map designs that will encourage us to ask questions for which answers can be found in the database.

We are likely to encounter some problems in automating a mapping process based on traditional design goals. These goals may not be as critical as previously thought when emphasis shifts to the map as database interface. Maybe we can rethink the issue of place names on maps, for example, and greatly reduce map clutter in the process. The same might be true of line generalisation, class intervals and other traditional mapping issues. We must be willing to challenge all design assumptions associated with printed maps if we are to optimise the design of the new interactive map form.

[. . .]

GIS – ideal and practice

To this point, much has been said about the relation between cartography and GIS technology. For purposes of discussion, an ideal geographic information system was often assumed. At present, there is no operational geographic information system with the properties of the ideal system outlined previously. Current geographic information systems tend to be either object-based or cell-based. Only some provide a means for integrating geometry/line (vector) and image (raster) data. We will have to go well beyond these GIS implementations if we are to achieve anything close to the ideal.

Currently, no single geoprocessing technology can provide the scope and depth of information needed to make complex environmental decisions. Surveying, remote sensing, image processing, photogrammetry, engineering and construction, surface modelling, computer-aided design and drafting (CADD), automated mapping, spatial analysis, statistical mapping, environmental process modelling (Hopkins 1977) and GIS technologies must all be integrated when tackling an environmental issue. Currently this can only be done with considerable labour and frustration. To achieve the desired goal, our current geographic information systems will have to be extended either to incorporate these geoprocessing technologies directly, or to make provision for ready access to them through the GIS interface. In the meantime, experts in geoprocessing technology integration will be very much in demand.

[. . .]

Conclusion

When the day of the technocrats has passed, we will still need to assess the human side of GIS technology. What good can GIS tools do for our fellow human beings? How can they be used to enhance human thought, communication and mobility? New map topics, new map forms, new ways of making maps and new ways of using maps can all be anticipated. These and related topics need to be explored as we consider the impact of GIS technology on cartography and explore how cartography can enhance the effectiveness of geographic information systems.

There will surely be moments when the pace and magnitude of change caused by information age advances are so overwhelming that the independent identity of cartography seems threatened. When this happens, we will do well to contemplate the secret of cartography's success through history. Those who predict the demise of cartography presuppose its rigidity. They could not be more wrong; flexibility and adaptability are the field's greatest sources of strength. The genius of cartography is that every few decades it is invigorated by adapting to technological advances or shifts in the way people view the environment. Each time this happens, the principles of the discipline can be rewritten. Currently, information systems supported by electronic technology are forcing a creative transformation of cartography as we know it. Fortunately, this promises life, not death.

In spite of this promise, there is cause to be wary. There is much to be lost by overselling GIS technology. Admittedly, the ability to generate graphic iterations easily encourages the cartographer to explore alternative design concepts before making final design decisions. Furthermore, a cartographer is likely to stumble upon new ideas, variations, colour schemes and so forth while refining initial design concepts and perfecting the final artwork. But the role of GIS tools in the conceptual process is clearly limited. They may have wonderful, almost magical powers to offer, but in the foreseeable future at least they will not replace the human element. Ideas come from people. GIS technology holds less promise as a creative aid than it does as an idea refinement and production tool. Until geographic information systems can read our minds, it will remain so.

References

Brand, S. (1987) *The Media Lab: Inventing the Future at MIT*, Viking, New York.

Bunge, W. (1962) *Theoretical Geography*, C.W.K Gleerup, Lund.

Chrisman, N.R., Cowen, D.J., Fisher, P.F. *et al.* (1989) Geographic information systems, in *Geography in America* (eds G.L. Gaile and C.J. Willmott), Merrill Publishing Co., Columbus, OH.

Davis, W. (1986) The origins of image making. *Current Anthropology*, **27** (3), 193–216.

Dreyfus, H.L. and Dreyfus, S.E. (1986) *Mind over Machine: The Power of Human Intuition and Expertise in the Era of the Computer*, The Free Press, New York.

Goodchild, M.F. (1988) Stepping over the line: technological constraints and the new cartography. *The American Cartographer*, **15** (3), 311–319.

Hopkins, L.D. (1977) Methods of generating land suitability maps: a comparative evaluation. *Journal of American Institute of Planners*, **43** (4), 386–398.

Keates, J.S. (1982) *Understanding Maps*, John Wiley & Sons, Inc., New York.

Kimerling, A.J. (1989) Cartography, in *Geography in America* (eds G.L. Gaile and C.J. Willmott), Merrill Publishing Co., Columbus, OH.

Maling, D.H. (1989) *Measurements from Maps: Principles and Methods of Cartometry*, Pergamon Press, Oxford.

Muehrcke, P.C. (1981) Whatever happened to geographic cartography? *The Professional Geographer*, **33** (4), 397–405.

Pagels, H.R. (1988) *The Dreams of Reason: The Computer and the Rise of the Sciences of Complexity*, Simon and Schuster, New York.

Robinson, A.H. (1952) *The Look of Maps*, University of Wisconsin Press, Madison, WI.

Robinson, A.H. and Petchenik, B.B. (1976) *The Nature of Maps: Essays Toward Understanding Maps and Mapping*, University of Chicago Press, Chicago.

Rucker, R. (1987) *Mind Tools: The Five Levels of Mathematical Reality*, Houghton Mifflin Company, Boston.

Tobler, W.R. (1976) Analytical cartography. *The American Cartographer*, **3**, 21–31.

White, R. (1989) Visual thinking in the Ice Age. *Scientific American*, **261** (1), 92–99.

Further reading

Fisher, P.F. (1998) Is GIS hidebound by the legacy of cartography. *The Cartographic Journal*, **35**, 5–9. [Argues the relation of GIS and cartography has to change but that map design skills are still essential in GIScience.]

Krygier, J. and Wood, D. (2005) *Making Maps: A Visual Guide to Map Design for GIS*, Guilford, New York. [An effective design guide for producing better maps using GIS tools and techniques.]

Pickles, J. (1995) *Ground Truth: The Social Implications of Geographic Information Systems*, Guilford, New York. [An edited volume of different critiques of cartographic rationality implied by many GIS practitioners, along with much relevant discussion on the politics of digital mapping.]

Schuurman, N. (2004) *GIS: A Short Introduction*, Blackwell Publishing, Oxford. [The best focused introduction to the principles underpinning contemporary GIS.]

Wright, D., Goodchild, M. and Proctor, J. (1997) Demystifying the persistent ambiguity of GIS as 'tool' versus 'science'. *Annals of the American Association of Geographers*, **87**, 346–363. [This paper explores the contested role of GIS and seeks to highlight possible ways forward.]

See also

- Chapter 1.11: Exploratory Cartographic Visualisation: Advancing the Agenda
- Chapter 2.5: Automation and Cartography
- Chapter 2.6: Cartographic Futures on a Digital Earth
- Chapter 2.11: Extending the Map Metaphor Using Web Delivered Multimedia
- Chapter 2.12: Imaging the World: The State of Online Mapping
- Chapter 3.3: Cartography as a Visual Technique
- Chapter 5.9: Affecting Geospatial Technologies: Toward a Feminist Politics of Emotion

Chapter 2.8

Remote Sensing of Urban/Suburban Infrastructure and Socio-Economic Attributes

John R. Jensen and Dave C. Cowen

Editors' overview

Jensen and Cowen's paper discusses the role of remote sensing technologies for mapping different aspects of the built environment. Written at the end of the 1990s, it details how developments in satellite imaging systems could augment and substitute for conventional on-the-ground surveys and urban cartography. Its focuses on the importance of considering both the spatial scale and temporal resolution of different imaging technologies and what aspects of urban environments they can effectively captured. It is clear from their discussion, and subsequent developments, that remote sensing technologies have become a vital component in cartographic production and are routinely used in a range of mapping applications.

Originally published in 1999: *Photogrammetric Engineering & Remote Sensing*, **65** (5), 611–622.

Introduction

Urban landscapes are composed of diverse materials (concrete, asphalt, metal, plastic, glass, shingles, water, grass, shrubs, trees and soil) arranged by humans in complex ways to build housing, transportation systems, utilities, commercial buildings and recreational landscapes (Welch 1982; Swerdlow 1998). The goal of this construction is usually to improve the quality-of-life. A significant number of professional businessmen and women and public organisations require up-to-date information about the city and suburban infrastructure. For example, detailed urban information is required by:

- city, county and regional planning agencies and councils of governments that legislate zoning regulations to improve the quality-of-life;
- city and state Departments of Commerce to stimulate development;
- Tax Assessor offices that maintain legal geographic descriptions of every parcel of land, assess its value and levy a tax mileage rate;
- Departments of Transportation that maintain existing facilities, build new facilities and prepare for future transportation demand;
- private utility companies (water, sewer, gas, electricity, telephone, cable) that attempt to predict where new demand will occur and plan for the most efficient and cost effective method of delivering services;
- Public Service Commissions that insure that utility services are available economically to the public;
- Departments of Parks, Recreation and Tourism who improve recreation facilities and promote tourism;

The Map Reader: Theories of Mapping Practice and Cartographic Representation, First Edition. Edited by Martin Dodge, Rob Kitchin and Chris Perkins.

- Departments of Emergency Management and Preparedness who plan for and allocate resources in the event of a disaster;
- private real estate companies attempting to find the ideal location for industrial, commercial and residential development; and
- residential, commercial and industrial developers.

The urban/suburban land these professionals manage or develop is of significant monetary value. Therefore, it is not surprising that city, county, state and federal agencies as well as private companies spend millions of dollars each year obtaining aerial photography and other forms of remotely sensed data to extract the required urban information. Much of the required information simply cannot be obtained through *in situ* site surveys.

Temporal, spectral and spatial characteristics of urban attributes and remote sensing systems

Many of the detailed urban/suburban attributes that businesses and public agencies require are summarised in Table 2.8.1. This paper reviews how remotely sensed data may be of value for collecting information about these attributes. To remotely sense these urban phenomena, it is first necessary to appreciate the urban attributes' temporal, spectral and spatial resolution characteristics.

Urban/suburban temporal considerations

Three types of temporal resolution should be considered when monitoring urban environments using remote sensor data. Firstly, urban/suburban phenomena progress through an identifiable developmental cycle much like vegetation progresses through a phenological cycle. For example, Jensen and Toll (1983) documented a ten-stage single-family residential housing development cycle at work in suburban Denver, Colorado, that progressed from (1) range land to (10) fully-landscaped residential housing, often within one year. The image analyst must understand the temporal development cycle of the urban phenomena. If it is not understood, embarrassing and costly interpretation mistakes can be made.

The second type of temporal resolution is how often it is possible for a remote sensor system to collect data of the urban landscape, for example every eight days, every 16 days or on demand. Generally, satellite sensors that can be pointed off-nadir (e.g. SPOT HRV) have higher temporal resolution than sensors that only sense the terrain at nadir (e.g. Landsat Thematic Mapper). Orbital characteristics of the satellite platform and the latitude of the study area also impact the revisit schedule. Remote sensor data may be collected on demand from sub-orbital aircraft (airplanes, helicopters), weather conditions permitting. Up-to-date remote sensor data are critical for most urban/suburban applications.

Finally, temporal resolution may refer to how often land managers/planners need a specific type of information. For example, local planning agencies may require population estimates every five to seven years in addition to the estimates provided by the decennial census. The managerial temporal resolution requirements for many important urban applications are summarised numerically in Table 2.8.1 and graphically in Figure 2.8.1.

Urban/suburban spectral considerations

Most image analysts would agree that, when extracting urban/suburban information from remotely sensed data, it is more important to have high spatial resolution (often $\leq$5 by 5 m) than high spectral resolution (i.e. a large number of multispectral bands). For example, local population estimates based on building unit counts usually require a minimum spatial resolution of from $\leq$0.25 to 5 m to detect, distinguish between and/or identify the type of individual buildings. Practically any visible band (e.g. green or red) or near-infrared spectral band at this spatial resolution will do. Of course, there must be sufficient spectral contrast between the object of interest (e.g. a building) and its background (e.g. the surrounding landscape) in order to detect, distinguish between and identify the object from its background.

While high spectral resolution is not required, there are still optimum portions of the electromagnetic spectrum that are especially useful for extracting certain types of urban/suburban information (Table 2.8.1). For example, USGS Level III land cover is best acquired using the visible colour (0.4–0.7 μm; V), near-infrared (0.7–1.1 μm; NIR), middle-infrared (1.5–2.5 μm; MIR) and/or panchromatic (0.5–0.7 μm) portions of the spectrum. Building perimeter, area and height information is best acquired using black-and-white panchromatic (0.5–0.7 μm) or colour imagery (0.4–0.7 μm). The thermal infrared portion of the spectrum (3–12 μm; TIR) may be used to obtain urban temperature measurements. Active microwave sensors may obtain imagery of cloud shrouded or tropical urban areas (e.g. Japanese JERS-1 L-band, Canadian RADARSAT C-band and European Space Agency ERS-1, 2 C-band).

[...]

Table 2.8.1 Urban/suburban attributes and the minimum remote sensing resolution required to provide such information

Attributes	Minimum Resolution Requirements		
	Temporal	Spatial	Spectral*
Land Use/Land Cover			
L1–USGS Level I	5–10 years	20–100 m	V–NIR–MIR–Radar
L2–USGS Level II	5–10 years	5–20 m	V–NIR–MIR–Radar
L3–USGS Level III	3–5 years	1–5 m	Pan–V–NIR–MIR
L4–USGS Level IV	1–3 years	0.25–1 m	Panchromatic
Building and Property Infrastructure			
B1–building perimeter, area, height and cadastral information (property lines)	1–5 years	0.25–0.5 m	Pan–Visible
Transportation Infrastructure			
TI–general road centreline	1–5 years	1–30 m	Pan–V–NIR
T2–precise road width	1–2 years	0.25–0.5 m	Pan–V
T3–traffic count studies (cars, airplanes, etc.)	5–10 min	0.25–0.5 m	Pan–V
T4–parking studies	10–60 min	0.25–0.5 m	Pan–V
Utility Infrastructure			
U1–general utility line mapping and routing	1–5 years	1–30 m	Pan–V–NIR
U2–precise utility line width, right-of-way	1–2 years	0.25–0.6 m	Pan–Visible
U3–location of poles, manholes, substations	1–2 years	0.25–0.6 m	Panchromatic
Digital Elevation Model (DEM) Creation			
Dl–large-scale DEM	5–10 years	0.25–0.5 m	Pan–Visible
D2–large-scale slope map	5–10 years	0.25–0.5 m	Pan–Visible
Socio-Economic Characteristics			
S1–local population estimation	5–7 years	0.25–5 m	Pan–V–NIR
S2–regional/national population estimation	5–15 years	5–20 m	Pan–V–NIR
S3–quality of life indicators	5–10 years	0.25–30 m	Pan–V–NIR
Energy Demand and Conservation			
El–energy demand and production potential	1–5 years	0.25–1 m	Pan–V–NIR
E2–building insulation surveys	1–5 years	1–5 m	TIR
Meteorological Data			
M1–weather prediction	3–25 min	1–8 km	V–NIR–TIR
M2–current temperature	3–25 min	1–8 km	TIR
M3–clear air and precipitation mode	6–10 min	1 km	WSR–88D Radar
M4–severe weather mode	5 min	1 km	WSR–88D Radar
M5–monitoring urban heat island effect	12–25 h	5–30 m	TIR
Critical Environmental Area Assessment			
C1–stable sensitive environments	1–2 years	1–10 m	V–NIR–MIR
C2–dynamic sensitive environments	1–6 months	0.25–2 m	V–NIR–MIR–TIR
Disaster Emergency Response			
DE1–pre-emergency imagery	1–5 years	1–5 m	Pan–V–NIR
DE2–post-emergency imagery	12 h–2 days	0.25–2 m	Pan–V–NIR–Radar
DE3–damaged housing stock	1–2 days	0.25–1 m	Pan–V–NIR
DE4–damaged transportation	1–2 days	0.25–1 m	Pan–V–NIR
DE5–damaged utilities, services	1–2 days	0.25–1 m	Pan–V–NIR

*Spectral bands: Pan = Panchromatic; V = visible colour; NIR = Near infrared; MIR = Middle infrared; TIR = Thermal infrared.

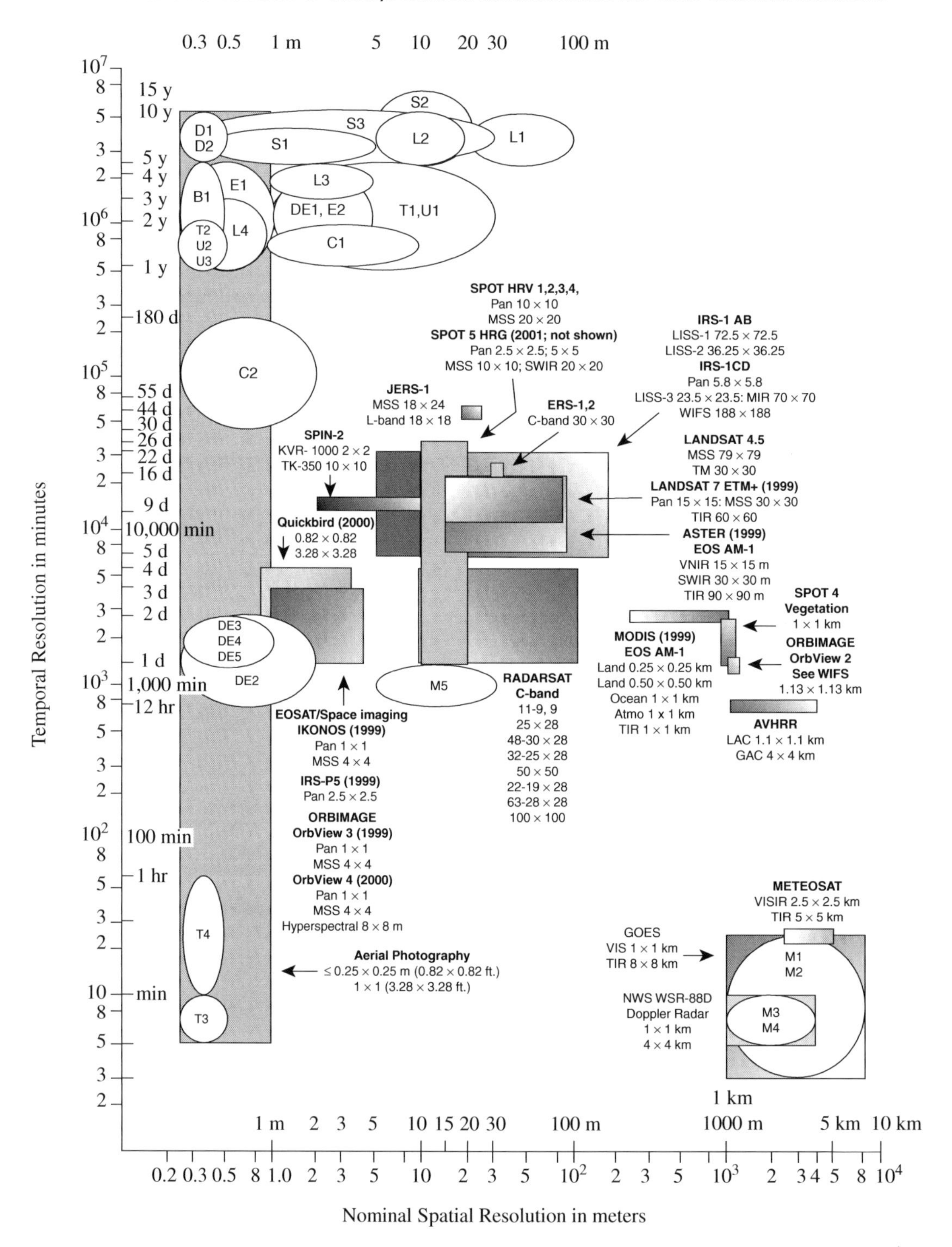

Figure 2.8.1 Subjective spatial and temporal resolution requirements for urban/suburban attributes overlaid on the spatial and temporal resolution capabilities of current and proposed remote sensor systems. (Refer to Table 2.8.1 for urban/suburban attribute codes.) Information presented in this type of diagram will constantly change due to (a) the development of new remote sensing instruments and their associated temporal and spatial resolutions, and (b) the user community continuously redefines existing data requirements and identifies new attributes to be collected.

Evaluation of urban/suburban attributes' spatial and temporal requirements and the availability of remote sensing systems to provide such information

The relationship between temporal and spatial data requirements for selected urban/suburban attributes and the temporal and spatial characteristics of available and proposed remote sensing systems is presented in Figure 2.8.1.

Land use/land cover

The term *land use* refers to how the land is being used. *Land cover* refers to the biophysical materials found on the land. For example, a state park may be used for recreation but have a deciduous forest cover. One method of organising land use/land cover information is to use a classification system. The most comprehensive hierarchical classification system for urban/suburban *land use* is the *Land-Based Classification Standard* (LBCS) under development by the American Planning Association (1998) that updates the 1965 Standard Land Use Coding Manual (Urban Renewal Administration 1965), which is cross-referenced with the 1987 Standard Industrial Classification (SIC) Manual (Bureau of the Budget 1987) and the updated North American Industrial Classification Standard (NAICS). The LBCS requires extensive input from *in situ* site surveys, aerial photography and satellite remote sensor data to obtain information at the *parcel* level on the following five characteristics: activity, function, site development, structure and ownership (American Planning Association 1998). The system does not provide information on land cover or vegetation characteristics in the urban environment because it is relying on the Federal Geographic Data Committee 'Standards' on these topics. The LBCS is not complete at this time. Therefore, the following discussion will focus on the use of the 'land use and land cover classification system for use with remote sensor data' developed by the US Geological Survey (Anderson *et al.* 1976).

[...] Generally, USGS Level I classes may be inventoried effectively using sensors with a nominal spatial resolution of $\geq$ 20–100 m such as the Landsat Multispectral Scanner (MSS) with 79 by 79 m ground resolution, the Thematic Mapper (TM) at 30 by 30 m, SPOT HRV (XS) at 20 by 20 m and Indian LISS 1–3 (72.5 by 72.5 m, 36.25 by 36.25 m, 23.5 by 23.5 m, respectively). [...]

Sensors with a minimum spatial resolution of approximately 5–20 m are generally required in order to obtain Level II information. The SPOT HRV and the Russian SPIN-2 TK-350 are the only operational satellite sensor systems providing 10 by 10 m panchromatic data. RADARSAT provides 11 by 9 m spatial resolution data for Level I and 11 land cover inventories even in cloud-shrouded tropical landscapes. Landsat 7 with its 15 by 15 m panchromatic band may be launched in 1999. More detailed Level III classes may be inventoried using a sensor with a spatial resolution of approximately 1–5 m (Welch 1982; Forester 1985) such as IRS-1CD pan (5.8 by 5.8 m data resampled to 5 by 5 m) or large-scale aerial photography. Future sensors may include EOSAT Space Imaging IKONOS (1 by 1 m pan and 4 by 4 m multispectral), EarthWatch Quickbird (0.8 by 0.8 m pan and 3.28 by 3.28 m multispectral), Orbview 3 and 4 (1 by 1 m pan and 4 by 4 m multispectral) and IRS P5 (2.5 by 2.5 m). The synergistic use of high spatial resolution panchromatic data (e.g. 1 by 1 m) merged with lower spatial resolution multispectral data (e.g. 4 by 4 m) will likely provide an image interpretation environment that is superior to using panchromatic data alone (Jensen 1996).

Level IV classes and building and cadastral (property line) information are best monitored using high spatial resolution panchromatic sensors, including aerial photography ($\leq$0.25 to 1 m) and, possibly, the proposed EOSAT Space Imaging IKONOS (1 by 1 m), Earthwatch Quickbird pan (0.8 by 0.8 m) and Orbview-3 (1 by 1 m) data.

Urban land use/land cover classes in Levels I through IV have temporal requirements ranging from 1 to 10 years (Table 2.8.1 and Figure 2.8.1). All the sensors mentioned have temporal resolutions of less than 55 days, so the temporal resolution of the land use/land cover attributes is satisfied by the current and proposed sensor systems.

Building and cadastral (property line) infrastructure

In addition to fundamental nominal scale land use and land cover information (i.e. identifying whether an object is a single-family residence or a commercial building), transportation planners, utility companies, tax assessors and others require more detailed information on building footprint perimeter, area and height; driveways; patios; fences; pools; storage buildings; and the distribution of landscaping every one to five years. These building and property parameters are best obtained using stereoscopic (overlapping) panchromatic aerial photography or other remote sensor data with a spatial resolution of $\leq$0.25–0.5 m (Jensen 1995; Warner 1996). For example, panchromatic stereoscopic aerial photography with 0.3 by 0.3 m spatial resolution was used to extract the building perimeter and area information for the single-family residential area in Figure 2.8.2. With this type of data, each building footprint, patio, outbuilding, tree, pool, driveway, fence and contour may be extracted. In many instances, the fence lines are the cadastral property lines. If the fence lines

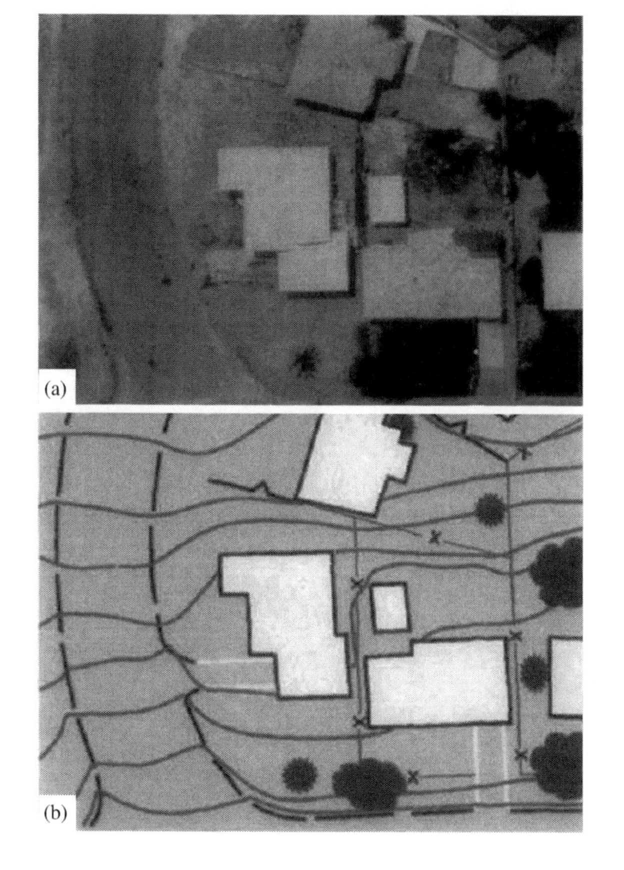

Figure 2.8.2 Planimetric cadastral information of a single-family residential area extracted photogrammetrically from panchromatic stereoscopic vertical aerial photography. Building footprints are in black, fence lines with x's in black, driveways in white, shrubs and trees in black, two-foot contours in continuous black lines and highway right-of-way in dashed lines.

are not visible or are not truly on the property line, the property lines are located by a surveyor and the information is overlaid onto an orthophotograph or planimetric map database to represent the legal cadastral (property) map. Many municipalities in the United States use high spatial resolution imagery such as this as the source for some of the cadastral information and/or as an image backdrop upon which surveyed cadastral and tax information are portrayed.

Detailed building perimeter, area and height data can be extracted from high spatial resolution ($\leq$0.25 to 0.5 m) stereoscopic imagery (Jensen *et al.* 1996). Such information can then be used to create three-dimensional displays of the urban terrain that we can walk through in a virtual reality environment if desired [...]. For example, Figure 2.8.3 depicts (1) a large-scale vertical aerial photograph of downtown Columbia, South Carolina, (2) a digital elevation model (DEM) of the same area extracted from the stereoscopic photography depicting the height of every building, and (3) the orthophotograph draped over the DEM, creating a virtual reality representation of a major street. Architects, planners, engineers and real estate personnel are beginning to use such information for a variety of purposes.

EOSAT Space Imaging (1999), OrbView (1999) and Earthwatch (2000) plan to provide fore–aft stereoscopic images from satellite-based platforms with approximately 0.8–1 m spatial resolution. Ridley *et al.* (1998) conducted a feasibility study and found that simulated 1 by 1 m satellite stereoscopic data could 'have potential for creating a national 3D building model if the processes were automated, which would produce a much cheaper source of building heights.' The accuracy of the building maximum heights (z) ranged between 1.5 and 3 m RMSE. Thus, the use of such imagery may not obtain the detailed planimetric (perimeter, area) and topographic detail and accuracy (building height and volume) that can be extracted from high spatial resolution stereoscopic aerial photography ($\leq$0.25–0.5 m). Research is required using real 1 by 1 m stereoscopic data obtained from satellite platforms (Ridley *et al.* 1998).

Transportation infrastructure

Engineers often use remote sensor data to: (1) update transportation network maps; (2) evaluate road, railroad and airport runway and tarmac condition; (3) study urban traffic patterns at choke points such as tunnels, bridges, shopping malls and airports; and (4) conduct parking studies [...]. One of the more prevalent forms of transportation data is the street centreline spatial data (SCSD). Three decades of practice have proven the value of differentiating between the left and right sides of each street segment and encoding attributes to them such as street names, address ranges, ZIP codes, census and political boundaries, and congressional districts. SCSD provide a good example of a national framework spatial data theme by virtue of their extensive current use in facility site selection, census operations, socio-economic planning studies and legislative redistricting (NRC 1995).

Road network centreline updating in rapidly developing areas may be performed every 1–5 years and, in areas with minimum tree density (or during the leaf-off season), can be accomplished using imagery with a spatial resolution of 1–30 m [...]. If more precise road dimensions are required, such as the exact centre of the road and the width of the road and sidewalks, then a spatial resolution of $\leq$0.2–0.5 m is required (Jensen *et al.* 1994). Currently, only aerial

Figure 2.8.3 (a) Panchromatic aerial photograph of Columbia, South Carolina (original at 1: 6000 scale). (b) Digital elevation model (DEM) derived from a stereo-pair and portrayed as a shaded-relief model. (c) An orthophotograph of the area draped over the DEM and displayed in a three-dimensional perspective projection. (d) The DEM and a viewshed model were used to identify the optimum building on which to place a cellular phone transceiver.

photography can provide such planimetric information (Figure 2.8.2).

Road, railroad and bridge conditions (cracks, potholes etc.) are routinely monitored both *in situ* and using high spatial resolution remote sensor data. [. . .] Careful inspection of high spatial resolution imagery ($\leq$0.25–0.5 m) by a trained analyst can provide significant information about the condition of the road and railroad [. . .]. Traffic count studies of automobiles, airplanes, boats, pedestrians and people in groups require very high temporal resolution data ranging from 5 to 10 minutes. Even when such timely data are available, it is difficult to resolve a car or boat using even 1 by 1 m data. This requires high spatial resolution imagery from $\leq$0.25–0.5 m. Such information can only be acquired using aerial photography, digital cameras, or video sensors that are (1) located on the top edges of buildings looking obliquely at the terrain or (2) placed in aircraft or helicopters and flown repetitively over the study areas. When such information is collected at an optimum time of day, future parking and traffic movement decisions can be made. Parking studies require the same high spatial resolution ($\leq$0.25–0.5 m) but slightly lower temporal resolution (10–60 minutes). Doppler radar has demonstrated some potential for monitoring traffic flow and volume.

Utility infrastructure

Urban/suburban environments are enormous consumers of electrical power, natural gas, telephone service and potable water (Haack *et al.* 1997). In addition, they create great quantities of refuse, waste water and sewage. The removal of storm water from urban impervious surfaces is also a serious problem [...]. Automated mapping/facilities management (AM/FM) and geographic information systems (GIS) have been developed to manage extensive right-of-way corridors for various utilities, especially pipelines [...]. The most fundamental task is to update maps to show a general centreline of the utility of interest, such as a power line right-of-way. This is relatively straightforward if the utility is not buried or obscured trees and if 1–30 m spatial resolution remote sensor data are available. It is also often necessary to identify prototype utility (e.g. pipeline) routes [...]. Such studies require geographically extensive imagery such as SPOT (20 by 20 m) or Landsat Thematic Mapper data (30 by 30 m) that have relatively large scene dimensions. The majority of the actual and proposed rights-of-way may be observed well on imagery with 1–30 m spatial resolution obtained once every one to five years. But, when it is necessary to inventory the exact location of transmission tower footpads, utility poles and manhole covers, the true centreline of the utility, the width of the utility right-of-way and the dimensions of buildings, pumphouses and substations, then it is necessary to have a spatial resolution of from $\leq$0.25 to 0.6 m [...]. Ideally, new facilities are inventoried every one to two years.

[...]

Socio-economic characteristics

Selected socio-economic characteristics may be extracted directly from remote sensor data or by using surrogate information derived from the imagery. Two of the most important attributes are (a) population estimation and (b) quality-of-life indicators.

(a) Population estimation

Knowing how many people live within a specific geographic area or administrative unit (e.g. city, county, state, country) is very powerful information. [...] Population estimation can be performed at the local, regional and national level based on (10) counts of individual dwelling units, (11) measurement of urbanised land areas (often referred to as settlement size) and (12) estimates derived from land use/land cover classification [...].

Remote sensing techniques may provide population estimates that approach the accuracy of traditional census methods if sufficiently accurate *in situ* data are available to calibrate the remote sensing model. Unfortunately, ground-based population estimations may be inaccurate [...]. In many instances in developing countries remote sensing methods may be superior to ground-based methods.

The most accurate remote sensing method for estimating the population of a local area is to count individual dwelling units based on the following assumptions:

- the imagery must be of sufficient spatial resolution to identify individual structures even through tree cover and whether they are residential, commercial or industrial buildings;
- some estimation of the average number of persons per dwelling unit must be available;
- some estimate of the number of homeless, seasonal and migratory workers is required; and
- it is assumed that all dwelling units are occupied, and that only *n* families live in each unit (calibrated using *in situ* investigation).

This is usually performed every five to seven years and requires high spatial resolution remotely sensed data ($\leq$0.25–5 m). For example, individual dwelling units in a section of Irmo, South Carolina, were extracted from 2.5 by 2.5 m aircraft multispectral data (Cowen *et al.* 1995). Correlation of the remote sensing derived dwelling unit data with US Bureau of the Census dwelling unit data for the 32 census block area yielded an $r^2 = 0.81$ (correlation coefficient of 0.91). These findings suggest that the new high spatial resolution panchromatic sensors may provide a good source of information for monitoring the housing stock of a community on a routine basis. This will enable local governments to anticipate and plan for schools and other services with data that have a much more frequent temporal resolution than does the decennial census. These data will also be of value for real estate, marketing and other business applications (Lo 1995). Unfortunately, the dwelling unit approach is not suit able for a regional/national census of population because it is too time consuming and costly (Sutton *et al.* 1997). [...]

[...]

(b) Quality-of-life indicators

Lo and Faber (1998) suggest that adequate income, decent housing, education and health services, and good physical environment are important indicators of social well-being and quality-of-life. Evaluating the quality-of-life of a population on a continuing basis is important because it helps planners and government agencies involved with the delivery of human services to be aware of problem areas. In the past, most quality-of-life studies made use of census data to extract socio-economic indicators. Only recently have factor analytic studies documented how quality-of-life indicators (such as house value, median family income, average number of rooms, average rent and education) can be estimated by extracting the urban attributes summarised in Table 2.8.2 from relatively high spatial resolution (≤0.25–30 m) imagery [...].

Onsrud *et al.* (1994), Curry (1997) and Slonecker *et al.* (1998) point out that scientists must exercise wise judgment when using remotely sensed data to extract socio-economic and/or quality-of-life information so that they do not infringe on an individual's right to privacy. The misuse of the high spatial resolution remote sensor data will likely be the impetus for future restrictive legislation.

[...]

Table 2.8.2 Urban/suburban attributes that maybe extracted from remote sensor data using the fundamental elements of image interpretation and used to assess housing quality and/or quality-of-life

Position	Attributes
Site	*Building*
	– single or multiple family
	– size (sq ft)
	– height (ft)
	– carport or garage (attached, detached)
	– age (derived by convergence of evidence)
	Lot
	– size (sq ft)
	– front and back yards (sq ft)
	– street frontage (ft)
	– driveway (paved, unpaved)
	– fenced
	– patio, deck, pool (in-ground, above-ground),
	– out buildings (sheds)
	– density of buildings per lot
	– percentage landscaped
	– health of vegetation (e.g. NDVI greenness)
	– property fronts paved or unpaved road
	– abandoned autos
	– refuse
Situation	*Adjacency to Community Amenities*
	– schools
	– churches
	– hospitals
	– fire station
	– library
	– shopping
	– open space, parks, golf courses
	Adjacency to Nuisances and Hazards
	– heavy street traffic
	– railroad or switchyard
	– airports and/or flight path
	– freeway
	– located on a floodplain
	– sewage treatment plant
	– industrial area
	– power plant or substation
	– overhead utility lines
	– swamps and marsh
	– steep terrain

Critical environmental area assessment

Urban/suburban environments often include very sensitive areas such as wetlands, endangered specie habitat, parks, land surrounding treatment plants and the land in urbanised watersheds that provides the runoff for potable drinking water. Relatively stable sensitive environments only need to be monitored every one to two years using a multispectral remote sensor collecting 1–10 m data. For extremely critical areas that could change rapidly, multispectral remote sensors (including a thermal infrared band) should obtain ≤0.25–2 m spatial resolution data every one to six months.

Disaster emergency response

Recent floods (Mississippi River in 1993; Albany, Georgia, in 1994), hurricanes (Hugo in 1989, Andrew in 1991, Fran in 1996), tornadoes (every year), fires, tanker spills and earthquakes (Northridge, California, in 1994) demonstrated that a rectified, pre-disaster image database is indispensable. The pre-disaster data only need to be updated every one to five years. It should be high spatial resolution (1–5 m) multispectral data if possible.

When disaster strikes, high resolution (≤0.25–2 m) panchromatic and/or near-infrared data should be acquired within 12 hours to 2 days [...]. If the terrain is shrouded in clouds, imaging radar might provide the most useful information. Post-disaster images are registered to the pre-disaster images, and manual and digital change detection takes place (Jensen 1996). If precise, quantitative information about damaged housing stock, disrupted transportation arteries, the flow of spilled materials and damage to above-ground utilities are required, it is advisable to acquire post-disaster ≤0.25–1 m panchromatic and near-infrared data within one to two days (Jensen *et al.* 1998). Such information was indispensable in assessing damages and allocating scarce clean-up resources during Hurricane Hugo, Hurricane Andrew, Hurricane Fran and the recent Northridge earthquake.

Observations

Table 2.8.1 and Figure 2.8.1 reveal that there are a number of remote sensing systems that currently provide some of the desired urban infrastructure and socio-economic information when the required spatial resolution is poorer than 4 by 4 m and the temporal resolution is between one and 55 days. However, very high spatial resolution data (≤1 by 1 m) is required to satisfy several of the data requirements. In fact, as shown in Figure 2.8.1, the only sensor that currently provides such data on-demand is aerial photography (≤0.25–0.5 m). EOSAT/Space Imaging IKONOS (1999) with its 1 by 1 m panchromatic data, OrbView 3 (1999) with its 1 by 1 m panchromatic data, and Earthwatch Quickbird with its 0.8 by 0.8 m panchromatic data (2000) may satisfy some of these urban data requirements, but not all. It may be necessary to develop higher spatial resolution (≤0.25–0.5 m) satellite remote sensor data to provide some of the detailed urban/suburban infrastructure and socio-economic information, or utilise aerial photography. None of the sensors can provide the 5–60 minute temporal resolution necessary for traffic and parking studies except for (1) repetitive aerial photography (very costly), or (2) the placement of digital or video cameras on the top edge of buildings to obtain an oblique view. [...]

References

American Planning Association (1998) *Land-Based Classification Standards: Draft Classification*, American Planning Association, Research Department, www.planning.org/lbcs.

Anderson, J.R., Hardy, E., Roach, J. and Witmer, R. (1976) A land use and land cover classification system for use with remote sensing data. USGS Professional Paper #964, Washington, DC.

Bureau of the Budget (1987) *Standard Industrial Classification Manual*, Government Printing Office, Washington, DC.

Cowen, D.J., Jensen, J.R., Bresnahan, G. *et al.* (1995) The design and implementation of an integrated GIS for environmental applications. *Photogmmmetric Engineering & Remote Sensing*, **61**, 1393–1404.

Curry, M.R. (1997) The digital individual and the private realm. *Annals of the Association of American Geographers*, **87** (4), 681–699.

Forester, B.C. (1985) An examination of some problems and solutions in monitoring urban areas from satellite platforms. *International Journal of Remote Sensing*, **6**, 139–151.

Haack, B., Guptill, S., Holz, R. *et al.* (1997) Urban analysis and planning, in *Manual of Photographic Interpretation*, American Society for Photogrammetry & Remote Sensing, Bethesda, MD, pp. 517–553.

Jensen, J.R. (1995) Issues involving the creation of digital elevation models and terrain corrected orthoimagery using soft-copy photogrammetry. *Geocarto International*, **10**, 1–17.

Jensen, J.R. (1996) *Introductory Digital Image Processing: A Remote Sensing Perspective*, Prentice-Hall, Saddle River, NJ.

Jensen, J.R. and Toll, D.L. (1983) Detecting residential land-use development at the urban fringe. *Photogmmmetric Engineering & Remote Sensing*, **48**, 629–643.

Jensen, J.R., Cowen, D.C., Halls, J. *et al.* (1994) Improved urban infrastructure mapping and forecasting for BellSouth using remote sensing and GIS technology. *Photogrammetric Engineering & Remote Sensing*, **60**, 339–346.

Jensen, J.R., Huang, X., Graves, D. and Hanning, R. (1996) Cellular phone transceiver site selection, in *Raster Imagery in Geographic Information Systems* (eds S. Morain and S. Baros), OnWard Press, Santa Fe, NM, pp. 117–125.

Jensen J.R., Halls, J. and Michel, J. (1998) A systems approach to environmental sensitivity index (ESI) mapping for oil spill contingency planning and response. *Photogrammetric Engineering & Remote Sensing*, **64** (10), 1003–1014.

Lo, C.P. (1995) Automated population and dwelling unit estimationfrom high-resolution satellite images: a GIS approach. *International Journal of Remote Sensing*, **16**, 17–34.

Lo, C.P. and Faber, B.J. (1998) Interpretation of Landsat Thematic Mapper and census data for quality of life assessment. *Remote Sensing of Environment*, **62** (2), 143–157.

NRC (1995) *A Data Foundation for the National Spatial Data Infrastructure, Mapping Science Committee*, National Research Council, Washington, DC.

Onsrud, H.J., Johnson, J.P. and Lopez, X.R. (1994) Protecting personal privacy in using geographic information systems. *Photogrammetric Engineering & Remote Sensing*, **60** (9), 1083–1095.

Ridley, H.M., Atkinson, P.M., Aplin, P. *et al.* (1998) Evaluating the potential of the forthcoming commercial US high-resolution satellite sensor imagery at the Ordnance Survey. *Photogrammetric Engineering & Remote Sensing*, **63** (8), 997–1005.

Slonecker, E.T., Shaw, D.M. and Lillesand, T.M. (1998) Emerging legal and ethical issues in advanced remote sensing technology. *Photogrammetric Engineering & Remote Sensing*, **64** (6), 589–595.

Sutton, P., Roberts, D., Elvidge, C. and Meij, H. (1997) A comparison of nighttime satellite imagery and population density for the continental United States. *Photogmmmetric Engineering & Remote Sensing*, **63**, 1303–1313.

Swerdlow, J.L. (1998) Making sense of the millennium. *National Geographic*, **193**, 2–33.

Urban Renewal Administration (1965) *Standard Land Use Coding Manual: A Standard System for Identifying and*

Coding Land Use Activities, Department of Commerce, Washington, DC.

Warner, W.S., Graham, R.W. and Read, R.E. (1996) *Small Format Aerial Photography*, Wittles Publishing, UK.

Welch, R. (1982) Spatial resolution requirements for urban studies. *International Journal of Remote Sensing*, **3**, 139–146.

Further reading

Baker, J.C., O'Connell, K.M. and Williamson, R.A. (2001) *Commercial Observation Satellites: At the Leading Edge of Global Transparency*, RAND-ASPRS, Santa Monica, CA. [A policy analysis of the impact of overhead imagery on democratic scrutiny and international politics.]

Jensen, J.R. (2000) *Remote Sensing of the Environment: An Earth Resource Perspective*, Prentice-Hall, Upper Saddle River, NJ. [One of the leading scholarly texts discussing the technologies and applications of remote sensing.]

Liverman, D., Moran, E.F., Rindfuss, R.R. and Stern, P.C. (1998) *People and Pixels: Linking Remote Sensing and Social Science*, National Academy Press, Washington, DC. [This report considers how satellite imaging could be better exploited to understand people and social activities.]

Stoney, W.E. (2005) Markets and opportunities. *Earth Imaging Journal*, **2** (1), 10–14. [A useful review of the numerous different satellite imaging systems and their capabilities.]

Williams, D.L., Goward, S. and Arvidson, T. (2006) Landsat: yesterday, today, and tomorrow. *Photogrammetric Engineering and Remote Sensing*, **72**, 1171–1178. [This paper considers the significance of the Landsat satellite imaging system, that has been vital to environment mapping, and how it could develop further.]

See also

- Chapter 2.3: Manufacturing Metaphors: Public Cartography, the Market, and Democracy
- Chapter 2.4: Maps and Mapping Technologies of the Persian Gulf War
- Chapter 2.12: Imaging the World: The State of Online Mapping
- Chapter 5.8: Cartographic Rationality and the Politics of Geosurveillance and Security

Chapter 2.9

Emergence of Map Projections, from *Flattening the Earth: Two Thousand Years of Map Projections*

John P. Snyder

Editors' overview

This extract is taken from Snyder's comprehensive book on the historical development of map projections. These are an often overlooked but crucial element of cartographic technology. They enable the production of flat maps of the three-dimensional globe that have spatial properties required for different contexts, scales and user tasks. Snyder traces how technical development in projections depends, in part, on new mathematical ideas and computational means, but have also been driven by wider social demands (ocean navigation, trade, war, land surveys and so on), along with an abiding concern for aesthetically pleasing and elegant solutions to an intractable problem (the 'perfect' projection). The twentieth century has seen a rapid growth in the number and type of map projections with the flexibility of computer processing of coordinates making wider choices practicable; yet relatively few projections are in widespread use and an understanding of appropriate choices remains limited.

Originally published in 1993: Excerpts from Chapters 1 and 4, in John P. Snyder *Flattening the Earth: Two Thousand Years of Map Projections*, University of Chicago Press, Chicago, pp. 1–4, 155–156, 274–277.

Emergence of map projections: classical through renaissance

For about two thousand years, the challenge of trying to represent the round earth on a flat surface has posed mathematical, philosophical and geographical problems that have attracted inventors of many types. Of course, the use of maps predates this period. The contemporary process of mapmaking, however, had a slow beginning because exploration of the earth as a whole is a relatively recent historical development.

Although the true shape of the earth was in doubt for centuries, a number of scholars were convinced the earth was spherical well before Claudius Ptolemy wrote his monumental *Geography* about AD 150 (Dilke 1987). Even scholars who considered the earth flat believed the skies were hemispherical. Thus it was soon apparent that preparing a flat map of a surface curving in all directions led to distortion. That distortion may take many forms – shape, area, distance, direction and interruptions or breaks between portions. In other words, a flat map cannot correctly represent the surface of a sphere.

The Map Reader: Theories of Mapping Practice and Cartographic Representation, First Edition. Edited by Martin Dodge, Rob Kitchin and Chris Perkins.

A globe also has drawbacks in spite of its basic freedom from distortion. A globe is bulky, of small scale and awkward to measure; less than half its surface can normally be seen at one time. As Ptolemy stated in his *Geography*:

> 'it is not easy to provide space large enough (on a globe) for all of the details that are to be inscribed thereon; nor can one fix one's eye at the same time on the whole sphere, but one or the other must be moved, that is, the eye or the sphere, if one wishes to see other places.
> In the second method wherein the earth is represented as a plane surface there is not this inconvenience. But a certain adjustment is required in representing the earth as a sphere in order that the distances noted therein may be shown on the surface of the globe congruent, as far as possible, with the real distance on the earth.' (Translation in Stevenson 1932: 39.)

The systematic representation of all or part of the surface of a round body, especially the earth, onto a flat or plane surface is called a map projection. Literally an infinite number of map projections are possible, and several hundred have been published. The designer of a map projection tries to minimise or eliminate some of the distortion, at the expense of more distortion of another type, preferably in a region of or off the map where distortion is less important.

Historically, one of the first steps in preparing a map is to lay down the graticule or net of meridians and parallels according to the selected map projection. The system of meridians and parallels, or lines of longitude and latitude, respectively, was developed near the beginning of the history of map projections. The history of the development of longitude and latitude is a long and arduous one [...] but the development of map projections as mathematical or geometrical designs continued in spite of the lack of accuracy of terrestrial measurements. The designer could treat meridians and parallels as if they represented precise coordinates, placed as straight lines or smooth curves on the map, even if the related geographical features were subject to misplacement.

In addition to longitude and latitude, terminology in this history involves frequent use of the type of projection. There have been several proposed classifications of map projections (d'Avezac-Macaya 1863: 473–485; Lee 1944; Goussinsky 1951; Tobler 1962; Calderera 1910). One of the most common classifications [...] is based on association, at least conceptually, with a developable surface (d'Avezac-Macaya 1863: 474–478). Such a surface may be laid flat without distortion. A cylinder or cone may be so developed, but not a globe. The most common projections may be conceptually and, in some cases, geometrically projected onto a cylinder, cone or plane placed tangent or secant to the globe, hence the categories *cylindrical* and *conic* projections. Because directions, or azimuths, are shown correctly from the centre of vertical projections onto a tangent plane, the latter projections are usually called *azimuthal* rather than planar. Many projections do not fall into these categories, leading to additional classes, some with related names like *pseudocylindrical* and *pseudoconic*.

Claudius Ptolemy and a handful of others, within two hundred years before or after the beginning of the Christian era, had provided detailed instructions concerning some methods of map projection, but many of the types of maps dominant before the Renaissance had bases that were philosophical rather than mathematical. These included world maps or *mappaemundi*, as they are called, such as symbolic T-O maps and others on which the landmasses were neatly fitted into a circle (Woodward 1987). (The T-O design is exemplified in Figure 2.9.1, which depicts the known continents separated by the Mediterranean Sea shaped like a T and surrounded by an O-shaped ocean.)

Whether the earth was spherical or flat was not a factor in these maps. They represented no projection as such. For over a thousand years the works of Ptolemy, both maps and technical exposition, were stagnant classics having little meaning as foundations from which to develop other approaches to projection. Occasionally there were other innovations in projection during the centuries prior to the Renaissance, but the subject was quiescent for a millennium under the most generous evaluation.

The transition from the Middle Ages to the Renaissance entailed a basic change in the concept of map projection. During the Renaissance (for our purposes, 1470–1669), interest in map projection began to expand. At last, mapmakers began to build on the work of their predecessors and to innovate, with their work often in turn replaced by

Figure 2.9.1 The oldest existing printed map known, a T-O map by Isidore of Seville, in Etymologiarum, 1472. (Reproduced from Woodward 1975: 9.)

still newer projections a few decades later. This development proceeded exponentially from the early fourteenth century to the present time.

The attraction toward scientifically sound map projection was prompted by various factors. Fundamental was the desire for knowledge itself, whether philosophical or scientific. Many of those contributing to the advancement of map construction, and specifically map projections, were renowned philosophers or ordained clergymen. The modern term *Renaissance man* originated from a multiskilled approach to learning often true of the cultural and scientific leaders of the sixteenth and seventeenth centuries. The study of map projections is now frequently termed mathematical cartography, but the development of appropriate mathematics was often contemporary with the development of map projections both during and after the Renaissance. Astronomy was also increasingly mathematical, and star maps utilised projections, so the fields of mathematics, astronomy and map projections often attracted the same individuals.

The development of map projections was also spurred by pragmatic considerations. Although projections were used to construct star maps of the apparently dome-shaped heavens long before the Middle Ages, the greater geographical awareness during the late Middle Ages and Renaissance led to an increased demand for flat maps of the globe.

The dozen or so different projections developed during the one-and-a-half millennia prior to the Renaissance were augmented by another ten or so general types of new or improved projections developed during the next two centuries. Most were pursuits of the classic goal of reducing the apparent distortion of the map, at least for the particular region being portrayed, but other projections provided a special type of information, such as Gerardus Mercator's projection for navigation. Still others served to remind the map user of the globular shape of the world. Most of the older as well as the newer projections were used for the cartography of the Renaissance.

Delineation of the use of the projections during the Renaissance must be incomplete, but some descriptions are available to us. The modern use of mathematics to prepare maps based on these projections in their ideal form is well known, but the actual techniques used in the fifteenth and sixteenth centuries, when trigonometric and algebraic techniques were still emerging, were often quite different. The draftsman of the Renaissance could work with elementary tools with varying origins. Compasses were known in ancient Egypt (use by a carpenter is mentioned in the Old Testament), and preparing a design with ruler, compasses and a lead or tin stylus was mentioned by Theophilus the Monk about 1100 (Dickinson 1949). Bronze tools from AD 79 (Pompeii and Herculaneum) include fixed proportional compasses. An extant set of related instruments from about 1618 includes pairs of compasses with a removable divider point, stylus, fluted pen and/or crayon holders, in addition to a square, a sector and other items (Dickinson 1949).

The mathematicians, geographers, physicists, clergymen, astronomers, mapmakers and others who have offered their solutions to the problem of flattening the globe have done so modestly or arrogantly. No one has found the ideal map projection, unless the requirements set are arbitrarily restrictive. The search continues, as the ever-escalating volume of literature asserts. Literature about the history of map projections has been much more limited. Aside from detailed treatises about early projections, the only lengthy study is an important but unillustrated paper in French (d'Avezac-Macaya 1863). Over three-fourths of published projections have appeared since then.

This chapter relies heavily on a few classic nineteenth and twentieth century treatises on the history of map projections, especially the paper by d'Avezac-Macaya (1863) as well as work by Nordenskiold (1889) and Keuning (1955). There are some contemporary descriptions of specific projections; although Keuning states that 'no textbook is known from the XVIth or the XVIIth century, in which they [projections] are treated', *Geographia generalis*, prepared by the short-lived Bernhard Varenius (1622–1650), fills much of that role (Keuning 1955: 3).

In very recent times, computers have been used not only to plot the meridians and parallels, but also to place geographic information in proper positions on a map projection. Until this era, the ease of plotting meridians and parallels by geometric means was of considerable importance, since other geographical data were often plotted by interpolation between these previously drawn lines. This was especially true in the years preceding and during the Renaissance. Furthermore, since maps were treated as works of art portraying the philosophy and beauty of the earth as well as providing technical information, the very shapes of meridians and parallels were important in implying the sphericity of the planet.

For many of the projections identified below, the mapmaker seldom gave us a name or construction technique. Instead, projections have been determined by measurements on and observations of the map itself.

[...]

Map projections of the twentieth century

The acceleration of map projection development in the nineteenth century was catalysed by the development and application of mathematical principles. The calculus, least squares and complex algebra opened up challenging approaches to mapping the increasingly explored and measured face of the earth. Reuse of old concepts also flourished, and the most used new and old projections generally continued to be the simplest to construct.

The twentieth century continued this mixture of complicated theory and simple, frequently naïve approaches to 'new' projections that were claimed in some cases to be not merely good but the best possible projection of the world without restriction. Inevitably, many of the new map projections were intended for world maps, since this could result in products visibly different to the viewer. Improvements in projections for medium- and large-scale maps, however, produced changes that could be detected only by careful measurement. In some cases, advocacy of new projections appeared motivated by desire for self-promotion, but this is a subjective charge occasionally confused with zeal (for example, Maling 1974 focused on variations of the cylindrical equal-area projection by Behrmann, Edwards and Peters).

Concurrent with the actual development of new projections during the twentieth century has been a proliferation of grid systems, used in almost every country as a means of locating a point by its rectangular coordinates according to an official ellipsoidal map projection, rather than by just its latitude and longitude. This method was formally used for systems beginning in central Europe in the late nineteenth century, although square or rectangular grids had originated with early Chinese maps and portolan charts (Campbell 1987; Thrower 1972: 24). At the same time, the number of projections used for most large-scale mapping dwindled to two or three, in particular the transverse Mercator and the Lambert conformal conic, although in a few countries the problem of map revision was not considered worth the effort of changing from such nonconformal projections as the Bonne or the azimuthal equidistant projection.

The more complicated projection developments were in some cases carried out like the proverbial reason for mountain climbing – because the challenge was there. Beyond the achievement of placing the world map conformally in a triangle, for example, there is little to recommend such a projection. The inventors of many of these mathematical novelties generally did not promote their work beyond modest scientific publication. Others applied older spherical projections, like the Albers equal-area conic, to the ellipsoid, so that they could better meet the needs of twentieth century precision.

The fact that, in spite of the many journals and diverse book publishers, very few projections have been independently duplicated is testimony to the great number of possibilities. The percentage of projections reinvented is about 5%, both before and after 1900.

The greatest new non-cartographic boon to twentieth century map projection development has been the modern computer. Older projections that lay dormant due to complexity could now be plotted quickly by persons following elementary instructions after programming, as they required only moderate proficiency in mathematics. Projections can now be developed using iterative solutions to complicated mathematical requirements, because the computer is so fast that an expert mathematician's analytical solution may be unnecessarily efficient, if indeed it is even possible in some cases.

The modern computer, however, is quite recent, and desktop crank calculators and logarithm tables were the fastest forms of calculation to more than two or three significant digits for innovators like Ernst Hammer (1858–1925) and Hans Maurer (1868–1945) of Germany, and Oscar S. Adams (1874–1962) of the US Coast and Geodetic Survey, who flourished during the early years of a century that saw an almost incomprehensible scope of changes.

[. . .]

Conclusions and outlook

The twentieth century produced a great mixture of map projections ranging from trivial to important, depending on one's viewpoint. With an increase in the number of cartographers and related professionals (as well as amateur cartophiles) and of journals available for presenting ideas, the numbers of map projections and books and papers about them continued to rise exponentially.

Sociological factors such as these can be used to explain the quantitative expansion of any continuing scientific field, but what of the quality? Employing relative usage as an index to quality is generally inappropriate here as in other consumer fields, because of political and promotional factors. The very different Gall orthographic and Robinson projections are no better or worse now than they were in 1855 and 1974, respectively, when they were first announced to the scientific public. However, in 1973 via Arno Peters (independently as his own) and in 1988 via the National Geographic Society (with full credit), these projections jumped from relative obscurity to substantial usage by not only their sponsors but others as well. Also, the nationalities of Gall, O.M. Miller, Ginzburg and Baranyi undoubtedly influenced the countries in which their world map projections are heavily used.

Claims that one projection shows less distortion than any other have been quoted in this work, and most are probably true using certain criteria stated or not stated by and varying with the inventor. These artificially restricting criteria present others with the same problems in choosing the best projection for a given application. The best equal-area projection for a very long narrow strip following a parallel of latitude is almost certainly the Albers equal-area conic, but almost any actual region is not this or any other idealised shape, and the property required may not be clear.

The choice of projection for a region –province, continent or ocean – should be governed by the shape and extent of the region, the purpose of the map, the likelihood that the map would be one of a series covering several regions or several characteristics of the same region, and the possible need for the maps of the set to fit together.

The selection of a world map projection is more subjective, and the choice generally makes much more of a visual difference. Mathematical analyses or minimisation of error is much less helpful in providing a visually satisfactory product, in part because of the diversity of land and water, polar and equatorial portions (Robinson 1986, 1988). For an overall view of the world, it would seem that for the model the designer of a flat map should draw chiefly upon the area, shape and distance relationships on the globe. Instead, the familiar Mercator projection was used as the model for the commonly seen Gall stereographic, modified Gall, Van der Grinten, Miller cylindrical and The Times projections – retaining the general appearance, but of course with modifications reducing the area distortion. Even the numerous other cylindrical and pseudocylindrical projections have the property in common with the Mercator that all points at the same distance from the equator on the globe are shown at the same perpendicular distance on the map, a desirable property closely related to climate which is not true of azimuthal, modified azimuthal and oblique and transverse projections in general. On the other hand, the latter types emphasise the curvature of meridians and parallels and the roundness of the globe, and azimuthals can better relate other locations to the centre. The mutual exclusiveness of many of these properties provides impetus to the ongoing developments in the field.

The modern computer has been an incentive in developing special low error projections, but to date these have tended to remain academic rather than become the basis of cartographic application. Those capturing most attention are still those that look distinctive, and pre-computer age world maps are the more popular twentieth century projections. This dichotomy of scientific analysis versus acceptability is true of many fields, however, and it helps give human interest to a subject that can otherwise be especially esoteric.

The current reluctance to use projections that are more mathematically complicated, in spite of the need to program the formulas only once for a given type of computer, will helpfully dissipate for medium- and large-scale maps. The increasing availability of both forward plotting (lat/long to *x, y*) and inverse data gathering (*x, y* to lat/long) formulas in a published algorithmic form for numerous projections at least since 1982, as well as in numerous computer software programs, is reducing this reluctance, but working with projections still strikes fear in the hearts of many trained cartographers and geographers because of the mathematical aspects. Just as the available instructions for computer-operated home appliances seem beyond the ability of many adults to follow, even though they need no knowledge of the computer program itself, so map projection programs require more than the routine instruction 'enter the standard parallels'.

It is more important that cartographers, for whom use of map projections is only one of many skills involved in map work, be brought closer to the projections already available than that still more projections be developed. [. . .]

In practice, the field of map projections will undoubtedly continue in various directions. More projections will be developed by both mathematically and educationally oriented individuals, but more of the latter will try to make the subject digestible to others. The more esoteric work of the past and present will be occasionally picked up by a mapmaker, but generally the accepted choices of the past will dominate future usage as they have dominated the present. The increased emphasis on the computer will not merely make it easier to prepare the printed map; it will also remove the need for many printed maps as Mark Monmonier (1985: 179) has stressed: 'A prime casualty of cartography's Electronic Transition will be the attitude that the map is a printed product'. Computers still take a few minutes to plot a map on a different projection, unless the map is already in computer storage, but this time is being reduced so that a series of maps will soon be able to appear rapidly on the screen with varying properties for a set of requirements, and the deficiencies of one projection can be shortly offset with another view having different advantages and shortcomings. Distances can be calculated and windowed on the terminal, offsetting scaling problems.

It has been said at various times that all worthwhile map projections have already been developed. The same was said decades ago of inventions in general. As Robinson and Sale (1969: 202) have written, 'It may reasonably be asserted that at present cartographers need to devote little time to devising new projections but rather would do better to become more proficient in selecting from the ones available. On the other hand', they added, 'if a new and particular use of maps requires a special type of projection, undeveloped as yet, such a projection might well be worth the time and effort spent in devising it'. The subject remains at least as alive as it was in the past.

References

Calderera, G. (1910) Sulla classificazione delle proirzioni geografiche. *Rivista Geografica Italiana,* **17** (9), 473–487.

Campbell, T. (1987) Portolan charts from the late thirteenth century to 1500, in *The History of Cartography, Volume One, Cartography in Prehistoric, Ancient, and Medieval Europe and the Mediterranean* (eds J.B. Harley and D. Woodward), University of Chicago Press, Chicago, pp. 371–463.

d'Avezac-Macaya, M.A.P. (1863) Coup d'oeil historique sur la projection des cartes de géographie. *Bulletin de la Société de Géographie*, **5** (5), 257–361, 438–85.

Dickinson, H.W. (1949) A brief history of draughtsmen's instruments. *Transactions of the Newcomen Society*, **27**, 73–84.

Dilke, O.A.W. (1987) The culmination of Greek cartography in Ptolemy, in *The History of Cartography, Volume One, Cartography in Prehistoric, Ancient, and Medieval Europe and the Mediterranean* (eds J.B. Harley and D. Woodward), University of Chicago Press, Chicago, pp. 177–200.

Goussinsky, B. (1951) On the classification of map projections. *Empire Survey Review*, **11** (80), 75–79.

Keuning, J. (1955) The history of geographical map projections until 1960. *Imago Mundi*, **12**, 1–24.

Lee, L.P. (1944) The nomenclature and classification of map projections. *Empire Survey Review*, **7** (51), 190–200.

Maling, D.H. (1974) Personal projections. *Geographical Magazine*, **46** (11), 599–600.

Monmonier, M.S. (1985) *Technological Transition in Cartography*, University of Wisconsin Press, Madison, WI.

Nordenskiold, A.E. (1889/1973) *Facsimile-Atlas to the Early History of Cartography with Reproductions of the Most Important Maps Printed in the XV and XVI Centuries*, Dover Publications, New York.

Robinson, A.H. (1986) *Which Map is Best? Projections for World Maps*, American Congress on Surveying and Mapping, Falls Church, VA.

Robinson, A.H. (1988) *Choosing a World Map: Attributes, Distortions, Classes, Aspects*, American Congress on Surveying and Mapping, Falls Church, VA.

Robinson, A.H. and Sale, R.D. (1969) *Elements of Cartography*, 3rd edn, John Wiley & Sons, Inc., New York.

Snyder, J.P. and Steward, H. (1988) *Bibliography of Map Projections*, US Geological Survey, Washington, DC.

Stevenson, E.L. (1932) *Geography of Claudius Ptolomy*, New York Public Library, New York.

Thrower, N.J. (1972) *Maps and Man*, Prentice-Hall, Englewood Cliffs, NJ.

Tobler, W.R. (1962) A classification of map projections. *Annals of the Association of American Geographers*, **52** (2), 167–175.

Woodward, D. (1975) *Five Centuries of Map Printing*, Chicago University Press, Chicago.

Woodward, D. (1987) Medieval mappaemundi, in *The History of Cartography, Volume One, Cartography in Prehistoric, Ancient, and Medieval Europe and the Mediterranean* (eds J.B. Harley and D. Woodward), University of Chicago Press, Chicago, pp. 286–370.

Further reading

Crampton, J.W. (1994) Cartography's defining moment: The Peters Projection controversy, 1974–1990. *Cartographica*, **31**, 16–32. [A critical and discursive analysis of the significance of the development and promotion of the Peters projection for cartographic practice and wider society.]

Iliffe, J. and Lott, R. (eds) (2008) *Datums and Map Projections: for Remote Sensing, GIS and Surveying*, 2nd edn, Whittles Publishing, Dunbeath. [A current practical guide to the use of projections across the spatial sciences.]

Monmonier, M.S. (2004) *Rhumb Lines and Map Wars: A Social History of the Mercator Projection*, University of Chicago Press, Chicago. [A rich historical treatment exploring the diverse contexts in which the iconic Mercator projection has been employed since its invention.]

Stewart, J.Q. (1943) The use and abuse of map projections. *Geographical Review*, **33**, 588–604. [An early academic guide to the appropriate deployment of projections.]

See also

- Chapter 1.12: The Agency of Mapping
- Chapter 2.2: A Century of Cartographic Change
- Chapter 2.5: Automation and Cartography
- Chapter 2.6: Cartographic Futures on a Digital Earth
- Chapter 2.11: Extending the Map Metaphor Using Web Delivered Multimedia
- Chapter 3.7: Area Cartograms: Their Use and Creation

Chapter 2.10

Mobile Mapping: An Emerging Technology for Spatial Data Acquisition

Rongxing Li

Editors' overview

Li's paper from 1997 considers how geographic positioning and spatial sensing technologies can be integrated on a single platform (like a truck) with the potential to create an innovative mobile mapping system. Such a system working in an largely automated and highly accurate fashion can speed up and reduce the costs of capture of cartographic data in comparison to conventional methods (especially surveying), with particular relevance to applications in terms of mapping highways and utility infrastructure. Mobile mapping also opens up new ways to represent space, particularly with extensive ground-level georeferenced imagery (which has been realised more recently as consumer services by Google Street View).

Originally published in 1997: *Photogrammetric Engineering & Remote Sensing,* **63** (9), 1085–1092.

Introduction

Large-scale spatial data and associated attributes in geographical information systems (GIS) are in high demand in order to generate new databases and to update existing databases in applications such as transportation, utility management and city planning. The acquisition of such information is mostly realised by aerial photogrammetry or terrestrial surveying using, for example, total stations. In the former case, aerial photographs may not provide sufficient detailed information regarding planimetric and horizontal object features, for example, manholes, positions of road curbs and centre lines, building facades and so on, because of the photo scale and perspective projection geometry. In the latter case, object features can be measured very accurately. However, the operational time of the field survey and associated costs are major concerns and sometimes render a project impossible. Furthermore, a road or utility survey along a highway may be costly or impractical if the traffic has to be stopped or detoured for a long period. However, a mobile mapping system, equipped with a mobile platform, navigation sensors and mapping sensors, is capable of solving the above problems. The mobile platform may be a land vehicle, a vessel or an aircraft.

Generally, navigation sensors, such as Global Positioning System (GPS) receivers, vehicle wheel sensors and inertial navigation systems (INS), provide both the track of the vehicle and position and orientation information of the mapping sensors. Objects to be surveyed are sensed directly by mapping sensors, for instance, charge coupled device (CCD) cameras, laser rangers and radar sensors. Because the orientation parameters of the mapping sensors are directly supplied by the navigation sensors, complicated computations such as photogrammetric triangulation are reduced or avoided. Spatial information regarding the objects is extracted directly from the georeferenced mapping sensor data by integrating navigation sensor data. Advantages of such a system include:

- Increased coverage capability, rapid turnaround time and, thus, improved efficiency of the field data acquisition.
- Integration of various sensors so that quality spatial and attribute data can be acquired and associated efficiently.

The Map Reader: Theories of Mapping Practice and Cartographic Representation, First Edition. Edited by Martin Dodge, Rob Kitchin and Chris Perkins.

- Simplified geometry for object measurements supported by direct control data from navigation sensors.
- Flexible data processing scheme with original data stored as archive data and specific objects measured at any time.
- Strongly georeferenced image sequences which provide an opportunity for automatic object recognition and efficient thematic GIS database generation.

Mobile mapping technology has been researched and developed since the late 1980s. It was inspired by the availability of GPS technology for civilian uses. [...]. The mobility of the mapping systems has always been a concern. It requires high resolution mapping sensors covering large areas, and high accuracy navigation sensors determining the positions and orientations of the vehicle. [...]

Land-vehicle-based mobile mapping systems result in, amongst others, (1) close distances between the systems and the objects to be surveyed, (2) no ground control and no triangulations across images exposed at different time, and (3) completely digital processing. These systems are designed mainly for mapping purposes. [...] Processing of the vast amount of mobile mapping data is subsequently a very important task. So far, there is no common commercial software capable of handling the data from different mobile mapping systems. Automation of the procedures of mobile mapping data processing has not been extensively researched because most efforts seem to have been made in the development of the data acquisition systems. However, the automation of processing such large observation databases is of great importance. Automatic matching of corresponding image points in an image sequence was reported by Li *et al.* (1994b) and Xin (1995). Extraction of road centrelines and curb lines from mobile mapping image sequences was researched by He and Novak (1992), Tao *et al.* (1996) and Li *et al.* (1996a). Three-dimensional coordinates in the object space calculated from the mobile mapping image sequences can be optimised by considering both precision and reliability (Li *et al.* 1996b). [...].

Mobile mapping technology

Spatial and time referencing in mobile mapping systems

The position of the moving platform changes dynamically along a predefined track to acquire mapping data of objects within its field of view. Two critical issues are (1) how to determine the dynamic positions of the platform itself at any time and (2) how to further derive the spatial information of the objects of interest in the field of view from the platform. Without appropriately defined spatial and time reference systems, it is difficult to describe relationships between the objects, the sensors, the platform and the world. The track of the vehicle is usually defined in a global coordinate system (Figure 2.10.1). Depending on the geographical extent of a project and the application requirements, this global coordinate system may be, for example, the Universal Transverse Mercator (UTM) projection system, a state plan coordinate system or a 3D Cartesian coordinate system. The platform coordinate system (X_v, Y_v, Z_v) is defined on the vehicle and used to integrate various sensors such as the GPS receiver, INS unit, laser device and cameras. An individual sensor, say the *i*th sensor, has its own local coordinate system (x_i, y_i, z_i), which is related to the platform coordinate system. The platform coordinate system is further referenced to the

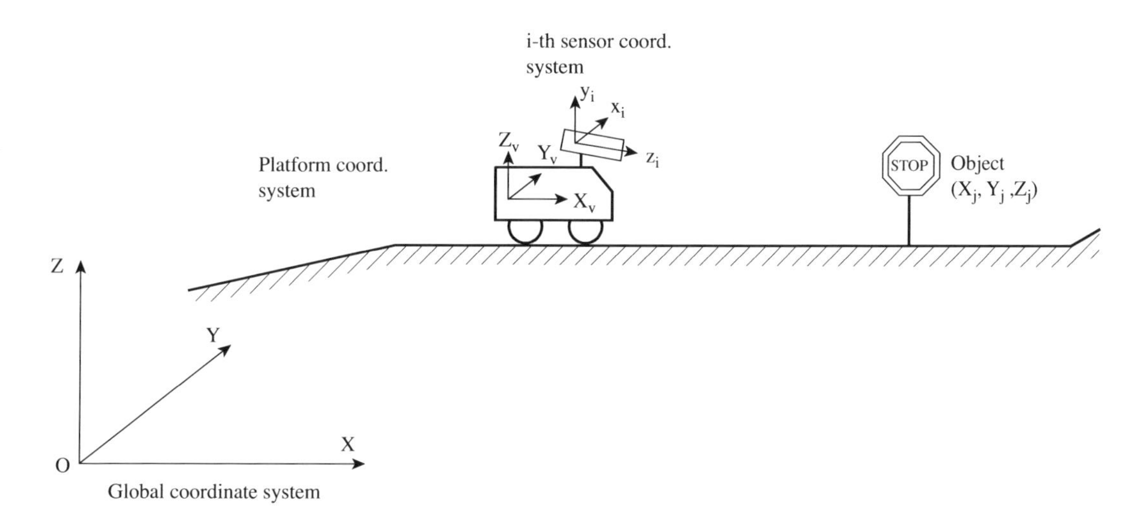

Figure 2.10.1 Relationships between objects, the platform and the world.

global coordinate system. For example, the object to be surveyed is a traffic sign which is within the field of view of the *i*th sensor (a camera in Figure 2.10.1). The objective is to derive the coordinates of the traffic sign in the global coordinate system (X_i, Y_i, Z_i). This can be realised by detecting the object using a single or multiple sensors and calculating its coordinates in a local coordinate system. By means of the platform coordinate system, the coordinates of the object in the local sensor coordinate system are transformed to the global coordinate system.

The mathematical models for calculating the object coordinates in the local coordinate system depend on the types of sensors. This will be discussed in a later section. The determination of transformation parameters between the local sensor coordinate systems and the platform coordinate system is usually carried out in a system calibration procedure. This procedure requires a special setting of precisely surveyed and well distributed control points, and has to be performed periodically, or whenever there is a change in the relationship between the sensors (Moffitt and Mikhail 1980; El-Sheimy and Schwarz 1993).

A time reference system is used to track the kinematic positions of the platform. It is also used in the synchronisation of the sensors to incorporate signals from various sources taken at different epochs. The signals are both spatially and temporally referenced to derive the global spatial information of the object detected by the sensors. The time reference system used in mobile mapping is dependent on the sensors used and the accuracy required. GPS time signals, related to Universal Coordinated Time (UCT), are available in received GPS data (Leick 1995). An INS, however, provides more frequent time updates (El-Sheimy and Schwarz 1993). Once a time system is chosen as the time reference system, signals from different sources can be integrated and the kinematic platform locations can be determined.

[...]

Extraction of spatial information

Derivation of spatial information

Spatial information from a direct measurement system is primarily obtained from the positions of the receiver. In this paper, methods for deriving spatial information from indirect measurement systems are discussed. If CCD cameras are employed (Figure 2.10.2), objects to be measured are identified in a pair of stereo images acquired by the cameras. The objective is to measure the objects in the images and to derive their positions in the global coordinate system. Suppose that a point, *J*, appears in the left and rights images with its image points as *j* and *j'*, respectively. Their coordinates in the images are measured as (x_j, y_j) and ($x_{j'}$, $y_{j'}$). According to photogrammetric principles (Moffitt and Mikhail 1980), the three vectors $\Delta\mathbf{r}_j^{C1}$, $\Delta\mathbf{r}_j^{C2}$, and **B** must lie on the same plane. This coplanarity condition can be expressed as a scalar triple product equation:

$$(\Delta\mathbf{r}_j^{C1} \times \Delta\mathbf{r}_j^{C2})^*\mathbf{B} = 0$$

This equation can be elaborated to a function of the observations (x_j, y_j) and ($x_{j'}$, $y_{j'}$), the calibration parameters, and the unknown coordinates of point *J* in the platform coordinate system. The calibration parameters include focal lengths of the cameras, the rotation matrices $\Delta\mathbf{M}_V^{C1}$ and $\Delta\mathbf{M}_V^{C2}$ from the local camera coordinate systems to the platform coordinate system, and the coordinates of the camera exposure centres in the platform coordinate system. [...]

Because the images are taken in sequence, one object point is usually coved by multiple images, which may form stereo image pairs with combinations of images taken at different epochs. A simple way to improve the accuracy is to measure the same point in all possible images for

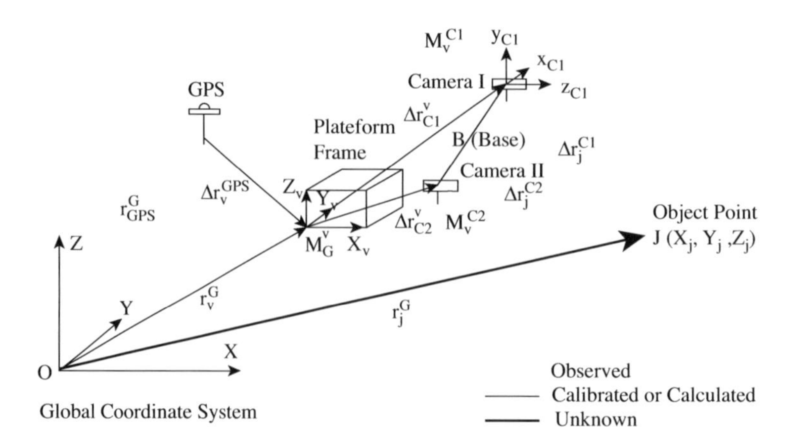

Figure 2.10.2 Mapping by indirect measurements.

calculating the coordinates of the point. This simultaneous approach was proven inefficient in comparison to a sequential transformation algorithm based on Givens transformation (Gruen and Kersten 1995). Further research by Li *et al.* (1996a) led to an algorithm for selecting an optimal image pair horn an image sequence for photogrammetric point intersection using Kalman filtering. This algorithm optimises both precision and reliability. Image matching techniques automate the procedure for measuring point and line features. Although not all features can be measured automatically at the moment, productivity can be improved by the matching techniques, considering the sizes of large databases to be generated.

Points, arcs and polygons are three primary spatial components of a GIS database. With points derived from the image sequences as discussed above, the other two components can easily be generated by an extended point measurement procedure (Li *et al.* 1994a, 1994b).

Derivation of attribute information

Attribute information is another important data category in a GIS database. It is associated with spatial features to describe additional non-spatial characteristics of objects. Some systems collect attribute data, such as recorded voices, indicating the wildlife species found in a location, or video images of road surfaces, together with the spatial information acquired, either as points for the species, or as 'continuous' lines of the road surfaces in video images. In an indirect measurement system, interactive interpretation of the image sequences is mostly used to derive the attributes. This requires that the mobile mapping software be designed and implemented in such a way that the attributes can be associated with the spatial features during the measurement procedure (Li *et al.* 1994b).

An ideal way to generate the attribute information is to extract the information from the image sequences automatically (He and Novak 1992; Li *et al.* 1996b; Tao *et al.* 1996). Currently, there are some very limited cases where this automation may be performed successfully. Firstly, text information in images, for example, street names on road signs and texts of traffic signs, can be extracted by pattern recognition methods. Depending on the orientation of the sign and of the cameras, the projection of the text on the image may be distorted or even invisible. Robust text recognition algorithms which consider the field conditions should be developed. Secondly, some specific types of objects, such as manholes and fire hydrants, have a symmetric geometry in the horizontal plane and their projections in the images are less distorted. Because the geometry of these objects is known, artificial images can be generated. Then these images can be matched with the object features in the real image sequences in order to automatically extract the attribute information. Finally, a general classification scheme, such as that used in the classification of satellite remote sensing images, is not appropriate for the mobile mapping images, because of the rapidly changing scales of the objects in the images and because of the extremely detailed and diverse information involved. The number of object classes in such an image is much higher than that of satellite imagery.

A semiautomatic procedure is practical for GIS database production. For each spatial feature, its geometry can be measured either interactively or automatically using image matching techniques. The attribute is then determined automatically if possible. Otherwise, an interpretation by the operator is necessary. In the current stage, this is still the most effective and reliable method for attribute derivation from the mobile mapping data.

Accuracy

The accuracy of the coordinates of a point in the global coordinate system derived from the mobile mapping data is one of the important quality measures. The errors involved are in turn contributed by those from the system components. The following discussion focuses on three error categories in mobile mapping systems (Table 2.10.1): control data errors, mapping sensor errors and the overall system error, based on a typical configuration with a land vehicle, DGPS, INS and CCD cameras.

Control data refer to observations from the navigation sensors used to derive the position and orientation parameters of other mapping sensors such as cameras. Thus, the accuracy of these observations directly affects the estimated positions and attitudes of the mapping sensors. Furthermore, it influences the point positions derived from the mapping sensor data. According to recent reports by Pottle (1995), El-Sheimy (1996) and Bossler and Toth (1996), if DGPS and post-processing techniques are employed, the accuracy of the dynamic platform positions can be determined in the range 6–20 cm. The speed of the vehicle is controlled under 70 km/hour. Angular parameters derived from an affordable strap-down INS are accurate to within 1–6 minutes if the time-dependent drift is corrected (Schwarz and El-Sheimy 1996; Hock *et al.* 1995). This angular error would result in a linear error of 1.5–8.7 cm at a distance of 50 m from the system. Inexpensive approaches using gyroscopes provide the angular parameters with an accuracy of 0.75–2.0 degrees (Cosandier *et al.* 1993). Experiments calculating the attitude information from data acquired by multiple GPS receivers on a vehicle gave an accuracy of 0.3–0.5 degrees with an average baseline between the receivers of 1.7–1.0 m (Lu 1995; Sun 1996). It is obvious that, at the present time, the angular parameters calculated from the observations of gyroscopes and multiple GPS receivers are not satisfactory for high accuracy control of the mapping sensor.

Table 2.10.1 Contribution of individual errors and the overall accuracy of the systems

	Error Type	Error	Conditions
Control Data	Platform positions from DGPs (dynamic)	6–20 cm	Speed of the vehicle <70 km
	Angular parameters from INS	1–6 minutes	Strapdown INS
	Angular parameters from gyroscopes	0.75–2 degrees	Vertical, heading and rate gyroscopes
	Angular parameters from multiple GPS antennas	0.3–0.5 degrees	The average baseline between 1.7 and 1.0 m
	Synchronisation	1–2 ms	1.9–3.9 cm at a vehicle speed of 70 km/h
Mapping Sensor Data	Calibration	2–5 mm	Well established control field
	Target positions in images	0.05 pixel	Image metrology techniques
	General objects in images	0.5 pixel	Zoom function
	Lens distortion	1.5 pixels	If not corrected
Overall System	Points in the object space	30–50 cm	DGPS, INS and medium level cameras

The synchronisation of the sensors gives an error of about 1–2 ms, which translates to a linear error of 1.9–3.9 cm at a vehicle speed of 70 km/h (Schwarz and El-Sheimy 1996).

Errors contained in the mapping sensor data include system calibration errors, image measurement errors and lens distortion errors (if not corrected) in the photogrammetric processing (Table 2.10.1). After a system calibration with a well established control field and an appropriate procedure (El-Sheimy and Schwarz 1993), the calibration errors can be as small as 2–5 mm. In the image space, locations of objects with clear image features and a symmetric geometry, such as marked targets used in system calibrations, can be determined at a subpixel level, for example 0.05 pixel, using image metrology techniques (Cosandier and Chapman 1992; Li 1993). Measurements of general objects can be performed with an accuracy of 0.5 pixel if a zoom function is employed. The lens distortion is usually modelled and corrected in the photogrammetric processing. If not corrected, this error may reach 1.5 pixels in the image space.

The errors caused by imprecise image measurements were reported to be 4 cm for objects within 50 m from the cameras, while the effect of the lens distortion reaches 12.7 cm if not corrected (Schwarz and El-Sheimy 1996). An estimate of the overall system accuracy depends on the accuracy of the individual components of the system. Taking the positional accuracy of an object point in the object space as the measure of the overall system accuracy, some systems have reached 30–50 cm (both horizontal and vertical) if the objects are 30–60 m from the system (Schwarz and El-Sheimy 1996; Bossler and Toth 1996; Hock *et al.* 1995). This overall system accuracy is important for defining application areas of the system, and also for evaluating the cost effectiveness of the system configuration. In practice, the images may have a sufficient resolution for identifying relatively small objects such as fire hydrants. The positions can be determined to an accuracy of 30–50 cm, for example. However, the positional errors may exceed the diameter of the fire hydrants. The important differences between resolution and accuracy should be distinguished. A relatively expensive, hence important, component in the system is the INS. Systems developed for applications not requiring highly accurate spatial data may be able to use alternatives such as gyroscopes or multiple GPS receivers to reduce the system costs.

Application considerations

Highway application

Because it employs dynamic data acquisition, mobile mapping technology can be directly used in highway related applications, such as traffic sign inventory, monitoring of speed and parking violations, generation of road network databases, and road surface condition inspection when laser technology is jointly applied. There are several advantages to using mobile mapping technology in highway applications. The data acquisition is performed without blocking traffic, assuming that traffic velocity is less than, for example, 70 km/h. The information obtained is diverse – a single collection can be used for multiple purposes. Moreover, because data can be both collected and processed in a short period, frequent and repetitive road surveys and database updating are both possible and affordable.

Objects along roads and highways, for example, traffic signs, light poles, bridges, road centrelines and so on, are usually represented as clear image features in the image sequences. Therefore, they can be identified easily and measured interactively to build a spatial database. In order to extract a road centreline, an image sequence of the road is needed. Each image pair supplies a segment of the centreline. The successive road segments from the sequence are measured continuously and combined to produce the entire road centreline.

Road centrelines and separating lines between lanes are painted in white or yellow. They have solid or dashed line patterns. Based on road surface and weather conditions, the quality of the lines in the image sequence varies. Manual extraction of the lines by the operator is relatively time consuming, although it is superior to other traditional surveying methods. Automation of this procedure has been researched. From the image sequence, centreline features are enhanced and extracted automatically. The corresponding 3D centreline segments are then generated in the object space (He and Novak 1992). Another approach defines a 3D centreline model in the object space as a physical Snake model (Tao *et al.* 1996). The Snake model is optimised to adjust the centreline shape using image features of the centreline as internal constraints, and using geometric conditions derived from other sensors of the system (GPS and INS) as external constraints. This method for automatic centreline extraction and reconstruction is reliable for different road conditions and line patterns. On the other hand, road curb lines, as opposed to painted centrelines, are projected onto the images based on their geometric shapes and material types. Therefore, curb lines can be more difficult to extract and identify automatically. Currently, curb line databases are built using semiautomatic approaches. The system provides the user with projected curb lines in a stereo image pair. The user is asked to confirm if the line pair suggested by the system is correct. Sometimes, because of image quality, the system cannot provide the data. In this case, the operator is asked to digitise the curb segments manually.

It is often required, in road surface condition inspection, that road surface cracks be located and measured. If the system is equipped with laser sensors, the depths between the sensors and the road surface are available as relative measurements. If the control data from the GPS and INS are available, the depth data derived from the laser sensors can be integrated into the global reference system and used to generate a digital road surface model (DRSM). The DRSM provides a geometric description of the road surface. This can be used to detect the locations, sizes and shapes of the road surface cracks. If the digital image sequence of the same road is processed and georeferenced, each grid point of the DRSM can be assigned a grey scale from the corresponding pixel of the image sequence. In this way, the images appear to be draped onto the DRSM, and a 3D road surface image is generated. By observing this 3D image using 3D visualisation tools, road surface conditions, including cracks, can be illustrated more efficiently. On the other hand, the DRSM can also be built from the stereo image sequence along the road. Corresponding road surface points appearing in a stereo pair can be measured by digital image matching techniques. Because one point may appear in more than one successive stereo pair, a point covered by an obstacle, such as a vehicle, in one stereo pair may be visible in the preceding or subsequent pairs. A gridding procedure is applied to calculate grid points of the DRSM. In comparison to the digital matching method using stereo image sequences, laser sensors generate more reliable surface models, because the depths are always measured directly. However, in the model built by the matching-based method, there may be areas with points that are not measured, but interpolated, because the areas are invisible or do not have sufficient image texture for image matching.

Application to facility mapping

Another useful application of mobile mapping technology is in the area of facility mapping. High voltage power transmission lines can be photographed by the mobile mapping system, and their positions can be measured from the image sequences. There are a number of important parameters which can be calculated from the mobile mapping data, for instance, the positions of the poles and/or towers, the positions of the insulators on each line which support the suspending transmission line segments, and the lowest points of the suspending line segments. In order to capture these desired transmission line features, the cameras have to be oriented somewhat upwards to aim at the towers and line segments. Consequently, a large portion of the resulting images contains the sky as background. This makes it easy to distinguish the targets from the background in the images because of the high contrast between them. Based on the same reason, automatic extraction of the transmission line segments in the image space is also possible. However, if epipolar geometry (Moffitt and Mikhail 1980; Li 1996) is used to determine the line segments in the object space, the 3D points along the line segments to be measured are dependent on the intersections between the epipolar lines and the line segments in the images (Figure 2.10.3). In case I, two cameras are forward looking, with an upper-left angle. For the *k*th image pair, the epipolar lines formed by $\text{Image}_{\text{left},k}$ and $\text{Image}_{\text{right},k}$ are almost parallel to the transmission lines. The intersections thus obtained are of low accuracy, considering the effect of both the intersection accuracy and errors of the epipolar lines caused by imperfect orientation parameters. This will consequently affect the accuracy of the 3D line segments in the object space derived thereby. There are four options for solving this problem. (1) In case I of Figure 2.10.3, if an appropriate overlapping area is available, subsequent images such as $\text{Image}_{\text{left},k}$ and $\text{Image}_{\text{right},k+1}$ are used to form a stereo pair so that the epipolar $\text{line}_{(\text{left},k)(\text{left},k+1)}$, has abetter intersecting angle with the transmission line. (2) In case II of

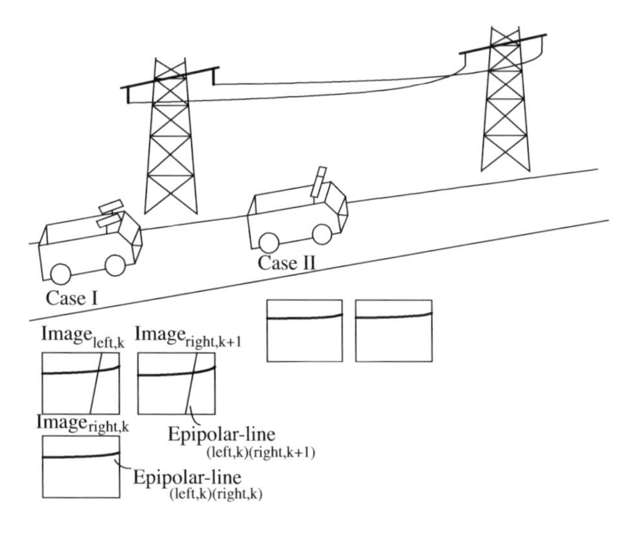

Figure 2.10.3 Camera orientation and epipolar geometry for determination of transmission lines.

Figure 2.10.3, the cameras are oriented toward the left side. This will also result in effective intersecting angles. (3) A hardware-based stereo viewing system can be used, for example, using a polarised system with a stereo glass and a special monitor. In this case, a 3D stereo model can be reconstructed. The operator is then able to view the transmission line in 3D and measure points along the line three dimensionally. (4) Better results can also be achieved by combining data acquired by laser ranging by a helicopter if available (Krabill 1989).

Some objects, such as fire hydrants and manholes, have symmetric geometric shapes and are less dependent on the camera positions and orientations. Thus, they appear similar in images. Based on the geometric shapes of the objects, simulated lighting sources and material characteristics, artificial images can be generated. These artificial images can then be compared with the image features in the sequences. A matching procedure between the artificial images and real images is performed. In this way, both the geometric information and attributes of fire hydrants and manholes appearing in the image sequences can be extracted efficiently.

A perspective on mobile mapping

One of the criteria used to measure the quality of the system is the accuracy of the location of objects measured using the acquired data. This accuracy is strongly influenced by the control data collected by the navigation sensors and the quality of the mapping sensors. The components of mobile mapping systems function efficiently and reliably as individual and independent systems in various applications. A mobile mapping system requires that these components work cooperatively.

Another factor to consider is cost. A highly accurate strap-down INS is currently the most costly component in the system. Low level INS and gyroscopes may be used, but they do not supply the same quality angular parameters which can be employed, for example, in camera orientation. The high cost of the INS make the entire system relatively expensive. If high accuracy is not essential, low cost gyroscopes can be used in order to make the system more affordable. The image resolution of the cameras has a great influence on the accuracy of the photogrammetric intersections of object points. This is especially critical because the physical baseline of the cameras is limited by the dimension of the land-based vehicle. For an object far from the cameras, the intersecting angle is small, and a one pixel error in the image will result in a large along-track error. High resolution cameras up to 4096 pixels by 4096 pixels will be available but will be rather expensive. In addition, the images acquired by high resolution cameras occupy a large memory and storage space. If multiple high resolution colour cameras are employed, the problems of efficient image data transmission and medium storage during the data acquisition, efficient object measurement and attribute extraction, and data archiving will be addressed.

GPS should provide 'continuous' and consistent data for absolute positional control. However, there are cases where GPS signals are blocked by high-rise buildings, tunnels and other objects. An integration of GPS signals at the last point before the signal blocking and the first point of signal recovery with INS trajectories bridges the gap of the control data. On the other hand, there are situations in which GPS signals are only blocked for a very short period, affecting one or two exposure stations. If not detected and corrected, these errors will distort both the positions of the camera exposure centres, and the object points derived from the image sequences. Therefore, an automated systematic quality checking procedure should be implemented to examine the GPS and INS data. If such an inconsistency exists, positions of a point derived from different image pairs of different exposure stations will show a large difference. Otherwise, they should have the same position within a certain tolerance. This quality checking procedure guarantees the quality of the control data used for camera orientation.

[...]

Efforts in mobile mapping research and development have been made by researchers and engineers over the last decade. The technology has evolved to such a degree as to allow mapping and GIS industries to use it for high flexibility in data acquisition. More information can be gained in less time, and with less effort, while achieving high productivity. [...]

References

Bossler, J.D. and Toth, C.K. (1996) Feature positioning accuracy in mobile mapping: results obtained by the GPSVan. *International Archives of Photogrammetry and Remote Sensing*, **31** (Part B2), 139–142.

Cosandier, D. and Chapman, M.A. (1992) High precision target location for industrial metrology. *Videometrics*, **1820**, 111–122.

Cosandier, D., Chapman, M.A. and Ivanco, T. (1993) Low cost attitude systems for airborne remote sensing and photogrammetry. *The Canadian Conference on GIS*, Ottawa.

El-Sheimy, N. (1996) A mobile multi-sensor system for GIS applications in urban centers. *International Archives of Photogrammetry and Remote Sensing*, **31** (Part B2), 95–100.

El-Sheimy, N. and Schwarz, K.-P. (1993) Kinematic positioning in three dimensions using CCD technology. *Proceedings of IEEE Vehicle Navigation Information Systems (VNIS)*, pp. 472–475.

Gruen, A. and Kersten, T.P. (1995) Sequential estimation in robot vision. *Photogrammetric Engineering & Remote Sensing*, **61** (1), 75–82.

He, G. and Novak, K. (1992) Automatic analysis of highway features from digital stereo images. *International Archives of Photogrammetry and Remote Sensing*, **29** (Part B3), 119–124.

Hock, C., Caspary, W., Heister, H. *et al.* (1995) Architecture and design of the kinematic survey system KiSS. *Proceedings of the Third International Workshop on High Precision Navigation*, Stuttgart, Germany, pp. 569–576.

Krabill, W.B. (1989) GPS applications to laser profiling and laser scanning for digital terrain models. *42nd Photogrammetric Week*, Stuttgart, Germany, pp. 329–340.

Leick, A. (1995) *GPS Satellite Surveying*, 2nd edn, John Wiley & Sons, Inc., New York.

Li, R. (1993) Building octree representations of 3D objects in CAD/CAM by digital image matching techniques. *Photogrammeteric Engineering & Remote Sensing*, **58** (12), 1685–1691.

Li, R. (1996) Design and implementation of a photogrammetric geo-calculator in a Windows environment. *Photogrammetric Engineering & Remote Sensing*, **62** (1), 85–88.

Li, R., Schwarz, K.-P., Chapman, M.A. and Gravel, M. (1994a) Integrated GPS and related technologies for rapid data acquisition. *GIS World*, April, 41–43.

Li, R., Chapman, M.A., Qian, L. *et al.* (1994b) Rapid GIS database generation using GPS/INS controlled CCD cameras. *Proceedings of the Canadian Conference on GIS*, pp. 465–477.

Li, R., Chapman, M.A., Qian, L. *et al.* (1996a) Mobile mapping for 3D GIS data acquisition. *International Archives of Photogrammetry and Remote Sensing*, **31** (Part B2), 232–237.

Li, R., Chapman, M.A. and Zou, W. (1996b) Optimal acquisition of 3D object coordinates from stereoscopic image sequences. *International Archives of Photogrammetry and Remote Sensing*, **31** (Part B3), 44–52.

Lu, G. (1995) Development of a GPS multi-antenna system for attitude determination. Unpublished PhD thesis. Department of Geomatics Engineering, University of Calgary.

Moffitt, F.H. and Mikhail, E.M. (1980) *Photogrammetry*, Harper & Row Publishers, New York.

Pottle, D. (1995) New mobile mapping system speeds data acquisition. *GIM*, **9** (9), 51–53.

Schwarz, K.-P. and El-Sheimy, N. (1996) Kinematic multi-sensor systems for close range digital imaging. *International Archives of Photogrammetry and Remote Sensing*, **31** (Part B3), 774–785.

Sun, H. (1996) Integration of Multi-Antenna GPS System with CCD Camera for Rapid GIS Data Acquisition. ENGO 661 Course Report, Department of Geomatics Engineering, University of Calgary.

Tao, C., Li, R. and Chapman, M.A. (1996) A model driven approach for extraction of road line features using stereo image sequences from a mobile mapping system, *ASPRS/ACSM Proceedings*, Baltimore, MD

Xin, Y. (1995) Automating geostation: a softcopy system. Unpublished Masters thesis. Department of Geomatics Engineering, University of Calgary, Canada.

Further reading

Gartner, G., Bennett, D.A. and Morita, T. (2007) Towards ubiquitous cartography. *Cartography and Geographic Information Science*, **34** (4), 247–257. [A discussion of the evolution of cartography with widespread mobile access and locational awareness.]

Grejner-Brzezinska, D.A., Li, R., Haala, N. and Toth, C. (2004) From mobile mapping to telegeoinformatics: paradigm shift in geospatial data acquisition, processing, and management. *Photogrammetric Engineering & Remote Sensing*, **70** (2), 197–210. [A paper looking beyond mobile mapping towards a new concept of telegeoinformatics.]

Kennedy, M. (2002) *The Global Positioning System and GIS: An Introduction*, 2nd edn, Taylor & Francis, New York. [The leading scholarly text discussing the technical issues of GPS and its various applications.]

Parks, L. (2001) Plotting the personal: Global positioning satellites and interactive media. *Cultural Geographies*, **8** (2), 209–222. [A consideration of the social effects of GPS tools and knowledge of location.]

Thielmann, T. (2010) Locative media and mediated localities. *Aether: The Journal of Media Geography*, **5a**, 1–18. [An insightful review of a range of literature on new media and its relation to mapping socio-spatial activities.]

See also

- Chapter 1.12: The Agency of Mapping
- Chapter 2.12: Imaging the World: The State of Online Mapping
- Chapter 3.11: Egocentric Design of Map-Based Mobile Services
- Chapter 3.12: The Geographic Beauty of a Photographic Archive
- Chapter 5.8: Cartographic Rationality and the Politics of Geosurveillance and Security

Chapter 2.11

Extending the Map Metaphor Using Web Delivered Multimedia

William Cartwright

Editors' overview

Cartwright's paper considers how the map can be made interactive through the application of multimedia technologies. Such software technologies give considerable creative potential to author cartographic products that are dynamic and responsive to users when delivered as digital new media. Cartwright outlines range of metaphoric frameworks to think about how new kinds of multimediated mapping could work, with different modes of information design, game-like interfaces and exploiting hyperlinked navigation offered by the Web.

Originally published in 1999: *International Journal of Geographical Information Science*, **13** (4), 335–353.

Introduction

[...]

The mapping industry has always striven to create products by adapting available technology. Some technological adaptations have worked, others have not. Those technological advancements that proved to be useful, and hence changed the way in which maps were produced, delivered or consumed, have been embraced by the profession and map users alike. They have become part of what mapping (in its broadest sense) now means. Some technologies, whilst proven to be technologically sound, have not met with the success expected. For example, when video discs were introduced they were looked upon as devices that heralded the introduction of multimedia, including the use of multimedia by the mapping sciences. However, whilst the technology and its applications were technically sound, the limited distribution of user video disc players and associated hardware produced a stillborn product. Also, some of the ideas behind the application of these products were not really that different than those employed since the printing press harnessed cartographers to thinking in terms of print-derived specifications (Cartwright 1995). Digital information, for the most part, has done little to promote fundamentally different cartographic products, with most work being directed towards the emulation of the products produced in the manual era. We still think of a map as a visual product where graphic appearance holds priority over content in information quality and resolution. [...]

Today's mapping systems need to provide knowledge to users, as well as data and information. Technology now provides the means of publishing information-rich geographical information products, from which knowledge can be gleaned from a generous delivery mechanism. The provision of different access metaphors may enable these computer-supported information-rich products to aid in imparting knowledge in ways that the user feels more comfortable with, and thus transfer information more effectively. Just continuing with the use of the metaphors that have been successful in the past cannot guarantee continued user acceptance of the format or the delivery mechanism. Users of conventional map displays need no longer be restricted by displays that are presented in isolation and in formats that are not attuned to their preferred mode of usage. By delivering geographical

The Map Reader: Theories of Mapping Practice and Cartographic Representation, First Edition. Edited by Martin Dodge, Rob Kitchin and Chris Perkins.

information in a timely manner, using presentations that the user can choose, and providing that information on equipment that emulates controls and procedures that the user is comfortable with or has already mastered, like using a video recorder, a better understanding of the information being presented may result as users can concentrate on the information being delivered rather than coping with mastering the access devices.

There exists a need to explore the applicability of multimedia metaphor extensions of geographical information products to providing users with a tool for enhanced information presentations that can be linked to 'real world' information. This would enable decisions to be made in concert with information obtained from the real world, rather than from an artificial, map-defined world, thus supporting the user's view of information. The hardware and software development related to multimedia packages and spatial information products like Geographical Information Systems have now reached a stage where the technological issues have largely been resolved. What needs to be investigated is the linking of powerful spatial information systems (that currently allow for the presentation of data in a spatially accurate and timely manner) to multimedia presentation packages. It is argued that the combination of metaphors used in conjunction with the map metaphor, delivered via a hybrid CD-ROM/Web package, would provide the means for users of contemporary geographical information to better understand the information being presented.

[. . .]

Geographical information provision/depiction using complementary metaphors

I [now] introduce an integrated set of metaphors intended to complement the traditional map metaphor for representation of geographic information. The fundamental contention made is that use of this integrated metaphor set will dramatically enhance access to and understanding of geographic information available through the WWW.

The proposed metaphor set adds nine metaphors to the traditional map metaphor for geographic information: the Storyteller, the Navigator, the Guide, the Sage, the Data Store, The Fact Book, the Gameplayer, the Theatre and the Toolbox. These metaphors are seen to offer access genres that are complementary to the use of maps and are, thus, intended to be used in conjunction with maps. The thrust of developing this metaphor set is to allow users to support their map use through multiple forms of information access. [. . .]

The Storyteller

The provision of a storytelling component in a comprehensive geographical information access system can offer users the simple option of being 'told' about the geography of a certain place or designated area. Storytelling can be used to give a sense of place, and can be implemented using electronic storylines. The interactivity of a multimedia environment allows storytelling to be nonlinear. Since the stories will need to be based on geographical fact, the stories presented and the choices that users are allowed to make will be guided by geographical realities. Geographical stories can be constructed according to one of Platt's (1995) electronic storytelling genres: user as observer, user as director, or user as actor.

Elements of the story can include text, pictures, animation, games and video. Even though the user appears to be in total control of the story, the author of the package will impose the real control and the user will control the package within the limits established by the author.

Whilst offering various forms of user control, Platt's methods are essentially linear storytelling approaches. Products offering more user choice about the type of story that could be delivered using multimedia, might be put together using one of Platt's main types of story structure. They can be constructed to allow the user to choose a linear or nonlinear mode as well as providing them with options concerning how they wish to become 'involved' in the story and how it should unfold.

An example of a nonlinear approach is to offer the user a pictorial landscape and, as the user moves the mouse over the screen image, hot spots will appear, offering the user the choice of being told a story about that particular part of the terrain. The advantage of using The Storyteller metaphor is that the user is able to experience a digitally produced story, told through the eyes of perhaps a panel of storytellers, each with their own expertise. This approach can provide a richer portrayal of the world than would be possible using a conventional product.

[. . .]

The Navigator

Maps are often used in a Web-based system as a tool to assist users in finding where information is located (Fabrikant and Buttenfield 1997). Some users who are inexpert at navigating with maps may need The Navigator's help. The Navigator metaphor would give support for users who were not good at interpreting abstract models, including maps. Information resources produced using this type of metaphor would enable the key points to be seen as the user moved through space by making use of key points, or landmarks, encountered as the user moves through space. Geographical resources could be viewed or experienced using 'key points' to indicate sites that should be visited, interpreted and consumed to give a correct interpretation of landforms, population patterns or communications systems.

Multimedia interpretations and presentations using the navigation theme could be built according to Hedberg and Harper (1992) navigation metaphor types: guide (for visual travel through an information 'landscape'); sequential navigation (where cues about where information was located would be made available sequentially); or a visual navigation that would provide a plan of the possible paths through the information landscape; or a hybrid metaphor that would combine appropriate component parts of all three. [...]

Navigation cues allow users to select the part of the current display about which they require more detailed or complementary information. Navigation cues can then 'lead' users to other information within the product or pertinent information outside the product that would be available on-line or locally. Multimedia products are able to provide this type of flexibility.

[...]

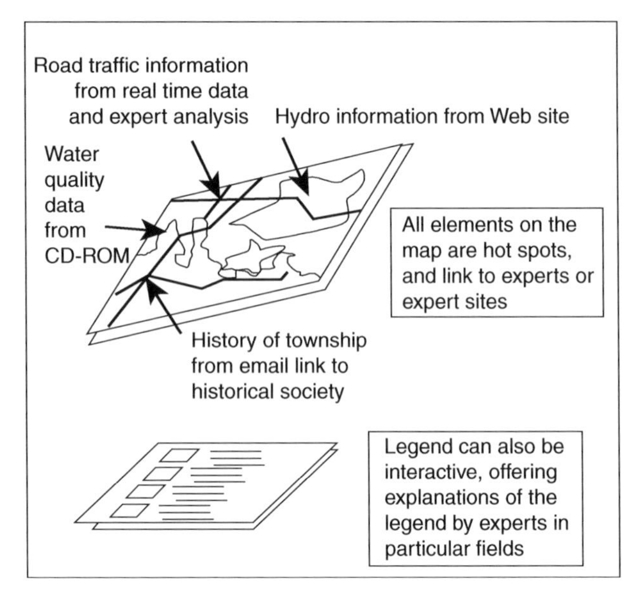

Figure 2.11.1 Map metaphor and Sage metaphor.

The Guide

The Guide metaphor assumes that the user has neither prior knowledge of the area being portrayed, nor the ability to effectively navigate through individual scenes or the entire package. The Guide allows the user to better appreciate and understand the information presented, since the Guide will have been to the location being studied before. The use of this metaphor allows multimedia products to focus on the important elements and to ignore the irrelevant. It can be used to build 'windows' that show the user where pertinent information resides. The Guide insures that pertinent information is automatically chosen for the user and a plan of viewing, one which suggests the best 'viewpoints' for specific information. When choosing the information that would be of interest to a particular user or useful in a particular application, the guiding descriptor would choose what and how to depict. It would consider the individual user's emotion, perception of the problem being solved, so as to construct a sense of place where the user can be placed to take full advantage of the geographical information landscape being depicted. The Guide navigates *for* the user and takes the user to places that the Guide computes as necessary information locations that must be visited by a particular user to properly comprehend the geographical information being provided.

The Sage

The Sage, as a metaphor, suggests access to expert advice or information that provides 'support' for decision making or information appreciation. The user can view the information immediately available on the multimedia presentation and 'hot links' can be made to experts in particular fields or to the author of the product. Using a topographical information set of layers, Figure 2.11.1 shows how hot links to experts could be used to provide answers to the sometimes overlooked 'why?' element of geographical information product interrogation. Information behind the surface map display could be further investigated and expert opinion sought as to why particular classifications were done or why selected information types were shown, users could also 'interrogate' the legend for more expert coverage and explanation. The Sage metaphor is similar to the use of 'clickable' hypermaps, as described by Kraak and van Driel (1996). [...]

The Data Store

The Data Store links to other information about the area under current investigation, without needing to display all of this information. This can provide a rich 'background' to a seemingly simple rendering of geographical information. Links can be made to information behind the display. For example, tools can be made using a multimedia package to allow the mapped information on a screen to form a 'bubble', thereby showing the user a glimpse of the types and amount of information that can be seen beyond the information shown on the surface depiction. In this example, interrogating the map surface for other 'deeper' or complementary information may provide the

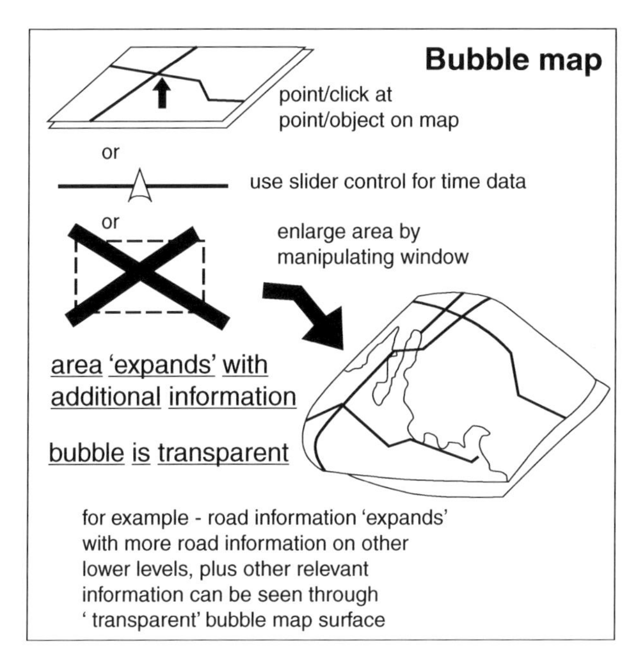

Figure 2.11.2 The concept of the 'Bubble map'. The user is able to manipulate the screen 'surface' and devise the most suitable interrogation method and view.

opportunity to display only that information needed for immediate information access via the map metaphor. Such an interrogation device is described as a 'Bubble map', whereby the user is able to manipulate the virtual map surface, to distort the type of view required and then to further explore and visualise the information stored 'below' the 'view surface' (Figure 2.11.2).

Where this concept differs from Monmonier's 'Nonlinear reprojection' (1977) and Snyder's 'Magnifying Glass' (1987) is the inclusion of interactivity. By incorporating multimedia 'sprites', multimedia elements that can be manipulated by the user, the user is able to 'peek' beneath the surface or 'look' more closer at the information by manipulating the multimedia presentation rather than re-computing mapping elements.

Obviously, this sort of treatment could be applied to any other multimedia element, allowing the user to experience the information displays in a fairly passive and non-exploratory manner or to 'dig' deeper for more information that supports the generalised depiction that was first available. The power of such a system is that the user could quickly browse the multimedia information package until the areas of immediate interest was found and then to more fully explore the information extraction possibilities. This could be applied to any multimedia element type.

The Fact Book

A Fact Book enables access to a plethora of facts, both about an area under study and other areas about which comparisons might have to be made. Multimedia elements constructed using a Fact Book metaphor will enable links to be made to facts from commercial or government information repositories. A possible way to implement the Fact Book metaphor is to link commercial, domestically consumed television broadcasts or on-line services to 'geographical location-responsive' resource providers. For example, in the near future if a family was at home watching the evening news on PC TV (van Niekerk 1997) and a place name or part of the world about which they were unfamiliar with appeared they could request more information from links provided. At the press of a button, hot spots would appear on the screen that, by pointing at the monitor and pressing the relevant buttons on the remote control, would provide information from a variety of resources (from within the local PC, from discrete local multimedia resources like CD-ROM or DVD-ROM, or from remote resources like Web-delivered Internet maps, facts and figures, or by optical cable delivered products). The television would immediately become the channel for immediate support information.

The Gameplayer

Multimedia elements developed under the Gameplayer metaphor will include things like a map-building game that allows users to learn the grammar of mapping. A geo video arcade game might also be utilised, where an educational package could be constructed to discover more about the 'stuff' that geography is made of. Other possible implementations of The Gameplayer metaphor involve simple photograph matching and information retrieval games that provide a novel approach to geographic information gathering. Such games might be designed to mirror expert problem solving processes.

[. . .]

Geographical information packages should be designed to access Gameplayer methods. Methods that could be built around Chesher and McCarthy's (1994) three genres of game are:

- Action: moving through a simulated landscape, avoiding environmental pitfalls.
- Puzzle/path: navigating through a virtual world where clues to navigation are uncovered by solving puzzles.
- Entertainment: where the user is provided with multimedia tools for building a comprehensive report on a

particular geographic location or region. Here, the multimedia resource is used to build another multimedia resource. If used in the context of an educational programme, this type of approach would enable a comprehensive resource to be constructed and developed by students themselves.

It is important to note here that perhaps the most important element of the Gameplayer metaphor is not the playing of games *per se*, but the use of gaming skills to explore geographical information.

The Theatre

Resources built along the guidelines of the Theatre metaphor are based around three elements: a stage, players and a script. (The concept of the Theatre is similar to ideas by Szegö 1987). The stage could be a four-dimensional geographical space (one that included time) that was the location where all of the play would take place. This stage should be dynamic and change over time and as a response to environmental and cultural forces. The players, either chosen by the user or pre-programmed by the product author, are the things that occur on the stage, or the elements of the landscape. The script controls both the stage and the players and can be 'written' by either the product author or, interactively, by the user.

When applied to geographical information users can 'build' the stage using multimedia tools, choose or define the players, then run the script (for pre-packaged scripts) or move from act to act using interactive controls, all with due consideration to geographical realities and natural and cultural controls. The play can incorporate any multimedia elements that best depicts the action, and the user would be able to direct the play and place players and multimedia 'sets' and special effects at will. The user will also probably be able to 'build' their own players by either selecting them from an electronic 'casting agency' or 'designing' them individually, so that they could undertake particular tasks or act in certain, predetermined or reactive, ways during defined acts. This is very much like the idea of 'agents' – since for the idea to work the 'actors' would need to include behaviours that vary with the context or situation they find themselves in. Building functionality into a computer using 'interface agents' has become an area of interest for research in human–computer interface design, building computer surrogates which possess a body of knowledge both about information being sought and how a particular user will use that information. Developers of multimedia systems are investigating the use of agents to move through the virtual space of information systems, relating huge pieces of information in short pieces of time. [...] Being able to more quickly locate precise and pertinent information via an agent makes for a more powerful geo-spatial information system.

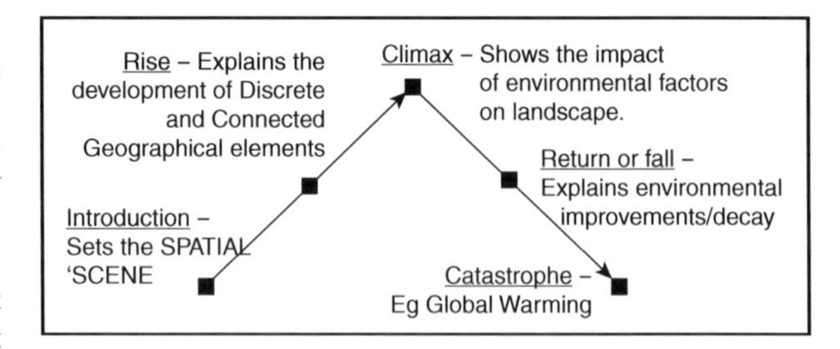

Figure 2.11.3 Shows how the Theatre metaphor might be applied to environmental mapping, after the lines proposed by Freytag.

Applying Freytag's Dynamic Anatomy of Theatre (Laurel 1993) – introduction, rise, climax, return/fall and catastrophe – to the presentation of geographical information spaces, a spatial story could be told (Figure 2.11.3).

By implementing Freytag's approach the understanding of the dynamic nature of geographical space might be improved, as time is compressed, the user is able to visualise all of the inputs into an environment. Alternatively, those from the Theatre background could use mapping to present a play differently. For example, Brenda Laurel developed an interactive multimedia installation, *Place Holder* (Rieser 1997), to illustrate how interactive narratives could be used. Participants were able to create their own stories based on improvised theatre and Virtual Reality (VR). They could alter their voices electronically to match mythical characters and fly through a recorded video landscape mapped onto a three-dimensional computer generated model.

The Toolbox

The Toolbox metaphor implies the use of tools that allow the user to make do or improvise, the way in which humans like to explore and discover. The word bricolage describes a completely different way of thinking than the 'generalist'/mathematical way. It is a 'use what you've got', 'improvise', 'make-do' way of problem solving, described by Claude Lévi-Strauss in his 1966 book *La pensée sauvage* (The Savage Mind) (Papert 1993). This approach provides the user with many ways of viewing information, as many multimedia tools would be provided. Standard ways of visualising geographical information would be complemented with measuring and assessing tools, with links to field surveys and expeditions and with tools of many other forms.

The Toolbox could offer 'real time' links to places in the world that are portrayed on the map by using Web cams (Cartwright 1997), where individual users 'on-line' could direct cameras to specific areas at a location or use a number of Web cams to make real time comparisons of different geographical locations. Information resources can be contacted and linked via the Internet, providing access to many different locations and laboratories and even exotic resources. Links are available to many sites around the world, from remote locations to international cities.

[...]

Discussion

Multimedia offers a different way than using conventional maps to extract pertinent facts that are current and complete. In spite of the potential, there is also a danger that in all of the hype about multimedia and what it offers, that this particular 'combination' of media types, all stored on some medium like CD-ROM or delivered via the Internet, may also eventually prove to be another 'sidetrack'. A hybrid CD-ROM/Web package promises to provide a mechanism that allows users to fully exploit the information on offer with contemporary communication and publishing systems.

Existing and future resources may well be interactive domestic televisions and telephones that provide interactive hyperlinked television mapping services, or enhanced mapping packages which 'link' to real expeditions via the use of CD-ROM or the Web, or the use of games interfaces, or hybrid tools like Web cams. At one time, actually producing products that had been envisaged was a major stumbling block. Now, we are hindered not by the lack of the means for production and delivery, but by thoughts and ideas about how new media might be applied. Once the current problems of lack of bandwidth and reasonable access to delivery technology in some parts of the world are rectified, this scenario of interactive enhanced mapping packages could be possible.

Viewed independently, these metaphors are not new and some of them have been tried before for representing geographical information, and/or the concepts underlying them have been used in areas outside the geographical sciences (usually to great effect). But, the integration of these metaphors in a multimedia environment designed to facilitate delivery and presentation of geographical information in a user-defined manner [...] could make available a powerful geographical interpretation tool.

Using multimedia to provide other map-complementing metaphors to explore geographical information allows that information to be made available in ways that best suit the user. If these metaphors are made available in a multimedia package they can provide a suite of artefacts that users can choose to explore and discover geographical information in their own manner. And, this offers the potential to be a much better method of exploring geographical information.

[...] To achieve the potential of this new 'rich' media, users need to be comfortable with altering and changing words, images and sounds. They must be able to quickly move from one media type to another. Historically, the inexpert users of paper maps were at a great disadvantage when trying to decipher the coloured lines, dots and shapes on a map that the designer–cartographer could perfectly comprehend. This type of problem is much magnified with multimedia maps and geographical information. Designers of these products need to be well aware of the usage skills of potential users in the plethora of media formats available with new media presentations.

Designers also need to develop skills to design products that allow users to understand the logic behind what is meant by user interfaces, to provide feedback that indicates if a procedure has been successful (even though the output may not be immediately apparent) and to design machine 'controls' so that they are positioned in a way that correlates with their effects. The designers of spatial products will not be able to merely apply design skills from paper maps or electronic maps and atlases to multimedia applications. New design methods and approaches need to be developed when applying multimedia for presenting spatial information. [...]

Today has been dubbed 'the Age of Access', where connectivity is driving towards the access of everyone to everyone, everything to everything, and everything to everyone. During the 1990s, knowledge has been promoted as one of the benefits of new technology and mapping. Geographical information products and systems of the 1990s need to be knowledge based. Dr Samuel Johnson (London 1775) said about knowledge and information that: 'Knowledge is of two kinds. We know a subject ourselves, or we know where we can find information'. Contemporary geographical information products can be employed to enable users to find information that enables them to enhance their knowledge of a particular subject. But, in the hands of an inexpert or novice user, such systems may prove to only provide a 'basket' of information, with no real way for understanding what is contained within that basket or its relevance. T. S. Eliot has said that knowledge should not be confused with information – there is a need for new media-enhanced geographical information products to provide the means of acquiring knowledge and not just voluminous amounts of information.

[...]

References

Cartwright, W.E. (1995) New maps and mapping strategies: contemporary communications. *SUC Bulletin*, **29**, 1–8.

Cartwright, W.E. (1997) New media and its application to the production of map products. *Computers and Geosciences*, **23**, 447–456.

Chesher, C. and McCarthy, P. (1994) When a floppy disk holds more than a CD-ROM, in *Proceedings of the Multimedia and Design Conference* (eds L. Maher, R. Coyne and S. Newton), University of Sydney, Australia, pp. 191–200.

Fabrikant, S.I. and Buttenfield, B.P. (1997) Envisioning user access to a large data archive. *Proceedings of GIS/LIS'97* 28–30 October 1997, Cincinati, OH.

Hedberg, J.G. and Harper, B. (1992) Creating interface metaphors for interactive multimedia. *Proceedings of the International Interactive Multimedia Symposium*, Mapping Sciences Institute, Perth, Australia, pp. 219–226.

Kraak M.-J. and van Driel, R. (1997) Principles of hypermaps, *Computers and Geosciences*, **23**, 457–464.

Laurel, B. (1993) *Computers as Theatre*, Addison Wesley, Reading, MA.

Monmonier, M. (1977) Nonlinear reprojection to reduce the congestion of symbols on thematic maps. *Canadian Cartographer*, **14**, 35–47.

Papert, S. (1993) *The Children's Machine: Rethinking School in the Age of the Computer*, Basic Books, New York.

Platt, C. (1995) Interactive entertainment. Who writes it? Who reads it? Who needs it? *Wired Magazine*, September, 144–149 195–197.

Rieser, M. (1997) Interactive narratives: a form of fiction? *Convergence – The Journal of Research into New Media Technologies*, **3**, 10–19.

Snyder, J.P. (1987) 'Magnifying glass' azimuthal projections. *American Cartographer*, **14**, 61–68.

Szegö, J. (1987) *Human Cartography. Mapping the World of Man*, Swedish Council for Building Research, Stockholm.

van Niekerk, M. (1997) Riding the PC TV revolution. *The Age*, 15 April, D1–D2.

Further reading

Cartwright, W., Peterson, M.P. and Gartner, G. (2007) *Multimedia Cartography*, 2nd edn, Springer., Berlin. [An edited volume bringing together many different perspectives about multimedia cartography.]

Dykes, J., MacEachren, A.M. and Kraak M.J. (2005) *Exploring Geovisualization*, Elsevier, London. [The 36 chapters report work from the research frontier relating to interactive mapping and cartographic exploration.]

Manovich, L. (2007) *The Language of Mew Media*, MIT Press, Cambridge, MA. [An influential work blending together ideas from visual studies, digital art and film theory to reconsider the nature of interactive media, such as mapping, in contemporary society.]

Pearce, M.W. (2008) Framing the days: place and narrative in cartography. *Cartography and Geographic Information Science*, **35** (1), 17–32. [Charts the potential of graphic strategies for story-telling about journeys in a map design and also as a novel explanatory framework for cartographic practice.]

See also

- Chapter 1.11: Exploratory Cartographic Visualisation: Advancing the Agenda
- Chapter 2.12: Imaging the World: The State of Online Mapping
- Chapter 3.7: Area Cartograms: Their Use and Creation
- Chapter 3.11: Egocentric Design of Map-Based Mobile Services
- Chapter 3.12: The Geographic Beauty of a Photographic Archive

Chapter 2.12

Imaging the World: The State of Online Mapping

Tom Geller

Editors' overview

Geller's short applied article reviews the range of major online mapping services that have emerged in the mid 2000s. The development and successful deployment of innovative web technologies by large corporations has delivered interactive mapping-on-demand to many millions of users daily. Significant software technologies have also opened up access to cartographic data allowing novel re-use in so-called map mash-ups. Consideration is also given to how flexible and open mapping technologies are re-engineering the ways states deliver cartography and also fostering virtual communities of practice to build new kinds of collaborative mapping. Given the rapid pace of technical and commercial developments in online mapping this article only provides a snapshot view of the key technologies and services available.

Originally published in 2007: *IEEE Computer Graphics and Applications*, March/April, 8–13.

Wherever computers are commonplace, free online mapping systems such as MapQuest and Yahoo! Maps are among the Web's most popular reference sites. In July 2005, a study by Hitwise showed the term *mapquest* as the second most popular online search term, with *maps* at fifth and *driving directions* at eighteenth. Consumer sites such as these compete to improve their interfaces and data relevance to better appeal to the browsing public. But innovation in online mapping isn't limited to these household names. Sites for the public good, such as those the United States' government manages, provide graphical access to data buried in such sources as the US Department of Agriculture Forest Service and the decennial census. Meanwhile, community mapping sites let the public provide local details, find nearby colleagues, and organise information in ways unknown 10 years ago.

Three trends dominate online mapping today:

- *Breadth:* The bringing together of more data sources, whether from public domain sources, community input, or extrapolation from multiple, diverse sources.
- *Depth:* Increasing the resolution and timeliness of those sources, particularly those relating to site photography.
- *Convergence:* Combining those sources in useful and easy-to-navigate ways.

[...]

Online mapping comes of age

[...]

Computerised mapping continued to develop into the 1980s but remained mostly the province of large organisations with specialised needs, mostly hidden from the public at large. One notable exception was the American Automobile Association's TripTiks. The AAA had provided its members with these simplified roadmaps since 1937, but in 1993 they contracted with R.R. Donnelley & Sons to

The Map Reader: Theories of Mapping Practice and Cartographic Representation, First Edition. Edited by Martin Dodge, Rob Kitchin and Chris Perkins.

computerise the system, eventually enabling complex routing and on-demand map printing. An R.R. Donnelley offshoot increased online mapping's visibility tremendously with its February 1996 launch of a system that remains a household name – MapQuest.com.

MapQuest.com certainly wasn't the first Web site to allow visitors to browse maps. Yahoo!, for example, featured city maps on its search site prior to MapQuest.com's launch. But MapQuest.com's launch made a huge impression and moved online mapping into the public sphere. Since then, both commercial and governmental mapping sites improved slowly but steadily until 2004, when an explosion of activity led to the fierce technological competition that continues today.

Online maps enhance consumer level search sites

The three most popular Internet search sites – Google, Yahoo! and MSN – have all adopted the view that mapping is a natural extension of searching. As a result, each has heavily funded mapping efforts and recent development shows the fruits of such investment. For example, Google acquired Keyhole in October 2004, leading to the release of Google Maps (originally Google Local) and Google Earth the next year. Yahoo! reworked its map interface in November 2005 and at the same time substantially expanded its developer tool suite for Yahoo! map integration into other Web sites. Microsoft has acquired low-altitude aerial photography on a massive level and incorporated it into Virtual Earth (part of Microsoft's Live Search). Despite these efforts by search-centric companies, the market-leading online mapping site as of this article remains MapQuest.com, now owned by America Online. Hitwise places MapQuest's market share as 56.2%, with Yahoo! Maps in second place at 19.6%; while Nielsen//NetRatings places MapQuest in the lead and Google Maps in second place.

The big four online map sites – Google Maps, Yahoo! Maps, MapQuest.com, and Virtual Earth – all count the following among their features:

- click-and-drag interface for navigation;
- driving directions between two points;
- zoom ability, ranging from planet to street level resolution;
- integration with a local businesses database, so a visitor can, for example, find and map all Chinese restaurants in a given neighbourhood;
- satellite view, showing landscape photographs from overhead (some such pictures are actually taken from airplanes); and
- access to several layers of information, such as roads, traffic and locations of sponsor storefronts.

All four sites provide APIs and other tools aimed at encouraging Web developers to incorporate the maps into their own sites. The sites have fairly liberal usage guidelines, and there's no cost to the developer for most non-commercial applications. One interesting result has been the rise of comparison sites that let you see how maps from these providers look side by side (for example, http://www.jonasson.org/maps/ and http://sidebyside.ning.com). Hundreds of other mash-up sites exist, combining ready-made maps with public domain data (such as voting records), proprietary data (such as store locations) or visitor-supplied information (such as street level photographs). [...]

Each mapping system's API is similar in that they all allow implementation through JavaScript; require the developer to register for a key; and include support for scaling, routing, map points and integrated controls – that is, developers don't need to add anything to permit user movement around the map. [...] According to lists maintained at http://programmableweb.com, Google Maps' API is by far the most popular choice for mash-ups, accounting in December 2006 for 722 of those listed, followed by Yahoo! Maps (64), Virtual Earth (55) and MapQuest (2).

[...]

3D maps: from the satellite to the street

If you define *online* site to include those accessed by programs other than a Web browser, the feature set expands considerably with Google Earth and Microsoft's Virtual Earth 3D, both of which are downloadable applications that frame the mapping data in a flyover format. [...]

Both Google Earth and Virtual Earth 3D depict the world primarily through a combination of satellite and aerial photography. The resolution varies depending on available sources and locations, with more densely populated areas tending to have higher resolution photography [...]. Both draw from a multitude of sources for their images, among them Sanborn, Europa Technologies, Navteq, TerraMetrics, NASA, Bluesky and Digital Globe. Both Google Earth and Virtual Earth 3D tell you in small type at the bottom of the screen the source of the imagery you're viewing.

Google Earth and Virtual Earth 3D differ in their use of viewing modes. Google Earth has, essentially, one seamless mode: you fly in and out, change your viewing angle, and move around the planet without interruption. Changes in views occur by adding or subtracting layers – for example, you could see geometric models of buildings by turning on the 3D buildings layer. By contrast, Virtual Earth 3D has distinct 2D and 3D views, each with its own features and idiosyncrasies. The 2D view gives you access to Microsoft's vast collection of bird's eye views, which are angled photos taken by low-flying planes in major cities throughout the United States (pop-ups notify you if an area you are viewing is available in bird's eye views). The 3D view displays terrain and building models and lets you change your viewing angle. Both Virtual Earth 3D views let you choose only the road map view, with no satellite photos – something not available in Google Earth.

Both programs also incorporate data from numerous sources, presented as layers that the user can turn on and off. Google Earth's implementation includes far more sources and gives the user greater flexibility in choosing which layer to view. By default, both applications include the categories you'd expect in any online atlas – roads, borders, place names, geographic features – along with data extrapolated from their Web search databases, such as hotels and restaurants.

With its longer time in the market, Google Earth, in particular, has been able to encourage outside parties to create their own information layers for display. These are done in Keyhole Markup Language (KML), an XML grammar and file format that can describe one or more placemarks (such as building locations), paths (such as roads), image overlays (such as congressional district groupings) and polygons. KML files carry several attributes, depending on file type. For example, a placemark KML file contains its:

- name, which shows up in a places list to the left of the main window;
- description, HTML-enabled text that appears when you click on the placemark;
- location data, encoded as latitude and longitude;
- style, describing the placemark's appearance, including opacity;
- view, controlling the viewer's height above the mark along with the number of degrees from north and the tilt from absolute overhead; and
- altitude above the ground where the placemark icon is floating.

Some KML additions to Google Earth are made dynamic by storing them on a server and applying the refresh attribute, which directs the server to feed a new version of the file to the client at a predetermined interval. Examples of dynamic KML files include real time weather maps, flight arrivals, and – more whimsically – the path of Santa's sleigh on Christmas Eve.

Google Earth and Microsoft Virtual Earth both provide 3D building models, but they're implemented quite differently. Virtual Earth 3D's default building models come with applied facade textures taken from photographs. Google Earth provides, at base, a 3D buildings layer that shows urban structures as untextured, massed models. These are simple but effective: you can, for example, lower the view to ground level, change the angle of view to look toward the horizon, and drive around a city with the gray, blocky buildings whizzing past you.

Virtual Earth 3D's approach to 3D modelling is in line with Microsoft's top-down philosophy. The company invested heavily to create properties that adhere to its corporate vision and that are complete out of the box. Google's approach to 3D modelling is more on the open source model; the company created a framework, made it publicly available, and hoped that it would populate the planet with great works. The framework in this case is SketchUp – a point-and-click 3D modelling program optimised for creating buildings and street scenes – and the 3D Warehouse – a collection of models created by both Google and creative users. Some of these models appear as a layer within Google Earth. [. . .] Google began by modelling major landmarks, such as San Francisco's City Hall (Figure 2.12.1). [. . .]

Layered GIS and government-sponsored mapping sites

While Yahoo!, Google, Microsoft and others battle to become the consumer's choice for getting travel (and shopping) directions, other parties target the audience that measures a map not by its graphical presentation, but by its variety, depth, and combinability of data. [. . .]

True to GIS's origins with the Canada Land Inventory, many such applications concern land resources and uses, with data collected through government sponsorship. One example is the National Geographic Society's MapMachine, a Flash-based program that ESRI built. The United States government, long a leader in mapping data and technology through the US Geological Survey (USGS), continues the tradition by providing Web-based viewers for such data. In many cases, there are multiple ways of accessing the same information through government sources, and a casual visitor might find bewildering the depth of information as a whole. Three gateways to government mapping information are Seamless Data

Figure 2.12.1 (a) A mass model of San Francisco's City Hall as seen from the roof of the nearby public library compared to (b) one created in Google Sketchup and distributed through the Google 3D Warehouse. While the mass models are less visually appealing, Google Earth has far more of this type of file.

Distribution System (http://seamless.usgs.gov), Earth Resources Observation and Science (http://eros.usgs.gov), and GISData Map Studio (http://gisdata.usgs.gov).

Two broad-based, layered map viewers that the USGS created are the National Map (http://www. nationalmap.gov) and National Atlas of the US (http://www.nationalatlas.gov). The National Map (Figure 2.12.2a) is the main map viewer program for the USGS. Layers available vary depending on the current view – for example, the California roads layer becomes available only if part of that state is in the viewing window. The National Atlas of the United States (Figure 2.12.2b) is a USGS online program that continues work started in 1874, when the government published its first official atlas, the *Statistical Atlas of the United States Based on the Results of the Ninth Census 1870.* The United States government published the atlas until 1970, when its final short run quickly sold out and was not reprinted. The current National Atlas site, Map Maker, launched in 1997; it gives access to dozens of layers, each of which is also available as a raw-data download. Additionally, the site acts as a gateway for educational materials, such as articles and printable maps.

[...]

Community-driven mapping projects

Finally, there are those maps whose creation is driven neither by corporate or governmental interests, but rather by direct public contributions. These maps tend to fall into two categories: community data that's added to an existing map structure and projects created from scratch.

Among those built on existing map structures, many rest on the APIs provided by the consumer mapping sites described previously. Notable examples include the following:

- Housing Maps (www.housingmaps.com; Figure 2.12.3) was one of the earliest mash-ups to take advantage of the Google Map API. The site takes RSS feeds from the popular free classified-ad site Craig's List (www.craigslist.org) and maps each property for rent or sale in selected cities.

- Frappr (www.frappr.com) and Platial (www.platial.com) are both companies that provide a front-end interface for visitors to create maps with multiple locations marked. A Frappr or Platial member can either mark the locations themselves (for example, 'places I've lived') or let other members place markers on their maps (for example, 'people planning to attend next month's science fiction convention in Pasadena').

- Wikimapia (www.wikimapia.org) uses Google Maps to show areas that visitors have tagged and described. The entire set of tags is also available as a KML file for Google Earth.

Google Earth also has facilities for adding data via KML files and sponsors online discussion boards for trading in such markers, making the entire program a sort of community site.

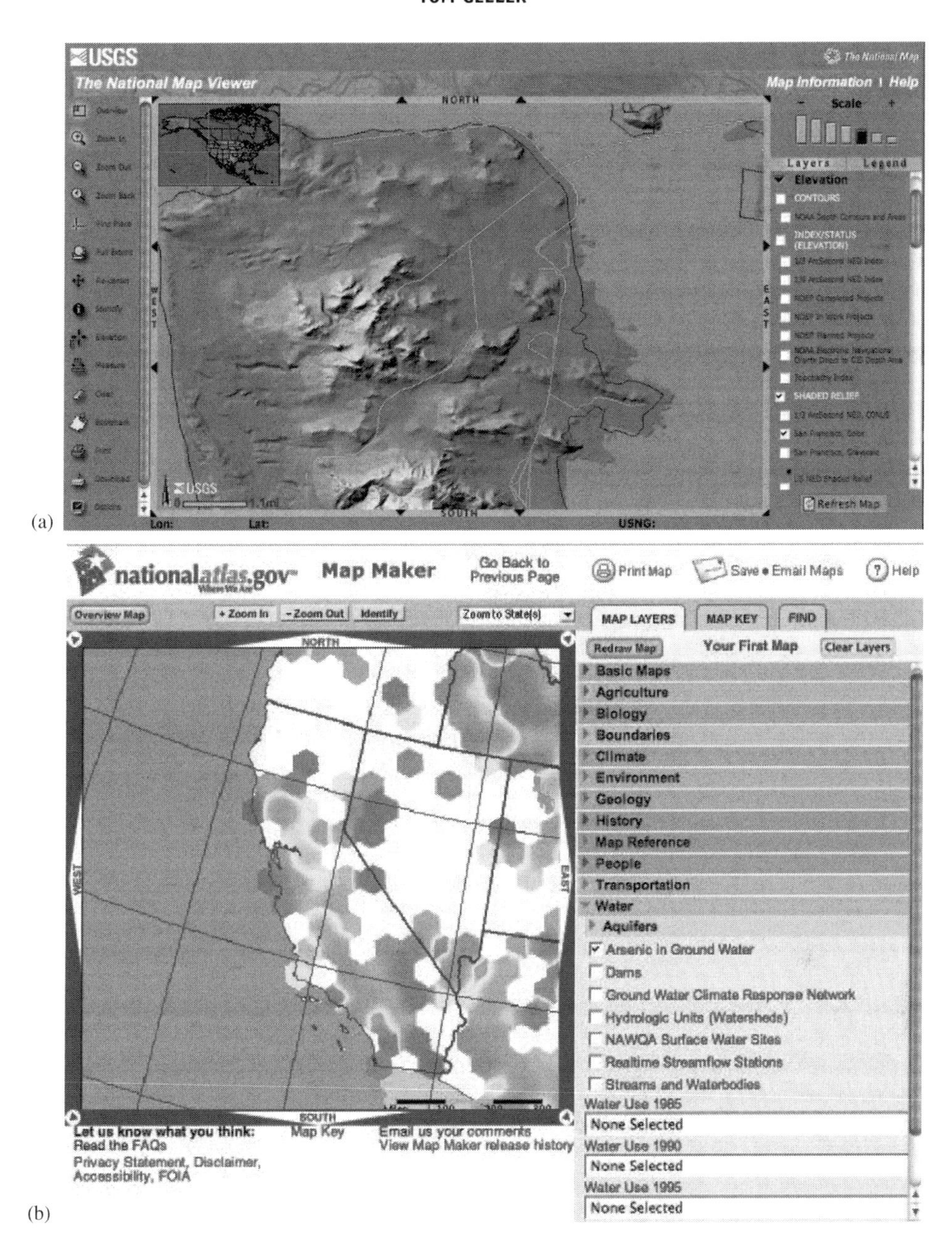

Figure 2.12.2 US Geological Survey map viewers. (a) The National Map presents a mapped interface to hundreds of layers of public domain data. Here we see how San Francisco's railways relate to the city's contours. (b) The interface to the National Atlas of the United States is typical of government hosted online mapping programs. A visitor selects one of the dozens of information layers via checkboxes, then hits a redraw map button to confirm the choice.

Finally, there are those sites that create both the map and data from scratch. CommonCensus Map Project (www.commoncensus.org; Figure 2.12.4) takes data from an ongoing survey to create maps showing geographic spheres of influence within the continental United States. The result is an irregularly updated, computer-generated set of maps demonstrating, for example, that people in Lowell, Massachusetts, consider themselves distinct from

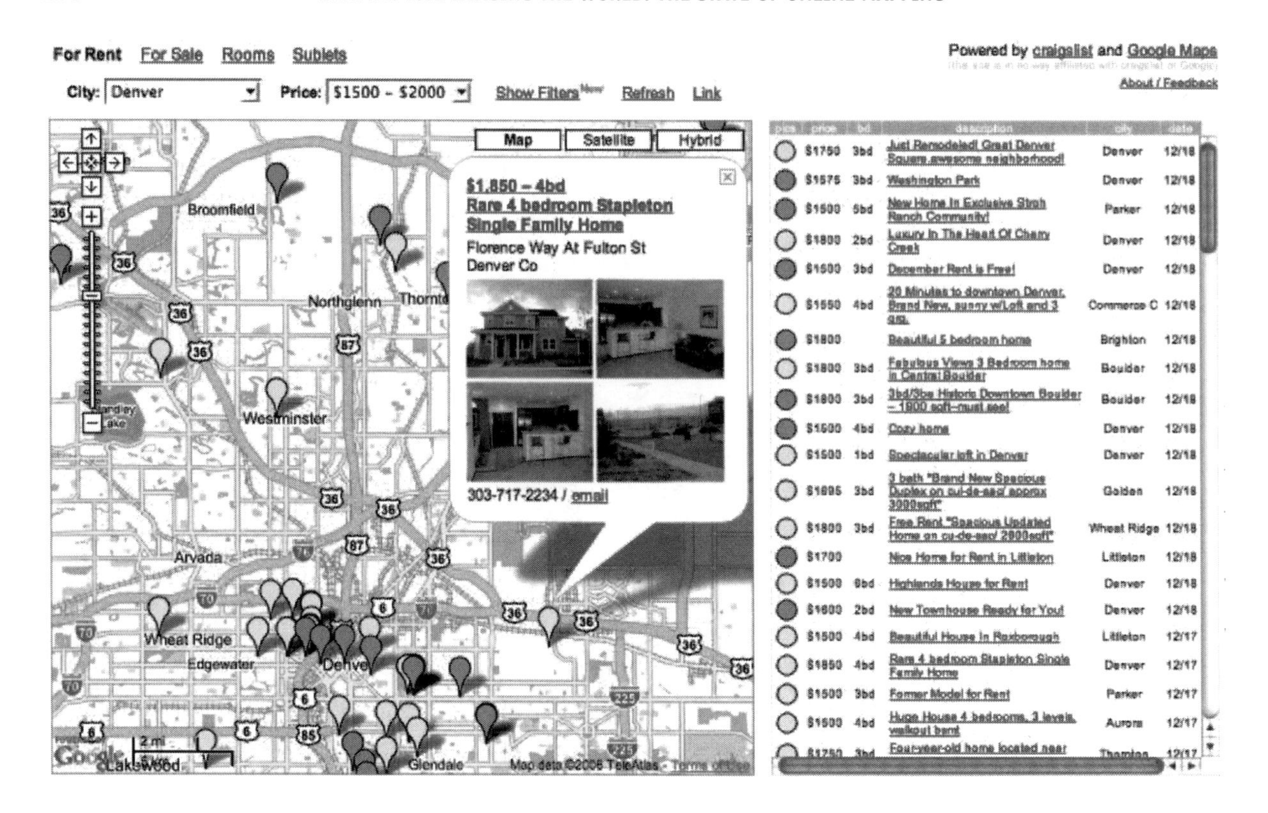

Figure 2.12.3 Housing Maps marries Google Maps and Craig's List, providing a visual guide to housing opportunities that's especially helpful to those unfamiliar with the text-only neighbourhood names found on Craig's List alone. Light grey pins point to listings with photos attached, while dark grey pins mark listings without photos.

Bostonians – and showing where, geographically, that distinction stops. OpenStreetMap (www.openstreetmap.org) notes that 'most maps you think of as free actually have legal or technical restrictions on their use, holding back people from using them in creative, productive or unexpected ways.' To remedy this, the (currently United Kingdom centric) project solicits visitors for GPS track logs in .gpx format, combining them to create maps that are bound by the comparatively liberal Attribution-ShareAlike license by Creative Commons (www.creativecommons.org).

[...]

Future directions

[...]

These consumer-oriented sites have introduced the utility of online maps to millions of people who would have been unlikely to discover the vast richness of GIS offerings. In return, the public has provided a sort of massive beta testing for this young field, leading to new interface paradigms and better access to data. As the public learns how the maps work (and vice versa), more data gets integrated into such systems. One sign of this convergence of GIS data and the consumer sphere is found in the December 2006 partnership between Google and NASA, with the stated goal of 'making the most useful of NASA's information available on the Internet,' such as 'real time weather visualisation and forecasting, high-resolution 3D maps of the moon and Mars, [and] real time tracking of the International Space Station and the space shuttle.'

That's not to suggest that consumer giants will swallow all public domain data and return it in a familiar map environment. For between the creation of information and its presentation lies the crucial middle – analysis. As such, there will always be a need for highly specialised, online mapping products to combine those data in unexpected ways and present them in formats appropriate to their audiences.

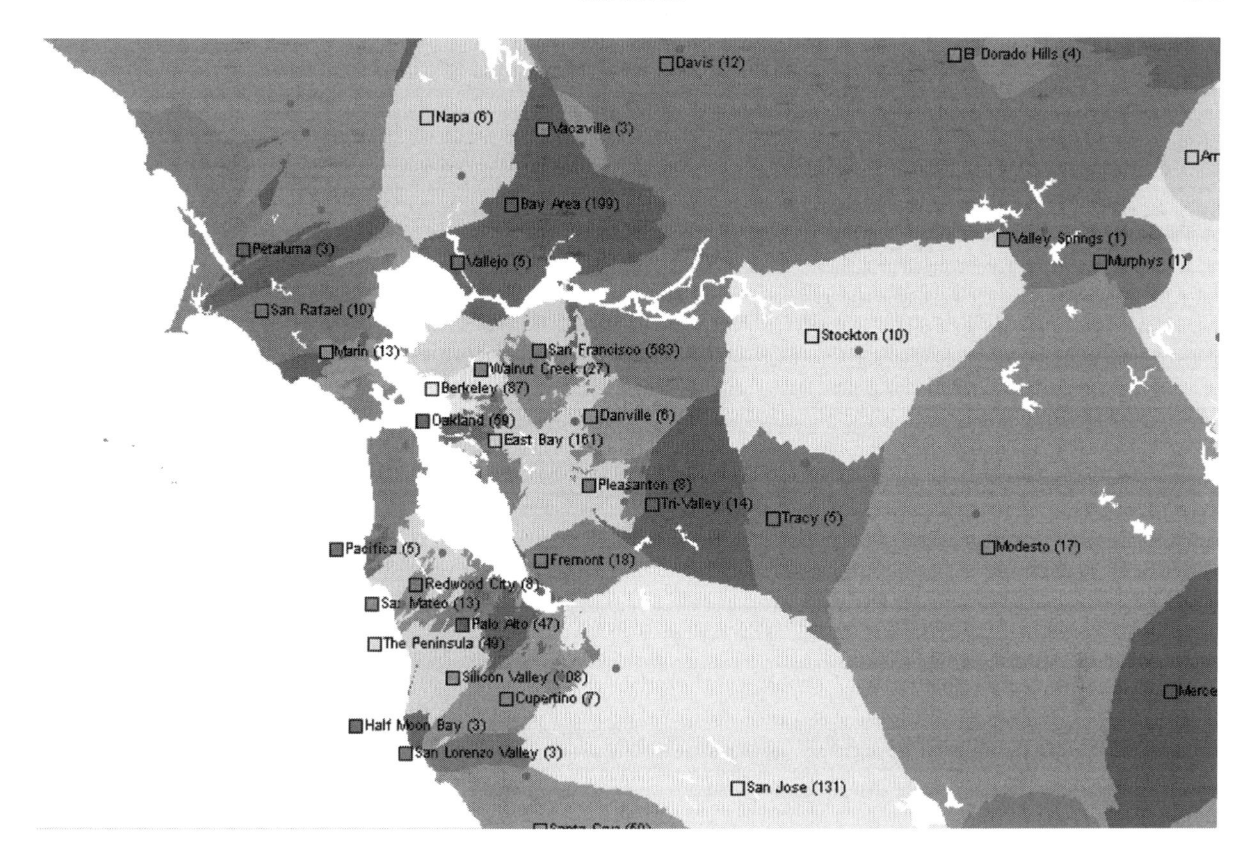

Figure 2.12.4 CommonCensus is a set of maps generated by computer from site visitors' responses to a survey. Numbers next to region names reflect the number of people who gave that response.

Further reading

Erle, S., Gibson, R. and Walsh, J. (2005) *Mapping Hacks: Tips & Tools for Electronic Cartography*, O'Reilly & Associates, Inc., Sebastopol, CA. [Perhaps the first published book giving ideas and examples for making map mash-ups.]

Miller, C.C. (2006) A beast in the field: the Google maps mashup as GIS/2. *Cartographica*, **41** (3), 187–199. [One of the first scholarly articles evaluating the wider significance of the Google Maps.]

Peterson, M.P. (2008) *International Perspectives on Maps and the Internet*, Springer, New York. [A useful edited volume documenting the diversity of online mapping.]

Taylor, D.R.F. and Caquard, S. (2006) Cybercartography: Maps and mapping in the information era. *Cartographica*, **41** (1), 1–5. [The introduction to a Cybercartographic theme issue in the journal that evaluates the significance of online and interactive mapping.]

Zook, M.A. and Graham, M. (2007) The creative reconstruction of the Internet: Google and the privatization of cyberspace and DigiPlace. *Geoforum*, **38** (6), 1322–1343. [A critical introduction to the social politics of Internet mapping which analyses some of the ways private capital has colonised virtual worlds and their cartographic representation.]

See also

- Chapter 2.6: Cartographic Futures on a Digital Earth
- Chapter 2.10: Mobile Mapping: An Emerging Technology for Spatial Data Acquisition
- Chapter 2.11: Extending the Map Metaphor Using Web Delivered Multimedia
- Chapter 3.10: Affective Geovisualisations
- Chapter 3.11: Egocentric Design of Map-Based Mobile Services
- Chapter 3.12: The Geographic Beauty of a Photographic Archive
- Chapter 4.10: Citizens as Sensors: The World of Volunteered Geography

SECTION THREE

Cartographic Aesthetics and Map Design

Chapter 3.1

Introductory Essay: Cartographic Aesthetics and Map Design

Chris Perkins, Martin Dodge and Rob Kitchin

Introduction

If there is one thing that upsets professional cartographers more than anything else it is a poorly designed map; a map that lacks conventions such as a scale bar, or legend, or fails to follow convention with respect to symbology, name placing and colour schemes, or is aesthetically unpleasing to the eye. In contrast, a well designed map not only follows conventions, but is beautiful to behold. It is perhaps no surprise then that cartography has often been called both a science and an art. A map is something that is crafted using scientific principles, which aims not only to faithfully represent the spatial relations of the world, but also to be aesthetically pleasing. Balancing these concerns is not straightforward and much research has been conducted to find map design principles that enhance both the communication and look of maps. In particular, such research gained prominence in the second half of the twentieth century after the publication of Arthur H. Robinson's monograph *The Look of Maps* in 1952 (excerpted here as Chapter 3.3).

This introductory essay explores some of the dimensions across which aesthetics and design matter, and delineates and explains how they are changing. Firstly, we consider some of the philosophical issues raised by focusing on different ways of understanding the design and 'the look' of the map. We then move on to consider the changing impacts of technology on map design and, in particular, upon the deployment of different kinds of thematic displays, before suggesting that technology alone offers only a partial means for explaining the deployment of changing visual techniques. We finish with a consideration of some of the practices and social contexts in which aesthetics and designs are most apparent, suggesting the subjective is still important in mapping and that more work needs to be undertaken into how mapping functions as a suite of social practices within wider visual culture. We conclude that earlier distinctions between artistic and scientific approaches to mapping may be rather unhelpful, and that that tensions between everyday practicalities and theoretical concerns are often overstated.

The nature of design and aesthetics

Robinson's work spelt out the need for a visual approach to cartography, grounded in a view of the discipline concerned above all else with communication. His research delineated many of the aesthetic factors that might be significant in effective map design. The resulting Robinsonian conceptualisation of cartography was strongly imbued with a functionalist rhetoric. Here, the primary role of the cartographer was to encode information in an optimal map design, such that the map reader would be better able to receive the cartographic message (Robinson and Petchenik 1976, excerpted as Chapter 1.3). For Robinson,

The Map Reader: Theories of Mapping Practice and Cartographic Representation, First Edition. Edited by Martin Dodge, Rob Kitchin and Chris Perkins.

aesthetic concerns were narrowly defined in distinctly normative terms: art had a purpose and the purpose was to raise the communicative efficiency of the map. Robinson argued treating maps as art could lead to arbitrary design decisions and that mapping needed to be based upon an objective application of best design practice.

Robinson posited that the process of map design can be broken down into sequences of different encoding and decoding operations. Visual matters play little role in data collection: it is in the abstraction, generalisation and symbolisation of information that design becomes important. Generalisation is itself often still a matter of aesthetics and compromise: the look of the map dictates what works best when considering how much simplification is required and may be particularly significant when maps depict specialist variables (Jenks 1963, excerpted as Chapter 3.4). Maps comprise combinations of line work, symbols, lettering and colours. These are all deployed through metrics that represent and control space: maps are projected, sometimes gridded, and usually uniformly scaled. Map design and projection choice inevitably impacts on the look of a map, a fact exploited by all the protagonists in the 'map wars' over the Peters projection (an equal area map that displayed the boundaries of countries in proportion to the size of their relative land mass – which looks distinctly different to the more common Mercator projection). Indeed, it was the unconventional look of the map that initially sparked the controversy (Crampton 1994; Monmonier 2004).

The 'success' of a symbol clearly impacts on overall design quality: decisions need to be taken on matters such as placement, sizing, an appropriate measurement level, the choice of a qualitative or quantitative representation and iconicity. In addition, Robinson *et al.* (1995) spelt out what might be termed the more gestalt-like features of a design, which work together to create an impression, including legibility, visual contrast, figure-ground effects, visual hierarchy and balance, and, rather as an after-thought, what are termed contextual items, but which largely elide anything beyond the surface of the map artefact itself.

This Robinsonian orthodoxy pervaded the emergence of academic cartography in North America, and continues to be reflected in the narrative of cartographic textbooks. Compare, for example, the sixth and final edition of the discipline-defining *Elements of Cartography* (Robinson *et al.* 1995) with a recent text aimed at the North American market (Tyner 2010). Neither spends much time on the elements of cartography that are most aesthetic, and, where they do, the aesthetic is defined in scientific rather than artistic terms. The principles of cartographic design, based upon a scientific understanding of how visual cognition works, are set out in systematic fashion, with the aim of reducing the chances of 'inappropriate' design choices.

In contrast, a different approach to information design comes from the work of Jacques Bertin and, in particular, the influential text *La Sémiologie Graphique* (1967, excerpted as Chapter 1.2). Bertin defined what came to be known as visual variables: primitives that designers can vary in order to construct the various visual codes which come together in map symbols and indeed complete maps. Alan MacEachren (1994, 1995, excerpted as Chapter 3.6) and others have subsequently expanded on Bertin's work, integrating cognitive and semiotic approaches to develop an approach to cartography centred on scientific visualisation. This also led to a focus on mapping processes, rather than simply optimal map design.

The rise of critical cartography in the 1990s generated a number of challenges to supposed scientific approaches to map design. On the one hand, social constructivist approaches argued that map design was infused with ideological and subjective decisions, even if it was framed scientifically. On the other, there was a concern that a focus on power relations inherent in design issues would push the focus towards exploring how power was embedded in maps, thus relegating issues of 'good design' to the margins. Krygier (1996) suggested that these challenges, along with technological change, made it more possible to escape the art/science dualism, by encouraging a focus on mapping as a 'sense making process' encompassing both. So, a concern for the aesthetic in cartography (Kent 2006) may be expressed through science as well as through art; see for example the consideration by Dykes and Wood (2008, excerpted as Chapter 3.12), where the elegant simplicity and intellectual focus of a tree map reflects beauty, and where the science of information visualisation is shown to work best through artistic registers. While Huffman (1996) has explored ways in which design might still matter in the relativistic postmodern world.

Forms of mapping and aesthetics

As well as significantly shaping the approach to map design, Robinson's work also influenced the form of mapping undertaken, and by default the look of maps. *Elements of Cartography* first published in 1953, and running to six subsequent revised editions, elided topographic matters. Instead, thematic mapping based on quantitative data dominates the text. As a result, the distinction into thematic mapping, and topographic survey or general purpose mapping, became reified in the day-to-day practices of cartography as a profession: cartographers were most likely to be trained in the design of the former, not the latter. It is perhaps unsurprising then that most subsequent Anglo-American textbooks have also had very little to say about the design of topographic maps. And perhaps these trends are exacerbated in the real world production of maps, with a gradual retreat from state-funded national surveys in the face of increasing competition from commercialised and globalised map sources such as TeleAtlas (underpinning much of Google Maps coverage). So, maybe what has been termed the 'blandscape' of multinationally sourced and internet-served mapping will increasingly supplant the national design imaginary offered by printed topographic products (Kent 2009).

The profusion of thematic cartography over the last century certainly reflects a changing aesthetic. Examining the timeline of significant data visualisation techniques, constructed by Michael Friendly and his collaborators, one is struck by the diversity of techniques that have been invented across many disciplines (Friendly and Denis 2010). Academic cartographers deploy choropleths, dasymmetric and dot distribution maps, isarithmic maps, proportional symbol maps, and cartograms, along with more novel multivariate geovisualisations encompassing the animated and multimediated data displays (Slocum *et al.* 2008). However, in practice, very few of these techniques have been deployed very much, or very well. Technological shifts such as desktop mapping packages and online geovisualisation have facilitated an emerging and radically different aesthetic, but paradoxically the same shifts have encouraged the mass profusion of often poorly designed thematic map output, centring around the use of off-the-shelf GI defaults and a limited number of map types.

Notable amongst these techniques has been the choropleth map. First named in 1938 by J.K. Wright, the technique creates maps that depict an average value for each area. Areas allocated to the same class are shaded the same: data are classified. So the designer can change the number of classes, the classification algorithm and the nature of the shading variation or sequencing (Evans 1977). Many of these issues are related to data generalisation, a fact developed long ago by Jenks (1963, excerpted as Chapter 3.4). Choropleths have probably been more researched than any other cartographic technique: their inadequacies were well documented by Wright in 1938, and have been extensively researched by academic cartographers in the years since. The technique hides any variation within the spatial frame of each enumeration district and is very often used in an inappropriate manner. An unimaginable number of possible displays may be made from the same data (but all the evidence suggests most users are unaware of this wide range); and all too often the sampling frame, the spatial units themselves, are a given and not available for the user to change.

Nevertheless, choropleth's are a ubiquitous design of data display. Martin (2005) found that 60% of all maps published in leading public health journals (published between 2000 and 2004) was comprised of choropleth maps. This over reliance on choropleth mapping reflects in their seeming simplicity and ease of construction, but also the social roles into which the maps are enrolled. So the classification of space and people, which this kind of thematic display facilitates, has been a useful aesthetic of governance (Crampton 2004).

However the increasing dominance of uniform national map designs, and the development of a thematic tradition, may well be much less pervasive than is supposed. In central Europe, Eduard Imhof exerted significant influence on cartographic practice and training. His classic 1965 work, *Kartographische Geländerdarstellung* (Chapter 3.2, excepted from an English language translation first published in 1982) implicitly recognised the complex interrelationship between symbols and the affective and emotional power of an evocative map design. Imhof noted, for example, that there can be a striking synergy of interest between cartographers and artists in their imitative images of mountains. The Swiss cartographic design tradition has continued to be applied to the depiction of relief in topographic mapping, and some of the most spectacular and aesthetic maps have been produced under the influence of Imhof's ideas (for a recent overview of work in this field see Hurni *et al.* 2001).

The Dutch cartographic tradition also placed greater emphasis upon aesthetic issues in cartographic design (Kraak and Ormeling 2010), as did John Keates's work in the United Kingdom (Keates 1984, 1993, 1996). Other researchers continued to emphasise the role of subjective decision making and craft in producing aesthetically pleasing map designs (Wood 1993), including critiques of published topographic mapping from researchers such as Collier *et al.* (2003). Consequently, the survival of different visual styles and designs of topographic maps in the face of often considerable pressure towards standardisation suggests topographic surveys continue to reflect national cultural values with map designs continuing to embody aesthetic conceptions of landscape (Kent and Vujakovic 2009). See Colour Plate One, for historically minded instigation.

The role of technologies

The visual appeal of maps mirrors the age when the image was produced. At one level this aesthetic variation reflects technological change. In Woodward's (1987) monograph about art and the history of cartography, the focus is largely upon an era prior to print production and mass consumption, when individualistic and artistic imagery was self-evident in mapping that clearly reflected its unique, craft origins. The worlds of the artist and cartographer were the same until the gradual emerging trade of military surveying began to encourage separation, a process facilitated, in part, by the application of new technologies. In contrast, contemporary mapping could be scripted as scientific, in particular after the nineteenth century invention of the thematic map (Robinson 1982). This historical generalisation has recently been challenged by an emerging focus on practice, for example, Edney (1993, excerpted as Chapter 1.10), who argues against narrowly progressive readings of map history, and in Cosgrove's (2005, excerpted as Chapter 3.9) analysis, which suggests that even in the twentieth century the worlds of artists and cartographers saw a continuing and active cross fertilisation.

However, it is undeniable that automation of map-making procedures in the mid-twentieth century encouraged a professionalisation of mapping that separated the worlds of the scientific mapmaker from those of the map user. The user simply read the map, whilst the maker sought to follow best professional practice. Only after the emergence of collaborative cartography and the widespread diffusion since early 2000s of online mapping tools have distinctions between map users and makers become rather more blurred in a noted upsurge of DIY mapping. This has led to a concern amongst many cartographers that we are entering an age of poorly designed, DIY maps.

Indeed, two recent trends highlight a growing recognition of the need to continue to focus on map design. The first is an emerging focus on the design of 'expert systems' that take map designers using a desktop or online GIS through design options, highlighting strategies that work, and those that might be inappropriate. For example Harrower and Brewer (2003, excerpted as Chapter 3.8) explore how colour might be deployed in choropleth displays (see Colour Plate Four). Their web-based ColorBrewer interface guides an unskilled user through the complex design choices available, offering help with an appropriate choice of sequence, matching colour schemes to display media and supporting output of colour specifications for appropriate use. Similar systems have been designed to guide novice designers through lettering and scale options. A second strategy has been to encourage better map design by taking design skills beyond the traditional academy and cartographic audience to try to get at amateur mapmakers in other professions (Darke and Spence 2008), and, in particular, by offering 'training' in visualisation aimed at the GI community. Many cartographic design texts are now targeted at this cross-over user group (Brewer 2008; Krygier and Wood 2005).

Technological change also facilitates shifts towards different and more diverse thematic displays. Dorling (1996, excerpted as Chapter 3.7) charts changes in the cartogram as a map form. The cartogram rescales representational space, so that the size of an area reflects a value ascribed to it rather than its geographical extent. The rather ugly blocky appearance of early cartograms, along with difficulties in designing them and the problems of recognising the places being mapped, may have hindered its widespread adoption, but the popularisation of an algorithm that preserved shape whilst converting areas into values, was influential on the publication of subsequent cartograms (see Gastner and Newman 2004 for the algorithm; Dorling, Newman and Barford 2008 for recent applications of this in the form of a global atlas).

More radical design challenges are faced if the designer wants to animate a display. Monmonier (1990, excerpted as Chapter 3.5) illustrates some of the many possible techniques for representing change in mapping. In the twenty years since this paper the web in particular has allowed many of these techniques to become commonplace, and the moving power of a map is increasingly deployed to depict changing phenomena across different media (Cartwright 1999, excerpted as Chapter 2.11). An overview of the state of knowledge around the design of these displays is provided by Lobben (2008).

Geovisualisation offers an emerging research agenda that has seen the development of many novel approaches and data display techniques (Dykes *et al.* 2005; MacEachren and Kraak 1997, excerpted as Chapter 1.11). Notable amongst these techniques are approaches to information visualisation, where different dimensions of variation in data, without any necessary spatial dimension, are visualised (Skupin and Fabrikant 2003). For example Dykes and Wood (2009, excerpted as Chapter 3.12) deploy tree maps as a technique to represent geographic characteristics of a geo-referenced photographic archive (see Colour Plate Four).

Technical advances and new ways of representing data are then still being discovered and deployed. The creative impulse is important in this kind of process and the worlds and art and science are no longer separate, if indeed they ever really were in mapping. Cosgrove (2005, excerpted as Chapter 3.9) suggests an overlap between the world of popular cartography, and in particular in the making of three-dimensional pictorial media maps, and the concerns of artists, in the period around the second world war in the United States that belies claims of objective rule-based design. Not only do cartographers deploy creative energy to design their functional maps, modern artists also deploy the apparently objective and scientific map to say something about the world. The recent upsurge in mapping by modern artists, charted by Harman (2009), reflects a set of concerns about living in the world that mirror those of a designer searching for an elegant design decision. And it is in the situated contextual practice of mapping that these issues come to a head.

The contexts, politics and practice of design

Whilst maps have always been displayed in different ways and through different media, recently there has been multiplication in display formats and contexts in which the map operates. For example, the same map will be read in very different ways if it is printed, folded, projected, mounted *in situ* in a 'You are Here' format, displayed in an exhibition, deployed as a graphic in association with other printed materials, displayed on a television screen, or a web site, or on a small screen of a mobile device or satnav system. A significant trend has been an emerging focus on context-specific design, from innovative work on web map design at the start of the new millennium (Kraak and Brown 2001) to a burgeoning research field relating to ubiquitous, or mobile cartography. A good example of the need for context-sensitive design is provided by Meng (2005, excerpted as Chapter 3.11), who explores the specific contextual requirements that flow from designing a map for display on a small mobile device, where use is likely to be personal, placed and transitory.

Contextually-informed design focuses on more than the map. Instead it considers factors such as the size of the display area, the nature of lighting, the nature of user interaction, the degree to which use might be individual or collaborative, the extent to which a display might be immersive, and the degree to which a design is fixed or under a user's control. Very few of these has yet received sufficient attention from the design literature and it has recently been argued that usability engineering approaches will be needed to ensure map designs work effectively given the diversity of contexts in which mapping is deployed (Haklay 2010). Instead of artificially simplified experiments, multiple methodologies, including speak-aloud protocols, video coding, participant observation, interviews and questionnaires, are likely to be deployed during investigations of real world map and geovisualisation display scenarios. Ethnographies of design practice will begin to reveal what designers actually do, instead of shoe horning their practice into pre-established rule structures. And this kind of situated design is much more likely to reflect on the politics of the aesthetic process, instead of pretending that everything can be known by the appliance of neutral science.

What practicing cartographers actually say about their skills and craft may indeed be as revealing as edicts from the academy. In 1999, The British Cartographic Society Design Group investigated best practice in map design. They identified five core principles: 'concept before compilation'; 'hierarchy with harmony'; 'simplicity from sacrifice'; 'maximum information at minimum cost'; and 'engage the emotion to engage the understanding'. These reflect a continuing focus on qualities that are much more likely to be associated with art than science, with rather zen-like slogans, encouraging creativity, reflection and holistic thinking (British Cartographic Society 1999).

Designers have probably always realised the emotional power that can work through mapping. And technological change opens up the possibilities for this kind of active engagement with 'affect'. Aitken and Craine (2006, excerpted as Chapter 3.10) highlight that mapmakers have much to learn in our designs from film-makers, who have long appreciated that they are working in a dream factory, where products are designed to do so much more than convey information. The moving image has a particular capacity to move its audience, and especially when accompanied by music. The animated and multimediated possibilities of new geovisualisations may be particularly effective if they engage with Aitken and Craine's suggestions and if they implement some of the practices in the British Cartographic Society guidelines.

However, static fixed historical displays also have the capacity to engage emotions. Look at the stark red and black imagery of William Bunge's nuclear war atlas (Colour Plate Six) and imagine its impact in the fearful world of the cold war. Its persuasive angry agitprop style offers a passionate cry of protest against the insanity of mutually-assured-destruction and the arms race. Technology has facilitated a resurgence of this kind of bottom-up counter-map design (Peluso 1995, excerpted as Chapter 5.6), and Wiki mechanisms exist for sharing and developing best practice in this field (Goodchild 2007, excerpted as Chapter 4.10, for an exploration of the changes this brings, and the Cloudmade web site at http://maps.cloudmade.com/editor for an example of a user-controlled design interface). It remains to be seen how researchers' work can be incorporated into these new design worlds, and how tensions between researched and professional design practice and everyday design practice might be resolved.

References

Aitken, S. and Craine, J. (2006) Affective geovisualizations. *Directions: A Magazine for GIS Professionals*, 7 February, www.directionsmag.com. (Excerpted as Chapter 3.10.)

Bertin, J. (1967) *La Sémiologie Graphique*, Gauthier-Villars, Paris.

Brewer, C.A. (2008) *Designed Maps: A Sourcebook for GIS Users*, ESRI Press, Redlands, CA.

British Cartographic Society (1999) *Five Principles of Map Design*, British Cartographic Society, London.

Cartwright, W. (1999) Extending the map metaphor using Web-delivered multimedia. *International Journal of Geographical Information Science*, **13** (4), 335–353. (Excerpted here as Chapter 2.11.)

Collier, P., Forrest, D. and Pearson, A. (2003) The representation of topographic information on maps: the depiction of relief. *The Cartographic Journal*, **40**, 17–26.

Cosgrove, D. (2005) Maps, mapping, modernity: art and cartography in the twentieth century. *Imago Mundi*, **57** (1), 35–54. (Excerpted as Chapter 3.9.)

Crampton, J.W. (1994) Cartography's defining moment: the Peters projection controversy, 1974–1990. *Cartographica*, **31**, 16–32.

Crampton, J.W. (2004) GIS and geographic governance: reconstructing the choropleth map. *Cartographica*, **39** (1), 41–53.

Darke, G. and Spence, M. (2008) *Cartography: An Introduction*, British Cartographic Society, London.

Dorling, D. (1996) Area cartograms: their use and creation. *Concepts and Techniques in Modern Geography*, **59**, http://qmrg.org.uk/files/2008/11/59-area-cartograms.pdf. (Excerpted as Chapter 3.7.)

Dorling, D., Newman, M. and Barford, A. (2008) *Atlas of The Real World*, Thames and Hudson, London.

Dykes, J. and Wood, J. (2009) The geographic beauty of a photographic archive, in *Beautiful Data: The Stories Behind Elegant Data Solutions* (eds T. Segaran and J. Hammerbacher), O'Reilly, Sebastopol, CA, pp. 85–104. (Excerpted as Chapter 3.12.)

Dykes, J., MacEachren, A. and Kraak, M.-J. (2005) *Exploring Geovisualization*, Elsevier, Amsterdam.

Edney, M.H. (1993) Cartography without 'progress': reinterpreting the nature and historical development of mapmaking. *Cartographica*, **30** (2/3) 54–68. (Excerpted as Chapter 1.10.)

Evans, I.S. (1977) The selection of class intervals. *Transactions of the Institute of British Geographers, New Series*, **2** (1), 98–124.

Friendly, M. and Denis, D.J. (2010) Milestones in the History of Thematic Cartography, Statistical Graphics, and Data Visualization. www.datavis.ca/milestones/.

Gastner, M.T. and Newman, M.E.J. (2004) Diffusion-based method for producing density-equalizing maps. *Proceedings of the National Academy of Sciences*, **101** (20), 7499–7504.

Godchild, M.F. (1997) Citizens as sensors: the world of volunteered geography. *GeoJournal*, **69** (4), 211–221.

Haklay, M. (2010) *Interacting with Geospatial Technologies*, John Wiley & Sons, Ltd, Chichester, UK.

Harman, K. (2009) *The Map as Art: Contemporary Artists Explore Cartography*, Princeton Architectural Press, Princeton, NJ.

Harrower, M. and Brewer, C.A. (2003) ColorBrewerorg: an online tool for selecting colour schemes for maps. *The Cartographic Journal*, **40** (1), 27–37. (Excerpted as Chapter 3.8.)

Huffman, N. (1996) You can't get here from there: reconstructing the relevance of design in postmodernism, in *Cartographic Design: Theoretical and Practical Perspectives* (eds C. Wood and C. Keller), John Wiley & Sons, Ltd, Chichester, UK.

Hurni, L., Kritz, K., Patterson, T. and Wheate, R. (2001) Special Issue: ICA Commission on Mountain Cartography. *Cartographica*, **38** (1–2).

Imhof, E. (1982) *Cartographic Relief Presentation*, de Gruyter, Berlin. (Excerpted as Chapter 3.2.)

Jenks, G. (1963) Generalization in statistical mapping. *Annals of the Association of American Geographers*, **53** (1), 15–26. (Excerpted as Chapter 3.4)

Keates, J.S. (1984) The cartographic art. *Cartographica*, **21** (1), 37–43.

Keates, J. (1993) Some reflections on cartographic design. *The Cartographic Journal*, **30** (2), 199–202.

Keates, J.S. (1996) *Understanding Maps*, 2nd edn, Longman, Harlow, UK.

Kent, A.J. (2006) Aesthetics: a lost cause in cartographic theory? *The Cartographic Journal*, **42** (2), 182–188.

Kent, A.J. (2009) Cartographic blandscapes and the new noise: finding the good view in a topographical mashup. *Society of Cartographers Bulletin*, **42**, 29–37.

Kent, A.J. and Vujakovic, P. (2009) Stylistic diversity in European state 1:50 000 topographic maps. *The Cartographic Journal*, **46** (3), 179–213.

Kraak M.-J. and Brown, A. (2001) *Web Cartography: Developments and Prospects*, Taylor & Francis, New York.

Kraak, M.-J. and Ormeling, F. (2010) *Cartography: Visualization of Spatial Data*, 3rd edn, Pearson, Harlow, UK.

Krygier, J. (1994) Sound and geographic visualization, in *Visualization in Modern Cartography* (eds A. MacEachren and D.R.F. Taylor), Pergamon, New York, pp. 149–66.

Krygier, J. (1996) Cartography as an art and a science? *The Cartographic Journal*, **32** (6), 3–10.

Krygier, J. and Wood, D. (2005) *Making Maps: A Visual Guide to Map Design for GIS*, Guilford, New York.

Lobben, A. (2008) Influence of data properties on animated maps. *Annals of the Association of American Geographers*, **98** (3), 583–603.

MacEachren, A.M. (1994) *Some Truth with Maps: A Primer on Symbolization and Design*, Association of American Geographers, Washington, DC. (Excerpted as Chapter 3.6.)

MacEachren, A.M. (1995) *How Maps Work*, Guilford, New York.

MacEachren, A.M. and Kraak, M.-J. (1997) Exploratory cartographic visualization: advancing the agenda. *Computers & Geosciences*, **23** (4), 335–343. (Excerpted here as Chapter 1.11.)

Martin, S. (2005) Cartography, discourse and disease: how maps shape scientific knowledge about disease. Unpublished Masters thesis. Anthropology and Geography, Georgia State University, Atlanta.

Meng, L. (2005) Egocentric design of map-based mobile services. *The Cartographic Journal*, **42** (1), 5–13. (Excerpted as Chapter 3.11.)

Monmonier, M. (1990) Strategies for the visualization of geographic time-series data. *Cartographica*, **27** (1), 30–45. (Excerpted as Chapter 3.5.)

Monmonier, M. (2004) *Rhumb Lines and Map Wars: A Social History of the Mercator Projection*, The University of Chicago Press, Chicago.

Peluso, N. (1995) Whose woods are these? Counter mapping forest territories in Kalimantan Indonesia. *Antipode*, **27** (4), 383–405. (Excerpted as Chapter 5.6.)

Robinson, A.H. (1952) *The Look of Maps*, University of Wisconsin Press, Madison, WI. (Excerpted as Chapter 3.3.)

Robinson, A.H. (1953) *Elements of Cartography*, John Wiley & Sons, Inc., New York.

Robinson, A.H. (1982) *Early Thematic Mapping in the History of Cartography*, University of Chicago Press, Chicago.

Robinson, A.H. and Petchenik, B.B. (1976) *The Nature of Maps: Essays Toward Understanding Maps and Mapping*, Chicago University Press, Chicago. (Excerpted as Chapter 1.3.)

Robinson, A.H., Morrison, J.L., Muehrcke, P.C. *et al.* (1995) *Elements of Cartography*, 6th edn, John Wiley & Sons, Inc., New York.

Skupin, A. and Fabrikant, S.I. (2003) Spatialization methods: a cartographic research agenda for non-geographic information visualization. *Cartography and Geographic Information Science*, **30**, 99–119.

Slocum, T.A., McMaster, RB., Kessler, F.C. and Howard, H.H. (2008) *Thematic Cartography and Geo-Visualization*, 3rd edn, Pearson, London.

Tyner, J. (2010) *Map Design*, Guilford, New York.

Wood, M. (1993) The map user's response to map design. *The Cartographic Journal*, **30** (2), 149–153.

Woodward, D. (1987) *Art and Cartography*, Chicago University Press, Chicago.

Wright, J.K. (1938) Problems in population mapping, in *Notes on Statistical Mapping, with Special Reference to the Mapping of Population Phenomenon* (ed. J. Wright), American Geographical Society and Population Association of America, New York, pp. 1–18.

Chapter 3.2

Interplay of Elements, from *Cartographic Relief Presentation*

Eduard Imhof

Editors' overview

In this chapter extracted from the English translation of Imhof's classic 1965 work, *Kartographische Geländerdarstellung*, he focuses upon the depiction of relief in topographic maps and, more specifically, on aspects of the interplay between different elements of the map symbolisation. He establishes the conceptual, graphical and technical reasons for interplay in relation to design principles, before exemplifying the implications of possible combinations of relief depiction techniques, drawing in particular upon Swiss topographic mapping experience.

English translation originally published in 1982: Chapter 14 in Eduard Imhof, *Cartographic Relief Presentation*, de Gruyter, Berlin, 325–345.

The nature and effect of interplay

The necessity for and the careful development of good interplay

[...] The successful interplay of elements is just as important as the good design of the elements themselves. But this can only be achieved if each is treated from the outset as part of the final map as a whole, and is designed accordingly to achieve the desired effects in combination with other symbols. The relief image always contains many symbols for ground cover, names and height values. These must also be fused with the terrain representation to provide a clear and purposeful overall design. This means that the terrain representation and these other symbols must be adjusted mutually to achieve success. [...]

In the interplay of elements, both combination and coordination, must be considered, as must questions of concept and graphic design. [...] 'Combination' is the joining of several things to form a new, planned whole. 'Coordination' means their arrangement to fit into a framework or structure. In the cartographic context, combination also refers to the conception of what should make up the new image whereas coordination refers more to the graphic aspects, to the manner in which the individual item is placed within the total structure.

[...]

Conceptual, graphic and technical aspects of interplay

When considering the interplay of the elements in a map, three different approaches must be examined:

(a) Conceptual interplay: Within the context of the purpose and scale of the map the following should be considered: what is appropriate and practicable as map content, what can be added to greatest advantage, and which elements have dubious significance for the purpose of the map? [...] The selection of elements to be included in the content depends largely [...] on the purpose of the map. [...] For example, contours would be of no value on a small-scale road map, a

The Map Reader: Theories of Mapping Practice and Cartographic Representation, First Edition. Edited by Martin Dodge, Rob Kitchin and Chris Perkins.

number of well selected spot heights being more appropriate. To improve orientation within the terrain, simple relief shading and perhaps a few coloured layer tints would suffice. A map with limited but carefully selected content will be much more successful in transmitting its information to the reader than would a sheet overloaded with trivial details.

(b) Graphic interplay: Graphic chaos destroys and removes all value from any combination of elements, no matter how well selected. [...] The same or similar types of patterns laid one on top of another will mutually interfere with or destroy each other. [...] On the other hand, different styles of graphic elements superimposed on one another are much less disruptive and can even reinforce each other in their effects. Thus, patterns of lines or strokes combined with unbroken area tints produce a good effect, the colour background not interfering with the line image. If bands of line patterns must be superimposed on one another then one should be made as open, large-meshed and strongly coloured as possible, while the other is made fine-grained, dense and in a light tint.

(c) Technical interplay: A successful graphic image is more than a little dependent on the technical processes of drawing and reproduction. [...]

Consistent generalisation and good standardisation

Every map, no matter how large its scale, depicts the terrain in a simplified form. [...] Selection of objects to be placed on a map, and the simplifications made in the image are more or less established in many cases by conventions and idealised symbols. But even the best principles or rules cannot guide the hand of the cartographer with certainty. [...] If a topographic map is the subject, and in particular one which has been surveyed throughout by the same topographer, then obviously a high degree of uniformity in content and generalisation will result amongst all the elements of the image. This homogeneity is destroyed, however, when a number of collaborators are on the job. [...]

The danger of internal inconsistency is much greater in small-scale maps which are produced by transferring information graphically from basic maps at larger scales. The latter are quite often very varied in scale, content and graphic form, and must therefore be checked carefully before they are used. [...] Often many cartographers may be at work on the same project. The various stages of the work or of the elements of the image may even be separated in time and space. [...] One cartographer draws only the stream network, a second the contours, a third produces the rock drawing, a fourth the shading, a fifth the layer tints, a sixth the numerical annotations and text, and so on. [...] Often in modern production processes the overall image is seen for the first time only when combined press proofs or combined multicolour copies are available. [...] In these circumstances, it is imperative that the conceptual and graphic interplay of all elements are strictly guided from the very beginning and that the generalisation of various elements be well organised.[...]

Careful emphasis and restraint. Mutual relationships between things

[...] If equal visual weight is given to all elements none of them is allowed to achieve its full value. Important things should stand out graphically above the incidental information. This can be brought about by thick heavy lines, by the enlargement and emphasis of symbols, and by employing striking colours or colour contrasts. [...] One must be able to judge which detail in any situation is the most important.

A topographical map for use in the field should not be a brightly coloured painting but rather an exact metrically correct drawing which is rich in content. It may, however, be shaded lightly and given colour tints to improve its legibility. A wall map, however, which is observed from a certain distance and is designed to provide a good overall picture of a large area, is more like a painting which has been supplemented by lines and symbols. [...]

In spite of the deliberate importance placed on the main elements of the map they should not overpower the other less important symbols. The graphic expression of something should not get out of control, nor should it mislead the reader into seeing more in an image than was put into it in the first place. [...] Cartography is the art of moderation, of the careful balancing of things. In cartography, the law of scale is in control.

There are many maps in which individual elements stand out above the other symbols to an unwarrantable extent. [...] In many topographic maps at scales of 1:20 000–1:50 000, the symbols for small forms and ground cover, the detritus dot pattern, rock hachures for karst or knolls, wedge-shaped hachures for small terraces, and so on are overemphasised in contrast to the overall surface form represented by contours. One cannot see the wood for the trees. [...] The manuscripts for maps such as these are normally drawn by topographers who are accustomed to detailed observation of the terrain and to the relationships of things at scales of 1:5000 and 1:10 000. They thus carry these relationships into the smaller-scale maps without sufficient adjustment of form. [...]

The mutual relationships of the symbol content of maps change with the scale. The smaller the scale, the more should the nature of the surface and its cover be subdued in favour of its general form. Richly detailed plans contain the whole, gaily coloured pattern as seen through close observation of landscape. At the other extreme, the smallest-scale maps present merely the bare, bold relief of the terrain, and even when colour tints are used they tend only to provide information on elevation zones.

Overlapping discontinuities substitution

Cases of the intersection, crossing and overlapping of lines, symbols, letters and numbers are countless in every map. It is characteristic of poor map design that the bolder, richer or otherwise dominating distribution subdues or even obliterates the weaker secondary features.

The elements of terrain representation are often barely recognisable beneath the houses and streets of cities in large-scale topographical maps. This is not easily avoided. Design problems such as these can normally be eliminated in cases of overlapping by smaller and isolated symbols. [...] Single lines [...] can be drawn over others without mutual interference, provided that the angle of intersection is not too small. It conforms to normal visual experience where one imagines lines or zones of lines continuing without interruption, even when they are interrupted by other lines, small symbols or letters. The same happens with simpler rock hachures. If two graphic elements of the same colour cross or overlap, they should be separated. Small local symbols should be brought out by stopping the lines which cross them immediately before they touch the symbol; Figures 3.2.1 and 3.2.2 are poor techniques, Figure 3.2.3 provides a good technique.

If two overlapping linear elements are very detailed, complicated and perhaps in different colours, then the method just described of interrupting the lines is difficult to accomplish with satisfactory precision. In cases of this kind small symbols [...] should be printed in a strong colour, while the elements of the terrain itself, for example contours and slope hachures, are printed considerably lighter, in brown or grey for example. The line pattern in the weaker colour appears to flow without interruption beneath the bolder symbols, like a stream under a bridge.

In many cases, the more important and visually more striking element takes over as a substitute for the weaker. Classic examples of this are the omission of political boundaries along stream lines or sharp mountain ridges. Several good and poor examples are contrasted in Figures 3.2.4–3.2.8. Substitution is valid only [...] if the continuity of the interrupted line is lost on account of it. In this respect many maps leave much to be desired.

Displacement, narrow passes

In narrow valleys, gorges and ravines, along steeply sloping ocean shores, along sea coasts and so on, streams

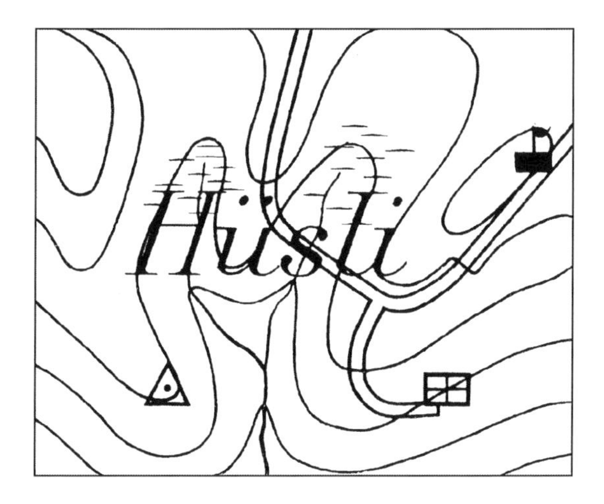

Figure 3.2.1 Poor combination.

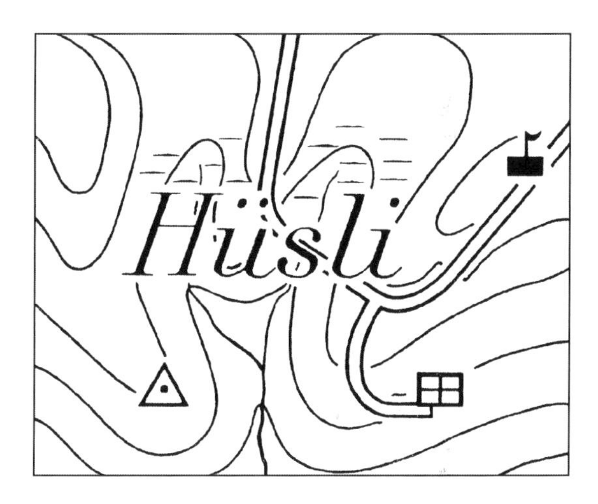

Figure 3.2.2 Badly planned combination.

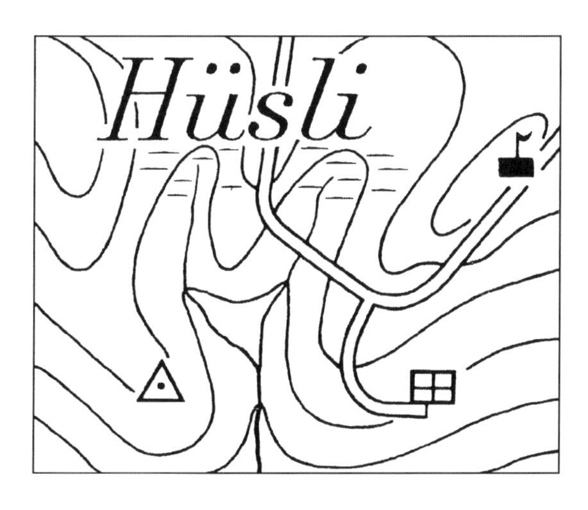

Figure 3.2.3 Good combination.

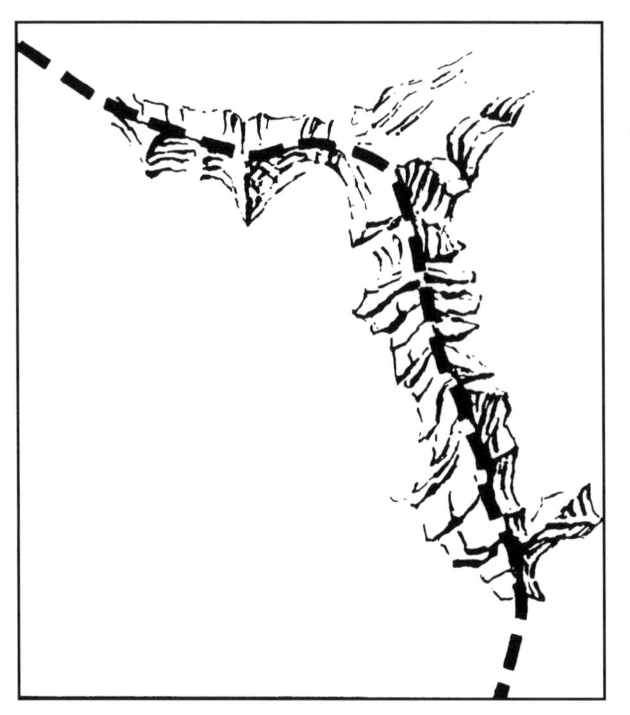

Figure 3.2.4 Poor solution. The boundary symbol destroys the clarity of the rocky ridge and gives it a dirty appearance.

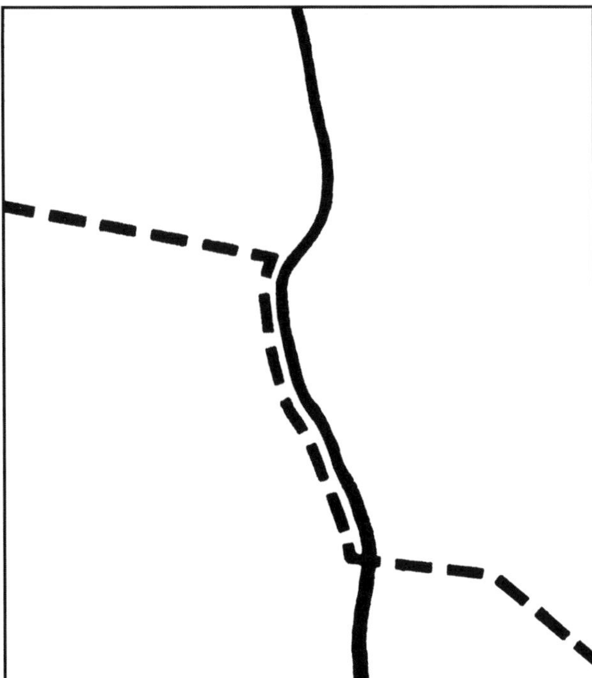

Figure 3.2.6 Poor solution. Unnecessary duplication of both the heavy stream line and the boundary line.

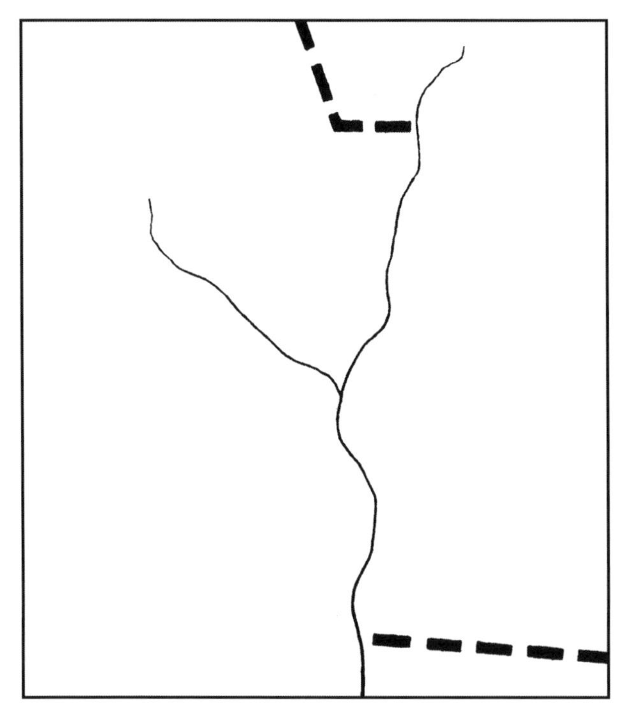

Figure 3.2.5 Poor solution. Such a weak stream line is an unsatisfactory substitute for the boundary symbol.

or shorelines, sloping roads, railways, settlements and often even cliff steps or sections of slopes are crowded together into very narrow spaces. [. . .] Every cartographer and all experienced map users know that individual topographic features in medium- and small-scale maps can be represented only in a greatly exaggerated form. At 1:100 000, for example, the corresponding reduction of an eight metre wide road would show as a double line only 0.08 mm wide. In order to maintain clarity, the line must be widened to about 0.8 mm, that is a ten times enlargement. [. . .]

In the Jura gorge near Moutier (Figure 3.2.9), the stream, railways and road take up a width of only 40 meters. In a topographical map at a scale of 1:200 000, however, they require a strip about 2.00 mm wide, even when they are placed as close together as possible, and this corresponds to 400 metres on the ground, an enlargement of ten times. The rock-covered slopes climb steeply upward on both sides of the stream. Therefore, if everything was depicted in its correct location, important sections of the terrain would be overlapped by the symbols mentioned earlier, and would become illegible (Figure 3.2.10). [. . .] Thus, the mapmaker has no other choice but to choose the lesser of two evils. Contour zones and rock drawing are pushed outward away from the symbol lines which bar their way. This positional error is then rectified in the uphill direction after the

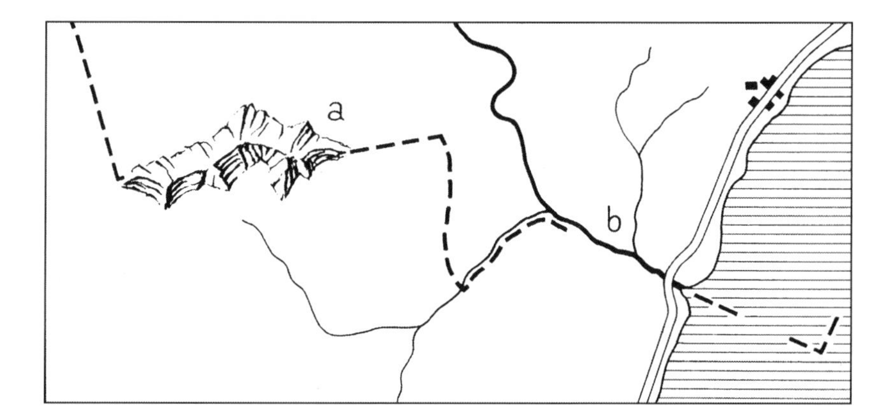

Figure 3.2.7 Good solutions. At 'a' the ridge line replaces the boundary symbol, at 'b', a bold stream line takes over this function. On lakes, it is often sufficient to indicate the initial sections and points of directional change in the boundary line.

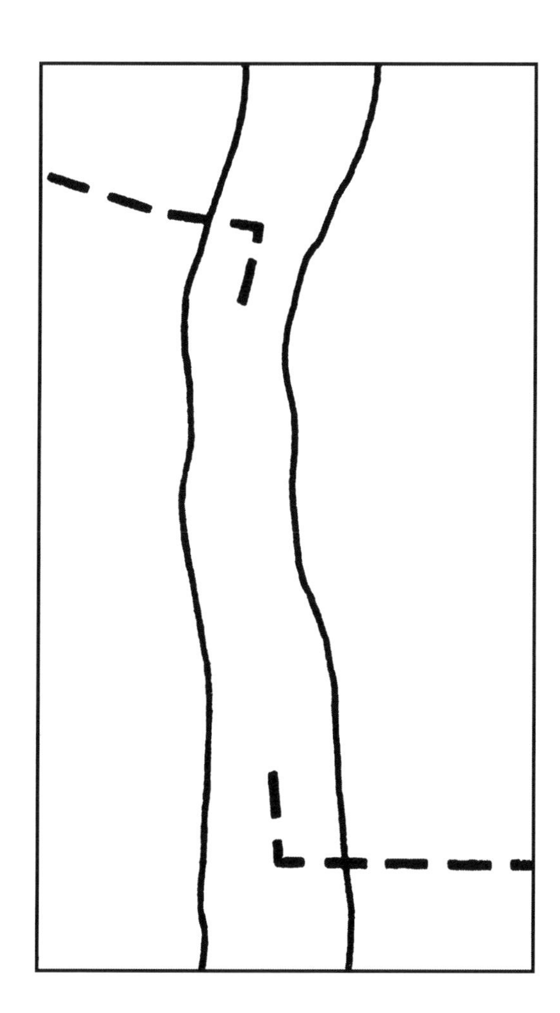

Figure 3.2.8 Good solution. Carrying the boundary symbol right along the stream is often unnecessary.

shortest distance possible and with as little change in form as possible (Figure 3.2.11). [...]

If the steep slope with its associated lines of communication borders on a lake [...] then the lake shore can hardly be displaced, since the outlines of lakes usually possess familiar shapes in maps, which one cannot allow to be distorted on account of a road or railway line.

Changes in tint value resulting from combination

The graphic expression of a line, a zone of lines, an areal colour and so on is changed when overlapping with other graphic elements and may even be changed by proximity alone. [...] Tonal mixtures, contrast effects and optical illusions occur. [...]

(1) A coloured land tone is modified by densely packed contours. A mixed tone of both elements results. [...]

(2) The colours of rock areas are similarly changed by close networks of black, grey, brown or reddish rock hachures. These in themselves convey the impression of a grey, brownish or reddish areal toning on white paper.

(3) Similar in nature are the changes of area colouring caused by debris dot patterns, dots for dunes or sand in maps of deserts, and in other maps, by small densely packed symbols for tundra, steppe, swamp or moor, as well as for cultivated land, vineyards an so on.

(4) In a three-dimensionally shaded relief map, the three-dimensional effect is often disturbed by the forested areas distributed over the map. To counteract this, one

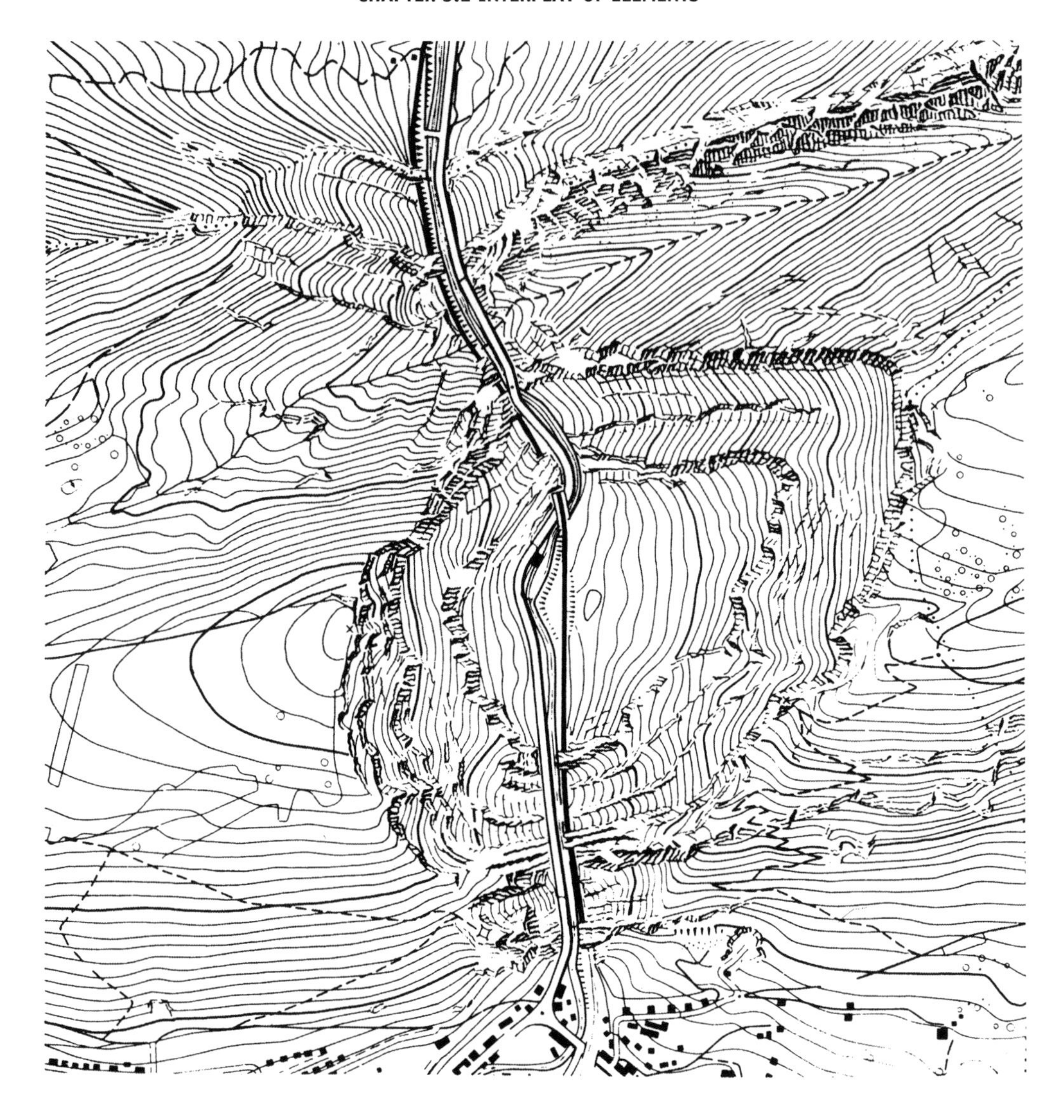

Figure 3.2.9 The gorge at Moutier in the Swiss Jura, scale 1:250 000. A very narrow pass with stream, railway and road.

can open or lighten the colour or the symbols used for forests, or one can combine the forest printing plate with the shading plate so that the forest plate also shows the relief.

Terrain representation and textual matter

All topographic maps normally contain several hundred, even several thousand, names and elevation values. In the interest of legibility they are printed black as a rule. This means that it is impossible to avoid completely disturbing the appearance and expressiveness of the terrain. However, there are many maps which are so thoughtlessly overloaded with names that the terrain can scarcely be recognised beneath the veil of letters. In order to reduce such interference to a minimum, the selection and arrangement of names must be undertaken with care, and the style and size of lettering must be well chosen. [. . .]

Combination of various elements of terrain representation

[. . .] When combining, one has always to keep the following two questions in mind:

- Which types of portrayal or elements should be combined in the case at hand?

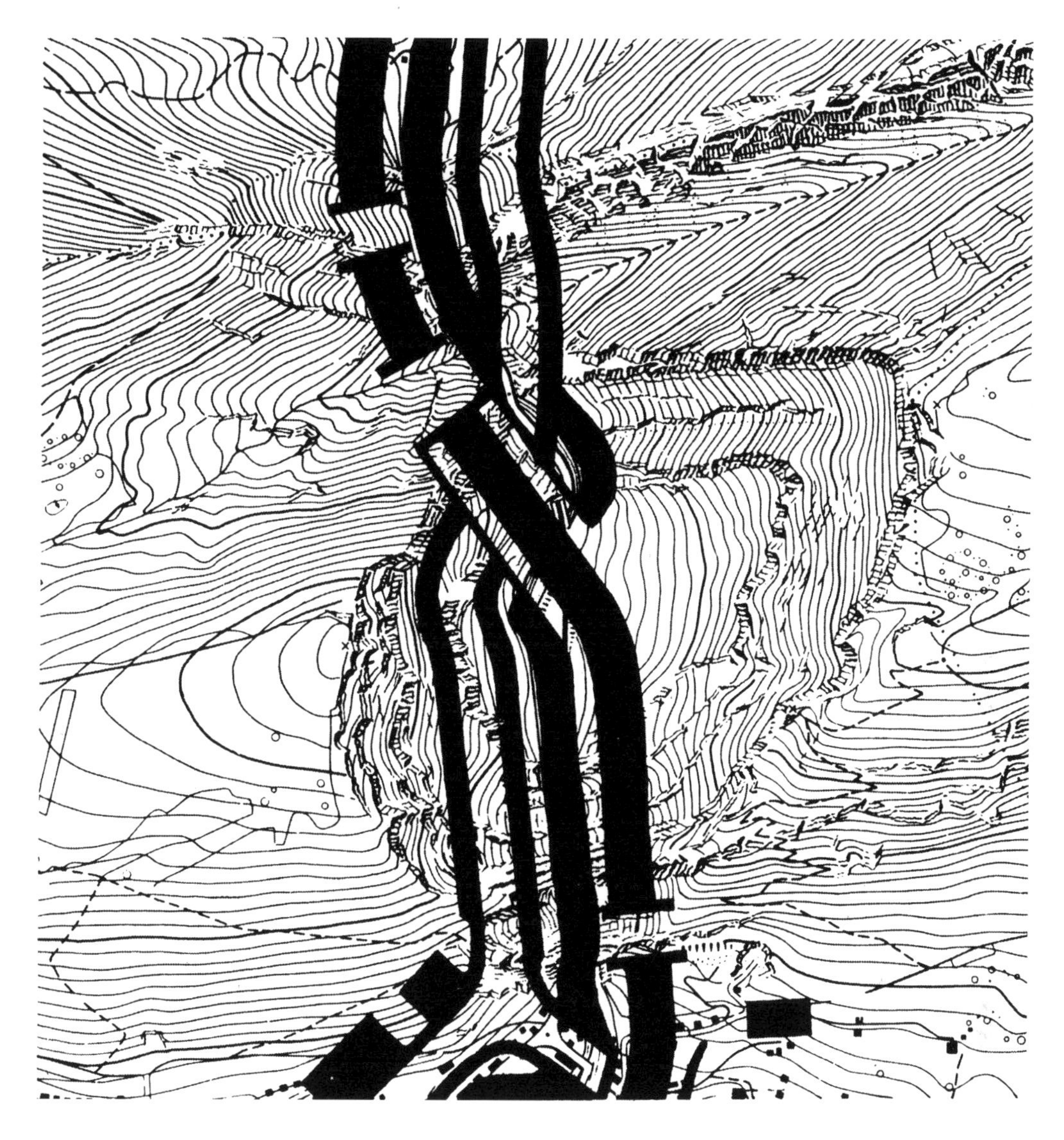

Figure 3.2.10 The same terrain at 1:25000. Stream, railway line and road simplified and widened for the 1:200 000 scale. A wide stretch of terrain is thus obscured.

- How should the individual element be depicted so that it provides strength of expression not only for itself but also in the combined portrayal, and taking other elements into account as far as possible?

Some elements give primarily metrical information, others serve mainly to provide expression and form for an image. One element allows small localised forms to be recognised and can be seen only in small areas. This group includes contours. Another element provides a good overall picture of large areas, for example hypsometric layer tints. One should always combine those elements which are as different or as contrasting as possible, both in the information which they contain and in their design characteristics: in other words, those symbols which supplement each other well in what they represent and are graphically compatible.

One should also consider the ways in which the colours of the combined elements may harmonize. Things which contrast in information content should also be contrasting in colour. Those things which are related in their meaning should not have clashing colours. The most important element, as far as the information is concerned, should stand out.

Figure 3.2.11 The same terrain at 1:200 000 enlarged to 1:25 000, with the displaced positions of some elements of the image.

There is an infinite number of possibilities of interplaying the elements. Some are found primarily in older maps, others in more recent maps. Many combinations are only suitable for larger-scale, others only for smaller-scale, maps. Possibilities also depend on the reproduction process, the number of available printing colours and so on. [. . .]

Combinations for large- and medium-scale maps

(1) Contours and slope- or shading-hachures. This symbol, introduced by early mapmakers, is still employed in the production of fine combined images but its employment is now no longer recommended. Contours present the metric form of terrain surface. Hachures, especially hachures constructed on the principle 'the steeper, the darker', attempt, in a less effective manner, to do the same thing and so there is little sense in combining them. [. . .] Contours in modern large- and medium-scale maps of mountain regions often twist and turn very sharply and suddenly. The stiff, regulated structures of slope and shadow hachures, however, are not sufficiently adaptable to these twists and bends. If both elements must be in agreement, the only solution is to adjust the forms to the hachure pattern by drastic generalisation. If such combinations were often successful in maps of the last century, it was simply because at that time contours were very inaccurately surveyed and smoothed out.

As a rule, contours are printed in strong dark brown or black ink, while hachures appear in medium brown, red–brown, or grey, and thus lose their shading effects. Such combinations of two dense patterns of strokes are not recommended, since they leave so little room for the rest of the map information.

(2) Rock depiction by means of contours, skeletal lines and hachures. Countless maps, even topographical surveys, suffer because of the poor interplay of these elements. The rock skeletal lines emphasise the sharp edges, fissures, runnels, terrace edges and so on. However, the rock contours, drawn separately, are often too extended and smooth so that it is impossible to achieve an accurate interplay [. . .]. The outcome is confusion. Rock edges or ridges, rock hachures and rock contours should reflect the same form down to the last detail. [. . .] A common deficiency of the photogrammetric plotting technique in rock, as well as in other terrain, is first to draw the contours, to smooth them out too much, and to leave the rock portrayal to the topographer or cartographer. A meaningful interplay and an adequate agreement between various elements can only be brought about if the main features of the rock are drawn first with a photogrammetric plotting machine and the contours fitted into this outline thereafter. If rock contours and rock hachures are not printed in the same colour, the printing results are often unsatisfactory in view of the complications in registering them properly. One of the least successful applications of rock contours is when rock skeletal lines run parallel to them, as would happen in the portrayal of horizontally layered rock. [. . .] The result of this combination is more often confusion than anything else.

(3) Contours and slope shading. Contours with equal vertical intervals, normally printed in black or brown, are frequently combined with grey or brown shading tones following the principle, 'the steeper, the darker'. This interplay is seldom enhancing, since these light–dark gradations attempt to repeat the metric portrayal expressed by the contours much less effectively. This combination is particularly unsuitable for narrow, steep-walled valleys in mountainous regions. [. . .] The technique is fairly successful, however, if used in maps of low plateau-like mountains and at large- and medium-scales.

(4) Contours with oblique hill shading or with combined shading. In medium-scale maps the combination of contours with oblique hill shading often produces outstanding results. The guilty cartographer then excuses his poor work by insisting that his contour pattern depicts the small forms well enough, the shadow tones being there merely to make the large forms more obvious, and that they should be allowed to stretch across the terrain in a broad, flat and generalised fashion. This unsatisfactory interpretation leads to a haziness in the contour image and to an ineffective, stylised approach to landforms. [. . .] The surface form as portrayed by the contours, and that which is presented through the play of light and shade, must harmonise with each other down to the finest detail. [. . .] It could be argued that a shading image which is too detailed and intricate introduces an untidy complexity into the map, making the overall impression less effective. If the shading is well 'worked over' and edited, however, this does not hold true. The careful depiction of small forms should in no way destroy the unity of large groups of forms. [. . .]

(5) Rock drawing and oblique hill shading. When combined with detailed hill shading, which fits exactly to the image, skeletal rock drawing can lead to excellent results. The two elements can supplement and support each other effectively, but only when the rock skeletal lines are drawn crisply and precisely (without unnecessary additional hachures) and printed in a strong brown, grey or violet. This is the only way in which it can maintain its definition clearly against the grey or blue–grey shading tones, and also continue to indicate form, combining itself with the relief shading into a unified whole. Examples of colour-separated originals are shown in Figure 3.2.12 (rock skeletal line) and Figure 3.2.13 (relief shading). Although such results are possible, many medium- and small-scale topographical maps suffer from poor attempts at just this combination. Rock hachures are often too dense and are printed in a colour so weak that they are obscured by the relief shading. Rock hachures which are too thick are normally unsuccessful on the illuminated surfaces of steep rock slopes; they darken them so much that the intended modelling effects never materialise.

(6) Shaded hachures and shading tones. This combination is also common, and is designed to improve the impression of relief in landforms. However, such overlapping of two elements, both of which have the same function, is not recommended.

(7) The landscape painting in plan view. Attempts to represent the landscape and provide a plan view portrayal which appears as natural as possible – an impressionist painting, for example – can be artistically stimulating when well executed. In cartography, however, plan view paintings are primarily of interest as experiments which might help one to answer the following questions: To what extent can the forms of natural phenomena be realised in the cartographical image, and what are the advantages and weaknesses of such imitative work? [. . .] To reproduce the complex play of light, shadow, colour and atmosphere in a map-like image demands artistic talent. But the technical reproduction aspects do not correspond to those normally used in cartography. Here, colour separations cannot be produced schematically, according to the topographic elements, but only according to the colours themselves. [. . .] What is achieved with such painting experiments? When successful, beautiful impressions, natural plan-view landscape pictures result – pictures which please the hearts of many viewers. But the end result is not a map. Paintings such as these are unable to provide, anywhere and in any respect, the topographic, conceptual and metric information which one expects from a map. [. . .] They are especially instructive in illustrating how ground cover colours and relief tones can be united without weakening the modelling effects and colour differentiations to any great extent. [. . .]

(8) Hill-shaded and coloured maps of medium and large scales, without contours. Relief shading can be combined with a layer tint system adjusted to give the aerial perspective effects seen in the landscape. The colour and shading tones here are as close to nature as possible, [. . .] most linear elements are lacking, which greatly increases the natural appearance, but reduces its power of expression and usefulness. Portrayals of this type are seldom found in maps of large and medium scale. They are generally prepared at the proof stage during the production of a contour map. [. . .] Shading tones should not be dull, mouse-grey and dirty looking, nor should they be stylised red–violet, or red–brown, but rather a hazy blue–grey. Toward the upper reaches, they should become more intensive and redder (more carmine), but toward the lower areas lighter and bluer. The bright blue used to induce aerial perspective in the hypsometric shading can be employed as a water tint at the same time. It may also be used as the first shading plate providing additional pale blue shadows, but care should be taken to restrain the blue from climbing too high up on the illuminated slopes. [. . .] The yellow tint should also be varied in intensity and made to conform with the various elevation layers as well as relief. In the

Figure 3.2.12 Mt Everest or Chomolongma. Rock drawing for a multi-coloured hill-shaded map at 1:100 000 scale. (Source: Imhof 1962.)

stronger shaded areas it is usually lightened or completely eliminated. [...] All rocky slopes on the light side possess the bright, yellowish-red toning of the highest layers. These tints are not used on permanent snow and glaciers. Finally, the colour transitions of the hypsometric tones are gradual. Shading tones should dominate the surface tones, they should not be allowed to be submerged by them. [...]

(9) Contours and rock portrayal combined with hill shading and colour tones. Equal vertical interval contours, usually differentiated by colour in the customary fashion, and rock skeletal lines or rock hachures, are combined with shading and colour tones. [...] The fairly demanding combination of reproduction and design is suitable for maps of medium and large scales [...]. At the larger scales, the forest may also be added by patterns of symbols or green area tints. At scales smaller than about 1:250 000 differentiation of contours by colour should be avoided except for glaciers. At scales smaller than 1:500 000 rock skeletal line drawing is omitted, since here it would not represent the form, no matter how meticulous the depiction. In this type of representation it is important to maintain a perfect reflection of form and the greatest possible harmony in the interplay of colours between the linear and areal relief elements. Contours and rock skeletal lines are designed to contain more metrical information for the map [...]. Rock hachures should be sufficiently faint and restrained on the illuminated slopes that they do not destroy the impression of three-dimensional form. Area tones should be transparent so as not to obscure the line

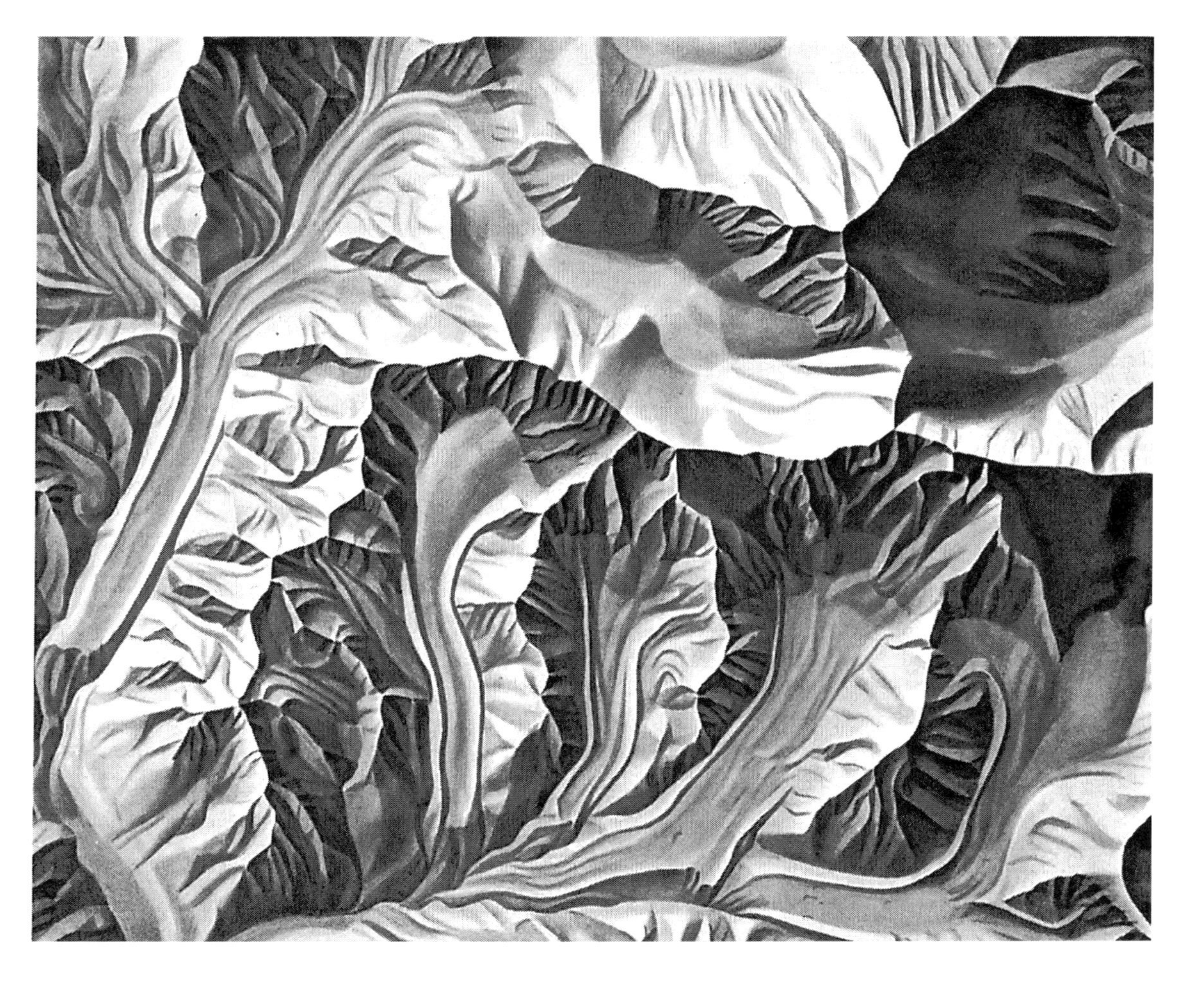

Figure 3.2.13 Mt Everest or Chomolongma. Shaded drawing for a coloured hill-shaded map with rock drawing and contours at 1:100 000 scale. (Source: Imhof 1962: 93.)

image. Dense cones of contours in a red–brown colour and dark brown rock outlines change the impression of area colours. [...] A medium, dull, bluish-brown or brown–violet is better suited for rock depiction than black, since such impure tones move them back from the black text and match them in with the area tones more neatly in a balanced picture. Dense, black patterns of lines and strokes would spoil the soft areal tones. The type of representation described here was developed many decades ago, primarily in Switzerland. It is often called the 'Swiss style,' but is in no way limited to Swiss or alpine terrain. It is suitable for all regions having a generally rough, ruggedly undulating terrain. [...] Since it combines the advantages of a metrical line scheme with the direct impression of naturalistic colours and shading tones, it is superior to all other types of representation.

[...] Earlier, strong brownish, reddish or yellow-reddish ground surface and illuminating tones dominated the usually pale shadow tones [...] such that the latter sank or were immersed in the dominant elements and were inaccurate. [...] Valley bottoms and flat land were therefore lighter than the adjacent illuminated slopes. Maps such as these were often ruined by too strong yellow and pink tints; also the hypsometric blue was delimited too crudely or abruptly, or even omitted altogether. A direct three-dimensional impression can only be produced by consistent oblique shading, by the priority of hill shading over the surface toning, by the most precise form of interplay between all elements and by removing from the map content as much unnecessary and disturbing secondary matter as possible.

Reproduction normally requires the following printing colours and printing steps:

(i) Strong, dull brown for settlements, communication networks, rock skeletal lines, contours on scree slopes and so on.

(ii) Strong blue for the stream network and for contours on glaciers.

(iii) Reddish-brown for contours in vegetated areas.

(iv) Blue–grey of medium strength to act as the richly variable main tone of relief shading, except for lakes.

(v) Grey–violet, medium strength, as a supplementary relief shading tone, except for lakes.

(vi) Light blue for hypsometric tints to reinforce shaded slopes and for lake surfaces.

(vii) Yellow for ground and light toning, apart from lakes and glaciers.

(viii) Pink, extremely weak, for rocks and high regions, except for lakes and glaciers.

(ix) Black for textual matter.

In addition to these nine printing colours, the following are often used:

(x) Violet, an additional supplementary relief shading tone

(xi) Green, for forest contours and tinted forested areas

(xii) Red, for railway lines and roads and so on, and in halftone tint for coloured bands along political boundaries.

The type of portrayal described here allows variations and also simplification and reduction in the number of printing colours. [...] The shading tones can be held back as far as possible and flat land and valley floors can be lightened an so on. [...]

Combinations for small-scale maps

At small scales, the forms to be portrayed are very flat as a whole and usually very finely detailed. With equal interval contour systems forms such as these cannot be brought out since the line patterns would be too coarse. [...] In place of combination with contour systems, systems involving layer tinting are normally used.

(10) Slope hachures produced according to the principle 'the steeper, the darker' and hypsometric tints. A poor obsolescent method. In addition to the deficiencies already mentioned under 'slope hachures', the following disadvantages should be noted. The slope hachure drawn to the principle 'the steeper, the darker' is a method of portrayal suitable for scales larger than about 1:200 000. [...] It cannot cope with the fine details of smaller scales. As stressed previously, however, layer tinting is particularly suited to small-scale maps, smaller than 1:200 000 for example. This means that there is a conflict in the suitability of both these techniques to different scales. The bold brown tones used formerly for higher elevation steps were produced partly by employing fairly coarse line screen patterns. These and the similar screen tint like hachures were mutually destructive. Both elements produce gradations of lightness but progressing in opposite directions and this leads to chaos when they are overlapped. A further weakness often lies in the lack of similarity in the forms produced by differing methods of generalisation.Shaded hachures and hypsometric tints. Despite certain weaknesses, this is a popular and, as far as content is concerned, quite a valuable combination for small-scale maps. It is frequently found in the physical maps of school atlases produced since the end of the nineteenth century. [...] Both shaded hachures and hypsometric tints are suitable for small-scale maps. Of course, the drawing or engraving of hachures involves a great deal of work, so that hachure images today, although technically more definite, must gradually give way to the more rapidly produced shading methods. (See combination No. 13–14 below.)

(11) Strong, brown layer tints composed of half-tone line screens and the delicate stroke structures of grey–brown, red–brown or grey hachures are also in conflict in these circumstances. Not infrequently, we find examples where they completely destroy one another. If the three-dimensional modelling effects of hachures are to be retained, then the layer tints must be given very delicate shades or printed as solid colour. Any line screen used in the area colour steps should be significantly finer in texture than that of the hachures. Furthermore, the printing colour brown or red–brown, used for the hypsometric tints, should be considerably lighter than that used for the hachures. Care should also be taken to avoid a contradiction between the generalisation of the hachure system and that of the layer tint outlines. If, as often happens, the contours bounding the layers are added to the hachures and colour tints, then the effect in mountainous areas is of a heavy build up of lines, unpleasant to the eye.

(12) Slope shading following the principle 'the steeper, the darker', combined with hypsometric tints. The same

applies here as applied in No. 10 for the combination of slope hachures and hypsometric tints. This combination is, therefore, poor and outmoded, despite the fact that it may often still be used since it is easy and fairly rapid to produce.

(13) Combined shading and traditional hypsometric tinting. [...] In the atlases and sheet maps of recent decades, the amalgamation of combined shading with traditional hypsometric tinting is often found. Combined shading is that which corresponds in its light–shade distribution to the shaded hachure, and hence to that type of shading without shadow on flat terrain. [...] Even combinations such as these are not completely satisfactory. Although the relief can be depicted in more detail with shading tones than with shaded hachures, it would nevertheless be obscured by strong brown or grey–violet elevation tints. [...] Here also, an inconsistent, imperfectly matching generalisation of relief tones and elevation can often create confusion. Recently, attempts have been made to overcome this interference with the relief impression by adjusting the layer tints in the same manner. [...] When this is done the lightest parts must be allowed to retain an adequate tone. Their combined printing then produces an image in which the light and shade tones of the relief are less disturbed and the elevation tints, even in the light areas, are just visible. [...] The contour outlines of layer tints are sometimes included in such combinations. A more definite delineation of colour in this way is usually obtained, however, only at the expense of clarity of the relief image.

(14) Oblique hill shading combined with hypsometric tints in small-scale maps. Hill shading under the assumption of oblique light can be combined with a sequence of hypsometric tints which provides the optimum effect of depth or three dimensions. The colour sequence, based on the idea of aerial perspective with increasing lightness with altitude, allows the shaded image to achieve its full effect and enables the desirable light–dark contrast to increase towards higher levels. [...] The lowland areas are greenish-grey; mountains of moderate altitude are light olive–brown; the highest, glacier-strewn elevations, however, have white bare effects, or, in the completed map, a violet–grey appearance, since the shading tones have no layer tints over them here. The detailed modelled forms are disturbed nowhere by the boundaries of the hypsometric tints. Maps such as these provide the ultimate in the impression of three-dimensional depth and form. [...]

(15) Relief shading combined with ground and vegetation colours in small-scale maps. More attempts have been made recently to combine, in small-scale maps, the three-dimensional impression of oblique shading with ground and vegetation colours. [...] The first examples of this type were maps of the United States [...] for the orientation of airline passengers. Instead of trying to depict elevation differences as they appear in nature, [...] the broader changes in colour in forested areas, farmland, plains and deserts were represented. Similar attempts were published and described later by Keates (1962). [...] At the small scale, however, the relief forms and the ground cover mosaic are so finely detailed and often have so little relation to one another that in certain areas great complexity and distortions of the relief are unavoidable. As a result of the flatness and spaciousness of the 'models', distinct aerial perspective hypsometric tints can scarcely be achieved by such combinations. All these difficulties lead to considerable generalisation in the concept and colours of the map and the introduction of conventional symbols. [...]

(16) Contours with equal vertical intervals, hachures and hypsometric tints. This combination of three elements is of little use. [...] Contours with equal vertical intervals belong to large- and medium-scale maps, hypsometric tints to the smaller scales. Maps of this type are untidy in design, the symbols employed working against one another. They are also uneconomical to produce. Only in exceptional cases can the method be usefully applied – such as in medium-scale maps of flat regions.

(17) Contours with equal vertical intervals, shading tones and hypsometric tints. [...] The combination of contours with equal vertical intervals with shading tones is very successful, but should be considered only for maps of large and medium scale. The addition of fixed hypsometric tints is seldom recommended, however, since the resultant banded effect would interfere with the detailed shading of small-scale maps to too great an extent. In general, the same untidiness and problems of scale as mentioned above, under No. 16, apply in the combination of these three elements. [...]

References

Imhof, E. (1962) *Schweizerischer Mittelschulatlas*, Orell Füssli Arts Graphiques, Zurich.

Keates, J. (1962) The small scale representation of the landscape in colour. *International Yearbook of Cartography*, **2**, 76–82.

Further reading

Collier, P., Forrest, D. and Pearson, A. (2003) The representation of topographic information on maps: the depiction of relief. *The Cartographic Journal*, **40**, 17–26. [An analytical comparison of effectiveness of different ways of representing relief.]

Hurni, L., Kritz, K., Patterson, T. and Wheate, R. (2001) Special Issue: ICA Commission on Mountain Cartography. *Cartographica*, **38** (1/2). [This theme issue provides a useful overview of the state of the art of mountain mapping, and its particular design challenges.]

Jenny, B. and Hurni, L. (2006) Swiss-style colour relief shading modulated by elevation and by exposure to illumination. *The Cartographic Journal*, **43** (3), 198–207. [A useful technical review historical development of relief shading and the presentation of new software solution.]

Kent, A.J. and Vujakovic, P. (2009) Stylistic diversity in European state 1:50 000 topographic maps. *The Cartographic Journal*, **46** (3), 179–213. [A systematic comparison of design strategies deployed in the publication of different European official 1:50 000 scale topographic maps.]

See also

- Chapter 1.2: Semiology of Graphics
- Chapter 1.7: Design on Signs/Myth and Meaning in Maps
- Chapter 3.4: Generalisation in Statistical Mapping
- Chapter 3.6: The Roles of Maps
- Chapter 3.8: ColorBrewer.org: An Online Tool for Selecting Colour Schemes for Maps

Chapter 3.3

Cartography as a Visual Technique, from *The Look of Maps*

Arthur H. Robinson

Editors' overview

This chapter forms the core argument in Robinson's PhD thesis, subsequently reworked and published as *The Look of Maps*. It justifies the need for a visual approach to cartography and establishes the limits of aesthetic and scientific approaches to mapmaking and reading, with a particular focus upon thematic maps. Robinson discusses the significance of lettering, colour and visual structure as the most significant factors in the design of mapping and establishes how treating cartography as a visual technique might lead to more effective maps.

Originally published in 1952: Chapter 3 of Arthur H. Robinson, *The Look of Maps*, University of Wisconsin Press, Madison, WI, 16–25.

Is cartography an art? If a map has a pleasant appearance does that make it a good map? These and many related questions constantly arise in discussions of the philosophy of cartography. To champion a categorical yes or no would be presumptuous. Yet the very frequency of the questions involving the relationship of art and cartography indicates that many students and practitioners have been unable to resolve the apparent, but perhaps unreal, incompatibility between science and art.

The assumption that effective cartographic technique and its evaluation is based in part on some subjective artistic or aesthetic sense on the part of the cartographer and map reader is somewhat disconcerting. For example, Raisz (1938: 2) claims that the 'effective use of lines or colours requires artistic judgment', and Wright (1942: 542) explains that the suitability of a symbol 'depends on the mapmaker's taste and sense of harmony.'

Throughout the literature there are numerous similar assertions regarding the assumed subjective aesthetic and artistic content of cartography. There is also a considerable tendency to define the subject as a kind of meeting place of science and art. This is exemplified by Eckert (1925: 670–677). He pleads for artistic imagination and intuition in cartographic portrayal and claims that the inter-action of such talents with scientific geography produces the aesthetic map. There is no question about the importance of imagination and new ideas, but it is equally important that significant processes be objectively investigated, whether it be the visual consumption of a graphic technique or a process in geomorphology. In order to understand the degree to which art enters into cartography, it is necessary to examine some of the fundamentals of each in order that the logic does not become simply a matter of semantics. It can perhaps best be approached by a comparison of the aims, techniques involved and the results accomplished by each activity.

Most scientific cartography is concerned with the dissemination of spatial knowledge. The aim of visual art is more difficult to express. Generally speaking it may be said to have two basic goals, the first of which is to provide aesthetic pleasure through visual (sensuous) stimuli. In many instances, and especially significant in cartography, this has taken the relatively crude form of ornamentation. This, so far as cartography is concerned, may well be considered a low form of art based on the uncritical and popular conception that anything graphic that is difficult is thereby artistic. Even assuming (and it is a difficult assumption) that the fancy borders, ornamental cartouches, curvaceous lettering and other decorative

The Map Reader: Theories of Mapping Practice and Cartographic Representation, First Edition. Edited by Martin Dodge, Rob Kitchin and Chris Perkins.

features so common on older maps (and still not uncommon) (Becker 1910) are a source of pleasure to a reader, it does not seem illogical to suggest that such 'art' does not add to the functional quality of the map. On the contrary, it may actually detract from it, for the attention may be drawn to these exhibitions of manual dexterity when it should be concerned with the data presented for consumption (Eckert-Greifendorff 1939). As with ornamentation, the use of colour on maps has also been held as irrefutable proof that there is aesthetic art in cartography.

With respect to the use of colour, ornamentation and other assumed elements of art, there seems to be considerable confusion between the creative motives involved and the effects produced. The fact that there is a 'joy of creation' in mapmaking is certainly no argument that there is art in cartography, for there is similar pleasure associated with any creative effort whether it be painting, composition of music or gardening. If a map is a functional object intended primarily to stimulate the intellectual aspects of our mental processes, the fact that it uses visual media to do so is beside the point. If the final composition also appears 'beautiful', that aspect is quite apart from its primary function and may, as previously pointed out, detract from its effectiveness as a map, since the aesthetic response may take precedence over or interfere with the intellectual response. That is not meant to imply the extreme, that something created for practical purposes must therefore not be pleasurable. Certainly any job, well done, especially a creative undertaking, provides pleasure both to its creator and observer. What is meant, on the contrary, is that any aesthetic stimuli which may be included in a map probably should be incorporated consciously and with full realisation of its effect on the other visual material.

The other aspect of art, the art which attempts to awaken various responses not necessarily of beauty, received little attention in cartography until the use of maps for propaganda purposes came into favour. It is difficult and perhaps unsatisfactory to attempt to separate the two aspects of art from the point of view of motivation because the techniques of the two are essentially similar, and certainly the motives are commonly combined or confused. In non-aesthetic art the aim may be any of a multitude of possibilities but a basic characteristic is the attempt to construct visual stimuli which will produce desired mental responses. It is well known that certain colours, shape combinations and line relationships produce predictable responses, including intellectual connotations such as simplicity, confusion, density, rhythm and balance. Recently two authors have considered its manifestation in cartography and, probably as a result of reaction to its use for propaganda purposes, have taken a negative approach. Wright (1942) illustrates the confusion that can occur between the visual and intellectual, and Speier (1941: 314) acknowledges that 'size, colour and design, can be made to serve propagandistic ends'.

There can be little doubt that if the use of visual techniques to stimulate predictable responses is accepted as within the field of art, then cartography includes artistic techniques. Such techniques obviously should be employed in the attempt to satisfy the functional requirements of a map, for a map is a graphic thing that, by any definition, cannot be visually sterile. The difficulty that arises in interpreting that fairly obvious conclusion is what proportion, if any, of the cartographer's artistic judgment, taste or sense of harmony should enter into his creation. Everyone has probably heard the fatuous statement (and a great majority of us have probably echoed it) 'I don't know what is good, but I know what I like'. If we proceed on the basis of that grotesque admission, we find ourselves spurning maps with red on them because we don't like red, we champion conventions because they are familiar, we make brightly coloured maps because they are brightly coloured, whether they convey the information or not. In short, our judgment of technique is based on convention, whim and fancy. What alternative bases can we adopt? There are two such. One is to standardise everything. Then all map readers, once they had learned the symbols and techniques would be presented only with varying combinations of familiar things. All cities always would appear as black circles and they would be named in Spartan Medium Italic type, and so on. The absurdity of such a proposition is, I hope, obvious.

The other alternative is to study and analyse the characteristics of perception as they apply to the visual presentation we call a map. They are complex to be sure, and even if the present state of knowledge concerning them is relatively scanty, logic will at least illuminate some of the broader aspects. Experience and research in other fields of visual presentation may assist. In order, however, to construct a basis for judgment it is necessary to admit that the scientific map is a functional object. Cartography is a technique, just as scientific writing or the language of mathematics, by which intellectual concepts are displayed for consumption. That from time to time a cartographer will include something for aesthetic purposes or will, by some technique or other, introduce a bit of levity on a map, does not alter the basic premise. It is only necessary that when such is done, it be done consciously and with full realisation of how such procedure will affect the resulting visual complex.

Cartography for educational purposes should be considered as no more and no less creative than writing. That master of functional writing, W.M. Davis (1911: 33), happily extended this thought to cartography when he wrote: 'It is well known that there are many geographical matters which are better presented pictorially, cartographically or diagrammatically than verbally. Hence, it is just as important to study the proper and effective use of various forms of graphic presentation, as it is to study the values of different methods, treatments, grades and forms of verbal presentation'. The ends desired thus dictate the means to be used to accomplish the purpose. The processes and functions which probably will occur between the eye and the mind of a reader must be predicted and analysed if the technique is to be properly evaluated. To do so is difficult to say the least. In the first place, there are not yet objective data upon which to base answers to many questions involving visual stimuli and mental response relationships. Part of a map is absorbed intellectually. That is to say, the visual shape and colour of an object may be, through convention, completely or partially submerged in its intellectual connotation, for example blue water, the dot-dash political boundary and familiar lettering. The remaining components of a map enter the brain as visual stimuli without intellectual meaning except through prior reference to a legend. Sometimes visual and intellectual qualities exist together in a single component, for example a large red USSR. The very adjacency of the visual forms within the map frame causes each to modify the intrinsic visual qualities of the other.

These stimuli have only recently been appreciated as having an important bearing, other than aesthetic, in the consumption of a map. True, Eckert (1908: 344) indicated some recognition of this nearly forty years ago when he noted that 'an artistic appearance, particularly a pleasing colouring, can deceive in regard to the scientific accuracy of a map'. But, except for colour, it was not until the employment of maps as tools of propaganda that the importance of visual relationships was generally appreciated. Most of the writers on cartography have touched upon this subject, usually implicitly, rarely directly, and visual relationships are generally considered to be artistic or psychological components which cannot be avoided, but which should be guarded against. Speier (1941: 313) echoed this sterile attitude when he wrote that 'the relationships of the different lines and areas we see, the shape of parts, the distribution of colours, symmetry or its absence all these are extraneous to the scientific purpose of a map.'

If visual relationships are to be utilized to accomplish a positive purpose, then it becomes necessary to establish principles for their employment. The complete evaluation of cartographic methodology therefore requires that, ideally, the visual and intellectual properties of all map data, techniques and media be analysed, as well as all the possible combinations of them. From the abstract point of view all the components of a map may be placed under two basic categories, (i) cartographic data and (ii) cartographic technique. As has been pointed out previously, it must be assumed that the selection of the data to be used on a map is primarily a function of the educational purpose to which the map is to be put and should, therefore, be determined without regard to cartographic technique. This, of course, can never be true in practice, since the range of techniques with which to convey geographic information is not unlimited, yet for purposes of evaluation of technique, consideration of content, wherever possible, must be removed. Its presence cannot fail to occasion responses which, consciously or unconsciously, will condition the evaluation of the technique. For this reason most preliminary advertising layouts are usually made up simply as organisations of shapes of varying value (brightness). The removal of these intellectual factors for purposes of evaluating technique is by no means as simple as it may appear at first consideration. As was noted, it is impossible in practice completely to separate data from technique, and since cartography is essentially a technique designed and existing for geographic data, it may be expected that data will play a large role in its method.

Cartographic data may be either quantitative or qualitative (Hettner 1910; Wright 1942, 1944). Quantitative data may be evaluated solely on bases of appropriateness and accuracy. The qualification of data involves problems such as its categorisation and generalisation, determination of ratios, and the choice of isograms and choroplethic limits. They lie closer to the substantive aspects than to the presentation. On the other hand, certain qualities of data, such as comparative visibility of shapes, colour relationships and a number of others, involve problems of visual evaluation. When considered from this point of view these aspects must also be included in the evaluation of technique. Consequently, it seems logical to separate cartographic method into two general categories: (1) substantive method and (2) visual method or technique. There is a third rubric in the complete classification of cartographic method, namely map reproduction. It acts primarily as a limiting factor on the other categories. One must always admit, however, that there is no clear and precise distinction between the categories. A map when analysed solely as a visual thing is essentially a group of more or less related items of all sizes, shapes and colours: the land–water relationship, the trim page, the map proper, the legend and title boxes, relative line width, the individual words, any massing of data, colour, or value areas, and so on. All of these may (and should) be considered merely as items varying in size, shape and colour, and bearing a direct visual relationship to one another. In the broadest classification of map components, these all may be grouped under the term 'design' and would so be considered if the map were being analysed purely as a graphic layout problem. But a map is more than that. Each of these elements of design has intellectual limitations which cannot be ignored: one is not able to organise the shapes freely, relative importance exists without regard to size and shape and utility takes precedence over the aesthetic. A map cannot, therefore, be evaluated solely on the basis of pure design, even though considerable insight into the problems of technique can be gained by so evaluating it.

The lettering on a map is the first element requiring evaluation. Names and words are the shapes or symbols most submerged in their intellectual connotations, but they do have visual form, and as such are significant in the overall organisation of the map. Their evaluation is based on two major aspects of their utility, the size and the design of the typeface. In addition, other important considerations in the methodology of map lettering include the appropriateness or suitability of the typeface, and the colour of print and background and its effect on legibility.

The organisation of the basic shapes within the map frame, that is to say, the structure of the map, has a significant bearing on the utility of the map. The controls which determine the possibilities of structural variation include elements of projection, balance, direction and many others.

The third element in the visual evaluation of a map is colour, used in the broad sense to include value (brightness) and intensity as well as hue. It is perhaps the most difficult of the cartographic techniques, since it is a significant element in both lettering and structure as well as in its restricted consideration. Its use with other elements markedly affects their visual effectiveness. It has, however, one use more or less peculiar to it, the portrayal of categories of either related or unrelated data. Different hues, values and intensities commonly appear in juxtaposition. It enters structural problems because its use changes the character of shapes. Thirdly, the colour seems to produce significant emotional and intellectual responses.

These three visual components of cartographic technique, lettering, structure and colour, encompass most of the aspects of a map capable of evaluation from the visual point of view.

References

Becker, F. (1910) Die kunst in der kartographie. *Geografische Zeitschritft*, **XVI**, 473–490.

Davis, W.M. (1911) The Colorado Front Range: a study in physiographic presentation. *Annals of the Association of American Geographers*, **1**, 21–83.

Eckert, M. (1908) On the nature of maps and map logic. *Bulletin of the American Geographical Society*, **40**, 344–351.

Eckert, M. (1925) *Die Kartenwissenschaft*, de Gruyter, Berlin.

Eckert-Greifendorff, M. (1939) *Kartografie*, de Gruyter, Berlin.

Hettner, A. (1910) Die eigenschaften and methoden der kartographischen darstellung. *Geografische Zeitschrift*, **16**, 75–82.

Raisz, E. (1938) *General Cartography*, McGraw-Hill, New York.

Speier, H. (1941) Magic geography. *Social Research*, **8**, 310–330.

Wright, J.K. (1942) Map makers are human: comments on the subjective in maps. *Geographical Review*, **32**, 527–544.

Wright, J.K. (1944) Cartographic considerations. *Geographical Review*, **34**, 649–652.

Further reading

Crampton, J. (2001) Maps as social constructions: power, communication and visualization. *Progress in Human Geography*, **25** (2), 235-252. [A critical examination of cognitive approaches to cartography placing Robinsonian visual thinking in relation to technological and epistemological change in cartography.]

Montello, D.R. (2002) Cognitive map design research in the twentieth century. *Cartography and Geographic Information Science*, **29** (3), 283-304. [Explores the seminal role of the *Look of Maps* in establishing the parameters for subsequent map design research.]

Robinson, A.H., Morrison, J.L., Muehrcke, P.C. *et al.* (1995) *Elements of Cartography*, 6th edn, John Wiley & Sons, Ltd, Chichester, UK. [For long the standard work on Robinsonian cartographic science, explores practical solutions for delivering map designs following principles established in this chapter.]

See also

- Chapter 1.3: On Maps and Mapping
- Chapter 1.7: Design on Signs/Myth and Meaning in Maps
- Chapter 2.2: A Century of Cartographic Change
- Chapter 3.6: The Roles of Maps
- Chapter 4.2: Map Makers are Human: Comments on the Subjective in Maps

Chapter 3.4

Generalisation in Statistical Mapping

George F. Jenks

Editors' overview

In this article Jenks considers factors affecting the successful communication of data values by cartographers using statistical mapping. He establishes the nature of the statistical surface, before exploring the implications for design of choropleth and isarithmic maps. The impact of different kinds of generalisation and classification strategies is considered in terms of the type visual pattern generated by changing the numbers of classes and the mathematical method of computing the breaks between classes. He concludes with a call for a more systematic consideration of the creative design choices and constraints limiting sensible outcomes.

Originally published in 1963: *Annals of the Association of American Geographers*, **53** (1), 15–26.

Among cartographers the term generalisation usually conveys the concept of selection, or simplification, or evaluation for relative significance, of landscape features which are to be mapped. The processes of generalisation, while creative and intellectual, depend upon visual comparisons between large-scale source materials and smaller scale compilations. Coastlines are smoothed, settlements are sifted for size or relative importance, patterns of drainage are simplified and so on. In each case, cartographic judgment and scientific integrity influence the cartographer in attempting to achieve a truthful representation of the landscape but his basic generalisations depend upon the fact that he can exploit his own and the map reader's experience in symbolisation.

Unlike landscape features, which can be seen by both the mapmaker and the map reader, abstract statistical distributions have no basis in direct observation. The patterns which are revealed by symbolisation do not create mental images which allow for critical comparisons and evaluations. In fact, the cartographer himself is often unaware of the nature or the degree of generalisation which he incorporates into his presentation. For example, how often does he analyse his representation in regard to the shape of the statistical surface being symbolised, or fully understand the variations involved in selecting different sizes of symbols, different numbers of classes to be shaded, or the results of manipulating and classing his data? Too often, he solves his symbolisation problems blindly, following precedents set by others or giving way to his own preconceived ideas and prejudices.

If we make the rather dangerous assumption that statistical cartographers possess a high degree of integrity and therefore desire to present data truthfully, then, why are statistical maps so often unreliable? The answer to this question appears to be that since it is difficult to visualise abstract forms, the cartographer makes a series of judgments without really understanding what effect these judgments will have upon the reader's interpretation of the distribution. In other words, would he make the same generalisation if he could visualise the statistical surface that he is trying to create?

This paper focuses attention upon three interrelated phases of statistical mapping: (1) the concept of the statistical surface; (2) generalisations resulting from the selection of numbers of classes; and (3) generalisations dependent upon the mathematics of classing data. In practice, the cartographer considers all three phases simultaneously, but to simplify exposition each will be treated

The Map Reader: Theories of Mapping Practice and Cartographic Representation, First Edition. Edited by Martin Dodge, Rob Kitchin and Chris Perkins.

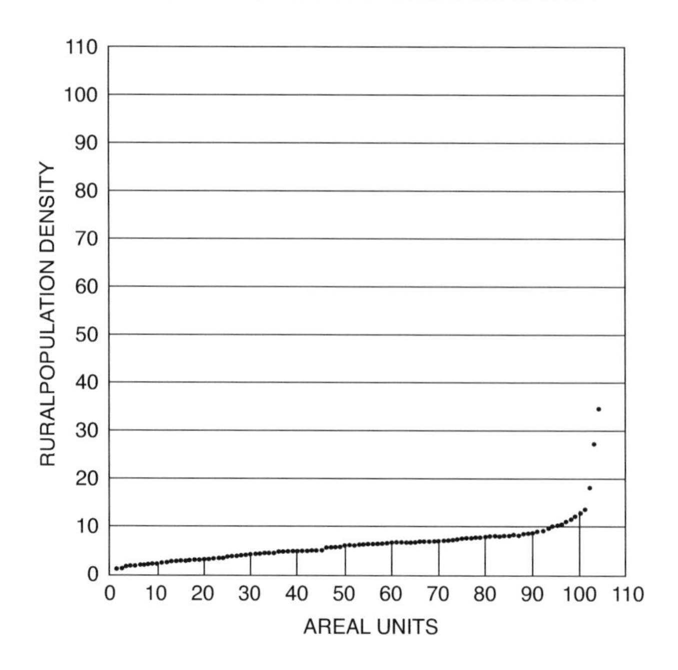

Figure 3.4.1 Graphic arrays are one of the analytic tools which aid cartographers in understanding the nature of mappable data. In this case, rural population densities for one hundred and five minor civil divisions were ordered and plotted by increasing value. [. . .] The curve shown here has a distinctive shape often found in population data and not infrequently in other data series which form the raw materials for statistical maps.

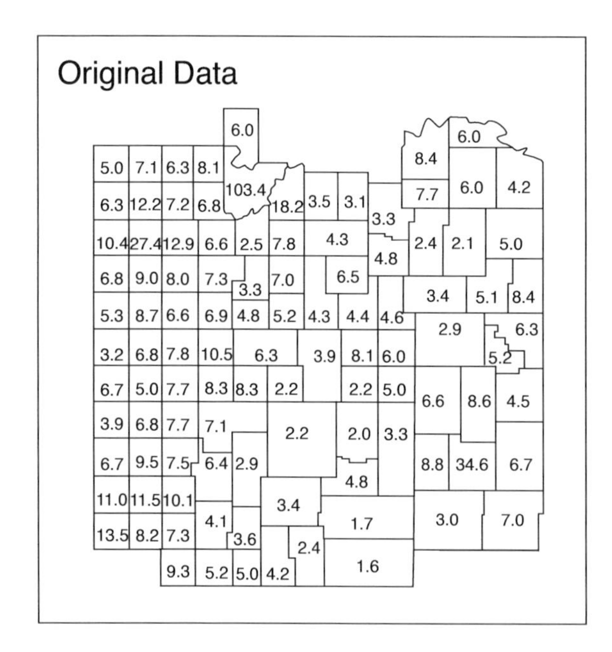

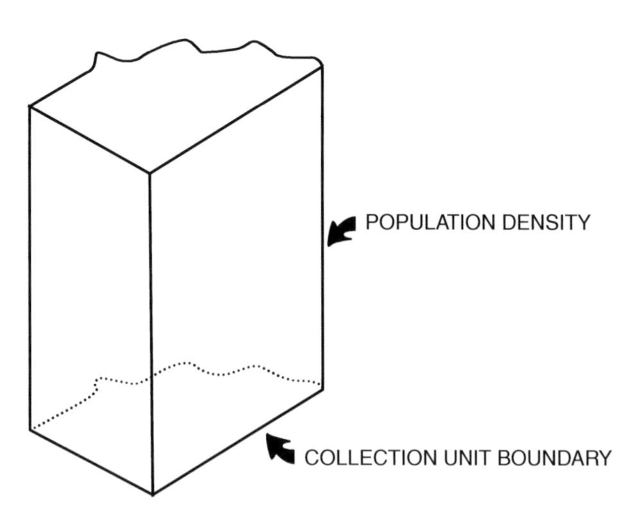

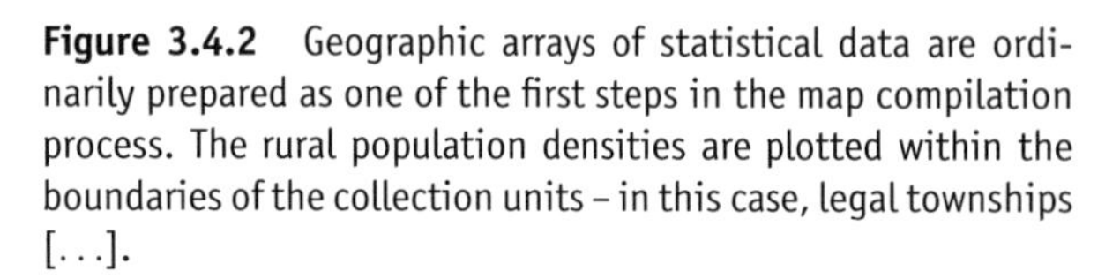

Figure 3.4.2 Geographic arrays of statistical data are ordinarily prepared as one of the first steps in the map compilation process. The rural population densities are plotted within the boundaries of the collection units – in this case, legal townships [. . .].

Figure 3.4.3 Since ratios pertain equally to the entire area of the collection unit a three-dimensional visualisation of population density would appear to be a prism. The top of the prism would be parallel to the base and the height would be scaled to the density value.

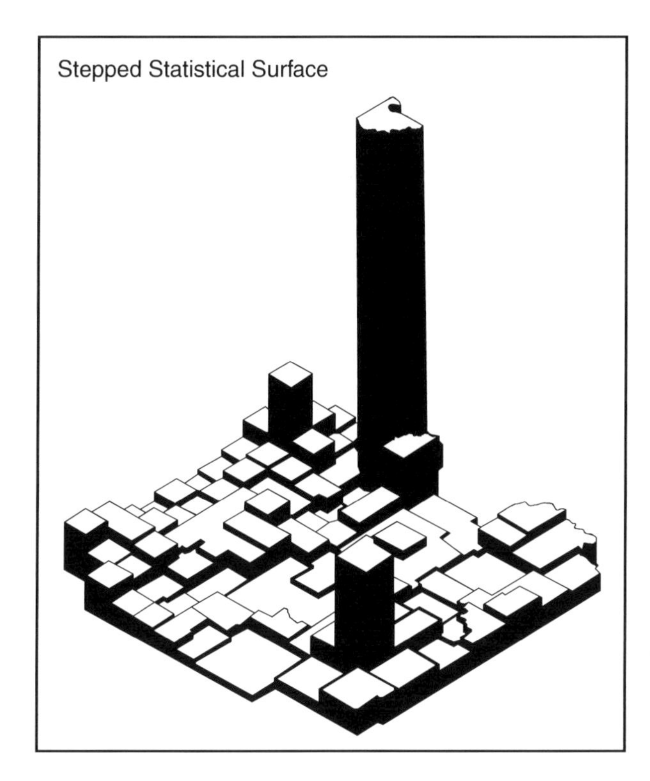

Figure 3.4.4 The stepped statistical surface is a composite of tangential prisms each of which represents a ratio for a single data collection unit. In this case the sample data are rural population densities and the height of each prism is proportional to these density ratios. When these data are symbolised on isometric block diagrams the cartographer is able to represent each density value exactly to scale and no generalisation by classing is necessary.

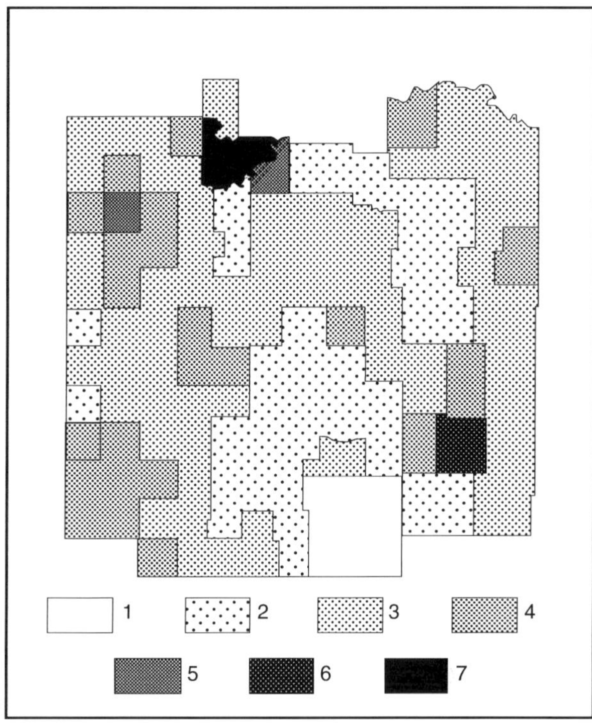

Figure 3.4.5 This is one of many different planimetric maps which can be made from the sample data. Choropleths and shading patterns are commonly used to symbolise data of this type and when there is considerable range in ratio values the data are classed into relatively few groups. Generalising by classing is necessary because the human eye cannot distinguish between grey (or colour) tones which have little contrast.

separately. The reader must be on guard not to lose track of the close interrelationships of all three phases in the map planning process, for if one concept is changed, the planner must automatically consider the effects of this modification upon the other two.

Further, the very nature of the map planning process, in which several cartographic problems are considered and solved simultaneously, complicates analytical presentation of the process. In the preceding paragraph three topics for discussion have been enumerated, but three-dimensional statistical surfaces are commonly represented by either choroplethic or isometric symbols and the solutions sought for one type of symbolisation need not pertain for the other. Thus, the author has found it advantageous to present this paper in three segments. The statistical surface concept is treated in the first part, since it is in this phase of planning that the symbolisation method is selected. The second and third sections treat choroplethic and isometric variations of statistical surfaces and in each of these sections problems relating *to* numbers of classes and to the mathematics of classing are discussed. [. . .]

The statistical surface concept

Most maps are two-dimensional representations of distributional phenomenon occurring at or near the surface of the earth but the conception of these phenomenon can be either two or three dimensional. For example, the average dot map can be conceived of only as a two-dimensional representation of a two-dimensional distribution because the data represented are individual or small groups of individual phenomenon which are located on the map by geographic coordinates. A contour map, on the other hand, must be conceived of as a two-dimensional representation of a three-dimensional phenomenon, since there is vertical as well as horizontal variation. Thus a

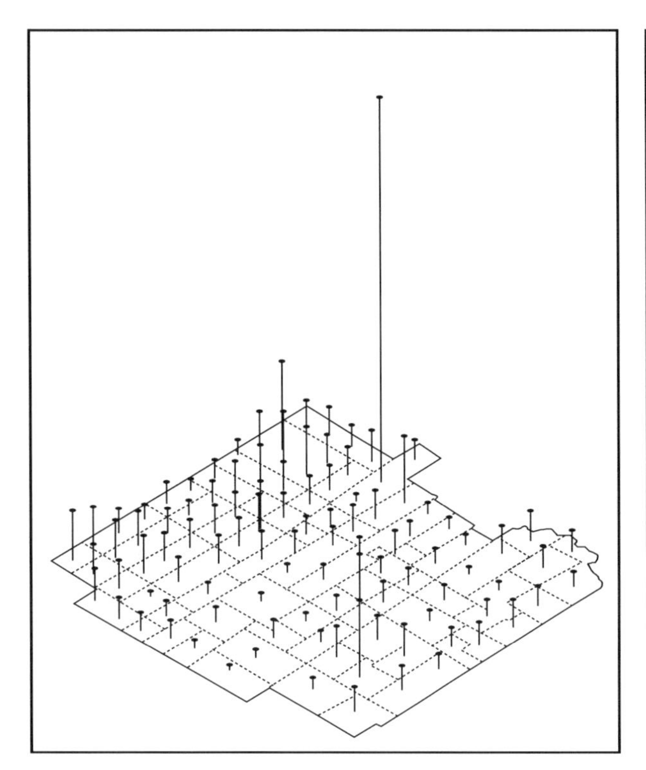

Figure 3.4.6 Population density values can be conceived as sample values in a smooth statistical surface. These sample values are visualised as elevated points above a datum plane. The relative heights of each point above the datum plane are scaled to the ratio value for the collection unit.

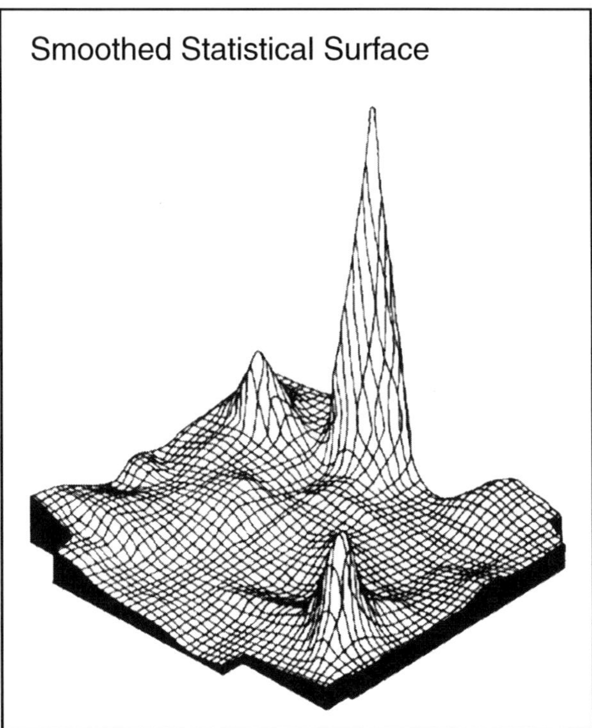

Figure 3.4.7 Smoothed statistical surfaces are inferred by interpolation between elevated control points. While not ordinarily used for enumerated data, smoothed surfaces are useful and necessary conceptions for statistical correlation. [...]

contour map represents the surface of a land volume which lies above or below a datum.

A wide variety of statistical data, when mapped, must also be conceived of as volumetric geographical quantities, since they vary in a third dimension over an area. Distributions such as those of precipitation, air pressure, population density, land use ratios and population potentials are phenomena of this type. Maps of these distributions must, therefore, be thought of as two-dimensional representations of statistical surfaces (Robinson 1961; Schmid and MacCannell 1955).

Visualisations of these statistical surfaces are more difficult to grasp than those for land surfaces because they are abstract and because they need not vary in a continuous manner. The land surface is real and experience tells us that we can proceed from one point to another without crossing a void. Some statistical surfaces are directly comparable, for example that of air pressure, since the data vary in a continuum and undulations in these surfaces can be thought of as 'troughs and ridges.' Other statistical surfaces do not fit this concept since there may be two high points separated by a void instead of a valley. Rural population density or value of farm products sold per farm are examples of discontinuous distributions of this type (James and Jones 1954: 10–11).

Discontinuities in statistical data can occur wherever the phenomena do not exist or wherever there are sharp breaks in the distribution. A case in point is rural population density, for there are voids in areas [...] where no rural people live. In addition, rural population densities need not be conceived of as continuously sloping statistical surfaces because high densities may occur in areas (the Nile valley) which are adjacent to very low density areas (the desert). Such distributions can be thought of as stepped statistical surfaces with each density value represented by a plane parallel to the datum.

The visual presentation of a statistical surface involves more than the mere classing of distributions into volumetric and non-volumetric categories. Data are collected by measuring values at points (weather data) and by enumerating for surfaces (census data), and the method of collection influences judgments relating to the representation of the statistical surface. Values measured at points are samples taken from an infinite number of such points, the totality of which form a smooth undulating surface. Enumerated data, however, require a different concept, since they may be thought of either as uniform values over a plane or as sample values for the centroid of the enumeration unit area (Robinson 1960: 182–190). The most common rationalisation of enumerated data is the former: a stepped surface, although this is changing with the increased utilisation of ratio and index values in geographic research.

To focus attention upon different concepts of statistical surfaces and related cartographic problems, a sample series of data has been processed into a variety of maps and statistical block diagrams. These data are rural population densities, by minor civil divisions, for eight central Kansas counties. All of these illustrations present different concepts of the same series of data, and they are presented to call attention to several variables which are introduced into statistical mapping during data processing. These data are arrayed graphically in Figure 3.4.1 and geographically in Figure 3.4.2

If one conceives of population density as a discontinuous volumetric distribution, the population density of each unit area would become prismatic in form, the base of each prism being contiguous with the surface of a civil division and the altitude proportional to the density value (Figure 3.4.3) (Schmid and MacCannell 1955: 229). A series of tangential prisms representing population density for numerous political divisions would form a stepped statistical surface (Figure 3.4.4). This stepped surface can be generalised and represented on a planimetric map using choroplethic symbolisation and area shadings (Figure 3.4.5). This representation is but one of a variety of different generalisations that could be made of the surface shown in Figure 3.4.4. Other variations are discussed in the following sections of this paper.

If population density is conceived of as a smooth statistical surface population density values are arbitrarily assigned to a point on the surface of the unit areas. Theoretically these control points should each be the centroid of the surface of the unit area, but in practice the visual centres of the collection areas are often used. The density values are then visualised as rising vertically above the datum and each becomes a point on the theoretical surface (Figure 3.4.6). This smoothly undulating surface can be represented, on a statistical block diagram, by a series of vertical profiles which are constructed parallel to the X and the Y axes (Figure 3.4.7). This surface, when plotted on a planimetric map, is symbolised by isarithmic lines Figure 3.4.8.

The cartographer [. . .] must make a series of decisions which pyramid into a finished structure. The foundation of the statistical map is the rationalisation of the statistical surface as either a stepped or a smooth surface, and this basic judgment influences all subsequent decisions. But [. . .] the final appearance or form of the map may be altered considerably by secondary and tertiary decisions. These secondary and tertiary judgments influence map interpretation because they determine those aspects of the map that are immediately visible. [. . .] Decisions which determine numbers of classes, isarithmic intervals, and class values form the statistical surface. For example, once a stepped statistical surface is decided upon, the choroplethic superstructure can be generalised by a few or many classes. Thus, the map reader may see either a simple or a complex version of the distribution although both maps are representations of a stepped statistical surface.

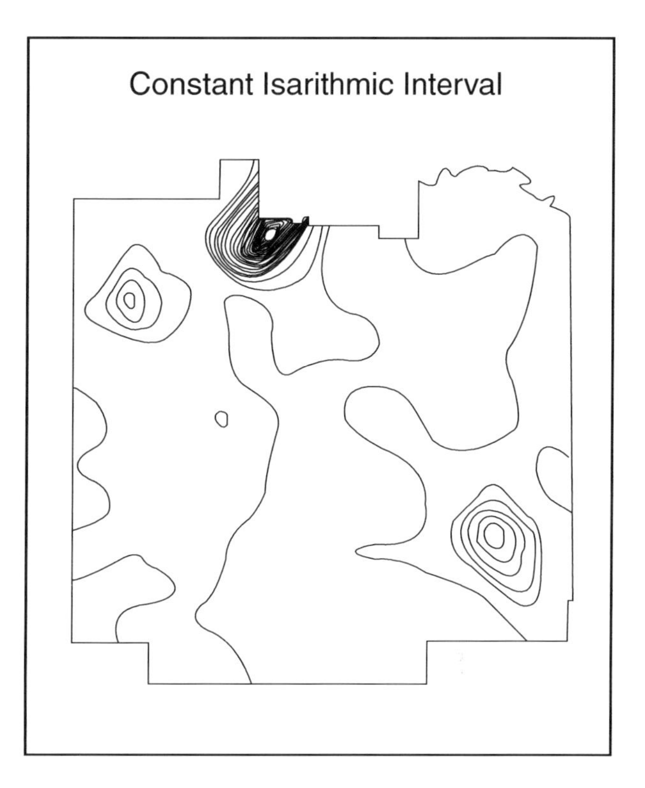

Figure 3.4.8 Smoothed statistical surfaces, like topographic surfaces, are usually symbolised with isometric or isarithmic lines. This representation was constructed using an isarithmic interval of five persons per square mile.

Variations in choroplethic statistical surfaces

Tabular statistical data collected for unit areas which are political or administrative divisions are often visualised as stepped statistical surfaces and represented by choroplethic symbolisation. When this is done, the cartographer has relatively few secondary and tertiary decisions to make, but these decisions influence map reader

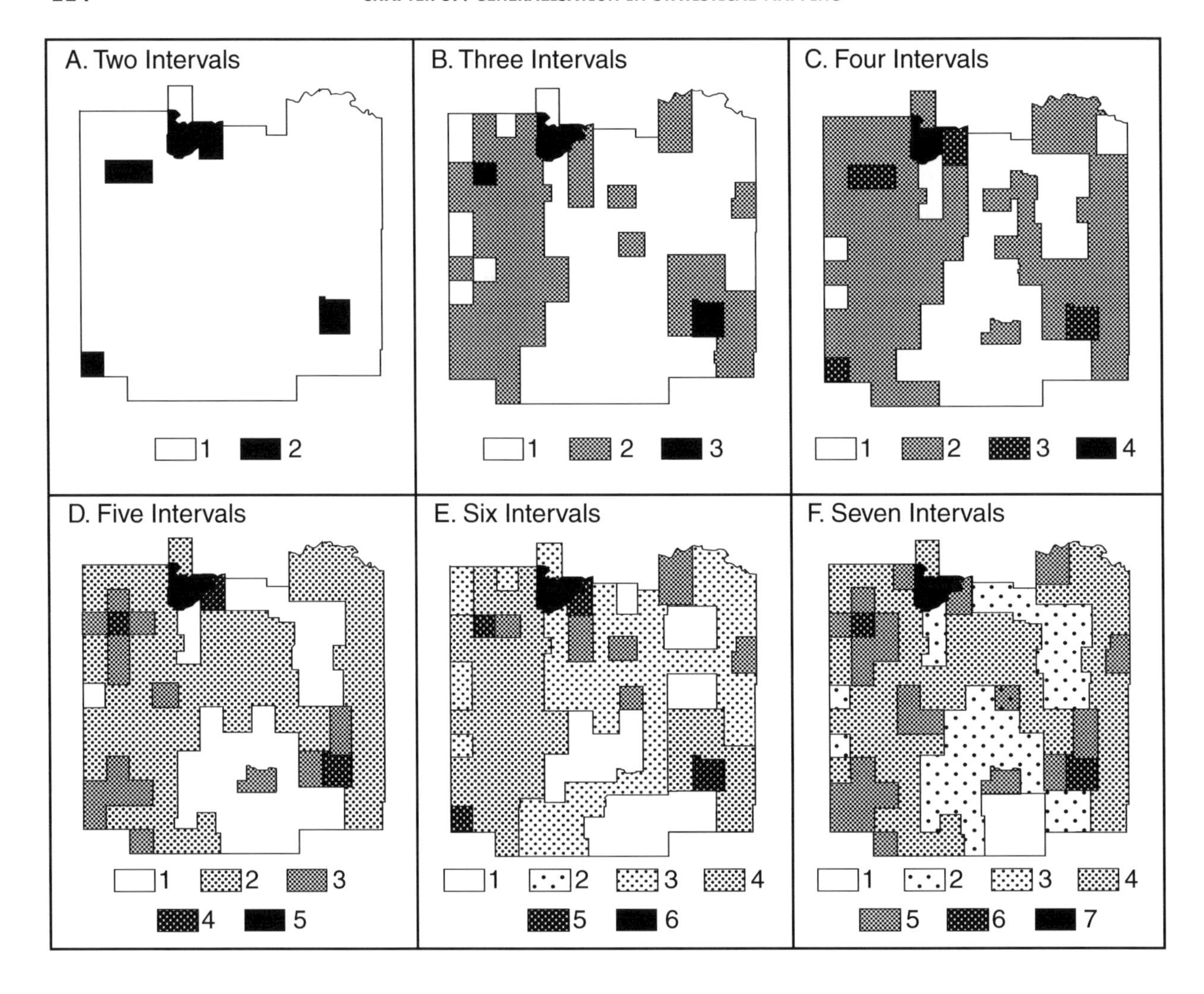

Figure 3.4.9 Different degrees of generalisation can be achieved in mapping statistical data by altering the number of classes. Here the sample data were generalised using geometric progressions and so differences in the maps are essentially due to differences in numbers of classes.[...]

interpretation to a very high degree. Two important secondary decisions involve the selection of numbers of classes and the choice of limits for these classes. A tertiary decision involves the choice of style, texture and value (darkness) of the areal shading patterns used. While these patterns modify the visual impact created by the map they are not directly related to data manipulation and thus not discussed in this paper (Robinson 1960; Williams 1960; Jenks and Knos 1961).

The selection of the number of classes and the determination of class limits are not two isolated cartographic decisions. Data must be processed for both of these aspects at one time, since changes in one will inevitably affect the other. There are, however, an almost infinite number of different combinations which could be determined and utilised. [...]

The average data series, used as raw material for a statistical map, covers a wide range of values which must be classed or grouped into relatively few categories. The degree of generalisation varies inversely with the number of categories, which can vary from two to more than ten. It is obvious that the minimum number of classes must be two, since no distribution is shown on a map covered uniformly by a single symbol. The upper limit in class numbers is determined, not only by the nature of the data, but by the fact that the human eye cannot distinguish between very slight differences in the value of shadings. The threshold of differentiation for

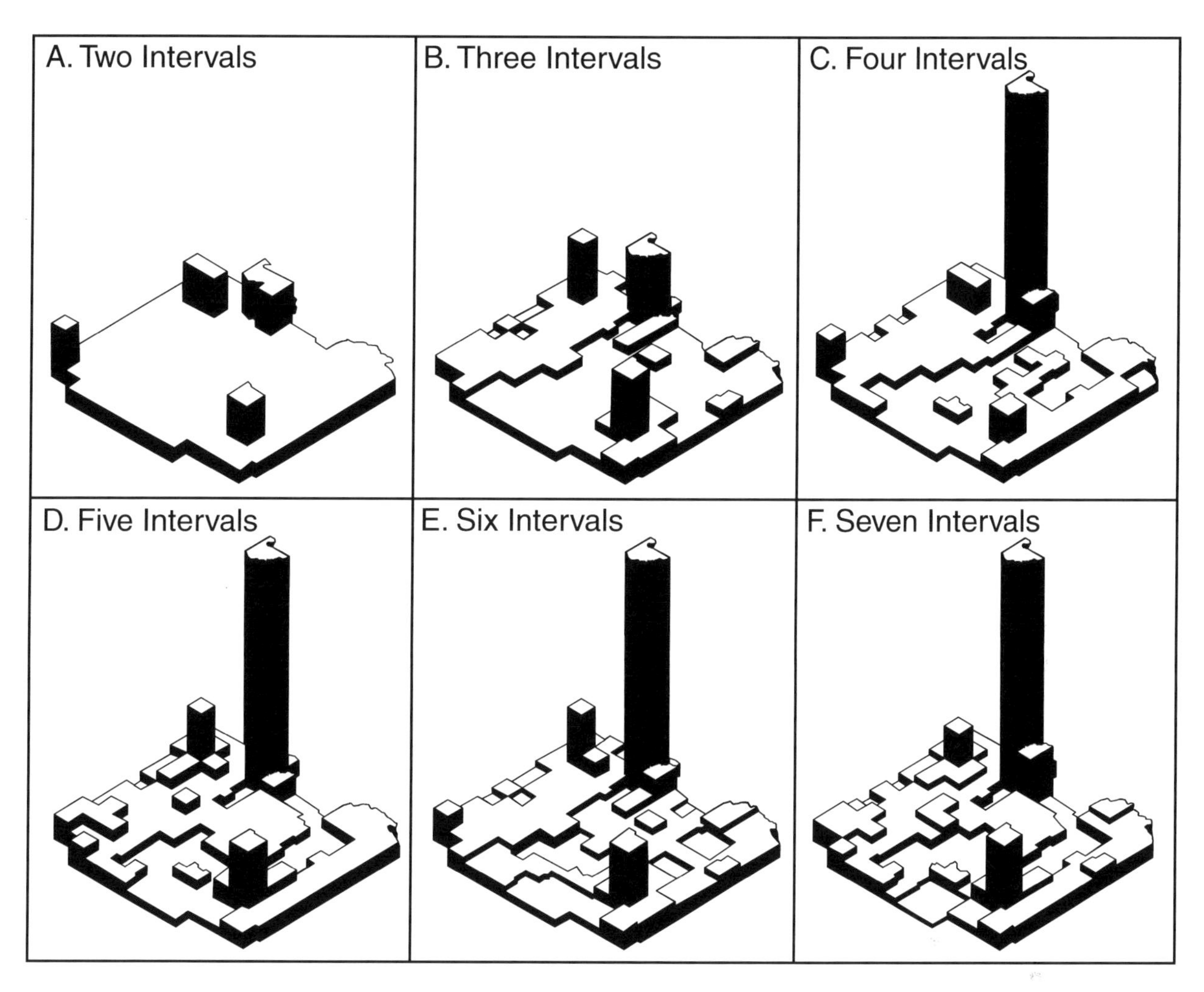

Figure 3.4.10 A comparison of these isometric blocks with that shown in Figure 3.4.4 will show that the degree of generalisation varies inversely with an increase in numbers of classes. More accurate generalisations could be achieved if more classes were used, but [. . .] seven or eight shadings seem to approach the differentiation limit of the average reader. The mean value for each class was used to determine the elevation of that class above the datum. [. . .] Since the data are highly skewed only one unit area falls into the largest class of maps C, D, E and F. This explains why these four maps have the tall column which is missing on the first two.

shadings is not known, either for colours or shades of grey, but it is believed that seven or eight shadings in a spectral sequence approaches the limit for average map readers (Jenks and Knos 1961).

Generalisation by numbers of classes is illustrated in Figure 3.4.9. The class limits for these maps are all based upon geometric progressions and the density of shadings were selected according to Williams' curve of the grey spectrum (Williams 1960). In addition, scale, pattern texture and drafting specifications were held constant. Thus, since the effects of these variables were minimized, differences are basically due to numbers of classes. The degree of generalisation in the statistical surfaces becomes readily apparent when the block diagrams of these maps (Figure 3.4.10) are compared with that of the original data (Figure 3.4.4).

Skilled cartographers usually follow a rational process in classing data, although they often utilise data processing procedures which are unexplained. An analysis of selected maps indicates that three conventional practices are followed and that a wide variety of others are used less frequently. The conventional class intervals are based upon rhythmic or equal steps, arithmetic and geometric progressions. [. . .] Less common, are intervals determined by frequency graphs, cumulative graphs and clinographs, or those determined from mathematical functions (MacKay 1955; Jenks and Coulson 1963).

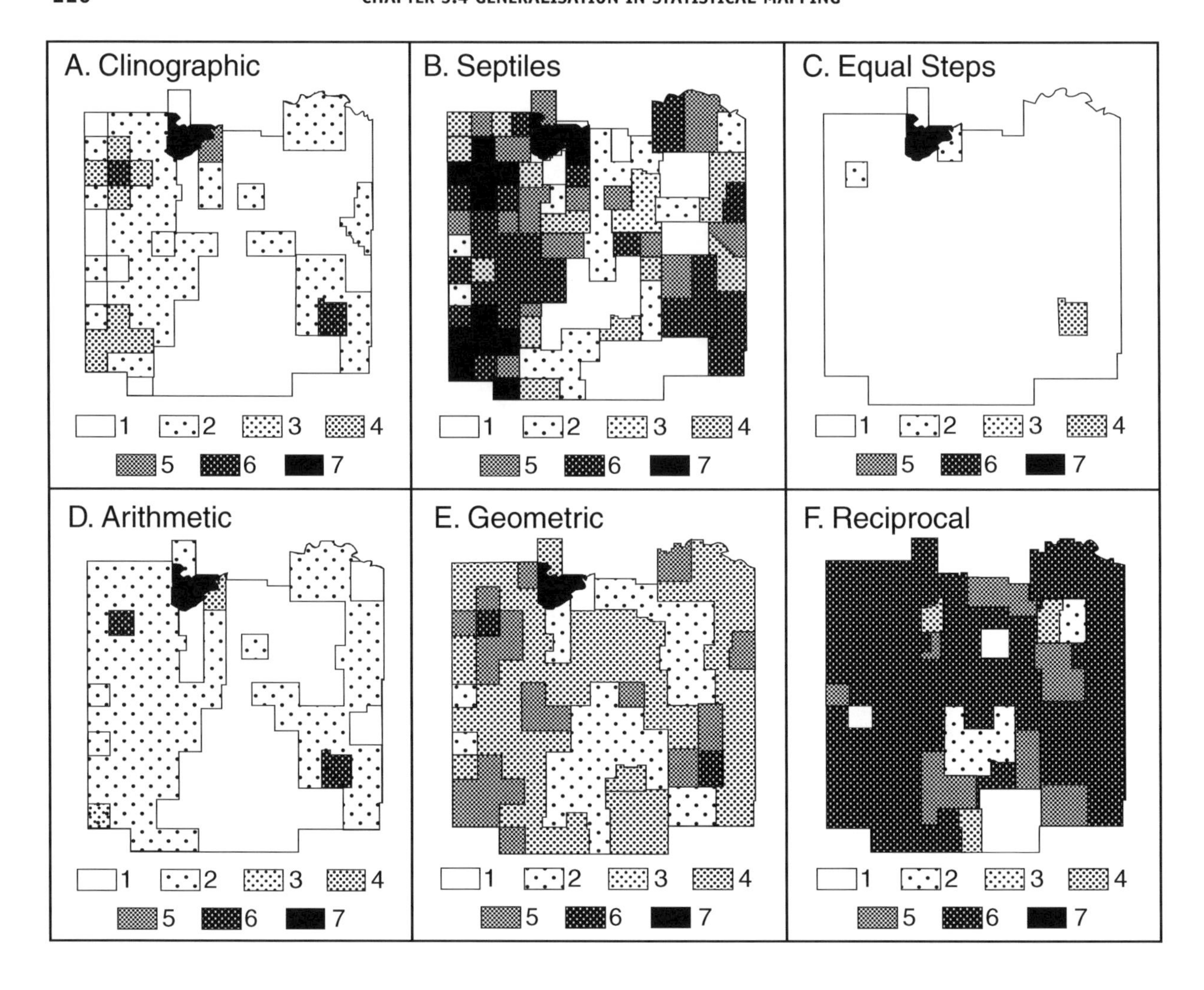

Figure 3.4.11 Different types of generalisation can be achieved by varying the method used in classing data for statistical maps. All of these maps were constructed with seven classes and the same shadings were used throughout. Therefore, differences can be attributed to the different mathematical systems used in classing. Break points were obtained from a clinograph of the data. The one hundred and five unit areas are divided into seven groups of fifteen unit areas each.[...]

Six maps are presented in Figure 3.4.11 which give six different generalisations of the sample data. All aspects of these maps were held constant except for changing class intervals and differences between them are solely due to that one factor. The relative degree and form of generalisation can be seen in the three-dimensional visualisations of these data in Figure 3.4.12, and by comparing these block diagrams with that of the original statistical surface (Figure 3.4.4).

The eleven different maps shown in Figures 3.4.9 and 3.4.11 are but a few of many different visualisations that can be constructed from one data series. All utilise choroplethic symbolisation and represent stepped statistical surfaces, and therefore all are closely related. Each, however, represents the data in a substantially different form. They are presented here to focus attention on the fact that the cartographer can control interpretation of a non-continuous distribution by the way in which he manipulates his data.

Variations in isarithmic statistical surfaces

Statistical data which are collected by measurement at specific geographic positions are ordinarily considered to be samples taken in a continuously undulating surface. The total form of the surface is then inferred by interpolating

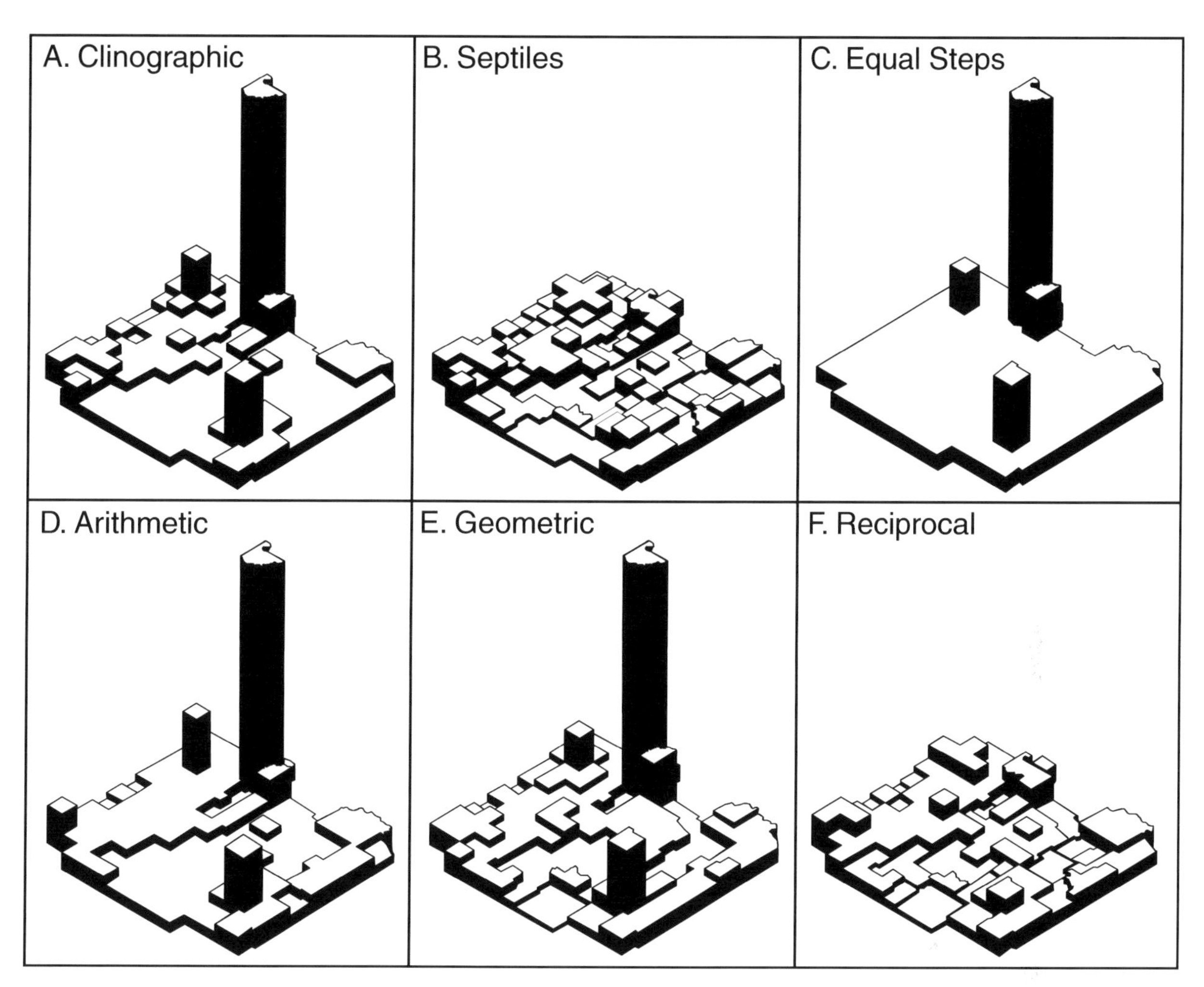

Figure 3.4.12 The form of generalisation achieved in the planimetric maps (Figure 3.4.11) is clearly demonstrated by these isometric blocks. Visual comparison of these blocks with that of the original data (Figure 3.4.4) shows that E is a better generalisation of the sample data than the others. However, with other sets of data one of the other methods of classing might give better results than a geometric progression. The mean value for each class was used to determine the elevation of that class above the datum.[. . .]

between these sample measurements. Of course, the correlation between the inferred surface and the actual surface will vary with the number, the relative positions and the degree of accuracy in measurement of these sample data (Blumenstock 1953; Alexander and Zahorchak 1943; Mackay 1953; Porter 1958). The correlation of the two surfaces will also vary with the skill and experience of the cartographer making the interpolations.

For purposes of the discussion which follows, let us assume that the sample data are sufficient in number, properly spaced and accurately measured. Let us further assume that the cartographic skill of the author is consistent throughout the presentation. Having made these assumptions, we can then concentrate on other variables and their relationship to the representation of a continuously undulating surface.

The choice of the isarithmic interval determines the degree of fineness or coarseness of the generalisation of a statistical surface. It can, therefore, be compared to the selection of numbers of classes in choroplethic mapping, since a large interval, like few classes, gives a gross generalisation. Selecting the isarithmic interval may be done by analysing the 'relative relief' and attempting to visualise the 'hills and valleys' of the statistical surface. [. . .] Thus, little relative relief would be represented by a small, and greater relative relief by a larger, isarithmic interval.

Figure 3.4.13 This isometric block presents a three-dimensional view of the isarithmic map in Figure 3.4.8. Compare the areas with lower population densities on this diagram with the diagram of the original data (Figure 3.4.4). It is readily apparent that much detail has been lost in these areas and that this generalisation is too coarse.

Figure 3.4.14 An isarithmic interval of two was used for areas with densities of less than ten persons per square mile and an interval of ten was used for the areas with higher densities. When two isarithmic intervals are used the map reader should be forewarned and the lines should be carefully labelled. When this is not done, as in this illustration, the form of the statistical surface can easily be misinterpreted.

Carrying this illusion a step farther, a decision might be reached, if the data suggest it, to use two different isarithmic intervals. This is rationalised in the same way that two contour intervals are sometimes used to represent adjacent level and mountainous topographic areas (Imhof 1961).

When a constant isarithmic interval is used to represent a continuous distribution, isarithmic patterns are directly comparable over the total mapped area. This aids the reader in interpreting the shape of the surface of the distribution, because he does not have to mentally shift from one vertical scale to another. Often, however, uniform isarithmic intervals obscure significant detail if the surface being mapped has great relative relief. Detail is lost either because a large interval over-generalises the flatter areas or because isarithms bleed together, creating amorphous blotches, when a smaller interval is used. The results of over-generalisation in flat areas are clearly shown in Figure 3.4.8, where the sample data were represented by an interval of five persons per square mile. This loss of detail is particularly striking when comparisons are made of the five interval isometric block (Figure 3.4.13) and the smooth surface shown in Figure 3.4.7.

When the sample data are represented by two different isarithmic intervals (two persons per square mile for the lower values and ten persons per square for the larger values) greater detail is shown in the flat areas without significant losses in the peaked areas (Figure 3.4.14). This compromise highlights the problem of interpretation however, since the large peak in the northern section of the map appears to have a steeper slope at its base than it has at higher elevations. Likewise, the peak in the southeast now appears to be almost butte-like in form, while in Figure 3.4.8 it appears to be a symmetrical pyramid. These 'gains and losses' are apparent when comparisons are made of isometric blocks in Figures 3.4.13 and 3.4.15.

If the cartographer wishes to employ area shading patterns with isarithms, to accentuate areal differentiation, he must make additional generalisations, since he

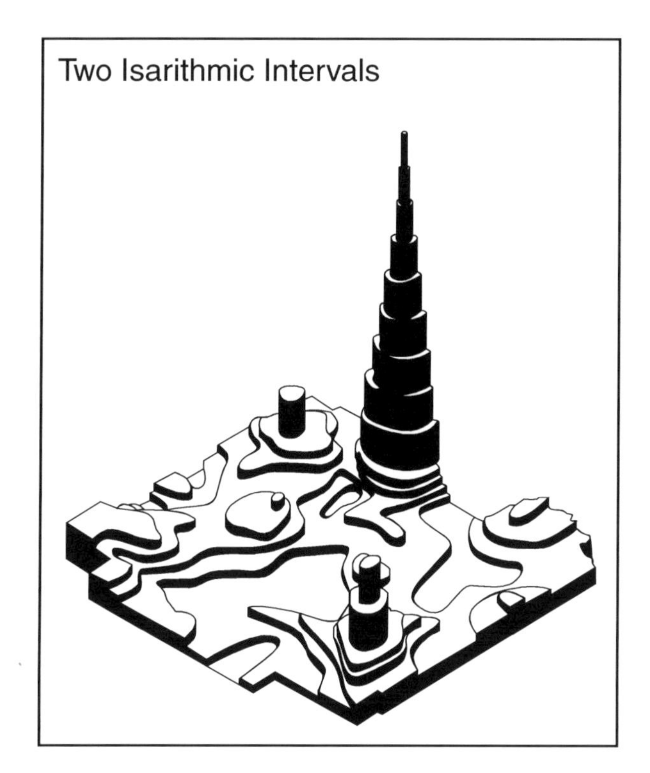

Figure 3.4.15 This generalisation more closely approaches the form of the original stepped statistical surface (Figure 3.4.4) than that shown in Figure 3.4.13. While this close correlation is apparent in the block diagrams, the planimetric presentation (Figure 3.4.14) is more difficult to interpret. For example, one could easily interpret the isarithmic pattern in the southeastern corner of the map in Figure 3.4.14 as representative of a butte-like form, but this feature has quite a different shape on the diagram above.

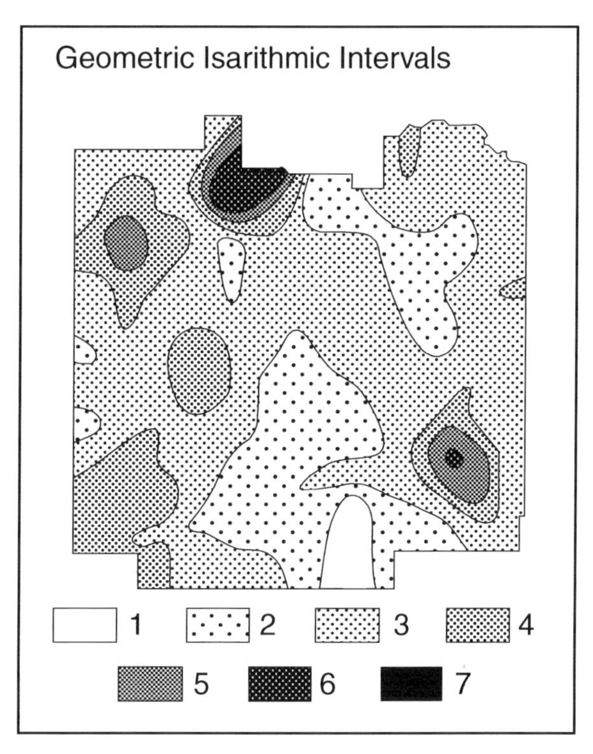

Figure 3.4.16 When shading is utilised to emphasise areal patterns on isarithmic maps the number of patterns is as limited as those used with choropleths. Here seven tones, ranging from white to black, are used between isarithms selected on a geometric progression. [...]

faces the same problems of categorisation that were discussed in the choroplethic section of this paper. Thus, if isarithms 2, 4, 8, 16, 32 and 64 are used to separate shadings on an isarithmic map of the sample data, the resultant map (Figure 3.5.16) corresponds exactly with the choroplethic map shown in Figure 3.4.9f. Comparisons of these two planimetric maps and their visualisations in isometric form (Figures 3.4.10f and 3.4.17) show this high degree of correspondence, but also demonstrate how the different concepts of the sample data influence the statistical surface.

Conclusions

The mapmaker can present to the map reader for interpretation only one of a multitude of different versions of a statistical distribution. In the creative process of developing this one map, the cartographer makes a series of simultaneous judgments involving his personal concept of the statistical surface, his concept of the most desirable degree of generalisation and his selection of a mathematical process for classing the data. These three judgments control and shape a generalised statistical surface, which is then symbolised to represent the abstract data. If the cartographer makes these judgments through rational processes he can transmit his concept of the distribution to the reader and the map reader is obligated to realise that this is a selected generalisation.

Map-makers are obligated, however, to create just as realistic representations of statistical data as they do of the earths surface. Too often we create statistical maps ‘blindly’, failing to recognise how a simple decision can affect interpretation of the final map. Although the decisions are simultaneous and are not separate logical steps, it is hoped that this analysis has established the need for recognition of the problems involved in mapping statistical surfaces and that more accurate maps may result.

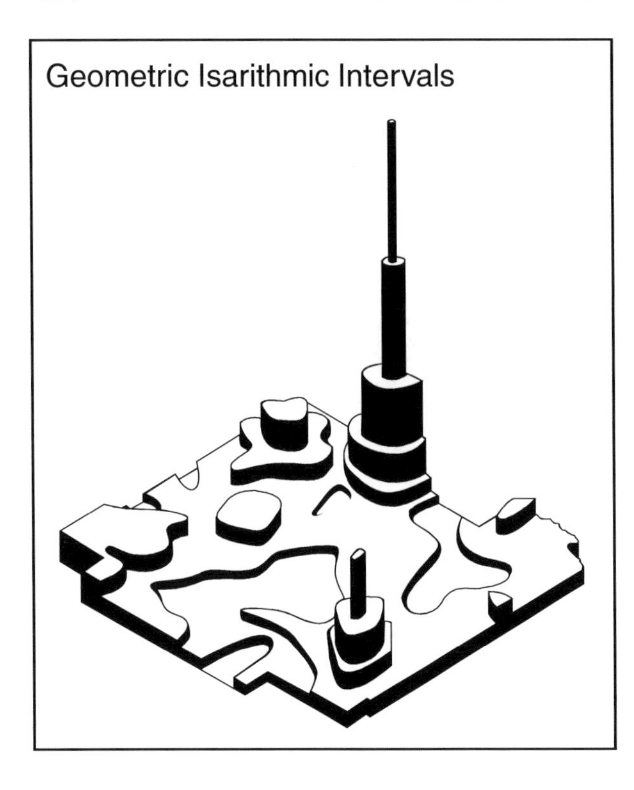

Figure 3.4.17 This block represents a three-dimensional view of the shaded isarithmic maps in Figure 3.4.16. It can, therefore, be directly compared with the block in 9F or 11E to give a clearer understanding of the two different concepts of representing statistical surfaces.

References

Alexander, J.W. and Zahorchak, G.A. (1943) Population-density maps of the United States; techniques and patterns. *Geographical Review*, **33**, 457–466.

Blumenstock, D.I. (1953) The reliability factor in the drawing of isarithms. *Annals of the Association of American Geographers*, **43**, 289–304.

Imhof, E. (1961) Isolinienkarten. *International Yearbook of Cartography*, **1**, 64–98.

James, P.E. and Jones, C.F. (1954) *American Geography: Inventory and Prospect*, Syracuse University Press, Syracuse, NY.

Jenks, G.F. and Coulson, M.R.C. (1963) Class intervals for statistical maps. *International Cartographic Yearbook*, **3**, 119–133.

Jenks, G.F. and Knos, D.S. (1961) The use of shading patterns in graded series. *Annals of the Association of American Geographers*, **51**, 16–34.

Mackay, J.R. (1953) The alternative choice in isopleth interpolation. *The Professional Geographer*, **4**, 2–4.

MacKay, J.R. (1955) An analysis of isopleth and choropleth class intervals. *Economic Geography*, **31**, 71–81.

Porter, P.W. (1958) Putting the isopleth in its place. *Proceedings of the Minnesota Academy of Science*, **25–26**, 372–384.

Robinson, A.H. (1960) *Elements of Cartography*, 2nd edn, John Wiley & Sons, Inc., New York.

Robinson, A.H. (1961) The cartographic representation of the statistical surface. *International Yearbook of Cartography*, **1**, 53–62.

Schmid, C.F. and MacCannell, E.H. (1955) Basic problems, techniques, and theory of isopleth mapping. *Journal American Statistical Association*, **50**, 220–239.

Williams, R.L. (1960) Map symbols: equal appearing intervals for printed screens. *Annals of the Association of American Geographers*, **48**, 226–253.

Further reading

Crampton, J.W. (2004) GIS and geographic governance: Reconstructing the choropleth map. *Cartographica*, **39** (1), 41-53. [An insightful reading of the politics of data classification and choropleth mapping.]

Evans, I.S. (1977) The selection of class intervals. *Transactions of the Institute of British Geographers, New Series*, **2** (1), 98-124. [Still probably the most comprehensive consideration of selection criteria for classification when using choropleth mapping.]

Monmomier, M. (1995) *How to Lie with Maps*, 2nd edn, University of Chicago Press, Chicago. [Chapter 10 of Monmonier's popular book provides an insightful discussion of the myriad ways statistical maps can be manipulated to produce widely different impression of reality.]

Slocum, T.A., McMaster, R.B., Kessler, F.C. and Howard, H.H. (2008) *Thematic Cartography and Geo-Visualization*, 3rd edn, Pearson, London. [A comprehensive 'state of the art' overview of many of the design principles underpinning approaches to thematic map design, with detailed discussion of selection, generalisation, classification and symbolisation of data.]

See also

- Chapter 3.2: Interplay of Elements
- Chapter 3.6: The Roles of Maps
- Chapter 3.8: ColorBrewer.org: An Online Tool for Selecting Colour Schemes for Maps
- Chapter 4.2: Map Makers are Human: Comments on the Subjective in Maps

Chapter 3.5

Strategies for the Visualisation of Geographic Time-Series Data

Mark Monmonier

Editors' overview

Monmonier's article reviews different ways of visualising geographical change, exploring graphic display technique for the simultaneous symbolic representation of time and space, and summarising these into a conceptual framework of potential use to cartographers, geographers and graphic designers. The techniques considered range from statistical diagrams, maps and video animations to interactive graphics systems with which the analyst might freely manipulate time as a variable. The display maybe spatial or non-spatial, single view or multiple view, static or dynamic.

Originally published in 1990: *Cartographica,* **27** (1), 30–45.

Graphic representations in time-attribute space

Time-series data traditionally have called for statistical diagrams like Figure 3.5.1, with time as the horizontal axis and a single variable as the vertical axis (Du Toit *et al.* 1986: 265–274). A separate trend line represents each place, and each trend line requires a label identifying the place represented. The label might be a place name, a directional abbreviation such as 'NE', or some other geographic metaphor.

A logarithmic scale for the vertical axis (Figure 3.5.2) is a useful modification that allows the slope of the trend line to portray relative rates of change. With an arithmetic scale the slopes portray only absolute change, not the rate of change. With Figure 3.5.1 the viewer should not compare slopes, whereas with Figure 3.5.2, he/she may validly interpret a steeper slope as representing a sharper rate of change than a more gentle slope.

The graphic might also focus attention on time periods, as in Figure 3.5.3, rather than upon sample points on the temporal continuum. In this case the vertical axis might show absolute change or the rate of change.

When the data include many places, symbols and labels readily overload the time-series graphic (Figure 3.5.4). Assigning each place a unique line symbol might alleviate graphic congestion but, as in Figure 3.5.5, a wide variety of qualitative line symbols yields a complex key and makes the graph difficult to read. Even when the number of places is not ridiculously large, criss-crossing trend lines are visually complex and require frequent references to a legend.

Various strategies can help the analyst cope with a plethora of places. As in Figure 3.5.6, a mean or median might represent all places for each time slice. Adding separate trend lines to show temporal trends for key places, as in Figure 3.5.7, can focus the viewer's attention on important departures from the average trend and provide meaningful comparisons for selected places. Figure 3.5.8 illustrates how the addition of so-called *error bars,* representing a standard deviation or the inter-quartile range, can portray variation throughout the region for each time sample as well as temporal trends in regional variation.

The Map Reader: Theories of Mapping Practice and Cartographic Representation, First Edition. Edited by Martin Dodge, Rob Kitchin and Chris Perkins.

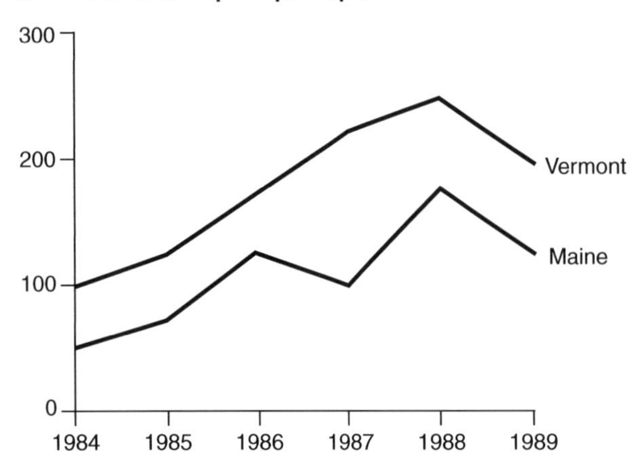

Figure 3.5.1 A typical time-series graph, with time scaled along the horizontal axis and the attribute scaled along the vertical axis. Two trend lines, both with a place-name label, illustrate geographic variation in an aspatial context.

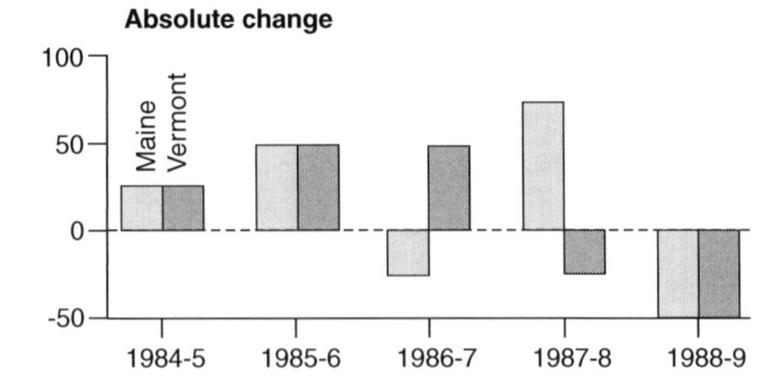

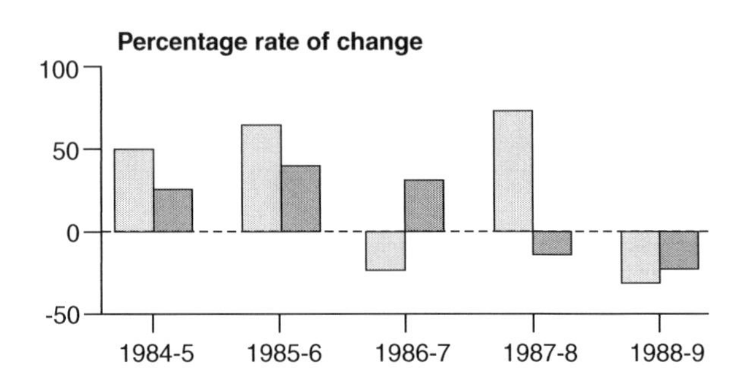

Figure 3.5.3 Time-series graphs can use bars, instead of lines, to focus on absolute change for time periods (above) and to focus on the rate of change (below). These graphs are based upon the data for Figures 3.5.1 and 3.5.2.

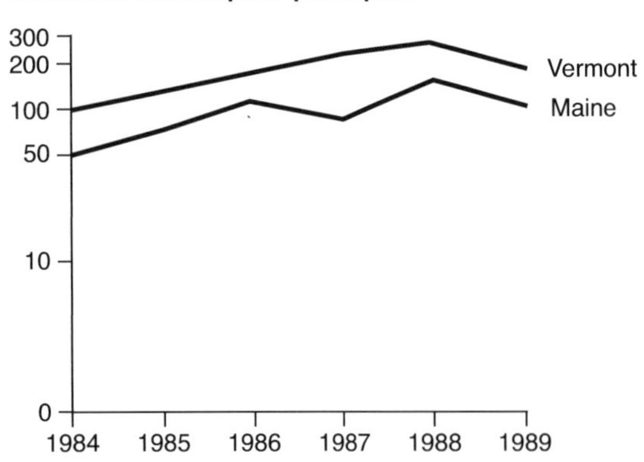

Figure 3.5.2 A logarithmic vertical scale replaces the arithmetic vertical scale of the diagram in Figure 3.5.1. This adjustment promotes comparison of the rate of change.

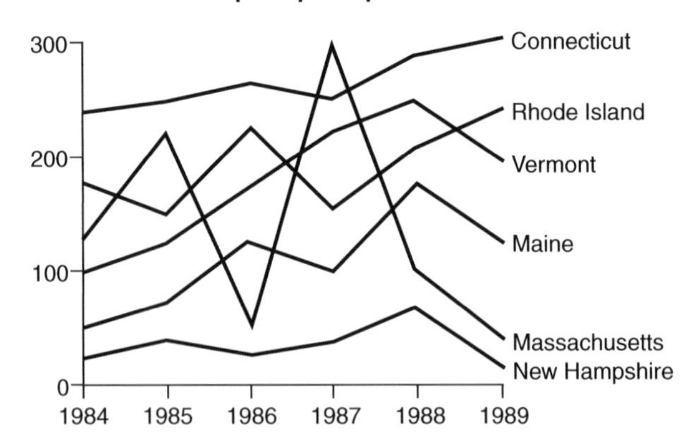

Figure 3.5.4 A time-series chart with many trend lines, each labelled with a place name.

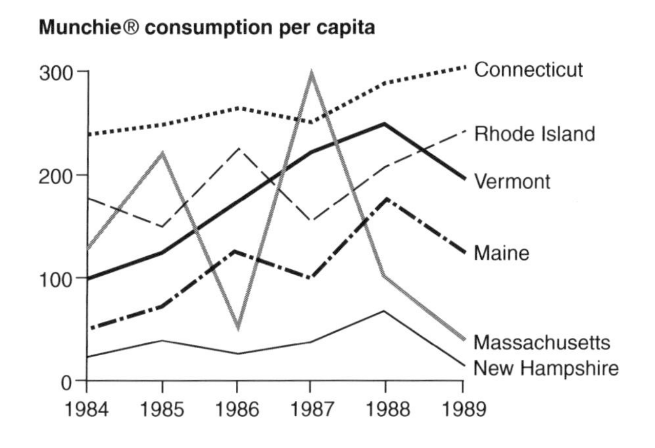

Figure 3.5.5 A time-series chart with many trend lines, each with a different patterned symbol.

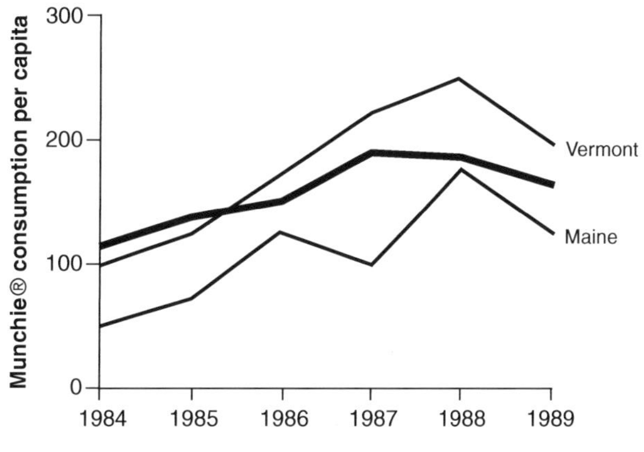

Figure 3.5.7 A time-series chart based on median values but including for comparison the trend lines for two significant places.

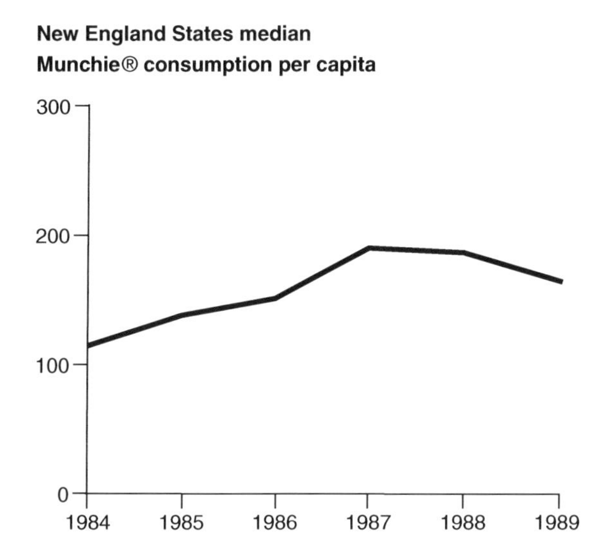

Figure 3.5.6 A time-series chart based upon the median value for all places for each recorded instant or period.

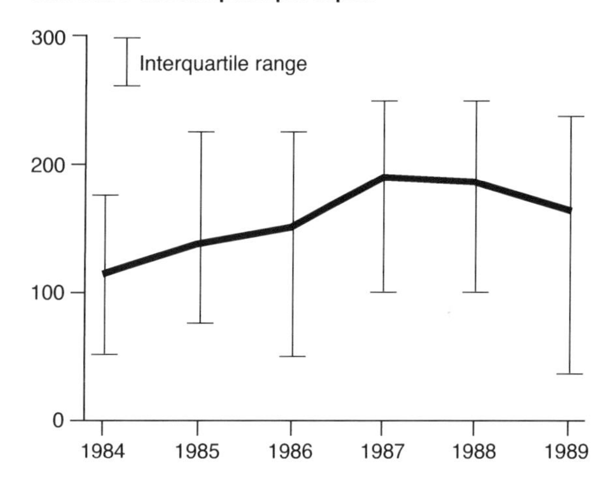

Figure 3.5.8 A time-series chart based on median values but including error bars to indicate each value's representativeness.

Graphic representation in geographic space

Thus far, place names and other geographic metaphors have provided the only link with the geographic space, and the phenomena have been univariate, not multivariate. For many applications, though, the analyst must cope with a set of places, each with its own data array (Figure 3.5.9), in which the rows, say, represent attributes and the columns represent instants or periods. To show relative location, he might also treat these data as sets of maps, perhaps organised as in Figure 3.5.10, with one set for each attribute subdivided by time sample. Historical atlases based on quantitative data commonly take this form, with each

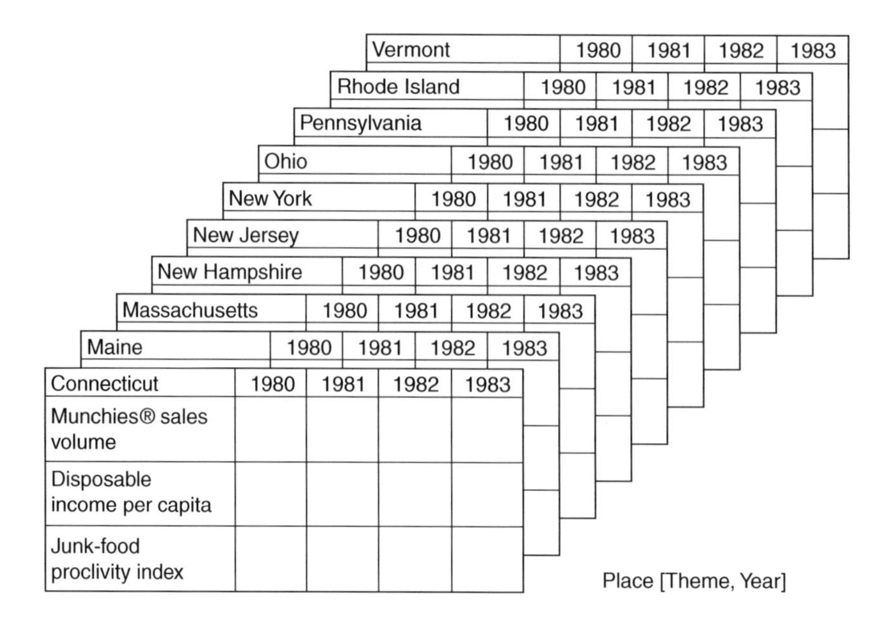

Figure 3.5.9 Data arrays, one for each place, with rows representing attributes and columns representing time periods.

attribute's set of maps arranged on a page or distributed over a sequence of adjacent pages.

If the number of instants of time is small, if the number of attributes also is small, and if the number of places is not too large, then a cartographic cross-classification array (Figure 3.5.11) might represent all the data in a single graphic (Monmonier 1979). Additional columns might even be inserted for periods between time samples, or additional rows might portray rates of change. The eye can readily slew from map to map, and the analyst can examine spatial and temporal trends simultaneously and even infer cross-correlation between variables. Yet for most data sets, small display screens or small pages render this approach unsuitable.

For a single variable observed for several instants or periods, individual cartographic point symbols, or *glyphs*, might portray a separate, spatially positioned series for each place. Figure 3.5.12 illustrates a few typical temporal glyphs. Although tiny time-series line graphs might be too visually complex for a map with many places, small clock-face, calendar, or framed time-line symbols could be useful for some applications.

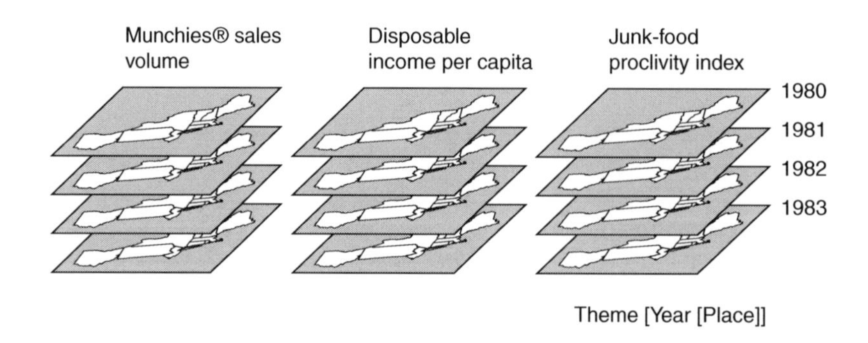

Figure 3.5.10 Stacks of maps grouped by attribute and within each attribute, by time.

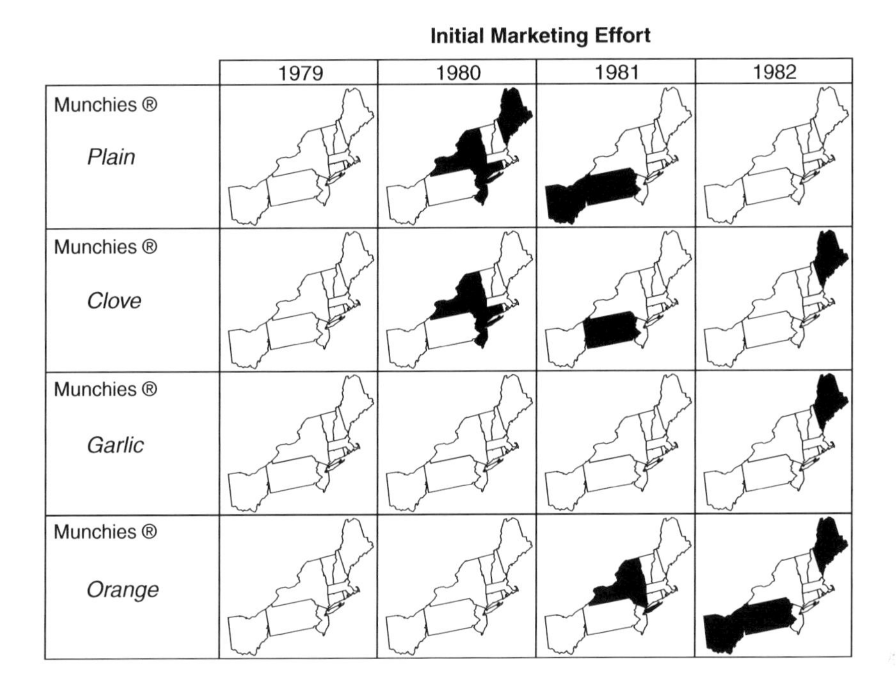

Figure 3.5.11 A cartographic cross-classification array, with rows representing attributes and columns representing time units.

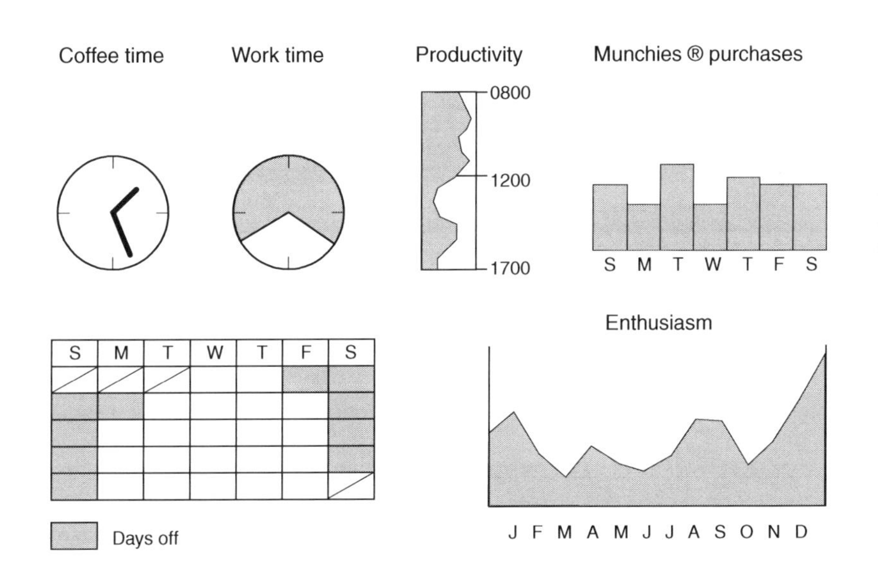

Figure 3.5.12 Typical temporal glyphs: the clock face, the calendar and the framed time-line symbol.

Another strategy is aggregation, perhaps the most severe example of which is the centre-of-population map used for decades by the US Bureau of the Census. As Figure 3.5.13 shows, for each census year, a single point symbol represents the centre of mass of the national population. By showing these symbols for successive censuses, the map demonstrates most dramatically the westward – and more recently, the south-westward –movement of the nation's population.

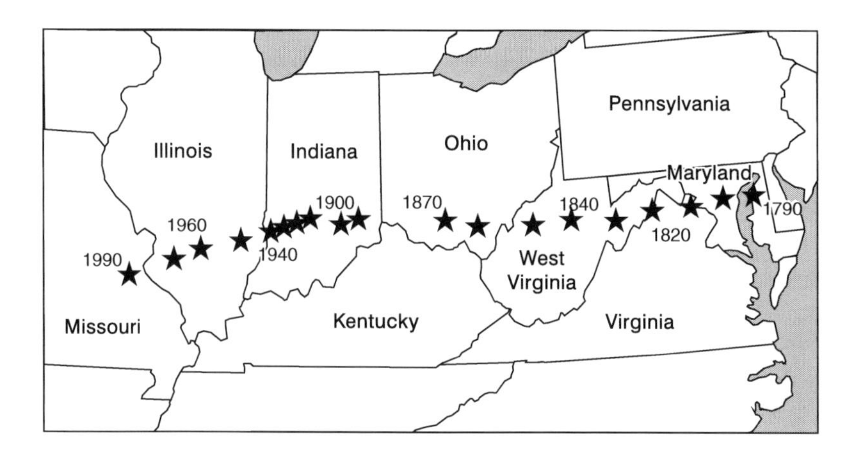

Figure 3.5.13 Map similar to the centre-of-population map used by the US Bureau of the Census to summarise the general westward shift of the United States population since the first census in 1790. (Adapted from Statistical Abstract of the United States, 1984, 104th edition, page 7.)

Other single-map generalisations include displays based upon single dates for each areal unit. A good example of this strategy is the county-unit map showing census year with peak population. This type of map can shock naïve viewers unaware that many counties had more people fifty years ago than they do today.

Another spatial variable well suited to a single-map portrayal is the time of first settlement. Isochronic lines for a polynomial trend-surface (Chorley and Haggett 1965) provide a concise generalisation of major thrusts in the advance of the settlement frontier, as in Figure 3.5.14, a county-level example for New York State. Canonical trend surfaces (Monmonier 1970) might be of use as well, to treat simultaneously the frontiers or innovation waves for several different ethnic groups or ideas.

Flow-linkage diagrams similar to Figure 3.5.15 offer a further generalisation of advancing settlement. These directed links, which attempt to reveal principal avenues of movement, focus the viewer's attention on corridors and

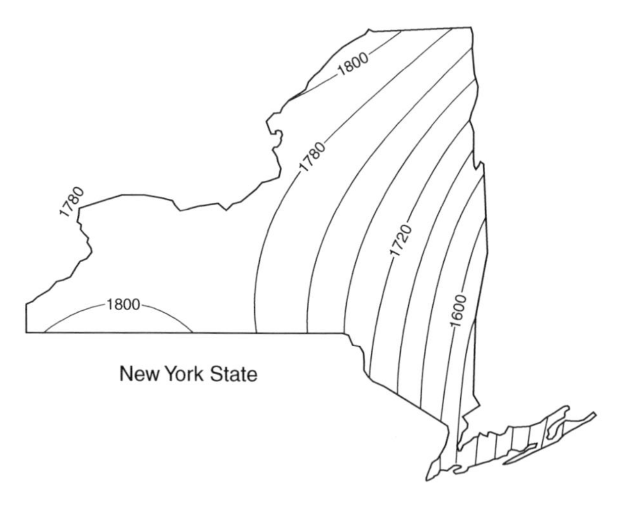

Figure 3.5.14 Quadratic polynomial trend surface showing general pattern of the time of first settlement of New York State counties.

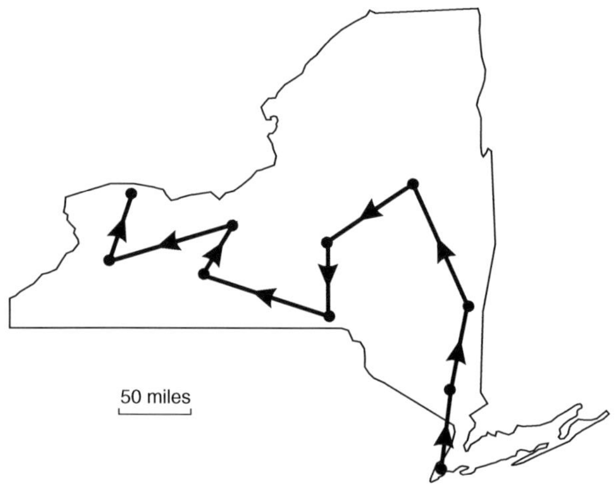

Figure 3.5.15 Flow-linkage trend diagram showing general pattern of the time of first settlement of New York State counties.

direction, not on the extent of settlement at particular times (Monmonier 1972).

Historical and cultural geographers frequently employ directional symbols to portray change over time. Arrow symbols are particularly useful in showing migration streams, the spatial diffusion of ideas, the migrations of tribes and refugees, and the advance and retreat of armies. Directional symbols can vary in size to represent relative magnitude, or vary in label, colour or pattern to represent a particular group or time period. Edward Tufte (1983: 40–41), in his widely acclaimed essay *The Visual Display of Quantitative Information*, lavishly praises Charles Joseph Minard's use of a variable width flow-line symbol on a map showing the declining size of the Napoleon's army in its abortive Moscow campaign of 1812: 'It may well be the best statistical graphic ever drawn.'

As a cartographic genre, these maps might be termed *dance maps*, after the choreography diagrams (Figure 3.5.16) used to teach the spatial mechanics of ballroom dancing to generations of students. These maps cover a period marked by several events, each described by map symbols describing a transition from one place to another. Dance maps are one of the three most common spatial-temporal displays.

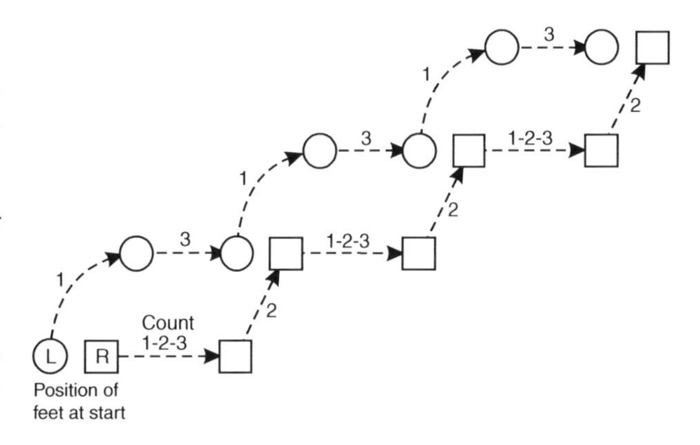

Figure 3.5.16 A prototypic dance map showing the woman's steps for the Hesitation Waltz (Source: Walker 1914: 46).

The second type is the *chess map* (Figure 3.5.17), so called because a separate map presents a snapshot for a discrete instant or period. Chess maps are juxtaposed so that the user can compare the pattern at time 1 with the pattern at time 2. The visual focus here is on the status of the phenomenon at these particular times, not on change *per se*. The chess map strategy would include a pair of choropleth maps representing the same population trait for 1980 and 1990, say, or a set of point-symbol maps showing the distribution of military bases for 1945 and 1985.

The third type (Figure 3.5.18) is called simply the *change map*. It refers to a single map on which symbols vary in value, size or some other appropriate visual variable to represent the direction, rate or absolute amount of change.

Multivariate data compression techniques, such as principal components analysis (Johnston 1978: 127–182) or canonical correlation (Davis 1986: 607–615), might prove useful in reducing the number of variables that

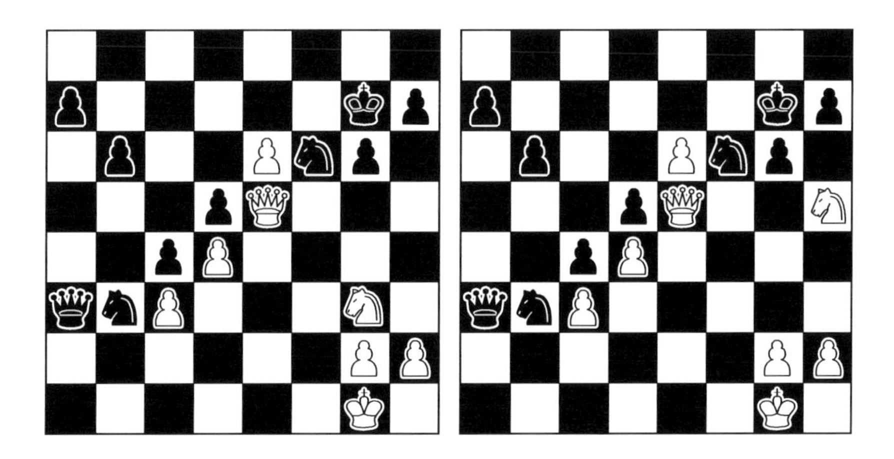

Figure 3.5.17 Chess maps are two or more geographic-space displays juxtaposed so that the viewer can compare spatial patterns for different times.

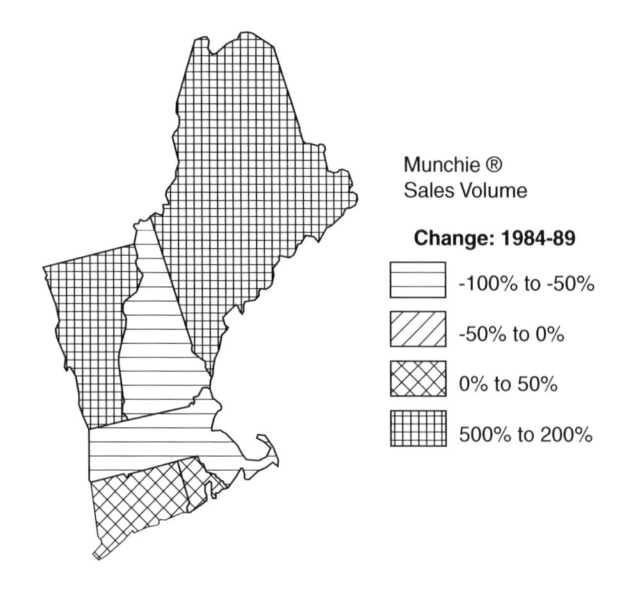

Figure 3.5.18 A change map showing rate of change by state.

need to be displayed. As Palm and Caruso (1972) have aptly demonstrated, labelling factors is often problematic, and map titles such as 'Factor 1' or 'Socioeconomic Status' can be confusing. Indeed, multivariate statistical methods are seldom suitable if the audience does not understand the underlying statistical and geometric principles. Yet for the analyst well grounded in both statistical and graphic analysis, multivariate methods – including classification techniques – can be highly effective for identifying redundant measures and for extracting summary maps from spatial-temporal data.

Hybrid representations with spatial and time-attribute axes

No multivariate analysis should be attempted without a prefatory univariate analysis. Statisticians advocating *exploratory data analysis* call for graphing the frequency distributions of each and every variable. Statistician John Tukey (1977: 56) warned that 'We have not looked at our results until we have displayed them effectively.' When the data are geographic, though, we need to display them in both the attribute space familiar to the statistician (Figure 3.5.19, right) and the geographic space that provides the necessary sense of place and relative location (Figure 3.5.19, left).

Treating just a single variable measured at two different times calls for an array of graphics (Figure 3.5.20) in both attribute space and geographic space. As a minimum, this array would include separate maps portraying the spatial variance at each of the two times, a map of change or the rate of change, three separate univariate histograms or cumulative frequency diagrams showing frequencies for each time as well as for change, and three separate scatter plots portraying in attribute space the bivariate relationships between the two sets of values and between each set of values and the rate or amount of change.

If the geographic space can be reduced to a single measure, such as distance from the equator or distance from the centre of town, a single hybrid graph might combine elements of both geographic and attribute space. The non-spatial axis might be the rate of change or even time itself. Graphic train schedules can employ this concept to reveal, at a glance, relative rates of travel

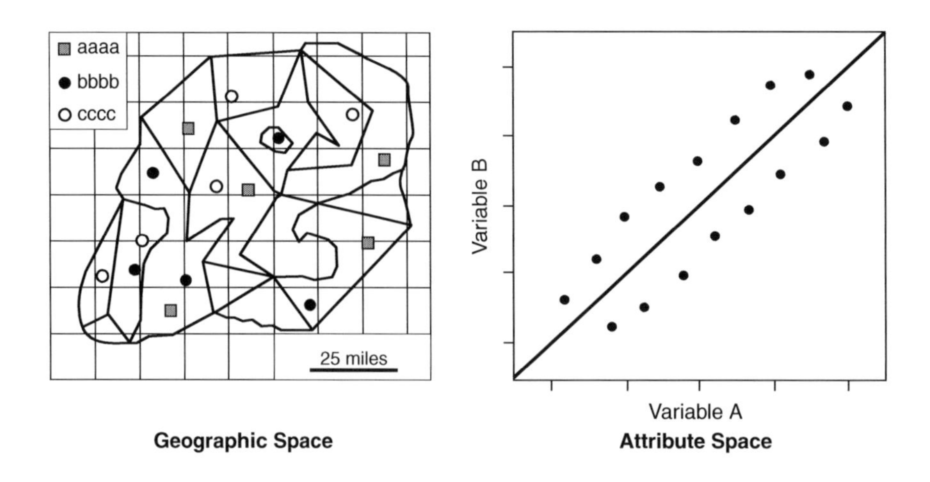

Figure 3.5.19 A comparison of attribute space and geographic space.

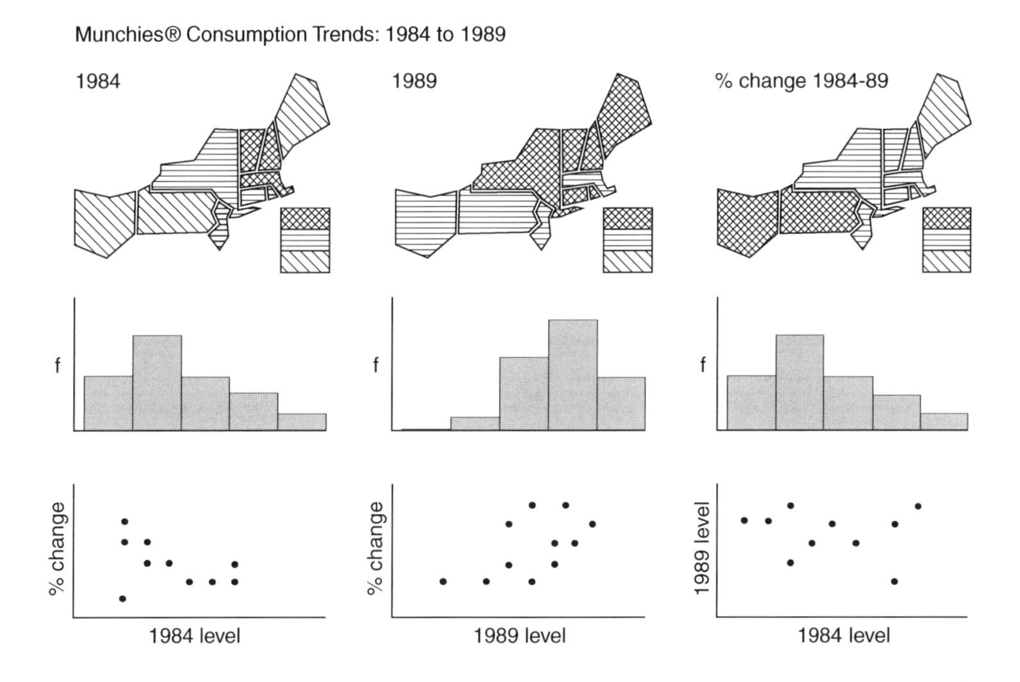

Figure 3.5.20 An array of graphics for exploring a single geographic attribute measured at two instants of time.

between stations and the length of stop-overs at terminals. As Figure 3.5.21 describes, each train starts at the top of the graph and moves downward and toward the right. Fast trains have a steep descent, and slow trains a comparatively gentle descent. A horizontal terrace represents lay-over time at a station. A graph with many trains, some fast and others slow, can show differences throughout the day in frequency, the relative advantages of express service and trains that cover only a portion of the route (Tufte 1983: 115–116). For commuter routes with generally frequent service and many stops, a graphic train schedule providing a quick overview might be more useful than a cumbersome numerical schedule consisting of lengthy tables and numerous footnotes.

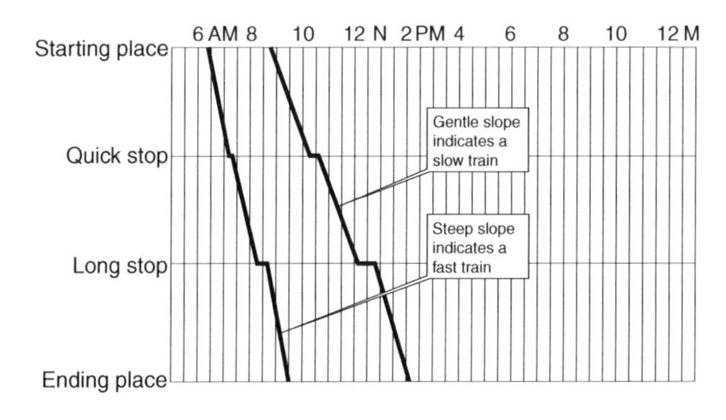

Figure 3.5.21 A graphic train schedule, with time scaled horizontally (from left to right) and distance shown vertically (from top to bottom).

Time as a visual variable

Computer graphics can elevate time to its proper place in graphic analysis. Graphics systems can do this in one of two ways. The first strategy is the interactive graphics system that allows the user to manipulate time freely, as with a *temporal scroll bar* (Figure 3.5.22), for moving between time periods. The viewer uses a mouse to point to the box in the scroll bar and to change the time displayed on the map by 'dragging' this box to the left (to select an earlier time) or to the right (to select a later, more recent time). A discrete temporal scroll bar (Figure 3.5.22, above) might address time in years or decades, whereas a continuous scroll bar might provide a graphic scale and reference time more precisely, for example, in days, hours, minutes or even seconds. As Carter (1988) notes, the map viewer often

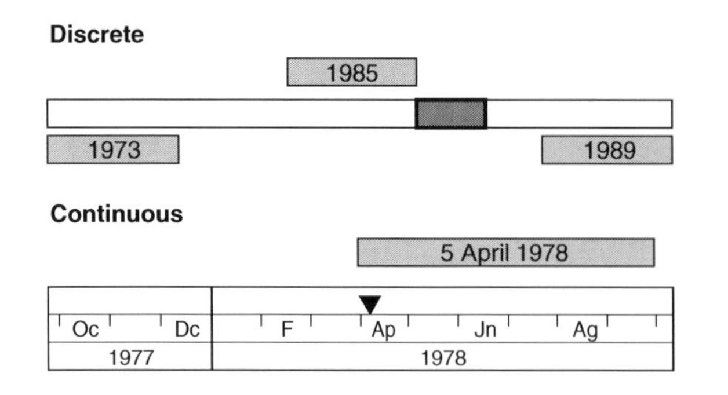

Figure 3.5.22 Temporal scroll bars can provide discrete (above) or relatively continuous (below) references to time.

needs to flip back and forth through a sequence of maps, and to move at his or her own pace. The second strategy is *animation graphics*, which provides a temporally ordered sequence of views so that the map becomes a scale model in both space and time.

Over the past three decades a number of geographic cartographers have addressed the uses of animation techniques for dynamic maps (Berlyant 1988; Moellering 1980; Thrower 1959, 1961; Tobler 1970), and in the 1980s videotex and microcomputers provided dynamic sequencing for the categories displayed on choropleth maps (Slocum *et al.* 1988; Taylor 1982). Dynamic maps can range in complexity and sophistication from a simple temporal sequence of complete maps, as might be shown in sequence with a single slide projector, to dynamic symbols that move across the map in the manner of video games. Other intriguing animation effects are possible, of course, including dramatic fades or dissolves of multiple views, progressive zooms, and rotating oblique views of statistical surfaces approached gradually, in the manner of an airplane circling an airport. Fading, fuzzy or blinking symbols might be particularly useful in dealing with data whose accuracy varies over time.

Another form of animation is the meaningful program or succession of views. In the 1970s statisticians addressed the problem of exploring graphically a database with a large number of variables with a technique called *projection pursuit* (Friedman and Tukey 1974; Huber 1985; Tukey and Tukey 1981). Projection pursuit leads the analyst to one or more potentially significant scatter plots by selecting a small number of optimum two-dimensional 'interesting' views chosen for their degree of clustering or 'clottedness'. The *grand tour*, a more recent elaboration of the projection pursuit model, seeks an optimal sequence of such interesting views (Asimov 1985; Buja and Asimov 1986). Both highly promising techniques are still largely experimental and little used.

An obvious extension, of course, is the addition of an optional geographic space viewport. This approach, called 'atlas touring' and currently under development, integrates maps and statistical graphics through the use of *graphic scripts*, composed using basic sequences called *graphic phrases* (Monmonier 1989b). A graphic phrase for the visual analysis of spatial-temporal data might, for example, partition the screen into four windows and generate an animated sequence of juxtaposed chess maps for pairs of individual years at the top of the display, a change map for the period in question at the lower right, and a time-series statistical diagram for the entire period of analysis at the lower left. Other graphic phrases might explore the spatial trends on a particular map or the spatial correlation among two variables. A map author might use several such graphic phrases to develop a graphic script examining the spatial-temporal trends, geographic trends and spatial co-variation of an electronic atlas in the form of a large spatial-temporal data set. Monitoring the script-writing behaviour of map authors should provide data to support the development of a still more advanced system – a system able to generate automatically a meaningful sequence of graphics that serve as a guided tour of an electronic atlas.

These techniques beg the question of how best to define and measure the meaningfulness of a map or scatter plot or the interest it might hold for a viewer. As statisticians associate inherent interest with the clottedness of a point cloud, geographic cartographers might regard as interesting a map that resembles a known set of regions, demonstrates a straightforward spatial trend, or otherwise exhibits a moderate to strong level of spatial autocorrelation, the geographic space equivalent of clottedness in attribute space. But for some distributions, a pattern that exhibits no discernable trend or lacks regional homogeneity might be more intriguing and noteworthy than one that meets such expectations.

High-interaction graphics avoid this issue by allowing the user to search the data for views he or she finds interesting. High-interaction graphics require a high-resolution display, a pointing device, a 'friendly' direct-manipulation graphics interface and a high-speed processor that allows the analyst to interact creatively with the display (Becker *et al.* 1987). The software must be flexible and the response time minimal. Historians and geographers have demonstrated the pedagogic utility of interactive systems that allow the user to juxtapose maps and frequency diagrams and to classify distributions portrayed on choropleth maps (Miller and Modell 1988).

Interactive systems for studying spatial-temporal data might incorporate *scatter plot brushing* (Figure 3.5.23, top)

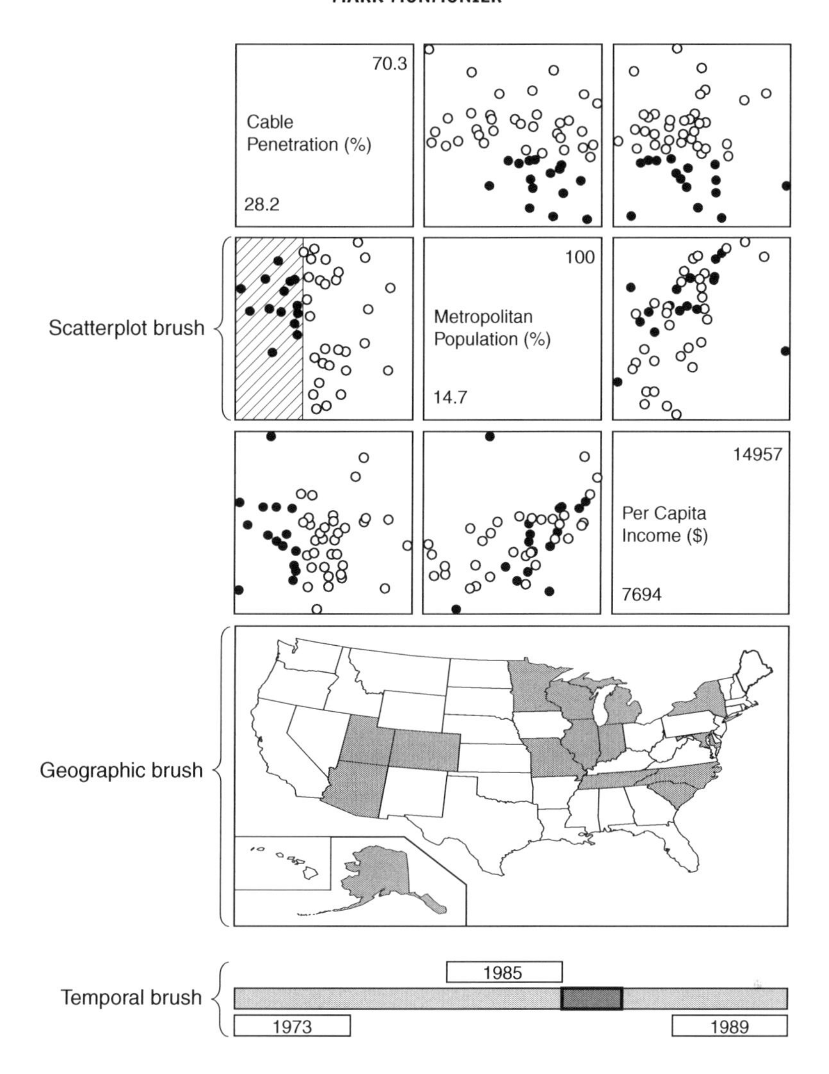

Figure 3.5.23 A scatter plot matrix with a scatter plot brush, a geographic brush and a temporal brush.

and a modification, the *geographic brush* (Figure 3.5.23, centre), with which the analyst highlights areas on the map and views them as well on the scatter plot matrix (Monmonier 1989a). The viewer might use a scatter plot brush to select a number of places by enclosing with a variable size rectangular frame the points that represent them in any of the nine scatter plots; the dots representing these places would be highlighted (darkened in this case) within the rectangular brush on the scatter plot in question and also highlighted in each of the other eight scatter plots as well as on the accompanying map. With a geographic brush, the analyst could select and highlight places by pointing to them on the map, by drawing a polygon around them, or by choosing one or more established regions, such as the Northeast, from a pull-down menu. For spatial-temporal data, addition of a *temporal brush* (Figure 3.5.23, bottom) is appropriate as well. The viewer might manipulate the time represented by the display by using the temporal scroll bar to call up the point clouds for some other time unit.

'Atlas touring' and geographic-temporal brushing can be complimentary. Data collected with an interactive system can be helpful in refining or tailoring definitions and measures of interest or meaningfulness. And a programmed series of views, which the analyst can interrupt at will, might serve as a useful introduction to the data – a graphic pump-primer, of sorts, to initiate analysis and encourage the generation and testing of hypotheses.

Concluding remarks

Table 3.5.1 is a conceptual framework that summarises the range of options for the graphic display of quantitative spatial-temporal data. It demonstrates that the principal tools available for the visualisation of spatial-temporal data are:

- multiple views, either in sequence or simultaneously in the same window;
- time scaling, as well as space scaling;
- interaction with the data and the display; and
- integration of maps and time-series graphics.

Cartographic research must turn away from its search for the single optimum map and begin to deal with sequences of maps and the need to integrate maps with statistical diagrams and text blocks containing definitions and other relevant information.

Table 3.5.1 A conceptual framework for the visualisation of geographic time-series data

Time-series graphs with place names or other geographic metaphors
Single static maps
◇ Temporal symbols
◇ Temporal aggregation
◇ Focused measurements (e.g. peak-year maps)
◇ 'Dance Maps' (movement)
◇ 'Change Maps' (rates, absolute change)
◇ Generalised maps focusing on transition and or diffusion
Multiple static maps (and graphs)
◇ 'Chess Maps' (juxtaposition)
◇ Cartographic cross-classification arrays (with maps and statistical graphics)
Single dynamic maps
◇ Sequenced symbols (accretion)
◇ Temporal sequences of views
◇ Symbols suggesting motion (pulsating directional symbols)
Multiple dynamic maps (and graphs)
◇ High-interaction graphic analysis
– Scatter plot brushing
– Geographic brushing
– Temporal brushing
◇ Programmed sequences of 'interesting' or 'meaningful' views
– Authored video animation
– 'Projection Pursuit' (not geographic)
– 'Grand Tour' (not geographic)
– 'Atlas Touring' (geographic)

References

Asimov, D. (1985) The grand tour: a tool for viewing multi-dimensional data. *SIAM Journal of Scientific and Statistical Computing*, **6**, 129–143.

Becker, R., Cleveland, W.S. and Wilks, A.R. (1987) Dynamic graphics for data analysis. *Statistical Science*, **2**, 355–395.

Berlyant, A.M. (1988) Geographic images and their properties. *Mapping Sciences and Remote Sensing*, **25**, 133–143.

Buja, A. and Asimov, D. (1986) Grand tour methods: an outline, in *Computer Science and Statistics: The Interface* (ed. D.M. Allen), Elsevier, New York, pp. 63–67.

Carter, J.R. (1988) The map viewing environment: a significant factor in cartographic design. *American Cartographer*, **15**, 379–385.

Chorley, R.J. and Haggett, P. (1965) Trend surface mapping in geographical research. *Transactions of the Institute of British Geographers*, **37**, 47–67.

Davis, J.C. (1986) *Statistics and Data Analysis in Geology*, 2nd edn, John Wiley & Sons, Inc., New York.

Du Toit, S.H.C., Steyn, A.G.W. and Stumpf, R.H. (1986) *Graphical Exploratory Data Analysis*, Springer-Verlag, New York.

Friedman, J.H. and Tukey, J.W. (1974) A projection pursuit algorithm for exploratory data analysis. *IEEE Transactions on Computers*, **C-23**, 881–890.

Huber, P.J. (1985) Projection pursuit. *Annals of Statistics*, **13**, 435–475.

Johnston, R.J. (1978) *Multivariate Statistical Analysis in Geography*, Longman, New York.

Miller, D.W. and Modell, J. (1988) Teaching United States history with the Great American History Machine. *Historical Methods*, **21**, 121–134.

Moellering, H. (1980) The real-time animation of three-dimensional maps. *American Cartographer*, **7**, 67–75.

Monmonier, M. (1970) A spatially-controlled principal components analysis. *Geographical Analysis*, **2**, 192–195.

Monmonier, M. (1972) Flow-linkage construction for spatial trend recognition. *Geographical Analysis*, **4**, 392–406.

Monmonier, M. (1979) An alternative isomorphism for the mapping of correlation. *International Yearbook of Cartography*, **19**, 77–89.

Monmonier, M. (1989a) Geographic brushing: enhancing exploratory analysis of the scatterplot matrix. *Geographical Analysis*, **21**, 81–84.

Monmonier, M. (1989b) Graphic scripts for the sequenced visualisation of geographic data. *Proceedings of GIS/LIS'89*, Orlando, Florida, pp. 381–389.

Palm, R. and Caruso, D. (1972) Labelling in factorial ecology. *Annals of the Association of American Geographers*, **62**, 122–133.

Slocum, T.A., Egbert, S.L., Prante, M.C. and Robeson, S.H. (1988) Developing an information system for choropleth

maps. *Proceedings of the Third International Symposium on Spatial Data Handling*, Sydney, pp. 293–305.
Taylor, D.R.F. (1982) The cartographic potential of Telidon. *Cartographica*, **19** (3–4), 18–30.
Thrower, N.J.W. (1959) Animated cartography. *Professional Geographer*, **11** (6), 9–12.
Thrower, N.J.W. (1961) Animated cartography in the United States. *International Yearbook of Cartography*, **1**, 20–30.
Tobler, W.R. (1970) A computer movie simulating urban growth in the Detroit region. *Economic Geography*, **46** (2), 234–240.
Tufte, E. (1983) *The Visual Display of Quantitative Information*, Graphics Press, Cheshire, CT.
Tukey, J.W. (1977) *Exploratory Data Analysis*, Addison-Wesley, Reading, MA.
Tukey, P.A. and Tukey, J.W. (1981) Preparation; pre-chosen sequences of views, in *Interpreting Multivariate Data* (ed. V. Barnett), John Wiley & Sons, Ltd, Chichester, UK, pp. 189–213.
Walker, C. (1914) *The Modern Dances: How to Dance Them*, Saul Brothers, Chicago.

Further reading

Harrower, M. (2004) A look at the history and future of animated maps. *Cartographica*, **39** (3), 33–42. [Traces the development of animated mapping and explores the potential of these kinds of displays.]
Peterson, M.P. (1995) *Interactive and Animated Cartography*, Prentice Hall, Englewood Cliffs, NJ. [The first systematic text bringing together research on different kinds of animated mapping.]
Lobben, A. (2008) Influence of data properties on animated maps. *Annals of the Association of American Geographers*, **98** (3), 583–603. [A useful critical examination of data properties, construction and cognition of different kinds of animated geovisualisations.]
MacEachren, A.M. (1996) *How Maps Work*, Guilford, New York. [Chapter 9 in MacEachren's book describes the graphic variables that are available for the representation of time in geovisualisation.]

See also

- Chapter 1.11: Exploratory Cartographic Visualisation: Advancing the Agenda
- Chapter 2.7: Cartography and Geographic Information Systems
- Chapter 2.11: Extending the Map Metaphor Using Web Delivered Multimedia
- Chapter 3.10: Affective Geovisualisations

Chapter 3.6

The Roles of Maps, from *Some Truth with Maps: A Primer on Symbolization and Design*

Alan M. MacEachren

Editors' overview

This excerpt from MacEachren's monograph describes map uses in geographic analysis and decision making. It suggests ways in which the specific use and user of a map are critical to proper symbolisation and design choices. Deploying examples from environmental issues, MacEachren contrasts roles of design in exploration, confirmation, synthesis and presentation of mapping in facilitating thinking in the private realm and communication in the public realm.

Originally published in 1994: Chapter 1 in Alan M. MacEachren, *Some Truth with Maps: A Primer on Symbolization and Design*, Association of American Geographers, Washington, DC, USA.

What is a map? What are maps for? When asked these questions, what comes to mind? For many people, the prototypical map is a highway map with the prototypical use being to facilitate travel from here to there. For others, a topo sheet is the first choice that comes to mind. Again, the typical use conceived of might be as a travel aid (for those who like to hike), or as a planning tool that allows delineation of drainage basins or provides a base upon which potential highway locations can be plotted. In all of these roles, the map is primarily a presentation device. It presents an abstract view of some portion of the world with an emphasis on selected features such as roads or terrain. When most map users, even trained cartographers, approach the task of map symbolisation and design, they typically assume a presentation goal for which a single map must be selected. This leads to questions about how to generate the *optimal map* for the task at hand.

The goal of producing an optimal map is not equally applicable to all maps. David DiBiase (1990) recently developed a graphic model of the range of uses to which maps and other graphics might be put in scientific research (Figure 3.6.1). This basic model is relevant, not only to science, but to applied geographical analysis and spatial decision making with a GIS. In this graphic depiction of the role of maps and graphics, we see that presentation is at one extreme of a range of map applications from exploratory data sifting in search of patterns that should be investigated to the presentation of data analysis results or of a plan for cleaning up a toxic spill. The concept of selecting an optimal map, although possibly relevant for presentation, becomes less relevant (and perhaps even counter-productive) as we approach the exploration end of the continuum.

In DiBiase's model, a critical distinction is made between 'visual thinking' and 'visual communication'. Maps can be used for both, but the goals of symbolisation and design will differ. For visual thinking, maps are tools that can prompt insight, reveal patterns in data and highlight anomalies. The goal is to help us notice something, such as a relationship between the location of industry and

The Map Reader: Theories of Mapping Practice and Cartographic Representation, First Edition. Edited by Martin Dodge, Rob Kitchin and Chris Perkins.

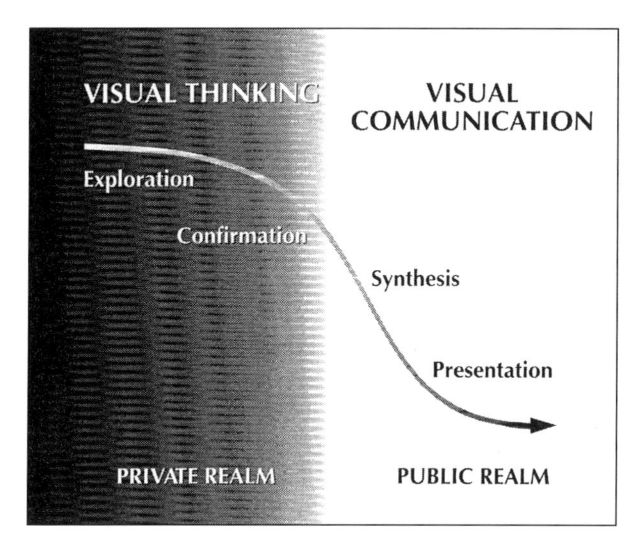

Figure 3.6.1 The functions of graphics in the research sequence. From the investigator's point of view, graphics change from reasoning tools to communication tools as an investigation expands from a private to a public endeavour. The downward slope of the curve suggests the hypothesis that visual thinking involves higher order cognitive tasks (for the investigator) than does visual communication.

incidence of health problems in a population. Symbolisation and design decisions made by the map author must facilitate our ability to notice the unexpected. In contrast to this pattern-seeking goal, the goal of visual communication is to make a point, to communicate what we think we know. If a comprehensive analysis using maps, health records, statistics, modelling and so on convinces you that there is a link between the geographic distribution of particular pollutants and a specific form of cancer, then a map can be designed to communicate this conclusion to the person formulating long term policy decisions.

Visual thinking and communication are not mutually exclusive map goals, but rather two ends of a continuum along which gradually changing goals lead (or should lead) to differences in design and symbolisation. At the visual thinking end of the continuum, for example, the map already has the analyst's attention. There is less need for visual 'hooks' that draw a reader to the illustration. In the middle of a long report, however, such hooks, in the form of dramatic colours, high contrast, iconic symbols, drop shadows and so on may serve an important role in drawing enough attention to the map for it to have an impact.

In the four sections that follow, examples are provided of map-based exploration, confirmation, synthesis and presentation. To provide continuity, all examples deal with environmental issues.

Exploration

Early in any investigation that involves the use of GIS to manipulate and examine spatial data, maps can be a particularly important exploratory tool. Maps and graphics at this stage often go undocumented because they remain in the private realm of an individual researcher or small group of researchers. The representations used are often schematic and generally not considered to be of 'publishable' quality. One such exploratory mapping endeavour that was not intended for a publication or report is described in a book chapter on visualisation in geographic research. The relevant passage is quoted below (MacEachren *et al.* 1992: 105–106):

> 'A fascinating example of the process of exploratory visualisation at work was provided by a group of colleagues. As a part of an evolving research effort dealing with issues of global climate, Dan Leathers, Brent Yarnal and Michael Palecki produced a choropleth map, originally hand drawn with coloured pencils, depicting correlations between US winter temperatures and the Pacific North American (PNA) teleconnection, a zonal index of the mid-tropospheric circulation over North America.'

The map presented an intriguing and unexpected pattern (Figure 3.6.2). A high PNA index indicates the presence of a 'deep, strong Aleutian low extending from the surface to the upper troposphere in the north-eastern Pacific, an upper level anticyclone centred over western Canada and a deeper than normal upper level trough located over the south-central United States' (Yarnal and Diaz 1986: 197–198). Negative values on the index indicate absence of this pattern, that is a jet stream with no dominant troughs or ridges. The choropleth map demonstrated a strong correlation between the PNA index and temperature for both the Northwestern and the Southeastern United States, with the Northwest being warmer than normal and the Southeast cooler.

In trying to explain this pattern, the researchers considered several possible explanations and connections to global circulation patterns. Going against the conventional wisdom that strong PNA patterns are a winter phenomenon and have little relationship to weather during other seasons, they found a similar relationship between the PNA index and temperature on an annual basis. A comparable relationship was identified for precipitation. Once discovered, the relationship was determined to have a strong physical basis, with topography exerting a blocking influence that reduces the association for the diagonal band from the Great Lakes to the Southwest. Working from the initial discovery, Leathers, Yarnal and Palecki produced a series of maps, did some numerical analysis and eventually demonstrated links among strong PNA patterns, El Nino events (warmer than normal ocean temperatures in the eastern Pacific) and polar circulation patterns. Their

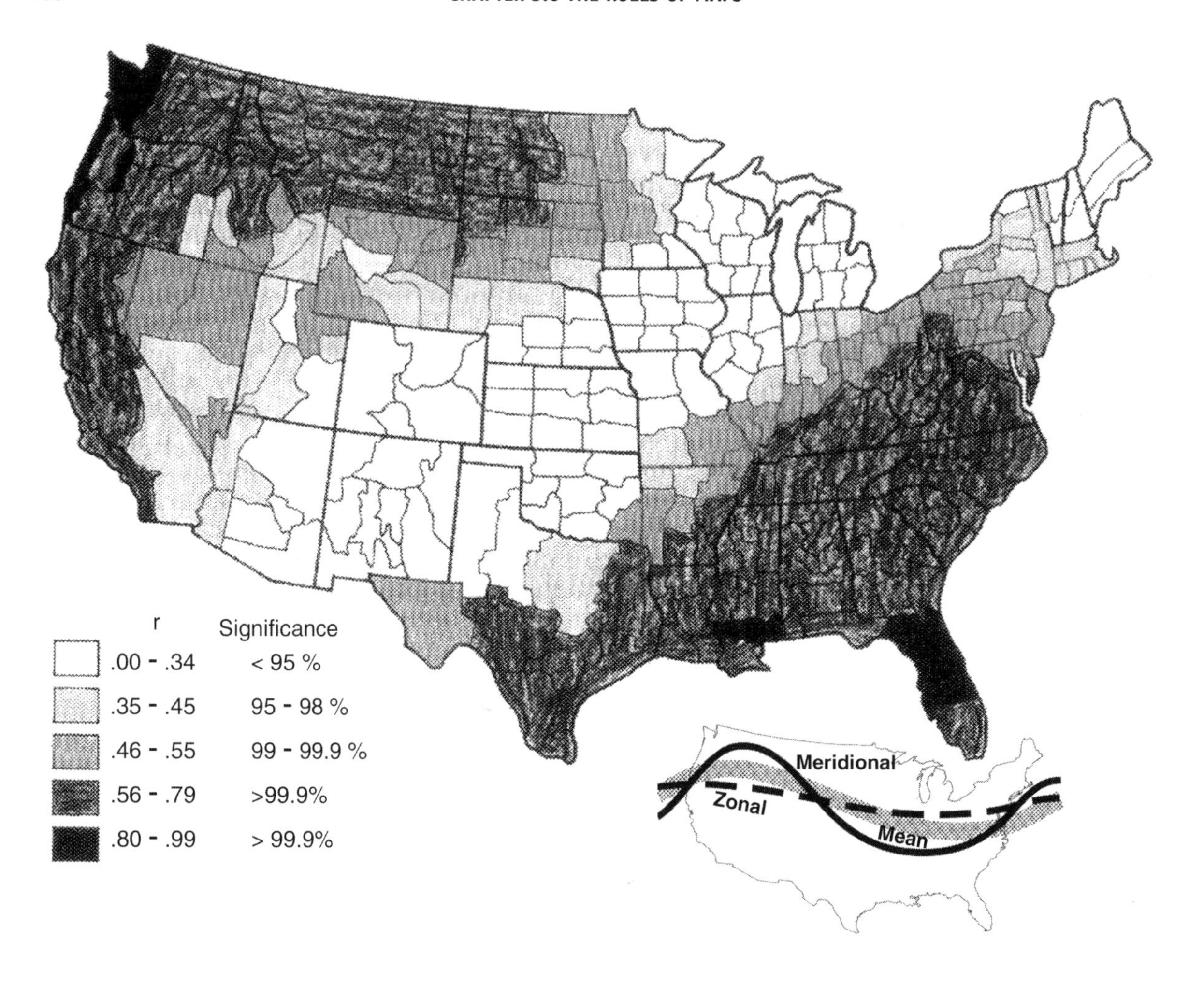

Figure 3.6.2 Correlation between the Pacific North American teleconnection index and mean winter US temperature. This choropleth map (the original was hand drawn using coloured pencils) represents values for each of the 344 climatic divisions of the conterminous US. The inset illustrates the mean position of the tropospheric flow in comparison to meridional and zonal flows (represented by positive and negative PNA indices, respectively).

findings have significant implications for long term weather prediction as well as for the regional impact of changes in global climate.

Confirmation

As an environmental investigation progresses, assumptions are made, questions are posed and hypotheses are generated. Maps and other visualisation tools at this stage of investigation are often used as a way to confirm initial suspicions. Separately, or in combination with other graphics, they can show the outcome of a modelling effort, or draw attention to anomalies by combining or transforming the original observations. These graphic expressions all play a role in confirming or countering hypotheses that have been posed. It is often the visual display of anomalies or residuals (cases that remain unexplained when a dependent variable is regressed against one or more independent variables) that leads to the greatest insights.

One interesting example of a confirmatory use of map-based graphics that turned up an unexpected relationship is provided in a video produced by the Deasy GeoGraphics Laboratory at Penn State University (DiBiase *et al.* 1991). The video uses animated maps and graphics to explore the results of predictions of various global climate models

(GCMs) about climate change and its impact on Mexico. The video was designed to support research by Liverman and O'Brien (1991) dealing with reliability of model estimates and what it means for agricultural planning in marginal areas.

Most results of GCM analysis have suggested that temperatures will increase if atmospheric carbon dioxide were to double. An animated map was produced on which histograms linked to thirteen Mexican cities showed means and variances for five different GCM predictions of temperature and precipitation throughout a year. It was clear from the animation that all models predict increases in temperature for a doubling of carbon dioxide scenario. The predicted increases are not just for the overall mean, but for every individual station. By itself, this result is a confirmation of expectations concerning global warming. In spite of this apparent confirmation, however, uncertainty is evident when the GCMs are compared. It is immediately clear that the models differ from one another, that these differences vary through the year, and that among-model variation is quite high during selected months.

To get a more complete picture of these model differences, data for individual cities were re-expressed. The re-expression, in the form of reordering of the data, is used to focus directly on intra-model variability. Temperature data for Puebla were re-expressed by transforming data from a chronological sequence to a sequence ordered from months in which model predictions agree most closely to months with high among-model variation (Figure 3.6.3). To enhance the important feature of this re-expression (the variability or uncertainty among models), duration of each scene in the animation was controlled by the among-model variability (i.e. scene duration increased as among-model differences increased). What resulted was a clear picture of highest disagreement (or greatest uncertainty) in the model predictions during the spring planting season – exactly when the greatest certainty is required. This planting season uncertainty had not been noticed until data were re-expressed for this animation. In this application, therefore, the multiple views of data undertaken as part of the confirmatory stage of analysis leads to the conclusion that current GCMs may be unreliable (or differ substantially in their reliability) for making estimates concerning potential impacts of increased atmospheric carbon dioxide on agriculture in marginal areas.

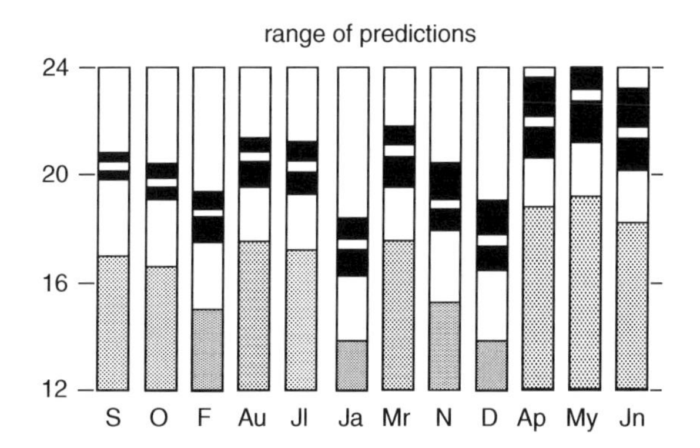

Figure 3.6.3 This set of figures compares predicted and observed temperatures for Puebla, Mexico. The grey bar represents monthly mean observed temperatures. The black zone symbolises the range of five GCM predictions for a two-times-CO_2 scenario. The white line across this zone is the mean of the model predictions. The scenes are ordered from months in which predictions vary the least to those varying the most.

Maps and other graphics can depict both patterns and anomalies, but are subject to their own data processing uncertainties. As a result, maps can mislead a user into seeing patterns that are not there, or missing patterns that are. Maps, therefore, can only be one part of a confirmatory stage of data analysis in which various analytical tools are brought to bear on the problem at hand.

Synthesis

As an analyst gains confidence in a perspective on an environmental issue, maps provide a tool through which ideas can be synthesised. They allow the analyst to produce a coherent but abstract statement concerning patterns and relationships being uncovered. Each stage in data processing and display represents a process of abstraction. Key features from the next lower level of analysis are highlighted and local details, random noise and so on are filtered away in an effort to get at the essence of a situation. The goal of synthesis is to make up for the lost detail by gaining perspective (i.e. stepping back from the data and the specifics to the point where the 'big picture' becomes apparent). Synthesis is an activity in which an expert makes informed decisions about what to emphasise, what to suppress and which relationships to show.

A typical example of a map produced at this stage of analysis can be found in the recent EPA Region 6 Comparative Risk Project Ecological Report (EPA 1990). For this project, a team of investigators used GIS and modelling techniques to evaluate 'the residual risk posed from 22 environmental problem areas identified by EPA Headquarters and the Regional Comparative Risk Project (RCRP) directors' (EPA 1990: 1). For 15 of the 22 problems, there was sufficient potential risk in Region 6 and sufficient data to carry out a complete evaluation. Using a mathematical model, an ecological risk index was calculated for each of these 15 problems and individual maps were prepared (Figure 3.6.4). These

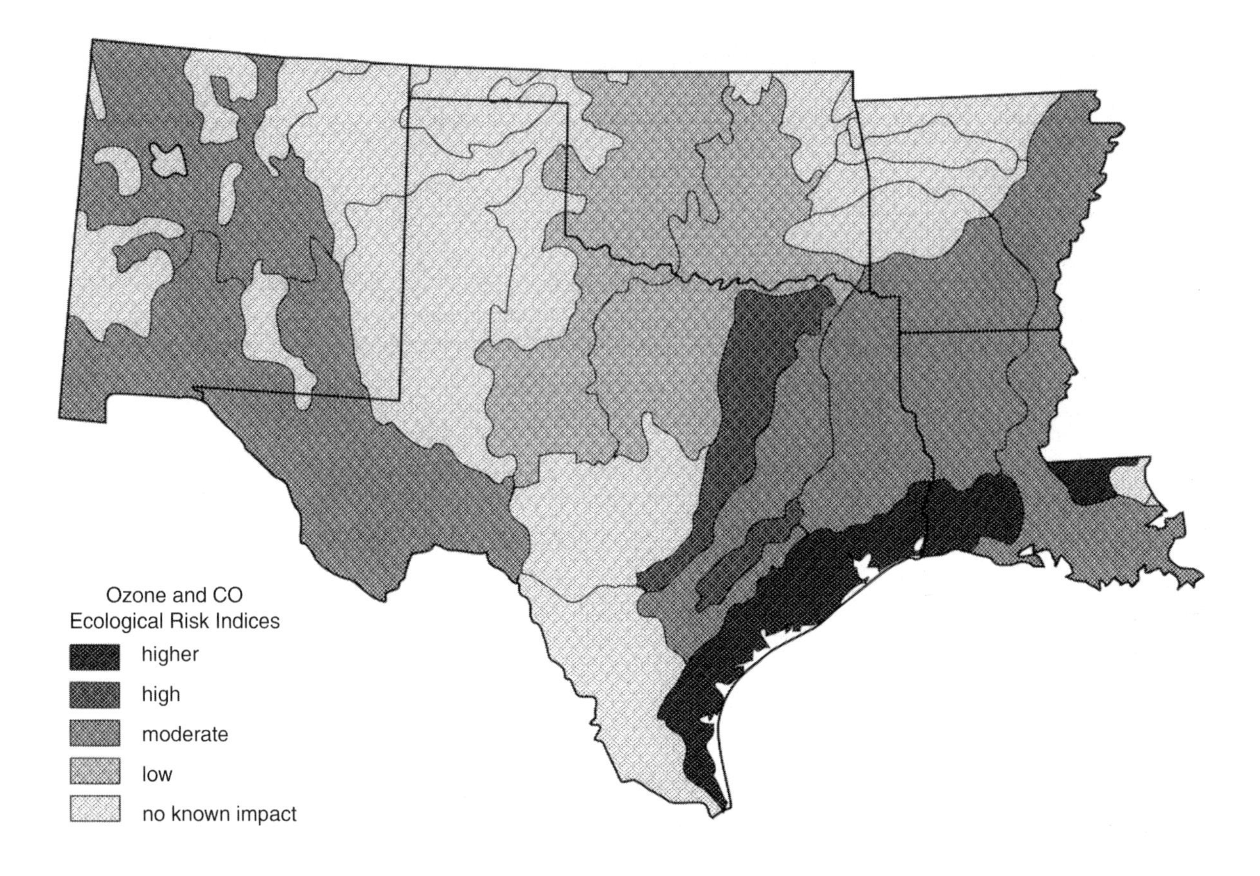

Figure 3.6.4 Map of ecological risk due to ozone and CO for EPA Region 6. After map 10 in Region 6 Comparative Risk Project Ecological Report (EPA 1990).

individual maps can be used in exploratory analysis to look for geographic patterns and anomalies. For example, it is apparent that ecological risk due to ozone and CO is particularly high along the Gulf Coast of Texas, but much of the Louisiana Gulf Coast is at only moderate risk.

A basic type of cartographic synthesis is used by the authors of the Ecological Report to address the overall implications of the 15 individual risks. An aggregate, or cumulative ecological risk index [ERI] is calculated by summing the individual risks for each ecoregion. The resulting map dramatically demonstrates the high ecological risk along the entire Gulf Coast and Mississippi River Valley, and the gradual decline in risk from the east to the west of the region (Figure 3.6.5). This synthesis is, of course, only one of many data/map syntheses that might be used to summarise the analysis. Different maps would result, for example, if the individual risks were not treated as equally serious (as unweighted summing of index values assumes). In addition, spatial filtering of the individual values might be used to account for interrelations among the adjacent ecoregions (i.e. spatial autocorrelation) or to filter out the impact of fixed ecoregion boundaries when considering phenomena that recognise no bounds (e.g. toxic air pollutants).

Presentation

Once you are convinced of a location (e.g. of toxic waste), an attribute at a location (e.g. alkalinity of surface water), or a relationship (e.g. between an industry and a health hazard) and you want to convince someone else of your assessment, a map is an ideal presentation device. A well designed map is convincing because it implies authenticity. People believe maps. It is this general acceptance of maps (in contrast to suspicion of statistical analysis) that makes maps a powerful presentation device, but one that must be constructed and used with care. For discussion of the negative consequences of this general acceptance

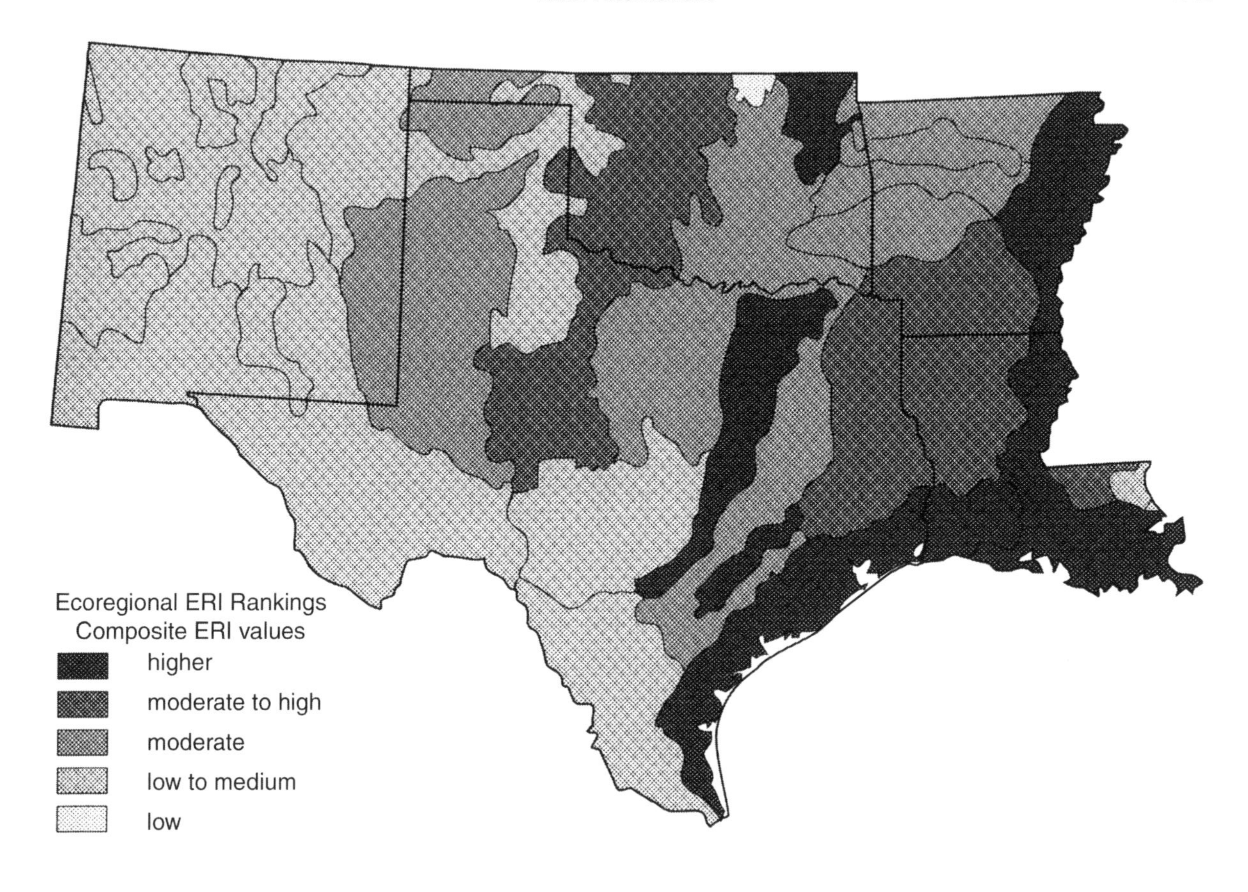

Figure 3.6.5 Ecoregional ERI rankings map that depicts the cumulative risk due to the 15 environmental risks assessed for region 6. After map 17 in Region 6 Comparative Risk Project Ecological Report (EPA 1990).

of truth and objectivity in maps see Harley (1989), Monmonier (1991) and Wood (1992).

An outstanding presentational map was developed at the US EPA by Omernik and Powers (1983) to illustrate the regional patterns of mean alkalinity of surface water in the conterminous US (Figure 3.6.6). 'The map was developed from mean annual total alkalinity values of approximately 2500 streams and lakes and from the apparent relationships of these data with land use and other macro-watershed characteristics, such as soil type and geology' (Omernik and Powers 1983: 133). Like the EPA Region 6 aggregate risk map discussed above, Omernik and Power's alkalinity map could be considered a synthesis. They, in fact, state that the map combines measurement and expert knowledge in ways similar to that used to fill in detail of precipitation patterns in areas of scarce data. Once they were satisfied with their synthesis, however, they moved on to the presentational stage. The emphasis on presentation is apparent in the goals they identify for the map and by its publication as a map supplement in a major geographic journal.

The authors cite three specific goals for their map: (1) provide a national perspective on the extent of the problem, (2) provide logic or rationale for selecting geographic areas for more detailed studies, and (3) allow more accurate regional economic assessments of acid-precipitation impacts on aquatic resources. Specifically, the map is intended to provide a picture of the sensitivity of water to acidification. Sensitivity (as indicated by alkalinity) is partitioned into six categories. Care was taken when selecting category breaks to avoid depicting either a worst case or a best case condition. [. . .] As the authors state, their intent was 'to show what one might expect to find in most surface waters most of the time' (Omernik and Powers 1983: 133). For analysis, particularly if working at a more local scale, the worst case and best case scenarios might prove valuable in planning responses to environmental disasters. The middle ground, however, is more appropriate for a

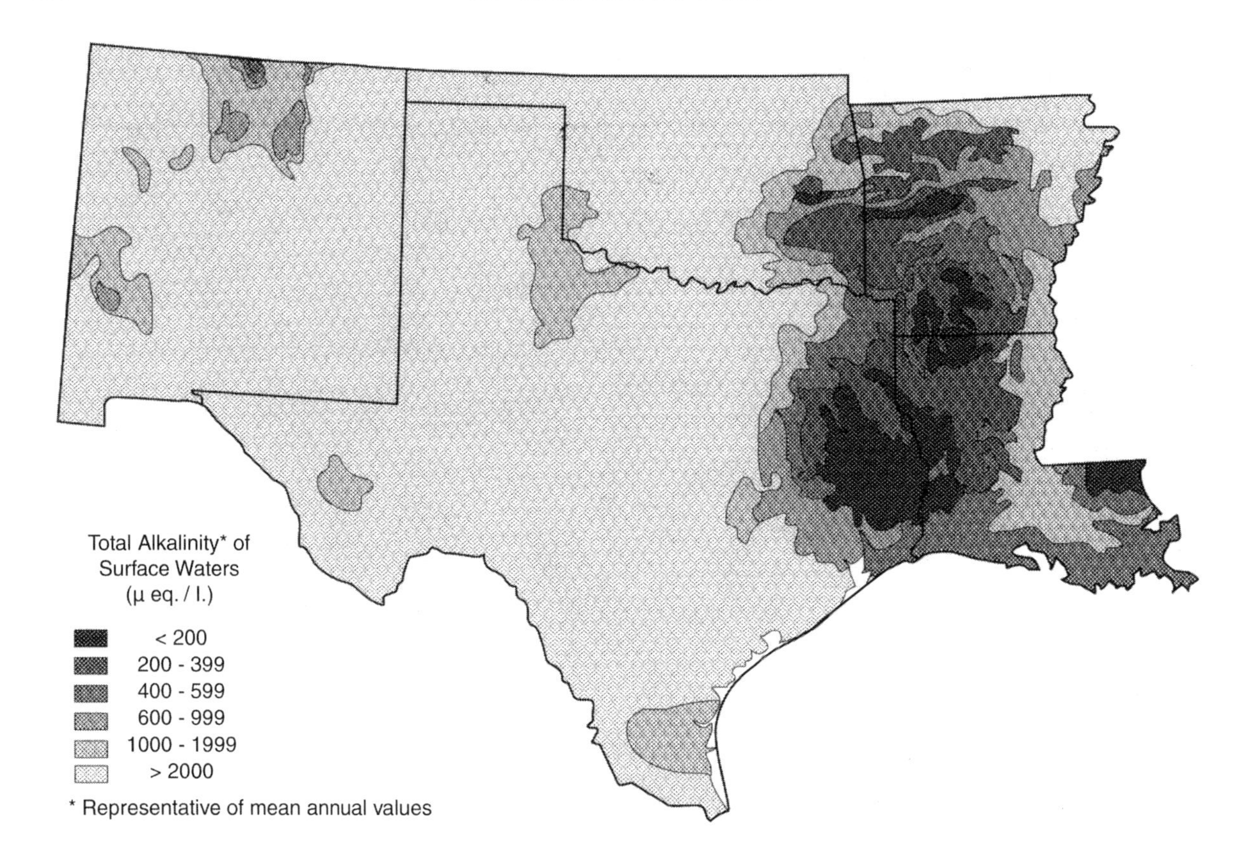

Figure 3.6.6 Reduced scale reproduction of the EPA Region 6 portion of Total Alkalinity of Surface Waters (Omernik and Powers 1983). Original map used area fills of reddish-brown through orange to yellow.

presentational graphic intended to guide broad national scale policy formulation.

Summary

In this chapter, a model of the role of maps and graphics in geographic analysis and decision making is presented. This model suggests a distinction between maps as visual thinking tools (used in early exploration of data sets and confirmation of patterns noticed) and maps as visual communication tools (that facilitate synthesis and presentation of results and conclusions). This distinction is particularly critical to design and symbolisation of maps in a GIS environment because GIS provides tools that facilitate multiple views of data and flexible map design/symbolisation choices. Although some design and symbolisation principles apply consistently across the entire exploration to presentation spectrum, the difference in goals (i.e. noticing the unexpected versus communicating the known) requires different strategies. [...]

References

DiBiase, D. (1990) Visualization in the earth sciences. *Earth and Mineral Sciences Bulletin*, **59** (2), 13–18.

DiBiaise, D., Krygier, J.B., Reeves, C. *et al.* (1991) *An Elementary Approach to Cartographic Animation*, Deasey GeoGraphics Laboratory, Penn State, PA.

EPA (1990) Region 6 Comparative Risk Project Ecological Report. United States Environmental Protection Agency, Dallas, TX.

Harley, J.B. (1989) Deconstructing the map. *Cartographica*, **26** (2), 1–20.

Liverman, D.M. and O'Brien, K. (1991) Global warming and climate change in Mexico. *Global Environmental Change*, **1** (4), 351–364.

MacEachren, A., Buttenfield, B., Campbell, J., *et al.* (1992) Visualization, in *Geography's Inner Worlds* (eds R. Abler, M. Marcus and J. Olson), Rutgers University Press, New Brunswick, NJ.

Monmonier, M. (1991) *How to Lie with Maps*, Chicago University Press, Chicago.

Omernik, J.M. and Powers, C.F. (1983) Total alkalinity of surface waters – a national map. *Annals of the Association of American Geographers*, **73** (1), 133–136.
Wood, D. (1992) *The Power of Maps*, Guilford, New York.
Yarnal, B. and Diaz, H.F. (1986) Relationships between extremes of the Southern Oscillation and the winter climate of the Anglo-American Pacific coast. *International Journal of Climatology*, **6**, 197–219.

Further reading

Dykes, J.A., MacEachren, A.M. and Kraak M.J. (2005) *Exploring Geovisualization*, Elsevier, London. [A wide ranging edited collection of papers on the possibilities opened out by a wider appreciation of the links between design and use in geovisualisation.]

MacEachren, A.M. (1995) *How Maps Work*, Guilford, New York. [A systematic evaluation linking semiotic and cognitive approaches to mapping and boring links between design and use.]

See also

- Chapter 1.11: Exploratory Cartographic Visualisation: Advancing the Agenda
- Chapter 2.11: Extending the Map Metaphor Using Web Delivered Multimedia
- Chapter 3.2: Interplay of Elements
- Chapter 3.3: Cartography as a Visual Technique
- Chapter 3.5: Strategies for the Visualisation of Geographic Time-Series Data
- Chapter 3.10: Affective Geovisualisations

Chapter 3.7

Area Cartograms: Their Use and Creation

Daniel Dorling

Editors' overview

Dorling re-energised the deployment of the cartogram as a design following his PhD research in the early 1990s, and in an ongoing series of atlases and publications in the twenty years since. In this excerpt from a short booklet he situates cartogram techniques, contrasting and explaining physical, mechanical and algorithmic methods used in their construction. He also explores the popularity of the technique in political cartography, epidemiological mapping, the mapping of mortality and the mapping of commuting and migration flows, before considering the explanatory power of the cartogram as a device for presenting complex socio-economic and political data.

Originally published in 1996: *Concepts and Techniques in Modern Geography 59*, Quantitative Methods Study Group, Institute of British Geographers.

Introduction

Maps are called cartograms when distortions of size, and occasionally of shape or distance, are made explicit and are seen as desirable. Typically places on a cartogram are drawn so that their size is in proportion to their human populations. However, their size could be made proportional to any measurable feature. Conventional maps can be seen as land area cartograms – as places on them are usually drawn in proportion to their land areas – although this is often not the case for features on these maps. [...] Cartograms are produced for a variety of purposes. They can be used, like the London Underground map, to help people find their way. In atlases they are often used for their ability to shock; cartograms where area is drawn in proportion to the wealth of people living in each place show a dramatic picture. A major argument for the use of equal population cartograms in human geography is that they produce a more socially just form of mapping by giving people more equitable representation in an image of the world. In research, cartograms are increasingly used to provide alternative base maps upon which other distributions can be drawn to see, for example, whether the incidence of a particular disease is spread evenly over the population. Cartograms can also be used like conventional maps where different areas are shaded with different colours to show variation over space. For instance, to be able to map both the absolute and relative concentration of elderly people across the population a cartogram is required. [...]

It may be helpful to begin with a simple example. Figure 3.7.1 gives population statistics and a map for a fictional island. Its area is 20 square kilometres and 100 people live on the island, which is divided into three districts: the Farm, Town and City. The figure shows that the mean population density of the island is five people per square kilometre, although most of its population live in the City at a density of 15 people per square kilometre. The equal area map of the island has been simplified to fit on a grid. The cartogram of the island was drawn by hand, starting with the City and trying to alter the shape of the island as little as possible while making the area of each district proportional to its population. Note that the Town still separates the Farm and City. The impact of this transformation is illustrated by using two pictograms in which icons are placed on both the map and cartogram to show the distribution of the population. The icons differentiate people depending on whether they are in work, but even on the cartogram it is difficult to

The Map Reader: Theories of Mapping Practice and Cartographic Representation, First Edition. Edited by Martin Dodge, Rob Kitchin and Chris Perkins.

Vital Statistics	Farm	Town	City	Total
Area	10	6	5	2060
People	10	30	60	100
Density	1	5	15	5
Working	5	10	15	30
Dependency	50%	33%	25%	30%

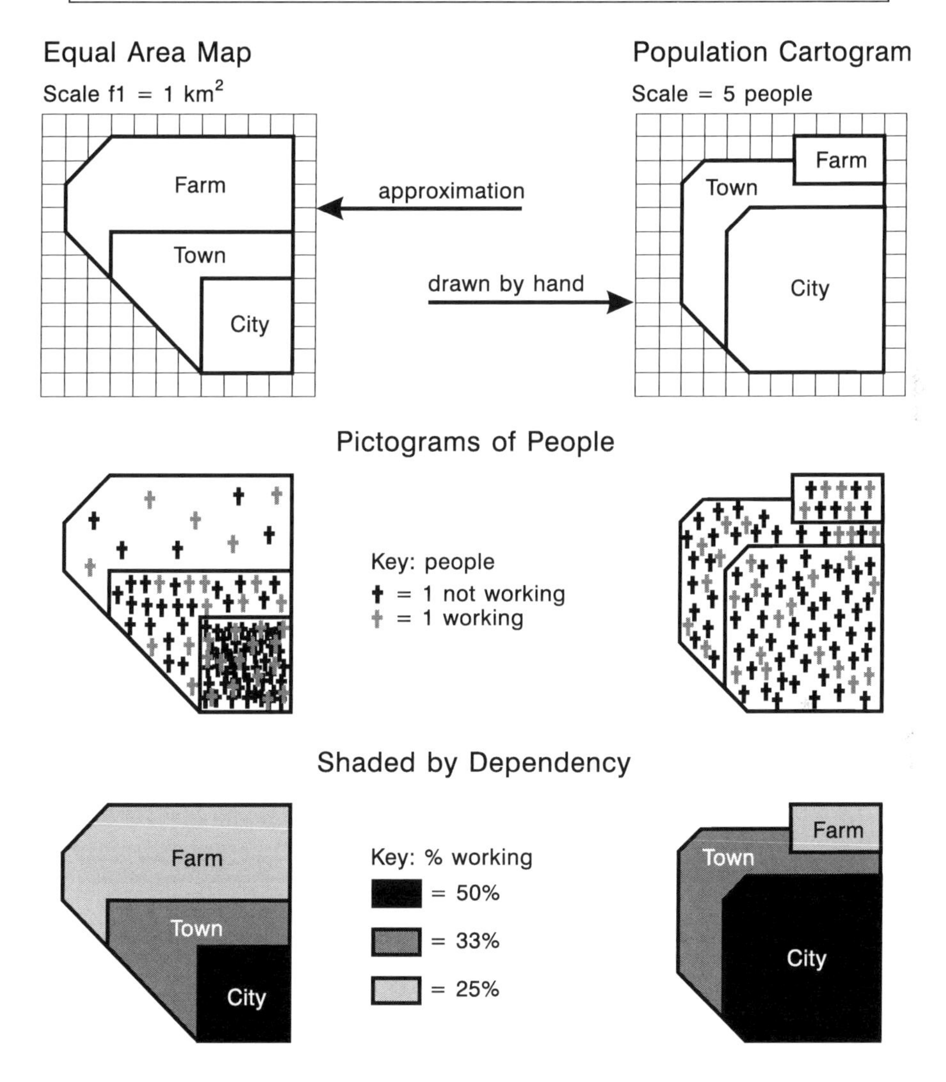

Figure 3.7.1 Example of a cartogram created and shaded by hand for a fictional island.

tell where workers are more numerous from these icons. The final pair of equal area map and population cartogram in the figure has each district shaded by the proportion of the population who are working (the dependency rate). Clearly the map and cartogram give very different impressions. One shows that on most land many people are working while the other shows that many people live in areas where most people do not work. [...]

People are unevenly distributed over the surface of the earth. This has always been so and is true no matter how closely these geographical patterns are studied. It appears that the ways in which people organise their lives require them to be spread unevenly over space. This presents a problem for looking at how their lives are organised spatially as, if conventional maps are used at any level of detail, the maps will always highlight the uneven distribution of people over the land rather than the differences between groups of people, which are of more interest. [...] Most cartograms are designed to make visualising the detailed human geography of people possible and sometimes even to make it easy.

Ever since the first maps were drawn, cartographers have had to decide how they will distort the shape of the earth to best show what they wish others to study. A frequent criticism of cartograms is that even cartograms based upon the same variable for the same areas of a country can look very different. There is nothing new about that in cartography. [...] Mapping involves making compromises between conflicting goals which result in the variety of views that we have of the world. Inevitably, they alter how we see different parts of this world. [...] With cartograms, distortion is not seen as unfortunate, but is deliberately used to convey information.

Methods

There are numerous methods of creating cartograms. [...] Each method produces useful maps which could not have been created by the alternative techniques. In general, manual methods are useful when only a few areas are being represented, while computer algorithms have to be used to produce area cartograms of many places. Mechanical methods provided a compromise between these two options in the past. [...]

Physical accretion models

The term 'physical accretion models' was first introduced by John Hunter and Johnathan Young. [...] They attempted to formalise the approach taken to produce the cartograms of British general election results drawn in *The Times* newspaper in both 1964 and 1966 (Hunter and Young 1968). The term physical accretion model describes the technique of constructing cartograms by using wooden tiles and rearranging them by hand. In the case of the old counties of England and Wales, 9214 wooden tiles were painted 62 different colours and were then manipulated to create a cartogram which was 2.4 square metres in size before being copied onto paper. The manipulation alone took 16 hours and involved attempting to maintain the shape of prominent features, such as estuaries, and preserving contiguity, as well as making area almost exactly proportional to population. In constructing that cartogram Hunter and Young began with London and proceeded northwards. [...]

Hunter and Meade (1971) discuss three-dimensional population models and how these cartograms can be constructed in the classroom. Many thousands of school children may have used this method and hundreds of undocumented cartograms could thus have been produced!

Mechanical methods

Only a year after John Hunter and Melinda Meade produced their series of cartograms created by combining wooden blocks, a paper was published [...] which illustrated a mechanical method of making population cartograms. The [...] remit was to create a map of Canada in which the divisions of the 1966 Census were drawn with their areas in proportion to their populations (Skoda and Robertson 1972). Furthermore, this map had to attempt to approximate the actual shape of places, maintain contiguity, include lines of longitude and latitude and the census tract boundaries of the twelve urban areas with populations in excess of 200 000 people. There were 266 census divisions to be included for the map of all of Canada and a further 1408 census tracts for the large cities. [...]

To create the isodemographic map of Canada, Skoda and Robertson developed a mechanical model in which 265 000 steel ball bearings, each with a diameter of one eighth of an inch, and each representing 140 people on the main map (and 70 people on the separate city maps) were poured into an equal land area map of Canada constructed from brass hinged aluminium strips. The ball bearings were weighed rather than counted and the whole model was assembled on several five metre two-plywood boards surfaced with Formica, which allowed the balls to move easily within the hinged metal district boundaries. [...] As the ball bearings were pressed down to form a continuous layer, correct contiguity was maintained – while a population cartogram was created. Transparent perspex sheets were used to apply an even pressure to all the ball bearings across the map and to prevent spillage. The authors had some licence in the design process and used this to attempt to keep the final result as conformal (shape preserving) as possible, [...] and the model resulted in a very accurate isodemographic map of Canada and particularly detailed population cartograms being produced of the major cities. [...]

The rising interest in urban geography in the 1960s and early 1970s was cited as one reason why this cartogram might be used by other researchers (Skoda and Robertson 1972: 25). [...] The way in which cartograms can radically change people's views of the world was being explored in some detail. In practice, however, there is little evidence that this cartogram was ever used to map social or economic variables. Just five years after the isodemographic map was printed, Canadian cartographers were publishing maps of general elections by using conventional equal land area projections (Coulson 1977). The one factor which advocates of cartograms have consistently underestimated has been the hostility to changing from conventional projections, and the more a cartogram is needed, such as in mapping aspects of the unevenly distributed human geography of Canada, the more unconventional that cartogram will look.

Competing cartogram algorithms

A year after Skoda and Robertson released the details of their mechanical model for constructing cartograms, a method which used a computer program was published (Tobler 1973). [...] To summarise Tobler's method, an estimate was made of the population living in each cell of a grid which had been laid over the United States of America. [...] The position of each vertex in the grid was then repeatedly altered according to the populations in, and the areas of, the four cells which surrounded it, until there was little improvement in the accuracy of the cartogram to be achieved by further iterations. [...]

Although the cartogram [...] was one of the first which had been created by computer to be published, Tobler had been working on algorithms to produce cartograms for several years before publishing his results (Tobler 1961) and has produced numerous examples since 1973. [...] Slight alterations to one algorithm can result in very different cartograms being drawn of the same places, even when the same statistics are used. All other computer algorithms for producing cartograms suffer from the same problem. All computer algorithms appear to produce differently shaped cartograms of the same distribution, often because of differences in the type of data being input, as well as due to differences between the algorithms themselves.

The propensity of slight alterations to a basic algorithm to produce very different end maps has resulted in many researchers claiming to have produced better algorithms since Tobler's first suggestions were published. [...] For example Guseyn-Zade and Tikunov (1993, 1994) published details of a 'new technique for constructing continuous cartograms'. Although there is a vast improvement in the speed of convergence of this method over others and the results may be more pleasing visually, this algorithm can be viewed as an improved version of the first implementation of a computer algorithm made by Tobler over thirty years ago. This is because all the methods described here have two common features. Firstly, they attempt to produce a numerical approximation to a pair of equations which cannot be solved by analytical means, and they attempt to do this through many small adjustments to the vertices of a digitised map (Tobler 1973: 216). Secondly, they cannot yet be used to produce cartograms as detailed as those produced by manual or mechanical methods.

Cellular automata cartograms

In 1968, John Conway, a mathematician at the University of Cambridge, invented the 'Game of Life', which was later credited with popularising the study of Cellular Automata (Toffoli and Margolus 1987). [...] A variant of the game can be developed to grow cartograms. The method would formally be called a cellular automata machine. [...] A grid of cells is set up in which each cell is assigned to one of the regions to be represented on the map, which is to have its area, eventually, in proportion to its population. To achieve this, regions represented by too few cells have to gain cells from regions represented by too many. [...] The checkerboard method of updating the cell values referred to in the algorithm is a standard technique (Toffoli and Margolus 1987: 118) and the stipulation that cells should be most prone to change state if they have few neighbours of their own kind, and less likely to change if they have many neighbours, causes the algorithm to simulate a process called 'annealing'. Cells on the corners of regions are most prone to be replaced by the cells of other regions and so the regions as a whole tend to form simpler shapes as the lengths of the boundaries drawn between them are minimised. However, under this implementation that advantage is partly offset by not altering cells if they are originally assigned to the sea (region '0'). Thus the shape of the coastline is maintained while contiguity is assured and a correct cartogram is produced. [...]

A major problem with preserving the coastline is that the complexity of boundaries within the cartogram is increased. [...] To overcome this problem the algorithm could be altered to allow cells in the sea to change state, but that might result in the loss of any coastline features, which are useful for identifying areas on the cartogram. A simpler solution is to transform the original base map to form a pseudo-cartogram, Here lines of latitude and longitude have been squeezed together or pulled apart to approximate an equal population cartogram, while those lines remain straight and parallel. When the algorithm is applied to this base a more visually acceptable solution is reached. Because this cartogram is continuous, every region is still connected to its original neighbours and no others and so this cartogram can be used to re-project a grid of equal population sized 'squares' onto the map of Britain. [...] This method of projection [...] does not result in a conformal map, nor does it necessarily produce a simple solution, but the final result always maintains contiguity.

Circular cartograms

Cartograms which can be shown to have good mathematical properties, such as approximating a conformal projection, are not necessarily the best options for cartographic purposes. If, for instance, it is desirable that areas on a map have boundaries which are as simple as possible, why not draw the areas as simple shapes in the first place? [...] The antecedents of this approach for making cartograms can be seen in work by Härö (1968: 456) [...] Johnston *et al.* (1988: 340) and Howe (1970) also show examples of cartograms which define areas as simple shapes for use as a base map. There is a long tradition of mapping with circles as symbols in cartography and the advantages of using this simple shape are well known (Dent 1972). [...] The metaphor for this algorithm is [...] a development of the programs which simulate the orbits of stars and planets.

Each region is drawn as a circle with its area in proportion to its population and is then treated as an object in a gravity model which is repelled by other circles with which it overlaps, but is attracted to circles which were neighbouring regions on the original map. Forces akin to forces of gravity are calculated, including velocity and acceleration, and the whole process is acted out as if it were occurring in treacle, to avoid any circles moving too quickly. [...] If this algorithm is left to run on a normal map for a few hundred iterations it eventually produces a solution in which no circles overlap and as many as possible are still in close contact with their original geographic neighbours. Some of the original coastline features are even retained [...] but a true continuous cartogram is not produced, as this is not possible where each place is drawn as a regular shape. [...] This cartogram has the advantage over more sophisticated forms of showing quite clearly how the populations of areas vary, without having to compare very complex shapes in more rural areas. The dominance of the major cities in the population structure of Britain is emphasised [...] which uses a functional rather than an administrative definition of cities. As the number of areas is increased the proportion of contacts which break topology decreases. This is one of the major advantages of this method of making cartograms – the picture becomes more accurate as more detail is added. [...]

A major problem with manual, mechanical and computer-based methods of producing continuous area cartograms is that the time this takes often escalates exponentially as the total number of individual areas to be represented rises. Computer algorithms are also prone to producing bottlenecks, which prevent the program progressing to an accurate solution when there are many area to be mapped. The time taken by the circular algorithm described here to converge to a sufficiently accurate result is in almost direct proportion to the number of areas being mapped [...] The maximum number of areas it has been used to map so far exceeds one hundred thousand (Dorling 1993).

Applications

Political cartography

[...]

There are distinct subject areas where the use of cartograms is particularly common. The most obvious, perhaps, is in mapping the results of elections (Upton 1994). To map the outcome of political elections on a traditional map may appear foolhardy, although this has not prevented entire atlases of election results being produced using equal land area maps (see Waller 1985, and, for a defence of the method, Kinnear 1968: 12). [...]

Use of cartograms often, of course, dramatically alters the impression of which party has won an election [...] but one aspect of political cartography which makes the use of cartograms difficult is that of frequent boundary changes. In Britain the boundaries of constituencies were altered many times before 1955 and there were large-scale redistributions of seats after the 1970, 1979 and 1992 general elections. When cartograms are drawn by computer this need not present a problem, but political cartograms drawn by hand can quickly become obsolete (Dorling 1994). A popular method for designing cartograms to study British general election results is to have each constituency drawn as a square. [...] Although there are cartographic advantages to raising simple symbols [...] their use in cartograms of Britain has often meant that the constituencies of London had to be shown as an inset (Johnston *et al.* 1988). [...] Interest in cartograms, in at least both Britain and America, tends to surge around the time of national elections (particularly when urban-based parties win and also just after population censuses have been taken). [...]

Early epidemiology

[...] Many doctors have been struck by the idea that they could learn more about disease through mapping. The earliest example often quoted is of a map of the incidence of cholera in London (Snow 1854), from which map a pump in Broad Street was identified as being at the centre of the clusters of cases of cholera in the 1848–1849 epidemic. The pump handle was removed. A cartogram of the area has yet to be constructed from the 1851 census of population data, so whether that pump was actually at the centre of a real cluster is still unknown. The first cartogram of London known to this writer is an 'epidemiological map' which was produced by a doctor working for the then London County Council Department of Public Health (Taylor 1955). The cartogram contained crosses, drawn in the borough rectangles, to show the incidence of polio during the 1947 epidemic. Because the rectangles were each drawn with the same height, their widths, as well as their areas, are proportional to population. The borough with the highest rate of polio and, hence, the tallest column of crosses in the figure was Shoreditch. Almost exactly one hundred years separates the two London epidemics, which were first drawn on a map and cartogram, respectively. Cartograms showing distributions within countries came later.

A claim was made to have produced the first cartograms showing national disease distributions only a decade after the crude cartogram of London was first drawn. The nation was Scotland and a separate cartogram was constructed by hand for each of eight age–sex groups (Forster 1966: 165). The author of these cartograms concluded that a national series of cartograms should be produced for each age–sex group for use in epidemiological studies in Britain. This was never done.

Mapping mortality

A National Atlas of Disease Mortality in the United Kingdom was published in 1963 under the auspices of the Royal Geographical Society; the atlas contained no cartograms. [...] Fortunately, a revised edition was published a few years later which made copious use of a 'demographic base map' (Howe 1970: 95). It is interesting to note that when the revised edition was being prepared, the president of the Society was Dudley Stamp who believed that 'The fundamental tool for the *(sic)* geographical analysis is undoubtedly the map or, perhaps more correctly, the cartogram' (Stamp 1962: 135). In the cartogram, which was used in the revised national atlas, squares were used to represent urban areas while diamonds were used to show statistics for rural districts. No attempt was made to maintain contiguity, but a stylised coastline was placed around the symbols, which were all drawn with their areas in proportion to the populations at risk from the disease being shown on each particular cartogram.

[...] What differentiates medical uses of cartograms most from political cartography is the mapping of clusters – individual cases of a disease or death which together might possibly be connected. [...] This process has been used to illustrate how cases of Salmonella food poisoning occurring in Arkansas in 1974 were not unduly clustered in Pulaski county (Dean 1976). Cartograms have been shown to be useful in determining that certain types of disease which appear clustered are in fact quite evenly distributed across the population. In more recent years researchers have turned their attention to trying to develop cartograms upon which actual, rather than illusory, clusters of disease can be identified.

The major problem with using population cartograms to identify clusters of disease is that the choice of which areas are closest to which on a cartogram can be quite arbitrary. For instance, if the same set of incidences of one particular disease were plotted on three different cartograms of America, then different parts of the country may appear to have dense clusters of cases depending on which cartogram was chosen. This would be true regardless of whether the clusters were to be identified by eye or by statistical procedures. [...] The proposition that there is no single 'true answer' as to whether a disease is clustered does not go down too well in some circles. Because of this problem a group of researchers at Berkeley developed a computer algorithm for identifying incidences of disease (Selvin *et al.* 1984). The algorithm they developed was not very different from other 'continuous transformation' cartogram algorithms and was still slow enough to warrant testing on a Cray supercomputer! (Selvin *et al.* 1988: 217). However, the authors of the algorithm do claim that their program would result in a unique transformation given an infinite amount of input data. [...] The algorithm was used to [...] create a cartogram of San Francisco county, upon which apparent clusters of disease were shown to be false (Selvin *et al.* 1988: 217). However, application of the method to another Californian County did provide evidence of some clustering of high cancer rates near oil refineries (Selvin *et al.* 1987).

Transformed flows

The term 'flows' covers a wide range of subjects. For instance, workers can be seen to 'flow' between different states of employment and votes are said to 'flow' between political parties. There are flows in medical mapping, such as were used to plot the movement of the influenza epidemic which spread across England and Wales in 1957 (Hunter and Young 1971). [...]

Commuting flows are amongst the most simple to map because people usually do not commute very far to work in a day so the flows tend to be short. [...] Which flows are shown depends on how a flow is defined to be significant and on the nature of the areas which are used to count the flows. In the centre of London there are many very small wards and hence fewer strong patterns can emerge. The final results of flow mapping depend both on the nature of the flows being mapped and on the way in which those flows are mapped.

[...] Although researchers have considered creating cartograms in which distance is proportional to time, few travel-time cartograms have been created (Angel and Hyman 1972). Almost all travel-time cartograms which do exist show travel time only from a single point using the fastest method of travel. [...]

A cartogram could be created in which travel-time distances between all points was shown; this would result in a complex surface being drawn but it should not be impossible to achieve. On this surface cities would appear as mountains, as commuters fight their way through the rush hour to enter them. The deep gorges of inter-city railway lines would cut through them, or perhaps appear as tunnels coming out of the mountains. The surface could be drawn upon the two-dimensional base of an equal population cartogram and so both time and population could be measured from its geometry. And, if a different surface were drawn for different times of the day or for different days of the year, those surfaces could be combined as a space-time volume in which distance is proportional to hours and volume is proportional to human lives. [...]

Conclusions

[...] I have attempted to illustrate how the use of cartograms is widespread in academic and popular writing and how cartograms are not as difficult to create as many readers might think. No commercial geographical information system exists which will create cartograms from scratch, but that is largely due to a perceived lack of demand and, perhaps, due to an inflated view of how difficult it is to implement computer programs to create cartograms. [...] In this final section some of the more complex uses to which cartograms can be put are illustrated [...] to show how the use of cartograms can shed light on complex spatial processes.

The evolving structure of British social geography, looking in detail at unemployment, voting and social segregation, [can be] visualised using cartograms. [...] Between 1955 and 1970 there were 630 constituencies in the country. [...] At this time seats where Labour came second were almost all won by the Conservative party and *vice versa.* However, the following six general elections saw a very different pattern emerge. For the general election of February 1974 the number of parliamentary constituencies was increased to 635. It increased again to 650 in 1983 and to 651 in 1992. [...] The rise and geographical spread of both the Liberals and the Nationalists in this period are quite striking (the Social Democrat Party are included with the Liberals here). Cartograms show that although the last four general elections were all won by the Conservative party, their share of seats in which they came second was falling over time. Most importantly, the way in which the main opposition parties had established clear geographical territories by the aftermath of the last general election is also made evident (Dorling 1994). The geographical evolution of political preferences over eleven general elections can be shown on a single A5 page if seats are drawn in proportion to the number of people who can vote.

[...] The longest series of unemployment rates for small areas extends back to 1978 for places known as 'amalgamated office areas' (which are groups of unemployment benefit offices). There are just over 850 of these areas and the number of people claiming unemployment benefit in each of them was recorded by the Department of Employment in almost every month over the sixteen years. [...] In this illustration the changing spatial distribution of unemployment in Britain is depicted using these statistics. The shading of each place at each time reflects its unemployment rate then, compared to its average unemployment rate over the whole period (for more details see Dorling 1995b). [...] The worst period for Devon, Cornwall and Wales was 1978–1979, for Scotland 1978–1979 and 1988–1989, but for the south east 1991–1993. For the latter region, the maps clearly show how the blight of unemployment took hold in central London in 1987 and spread out across south-east England in subsequent years. Compare this pattern to the rise in Conservative party losses in London at the last general election and the return of the Labour party as the main political contender between the capital and the midlands. The use of cartograms allows possibly unexpected connections to be seen between different spatial processes where the importance of events is weighted by how many people are involved. [...]

It would be misleading to think of the north of the country as economically prosperous in static terms at the start of the 1990s. Although the south did experience particularly high rates of unemployment then, unemployment is not the only way in which loss of employment is expressed. Increasingly, levels of permanent sickness and early retirement rose in Britain during the last decade and, in places, these factors resulted in more people of working age being out of employment than did unemployment alone.

A cartogram of over ten thousand wards in Britain (part-postcode sectors in Scotland) allocates [wards] to one of sixteen categories depending on the proportion of people of working age who were either unemployed or were retired or permanently sick in that ward. Thus over twenty thousand statistics are used [...], two for each ward resulting in a bivariate shading scheme. [...] Almost twice as many statistics are being used as in the last [example]. [...] In some wards over 10% of the working population was unemployed and over 10% was prematurely retired or permanently sick. [...] A very clear north/south and urban/rural divide still separates the people of Britain when shown at this level, despite the equalizing effects of the last recession. The pattern for each of the other fifteen categories of ward [...] also tells a story about the spatial social structure of this country [...] (Dorling 1995a; Dorling and Woodward 1996).

The relative levels of four groups of people within each ward [can also be shown in cartograms]. [...] The four groups are: people who were working in managerial or professional jobs, [...] people in supervisory or artisan type jobs, [...] people in semiskilled or strictly supervised jobs [...] and people looking for work who had not been employed in the last ten years [...]. Wards where a disproportionate share of the workforce belongs to a single one of these groups are given its 'primary' colour. Wards where two groups are disproportionately represented are shaded by the mix of those colours so that a ward with many managers and artisans would be shaded light green, and so on. An excess of workers with no occupation in an area (the long-term unemployed and people in the armed forces) changes the shade from light to dark. This mixing results in fourteen possible categories of ward [...]

If different levels of each group were distinguished, rather than just over- or under-average levels being used, then many hundreds of shades would be needed; but already with this minimal number the pattern which is revealed is complex. Around almost every city in Britain are localities containing disproportionate numbers of single groups. A majority (61%) of the population live in areas where either one or two of the three groups in employment are over-represented and the other two or three groups are under-represented. [...] This cartogram contains a detailed picture of the way in which people in Britain are spread over the country according to the kind of work they do or have done. On a conventional map the red, purple and black would appear to occupy very little space. What is unusual here is the level of detail which is included in this series of colour cartograms. However, these show detail which is still nowhere near the limit possible with high quality colour printing and computer animation.

Population cartograms are, almost by definition, economical with space, but it is still surprising to see how much land area in a country like Britain is sparsely populated, and hence how much detail can be afforded to the cities when the countryside is shrunk. Whatever you choose to use cartograms for, from studying participation in elections, to the spread of a disease, or the social structure of a country, the very different perspectives they show are likely to alter the way you imagine the processes behind these patterns to be operating. In human geography, mapping with population cartograms changes the perspective from concentrating on the populations of the countryside to the inhabitants of towns and cities where most of the people live. The last cartogram to appear in *The Times* newspaper was drawn on 4 April 1966 following the Labour victory, which was orchestrated from the cities. When the Conservative party regained power at the following general election the newspaper reverted back to a conventional projection, which made the result appear like a landslide for the party of the right (*The Times*, 22 June 1970). Claiming space on maps is as much a political process as a technical one. People who are seen not to matter are often neither counted nor shown in studying society.

References

Angel, S. and Hyman, G.M. (1972) Transformations and geographic theory. *Geographical Analysis*, **4** (4), 350–367.

Coulson, M.R.C. (1977) Political truth and the graphic image. *The Canadian Cartographer*, **14** (2), 101–111.

Dean, A.G. (1976) Population-based spot maps: an epidemiologic technique. *American Journal of Public Health*, **66** (10), 988–989.

Dent, B.D. (1972) A note on the importance of shape in cartogram communication. *The Journal of Geography*, **71** (7), 393–401.

Dorling, D. (1993) Map design for census mapping. *The Cartographic Journal*, **30** (2), 167–183.

Dorling, D. (1994) Bringing elections back to life. *Geographical Magazine*, **66** (12), 20–21.

Dorling, D. (1995a) *A New Social Atlas of Britain*, John Wiley & Sons, Ltd, Chichester, UK.

Dorling, D. (1995b) The changing human geography of Britain. *Geography Review*, **8** (4), 2–6.

Dorling, D. and Woodward, E. (1996) Social polarisation 1971–1991: a micro-geographical analysis of Britain. *Progress in Planning*, **45** (2), 67–122.

Forster, F. (1966) Use of a demographic base map for the presentation of areal data in epidemiology. *British Journal of Preventative and Social Medicine*, **20**, 165–171.

Guseyn-Zade, S.M. and Tikunov, V.S. (1993) A new technique for constructing continuous cartograms. *Cartography and Geographic Information Systems*, **20** (3), 167–173.

Guseyn-Zade, S.M. and Tikunov, V.S. (1994) Numerical methods in the compilation of transformed images. *Mapping Sciences and Remote Sensing*, **31**, 66–85.

Häro, A.S. (1968) Area cartogram of the SMSA population of the United States. *Annals of the Association of American Geographers*, **58**, 452–460.

Howe, G.M. (1970) *National Atlas of Disease Mortality in the United Kingdom*, revised and enlarged edn, Thomas Nelson and Sons, London.

Hunter, J.M. and Meade, M.S. (1971) Population models in the high school. *The Journal of Geography*, **70**, 95–104.

Hunter, J.M. and Young, J.C. (1968) A technique for the construction of quantitative cartograms by physical accretion models. *Professional Geographer*, **20** (6), 402–407.

Hunter, J.M. and Young, J.C. (1971) Diffusion of influenza in England and Wales. *Annals of the Association of American Geographers*, **61** (4), 637–653.

Johnston, R.J., Pattie, C.J. and Allsopp, J.G. (1988) *A Nation Dividing? The Electoral Map of Great Britain 1979–1987*, Longman, London.

Kinnear, M. (1968) *The British Voter: An Atlas and Survey Since 1885*, BT Batsford, London.

Selvin, S., Merrill, D., Sacks, S. *et al.* (1984) Transformations of maps to investigate clusters of disease. Lawrence Berkeley Lab Report LBL-18550.

Selvin, S., Shaw, G., Schulman, J. and Merrill, D. (1987) Spatial distribution of disease: three case studies. *Journal of the National Cancer Institute*, **79** (3), 417–423.

Selvin, S., Merrill, D., Schulman, J. *et al.* (1988) Transformations of maps to investigate clusters of disease. *Social Science and Medicine*, **26** (2), 215–221.

Skoda, L. and Robertson, J.C. (1972) Isodemographic map of Canada. Geographical Paper No. 50, Lands Directorate, Department of the Environment, Ottawa, Canada.

Snow, J. (1854) *On the Mode of Communication of Cholera*, 2nd edn, Churchill, London.

Stamp, L.D. (1962) A geographer's postscript, in *Taxonomy and Geography* (ed. D. Nichols), The Systematics Association, London, pp. 153–158.

Taylor, I. (1955) An epidemiology map. *Ministry of Health Monthly Bulletin*, **14**, 200–201.

Tobler, W.R. (1961) Map transformations of geographic space. Unpublished PhD thesis. Department of Geography, University of Washington.

Tobler, W.R. (1973) A continuous transformation useful for districting. *Annals of the New York Academy of Sciences*, **219** (9), 215–220.

Toffoli, T. and Margolus, N. (1987) *Cellular Automata Machines*, MIT Press, Cambridge, MA.

Upton, G.J.G. (1994) Picturing the 1992 British general election. *Journal of the Royal Statistical Society Series A*, **157** (2), 231–252.

Waller, R. (1985) *The Atlas of British Politics*, Croom Helm, London.

Further reading

Dorling, D., Newman, M., and Barford, A. (2009) *Atlas of The Real World*, Thames and Hudson, London. [Copious number of world cartograms and statistical explanations from Dorling and colleague's *WorldMapper* project that links the Gastner and Newman algorithm to UN demographic datasets to reveal striking cartograms of global inequality – see also www.worldmapper.org/about.html.]

Gastner, M.T. and Newman, M.E.J. (2004) Diffusion-based method for producing density-equalizing maps. *Proceedings of the National Academy of Sciences*, **101** (20), 7499–7504. [The publication of significant new algorithm for the automated generation of density-equalizing but shape-preserving cartograms.]

Raisz, E. (1934) The rectangular statistical cartogram. *Geographical Review*, **24**, 292–296. [One of the earliest academic papers describing the construction and utility of cartograms.]

Tobler, W. (2004) Thirty-five years of computer cartograms. *Annals of the Association of American Geographers*, **94**, 58–73. [A critical and wide-ranging overview of the developmental history of cartograms from the most influential figure in their invigoration during the early days of automated cartography.]

See also

- Chapter 2.5: Automation and Cartography
- Chapter 2.9: Emergence of Map Projections
- Chapter 3.4: Generalisation in Statistical Mapping
- Chapter 3.8: ColorBrewer.org: An Online Tool for Selecting Colour Schemes for Maps
- Chapter 3.12: The Geographic Beauty of a Photographic Archive

Chapter 3.8

ColorBrewer.org: An Online Tool for Selecting Colour Schemes for Maps

Mark Harrower and Cynthia A. Brewer

Editors' overview

Choosing effective colour schemes for thematic maps is surprisingly difficult. ColorBrewer is an online tool designed to take some of the guesswork out of this process by helping users select appropriate schemes for their specific mapping needs by considering: the number of data classes; the nature of their data (matched with sequential, diverging and qualitative schemes); and the end-use environment for the map (e.g. CRT, LCD, printed, projected, photocopied). ColorBrewer contains 'learn more' tutorials to help guide users, prompts them to test-drive colour schemes as both map and legend, and provides output in five colour specification systems.

Originally published in 2003: *The Cartographic Journal*, **40** (1), 27–37.

Introduction

Colour plays a central role in thematic cartography. Despite this, using colour effectively on maps is surprisingly difficult and often exceeds the skill and understanding of novice mapmakers. On the one hand, a 'good' colour scheme needs to be attractive. On the other hand, the colour scheme must also support the message of the map and be appropriately matched to the nature of the data. Moreover, colour schemes that work for one map (e.g. a choropleth map of income) will not necessarily work for another (e.g. a map of dominant commercial sector by county). Relying on the same colour scheme for all thematic mapping needs is a mistake. Unfortunately, many novice mapmakers have become conditioned to use the default colour schemes built into commercial mapping and GIS packages, and sometimes are even unaware that they can change those default colours. If colour schemes are not carefully constructed and applied to the data, the reader may become frustrated, confused, or worse, misled by the map.

An understanding of how to manipulate the three perceptual dimensions of colour (hue, saturation and lightness) is required to create attractive and logical colour sequences for thematic maps. These dimensions are three of the visual variables or basic 'building blocks' of graphics (Bertin 1983). Nearly a century of colour theory development combined with perceptual testing has allowed us to better understand how individuals perceive and understand colour. From a semiotic perspective, colour works as a 'sign vehicle' when the map reader is able to understand that a specific colour on the map represents or *stands for* something in the real world and that multiple occurrences of that colour stand for the same kind of thing (MacEachren 1995).

The problem

Federal agencies, such as the US Census Bureau, National Center for Health Statistics and National Cancer Institute, produce numerous maps for both internal and external use (Brewer and Suchan 2001; Pickle *et al.* 1996; Devesa *et al.* 1999). As collectors and distributors of geospatial data, these agencies make extensive use of thematic maps – especially choropleth maps –to communicate facts to the

The Map Reader: Theories of Mapping Practice and Cartographic Representation, First Edition. Edited by Martin Dodge, Rob Kitchin and Chris Perkins.

public and to facilitate exploratory data analysis by their own researchers. The individuals who produce these maps rarely have the time to worry about carefully crafting individual colour schemes every time they make a map. Although most GIS software incorporates default colour schemes for thematic maps, it provides no guidance on how to best use these default colour schemes, and many of these schemes are simply unattractive. More importantly, these default schemes are not appropriate for all mapping tasks.

Given the importance of maps in the activities of these agencies, and the central role colour plays on maps, poor use of colour is a concern. Agencies, and more generally novice mapmakers, would benefit from tools that could quickly guide them through the colour selection process and let them 'test drive' a colour scheme and see it on a map before committing to it. Such tools would let them work more *quickly* and *confidently* by reducing the chance of misapplied colour schemes on thematic maps.

The solution: www.ColorBrewer.org

This paper outlines the development of ColorBrewer, an online tool that helps users identify appropriate colour schemes for maps. The reader is encouraged to try ColorBrewer online at www.ColorBrewer.org. The impetus for this work came from watching individuals that produce thematic maps at federal agencies struggle with selecting good colour schemes for maps. ColorBrewer is designed to take some of the guesswork out of this process. The system suggests possible colour schemes by prompting the user to identify how many data classes they have and what kind of colour scheme best suits their data and map message (i.e. sequential, diverging or qualitative; explained below).

ColorBrewer also provides guidance for varied display environments – CRT, laptop, colour laser print, photocopy and LCD projector – and warns where and when a colour scheme might fail. It is important to remind users that the success of a colour scheme is, in large part, a product of the display medium. For example, colours that are easily differentiated on a cathode-ray tube (CRT) monitor may look indistinguishable on a laptop LCD display. As anyone who has made colour prints knows, startling shifts in the appearance of colours can occur when moving from electronic displays to printed displays. Compounding this, many of the maps produced by federal agencies must work across multiple media (e.g. both as an electronic display and as a paper report) and designing truly robust colour schemes that work well in multiple environments is difficult.

In addition to simply suggesting colour schemes, ColorBrewer allows the user to 'test drive' each of the colour schemes to see how well they perform as both an ordered legend and as a choropleth map, in order to give the user a better sense of how an individual scheme will function when it is applied to a real map. Differentiating colours within a complex distribution – especially when additional map information such as place names and line-work is overlaid – is a harder perceptual task than seeing differences in simple and logically ordered data legends. Put another way, just because a map reader can see subtle differences between the individual colour patches in a legend does not mean they will be able to recognise those same differences on the map (Brewer 1997).

The appearance of colours can change with the presence or absence of enumeration unit borders, and the colour of those borders. To accommodate this, ColorBrewer allows users to turn enumeration borders on and off, as well as change their colour. Our goal was not to suggest what border colours to use, but to remind the user that this is yet another component to map design. Since thematic maps often contain additional base information, similar functionality has been built in with an overlay of roads and cities to demonstrate the effects of additional line-work and point symbols on the appearance of the colours. The colour of these overlays can also be changed. Though the examples we have chosen are highways and cities, they should give users a good idea of how other line-work, points and text will function on their map. Some colour schemes make it easier to read base information than others, and this legibility may be more important for some map purposes than others.

Not all maps are displayed with a white background. Black backgrounds, for example, have become common in on-screen geovisualisation environments. Since the appearance of a specific colour is influenced by the colours that surround it, ColorBrewer allows users to adjust the colour of the background display to determine how well different colours function against different backgrounds. In short, the ColorBrewer system:

- asks the user to specify the number of classes and type of colour scheme;
- displays a selection of colour schemes for the user to choose among;
- presents the selected colour scheme on a map and in a legend display;
- displays guidance for each scheme on the potential for good results in different media;
- suggests whether a selected scheme will accommodate colour-blind people;
- allows the user to change colours of enumeration unit borders (or remove them);

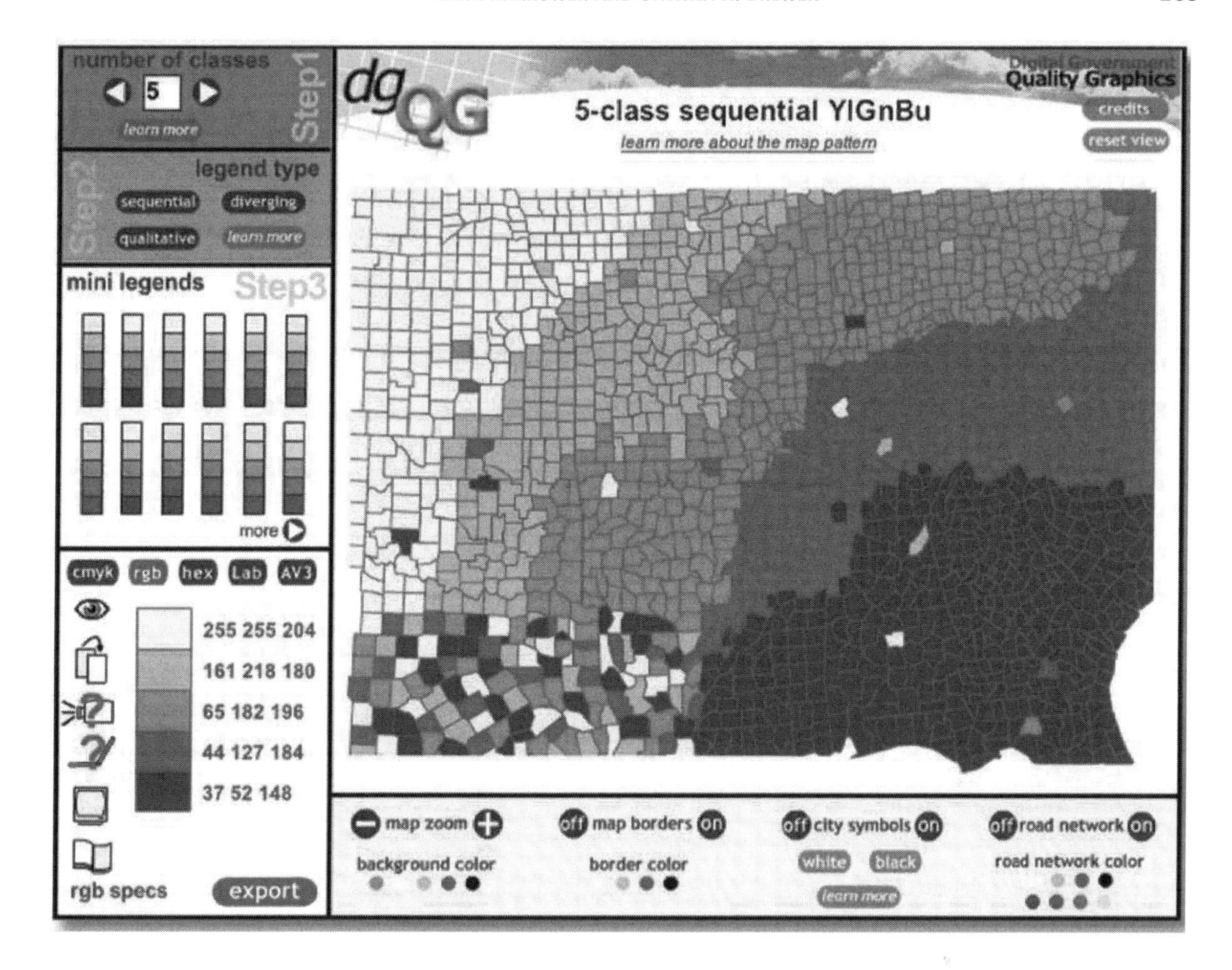

Figure 3.8.1 ColorBrewer contains 35 colour scheme sets that can be used to create legends for maps ranging from 3 to 12 classes. The sets are organised into three kinds of colour scheme: (a) sequential, (b) diverging and (c) qualitative. (See Colour Plate Four.)

- displays an overlay of sample line, point and text base information;
- displays a variety of map background colours.

Beyond our description of system basics, a more in-depth discussion of the rationale behind and implementation of ColorBrewer follows in the next section.

Kinds of colour schemes

A total of 35 colour scheme 'sets' are contained in ColorBrewer and they are divided into three groups: qualitative, sequential and diverging. These sets have been designed to produce attractive colour schemes of similar appearance for maps ranging from three to twelve classes. Not all colour schemes can be expanded to 12 classes, and many of the sequential scheme sets, for example, stop at nine classes because further divisions within the scheme are perceptually unreliable.

Colour can be used to imply categorical differences (e.g. forest, city, marsh) or to imply ordered differences (e.g. population density rates). [...] The kinds of data appropriately matched to these types of schemes are outlined in the following sections (Figure 3.8.1; see Colour Plate Four).

Sequential colour schemes

Sequential colour schemes imply order and are suited to representing data that range from low-to-high values either on an ordinal scale (e.g. cold, warm, hot) or on a numerical scale (e.g. age classes of 0–9, 10–19, 20–29, etc.). Lightness steps dominate the look of these schemes, usually with light colours for low data values and dark colours for high values. 'Dark equals more' is a standard cartographic convention. Sequential schemes can be either single hue (e.g. same blue, with different lightness and saturation levels) or multi-hued (e.g. light yellow through dark green). ColorBrewer includes 12 multi-hued sequential schemes and six single-hued schemes. We included more perceptually-graded multi-hue sequential colour schemes in ColorBrewer for two reasons: (1) they provide better colour contrast between classes and (2) they are more difficult to create than single-hue schemes because all three dimensions of colour are changing simultaneously. Moreover, since the default sequential colour schemes in commercial mapping and GIS packages are usually single hued, we felt the inclusion of multi-hued schemes would help novice map makers who wished to use more sophisticated colour schemes.

Diverging colour schemes

Diverging colour schemes should be used when a critical data class or break point needs to be emphasised. The break or class in the middle of the sequence is emphasised by a hue and lightness change and should represent a critical value in the data, such as the mean, median or zero. For example, a choropleth map of poverty rates might be designed to emphasise the national rate (midway through the range of rates shown on the map), so that places above and below the national rate are shown with different hues and, thus, have similar visual emphasis. Diverging schemes are always multi-hue sequences and, because of the way in which lightness is varied, do not make good black and white photocopies or prints (which only capture differences in lightness). Although we have designed the diverging schemes to be symmetrical (e.g. equal number of colours on either side of the middle break point), designers may need to customise schemes by moving the critical break/class closer to one end of the sequence. For example, a map of population change might have two classes of population loss and five classes of growth, requiring a scheme with only two colours on one side of a zero-change break and five on the other. To construct an asymmetrical scheme, the user should choose a ColorBrewer scheme with more colours than they need and omit colours (as needed) from one side of the scheme.

For this project both the sequential and diverging schemes were constructed from conceptual arcs across the outer shell of colour space. Colours were chosen that are well saturated and organised in orderly lightness sequences. The scheme was not designed using perceptual colour specifications, but knowledge of the relationships between CMYK colour mixture (cyan, magenta, yellow and black) and perceptually ordered colour spaces were used, such as Munsell (Brewer 1989), to design the schemes. The diverging schemes generally arc over the top of perceptual colour space (with white or light colours in the middle of the arc). The multi-hue sequential schemes include more hue change through the middle of the ranges and more lightness change at the ends of the schemes. ColorBrewer qualitative schemes generally maintain useful hue contrast with similar lightness and saturation for most colours (the exceptions are obviously the Paired and Accents schemes which intentionally include lightness and saturation differences).

Qualitative colour schemes

Qualitative colour schemes rely primarily on differences in hue to create a colour scheme that does not imply order, merely difference in kind. Since there is no conceptual ranking in nominal data it is inappropriate to imply order when depicting these data with colour (for example, by using a light-to-dark single-hue sequence). Qualitative schemes work best when hue is varied and saturation and lightness are kept or nearly constant. We do not recommend arbitrarily using strong 'neon' colours (i.e. high saturation) and pastel colours (i.e. light and low saturation) in the same qualitative colour scheme because these variations in saturation might imply order.

In addition to standard qualitative schemes, we offer two subcategories that do not maintain consistent lightness: Paired and Accents. Accent schemes allow the designer to customise qualitative maps by accenting small areas or important classes with visually stronger colours. A small number of colours that are more saturated, darker or lighter than others in the scheme are offered as part of the Accent schemes. These accent colours should be used for classes that need emphasis for a particular map topic. Designers must beware of unintentionally emphasising classes when using these qualitative schemes. Paired schemes present a series of lightness pairs for each hue (such as light green and dark green). Often a qualitative map will include classes that should be visually related, though they are not explicitly ordered. For example, 'coniferous' and 'broadleaf' woodland would be suitably represented with dark and light green land-cover classes. Although designers will probably not need to use for an entire Paired scheme, these pairs can be combined with other qualitative schemes to build a custom scheme for a particular map. Qualitative schemes are, thus, more flexible than either diverging or sequential for which an implied order in the colour sequence is maintained.

Number of data classes

Choosing the number of data classes is an important part of thematic map design. Although increasing the number of data classes on a thematic map will result in a more

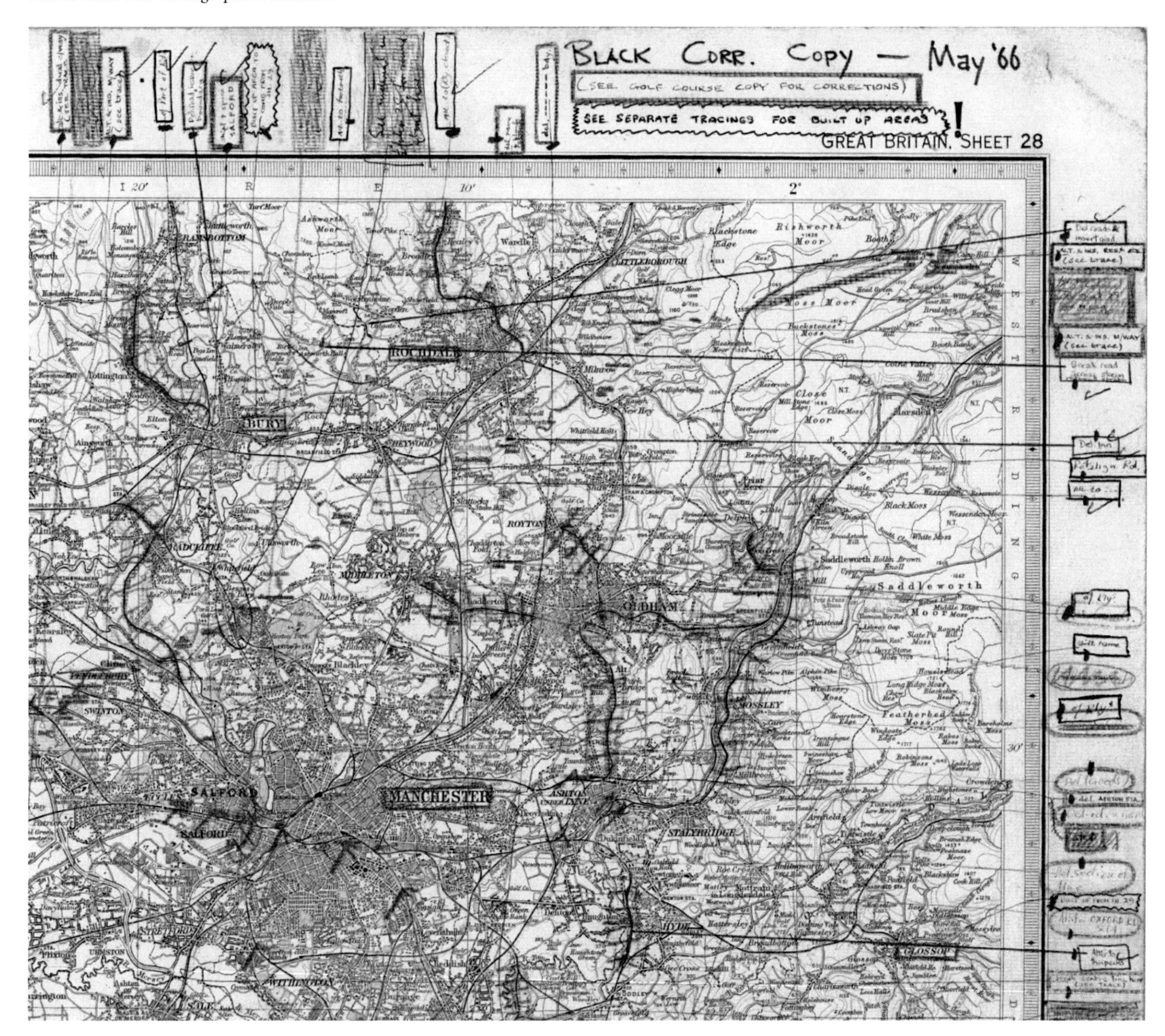

"Black Correction Copy" - Bartholomew & Sons 'Half-Inch' Merseyside map (Sheet No. 28), 1966.

The medium scale topographic map, ubiquitously produced in many nations by state agencies and commercial firms, has been a powerful 'default' view of territory throughout much of the twentieth century. Purchased by a public with increasing leisure time these maps come to signify important aspects of the landscape itself, and the map's sheet-based division of space becomes a container of administrative activities. It is easy to forget how much unseen cartographic labour goes into the production of such everyday maps.

To illustrate something of the emergent processes of cartographic production here are two excerpts for the Greater Manchester area of the 1966 Bartholomew & Sons 'Half-Inch' map of Merseyside (Sheet No. 28). On the left is one of a series of the working drafts, never normally seen by the public, which has been messily annotated by the cartographer with corrections, deletions and additions relating to changing transport infrastructure.

On the right is the finished, published sheet with its precise representation of the territory, packed with geographic information, plotted over a layer-coloured relief backdrop. This working draft and a wealth of other information about the cartographic production process and commercial practices of leading map making firm is available at the *Bartholomew Archive*, a unique resource held the National Library of Scotland in Edinburgh.

Further reading:

- *Bartholomew Archive* details and searchable catalogues at www.nls.uk/bartholomew
- Fleet, C. and Withers, C.W.J. (2010) Maps and map history using the Bartholomew Archive, National Library of Scotland. *Imago Mundi*, **62** (1), 92–97.
- Hodson, Y. (1999) *Popular Maps: The Ordnance Survey Popular Edition One-Inch Map of England and Wales, 1919–1926*, The Charles Close Society, London.
- Nicholson, T. (2000) Bartholomew and the half-inch layer coloured map 1883–1903. *The Cartographic Journal*, **37**, 123–45.

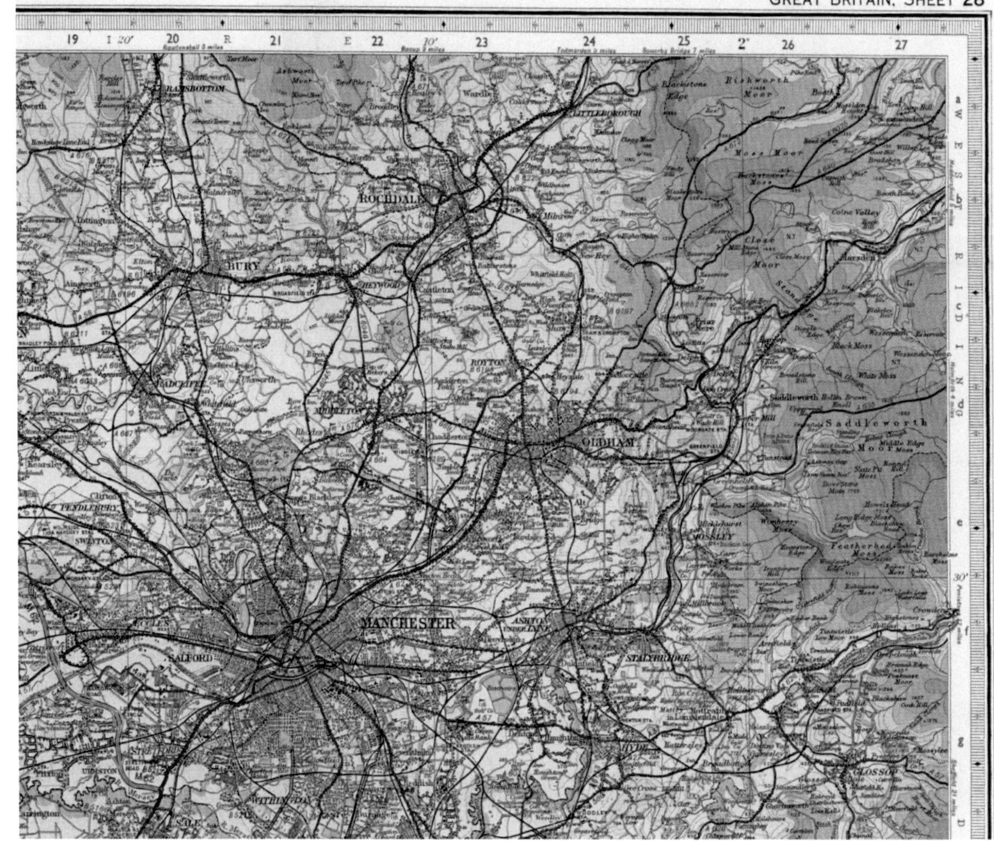
GREAT BRITAIN, SHEET 28
RAMSBOTTOM
LITTLEBOROUGH
ROCHDALE
BURY
HEYWOOD
RADCLIFFE
ROYTON
MIDDLETON
OLDHAM
Saddleworth
PENDLEBURY
SWINTON
MANCHESTER
SALFORD
ECCLES
ASHTON UNDER LYNE
STALYBRIDGE
HYDE
GLOSSOP
MOSSLEY
WITHINGTON
Marsden
Black Moss
Featherbed Moss

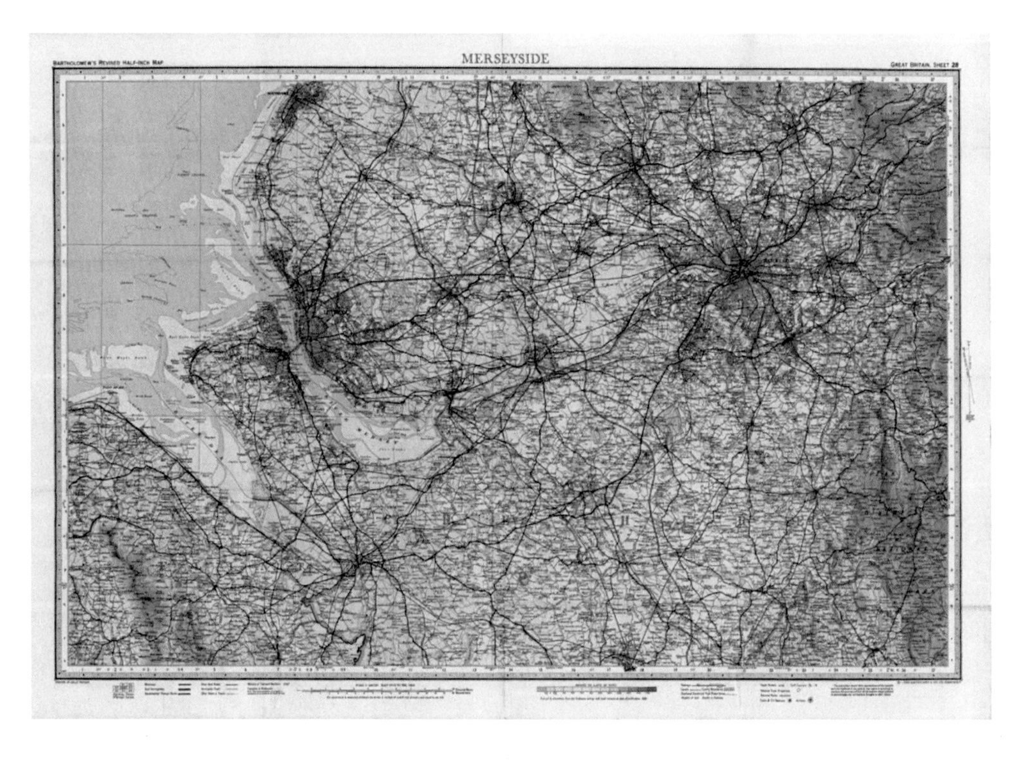
BARTHOLOMEW'S REVISED HALF-INCH MAP
MERSEYSIDE
GREAT BRITAIN SHEET 28

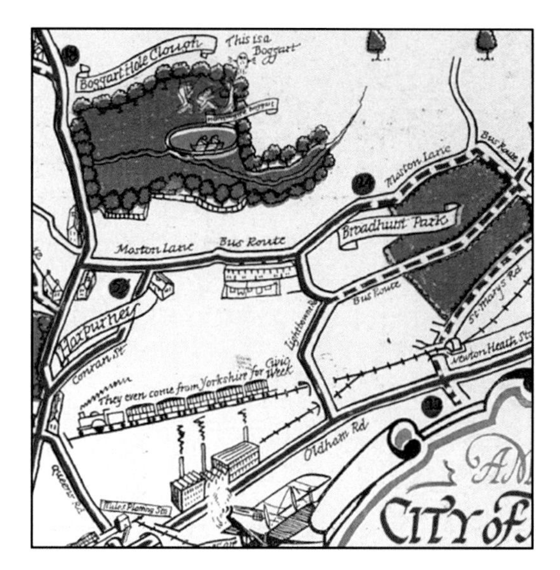

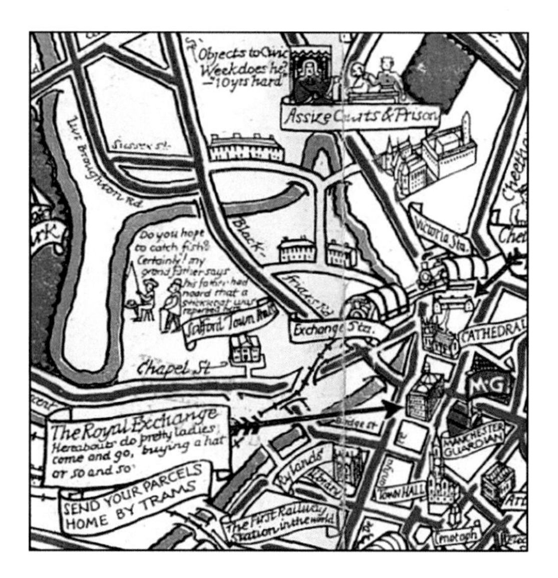

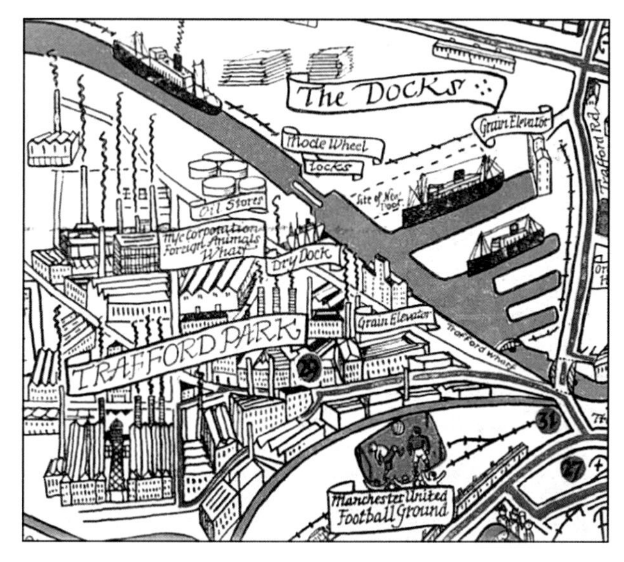

"A Map of the City of Manchester in the Year of its First Civic Week, A.D. 1926"

This free map was published in a special supplement to the Manchester Guardian newspaper to promote a major cultural festival in 1926, highlighting 'the ample means provided by the City Fathers by which the Citizens may transport and disport themselves in public vehicles'. It is a fine exemplar of pictorial cartographic design that playfully blends together utilitarian information with cartoon details of the idiosyncrasies of place and local history. The hand-drawn aesthetic, bright colours, pictographic symbols and heraldic motifs are consciously employed to make the map appear to be amateur and rather old-fashioned in marked comparison to the cleanliness of professionally printed topographic maps (see Colour Plate One). But it is clearly carefully crafted and well designed, with considerable thought around content and a skilful layout from a practiced cartographer or graphic designer (who remains unknown to us, their initials are *WM*).

The map has a depth of cartographic detail with narrative layers to be read through close inspection. Much whimsical humour is evident, for example the hunting 'Boggarts' and a dig at Yorkshire folk (see excerpt top left). Facets of city culture and the connections to people and events are portrayed, with caricatured figures speaking aloud (see excerpt top right) - such social life is strictly off limits on normal street maps. Whilst the focus is on leisure and pleasure with the verdant green of parks and gardens standing out, the map does not necessarily try to deny the industrial nature of Manchester, with numerous factories, docks and smoking chimneys all charmingly charted (see excerpt above middle).

Further reading:

- *Mapping Manchester: Cartographic Stories of the City*, www.mappingmanchester.org
- Cosgrove, D. (2005) Maps, mapping, modernity: art and cartography in the twentieth century. *Imago Mundi*, **57** (1), 35–54. (Excepted as Chapter 3.9.)
- Holmes, N. (1991) *Pictorial Maps*, Watson-Guptill Publications, New York.

(Courtesy of Manchester City Library and Archives)

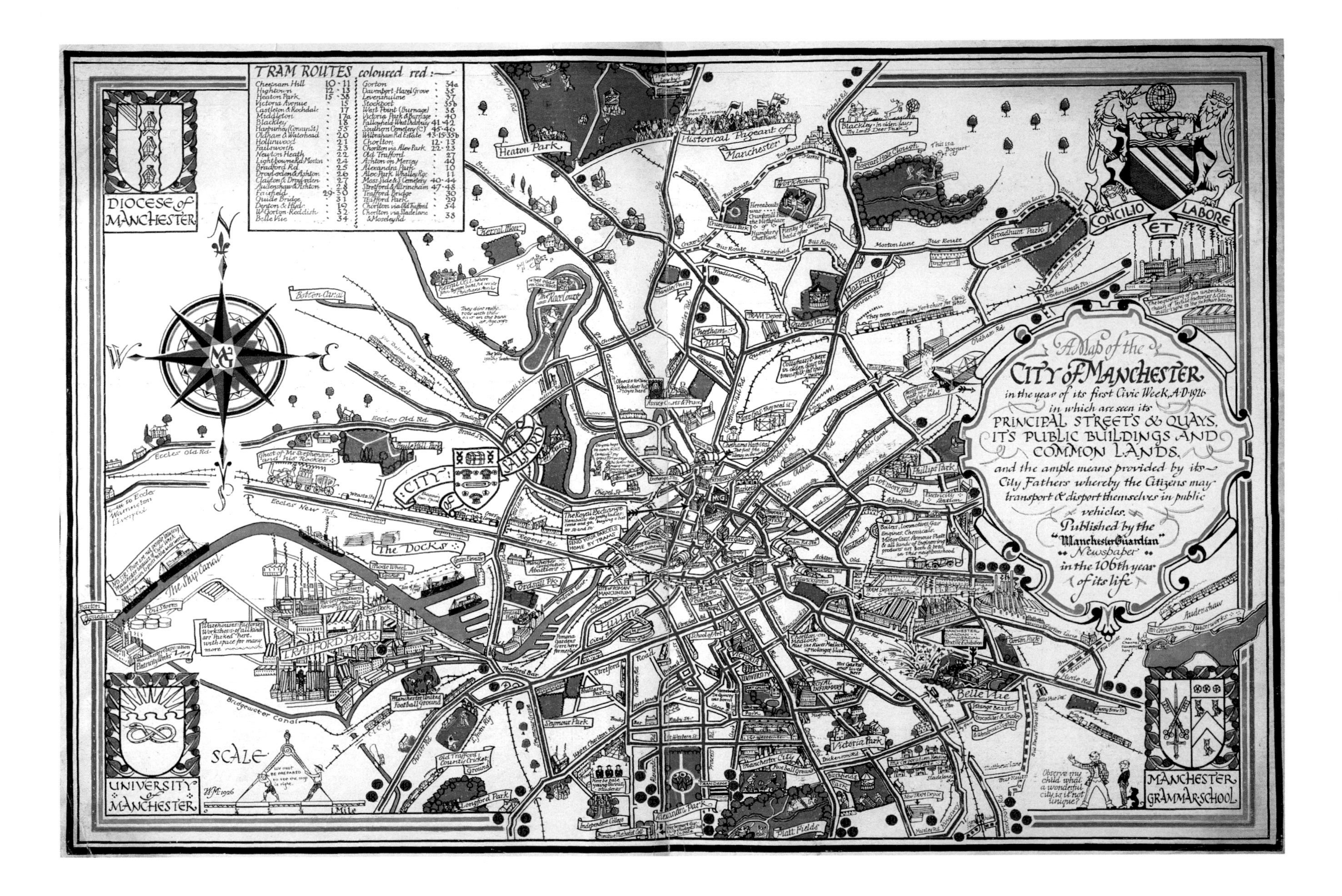
A Map of the
CITY of MANCHESTER
in the year of its first Civic Week, A.D. 1926
in which are seen its
PRINCIPAL STREETS & QUAYS,
IT'S PUBLIC BUILDINGS AND
COMMON LANDS,
and the ample means provided by its City Fathers whereby the Citizens may transport & disport themselves in public vehicles.
Published by the
"Manchester Guardian"
Newspaper
in the 106th year of its life
TRAM ROUTES coloured red:—
Cheetham Hill 10·11
Hightown 12·13
Heaton Park 15·38
Victoria Avenue · 15
Castleton & Rochdale · 17
Middleton · 17a
Blackley · 18
Harpurhey (Conran St) · 55
Oldham & Waterhead · 20
Hollinwood · 21
Failsworth · 23
Newton Heath · 22
Lightbowne Rd. Moston · 24
Bradford Rd. · 25
Droylsden & Ashton · 26
Clayton & Droylsden · 27
Audenshaw & Ashton · 28
Fairfield 29·30
Guide Bridge · 31
Denton & Hyde · 19
W. Gorton-Reddish · 32
Belle Vue · 34
Gorton · 34a
Davenport-Hazel Grove · 35
Levenshulme · 37
Stockport · 35b
West Point (Burnage) · 38
Victoria Park & Burnage · 40
Fallowfield-West Didsbury 41·42
Southern Cemetery (C) 45·46
Wilbraham Rd Estate 43·15·35b
Chorlton 12·13
Chorlton via Alex Park 22·23
Old Trafford · 27
Ashton on Mersey · 49
Alexandra Park · 10
Alex Park Whalley Rge · 11
Moss Side & S. Cemetery 40·44
Stretford & Altrincham 47·48
Trafford Bridge · 30
Trafford Park · 29
Chorlton via Old Trafford · 54
Chorlton via Slade lane & Moseley Rd · 38
DIOCESE of MANCHESTER
CONCILIO ET LABORE
UNIVERSITY of MANCHESTER
MANCHESTER GRAMMAR SCHOOL
CITY OF SALFORD
Heaton Park
Historical Pageant of Manchester
Workhouse
Broadhurst Park
Queens Park
Philips Park
The Docks
The Ship Canal
TRAFFORD PARK
Manchester United Football Ground
Old Trafford County Cricket Ground
Longford Park
Seymour Park
Alexandra Park
Platt Fields
Victoria Park
Belle Vue
Manchester City Football Ground
Royal Infirmary
Bolton Canal
Bridgewater Canal
Eccles New Rd.
Eccles Old Rd.
Oldham Rd
Bury New Rd.
Bury Old Rd.
The Race Course
Observe my child what a wonderful city, is it not unique?
SCALE
1 Mile
WM 1926

Colour Plate Five: Visualising the Efforts of Volunteer Cartographers

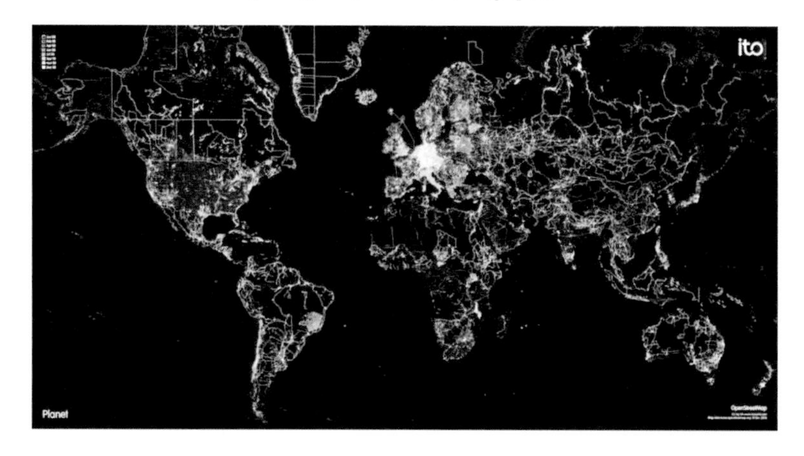

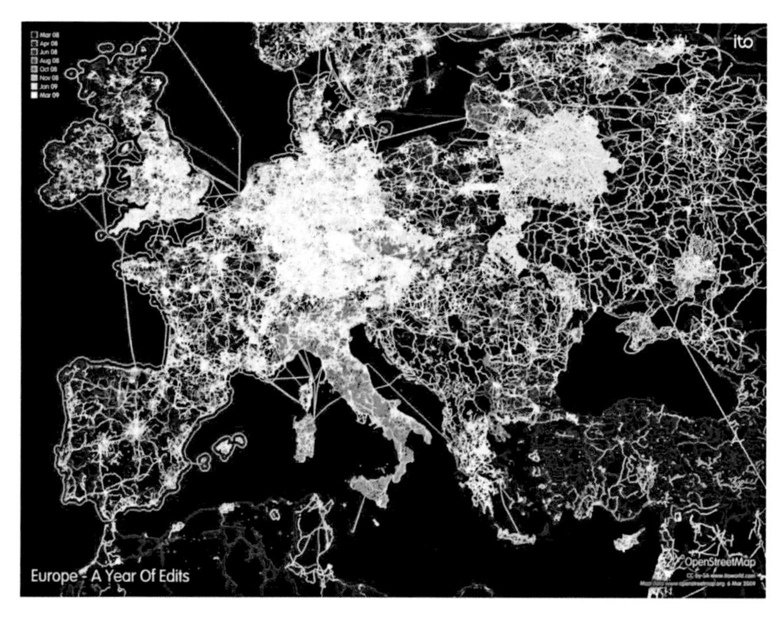

Images from "OSM Year of Edits", 2008

The nature of cartographic production is currently undergoing significant changes, with a radical challenge to established state agencies and large commercial vendors from a new wave of amateur mapmakers and volunteer cartographers who are empowered by Web2.0 technologies. The images here are from a short animated geographic visualizations created by ITO World to show the growth in one of the leading volunteer mapping projects called *OpenStreetMap* (OSM).

OSM exploits the capacities of Web2.0 to foster online collaborative mapping, with a system to share cartographic data based on a wiki model (in which everyone can edit everyone else's data). The resulting worldwide map data is open-source in that it is free to use. The wiki and open-source nature of OSM have side-effects in exposing details on how cartography comes into being – such as the authorship and date of creation and editing – which itself can be mapped. This information has been exploited to map the editing efforts of volunteers contributing to OSM. Areas visible have been added or changed in the year, with the brighter features being the more recently edited. Whilst the extensive areas of the world ablaze with volunteer mapping activity stand out, much also remains black and blank – devoid of activity. The patterns are evidently fragmented, reflecting the unpredictable and undirected nature of volunteer projects with people busy mapping their local patch whilst other areas are vacant awaiting a volunteer to get the OSM bug. Despite rapid growth in OSM since its inception in 2004 there are still large blanks in coverage; even in the brightest areas – such as part of Europe – where editing activity is at its most intense, the map is far from complete (here the intensity of data display tends to mask what is missing). Still OSM could well represent one of the major future routes for cartography and, potentially, quite radically change what aspects of space get mapped.

Further reading:

- The *OpenStreetMap* project at www.openstreetmap.org
- Goodchild, M. (2007) Citizens as sensors: the world of volunteered geography. *GeoJournal*, **69** (4), 211–21. (Excerpted as Chapter 4.10.)
- Haklay, M. (2010) How good is volunteered geographical information? A comparative study of OpenStreetMap and Ordnance Survey datasets. *Environment and Planning B: Planning and Design*, **37** (4), 682–703.
- The 'OSM Year of Edits' video produced by ITO World can be viewed at www.vimeo.com/2598878

(Courtesy of Peter Miller, ITO World, www.itoworld.com)

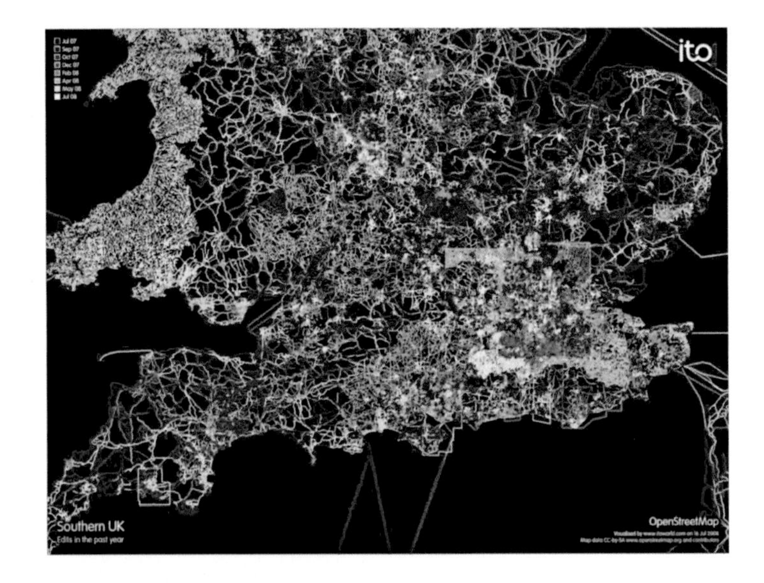

ito
Southern UK
Edits in the past year
OpenStreetMap

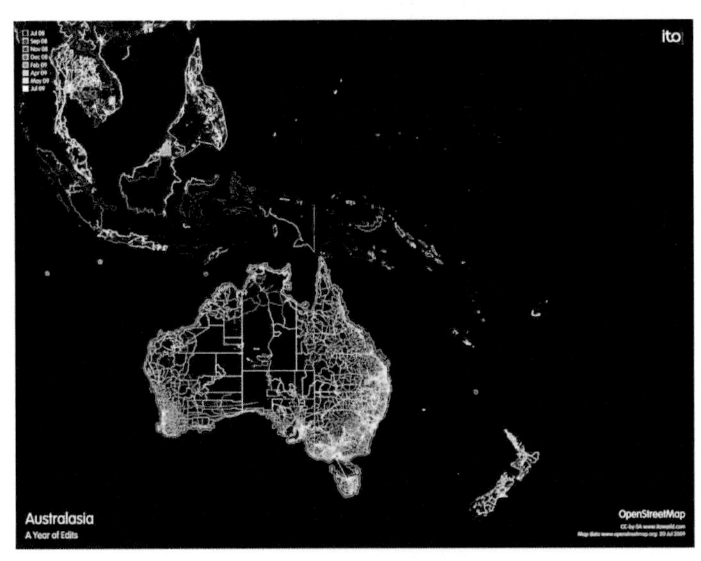

ito
Australasia
A Year of Edits
OpenStreetMap

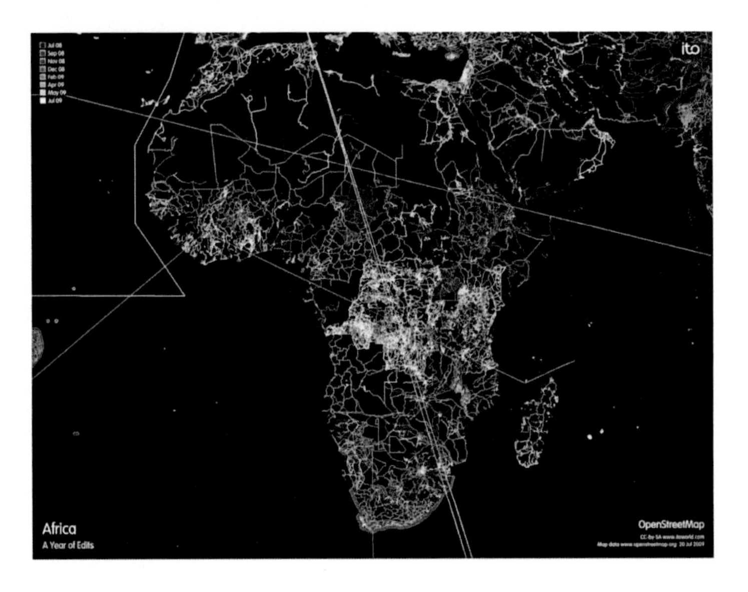

ito
Africa
A Year of Edits
OpenStreetMap

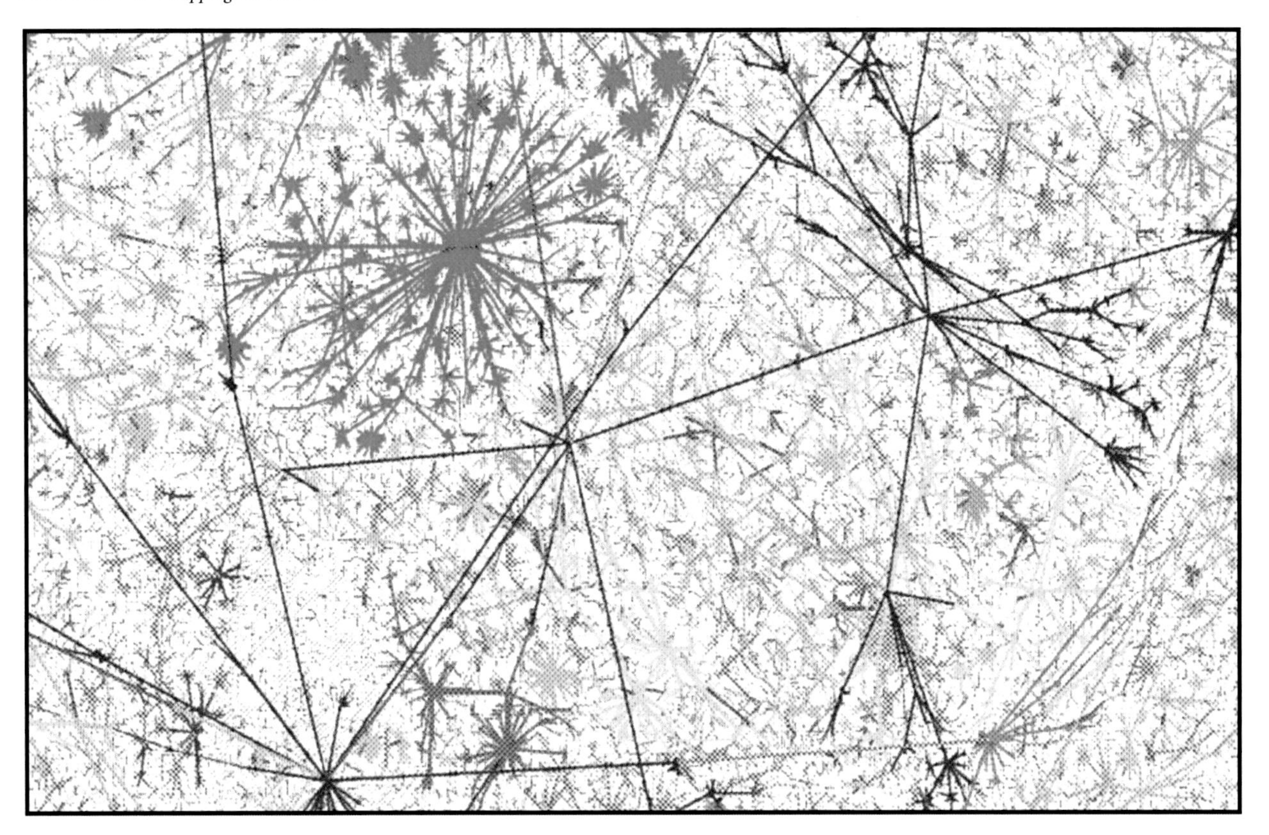

"Internet connectivity graph", December 2000.

In terms of attempts at mapping the virtual connections of cyberspace, the striking graphs produced by Bill Cheswick and Hal Burch in their 'Internet Mapping Project' provide some of most evocative representations. They have been variously described as looking like a peacock's wing, a human lung or a coral reef. As a visual mapping it has distinctly organic aesthetics, with fractal-like complexity we associate with natural forms.

While their work is not cartographic in strictly geographic terms, it is a spatial mapping that seeks to displays the topology of thousands of interconnected networks to provide an overview of the core of the Internet in a single snapshot. They map the Internet in an abstract space: as Cheswick once stated: "We don't try to lay out the Internet according to geography . . . The Internet is its own space, independent of geography." The topology data are surveyed by using the Internet to measure itself on a daily basis, charting the routes to a large number of end points (usually Web servers). The striking example shown here is a spatialization of data gathered on 11 December 2000, representing connections between nearly 100,000 internet nodes.

Maps are rendered using customised graph drawing software which takes many hours to lay out the final image. The algorithm nevertheless uses simple rules, with forces of attraction and repulsion jostling the nodes into a stable, legible configuration. The end result is a static image, but there are many permutations in the algorithm to generate different layouts and colour-codings of the links according to different criteria (such as network ownership or nationality). Its important also to realise that the pattern apparent in the graph is not a 'natural' projection of the internet – it has no inherent visual structure: there is no up and down, or east and west – but an artefact of the layout algorithm. In the example shown, links have been colour-coded according to the network owner, seeking to highlight who "owns" the largest sections of the Internet.

Further reading:

- Details at www.cheswick.com/ches/map/
- Burch, H. and Cheswick, B. (1999) Mapping the Internet. *Computer*, April, 97–102.
- Dodge, M. and Kitchin, R. (2000) *Mapping Cyberspace*, Routledge, London.
- Harpold, T. (1999) Dark continents: a critique of Internet metageographies. *Postmodern Culture*, **9** (2), http://muse.jhu.edu/journals/pmc/v009/9.2harpold.html.

(Courtesy of Bill Cheswick and Lumeta Corporation, www.lumeta.com)

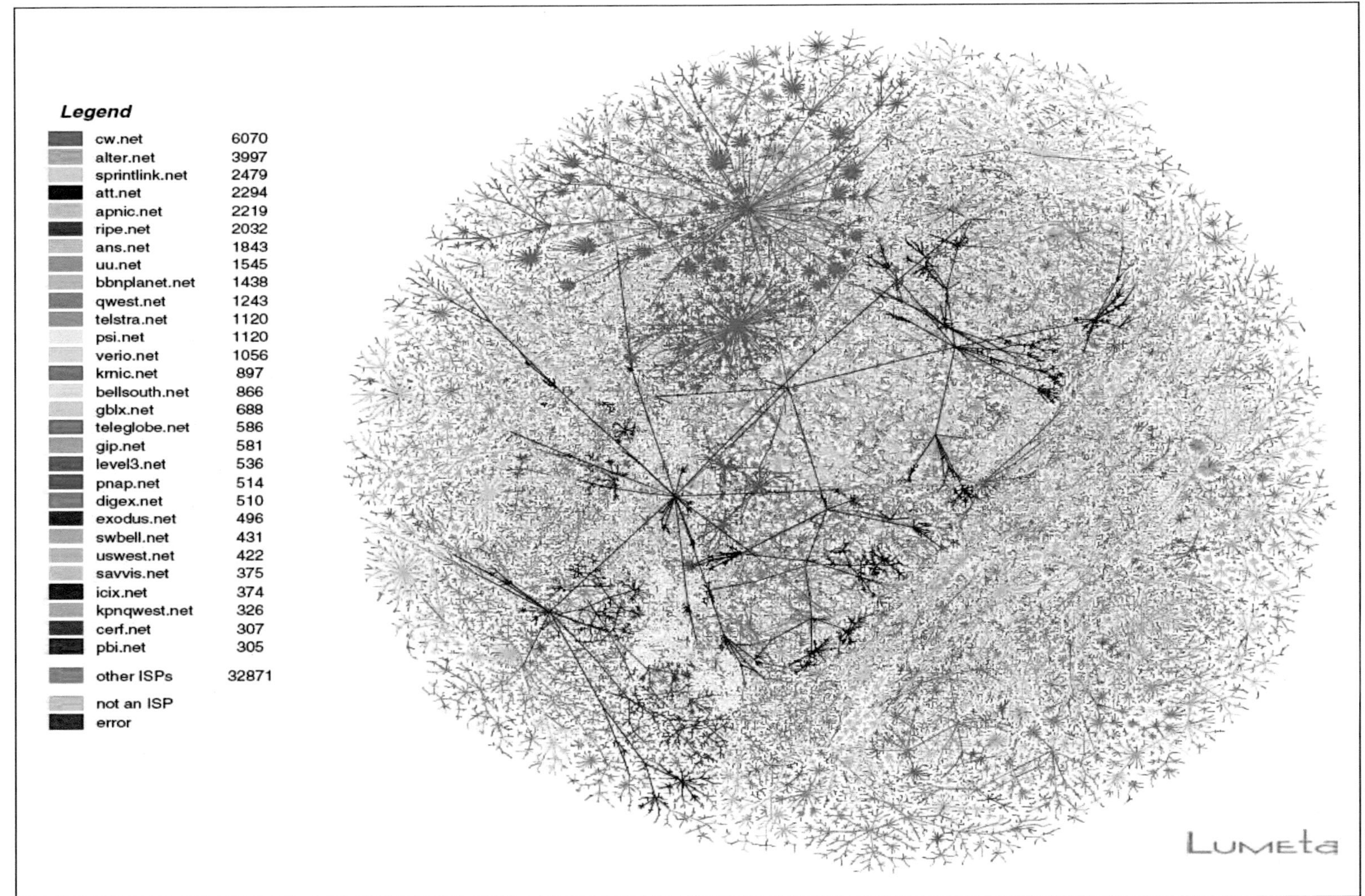

Legend
cw.net 6070
alter.net 3997
sprintlink.net 2479
att.net 2294
apnic.net 2219
ripe.net 2032
ans.net 1843
uu.net 1545
bbnplanet.net 1438
qwest.net 1243
telstra.net 1120
psi.net 1120
verio.net 1056
krnic.net 897
bellsouth.net 866
gblx.net 688
teleglobe.net 586
gip.net 581
level3.net 536
pnap.net 514
digex.net 510
exodus.net 496
swbell.net 431
uswest.net 422
savvis.net 375
icix.net 374
kpnqwest.net 326
cerf.net 307
pbi.net 305
other ISPs 32871
not an ISP
error
Lumeta

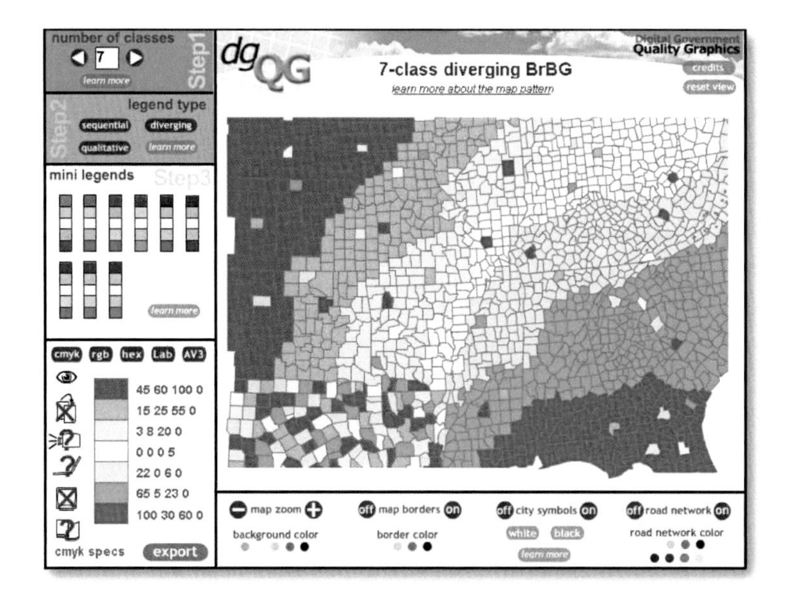

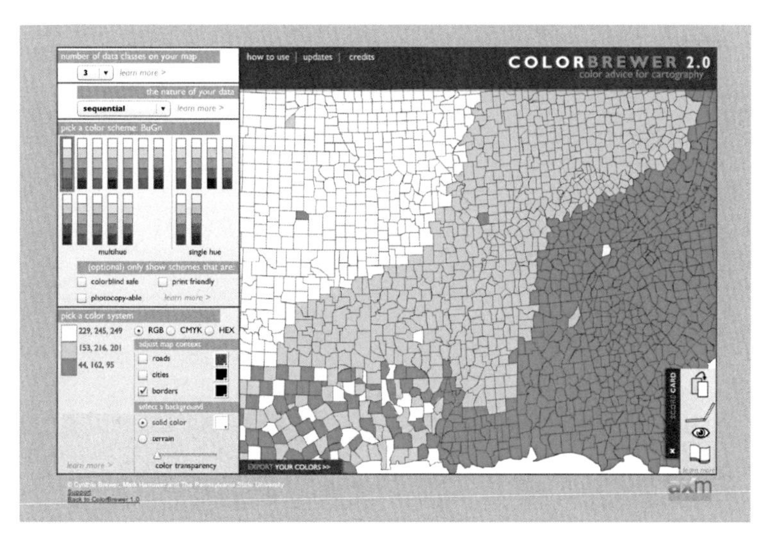

The ColorBrewer Tool

While it has become very much easier and quicker to create complex thematic maps from statistical data in mapping software, GISs and now web services, these tools often provide limited assistance to users on choosing appropriate designs for different tasks and display media. In the most ubiquitous thematic approach, the choropleth map, with areas shaded and classified according to data values, the choice of colour is often crucial to the effectiveness of the representation. The choice of colours and their sequencing can affect both the functional quality of the map (its legibility and 'efficiency' of information communication) and also the subjective interpretation of spatial patterns displayed (certain colours and combinations strongly connote meanings). When such choropleth mapping is deployed in policy documents or to convey health statistics, for example, an accurate understanding of spatial patterns can be essential.

Shown above are screenshots of an innovative online guidance system – ColorBrewer – that is designed to help users make more informed choices of colour for choropleth mapping situations. It was developed by Mark Harrower and Cynthia Brewer over several years and iterations (top image: shows version 1; bottom image: current version 2). ColorBrewer has three key features: interactive trial-and-error experimentation, specific guidance for the effectiveness of sequences according to different display media and the ability to output colour specifications for actual use.

Further reading:

- Try the current *ColorBrewer* application at http://colorbrewer2.org/
- Brewer, C. (2008) *Designed Maps*, ESRI Press, Redlands, CA.
- Harrower, M. and Brewer, C. (2003) ColorBrewer.org: an online tool for selecting colour schemes for maps. *The Cartographic Journal*, **40** (1), 27–37. (Excerpted as Chapter 3.8.)
- Jenks, G. (1963) Generalization in statistical mapping. *Annals of the Association of American Geographers*, **53** (1) 15–26. (Excerpted as Chapter 3.4.)

(Courtesy of Mark Harrower, Axis Maps)

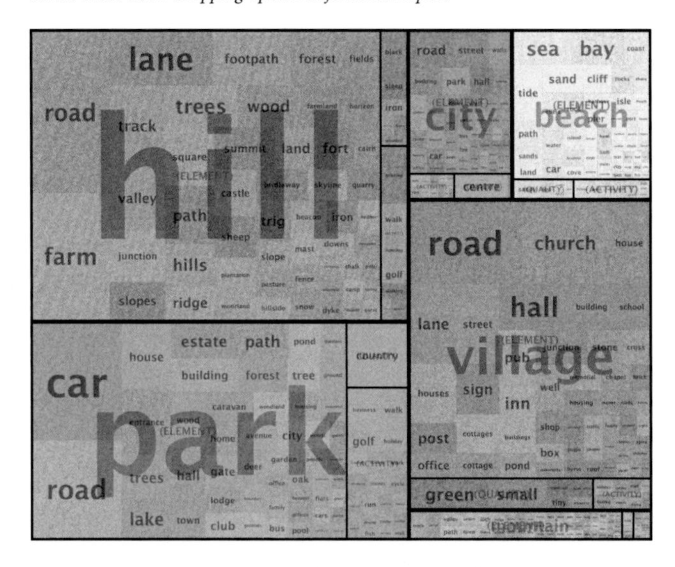

Treemaps of Geograph Archive Images

The application of mapping to display, analyse and understand the structure of large corpuses of information has been a growth area of scholarship and innovation in applied design over the two decades. Under the banner of 'information visualization' a number of new ways to summarise semantic structures using spatial representations have been developed (including huge graph displays, see Colour Plate Two). Shown above are examples of one such spatial mapping to show information structures, here applied to summarise the latent patterns of tags of photographs of places in a large online archive. Created by Jason Dykes and Jo Wood, two leading innovators in cartography and geovisualization, these 'treemaps' are deployed to help people to understand the million plus voluntarily submitted and georeferenced photographs on the *Geograph* web site. These two images reveal the beautiful potential of complex information visualization software to reveal patterns otherwise hidden in massive datasets.

The top image shows a treemap of terms occurring in *Geograph* titles and comments for six selected scene types. The sizes of the nodes represent frequency of terms. Colours emphasize the scene type/facet/descriptor hierarchy with an inherited random scheme. Layout maintains square shapes amongst the nodes.

The image below is a treemap of the same data following a more sophisticated analysis. It also shows terms occurring in *Geograph* titles and comments for six selected scene types. Node sizes represent term occurrence, and colours represent absolute spatial locations with a CIELab colour scheme. Displacement vectors show absolute locations of non-leaf nodes.

Further reading:

- Explore and contribute to the *Geograph* online archive at www.geograph.org.uk
- Dykes, J. and Wood, J. (2009) The geographic beauty of a photographic archive, in *Beautiful Data: The Stories Behind Elegant Data Solution* (eds T. Segaran and J. Hammerbacher), O'Reilly, Sebastopol, CA, pp. 85–104. (Excerpted as Chapter 3.12.)
- Shneiderman, B. (1992) Tree visualization with tree-maps: 2-d space-filling approach. *ACM Transactions on Graphics*, **11**, 92–99.
- Skupin, A. and Fabrikant, S. (2003) Spatialization methods: a cartographic research agenda for non-geographic information visualization. *Cartography and Geographic Information Science*, **30**, 99–119.

(Courtesy of Jason Dykes, City University, London)

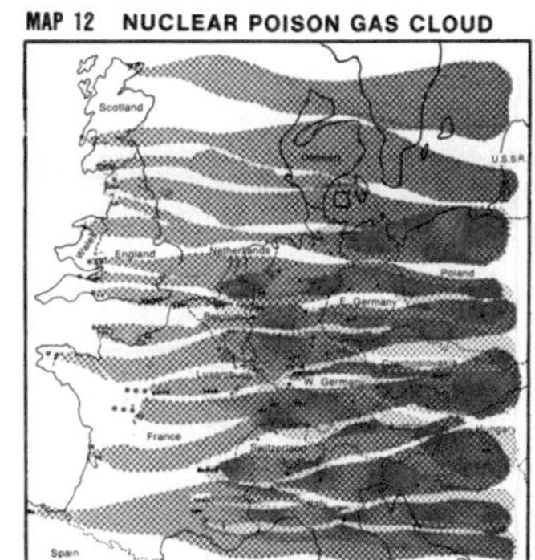

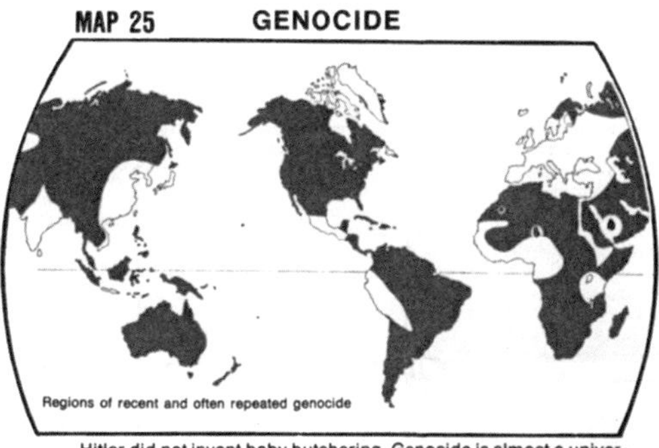

"The Nuclear War Atlas" 1982 (58 x 86 cm sheet; Society for Human Exploration)

Cartography is a potent weapon in war but can also be redeployed to protest against the waging of war and the logics of military doctrine. In this example of protest cartography the anti-war rhetoric is evident throughout the 28 maps arrayed across the sheet, with their purposeful design to highlight the dangers of nuclear weapons and to directly challenge the spatial strategies of nuclear warriors.

The poster sized *Nuclear War Atlas* was produced in the early 1980s – a time of increasing Cold War tension – under the direction of radical geographer William Bunge. In his overtly political work he used the scholarly methods of spatial analysis, quantitative measurement and statistical cartography to highlight social inequalities and seek political justice. In this mapping artefact, the reader is bombarded with an intense collage of cartography, drawn in starkly effective red and black, at differing scales and using different projections. The geographical effects of nuclear weapons and concomitant devastating impacts on humanity are plotted scientifically to affect the reader. The polemical titles mask what are a set of maps packed with information and ideas – undoubtedly propagandist – but based on evidence and thought. Blast effects on cities such as Chicago are mapped with graphic pictograms of death at varying distance from the epicentre (see excerpt top left). The spatial patterns of lethal radiation poisoning are plotted in ominous red clouds blown by the prevailing winds (see excerpt top right). The atlas also uses maps to build narrative connections between potential nuclear annihilation and other events that effect the future of humanity (excerpt above middle). The maps are all about death but they are also progressive call for action to insure life.

Further reading:

- Bunge, W. (1971) *Fitzgerald: Geography of a Revolution*, Schenkman Publishing Company, Cambridge, MA.
- Bunge, W. (1988) *The Nuclear War Atlas*, Basil Blackwell, Oxford.
- Slavick, E. (2007) *Bomb After Bomb: A Violent Cartography*, Charta, New York.
- Wood, D. (2010) *Rethinking the Power of Maps*, Guilford, New York.

(Courtesy of the John Rylands University Library)

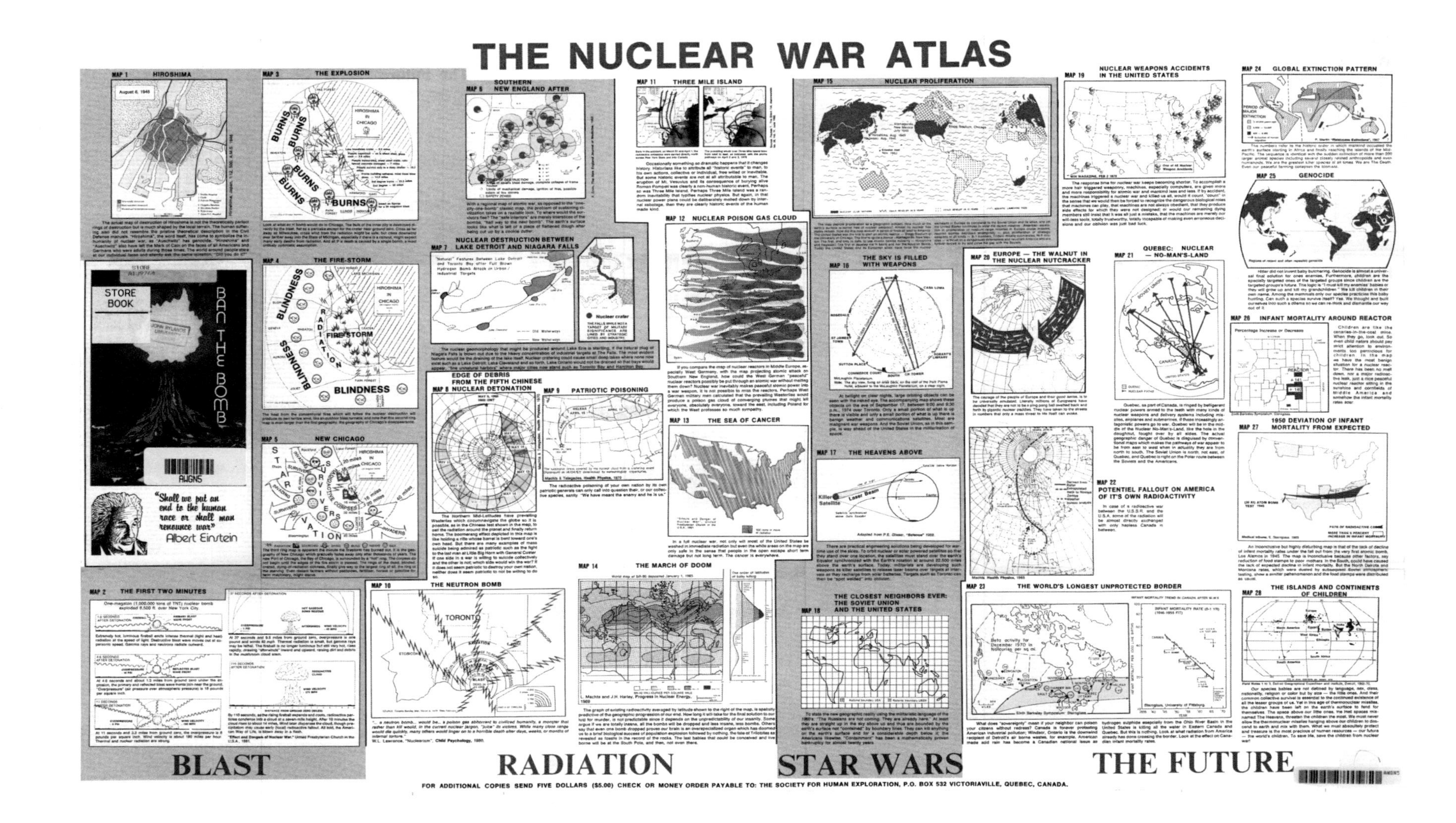
THE NUCLEAR WAR ATLAS
MAP 1 HIROSHIMA
August 6, 1945
MAP 3 THE EXPLOSION
BURNS
MAP 4 THE FIRE-STORM
BLINDNESS
FIRE STORM
MAP 5 NEW CHICAGO
MAP 6 SOUTHERN NEW ENGLAND AFTER
MAP 7 NUCLEAR DESTRUCTION BETWEEN LAKE DETROIT AND NIAGARA FALLS
MAP 8 EDGE OF DEBRIS FROM THE FIFTH CHINESE NUCLEAR DETONATION
MAP 9 PATRIOTIC POISONING
MAP 11 THREE MILE ISLAND
MAP 12 NUCLEAR POISON GAS CLOUD
MAP 13 THE SEA OF CANCER
MAP 15 NUCLEAR PROLIFERATION
MAP 16 THE SKY IS FILLED WITH WEAPONS
MAP 17 THE HEAVENS ABOVE
Killer Satellite
Laser Beam
MAP 19 NUCLEAR WEAPONS ACCIDENTS IN THE UNITED STATES
MAP 20 EUROPE — THE WALNUT IN THE NUCLEAR NUTCRACKER
MAP 21 QUEBEC: NUCLEAR — NO-MAN'S-LAND
MAP 22 POTENTIEL FALLOUT ON AMERICA OF IT'S OWN RADIOACTIVITY
MAP 24 GLOBAL EXTINCTION PATTERN
MAP 25 GENOCIDE
MAP 26 INFANT MORTALITY AROUND REACTOR
MAP 27 1950 DEVIATION OF INFANT MORTALITY FROM EXPECTED
STORE BOOK
BAN THE BOMB
"Shall we put an end to the human race or shall man renounce war?"
Albert Einstein
MAP 2 THE FIRST TWO MINUTES
MAP 10 THE NEUTRON BOMB
TORONTO
MAP 14 THE MARCH OF DOOM
MAP 18 THE CLOSEST NEIGHBORS EVER: THE SOVIET UNION AND THE UNITED STATES
MAP 23 THE WORLD'S LONGEST UNPROTECTED BORDER
MAP 28 THE ISLANDS AND CONTINENTS OF CHILDREN
BLAST
RADIATION
STAR WARS
THE FUTURE
FOR ADDITIONAL COPIES SEND FIVE DOLLARS ($5.00) CHECK OR MONEY ORDER PAYABLE TO: THE SOCIETY FOR HUMAN EXPLORATION, P.O. BOX 532 VICTORIAVILLE, QUEBEC, CANADA.

'information rich' map by decreasing the amount of data generalisation, if the cartographer uses too many data classes they may compromise map legibility – more classes require more colours that become increasingly difficult to tell apart. The maximum number of data classes ColorBrewer supports is 12, although only a few of the colour schemes can be divided into this many steps and remain differentiable. As a general rule of thumb, cartographers seldom use more than seven classes on a choropleth map. Isoline maps, or choropleth maps with very regular spatial patterns, can safely use more than seven data classes because similar colours are seen next to each other, making them easier to distinguish. The appearance of a map distribution will also be less varied among different classification algorithms (for example, quantile versus equal intervals) with more classes (Muller 1976).

ColorBrewer is not a data analysis tool and will not tell cartographers how many data classes they should use for a given mapping project (cf. MacEachren 1994; Slocum 1999 for summaries of a method for objectively calculating a suitable number of classes). Rather, if a cartographer knows how many data classes they would like to use, ColorBrewer will suggest appropriate colour schemes. Because the main map in ColorBrewer is designed as a diagnostic tool for evaluating the robustness of different schemes in different display contexts, the system is designed to *dissuade* cartographers from attempting to use too many data classes (i.e. colours). Furthermore, a ten-class scheme that is reliable on a CRT display will most likely fail on a laptop LCD display because of different contrast characteristics. Thus, the number of data classes a designer should use for a specific map is a product of the data characteristics, the intended message of the map, the target audience and (importantly) the display medium. A discussion of how this diagnostic map was constructed and how it works follows.

The map as a diagnostic tool

A well-known problem with choropleth maps is *simultaneous contrast* (Brewer 1997). For example, a single enumeration unit (e.g. county) of medium lightness that is surrounded by dark enumeration units will appear lighter than it actually is. Thus, the map reader will not be able to accurately match these outlying enumeration units with colours in the legend because they will appear lighter on the map than they do on the legend. A related problem involves the similarity in appearance of all light-coloured outliers surrounded by dark areas. For example, two different light yellows side-by-side would be easily differentiated. When physically separated and surrounded by dark colours, they will likely appear to be the *same* colour. As a general rule, the more complex the spatial patterns of the maps, the harder it will be to distinguish slightly different colours. To illustrate this, the maps in ColorBrewer present colours as both random distributions and well-ordered distributions (sequential banding). All other things being equal, a colour scheme will fail to be fully differentiable in the random portion of the map before it fails in the ordered portion. [...]

There are two ways the 'diagnostic map' in ColorBrewer can be used to evaluate a colour scheme. Firstly, can you clearly see every colour in the random section of the map? For example, if you have chosen a ten-class colour sequence, make sure that you are able to clearly see ten *unique colours* in this random section. Secondly, within each large band of colour on the map, we placed one polygon filled with each of the other map colours (the outliers). For example, if you have selected a five-class map, there will be four outliers per band, demonstrating the appearance of all map colours with each as a surrounding colour. [...]

Colour specifications: output from ColorBrewer

Once the user has identified a colour scheme they wish to use, ColorBrewer can display the numerical specifications of that scheme in five different colour specification formats: CMYK, RGB, hexadecimal, Lab and AV3 (ArcView 3.x HSV). [...] Red–green–blue (RGB) is fundamental for specifying onscreen colours (since these additive primaries are used to produce emitted colour). Cyan–magenta–yellow–black (CMYK) percentages are the standard ink specification for printing. Hexadecimal colour specifications are used to define colours on Web pages and in Macromedia Flash. Hexadecimal colour specifications are RGB specifications in base 16. Lab ('L' for lightness, 'a' for red–green, and 'b' for yellow–blue axes) is a perceptually scaled colour system that is available in some graphics and mapping packages, so we have included it in the hope of moving toward perceptual specifications for perceptually ordered colours. [...]

The colour specifications in ColorBrewer should never be treated as ironclad guarantees, since colour reproduction (whether onscreen or in print) is an inexact science. As anyone knows who has done production print work before, we cannot guarantee that our CMYK percentages will produce exactly the same colours on the printed page that you see on screen. It is best to think of these CMYK ink percentages as a good starting point but, because printers vary, some 'tinkering' with the exact ratio of inks will be necessary to create a satisfactory final printed product. Colour laser print quality is difficult to control and often produces marked shifts in colour appearance from screen to paper. From our own experiences, cyan pigments seem to be hard to control and light yellows, for example, can become light greens even on expensive colour laser printers. Worse, colour laser output can change over time (as toner amounts change and imaging drums degrade), and even the kind of paper used will influence the appearance of a colour map. Therefore, it should surprise no one that many trial prints may be required before the colours look right. Even if a designer has no intention of printing his or her maps, the situation is only slightly better with digital displays since computer monitors are calibrated differently, and cross-platform differences (e.g. from Mac to PC) can change the appearance of colours, although these changes are somewhat more predictable.

Guidance from usability icons

In an effort to help the user navigate the colour reproduction process, we have tested each of the permutations of every colour scheme in ColorBrewer across multiple display types, platforms and printed output. We systematically distilled these qualitative differences into 'use guidelines' based on media and display environments. These guidelines are included with ColorBrewer and represent a time saving function of the tool for novice and expert cartographers alike. In other words, we offer to steer users away from potential problems because we have tried every possible combination of every colour scheme in multiple display environments.

We examined each of the colour scheme sets using two CRT screens (Mac and PC), two laptop LCD screens (old and new), two LCD projectors (old and new) and prints from a Tektronix colour laser printer and a black-and-white laser printer. We evaluated the schemes by deciding whether we could differentiate all of the colours on the multiple backgrounds offered in the ColorBrewer map display. If we had difficulty with the contrast in a scheme, the icon was marked with a red 'X.' If we saw a difference on one display but not another, or if the difference was weak but visible, the usability icon was marked with a red '?' [...] These approximate evaluations were made using both a theoretical understanding of colour-blind confusions throughout colour space (Olson and Brewer 1997) as well as an evaluation of the schemes by an individual with red–green colour impairment, although we would like to see more thorough testing of the schemes to confirm that they accommodate a wide range of colour vision impairments. Generally, the hue pairs for diverging schemes were selected to accommodate colour-blind readers, with the exception of red–yellow–green and full spectral schemes. Almost all of the qualitative schemes are difficult for colour-blind readers and all of the sequential schemes are useful to them, because they include visible lightness differences.

Software development: Flash 5

The list of functional requirements for ColorBrewer was long: it needed to be (1) Web-based, (2) download quickly, (3) run well on 'trailing-edge' computers, (4) work across different computer platforms (e.g. Mac, Windows, UNIX), (5) allow for a high degree of user interactivity and (6) require little training time to use. The amount of control that users have in ColorBrewer is less than a full GIS package, but somewhat more than most multimedia Web content. The success of ColorBrewer depended upon providing visitors with enough flexibility and power to quickly find and 'test drive' many colour schemes, without overwhelming them with a complicated interface. Few Web technologies could fulfil all of our requirements. Although ColorBrewer could have been built in JAVA, this required a level of programming expertise beyond our abilities and would require that every component be built from the ground up. Instead, Macromedia Flash 5 was selected as it offered the right combination of graphic design capabilities, programming flexibility and rapid development.

Although originally conceived of as a tool for creating simple non-interactive animations for Web pages, Flash has matured into a powerful tool for creating dynamic and interactive Web material. Flash content has become the *de facto* standard for dynamic multimedia online. Because they are vector-based, Flash files are remarkably small and, hence, download quickly. The speed with which online applications download is directly proportional to their success with the public. Flash is attractive because it is compatible with all major Web browsers (using the Shockwave plug-in) and runs well on lower-end computers. Moreover, current estimates are that over 97% of Web users have the Flash plug-in already installed (Macromedia 2002). Those who need the Flash plug-in can find it for free on the Macromedia website (www.macromedia.com) and the introductory screen for ColorBrewer offers a link to the site. Flash's object-oriented programming language, called Actionscript, allowed us to add functionality such as zooming, panning, pull-down menus and 'pop-up' help windows. [...]

Learning an interface consists of at least two critical steps: knowing *what* the buttons do and knowing *the order* in which to use them. It never ceases to amaze us that software engineers often hide important interface controls or options deep within the interface. By labelling interface sections in ColorBrewer Step #1, Step #2, and so on, users should be able to quickly load colour schemes and understand the sequence of actions that are required to load those schemes. Simply put, the most important controls are highest in the visual hierarchy of the interface. To further assist users, some interface controls appear only once they are needed. For example, the Step 3 window that displays the mini colour legends is blank until the user has completed Steps 1 and 2 (Figure 3.8.1). Thus, the user is 'directed' through the interface rather than left to wonder what button to click next.

ColorBrewer has a footprint of 800 × 600 pixels so that it will fit on smaller computer monitors typical of laptops and older machines. That is a very limited space in which to fit all of the interface elements, especially since we wanted to devote as much screen space as possible to the map. One solution to this problem was to build *zooming* and *panning* capabilities into the main map so that users can enlarge areas of interest or produce polygon sizes more similar to their intended application. Another solution was to use 'pop-up' windows for non-critical interface controls. For example, a number of 'learn more' buttons are embedded throughout the interface. These concise help files only appear on demand and provide explanations of how to use ColorBrewer, as well as some of the theory behind colour use and map representation. Importantly, these help buttons are context specific and the material they display is directly related to the portion of the interface from which they were initiated. This allows the user to retrieve the information they need with one button click, rather than force them to navigate through an online manual or interact with onscreen wizards. For the expert, or return user, these help features can be ignored and do not inhibit their work.

The Actionscript code that makes the system run has been optimised so that the controls will react almost instantly to user requests. All other things being equal, the faster a system reacts to user input, the more it will inspire confidence and encourage users to explore its functionality. Audio feedback is also built into the system to reinforce the message that clicking on buttons initiates actions. Consistent design elements, such as rollover states for buttons (that change colour or grow when they are touched by the cursor), further reinforce that the interface is 'hot' and at the control of the user. Consistent design also decreases the amount of time it takes the user to learn an interface because things always behave the same way.

[...]

Conclusion

ColorBrewer.org has been online since August 2001. Since that time, we have received numerous unsolicited emails from satisfied users around the globe. Judging from their reactions, ColorBrewer appears to be successful: they have told us that the system is easy to use, attractive and has been a tremendous help in their own design work. Not inconsequentially, ColorBrewer is available for free, 24 hours a day, to anyone with a Web connection. ColorBrewer is an online tool that is designed to help mapmakers select effective colour schemes for thematic maps. ColorBrewer has been designed to fill existing gaps in the usability of GIS and graphics software by providing not only attractive colour schemes, but also recommendations on how to use those colour schemes most effectively. As such, it has found an audience with both map designers who rely on it to speed their work and college-level educators who use it in their classroom to demonstrate important concepts in colour use and design. Some of the valuable features in ColorBrewer include: (1) multi-hued perceptually ordered colour schemes; (2) 'learn more' features which advise designers on some of the theory behind colour use and map making; (3) the ability to check how well different colours schemes will hold up when additional map information is present, or the colours of the background and enumeration unit borders are changed; (4) the ability to check how well different colours schemes will perform when the mapped patterns are both ordered (e.g. banded) and complex (e.g. heterogeneous distributions); and (5) colour output specifications in five commonly used colour systems (CMYK, RGB, hexadecimal, Lab and ArcView 3.x HSV).

Perhaps the most valuable feature in ColorBrewer is the use guidelines for every colour scheme. In total, 385 unique colour schemes have been evaluated across different computer platforms and monitors (i.e. Mac and PC, projected LCD, laptop LCD and CRT), for possible colour-blind confusions, as well as in printed formats (both as colour and black-and-white prints). Given the importance of designing maps for the medium in which they will be used, these guidelines will help cartographers avoid selecting colour schemes that look attractive in their *design medium* (usually a CRT monitor) but fail in their *display medium.*

References

Bertin, J. (1983) *Semiology of Graphics*, University of Wisconsin–Madison, Madison, WI.

Brewer, C.A. (1989) The development of process-printed Munsell charts for selecting map colours. *The American Cartographer*, **16**, 269–278.

Brewer, C.A. (1997) Evaluation of a model for predicting simultaneous contrast on colour maps. *The Professional Geographer*, **49**, 280–294.

Brewer, C.A. and Suchan, T.A. (2001) *Mapping Census 2000: The Geography of US Diversity*, US Government Printing Office, Washington, DC.

Devesa, S.S., Grauman, D.J., Pennello, G.A. *et al.* (1999) *Atlas of Cancer Mortality in the United States, 1950–1994*, National Institutes of Health, Bethesda, MD.

MacEachren, A.M. (1994) *Some Truth with Maps: a Primer on Symbolization and Design*, Association of American Geographers, Washington, DC.

MacEachren, A.M. (1995) *How Maps Work*, Guilford, New York.

Muller, J.-C. (1976) Number of classes and choropleth pattern characteristics. *The American Cartographer*, **3**, 169–176.

Olson, J.M. and Brewer, C.A. (1997) An evaluation of colour selections to accommodate map users with colour vision impairments. *Annals of the Association of American Geographers*, **87**, 103–134.

Pickle, L.W., Mungiole, M., Jones, G.K. and White, A.A. (1996) *Atlas of United States Mortality*, National Center for Health Statistics, Hyattsville, MD.

Slocum, T.A. (1999) *Thematic Cartography and Visualization*, Prentice-Hall, Upper Saddle River, NJ.

Further reading

Brewer, C., MacEachren, A., Pickle, L. and Herrmann, D. (1997) Mapping mortality: evaluating color schemes for choropleth maps. *Annals of the Association of American Geographers*, **87**, 411–438. [Provides an effective summary of practical research into the use of colour in thematic maps.]

2010http://colorbrewer2.org/ [The current version of the expert guidance described in this article.]

See also

- Chapter 3.3: Cartography as a Visual Technique
- Chapter 3.4: Generalisation in Statistical Mapping
- Chapter 3.7: Area Cartograms: Their Use and Creation

Chapter 3.9

Maps, Mapping, Modernity: Art and Cartography in the Twentieth Century

Denis Cosgrove

Editors' overview

This article draws on theoretical work in Science Studies to explore relationships between artistic practices and cartography in the twentieth century. It suggests that a focus on mapping practices reveals ongoing and strong connections between cartography, art and modernity. These connections are fleshed out in case studies on avant-garde artistic encounters with mapping, and in particular in the work of Duchamps and Johns, and in the worlds of Surrealism, Situationism and Conceptual art. Parallel to these artistic expressions the world of popular mass media cartography of mid twentieth century Los Angeles also reveals a sustained engagement with artistic concerns, in particular in the oblique aerial work of Richard Edes Harrison and Charles Owens. It is argued that regarding the worlds of artists and cartographers as separate elides a continuing and active cross fertilisation.

Originally published in 2005: *Imago Mundi,* **57** (1), 35–54.

Among the many questions opened up within the history of cartography during the 'theoretical turn' since the early 1980s is that of understanding the relationships between mapmaking and artistic practice. [...] Both are concerned with technical questions of content selection and emphasis, medium, line, colour and symbolisation, and require similar decisions about form, composition, framing and perspective [...]. Unsurprisingly, the relationship between art and cartography came under scrutiny in the 1980s, as historians of cartography sought to bring new critical thinking to bear on the interpretation of maps and tried to broaden our understanding of mapping practices, not least the conventional historiography of cartography's evolution from art to science. The art/science debate was intimately bound to the more general critical turn within the history of cartography. [...]

Since then, critical interpretative and iconographic methods of art history have been widely applied to maps, and interest among contemporary artists in mapping themes has significantly increased. [...] Woodward (1987: 1) identified contemporary artistic interest in maps as a principal reason for scholars to 'explore the complex intermingling of art and science ... found in the map'. Most critical scholarship tends to focus on what were once regarded as decorative elements on maps, and the artists whose cartographical interests are routinely studied remain overwhelmingly drawn from the pre-modern period. [...] This article by way of contrast focuses on the relationship between art and science in general and within the context of twentieth century artistic engagement with mapping practices and with maps as art objects.

Art, science and cartography

[...] There is now a significant literature on the constitutive role of visual images, including maps, in the practices of science which undermines claims that science's 'analytical, independent and reportive' method makes no

The Map Reader: Theories of Mapping Practice and Cartographic Representation, First Edition. Edited by Martin Dodge, Rob Kitchin and Chris Perkins.

call on the persuasive role of aesthetics. Work published in the field of Science and Technology Studies has considerably reconfigured our understanding of how science works (Dalston and Galison 1992; Shapin 1998; Livingstone 2003). Science's nexus of instrumentation, mechanisation and objective representation strategies (such as quantification or photography) does not escape aesthetic appeal. Rather, science deploys the aesthetics of plain style within a broader persuasive strategy. Mathematics, language and illustration, including various forms of maps, play a major role in securing claims to the universality, authority and immutability of scientific knowledge [...] and help to secure claims about spatial relations and processes that are invisible to the individual observer and often based on evidence geographically removed from the site of map use. Numbers, lines, colours and key coding reinforce the thematic map's scientific status as an 'immutable mobile', to borrow the philosopher of science Bruno Latour's term for an instrument that preserves the meaning and truth claims of scientific observations as they circulate across space and time (Latour 1999). Latour regards an 'immutable mobile' as a vehicle for transferring through space scientific information gathered at a specific site in a way that preserves, through the unchanging nature of the vehicle, the validity of that information. The map is a paradigm example. [...] Latour (1998: 425) further points out that in the process of circulation by which scientific knowledge is universalised, 'information is never simply transferred, it is always radically transformed from one medium to the next ... it pays for its transport through a heavy price in transformations'. To achieve immutability (for example, by means of a scientific map), the information contained undergoes transformation, a process which, in principle, is not different from that of artistic production of spatial images.

The shared epistemology of art and science [...] is thus true for work from any historical period. It has remained obscured, however, by claims for representational objectivity developed within modern science. As Latour (1998: 423) points out, 'it was much more difficult to extirpate scientific activity from its epistemological past than to free art history from aesthetics. Once the two moves were completed, a vast common ground was opened and, in recent years, a flurry of studies has 'vascularised' the connection between visualisation in science and the visual arts'. These moves set the context in which critical deconstructions of scientific mapping projects have revealed the significance of culture, location, individual subjectivity and artistic imagination in the creation of maps as scientific instruments (Stafford 1984, 1994; Dalston 1998; Burnett 2000).

Latour's reference to the freeing of art history from aesthetics signals a second challenge to Woodward's framework from the perspective of art practice. The art historians who contributed to Woodward's *Art and Cartography* largely avoided considerations of aesthetics. [...] In fact, among artists themselves, aesthetics had been under siege for much of the twentieth century. Artists' embrace of the revolutionary political term avant-garde betrays a concern that self-consciously modern art should break with such bourgeois and commercial concerns as beauty and aesthetics. Their manoeuvre prejudices any attempt to study the quite considerable engagement of twentieth century artists with maps within the rubric applied by Woodward to pre-modern art and cartography. Indeed, just as Science Studies have directed attention away from the truth claims of science towards its specific and located practices, including its use of cartography, so artists' move directs us away from the map as art and towards the place of cartography within artistic practice. [...] That said, it is important to acknowledge that avant-garde ideas and practices by no means encompassed the whole of what goes under the rubric of Modern art in the twentieth century, and that Modernism has found many other artistic expressions, some of which also involved maps and mapmaking. Pictorial mapping exploded in the middle years of the twentieth century in such popular media as newspapers, magazines and film, where it was strongly influenced by some of the same cultural currents that gave rise to Pop Art.

Avant-garde art and cartography

By the end of the nineteenth century the practices of modern cartography were well in place. States had surveying and cartographical institutions producing topographical maps of their territory and of their colonial possessions; statistical mapping was a significant tool of bureaucracy and social regulation; and map use and interpretation were widely taught in schools. Cartography was a profession that embraced such scientific practices as classification, quantification and instrumentation to secure the truth of its visual records and representations. Later, photography and aerial survey reinforced the trust in mimesis. Early twentieth century Modern artists were similarly concerned with the relationships between vision and space, but their recognition of the complexities and subjectivities involved directed the attention of groups such as the Post-Impressionists, Fauvists and Cubists away from cartographical concepts and practices as objects of potential artistic interest. Crudely speaking, while cartographers were striving for methodological rigour in such matters as projection, scale, topographical representation and nomenclature, [...] art was dominated by a series of avant-garde movements whose intention was to criticise and subvert long-standing ideas and practices of representational art. That critical intent soon redrew the boundaries of art itself as a socially and politically engaged activity.

Closer examination of art interests and practices suggests some nuances to this familiar story. [. . .] Artistic experimentation occasionally generated images that have a strongly cartographical feel. Futurism in the late 1910s and in the 1920s, for example, was inspired precisely by the ways in which modern technologies had transformed the experience of space and time. Futurists responded enthusiastically to the same opportunities for seeing the land from an aircraft, which challenged and expanded cartography itself. Italian 'aeropaintings', for instance, conveyed the experience of speed and fractured vision which the view from an aircraft allowed (Mantura *et al.* 1990). Italian embrace of Modernism extended into cartography as geopolitical mapmakers employed its design strategies and typography to dramatise their cartographical arguments (Atkinson 1995). Brief consideration of two key twentieth century avant-garde artists permits closer examination of the complex connection with mapping and maps.

Duchamp and Johns

Marchel Duchamp (1887–1968) is considered one of Modernism's guiding spirits and a key influence over avant-garde art practices. His 'readymades' of the 1940s – common objects of everyday use only slightly modified, if at all, by the artist but turned into art by selection and relocation – radically transformed art. Housefield (1992) has recently argued that the readymades were strongly influenced by a cartographical impulse to represent actual spaces. Today, Duchamp's readymades are [. . .] treated as discrete objects, but they were originally gathered together in his New York studio and made visible as a collection of related objects. Housefield shows that each object can be connected, conceptually and to some extent formally, to a specific site in Paris [. . .], notes the popularity in Duchamp's youth of pictorial maps as guides to monumental Paris and [. . .] suggests that Duchamp's readymades engage analogy, humour and shifts in scale to translate elements of the human made urban landscape into the interior landscape of the studio. Such shifts and translations parallel the physical and conceptual transformations of landscapes into cartographic representations, or maps. Housefield closes by remarking, that although modern art does not always represent the landscape in immediately recognisable ways, the relationships with geography merit more consideration than they have received [. . .].

Examining relations between modern art and cartography is significant in view of Modernism's consistent desire to confront figurative and representational conventions. Modern painters displayed intense interest in the grid as a pictorial device. [. . .] But their response was to the grid's capacity to express 'the absolute autonomy of art – anti-natural, anti-mimetic, antireal' – rather than to its ability to frame and compose the spatial arrangement and scalar representation of material places and landscapes. 'Unlike perspective, the grid does not map out the space of a room or a landscape or a group of figures onto the surface of a painting. Indeed, if it maps anything, it maps the surface of the painting itself' (Krauss 1980).

Jasper Johns's encaustic and collage map (1963) may be read as an extension of his Modernist fascination with the painterly surface as much as a direct engagement with cartography. Johns's image looks like a crude map of the conterminous states of the United States, each labelled and rendered in a different colour. The wax-like encaustic is thickly layered, and the brush strokes are obvious. They draw attention to the unstable surface of the image as much as to the provisionality of meaning in the mapped space. Johns consciously juggles with cartographical conventions: varying the lettering and nomenclature of states, cutting place names off at the margins, and using a muted grey palette set off by occasional points of primary colour. The 'key' in the corner of the image, [. . .] refers only to itself, rather than the contents of the 'map'. Collage pieces, [. . .] preserved by the encaustic material used to depict the states, denote specific places and times. Johns is questioning the pretence of the map to be more than its surface (Yau 1996).

Surrealism and Situationism

Surrealism was the avant-garde Modern art movement that explicitly engaged cartography as practice rather than simply the map as image. Post-war advances in cognitive psychology challenged many of the assumptions about the transparency of representational images by emphasising the importance of individual and social perceptions. Ability to recognise and understand map images was learned and cultural rather than a function of the map's scientific objectivity and design clarity. Meanwhile Situationism, a second-generation Surrealist movement, stimulated intense interest in the map as a communicative device and in the subversive potentials of mapping practices. Situationism's conscious move beyond the art world of studios and galleries into the spaces of everyday life reinforced this concern with mapping as a means of engaging graphically and actively with material spaces. [. . .] Guy Debord sought to connect art practice directly to the geography of the city. His concept of psychogeography was part of a set of radical responses to rationalist and functionalist urban planning [. . .], which he believed to be destroying the social and psychological well-being of urban communities. [. . .] The connected practice of the urban dérive, or drift, intended to generate chance encounters and provocative interactions with other individuals, involved a kind of subversive survey of urban space that both stimulated and recorded 'transient passage through varied ambiances' (Wollen 1999: 30). [...] To illustrate these experiments, between 1955 and 1959, Debord and his Danish colleague, Asger Jorn, produced various collages bringing together map fragments, images and texts that captured urban space and experience in Paris and Copenhagen. [...] Like Duchamp, Debord's psychogeographical street maps of Paris drew upon popular pictorial maps. Debord explicitly used G. Peltier's Guide Tirade de Paris (1951) and his Vue de Paris à vol d'oiseau (1956). Peltier's totalising and commanding vision of the city derived in part from aerial photography. [...] Pictorial maps perfectly captured the distanciated spatial vision of mid-century urban planning in European cities.

Debord's dust jacket for the book illustrates the concept of the dérive as a mode of experiencing the city intimately from below. The mastering perspective of the bird's-eye view is broken into arbitrary fragments representing districts to be walked and in which encounters might take place. The red arrows connecting these spaces represent taxi rides or other less intimate connections across urban space.

Hermann Bőllmann's [...] axiometric image of Manhattan – the paradigm Modernist landscape – renders to great visual effect the synoptic, mastering gaze that Michel de Certeau, heir to the Situationist critique, later dissected in his The Practice of Everyday Life (De Certeau 1988). The Situationist response to the urban vision represented by such cartography was to cut the map of Paris or Amsterdam into 'islands' of urban space joined only by thick red arrows or dark ribbons that evoke the emotional connections made within and between such locales by the artist/mapmaker. Urban 'mapping' is thus transformed into a pictorial art practice. [...]

Conceptual art and beyond

In the late 1960s and 1970s, conceptual artists, for whom painting [...] had reached the limits of expression, also took an interest in mapping. Adopting a less activist relationship with social issues than the Situationists, conceptual artists focused on the idea of an artwork, on theoretical methodologies of documentation, on site and on performance. These concerns directed their engagement with cartography further towards the processes of mapping. [...] Their interests in documentation and site specificity, for example, not only signalled the value of the map as a mode of spatial representation but also recognised that mapping and its powerful visual codes made it a highly effective subject for creative manipulation (Wollen 1999; Curnow 1999). From its earliest stages maps played a role in conceptual art. [...] In Douglas Huebler's Site Sculpture Projects (1968), maps play a dual role of instantiating the site and documenting the projects, whose aim was to redefine sculpture. [...] Like Jasper Johns's earlier map, the conception of Huebler's project owes much to the 'mapped' nature of American social space, in which political and survey boundaries are strongly orientated to lines of latitude and longitude. [...] For Huebler, objective, scientific and even banal everyday aspects of the map became positive advantages in supporting the goal of divorcing art from visual pleasure.

[...] Conceptual and post-conceptual artists from the 1970s to the present have sustained a critical conversation with cartography. A recurrent theme in site-specific and performance art has been the use of interactions between people and things in material space to 'map out' locations, routes and journeys within cities. The language and practice of maps have been deployed to structure journeys and interactions, as in the topographical work of Land Artists discussed by Stephen Bann (1994). [...] Other artists have been attracted less to the survey and recording aspects of mapping than to the techniques and processes of map making and to ways of manipulating such apparently determinative elements as projection and scale. Agnes Denes's Isometric Systems in Isotropic Space–Map Projections (1974) projected the world map on to such mathematical figures as the cube, doughnut and snail shell. Lilla LoCurto and William Outcault's Selfportrait.map (2000) used a selection of spherical projections to explore the formal relationships between the globe and the human body and the technical problems of representing their curvilinear surfaces on the two dimensional plane. The New Zealand artist Ruth Watson has devoted numerous works since the mid-1990s to an exploration of the heart shaped cordiform projection. She makes images of the world centred on the South Pole rather than the North, creating them from photographs of the tongue's surface and metal pins (Lingua Geographica), gold chocolate wrapping paper (Take Heart), and displays them in locations closely connected with mapping and colonialism, such as the Stedelijk Museum in Amsterdam. [...] Further exploring the connections between cartography and imperialism, English artist Pat Naldi projected maps of unclaimed territories and a childhood map of the British Empire onto the walls of the British School in Rome in 2001. Laura Kurgan meanwhile has addressed the politics of contemporary mapping technologies, using SPOT satellite images taken during the Balkan wars of the 1990s to map out the locations of mass graves of ethnic cleansing in Kosovo. The critical and post-colonial aspects of Watson's, Naldi's and Kurgan's map images reflect an established radicalism in conceptual art's engagement with mapping. [...]

In recent years, the ideological stridency of some critical artwork has been paralleled by a more nuanced artistic use of the map. This reflects a recognition that deconstruction has successfully challenged the map's naturalising powers and our ability to acknowledge the visual appeal of mapped images, without necessarily evading their problematic connections with power and exploitation. Thus the Irish artist Kathy Prendergast has developed a series of works in her Atlas of Emotions (1999) which abstract from topographical maps of Canada and the United States place names that record European colonisation, such as the psychologically charged toponym 'Lost'. Re-mapping these words challenges assumptions that colonial exploration and cartography were straightforward [...]. They suggest the coexistence of cognitive dissonance in unfamiliar territory and affective relations with the earth's surface (Nash 1998). Japanese artist Satori Matoba skilfully dissolves maps of different but politically connected locations into each other. Her image of Pearl Harbor/Hiroshima (1998), for example, disrupts the taken for granted meanings of both places by merging the two topographical maps into a single, visually smooth cartographical surface.

The engagement of contemporary art with cartographical images and practice, sanctified by a series of major exhibitions in the 1990s, has thus developed into a field of intensive and continuing work, in which the map is the focus for widely diversified critical and graphic attention. The complex history of twentieth century relationships between cartography and the artistic avant-garde awaits a detailed, authoritative study. Such a study will need to engage – and indeed map – the highly complex and varied expression of Modernity in art and may reveal unexpected impacts of art theory and practice within cartography itself.

Popular culture, art and cartography

I have concentrated so far on avant-garde Modernism that was highly intellectualised and concentrated in a limited number of artistic centres, such as Paris and New York. But the avant-garde was not hermetically sealed from popular culture, and Modern art had a highly varied expression, extending well beyond the circle of self-consciously radical artists and comprising a more complex geography than just these two cities. For the more popular artistic expressions of Modernism, I draw on work in Los Angeles. [...] By the 1930s the film industry in Hollywood had attracted a huge range of artistic talent to southern California, generating innovation in many fields of graphic art, including line drawing, cartoons, comics, posters and other forms of commercial illustration. By the 1960s this work was beginning to have an impact on such avant-garde movements as Pop Art [...].

Before turning to the direct impact of Hollywood on popular cartography, it is important to recall the significance of the map in mass culture in the twentieth century. [...] Map images became ubiquitous in mass media; they were found in newspapers, on screen and in advertising, travel and tourism. By mid century, cartographical literacy and graphic communications had become so pervasive that scientific cartographers [...] found it impossible to control the output of cartographical images. The cartographers' very success in achieving scientific status for their work had given the map enormous authority. Faced with the volume of popular maps, professional cartographers were frequently reduced to impotent rage at the vulgarity and inaccuracy of what passed for a map (Pickles 1992). Those who produced maps in news journals and magazines rarely possessed the technical training of professional cartographers, nor would many have made a serious claim to any artistry for their work. Monmonier (1989: 14) describes twentieth century journalistic mapping as 'a distinctive cartographic genre... generally simple in content and symbolisation... Unfortunately', he states, 'news publishers tend to hire artists untrained in cartographic principles, and news maps sometimes reflect an ignorance of map projections or cartographic conventions'. Monmonier's evidence entirely supports such a conclusion, but his judgment betrays a lingering attachment to scientific cartography as the criterion by which all map images are to be evaluated.

[...] Placing popular map images within the context of modern culture's consistent erasure of canonical distinctions and hierarchies yields a richer understanding of their role and relations with both art and cartography, as Schulten (1998, 2001) has shown in her study of Richard Edes Harrison's wartime mapmaking. Harrison's 1944 atlas, Look at the World, contains some of the most graphically adventurous maps of the twentieth century. [...] Despite Harrison's lack of formal cartographical training, he nevertheless recognised the advantages of orthographic and azimuthal projections for conveying the scale and spatial relations of a world shrunken by powered flight. His foreshortened picture-maps illustrated geo-strategic relations as if viewed from the cockpit of an infinitely high-flying aircraft (Harrison 1944). Although Harrison constantly emphasised that his maps showed the 'true' nature of spatial relations in a world of air power, he was equally explicit about their persuasive function; they were intended to explain the first truly global conflict to citizens in a modern democracy through graphically dramatic images published in mass-circulation photo-journals such as *Life* and *Fortune*.

[...] At least two California daily newspapers published highly original, full-colour maps explaining the war to their readers in considerable and vivid detail. Neither scientific in their cartography nor consciously avant-garde in their art, the creators of these maps called upon the scientific authority that the map possessed in modern society and the graphic techniques developed in films and comic books to produce dramatic spatial images for a mass readership.

Pictorial war maps

From the 1930s into the 1950s, the Los Angeles Times and the San Francisco Examiner, devoted whole pages to what are best called pictorial maps. The maps were the work of the newspapers' staff artists, Charles Hamilton Owens and Howard Burke respectively. The pictorial style of two artists is similar. Each adopted axiometric mapping, often incorporating detailed cartographical studies, drawings and realistic landscape features such as buildings and vegetation into the main image. Blocks of text and labelling provide explanatory narrative. In Owens's case, a characteristic feature was to surround the map with pencil sketches dramatising its subject or promoting geographical recognition through iconic locational images. Charles Owens, who is the principal subject of this study, worked between 1910 and 1912 for the New York Herald and other New York newspapers and after that at the Los Angeles Examiner before joining the Los Angeles Times in 1918, where he stayed until his retirement in 1952. Howard Burke is known only for his work as staff artist at the San Francisco Examiner, where he was active in the same mid-century years as Owens. Owens was more prolific than Burke. From the time he joined the Los Angeles Times, he sketched maps alongside his drawings of court-room scenes, crime locations, sporting events and the myriad other items. Among Owens's earliest maps are those describing the fate of the German colonies in the Versailles negotiations of 1919 and the 1922 division of Ireland. His cartographical contribution peaked in a series of nearly two hundred full-page, colour war maps published on Mondays between February 1942 and August 1945. [...] Regardless of the scale of the area mapped, [...] a sharply curving horizon places the observer high in the heavens, as a witness to the earth's curvature, while swooping low over the details of topography and landscape. Owens's maps reflect the excitement and imaginative stimulus of what mid-century American political commentators called 'the airman's vision', a synoptic, global perspective (Cosgrove 2001). [...] A characteristic feature of Owens's maps is the sketched battle scene or landscape, rendered in pencil and charcoal and placed above the horizon line to frame the mapped space. These scenes give his cartography the intensity and immediacy, in the style of the war comics that remained popular into the 1950s. The most obvious source of Owens's vistas is the storyboard, with which visits to film studios had familiarised him. Map and narrative sketches often interpenetrate as the latter invades the former, producing dynamic effects reminiscent of films or newsreels.

Owens's use of the directional arrow, barbed 'front' and solid line barrier came from military mapping and geopolitics and enhanced the sense of dynamism within his maps. The influence of Hollywood is also apparent here: an opening sequence of dotted lines or arrows snaking across a map to set the story's context was a common device in the Hollywood films of the 1940s, such as in the opening frames of *Casablanca*, Owens's various techniques for constructing informative but dramatically suggestive maps have distinct Modernist echoes.

Collage was pioneered by Picasso, Braque and other Cubist painters in the years immediately preceding the First World War as a form of visual rapportage. For Owens, the technique captured well the strategic imperatives of the first air war. [...] The provisional nature of the image also enhanced its dramatic qualities. More conventional techniques, such as colour selection and tone, play a central role in Owens's maps, enhancing their urgency by speaking directly, often harshly, in primary blocks, or emphasising the gloomy drama of war landscapes in their margins.

These various techniques [...] reflect his immersion in a southern California visual culture that was Modernist without being avant-garde, a culture that emphasised speed, the moving image and dramatic graphic communication. Conservative in pictorial style and lacking contact with self-conscious art movements, Owens's mapping reveals the influence of popular and permeable Modernism that suffused twentieth-century southern Californian culture. [...].

Pictorial maps and popular culture

Charles Owens's cartography drew upon and contributed to a modernist visual and spatial culture in the making in 1920s and 1930s southern California. It was a culture closely associated with cars and aeroplanes, films and comic strips. [...] The Times publisher – Owens's employer and friend – was Harry Chandler, around whom orbited both local oligarchs, who shared Chandler's hotel and real-estate interests, and a bohemian intelligentsia associated with the city's entertainment and academic institutions. All shared the vision that Chandler trumpeted in his newspaper, of Los Angeles as a city of the future. [...] Owens's artistry contributed directly to the vision by illustrating many of the engineering, architectural and cultural projects that were to transform southern California's twentieth century landscape. [...] Many of his drawings incorporated bird's-eye landscapes and maps based on the artist's astonishing capacity to re-create topographies either from studies made during his frequent flights over the region or from memory.

During three decades, Charles Owens mapped southern California's spaces of modernity using the orthographic perspective that so fascinated artists such as Duchamp and Debord. But his vision grew out of such popular-cultural features as the automobile, aerial photography and film rather than any formal theory of the art of cartography. Early in Owens's career at the Los Angeles Times, he produced a set of landscape paintings to serve as full-colour cover illustrations for *Touring Topics*, the Automobile Club of Southern California's monthly magazine (Ott 2000) Owens's Auto Club illustrations pictured an iconic southern California of Spanish missions, Washingtonia palms, Joshua trees, Monterey pines on the California coast, always with a car placed prominently within the scene. The automobile turned California's coasts, mountains and deserts into scenery, landscapes for visual consumption, and Owens's illustrations in the Los Angeles Times mapped new routes as they were opened. The scenic mode of vision encouraged by the newspaper's Motoring and Outdoor sections, and by magazines such as Touring Topics, might be called a new 'scopic regime', connecting landscape scenery, consumption and speed (Jay 1993). [. . .] This new spatiality stimulated new cartography [. . .] [such as] novel ways of illustrating space, of which Owens and Burke were pioneers. [. . .]

The association of speed, space and new ways of seeing created by the automobile was even more pronounced in the experience of powered flight. [. . .] Since southern California was developing as an important centre of aircraft manufacture and air culture in the early twentieth century, the aerial view of landscape was particularly appropriate for negotiating its wide, often uninhabited, desert and mountainous terrain. From early in his career with the Los Angeles Times, Charles Owens used aircraft to [. . .] obtain new perspectives on the rapidly transforming landscape of the area. When he was unable to sketch or photograph personally from an aircraft, he would often base his orthographic landscape drawings and maps on commercial aerial photographs.

[. . .] If the automobile were both the generator of southern California's revolutionary form of sprawling, poly-centred urbanisation and the necessary instrument for negotiating modern landscape on the ground, the oblique air photo was particularly suited to capturing its spatial logic. [. . .] By the outbreak of war in 1941, Owens had perfected a cartography for mapping the scale and speed of aerial warfare in the Pacific.

Owens's connections with Los Angeles's iconic modern film industry were less intense than his associations with the automobile and aeroplane, but they were nonetheless close. [. . .] His studio sketches reflect a fascination with the moving image that is also apparent in Owens's infatuation with boxing. Owens took ringside seats at boxing matches in order to develop his ability to capture the body in motion and the drama of the fight. [. . .] Like driving and flying, the cinema offers a kinetic spatial experience characteristic of modernity, transforming the possibilities for representing space cartographically. The rapidly pencilled battle scenes on Owens's war maps echo the sparse, dramatic lines of the film storyboard and the ringside sketch.

Southern California's cultural modernity is captured in Owens's action dynamics with their emphasis on technologies of speed and mobility, in the montage format of their graphic architecture, in the collage techniques used in their construction, in the high perspective views, zoom techniques and angle shots that seem to borrow from aerial photography and film making as new ways of seeing and experiencing space and mobility. Owens's work is a modernist cartography for air-age America, not only in its mode of positioning the viewer and its ability to capture speed, but also in its geographical sweep over hemispheres and oceans. [. . .] Modern artists only marginally interested in the map were, in fact, perfectly suited to mapping the spatiality of a wholly twentieth century landscape in southern California and a contemporary space of war in the Pacific.

Cartographic art in the twentieth century

While the idea that cartography has progressed historically from art to science was effectively expunged in the late 1980s, epistemological distinctions between art and science in mapping have remained relatively unexamined, and twentieth century relations between the two practices are unexplored. Since the 1990s, thinking in Science and Technology Studies has tended to dissolve epistemological distinctions between art and science and to highlight the aesthetic role played by scientific images, including maps, in securing science's truth claims. At the same time, Modern artists have rejected aesthetics as the defining feature of their work, distinguishing art on the basis of imaginative, creative, provocative and exploratory practice, all features art shares with science.

[. . .] So in order to explore the relations between art and cartography in the modern period we should shift attention away from the map itself and towards the process of mapping and consider the ways in which maps are deployed in both scientific and artistic projects. During the twentieth century, when the divorce of art and science seemed most complete, we discover a continuous but complex conversation between art, science and cartography. It is best understood in terms of actual practices [. . .], rather than through universal definitions of 'art' and 'science'. Over the course of the century, avant-garde artists consistently distanced their work from aesthetics and used scientific and technological advances for a variety of imaginative, creative and critical ends. As mapping technologies and spatial referencing became even more important this artistic interest increased. [. . .]

One of the map types that attracted particular critical attention among twentieth century artists was the high-angle oblique pictorial map. [...] Such maps incorporated various aspects of modernity: the new spatial perspectives introduced by powered flight, the scale and logic of contemporary spaces that are comprehensible only from above, the synoptic vision of modern state power and twentieth century mass culture, of which tourism is an expression. While not necessarily constrained by scientific cartography, these maps drew on the authority that professional cartography has gained in modern society, thereby attracting the criticism of avant-garde artists. In contrast to avant-garde criticism, less theoretically informed popular artists such as Richard Edes Harrison and Charles Owens favoured synoptic pictorial cartography, using orthographic maps and bird's-eye views to educate a mass public about the novel spatial relations of modern warfare. Seeking to map the spatialities of modernity introduced by the automobile and the aeroplane, they turned to the technologies of aerial photography and cinema and techniques such as collage and montage. Owens transformed the pictorial map into a dynamic image of mid-twentieth century spatial relations for southern California and the theatre of war in the Pacific. But these more popular maps also relied for their success as artistic projects on cartographical literacy and the social authority of mapping achieved by professional cartographers. Scientific cartographic [...] practitioners fretted over the potential deceptions of creative, pictorial and decorative mapping [...]. Meanwhile, a significant strand within avant-garde and popular modern art nurtured and renovated the long historical connections between art and cartography. It did so in consciously political ways that, with hindsight laid the foundations for critical engagements today that challenge the stature of both.

References

Atkinson, D. (1995) Geopolitics and the geographical imagination in Fascist Italy. Unpublished doctoral thesis, Loughborough University of Technology.

Bann, S. (1994) The map as an index of the real: land art and the authentication of travel. *Imago Mundi*, **46**, 9–18.

Burnett, D.G. (2000) *Masters of All They Surveyed: Exploration, Geography – A British El Dorado*, University of Chicago Press, Chicago.

Cosgrove, D. (2001) *Apollo's Eye*, The Johns Hopkins University Press, Baltimore, MD.

Curnow, W. (1999) Mapping and the expanded field of contemporary art, in *Mappings* (ed. D. Cosgrove), Reaktion, London, pp. 253–268.

Dalston, L. (1998) *Wonders and the Order of Nature 1150–1750*, Zone Books, New York.

Dalston, L. and Galison, P. (1992) The image of objectivity. *Representations*, **40**, 81–128.

De Certeau, M. (1988) *The Practice of Everyday Life*, University of California Press, Berkeley, CA.

Harrison, R.E. (1944) *Look at the World: The Fortune World Atlas for World Strategy*, Fortune, New York.

Housefield, J. (1992) Marcel Duchamp's art and the geography of modern Paris. *Geographical Review*, **92** (4), 477–502.

Jay, M. (1993) *Downcast Eyes: The Denigration of Vision in Twentieth-Century French Thought*, University of California Press, Berkeley, CA.

Krauss, R. (1980) Grids, you say, in *Grids: Format and Image in Twentieth-Century Art*, Pace Gallery, New York.

Latour, B. (1998) How to be iconophilic in art, science and religion? in *Picturing Science, Producing Art* (eds C.A. Jones and P. Galison), Routledge, London.

Latour, B. (1999) *Pandora's Hope: Essays on the Reality of Science Studies*, Harvard University Press, Cambridge, MA.

Livingstone, D. (2003) *Putting Science in Its Place: Geographies of Scientific Knowledge*, University of Chicago Press, Chicago.

Mantura, B., Rosazza-Ferraris, P. and Velani, L. (1990) *Futurism in Flight: 'Aeropittura' Paintings and Sculptures of Man's Conquest of Space*, Aeritalia Societa Aerospaziale Italiana, London.

Monmonier, M.S. (1989) *Maps with the News*, Chicago University Press, Chicago.

Nash, C. (1998) Mapping emotion. *Environment and Planning D: Society and Space*, **16**, 1–9.

Ott, J. (2000) Landscapes of consumption: auto tourism and visual culture in California, 1920–1940, in *Reading California. Art, Image, Identity, 1900–2000* (eds S. Barron, S. Bernstein and I.S. Fort), University of California Press, Berkeley, CA, pp. 51–67.

Pickles, J. (1992) Texts, hermeneutics and propaganda maps, in *Writing Worlds: Discourse, Text and Metaphor in the Representation of Landscape* (eds T.J. Barnes and J.S. Duncan), Routledge, London, pp. 193–230.

Schulten, S. (1998) Richard Edes Harrison and the challenge to American cartography. *Imago Mundi*, **50**, 174–188.

Schulten, S. (2001) *The Geographical Imagination in America, 1880–1950*, University of Chicago Press, Chicago.

Shapin, S. (1998) Placing the view from nowhere: historical and sociological problems in the location of science. *Transactions of the Institute of British Geographers*, **23**, 5–12.

Stafford, B.M. (1984) *Voyage into Substance: Art, Science, Nature and the Illustrated Travel Account, 1760–1840*, MIT Press, Cambridge, MA.

Stafford, B.M. (1994) *Artful Science: Enlightenment, Entertainment and the Eclipse of Visual Education*, MIT Press, Cambridge, MA.

Wollen, P. (1999) Mappings: Situationists and/or Conceptualists, in *Rewriting Conceptual Art* (eds M. Newman and J. Bird), Reaktion, London.

Woodward, D. (1987) Introduction, in *Art and Cartography* (ed. D. Woodward), Chicago University Press, Chicago.
Yau, J. (1996) *The United States of Jasper Johns*, Zoland Books, Cambridge, MA.

Further reading

Harman, K. (2009) *The Map as Art: Contemporary Artists Explore Cartography*, Princeton Architectural Press, New York. [A profusely illustrated and rigorously edited overview of modern artistic engagement with mapping.]

Latour, B. (1987) *Science in Action. How to Follow Scientists and Engineers Through Society*, Open University Press, Milton Keynes, UK. [The original exposition of STS-based approaches to the construction of knowledge wherein mapping was first ascribed immutable and mobile characteristics.]

Nold, C. (2009) *Emotional Cartography: Technologies of the Self*. http://emotionalcartography.net/ [An e-book of contemporary essays by artists and cultural theorists on how emotional aspects of everyday life can be visualised using technology.]

Wood, D. (2010) *Rethinking the Power of Maps*, Guilford, New York. [Chapter 7 on 'Map art: stripping the mask from the map' provides a systematic, critical and insightful consideration of contemporary artistic mapping practices.]

See also

- Chapter 1.9: Visualisation and Cognition: Drawing Things Together
- Chapter 1.12: The Agency of Mapping
- Chapter 2.11: Extending the Map Metaphor Using Web Delivered Multimedia
- Chapter 3.10: Affective Geovisualisations
- Chapter 4.8: Refiguring Geography: Parish Maps of Common Ground
- Chapter 5.9: Affecting Geospatial Technologies: Toward a Feminist Politics of Emotion

Chapter 3.10

Affective Geovisualisations

Stuart Aitken and James Craine

Editors' overview

This article argues the case for a more emotional approach to the aesthetics of mapping and suggests that geovisualisation can and should learn from film and deploy its tools in ways that powerfully affect everyday lives.

Originally published in 2006: *Directions: A Magazine for GIS Professionals*, 7 February. www.directionsmag.com/article.php?article_id=2097

> 'By means of the film it would be possible to infuse certain subjects, such as geography, which is at present wound off organ-like in the forms of dead descriptions, with the pulsating life of a metropolis.'
>
> *Albert Einstein*

It may be argued that the very heart of geography and GIS is constituted in large part by the practice of looking and is, in effect, the stuff of images. If we agree, then geovisualisation is highlighted as an important disciplinary and practical endeavour. The period from 1985 witnessed increasing recognition of the power of articulate, moving images to intervene in the ongoing transformations of everyday geography, and yet there remains a reticence within the geovisualisation community to fully embrace the emotional power of cinema.

This essay is about the intensification of emotional life that is possible through moving spatial images. Our primary assumption is that while data visualised through GIS can be provocative, it is often joyless and over-calculated, with a tendency for the program to overwhelm the content. Even the best GIS-visualised data are often more interesting to think about than to experience, more interesting to create than to comprehend – it is most often not the product of a searching soul but of a highly computer-literate mind. And so we argue that although today's geovisualised data and digitally formatted movies may look different and may appear to have separate functions, they are both digital, pixelised, spatial data that engage users/consumers in similar ways. The problem for geographers and GIS specialists, we submit, is that cinematic landscapes, in which humans and human culture play a very important part, are a much more powerful producer of emotional geographies than static or animated cartographies, no matter how the data are visualised. If cinema is more concerned with engaging emotions than celebrating computerised visuals, then why should this not also be the case for geovisualisations? As neurologist Antonio Damasio (1994) has shown, emotions are a huge part of, and are not separate from, our intellectual reasoning.

A case for affect

Think back to the last time you lost yourself in a series of contours, shadings and cartographic symbols. Perhaps you have a treasured map at hand that will do the job for you right now (indeed, the boundlessness of imagination enables continuous return to the same map for fresh marvels). This process involves a particularly quirky engagement, pouring over the image to the extent that we may lose a conscious connection to our corporeality. The space between our conscious knowledge of our bodies and the borders of the map merge. Perhaps time disappears. We are lost to the task of imagining what it would be like to be in this place for the first time. We imagine tramping over the hills or along the streets depicted in the map. We pick up maps of exotic places and indulge our amazement at the contemporaneous

The Map Reader: Theories of Mapping Practice and Cartographic Representation, First Edition. Edited by Martin Dodge, Rob Kitchin and Chris Perkins.

heterogeneity of the planet, what Doreen Massey (2005: 15) calls 'spatial delight'. There may even be an imagining that projects knowledge of local weather, waves, tides, or of the ways mists curl around mountain tops, knowledge not directly accessible from the map.

And there is other knowledge not necessarily accessible from the maps which gives us pause. Two decades of warnings from cartographers such as Brian Harley, Denis Wood and Mark Monmonier highlight the inherent power of maps: the grotesque distortions of variety and uniqueness, the colonialism, the propaganda, the god-trick, the diminishing/submersion of certain subjects for others. Even with these admonitions and understandings, there is still wonder. We argue, and this is our central twist, that more so than cartography, cinema produces wonders that affectively engage and absorb us – they intensify our lives, if only momentarily. There is no reason why geovisualised data should not do the same.

Filmic spatial data

So, let us think for a moment about the mechanics behind the emotional wonders of moving images. Cinema is a particular kind of movement propagated by single picture frames passing rapidly in front of a projection light at a prescribed rate; in its digital form the movement is accomplished by changing hues and colours of the individual pixels that compose the viewing screen.

The active promotion of images in motion in the United States began on 13 June 1891 when *Harpers' Weekly* announced that Thomas Edison had invented a kinetograph, a combination of a moving picture machine and a phonograph. The 1902 Sears catalogue described the kinetoscope's ability to render a 'pictorial representation, not lifelike merely, but apparently life itself, with every movement, every action and every detail brought so vividly before the audience' (quoted in Denzin 1995: 16). What was revolutionary about the kinetoscope was its power over the image. By the 1920s Soviet film maker Sergei Eisenstein was breaking ground with kino-pravda (film truth). Eisenstein's notion of 'the-image-in-motion-over-time-through-space-with-sequence' was about filmic rhythm and the ways film time-space could be manipulated for fullest effect (Aitken 1991). Eisenstein's image-event was not about presenting real life, or even film as a mirror of real life. The focus of kino-pravda on filmic rhythms, sequences, framings, spectacles, jump cuts and montages was about the creation of illusion to the extent that we, the consumers of the images, would suspend our disbelief.

From cartography, DiBiase and his colleagues (1992: 206), closely paralleling the earlier thoughts of Eisenstein, extend spatial images into the realm of non-static, animated representations existing within two or three spatial dimensions plus a temporal dimension. Their dynamic variables of duration (the number of units of time that a scene is displayed), rate of change (a proportion formed by the magnitude of change in an attribute and the duration of scene) and order (the logic of chronological sequencing of scenes associated with a time-series data set) are used to form dynamic representations. This is precisely what Eisenstein did with kino-pravda, but his focus was on the emotional impact – he called it shock value – of the image-event.

Affecting the geovisual

Geographers learn from looking and the geovisualisation approaches established by DiBiase, but also Cartwright, MacEachren and others, offer a unique way of interpreting visual representations. Indeed, MacEachren's famous elaboration of semiotics to cartography is about a science and art that is closely connected to the study of film (Aitken and Craine 2005). Let us assume that geography must move past pure semiotics and cognitive theory in order to open up the discipline to fresh investigation. Further, geovisualisation must refuse to take vision for granted and should, instead, insist on 'problematising', theorising, critiquing and 'historicalising' the visual process. As a case in point, data exploration through the use of 'map metaphors', Cartwright (2004: 32) famously argues, is changed to 'new forms of multisensory and multimedia communication' where visual information gathering is augmented with other sensory stimuli. Unfortunately, Cartwright stops short of exploring the important dimensions of affect and absorption.

With the recognition of multimedia as a viable way to communicate geographic information, we can examine how new technologies provide new modes of data exploration and new forms of analysis and cognition through affect. To take just one example, Platt (1995) shows how 'storytelling' offers the user the option of engaging the geography of a particular place. Video gaming is a case in point, where filmic digital images are merged with usable/controllable products to create innumerable storylines. Imagine the young gamer deftly avoiding missiles and light-beams from enemy fighters and courageously engaging the enemy in LucasArt's Star Wars series as a Death Star brings its planet-destroying cannon on the unsuspecting world below. Her passion may for some border on addiction, but it is nothing if it is not self-absorbed. Do you recognise something important about the ways young people engage with these games? Watch her weave in and out of the canyons that comprise the surface of the Death Star. And like Luke Skywalker feeling the force, she successfully sends missiles down the postage stamp opening into the main reactor. Boom! We are treated to a spectacular conflagration. Next level and a new storyline.

She is now ramping down a hierarchy of scaled images towards the planet below. The images coalesce in a way that suspends her disbelief about a myriad of pixels changing hue. Indeed, she has no knowledge of the mechanics of digital animation. She is simply hurtling towards the surface to chase some of the renegade enemy fighters that escape the conflagration aboard the Death Star. Now she is down in some desert canyons, chasing more enemy fighters. You are impressed at this time with the way she negotiates scale. The pixels change hue to form digital images of dynamic landscapes, aerial photographs if you will, from the planetary scale down to just above the desert floor.

We'd like to finish with a thought about the ways film and animation contrive the stories of our lives. We believe that this is about the power of moving images to create spaces and affect societal change. And it is about the subtle pervasiveness of these images in contemporary culture. Einstein (in the opening epigram) was right when he suggested that geographers do not appreciate the power of the moving image. Nor do we know how to use that power to better the world in which we live through emotion-charged geovisualisation. Obviously, this power goes way beyond video gaming. As we are absorbed into mobile geographic images, it is possible also to be affected by the power of those images to highlight social injustice, geographies of AIDS, or the tragedies of the global sex trade. Affective geovisualisations are soulful; they tug at our hearts to the extent that we may be mobilised to action.

References

Aitken, S.C. (1991) A transactional geography of the image-event: the films of Scottish director, Bill Forsyth, *Transactions of the Institute of British Geographers*, **16**, 105–118.

Aitken, S.C. and Craine, J. (2005) Visual methodologies: what you see is not always what you get, in *Methods in Human Geography* (eds R. Flowerdew and D. Martin), Prentice Hall, Harlow, UK, pp. 250–269.

Cartwright, W. (2004) Geographical visualization: past, present and future development, *Journal of Spatial Science*, **49** (1), 25–36.

Damasio, A. (1994) *Descartes' Error: Emotion, Reason, and the Human Brain*, Penguin Books, New York.

Denzin, N. (1995) *The Cinematic Society: The Voyeur's Gaze*, Sage, London.

DiBiase, D., MacEachren, A., Krygier, J. and Reeves, C. (1992) Animation and the role of map design in scientific visualization, *Cartography and Geographic Information Systems*, **19** (4), 201–214.

Massey, D. (2005) *For Space*, Sage, London.

Platt, C. (1995) Interactive Entertainment, *Wired Magazine*, **3** (9). www.wired.com/wired/archive/3.09/interactive.html.

Further reading

Conley, T. (2007) *Cartographic Cinema*, University of Minnesota Press, Minnesota, MN. [Theoretical exploration of the relations between mapping and motion pictures.]

Kwan, M.P. (2007) Affecting geospatial technologies: toward a feminist politics of emotion. *The Professional Geographer*, **59**, 33–54. [Provides a critical and feminist argument for a more emotional engagement with geospatial technologies. Excerpted as Chapter 5.9.]

Lammes, S. (2008) Playing the world: computer games, cartography and spatial stories. *Aether: The Journal of Media Journal*, **3**, 84–96. [Discusses the mutable qualities of digital mapping deployed in video games.]

Nold, C. (2009) *Emotional Cartography: Technologies of the Self*, http://emotionalcartography.net/. [An e-book of contemporary essays by artists and cultural theorists on how emotional aspects of everyday life can be visualised using technology.]

See also

- Chapter 1.12: The Agency of Mapping
- Chapter 2.11: Extending the Map Metaphor Using Web Delivered Multimedia
- Chapter 3.9: Mapping, Modernity: Art and Cartography in the Twentieth Century
- Chapter 5.9: Affecting Geospatial Technologies: Toward a Feminist Politics of Emotion

Chapter 3.11

Egocentric Design of Map-Based Mobile Services

Liqiu Meng

Editors' overview

Meng's article focuses on egocentric design of maps for mobile applications, which are characterised by short-term, transient usage by individual users. It highlights the key importance of centring, redundant coding, continuously varying multiple levels of detail, the severe limits of display space, single window display, augmented focusing, orientation and affect. Her discussion concludes that a greater focus on usability studies is required in ongoing mobile map design research.

Originally published in 2005: *The Cartographic Journal*, **42** (1), 5–13.

Background

The widespread Internet access and the booming ubiquitous computing technologies have not only blurred the distinction between office and home, but substantially contributed to the increasing mobility of our life. Hand-held mobile devices that have already exceeded traditional PCs in number are rapidly evolving from toys to tools. Experiences and statistics hitherto have shown that maps remain the most popular communication language of spatial information also for mobile applications, apart from the fact that more and more location-based services are being integrated with the physical environments, especially urban areas where computer chips are nearly omnipresent. Being equipped with wireless hand-held maps, mobile people can be accurately and timely informed of the events taking place in their surroundings, get therefore well prepared for their tasks.

'Putting yourself in the world and the world on your palm', however, does not automatically lead to an improved mobility unless both worlds 'melt' together in your brain. The synchronous interactions with the reality and its map are conducted through effortful collaborations of sensory and motor organs of the user. In order to keep the attention of a mobile user on his interaction with the reality, the map should be used unobtrusively. This requires, on the one hand, an intuitively operable mobile device of a nearly invisible size, on the other hand, a pervasive visibility of map symbols necessary for their immediate comprehension. Such a paradoxical requirement makes the design of map-based mobile services a challenging research topic.

[. . .]

Internet maps

In addition to the usual map functions, an Internet map often serves as a metaphor to spatialise the information space and as a collaborative thinking instrument shared by spatially separated users. Its individual symbols can be treated as hyperlinks leading to various sorts of virtual places in cyberspace. These open-ended hyperlinks make an Internet map more powerful, at the same time, more fragile than a closed mapping system. [. . .]. This means that both the designer and the user are confronted with a

The Map Reader: Theories of Mapping Practice and Cartographic Representation, First Edition. Edited by Martin Dodge, Rob Kitchin and Chris Perkins.

cognitive overhead associated with the encoding and decoding of hypermedia information. Although the designer is able to make full use of audio-visual variables to distinguish the hyperlinks from other symbols and provide necessary navigational guide, he has little control over the linked contents and their design parameters. As soon as the user decides to click on a recognised hyperlink, he runs certain risk of invoking unexpected change of the web page appearance, losing his gained information from the vision field, getting irritated by erroneous links or cryptic message, landing nowhere or finding no way back. The erratic characteristics of hyperlinks are so far a major barrier that hampers the usability of Internet maps. Being bound to stationary computers an Internet map usually occupies the entire vision field of the user, thus demands their exclusive attention. Moreover, the unpredictable reaction time of hyperlinks always reminds the user of the 'thickness' of a web map.

Mobile maps

The realisation of wireless Internet access has finally brought web maps back to mobile environments where they are most needed. Along with the triumph of unification of two open-ended systems – the real and virtual world – however, cartographers are facing a number of more acute design constraints. The miniaturised display devices make mobile maps more personal services than their predecessors. Although the same mobile map can be shared by virtually networked and spatially separated users like a usual web map, it does not primarily act as a collaborative thinking instrument, rather a common memory to back up the group mobility. The contents and presentation styles of a mobile map need to be adapted to the actual requirements and cognitive abilities of individual mobile users. Often the short-term memory of a mobile user can accommodate far less information than the maximally allowed visual load on the display surface. Figure 3.11.1 illustrates the difference between a presentation based on empirically determined minimal graphic dimensions for screen resolution and a presentation with the most relevant information for a biker. Apart from technical issues such as network accessibility, positioning quality and transmission speed, the designer has the essential task to match in real time a 'meagre' map with the 'sharp' user requirements filtrated through a very narrow space-time slot.

Necessity and usability of egocentric maps for mobile applications

[...] From a designer's perspective the production of a geocentric map intended for a large target group might be cost effective because every user can get something useful from the map. In mobile usage context, however, a geocentric map can hardly remain usable due to the following

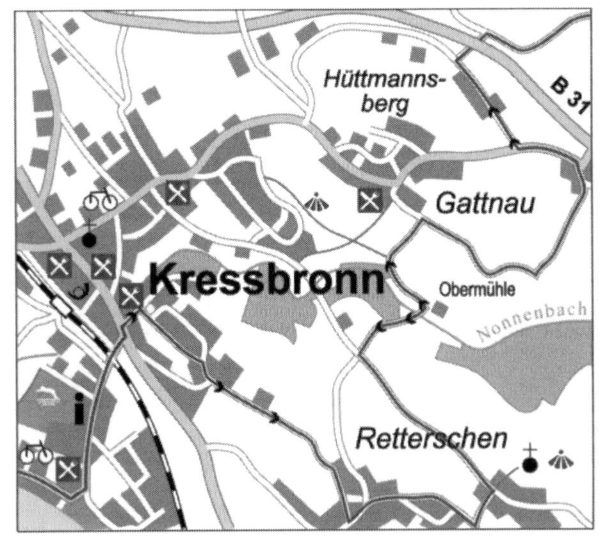

Figure 3.11.1 Presentation based on empirically determined minimal graphic dimensions for screen resolution (left); presentation with the most relevant information for a biker (right).

reasons: what a large target group needs is often much more than what is necessary for an individual user; the unnecessarily large amount of information claims much network capacity to get transmitted from server to client; the rendered graphics is likely cluttered on the small display, which makes the recognition of particular symbols difficult. Unlike a geocentric map, a mobile map is somewhat like a snapshot with highly selective information from an environment of a very limited scope. It should be able to:

- Support the user non-intrusively. The mobile environment badly needs the so-called 'subdued computing' (Ibach and Horbank 2004). The interaction with the map should take place intuitively and make the user feel like effortlessly fetching information from the ambient environment into which map has discreetly melted.

- Draw the user's attention immediately to their wanted information, which is either located at the most salient place within the vision field or visualised as a floating figure on a receding background.

Not only the technical factors, such as limited display size, energy supply and bandwidth of wireless network, but also non-technical factors, ranging from time-critical user tasks, constantly altering environments to volatile user emotions, force the designer to accommodate in the mobile map only the information that is instantly needed and effortlessly comprehensible. In this sense, the general postulate in conventional cartography 'Map use is an effortful process that needs training' is no longer valid for mobile applications. Whether a mobile map is immediately usable depends largely on how well it satisfies the egocentric needs of its intended user. Figure 3.11.2 shows an egocentric presentation of TomTom GO (www.tomtom.com) for a driver. The driver simply needs to follow the arrows intuitively embedded in the perspective view.

The usability of an egocentric mobile map can be judged by the degree to which both pragmatic and hedonistic requirements of its target user are satisfied. The pragmatic requirements are objective, easily measurable and purpose orientated. They concern the effectiveness and efficiency of map use. While the effectiveness has to do with questions such as whether the map contains information wanted by the user, whether the user really understands the meanings of map symbols and whether the interactive functions achieve the expected performance, the efficiency can be measured by the times spent on successfully finished actions such as searching, comparing, navigating and so on, the number of successfully finished actions within a unit time, the price of the service compared with its performance and so on. Hedonistic requirements are subjective, diffuse and independent of map purpose. They concern the user's emotion, such as joyfulness or irritation,

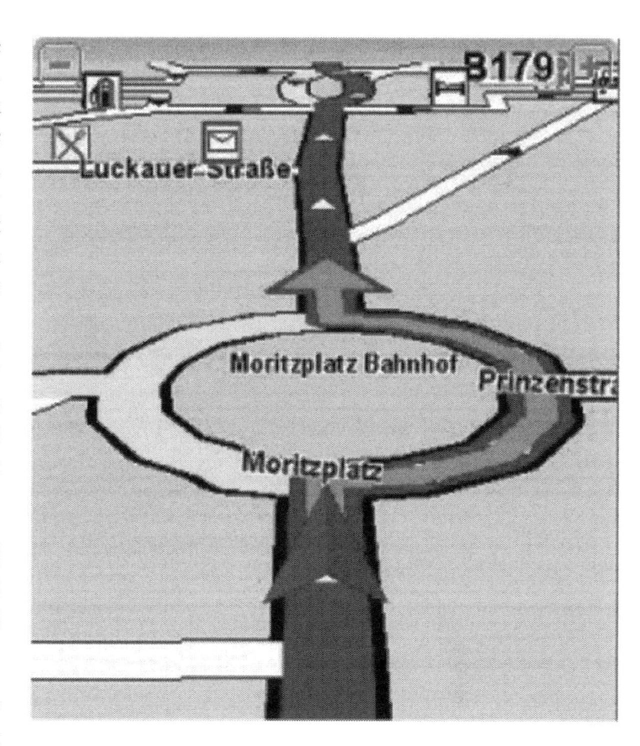

Figure 3.11.2 Screenshot of an egocentric presentation of TomTomGO (www.tomtom.com).

during the interaction with the map [...]. A well designed egocentric map is able to relax the user, extend their physical capability and raise his self-confidence. Often the hedonistic quality is correlated with the pragmatic quality. On the one hand, effective map contents tend to enhance the feeling of trust and safety. On the other hand, in a relaxed mood the user tends to be more efficient.

The ego centre of a mobile map user

The profile of a mobile user from which their ego centre will be derived is generally composed of a situation-independent part and a situation-sensitive part, where the situation is a function of space and time (Reichenbacher 2004: 77). Characteristics such as user goal, demographic data, personal preferences, habits, visual literacy, ability of spatial cognition, domain knowledge and computer experiences are relatively situation independent. Their values tend to remain stable throughout a map use session. Other characteristics such as user activity, information need and emotion are apparently situation sensitive. Depending on the mobile usage context, the ego centre is reflected in various aspects, such as (1) the current user location in the form of a point, a route or a region of varying size, (2) one or many locations that are currently of interest to the user, (3) data items that are currently of interest to the user, (4) actions or operations that are frequently performed by the user, (5) symbolisation styles preferred by the user, and (6) interaction modalities preferred by the user. In its most general form the ego centre can be expressed as a function of all these aspects. Detecting such an ego centre requires a continuous contact with the user in their mobile usage context.

As reported in Meng (2004), various approaches can be applied to gather implicit and explicit evidence about ego centres of an individual mobile map user. The usual outcome of these approaches consists of both objectively recorded user behaviour and subjectively expressed user opinions. In a post-processing stage, the following tasks need to be performed:

- Categorisation of user tasks of various granularity levels based on their recurring frequency.
- Identification of user stereotypes based on their interaction trajectories and personal constructs.
- Conversion of user statements into values or value ranges of individual design parameters.
- Enrichment of the user location with semantic attributes or events.
- Classification of user requirements based on their relative stiffness into 'must', 'should', 'could' and 'may'.

With regard to the transient nature of mobile environments, however, it is often too late to begin with the egocentric map design after the ego centre is precisely determined. In fact, the design process can take place at the same time as the process of user tracking, or as soon as some evidence of the ego centre has been determined.

Design patterns of egocentric mobile maps

The design process could lead to many alternative solutions, with each trying to highlight the contents of the ego centre in a distinctive manner. The following design patterns that have been partly practised in cartography, computer graphics and human–computer interface design are worthy to be (re) considered in the mobile usage context.

Centring

If the ego centre is a spatially concentrated location such as a building block, a street segment or a region within the walking distance, it will be placed around the optical or geometrical centre of the vision field. Such an intuitive graphic transform can be easily realised as soon as the location of the ego centre is explicitly defined by the user themselves or automatically determined by means of positioning techniques. A mobile map that is dynamically centred on the location of the user's actual interest [...] possesses, in addition to its functionality, a certain hedonistic quality because it implies that the system is working, thus makes the user feel relaxed.

Redundant encoding

The contents of ego centre can be visually differentiated from the peripheral information by using a graphic variable such as colour tone. [...] Adding the label, the ego centre is made even more salient. Though this redundant encoding method does not bring additional information, it helps to reinforce the visual acuity of the ego centre and at the same time allows users who miss one graphic variable (e.g. colour-blind users) to attend to the ego centre.

Continuously varying level of detail

By applying distortion techniques such as anamorphosis, multifocal projection, fisheye view, interactive magnifier [...], a spatially concentrated ego centre can be displayed as a blown-up circular or rectangular area containing fine but legible details while the peripheral area is progressively displaced, compressed and generalised. Since the enlarged ego centre is balanced by the reduced peripheral area, the overall display space remains constant. What the user sees is a detailed ego centre in continuous connection to an overview of its context.

Multiple levels of detail

A task-relevant location can be expressed at multiple discrete levels of details. For instance, it is possible to present the actual location of a city tourist at street level and in an overview. These two levels can be automatically switched over to each other, or they can be visualised simultaneously by embedding the higher level of details in a lower one. Figure 3.11.3 demonstrates one such example. Further, the overview can be treated as a movable piece that does not need to be laid out precisely (Tidwell 1999). The user can freely drag the overview to an adjacent place that will neither cause eye shift nor interfere with the actual ego centre. Moreover, the synchronized overview and detailed view allows the user to jump to a new ego centre whenever he wants.

Figure 3.11.3 A detailed route from Munich airport to Putzbrunn embedded in the overview of Munich city.

Space contraction

In the case that the ego centre is composed of spatially widely separated locations, its overall extension will be intentionally contracted to make the associated contents simultaneously visible within the same window. The level of details and the topological relations remain unchanged during the contraction. [...] This design pattern eliminates the need to use interactions such as scrolling and panning.

Single window with details on demand

The screen space of a mobile device is obviously not able to accommodate the task-relevant mapping contents in multiple windows adjacent to each other. Drilling through the tiled windows, on the other hand, may cause the abrupt change of the presentation or disorientation. For this reason, a single window with details on demand is a preferred design pattern. While the actually required spatial contents are made visible within the same window, the supporting information, such as legend, live picture and on-line help instructions, is kept hidden. The user is allowed to make the hidden information temporarily visible (or audible) at a selectable granularity level. For instance, if the user tries to click on a point of interest, its associated details will be broadcasted by an anthropomorphic agent or pop-up in a blown-up frame which has, in case of plain text, a manual or automatic scrolling function adapted to the average reading speed.

Augmented focusing

If the ego centre contains many salient but sparsely distributed spatial features, the mobile user with his limited memory load may still have difficulties attending to all of them simultaneously. Moreover, large jumps from one salient feature to another may cause the user to become confused or disoriented. Therefore, a fine tuning of the distinctions among these salient features based on their relevance to the actual user action is necessary. The one or two most relevant map features for the actual user action will be visually further enhanced with extra design elements, such as a bounding box, three-dimensional symbols, blinking, voice, large label and animated magnifier. Being termed as augmented focusing, this design pattern proves desirable for time critical tasks such as driving or walking along a complex route. Although the mobile user needs an overview of the entire route containing the starting point, the destination, a number of intermediate stations as well as landmarks, at each particular moment, however, they may be just attentive to the most recently passed station and the on-coming station with its one or two landmarks. [...]

Orientation gesture

Orientation plays a significant role for many navigation related tasks. Technically it is possible to align a two-dimensional map and its labelling dynamically with the actual moving direction of the user, frequent rotations, however, are ergonomically uncomfortable. The alternative solution, with a three-dimensional perspective scene viewed from the actual user location, as shown in Figure 3.11.1, may guide an individual user more intuitively. Yet three-dimensional perspective scene makes it hard for the user to judge the relative distances in addition to typical problems caused by occlusion. With regard to these known drawbacks, local orientation gesture such as arrows or animated objects integrated in the static scene proves an inexpensive and more flexible solution. Being guided by the orientation gesture the user does not have to change their cognitive orientation model established through his prior map reading exercises. Meanwhile, the availability of a consistent reference invokes the feeling of safety. Furthermore, such an egocentric map can be shared by a group of users moving in different directions within the same scene. Figure 3.11.4 illustrates a scene of group dating where the actual location and moving direction of

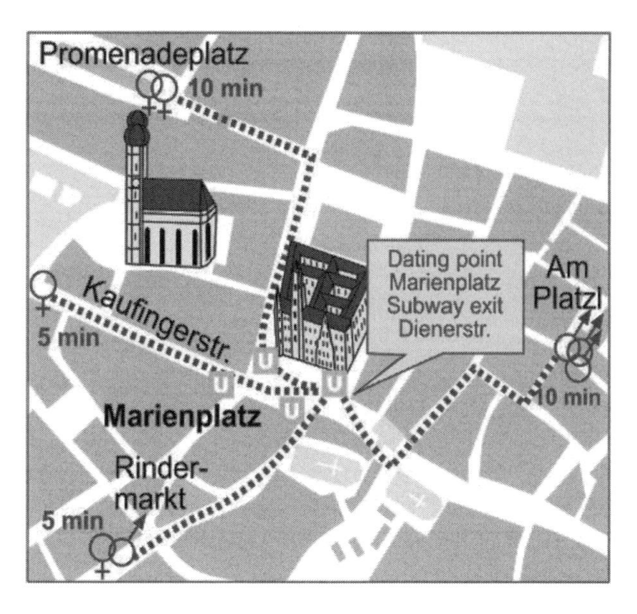

Figure 3.11.4 An egocentric map shared by eight people who are approaching from different locations to a dating point.

every group member are visually highlighted upon a grey background.

Affective emphasis

Many users do not consult a map at all unless they are disoriented or seized with a panic (Muehrcke 1992). Situations that cause troubles are often demanding situations involving many error-prone spatial decisions. Emotionally unstable users may behave entirely differently from what they would likely do in normal situations. Some may experience a memory blackout, while others may, on the contrary, show an increased performance of information processing. Even the same user may react differently to similar panics that occur on different occasions. The design pattern 'affective emphasis' has the goal to encode the spatial information with symbols that do not necessarily carry additional information, but can possibly invoke positive emotions, thus calm down a panic. [...]

Research challenges

The cartographic practice hitherto has very well taken the general physiological and psychological user capabilities into account but largely ignored the special user requirements, partly out of economical concerns. With the widespread mobile geoservices, there is a growing demand on personalised map services because most of the mobile tasks are sensitive to location and/or time and map use is always a personal experience. Unlike the geocentric maps that serve many purposes and many users in a balanced manner, an egocentric map is intended for an individual person for their task at hand. Therefore, no compromise is necessary. While the geocentric map design strives for the ultimate goal that sounds like 'what you see is what you want', the egocentric map design focuses on the goal that may sound in a reversed order 'what you want is what you see'. According to the motto 'if it doesn't feel right, who cares if it works' (Hassenzahl and Hofvenschiold 2003), mobile map services will not be accepted by their intended users until they satisfy both the pragmatic and hedonistic requirements. Egocentric mobile maps do not attempt to replace the geocentric maps. Rather they will serve as a more usable means to support the mobility.

One of the key issues in designing egocentric maps deals with the detection of ego centres by tracking dynamic user behaviour and constructing user profiles. What matters most in the mobile context are user's actual task, information need and cognitive ability. Activity theory based on the belief 'you are what you do' proves an efficient approach to user modelling [...]. Being driven by a certain task or goal, user behaviours such as interactions with the map and mobile trajectories in physical environments (Mountain 2004) give valuable cues on the time pressure, the information need and the way this need gets satisfied. However, activity theory has its limitations as soon as the actual cognitive ability of the users is concerned. It is generally known that the actual cognitive ability of a user is correlated with his actual emotion state, which tends to fluctuate more in mobile than in stationary environments, and more with time pressure than without. Bearing this fact in mind, it would be a better choice to embed personalised solutions in the mobile map than to screw up or down the mapped contents. By providing ready-to-work services instead of ready-to-get information in troublesome situations, the mobile map takes over an essential part of mental effort for information processing from the user, thus making the non-deterministic influence of the user's actual cognition ability less critical.

In the realisation of egocentric mobile maps on small display devices then, the designer typically has to find trade-offs:

- between the frequency of adaptation and the necessary consistency a mobile user would like to rely on;
- between the degree of adaptation and the degree of interaction;

- between the maximally allowed visual load on a mobile display device and the minimum amount of information required by the user for a certain moment;
- between the maximum number of visual signs a user can recognise within a certain time limit and the minimum number of information units he can efficiently remember; and
- between reusable and one-time design patterns.

This list is not intended to cover all the research questions involved in the egocentric design process. Rather, it tries to highlight some essential problems that are frequently encountered in the design practice.

References

Hassenzahl, M. and Hofvenschiold, E. (2003) If it doesn't feel right, who cares if it works? oder muss Software mehr als nur gebrauchstauglich sein? *Proceedings of the 1st Annual GC-UPA Track*, Stuttgart, Germany.

Ibach, P. and Horbank, M. (2004) Highly available location-based services in mobile environments. *International Service Availability Symposium*, Munich, Germany.

Meng, L. (2004) About egocentric geovisualization. *Proceedings of the 12th International Conference on Geoinformatics*, University of Gavle, Sweden.

Mountain, D. (2004) Exploring mobile trajectories: interactive approaches for spatio-temporal data EURESCO Geovisualisation Conference, Kolymbari, Greece.

Muehrcke, P.C. (1992) *Map Use: Reading, Analysis, and Interpretation*, 3rd edn, JP Publication, University of Wisconsin, WI.

Reichenbacher, T. (2004) Mobile cartography: adaptive visualization of geographic information on mobile devices. Unpublished PhD thesis, Technical University of Munich, Germany.

Tidwell, J. (1999) *Common Ground: A Pattern Language for Human–Computer Interface Design.* http://www.mit.edu/jtidwell/common_ground.html.

Further reading

Dobson, J. and Fisher, P. (2003) Geoslavery, *IEEE Technology and Society*, **22**, 47–52. [A well argued polemical piece that highlights the range of dangers of locative technologies and personal tracking in terms of individual privacy and democratic freedoms.]

French, R.L. (2006) Maps on wheels: The evolution of intelligent automobile navigation, in *Cartographies of Travel and Navigation* (ed. J.R. Akerman), University of Chicago Press, Chicago, IL. [An historical introduction to the development of mobile navigational aids.]

Raper, J.F., Gartner, G., Karimi, H. and Rizos, C. (2007) A critical evaluation of location based services and their potential, *Journal of Location Based Services*, **1** (1), 5–45. [A cogent and wide ranging review of locational awareness technologies and application with much of relevance to mapping.]

See also

- Chapter 2.10: Mobile Mapping: An Emerging Technology for Spatial Data Acquisition
- Chapter 2.11: Extending the Map Metaphor Using Web Delivered Multimedia
- Chapter 2.12: Imaging the World: The State of Online Mapping
- Chapter 3.12: The Geographic Beauty of a Photographic Archive
- Chapter 4.10: Citizens as Sensors: The World of Volunteered Geography
- Chapter 4.11: Usability Evaluation of Web Mapping Sites

Chapter 3.12

The Geographic Beauty of a Photographic Archive

Jason Dykes and Jo Wood

Editors' overview

During the 1990s and in the first decade of the new century innovative ways of representing information in a visual way have developed into a burgeoning research field. The possible dimensions of information visualisation extended beyond the terrain occupied by cartographic representations of geographic space. Meanwhile, data are increasingly being made and disseminated collaboratively. This final article in the section examines the interface between these two worlds, and in particular on the aesthetics of an encounter between The *Geograph* project, an online crowd-sourced archive of geographically referenced photographs, and ways in which geographic qualities of these data might be represented using information visualisation techniques. Dykes and Wood chart these attributes using a number of different 'treemaps'. They conclude that these data-dense graphics reveal aspects of beauty in the encounter which are at once creative, innovative and problem solving.

Originally published in 2009: Chapter 6 in *Beautiful Data: the Stories Behind Elegant Data Solutions* (eds Toby Segaran and Jeff Hammerbacher), O'Reilly, Sebastopol, CA, 85–104.

Photographs can be beautiful. It seems almost demeaning to consider something that can capture experience, kindle emotion and invoke the sublime as merely data. Yet once stored digitally, we can process a photograph's binary digits just as we might any other stream of numeric data. But we may go further. By collecting those photographic representations together, by describing them, by arranging them, we can create context. Here a new beauty emerges, something that is fed by the beauty of the images that comprises the collection, but which is so much more than the sum of its parts.

In this chapter we explore the beauty that emerges when we consider the geography of a photographic collection and we try to map that geography visually. By geography we mean the information that allows us to associate something with a place and location (two quite distinct concepts). And when we're dealing with data, there's a lot of geography about. Some estimates suggest that up to 80% of data are geographic (MacEachren and Kraak 2001). This information might be recorded directly through latitude and longitude coordinates, or indirectly though association with a postcode, a name or some other notion of place. That geography can be a useful way of organising, filtering and interpreting data.

Geography can be assigned to data in a number of ways. It may be part of the data collection process (for example, through satellite remote sensing). It may emerge during data query and interpretation (for example through location based services such as Google Local). Or it may be generated through more sophisticated spatio-temporal analysis as typified by the current interest in geovisual analytics (Andrienko *et al.* 2008). Here we will use as our starting point data that have been made geographic as part of the collection process through locations but that also contain more general descriptions of place. The *Geograph* archive contains over one million photographs that have been pinned to a specific latitude and longitude, either automatically by GPS enabled devices such as the iPhone, or by individuals who have manually located their photos on a map. Additionally, the geography of these

The Map Reader: Theories of Mapping Practice and Cartographic Representation, First Edition. Edited by Martin Dodge, Rob Kitchin and Chris Perkins.

photos has also been described by their owners as freeform text, perhaps by naming nearby locations, or by describing features or activities captured by the photo. There is complexity and subtlety here, and, as we shall see, beauty can emerge when we try to visualise this to enhance the interplay between location and descriptions of place.

By beauty we mean a characteristic of an entity that provides pleasure, meaning or satisfaction. Things of beauty may exhibit all sorts of qualities relating to aesthetics, truth, harmony, elegance, efficiency and sublimity. Where function is important, efficiency is a key trait. For example, when writing beautiful code, it is frequently the case that the code has a very specific purpose, such as sorting a list, solving a system of linear equations or performing a Fourier transform. The beauty of the code can result from the effectiveness in meeting that purpose (Kosara 2007). When dealing with beautiful data, such a purpose might not be quite so precise. Exploration of data is an important part of the scientific endeavour and can lead to insights, hypotheses to be tested and validation of prior theory. Beautiful data warrant exploration. They contain patterns, structures and anomalies that are not immediately apparent but emerge as reward for mining the hidden depths within. Here we build upon two long traditions for exploring data in a visual manner. The complexity of spatial relationships and the way in which they can be encoded with graphics means that visual methods have a long history in the geo-sciences, where they are widely used to support knowledge discovery and to communicate information. Cartography has developed robust techniques for showing geographic data visually. Over the centuries it has innovatively and successfully combined the objective rigour of science with more subjective skills of interpretation design and critique. Maps themselves can be things of beauty as well as their referents. Information visualisation encapsulates the process of visual exploration of data that may not have any geographic component, through the generation of graphs, charts and visual interaction. This chapter reports on approaches to exploring beautiful geographic data that combine elements of cartography and information visualisation.

Beauty in data – *Geograph*

We consider *Geograph* a beautiful data set for a variety of reasons. An engaging combination of valued data source, online community, game and motivation for exploring the countryside, this rapidly expanding archive of geo-referenced and annotated photographs of the landscape of the British Isles gives ample scope for considering beauty in the big picture and in myriad small details.

Originally conceived by Gary Rogers, and supported by Paul Dixon, Barry Hunter and a growing team of moderators, *Geograph* is an effort to collect 'Geographically representative photographs and information' (Geograph 2008) for each one square kilometre grid cell in Great Britain and Ireland. At the time of writing this process has involved over 8000 contributors collecting and documenting more than a million photographs with almost 90% of the 244 000 one square kilometre grid squares in Britain '*Geographed*'.

Contributors to *Geograph* are free to select their representative view, but the geographic features should be indicative of the typical human and physical geography in the grid cell – 'think what a child looking at a map in a geography lesson might find useful when trying to make sense of what the human and physical geographical features in a given grid square actually look like' (Geograph 2008, www.geograph.org.uk/article/Geograph-or-supplemental).

We see beauty in various aspects of *Geograph*:

- The objective of producing an open archive of purposefully selected and annotated images has an engaging blend of simplicity (the idea), complexity (the process – which is structured in a manner that also has some beauty – see below) and utility (the resource). The associated organisation and devotion of those running the project to maintain the collection and make it usable are admirable and impressive.

- There is beauty in the 'collective effort' approach used in generating the archive from the 'bottom up' and the way in which technology and person-power are used to achieve the aims. There is considerable beauty in the collaborative 'citizens as sensors' (Goodchild 2007) nature of the project that relies upon common understanding and broad cooperation with limited individual gain.

- The implementation and presentation of the idea in an engaging, accessible and stimulating web site that provides access to the information in so many ways has a considerable aesthetic and technical quality.

- The maps generated from *Geographs* at a range of scales to provide insights into the various processes involved in generating the collection are beautiful. These include a number of innovative cartographic representations and interactive features that provide access to this mass of information that change as the collection is updated. These include *Geograph* densities, distribution maps of contributors and their contributions and geo-referenced photomosaics.

- The individual contributions have aesthetic appeal as representations of human and physical landscape determined by those who inhabit and visit the locations that are depicted. This is formalised by the community, as candidates for the *Geograph* of the Year (GOTY) are chosen on a weekly basis from the photographs contributed.

Geograph is somewhat typical of the broad user-generated geo-referenced data sets that are becoming available but its quality, implementation and uniqueness make it especially beautiful in our view. There are real opportunities for exploring the *Geograph* data and generating new knowledge that may further accentuate its beauty. We provide some examples in the sections that follow that describe some of our visual exploration of the geographies represented in the collection.

Visualisation, beauty and treemaps

Before we delve into the *Geograph* archive itself, it is worth considering some of the motivations and visual techniques that might be usefully applied to exploring the collection. The cartographic tradition has understandably focused predominantly on static cartography that is reproduced using traditional media. Most of our work in the last 15 years has been interactive – utilising digital technologies for reconsidering maps and redeploying them as responsive interfaces that dynamically respond to queries to support exploration (Fisher 1998). We have aimed to ensure aesthetic quality in the smoothness of the interactions and satisfaction by developing informed views and interactions that provoke thought and discovery. However, some of our recent work has re-emphasised *data density* and focused back on the fundamental cartographic design decisions associated with generating layouts and symbolism that use space efficiently and effectively. This work has been partially motivated by hardware advances that have made processing and displaying large data-dense graphics more feasible, but is in effect a response to the need for new effective, informative and aesthetically pleasing methods for graphically representing the kinds of larger data sets that are increasingly available as we follow Tufte's (1983) advice to 'present many numbers in a small space'.

What is beauty in visual data exploration?

As we consider beauty to be associated with some stimulus that results in a positive perceptual experience involving pleasure, meaning or satisfaction it is an inherently subjective quality. In visualisation beauty is often in the eye of the developer or designer. Apposite calls have been made to the community to formally critique visualisation work in an effort to consider aesthetics more collectively (Kosara 2007). However, until a usable body of knowledge is developed we are reliant upon broad principles and rules of thumb when developing aesthetically pleasing graphics.

A number of these are used by Oram and Wilson (2007) and can be usefully applied to data visualisation. For example, Kernighan (2007) identifies characteristics of beautiful code that include compactness, elegance, efficiency and utility and informally quantifies compactness by indicating that 'Ideally the code would fit onto a single page'. Matsumoto (2007) deems code to be less than beautiful if it is difficult to understand and has used this principle to inform his approach to developing the Ruby programming language. But 'difficult to understand' should not be confused with complexity. In the context of visualisation a simple graphic that is easy to understand but shows very little data is not beautiful. Rather, beautiful data visualisation shows things that are complex but in a way that makes them easier to understand – and may aim to do so on a single 'page'.

In our case we endeavour to use space efficiently to show multiple (spatial and other) relationships by augmenting and synthesising existing approaches to cartography and information visualisation. We seek to do so in ways that are compact enough to fit onto a single (large) page or screen and sufficiently elegant to reveal both overall structure (*Gestalt*) and local detail (*details on demand*) concurrently.

One way we can do so under these constraints is by considering the idea of data/location ratio. Tufte (1983) conceived of the notion of the data/ink ratio – a heuristic that forces the designer of a graphic to consider what proportion of the ink on a page is used directly to represent data. The larger the ratio, the more efficient is the use of graphic symbolisation and the greater the depth of information that can be revealed. This metric of form and function might be considered as contributing to the beauty of a data graphic. Likewise, the data/location ratio is the degree to which the position of a graphical element on a page reflects characteristics of the data it represents. Traditional cartographic maps score highly in this respect, since the location of a symbol on the page usually identifies the geographic location of its referent. Many information graphics are perhaps less efficient in this regard, as are some maps – such as cartograms and schematics – often for good reason. We would argue that the efficient use of space is an important aspect of beautiful visualisation in that it supports the process of visual discovery of geographical pattern. This is increasingly so in the context of the volumes of data with a geographic component, such as large volunteered collections. In short, there is beauty in using space effectively.

Making treemaps beautiful – a geographic perspective

Treemaps are space-filling representations of hierarchies (Shneiderman 1992). Like many beautiful ideas, the treemap is based on an elegantly simple concept. An item of data is represented as a rectangle. If that data item itself contains a collection of other data items (a defining feature of any hierarchy), each is represented as a smaller rectangle that sits inside the 'parent' rectangle. In turn, these smaller rectangles can themselves contain even smaller ones that sit inside them and so on. The rectangles are arranged such that they fill the entire graphical space without any gaps. Each rectangle, or node, can be sized according to some characteristic of the data it represents. It can also be coloured in response to the data as well as being labelled in some meaningful way. All rectangles are visible – they do not overlap. There is elegance in the compactness of the representation (a single coloured and labelled rectangle can simultaneously show three or more independent characteristics of some data). The simple geometric form of each node (a rectangle) lends itself to the representation of large datasets, since a treemap can simultaneously show almost as many nodes as there are pixels on a screen. There is also some cognitive elegance in that the semantic containment relationships of the hierarchy are represented directly as geometric containments in the treemap (parent nodes enclose child nodes).

We saw opportunities for employing treemaps to represent large quantities of information recorded as geographical and thematic hierarchies. And by constructing new hierarchies we saw the possibility for exploring notions of place in the large numbers of records in the *Geograph* data set. However, treemaps have been widely critiqued for a number of reasons. Ironically, the aesthetic quality of treemaps has been criticised (Cawthon and van de Moere 2007), although we would argue this is more a function of implementation than design *per se*. More significantly, the arbitrary placement of nodes in a treemap significantly lowers the data/location ratio. Most existing treemap layout algorithms locate nodes in order to maximise their aspect ratios (making nodes as square as possible – important for aesthetics and size comparison tasks) and to aid readability (maximising horizontal linear continuity). Very few are concerned with using graphical location to represent some aspect of the data. This results in treemaps with linear discontinuities and arbitrary node placement that counters established best practice in cartography and statistical graphics whereby locations on the plane are regarded as the primary means of representing relational information (Bertin 1983). Arbitrary location of nodes within treemaps fails to take advantage of the 'first law of cognitive geography' (Fabrikant *et al.* 2002) whereby near things are regarded as more similar than distant things.

We saw scope for ordering nodes at all levels of a two dimensional treemap according to one or two dimensional orderings in the data (Wood and Dykes 2008). In a sense we address one of the key problems associated with the treemap, namely that the primary information carrying dimensions are not fully utilised, by mapping one (or more) data dimensions to them. We use space within the treemap to represent one dimensionally ordered or two dimensionally spatially arranged relationships in our data.

A geographic perspective on *Geograph* Term Use

The concept of place is a complex one that is not well described with simple latitude–longitude coordinate pairs. It is more than simply location in that it also says something about the nature of features that create a sense of place. It can rely on intangible, subjective and sometimes contradictory characteristics that traditionally are not well represented in digital data sets. Volunteered geographic information such as the personal descriptions of place available in *Geograph* gives us access to new and multiple perspectives that reflect a range of viewpoints and enable us to begin to consider alternative notions of place as we attempt to describe it more effectively.

Consequently, Ross Purves and Alistair Edwardes have been using *Geograph* as a source of descriptions of place in their research at the University of Zurich. Their ultimate objective involves improving information retrieval by automatically adding indexing terms to geo-referenced digital photographs that relate to popular notions of place. Their work involves validating previous studies and forming new perspectives by comparing *Geograph* to existing efforts to describe place and analysing term co-occurrence in the *Geograph* descriptions (Edwardes and Purves 2007). A popular approach in the literature involves identifying 'basic level' or 'scene types' through which place is described. These informative and summative descriptions of place have been traditionally derived through human subject testing. Such tests are notoriously difficult to organise and typically involve small numbers of participants making it hard to generalise from the results or repeat the experiments. Edwardes and Purves (2007) evaluated the way in which *Geograph* contributors use scene types and descriptions of them and found terms and their rankings to be similar to those reported in previous studies.

Having to an extent validated the terms used in the collection, we collectively identified opportunities for exploring the nature, structure and geography of some of these relationships in *Geograph*. In particular, we wished to understand the relationship between photographic content, photographic location and textual description of place as recorded in contributors' annotations of their photographs. A visualisation approach seemed appropriate and treemaps provided an opportunity for the exploration of these characteristics of *Geograph*.

The examples that follow document some of the ways in which spatial treemaps and other graphics were used in our exploration of *Geograph* as we developed our shared understanding of the collection and descriptions of place.

Representing the term hierarchy

The *Geograph* archive was processed in April 2008, at which time approximately 750 000 images had been geo-located and given a title and textual description. We focused on images that related to six scene types deemed particularly interesting through Ross and Alistair's analysis: beach, village, city, park, mountain and hill. For each of these scene types we then selected the most popular descriptive terms in each of three different facets: activities (predominantly verbs), elements (predominantly nouns) and qualities (predominantly adjectives). This resulted in a term co-occurrence hierarchy of four scene types, each containing three facets, each containing a number of agreed descriptors. A treemap reflecting this hierarchy contains nodes for each of the co-occurrences of our selected scene types with a popular descriptor and reveals some structure in the descriptions of place used in Geograph. The treemap shown in Figure 3.12.1 (see Colour Plate Four) uses an ordered squarified layout algorithm to optimise shape and locational consistency of nodes (Wood and Dykes 2008). It includes 144 000 leaf nodes – one for every time a scene type and descriptor co-occur in *Geograph*.

Each node is coloured using an inherited random scheme where scene types are randomly coloured and children (facets, descriptors and photos) inherit this colour with a minor mutation. Whilst the colours have no independent meaning this colouring scheme is used to emphasise the hierarchical structure of the classification. We can see variously and concurrently that hill is a more popular

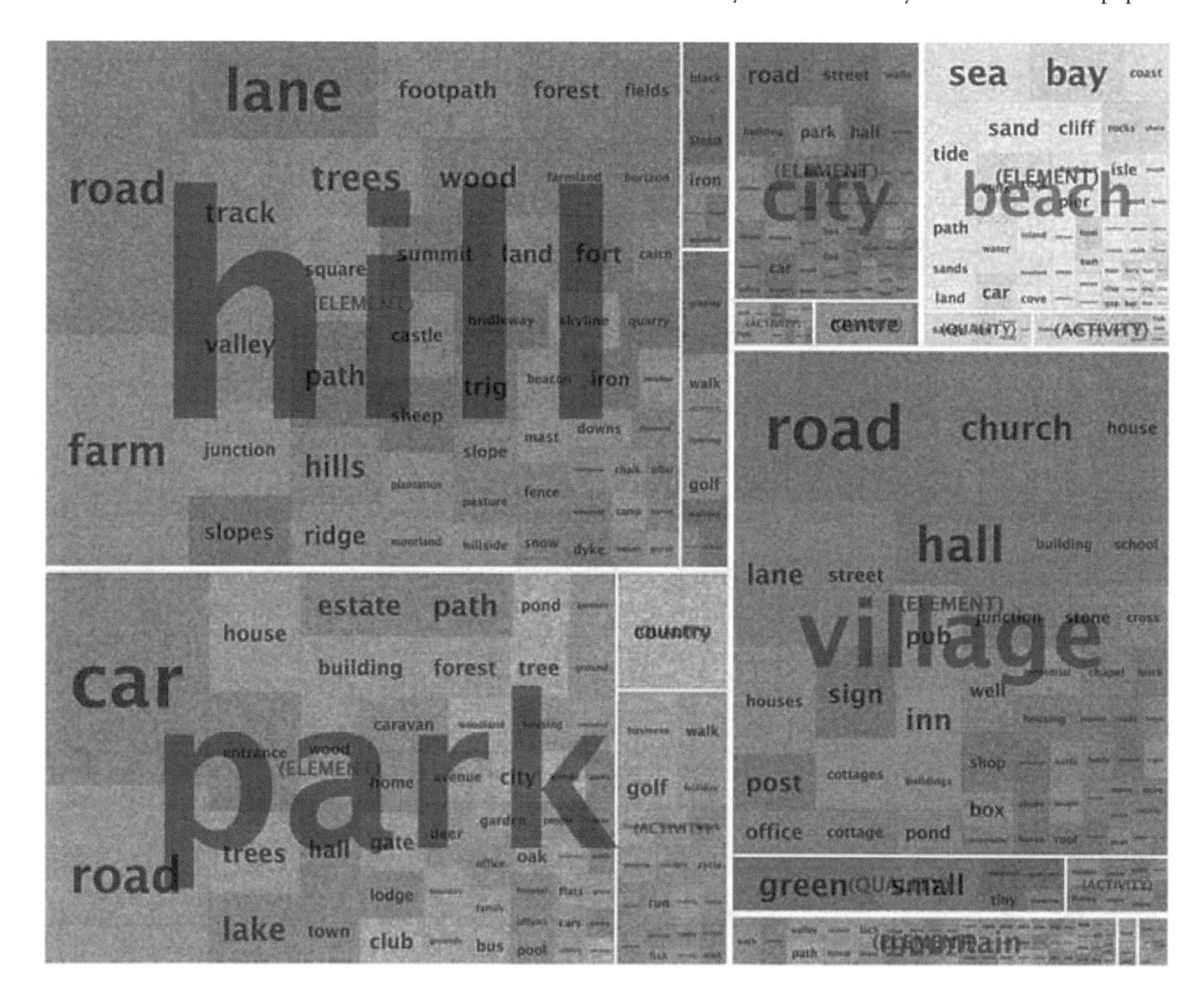

Figure 3.12.1 Treemap of terms occurring in *Geograph* titles and comments for six selected scene types. Node sizes represent term occurrence. Colours emphasise the scene type/facet/descriptor hierarchy with an inherited random scheme. Layout uses an 'ordered squarified' approach to maintain square shapes amongst nodes. (See Colour Plate Four.)

term than park, village, city, beach or mountain, as it occupies a larger area in the treemap. The elements facet is consistently more popular than qualities or activities across these scene types. The activity facet is particularly strongly associated with park. The descriptor road dominates four of the scene types, but not beach or mountain. Descriptors such as valley and path are used more frequently with hill than mountain although in relative terms these descriptors are more commonly related with mountain than hill. Loch is used more frequently with mountain than hill. Footpath is popular as a descriptor of hill but not mountain. One important aspect of the beauty of this kind of data-dense information is that many other relationships are displayed concurrently and might equally be described.

Representing absolute location with colour

While the treemaps described above provide some indication of how place is described, they say little about its relationship with location. We explored a couple of ways in which locational information may be added to the treemap. The first involved using colour to provide an indication of absolute location within the British Isles.

The challenge was to represent the two dimensions of photo location (easting and northing) with a colour that distinguished it from other photos in different locations. Most colour spaces are defined using three components (for example, red, green and blue or hue, saturation and value), so selecting just two components to represent eastings and northings is problematic. Additionally, most colour schemes are perceptually non-uniform; in other words the perceived similarity of two colours a fixed distance apart from each other varies across the colour space. We therefore chose to use the CIELab colour model that provides a more perceptually consistent colour gamut. By representing the eastings and northings of each photo's location with the a and b components of the colour space, we were able to produce a geographic map of colour where south westerly locations were coloured orange, south easterly locations green, north easterly blue and north westerly purple. Central locations tended towards brown and the degree of colour similarity between two nodes provided an indication of locational similarity of the photos they represent.

Some locational influences on descriptions of place are apparent from this view. For example, within hill the descriptors track, summit and cairn exhibit different geographies to downs, chalk, barrow and junction; mountain, beach and village have different locational characteristics; activities and qualities are geographically different within beach. The complexities associated with the relationship between location and place are also apparent.

Representing relative location with spatial treemaps

The use of colour to show location has some aesthetic appeal and provides some insight into location–place relationships, but it is limited in its effectiveness. In particular, it requires of the reader a memory of how colour is related to location. It also fails to use node position in any meaningful way. So instead, we can map the geographic location of each photo to node position within the treemap such that northerly photos appear towards the top of an enclosing node's space, westerly towards the left and so on. Because the treemap will fill the space with non-overlapping rectangles, we cannot provide an exact spatial mapping of location, but this form of layout does give an indication of relative location of nodes. If we are concerned with exploring the locational aspects of the place descriptors we can use CIELab colouring to emphasise absolute location, or we can represent some other aspect of the data with colour (such as term importance) whilst retaining a strong cartographic metaphor. [...]

Representing location displacement

Although we can identify some spatial patterns in our hierarchical structure, it is not always clear the degree to which nodes have been displaced from their true geographic location in order to be tessellated within the treemap space. CIELab colouring can give some indication of this displacement. [...] We can improve things further, however, by indicating graphically how a photo or group of photos has been 'moved' during the tessellation. Doing so follows the advice of Skupin and Fabrikant (2003) who recommend that cognitively plausible cartography should use appropriate methods for communicating this form of error.

Figure 3.12.2 superimposes collection of lines on the treemap. These join each node's treemap position to its geographic location – the longer the line, the greater the displacement. See, for example the displacement vectors associated with the quality and activity facets in beach that confirm their different geographies. The design goal behind these lines was to provide additional spatial context to the treemap while retaining the ability to explore the term hierarchy. Thicker lines are used to show scene type than facet displacement. Very thin lines can be used to show the displacement of individual *Geographs* as many hundreds of thousands of lines may be drawn. [...] This concurrent view of term hierarchy and both relative and absolute geography allows us to consider the geographies that we noted of track, summit and cairn and compare them with those of downs, chalk, barrow and junction simultaneously.

Figure 3.12.2 Spatial treemap of terms occurring in *Geograph* titles and comments for six selected scene types. Node sizes represent term occurrence and colours represent absolute spatial locations with CIELab scheme. Displacement vectors show absolute locations of non-leaf nodes (scene types, facets and descriptors). (See Colour Plate Four.)

This spatial arrangement may draw our attention to patterns, structures, anomalies and relationships that have not been previously considered and insights into the data. For example: bridleway, path and track have different geographies within hill; fishing and cricket are activities with different geographies in village; compare chapel and church in village, or golf with hill to golf with park. [...]

Beauty in discovery

We consider spatially ordered treemaps of large hierarchical data sets that have high data densities and high data/location ratios to be aesthetically pleasing and offer the figures presented in this chapter as candidate beautiful depictions of beautiful data. But things of beauty should be lucid, usable and ultimately satisfying as well as elegant. Matsumoto (2007) expresses this in the context of computer code through his belief that beautiful code should be readable. Maps make complex geo-spatial data readable in a fit-for-purpose manner that enables multiple spatial relationships to be determined and tasks such as navigation, geographic comparison and pattern detection to be achieved. Likewise, our efforts aim to make analytical sense of the geography of the language of *Geograph* through readable graphics. Can we read *Geograph*? We think that these and other graphics mean that we're well on our way to interpreting some of the complexity of this example volunteered geo-spatial data set.

The spatial relationships depicted in these views are dependent upon aspect ratios and the average locations of nodes within leaves. We may be interested in more precise geographies than those described here and the data-dense treemaps have been useful in selecting candidate facets and descriptors for more traditional mapping. Maps of term co-occurrence derived through the treemaps have enabled us to identify quantifiable spatial differences in term use. For example, we have found that valley is used with hill more frequently than might be expected in the south and south-west, whilst summit is used more frequently in the north. These relationships persist at different scales and when using alternative means of term selection from *Geograph*. They may reflect a bias in selecting particular aspects of the landscape to be recorded, regional geography, linguistic differences, personal preferences of contributors in particular places or likely a combination of these factors, but our reading of the treemaps draws attention to these characteristics in the light of others. After all, place is a subjective notion, one that is the product of many influences. Also note the rather rustic terms used with village here. These may not be due to mindful bias but may relate to selection of aspects of landscape that are aesthetically pleasing for *Geograph* – meaning that our reading of the data reinforces our contention that we are dealing with beautiful data. We intend exploring the geography of such terms further with maps and more formal spatial analysis. These and other aspects of the geography of user-selected descriptive terminology are being used by our colleagues in Zurich in developing an ontology for describing place.

Truth

'Truth' also has some role in beauty and aesthetic. Our views and the knowledge derived from them are incomplete, but these and other methods such as those discussed above help us establish (or at least move towards achieving establishing) possible truths ... against all the odds when the size, structure, diversity and complexity of data are considered. But who better to establish truth than a large group of volunteers working collectively? And how better to explore truths than through carefully designed graphics that begin to help us reveal structure and aid discovery?

Sublimity

We argue that the graphics presented here are useful, but complex and indeed that the complexity associated with geographies of term use and place in Britain have a sublime quality. Robert Kosara (2007) argues that pragmatic visualisation designed for data analysis should draw on graphics that are anti-sublime to aid efficient interpretation. He reasons convincingly that the enigmatic characteristics of sublime graphics are likely to have a detrimental effect on their interpretation in the visualisation process in terms of objectivity and speed.

In our experience, however, much of the applied interpretation of geographic information requires consideration of aspects of the geography that lie beyond the depicted data set and such judgements are likely to be subjective and less than immediate. We might term these enigmatic qualities. Kosara contends that sublime art 'cannot be easily readable' and 'must present enough of an enigma to keep an audience interested without being easy to solve'. In our abstract cartography we are trying to make multiple relationships readable in the wider context of these enigmatic and personal characteristics of space and place, which mean that the analysis of geographic information is difficult. We try to do so in a manner that keeps analysts interested in the data and enables them to generate ideas from it. The close relationships between science and design, data and interpretation, producer and consumer inherent in cartography give rise to a sublime quality in many maps that is viewed positively by many geographers as they interpret them. Berann's panoramas (Patterson 2000) and Imhof's (1965) relief maps are cases in point. [...]

As such, in the case of many maps we regard graphics that are based solely upon direct depictions of external spatial data as having subliminal qualities that, according to experience and cognition, may invoke a deep intellectual response that is beneficial to the viewer or analyst.

Graphics of *Geograph* also have a sublime quality in part because they reflect the sublime nature of the data and its collection. Indeed this sublimity is in part what makes *Geograph* and the graphics generated from it appealing, interesting, useful and ultimately beautiful. Kosara identifies the sublime as contributing an aesthetic quality to graphics but regards sublimity and efficiency as criteria that can be in conflict when considering beauty in data visualisation. It remains to be seen whether we are able to utilise the sublime qualities of the *Geograph* visualisation to our advantage in understanding the way in which language is used to describe places in Britain, but we feel that the approaches documented here have give us a start.

Reflection and conclusion

We have argued that the beauty in data lies in its depth. Beauty emerges as previously hidden structures and patterns are revealed. These patterns prompt new thoughts and questions about the data. They inspire. They encourage exploration. They provide insight.

Beautiful data encourage beautiful visualisation, for that also encourages exploration and rewards the viewer who explores. Visualisation is particularly apt for exploring geographic patterns, as centuries of cartography have demonstrated.

The examples we have explored show how we can use beautiful data such as *Geograph* to address the geographic context of *place* with a view to using such descriptions for information management and retrieval. Sophisticated, data-dense graphics with high data/ink and data/location ratios and aesthetic appeal are an important part of this process.

We broadly consider beauty to be a characteristic of an entity that provides pleasure, meaning or satisfaction. In terms of data and its representation, various aspects of *Geograph* have these qualities. Our visualisation of some facets of *Geograph* is innovative, creative and has a problem-solving basis. It has helped inform our colleagues' work and opened new analytical avenues. We would put forward these data-dense graphics of the people's descriptions of the human and physical geography of the British Isles in time and place as being as amongst the most beautiful that we have created.

References

Andrienko, G., Andrienko, N. and Dykes, J. *et al.* (2008) Geovisualization of dynamics, movement and change: key issues and developing approaches in visualization research. *Information Visualization*, **7**, 173–180.

Bertin, J. (1983) *Semiology of Graphics*, University of Wisconsin Press, Madison, WI.

Cawthon, N. and van de Moere, A. (2007) The effect of aesthetic on the usability of data visualization. *Proceedings of the 11th International Conference Information Visualization*, IEEE Computer Society, Washington, DC, pp. 637–648.

Edwardes, A. and Purves, R. (2007) A theoretical grounding for semantic descriptions of place, in *LNCS: Proceedings of 7th Intl Workshop on Web and Wireless GIS, W2GIS* (eds M. Ware and G. Taylor), pp. 106–120.

Fabrikant, S., Ruocco, M. and Middleton, R. *et al.* (2002) The first law of cognitive geography: distance and similarity in semantic space, *Proceedings of GIScience 2002*, Boulder, CO, pp. 31–33.

Fisher, P.F. (1998) Is GIS hidebound by the legacy of cartography? *The Cartographic Journal*, **35**, 5–9.

Geograph, (2008) *Geograph British Isles* – photograph every grid square. http://www.geograph.co.uk/.

Goodchild, M. (2007) Citizens as sensors: the world of volunteered geography. *GeoJournal*, **69**, 211–221.

Imhof, E. (1965) *Kartographische Geländedarstellung*, de Gruyter, Berlin.

Kernighan, B.W. (2007) A regular expression matcher, in *Beautiful Code: Leading Programmers Explain How They Think* (eds A. Oram and G. Wilson), O'Reilly, Sebastopol, CA, pp. 1–9.

Kosara, R. (2007) Visualization Criticism: the missing link between information visualization and art. *Proceedings of the 11th International Conference Information Visualization*, IEEE Computer Society, Washington, DC, pp. 631–636.

MacEachren, A.M. and Kraak, M.J. (2001) Research challenges in geovisualization. *Cartography and Geographic Information Science*, **28**, 3–12.

Matsumoto, Y. (2007) Treating code as an essay, in *Beautiful Code: Leading Programmers Explain How They Think* (eds A. Oram and G. Wilson), O'Reilly, Sebastopol, CA, pp. 477–481.

Oram, A. and Wilson, G. (2007) *Beautiful Code: Leading Programmers Explain How They Think*, O'Reilly, Sebastopol, CA.

Patterson, T. (2000) A view from on high: Heinrich Berann's panoramas and landscape visualization techniques for the US National Park Service. *Cartographic Perspectives*, **36**, 38–65.

Shneiderman, B. (1992) Tree visualization with tree-maps: 2-d space-filling approach. *ACM Transactions on Graphics*, **11**, 92–99.

Skupin, A. and Fabrikant, S. (2003) Spatialization methods: a cartographic research agenda for non-geographic information visualization. *Cartography and Geographic Information Science*, **30**, 99–119.

Tufte, E.R. (1983) *The Visual Display of Quantitative Information*, Graphics Press, Cheshire, CT.

Wood, J. and Dykes, J. (2008) Spatially ordered treemaps. *IEEE Transactions on Visualization and Computer Graphics*, **14**, 1348–1355.

Further reading

Cartwright, W., Gartner, G. and Lehn, A. (2008) *Cartography and Art*, Springer, Berlin. [An edited collection of papers exploring beauty in mapping and the relationship between artistic and mapping practice.]

Dodge, M. and Kitchin, R. (2001) *Mapping Cyberspace*, Routledge, London. [A discussion of the different ways information spaces have been measured and mapped.]

Dykes, J., MacEachren, A. and Kraak M.J. (2005) *Exploring Geovisualization*, Elsevier, London. [The 36-chapter volume covers a wide range of work from interactive mapping, cartographic exploration, and data visualisation research: the best overview of the field.]

Friendly, M. and Denis, D.J. (2001) *Milestones in the History of Thematic Cartography, Statistical Graphics, and Data Visualization*. http://www.math.yorku.ca/SCS/Gallery/milestone/. [A richly illustrated and growing online archive that traces the development of techniques and highlighting both innovations and outstanding examples of data graphics across diverse fields.]

See also

- Chapter 1.11: Exploratory Cartographic Visualisation: Advancing the Agenda
- Chapter 2.11 Extending the Map Metaphor Using Web Delivered Multimedia
- Chapter 2.12 Imaging the World: The State of Online Mapping
- Chapter 3.10: Affective Geovisualisations
- Chapter 4.10: Citizens as Sensors: the World of Volunteered Geography

SECTION FOUR

Cognition and Cultures of Mapping

Chapter 4.1

Introductory Essay: Cognition and Cultures of Mapping

Chris Perkins, Rob Kitchin and Martin Dodge

Introduction

Maps are produced and used by people; they are the product of the skills and abilities of individuals embedded in particular cultures and inherently reflect those skills and wider culture. It is now widely accepted that mapping is a cognitive and embodied activity, a set of processes that people engage with in order to make sense of, and connections with, the world 'out there'. For some, mapping is an essential ability; an intrinsic cognitive function of being human (Blaut 1991, excerpted as Chapter 4.4; Blaut *et al.* 2003). Regardless of whether mapping abilities are nativist or nurtured, however, maps exist in all human cultures, and reflect everyday subjectivities. Wright (1942, excerpted as Chapter 4.2) explored many of the dimensions of this subjectivity, highlighting that: 'the qualities of integrity, judgment, critical acumen, and the like are as much required in the interpretation of maps as in the preparation of them' (p. 543). Maps that emerge from these subjective and social processes are enrolled in a myriad of tasks, and therefore it is perhaps unsurprising that the links between people and cartographic practices have been understood in many different ways.

This section of the book focuses upon people, culture and mapping, and the diverse ways in which scholars have explored the relationship between maps, mapping, individuals and their social contexts and cultural meanings. The most obvious difference between the excerpts is between those scholars who focus upon individuals and their cognitive abilities to understand, produce and read maps, as against those who focus upon the cultural context within which maps are created and used, and the wider meanings associated with mapping as a whole.

Cognitive approaches to mapping

Over the past four decades environmental psychologists and behavioural geographers have investigated the relations between individuals and their world, and the individual processing of spatial information about that world. Their focus has often been upon the development of individual mapping skills – the ability to understand, process and create maps – or upon experimental investigation of particular and carefully controlled subsets of map reading tasks (Blades *et al.* 2002; Downs and Stea 1973a; Lloyd 2000, excerpted as Chapter 4.9). Research informed by psychological methods has often sought to establish best practice, either in cartographic design, or in delivering optimal map skills teaching, or in determining cognitive limits to perception (Montello 2002). Methods have tended towards controlled experiments, which simplify the complexity of real world cartographic practice, so as to build an incremental understanding of map use.

Cognitive approaches to cartography are grounded in a number of different traditions. On the one hand, there

The Map Reader: Theories of Mapping Practice and Cartographic Representation, First Edition. Edited by Martin Dodge, Rob Kitchin and Chris Perkins.

is a body of work centred around cognitive mapping. These scholars proceed on the basis that each individual possesses a 'cognitive map' of the world; that is a mental construct that allows them to process and synthesise spatial information and guides spatial decision and choice making. Initially popularised by Tolman's (1948) influential work on the spatial behaviour of rats, the concept remains a powerful influence on the field of spatial cognition (see Kitchin and Freundschuh, 2000, for an overview). Downs and Stea (1973b: 8) offer the following definition:

> 'cognitive mapping is a process composed of a series of psychological transformations by which an individual acquires, codes, stores, recalls and decodes information about the relative locations and attributes of phenomena in his everyday spatial environment'

Even the most ardent proponents of the idea recognised that the existence of a cognitive map is almost impossible to prove, and treated the concept as a useful tool with which to understand how people dealt with everyday navigational and spatial demands. The notion is firmly grounded in a view of sense-making as being concerned with transmission of information, in a process of communication.

Nonetheless, the relationship between cognitive mapping and cartography has been pursued at two levels. The first considers how maps can be used to improve a person's understanding of a place and how such information is integrated into a person's cognitive map (Butler *et al.* 1993; Lloyd 1993). The usual approach here is to compare the spatial knowledge of one group of individuals who learnt an area by traversing it with another group that learnt the street layout purely from a map (with studies showing that those who had access to the map had a more accurate and complete understanding). The second is to consider how individuals mentally engage with, understand, learn and memorise cartographic information, with the aim of determining how map design might be improved in ways that make maps easier to comprehend and use. Lloyd (2000, excerpted as Chapter 4.9), for example, explores how early psychophysical experimentation, and, in particular, eye movement studies, sought to make links between perception of particular graphic symbols or parts of maps, and cognitive activity. More nuanced experimentation emerged that was grounded in a more conceptual approach to mapping, brought together in an impressive synthesis by MacEachren (1995), and the tradition continues to this day, for example in recent fMRI research examining links between activity in different parts of the brain, and map reading or geovizualisation tasks (Lobben *et al.* 2009).

A second strand of work is grounded in ideas of mapping as visual communication and comes from map design scholarship, in particular the work of Arthur H. Robinson. Montello (2002) suggests Robinson's *The Look of Maps* (1952, excerpted as Chapter 3.3) had a profound influence on research into map design because of the conceptual framework of visual communication that it propounded, and which Robinson's students and colleagues successfully deployed during subsequent decades. This functional approach to mapping was justified by its theoretical grounding in a particular view of cognition, which encouraged researchers to focus on cartographic practice as being defined by map reading (also see Morrison 1976, excerpted as Chapter 1.4; Board 1972, excerpted as Chapter 1.6).

Together these bodies of work have encouraged a number of continuing research foci. Montello (2002) suggests cognitive approaches to map design research probably peaked in the late 1970s and early 1980s, before the rise of GIS, and before epistemological challenges from social constructivist thought, that came to question the validity of communication as a device for understanding mapping. (See introductory essays for Sections 1 and 2.) However a more recent resurgence is evidenced by the establishment of a new International Cartographic Association (ICA) Working Group on Map Use and Users, and the publication of theme issues in key cartographic journals (Fabrikant and Lobben 2009; Van Elzakker *et al.* 2008). The notable diversity of new display variables offered by geovisualisation is gradually being investigated (Nivala *et al.* 2008; excerpted as Chapter 4.11), although Fabrikant and Lobben (2009) are rather pessimistic about progress to date. Empirical user testing of digital map interfaces reveals them to be just as poorly designed as were many paper cartographic products investigated in the first wave of cognitive research into map designs forty years earlier (perhaps unsurprisingly given the relative lack of collaboration between cartographic researchers and system designers).

An ongoing second strand of research focuses upon map skills and how they are deployed by different groups of people. Here, the focus is not upon how map designs work, but rather upon map reading skills of different social groups. So, for example, children of different ages have received particular attention (see Wiegand 2006 for a useful review of this field). Gendered map use has been investigated (Gilmartin and Patten 1984). Mapping skills of different groups of disabled people have been observed and tested (Matthews and Vujakovic 1995; Ungar *et al.* 1997). Mapping abilities needed for specific kinds of leisure pursuit have been investigated (see Crampton 1992 on expert and novice orienteers).

Much of James Blaut's career focused on bringing together work of this kind in order to amass evidence for what he termed 'natural mapping' (Blaut 1991, excerpted as Chapter 4.4). He was seeking to establish the universal and human nature of mapping skills as a cognitive process, but also as a cultural universal (disputed by Downs and Liben 1991). Much of this work has a practical or normative rational. By defining existing map skills, ways of teaching improved map skills can be devised.

Cultural approaches to mapping

In contrast to cognitive approaches, anthropologists and cultural geographers have tended to focus more upon everyday cultural practices, than upon individual skills and cognition. They have emphasised the role of maps as part of a shared identity and explored the cultural processes through which mapping as a practice, or the cartographic artefact as an object, has come to hold particular meanings. The map is treated not as a neutral representation or functional communication device, but rather as a part of culture, with an influence upon other aspects of life (Perkins 2008). As such, attention has been less on the atomistic and functional elements of the map, and more with the map as an object as a whole, and its real world circulation and enrolling into different cultural contexts.

It is only in the last thirty years that cultural approaches to cartography have emerged as a key research field. The differences from cognitive approaches are best understood by referring to two papers that, from their titles at least, might well be grounded in an awareness of cognition. Reeves (1993, excerpted as Chapter 4.6) article, *Reading Maps*, explores the practices of map reading in the early modern European world, and the ways these changed over time in different cultural contexts, with almost no consideration of the individual cognitive processes implicit in those practices. Instead, she interprets changes in the *cultural* practices of map reading, through carefully chosen examples from fine art and literature, to reveal mutability, and the important social roles played by cartographic representations. Her methodology depends upon historical and literary scholarship, not controlled testing of human subjects. Women and men, for example, read mapping in particular ways, because of cultural roles and interplay of different media, not because their brains dictated a particular way of reading.

The second paper, by Orlove (1991, excerpted as Chapter 4.7), also focuses upon *Reading Maps*, and also reaches a strongly cultural conclusion, albeit deploying methodologies sourced from anthropology and indigenous knowledges. Here, the focus is upon the cultural politics of the reading process. Instead of an emphasis on the signs and symbols on the map and an investigation of what they signify, as if meaning is fixed in individual processing of information, Orlove focuses upon the social and cultural processes through which the map reading process comes to produce certain interpretations. Different social groups deploy mapping in ways that reinforce their own interpretations: instead of focusing upon the neutral fixed meaning in the mapping, the task of the researcher becomes one of unpacking the social processes around which meaning coalesces (often contested and political in the case of Lake Titicaca reeds; see also discussion in Chapter 5.1 on counter-mapping). There is a real world concern for exploring how maps are deployed, instead of a narrow focus upon the cognitive processes underpinning any reading (see Perkins and Gardiner 2003 for an examination of the limits of cognition). Orlove's paper was one of the first to adopt this kind of positioned and ethnographic approach for cartographic knowledge. It reflects a growing academic concern with indigenous mapping practices (Peluso 1995, excerpted as Chapter 5.6; Sparke 1998, excerpted as Chapter 5.7; Wood 2010) and with mapping as a process (Rundstrom 1991).

This emphasis upon mapping practices echoes moves across the social sciences towards post-constructivist thought. Anthropological ways of approaching cartographic practices now involve immersion and participatory approaches, instead of distance and objectivity. And the focus of study is increasingly mapping practice, instead of the fixed form of the map as a representational object. These trends can be seen in a number of practical initiatives and empirical studies (for example, Grasseni 2004 on the co-construction of ideas of landscape in Italian local mapping initiatives; Parker 2006 on the empowering potential of community mapping in Seattle; Perkins 2007 on the cultural context of community mapping initiatives in Britain). Crouch and Matless (1996, excerpted as Chapter 4.8) focus on the ambiguities of community-led local mapping initiatives using a Deleuzian reading of the Common Ground Parish Map Project, with case studies of how mapping speaks for but also responds to contested notions of place. The changing relationships that emerge from the interplay of aesthetics, politics and situated mapping are all embedded in cultural contexts and embodied practice that must be interpreted to gain real understanding of their meaning.

These trends towards community-based and local mapping reflect more than just intellectual fashion. They also indicate a significant democratisation of mapping, dating from the last decade of the twentieth century, but with roots that can be traced back to pioneering work by William Bunge in the 1960s and subsequently, (Colour Plate Six) and artistic encounters with mapping even earlier (Bunge 1971; Peluso 1995; Wood 2010 on the rise of 'counter-mapping', participatory GIS and artistic mapping; Pinder 1996 for an analysis of the Situationist artistic encounter with mapping). The rise of community-oriented mapping reflects power perhaps shifting away from the nation-state, towards other and everyday mappers (Geller 2007, excerpted as Chapter 2.12; Goodchild 2007, excerpted as Chapter 4.10).

One of the most interesting recent trends in this context has seen the rise of Web-facilitated 'crowd sourced' mapping. Instead of a centrally controlled and institutional authored cartographic product, the Web offers an infrastructure through which many people can collaborate in a shared, participatory endeavour (Surowiecki 2004; Sui 2008). These changes focus attention on the processes through which mapping emerges in complex technologically mediated systems and, together with other locative technologies (see Section 2), have been designated 'neogeography' (Haklay *et al.* 2008). Goodchild (2007, excerpted as Chapter 4.10) is one of the first to delineate the likely impacts of these trends on mainstream cartography. His notion of volunteered geographic information reflects the GI industry perception of the trend, in so far as it discusses the potential for using peoples' data, rather than the potential for crowd sourcing to create new mapping opportunities for people that are out of the control of 'old' institutions. Initiatives like OpenStreetMap offer a new model for people to become involved in making and deploying maps, and are already approaching the data quality of many state and commercially produced maps (Haklay 2010). (Colour Plate Five.)

Research has also started to address how people relate to maps and the mapping process in a cultural and emotional sense. As a part of visual culture, maps have a uniquely affectual role to play. They evoke emotions and carry inherent connotations with them. People tend to believe what they see on a map. The medium evokes an authority, making a link between places and things that happen in those places (Wood and Fels 2008, for example, provides a discussion of the ways in which mapping of nature evokes different feelings about the natures being represented). But mapping can also have different tones, evoking pleasure (Wood 1987), arousal (Nold 2009), ambivalence (Hawthorne *et al.* 2008) and humour (Caquard and Dormann 2008). The affect of mapping is an emerging research focus (Aitken and Craine 2006, excerpted as Chapter 3.10; Kwan 2007, excerpted as Chapter 5.9). Harley (1987, excerpted as Chapter 4.5) describes one map sheet within the frame of which various biographies have been played out. The people making the map, the histories of the place, and indeed personal biographies of the author are all charted and reflected in the collected artefact (Perkins 2008 gives an exploration of the motivations behind map collecting). Instead of a cognitive analysis of mechanisms of map reading, this kind of research is exploring the narrative potential of mapping (Pearce 2008).

Conclusions

Cognitive and cultural approaches to cartography provide rich insight into the relationship between people and maps. After a brief hiatus between the mid 1980s and mid 1990s, cognitive research into reading maps and geovisualisations has continued apace, providing insights into how people learn from maps and how maps might be designed to improve their legibility and comprehension. In addition, researchers from across the humanities and social sciences have created a detailed understanding of the role of culture in the production and reading of maps and their effects on the societies in which they are used. In the second decade of the twentieth century we can expect further refinement of ideas as rapid technological changes such as Mapping 2.0 (Crampton 2009) encourage an increasing focus upon the human processes of mapping, in all their cultural diversity, along with a progressive improvement in knowledge of how affective geovisualisations might work as crafted designs.

References

Aitken, S. and Craine, J. (2006) Affective geovisualizations. *Directions: A Magazine for GIS Professionals*, 7 February. www.directionsmag.com. (Excerpted as Chapter 3.10.)

Blades, M., Lippa, Y., Golledge, R. *et al.* (2002) Wayfinding by people with visual impairments: the effect of spatial tasks on the ability to learn a novel route. *Journal of Visual Impairment and Blindness*, **96**, 407–419.

Blaut, J.M. (1991) Natural mapping. *Transactions of the Institute of British Geographers*, **16** (1), 55–74. (Excerpted as Chapter 4.4.)

Blaut, J.M., Stea, D., Spencer, C. and Blades, M. (2003) Mapping as a cultural and cognitive universal. *Annals of the Association of American Geographers*, **93** (1), 165–185.

Board, C. (1972) Cartographic communication. *Cartographica*, **18** (2), 42–78. (Excerpted as Chapter 1.6.)

Bunge, W. (1971) *Fitzgerald: Geography of a Revolution*, Schenkman Publishing Company, Cambridge, MA.

Butler, D.L., Acquino, A.L., Hissong, A.A. and Scott, P.A. (1993) Wayfinding by newcomers in a complex building. *Human Factors*, **35** (1), 159–173.

Caquard, S. and Dormann, C. (2008) Humorous maps: explorations of an alternative cartography. *Cartography and Geographic Information Science*, **35** (1), 51–64.

Crampton, J. (1992) A cognitive analysis of wayfinding expertise. *Cartographica*, **29**, 46–65.

Crampton, J. (2009) Cartography: maps 2.0? *Progress in Human Geography*, **33**, 91–100.

Crouch, D. and Matless, D. (1996) Refiguring geography: Parish maps of Common Ground. *Transactions of the Institute of British Geographers*, **21**, 236–255. (Excerpted as Chapter 4.8.)

Downs, R.M. and Liben, L. (1991) Understanding maps as symbols: the development of map concepts in children, in *Advances in Child Development and Behavior* (ed. H.W. Reese), Academic Press, New York.

Downs, R.M. and Stea, D. (1973a) *Image and Environment: Cognitive Mapping and Spatial Behavior*, Aldine Press, Chicago.

Downs, R.M. and Stea, D. (1973b) Cognitive maps and spatial behavior: process and products, in *Image and Environment: Cognitive Mapping and Spatial Behavior* (eds R.M. Downs and D. Stea), Aldine Press, Chicago, pp. 8–26. (Excerpted as Chapter 4.3.)

Fabrikant, S.I. and Lobben, A. (2009) Cognitive issues in geographic information visualization. *Cartographica*, **44** (3).

Geller, T. (2007) Imaging the world: the state of online mapping. *IEEE Computer Graphics and Applications*, March/April, 8–13.

Gilmartin, P. and Patton, J.C. (1984) Comparing the sexes on spatial abilities: map-use skills. *Annals of the Association of American Geographers*, **74** (4), 605–619.

Goodchild, M.F. (2007) Citizens as sensors: the world of volunteered geography. *GeoJournal*, **69** (4), 211–221. (Excerpted as Chapter 4.10.)

Grasseni, C. (2004) Skilled landscapes: mapping practices of locality. *Environment and Planning D*, **22**, 699–717.

Haklay, M. (2010) How good is volunteered geographical information? A comparative study of OpenStreetMap and Ordnance Survey datasets. *Environment and Planning B: Planning and Design*, **37** (4), 682–703.

Haklay, M., Singleton, A.D. and Parker, C. (2008) Web mapping 2.0: the neogeography of the geoweb. *Geography Compass*, **2** (6), 2011–2039.

Harley, J.B. (1987) The map as biography: thoughts on Ordnance Survey, Six-Inch Sheet Devonshire CIX, SE, Newton Abbot. *The Map Collector*, **41**, 18–20. (Excerpted as Chapter 4.5.)

Hawthorne, T., Krygier, J. and Kwan, M.P. (2008) Mapping ambivalence: exploring the geographies of community change and rails-to-trails development using photo-based Q method and PPGIS. *Geoforum*, **39** (2), 1058–1078.

Kitchin, R. and Freundschuh, S. (2000) *Cognitive Mapping: Past, Present and Future*, Routledge, London.

Kwan, M.-P. (2007) Affecting geospatial technologies: toward a feminist politics of emotion. *The Professional Geographer*, **59** (1), 22–34. (Excerpted as Chapter 5.9.)

Lloyd, R. (1993) Cognitive processes and cartographic maps, in *Behavior and Environment: Psychological and Geographical Approaches* (eds T. Gärling and R.G. Golledge), North-Holland, Amsterdam, pp. 141–169.

Lloyd, R. (2000) Cognitive maps: encoding and decoding information, in *Cognitive Mapping: Past, Present and Future* (eds R. Kitchin and S. Freundschuh), Routledge, London, pp. 84–107. (Excerpted as Chapter 4.9.)

Lobben, A., Lawrence, M. and Olson, J.M. (2009) fMRI and human subjects research in cartography. *Cartographica*, **44** (3), 159–170.

MacEachren, A.M. (1995) *How Maps Work*, Guilford, New York.

Matthews, M.H. and Vujakovic, P. (1995) Private worlds and public places: mapping the environmental values of wheelchair users. *Environment and Planning A*, **27**, 1069–1083.

Montello, D.R. (2002) Cognitive map design research in the twentieth century. *Cartography and Geographic Information Science*, **29** (3), 283–304.

Morrison, J. (1976) The sience of cartography and its essential processes. *International Yearbook of Cartography*, **16**, 84–97. (Excerpted as Chapter 1.4.)

Nivala, A.M., Brewster, S. and Sarjakoski, L.T. (2008) Usability evaluation of web mapping sites. *The Cartographic Journal*, **45** (2), 129–138. (Excerpted as Chapter 4.11.)

Nold, C. (2009) *Emotional Cartography: Technologies of the Self*. http://emotionalcartography.net/.

Orlove, B. (1991) Mapping reeds and reading maps: the politics of representation in Lake Titicaca. *American Ethnologist*, **18** (1), 3–38. (Excerpted as Chapter 4.7.)

Parker, B. (2006) Constructing community through maps? Power and praxis in community mapping. *Professional Geographer*, **58** (4), 470–484.

Pearce, M. (2008) Framing the days: place and narrative in cartography. *Cartography and Geographic Information Science*, **35**, 17–32.

Peluso, N.L. (1995) Whose woods are these? Counter-mapping forest territories in Kalimantan, Indonesia. *Antipode*, **27** (4), 383–406. (Excerpted as Chapter 5.6.)

Perkins, C. (2007) Community mapping. *The Cartographic Journal*, **44** (2), 127–137.

Perkins, C. (2008) Cultures of map use. *The Cartographic Journal*, **45** (2), 150–158.

Perkins, C. and Gardiner, E.A.H. (2003) Real world map reading strategies. *The Cartographic Journal*, **40** (3), 265–268.

Pinder, D. (1996) Subverting cartography: the situationists and maps of the city. *Environment and Planning A*, **28** (3), 405–427.

Reeves, E. (1993) Reading maps. *Word and Image*, **9** (1), 51–65. (Excerpted as Chapter 4.6.)

Robinson, A. (1952) *The Look of Maps*, University of Wisconsin Press, Madison, WI. (Excerpted as Chapter 3.3.)

Rundstrom, R.A. (1991) Mapping, postmodernism, indigenous people and the changing direction of North American cartography. *Cartographica*, **28**, 1–12.

Sparke, M. (1998) A map that roared and an original atlas: Canada, cartography, and the narration of nation. *Annals of the Association of American Geographers*, **88** (3), 463–495. (Excerpted as Chapter 5.7.)

Sui, D.Z. (2008) The wikification of GIS and its consequences: or Angelina Jolie's new tattoo and the future of GIS. *Computers, Environment and Urban Systems*, **32** (1), 1–5.

Surowiecki, J. (2004) *The Wisdom of Crowds*, Little Brown, London.

Tolman, E.C. (1948) Cognitive maps in rats and men. *Psychological Review*, **55** (4), 189–208.

Ungar, S., Blades, M. and Spencer, C. (1997) Strategies for knowledge acquisition from cartographic maps by blind and visually impaired adults. *The Cartographic Journal*, **34**, 93–110.

Van Elzakker C., Nivala, A., Pucher, A. and Forrest, D. (2008) Use and user issues in geographic information processing and dissemination theme issue. *The Cartographic Journal*, **45** (2).

Wiegand, P. (2006) *Learning and Teaching with Maps*, Routledge, London.

Wood, D. (1987) Pleasure in the idea: the atlas as a narrative form. *Cartographica*, **24** (1), 24–45.

Wood, D. (2010) *Rethinking the Power of Maps*, Guilford, New York.

Wood, D. and Fels, J. (2008) *The Natures of Maps*, Chicago University Press, Chicago.

Wright, J.K. (1942) Map makers are human: comments on the subjective in maps. *Geographical Review*, **32** (4), 527–544. (Excerpted as Chapter 4.2.)

Chapter 4.2

Map Makers are Human: Comments on the Subjective in Maps

John K. Wright

Editors' overview

Written in 1942, at a time when cartography served the functional requirement of waging war, Wright's critical piece highlighted what has come to be an accepted orthodoxy fifty years later: that people and their subjectivities are important in the mapping process. At the time of publication Wright's paper must have seemed radical indeed. He highlights the mental and moral qualities that are important in aspects of map design and use, focusing in particular upon scientific integrity, authorial judgment, consistency, harmony, taste and progressiveness. Wright then discusses simplification and amplification before focusing in some detail on the subjectivities inherent in the mapping of quantitative data and their generalisation and classification.

Originally published in 1942: *The Geographical Review,* **32** (4), 527–544.

Like bombers and submarines, maps are indispensable instruments of war. In the light of the information they provide, momentous strategic decisions are being made today: ships and planes, men and munitions, are being moved. Maps help to form public opinion and build public morale. When the war is over, they will contribute to shaping the thought and action of those responsible for the reconstruction of a shattered world. Hence it is important in these times that the nature of the information they set forth should be well understood.

Maps are drawn by men and not turned out automatically by machines, and consequently are influenced by human shortcomings. Although this fact itself is self-evident, some of its implications are often overlooked. The trim, precise and clean-cut appearance that a well drawn map presents lends it an air of scientific authenticity that may or may not be deserved. A map may be like a person who talks clearly and convincingly on a subject of which his knowledge is imperfect. We tend to assume too readily that the depiction of the arrangement of things on the earth's surface on a map is equivalent to a photograph, which, of course, is by no means the case. The object before the camera draws its own image through the operation of optical and chemical processes. The image on a map is drawn by human hands, controlled by operations in a human mind. Every map is thus a reflection partly of objective realities and partly of subjective elements (Eckert 1908). No map can be wholly objective. Even a photograph taken vertically downward is subjective, in the sense that the photographer's choice determined the tract of country shown in it and also the time of day when the film was exposed and thus the aspect of the shadows appearing in the picture. Likewise, no map is wholly 'non-objective', as some forms of painting and sculpture are said to be. Even a map of an imaginary country is objective, in the sense that the mountains, roads, towns and so on that it pictures were suggested by corresponding objective things in the real world.

The maps produced by governmental surveys or made in the field by explorers are more or less directly copied from nature. As in the case of the memoirs and letters written by those who have participated in historic events, their quality is influenced by the experience and powers of observation of their makers. A topographic map drawn by a man familiar with geology and physiography is likely to be far more expressive of relief and drainage than one

The Map Reader: Theories of Mapping Practice and Cartographic Representation, First Edition. Edited by Martin Dodge, Rob Kitchin and Chris Perkins.

drawn by an untrained observer, just as the memoirs of an experienced statesman are likely to present a more truthful record of an international conference than those of some inexperienced journalist. Many maps, however, are not drawn from nature but are compiled from such documentary sources as other maps, surveyors' notes and sketches, photographs, travellers' reports, statistics and the like. As these sources are themselves man-made, the subjective elements they contain are carried over into the maps based on them. In the following paragraphs maps will be considered in the light of the effects on their sources and their compilation of certain mental and moral qualities: scientific integrity, judgment, consistency, progressiveness and their opposites.

Scientific integrity

Fundamental among these qualities is scientific integrity: devotion to the truth and a will to record it as accurately as possible. The strength of this devotion varies with the individual. Not all cartographers are above attempting to make their maps seem more accurate than they actually are by drawing rivers, coasts, form lines and so on with an intricacy of detail derived largely from the imagination. This may be done to cover up the use of inadequate source materials or, what is worse, to mask carelessness in the use of adequate sources. Indifference to the truth may also show itself in failure to counteract, where it would be feasible and desirable to do so, the exaggerated impression of accuracy often due to the clean-cut appearance of a map. Admittedly this is not always easy of accomplishment.

A map is not like a printed text, in which statements can be qualified with fine shades of meaning. One cannot, on the face of a map, cite the evidence used and discuss its validity. A town or a mountain must be shown in one place, even though three sources of apparently equal validity may locate it at three different places. Nevertheless, there are certain ways in which the mapmaker can, within limits, modify the definiteness of his commitment. If he is not sure of the courses of rivers or if contours are approximate form lines only and are not based on actual field surveys, he can at least show them by broken lines. He can introduce question marks or, as on marine charts, such challenging letters as P.D. (position doubtful) or E.D. (existence doubtful) alongside islands and shoals. Another device is to include in the margin of the map a small diagram of the region showing the character of the surveys and other sources on which the map is based – a 'relative reliability' diagram. This accomplishes to some extent what the careful historian or economist does with qualifying phrases in his text and appended critiques of his sources. Although this device is not often used at present, it may someday become a standard practice in cartographical scholarship where maps, especially maps covering considerable tracts of territory, are based on different source materials of varying validity.

Beware of maps prepared to substantiate a pet theory! There is a well known type of reasoning that begins with a theory, gathers statistics and other data that seem to support it, makes a map on the basis of the statistics, and finally 'proves' the theory by reference to the map. The dishonesty in such a procedure may be unconscious, but there is a large use of maps in propaganda with a view to conscious and deliberate deception in the service of special interests. The relative areas of different regions as disclosed on maps in railroad timetables are usually deliberately distorted so as to show particular railroad systems to best advantage. More subtle and dangerous is the type of deception found on maps designed for propaganda purposes – maps on which facts are played up or played down, omitted or invented, for nationalistic ends (Speier 1941).

Maps are not confined to the representation of a given state of affairs. They can be drawn to symbolise changes, or as blueprints of the future. They may make certain traits and properties of the world they depict more intelligible – or may distort or deny them. [...] They may reveal policies or illustrate doctrines. They may give information, but they may also plead. Maps can be symbols of conquest or tokens of revenge, instruments for airing grievances or expressions of pride. Speier (1941) analyses the ingenious manipulations of design and colour to be found on some of the 'geopolitical' maps issued in the interests of totalitarian propaganda. For example, on a map to illustrate the repatriation of Germans who had been living in Latvia, the German minority about to return to the fatherland is represented by a row of thirteen identical symbols, each standing for five thousand men. The symbols extend over the whole area of Latvia where it is widest, from Libau in the west to the eastern border. The size of the symbol is so chosen that the country seems to be populated by Germans, whereas, in point of fact, the German minority amounts to 3.7% of the total Latvian population. Speier exposes the more blatant forms of propagandistic maps intended to influence the masses.

Nationalistic bias may also reveal itself in detailed maps published in serious books and learned periodicals. Superficially these may seem to have been prepared according to sound scientific principles. They may accurately record accurate statistics, but with a symbolism devised to overemphasise the distribution of one people or language or religion or set of institutions. Assume, for example, that the frontier province of Pomeria, which formerly belonged to Sudia, was annexed by Nordia in the last war and that a recent census has shown that half of the population of Pomeria are Nordians and half Sudians. The Nordians are concentrated in the towns, the Sudians form the bulk of the rural population. On a detailed map of Pomeria in the Bulletin of the Sudian Geographical Society the areas where less than 10% of the total population is Sudian are left white and those where more than 90% is Sudian are shown in dark red, with intermediate gradations of lighter red to bring out intermediate percentages. Clearly most of the map would be red, giving an impression of a preponderant Sudian population. Eckert (1925: 145) discussed this particular type of deception, which he asserted had 'recently been propagated on certain non-German maps', notably 'maps showing the languages and the plebiscitary vote in Upper Silesia, which were drawn neither by German hands nor in the German scientific spirit.'

Judgment

Fully as important as scientific integrity in the making of maps, indeed, largely a function of scientific integrity, is judgment. This embraces critical acumen in the selection of source materials, discrimination in the use of techniques, taste in choice and arrangement of colours, symbols, lettering and so on., and, throughout, a feeling for consistency.

As maps are only rarely accompanied by relative-reliability diagrams, critiques of their sources, and explanations of the graphic techniques by which information is disclosed on them, it is not always possible to test the quality of the judgment that has gone into their construction. The essential accuracy of certain types of map, however, can be taken on faith. One hardly needs to question the basic reliability of the charts and topographic sheets issued by governmental institutions such as the United States Coast and Geodetic Survey (Patton 1927) and the United States Geological Survey, whose high standards are well known. On the other hand, it is not safe to take on faith the reliability of the average reference map, atlas map or statistical map. A general reference map can usually be spot-checked by comparison with good maps of the same region on larger scales, and a statistical map spot-checked against the sources if these can be identified and found. Whether or not such tests need to be carried out depends, of course, on the uses to which a map is to be put. A map exact enough for measuring the air-line distance from Boston to New York may be quite unsuitable for measuring the total length of the Maine coast line. In general, unless one has solid ground for confidence in the integrity and judgment of the makers, detailed information derived from maps should not be incorporated in other maps or used as a basis for conclusions and decisions of importance. The fallacy of over-refined inferences from maps should be carefully avoided. One can with good reason accept the general over-all picture that a map presents, but this does not mean that the map in its every particular presents the gospel truth.

Simplification and amplification

Although a map's reliability is no higher than that of its sources, it may be considerably lower unless good judgment is exercised in its compilation. In this process the 'raw' information provided by the sources is transformed by the cartographer. Two operations may be carried out. If the 'raw' information is too intricate or abundant to be fully reproduced to the scale of the map as it stands, it may be simplified and generalised. If it is too scanty, it may be amplified and elaborated. [...] For example, where the records yield scanty details, as for ancient times, the historian seeks to fill in the gaps on the basis of inference and conjecture. Where the information is superabundant, as for modern times, he selects. In both cases he does not copy the sources but gives a partly subjective interpretation. Frequently he must both amplify and simplify in the same study. So, too, one part of a map may be the result of amplification of the sources, another part the result of simplification.

For most general reference maps the sources are simplified and generalised to some extent by the removal of minor irregularities in coast lines, railroads, slopes and so on, as well as by the omission of many features: towns, hills, mountains, lakes and so on. If the process of simplification and generalisation is not carried out consistently, the results may be misleading. Sometimes one category of information is simplified to excess in relation to another. For instance, the omission of stream-lines where form lines or contours are reproduced in detail may give an erroneous impression of dry watercourses. This is an inconsistency found even on standard atlas maps. There is also inconsistency where specific information is more radically simplified on one part of a map than on another. Such inconsistency is sometimes necessary on even the best of general reference maps to avoid overcrowding; for example, where, in the more densely settled regions, towns much larger than those shown in the sparsely settled regions are omitted because there is no room for them. In such cases the mapmaker must use his discrimination in deciding what to omit, and his map, although it may gain in legibility and beauty, will lose in reliability, at least with regard to the specific distributions that are inconsistently shown.

Amplification of the information provided by the source materials is the addition of details regarding the exact location of which the sources fail to provide precise evidence. Amplification is, of course, wholly unjustifiable where there is no sound evidence of the existence of the details added, as where an unsurveyed river is shown with sinuosities in the sincere belief that they are more 'natural' than a straighter course would be. If, however, reliable observers report having crossed the river at several points, the careful cartographer need not omit it from his map because he has little to guide him in drawing its exact course. It is usually preferable to plot the river by conjecture, in the light of whatever evidence there may be. On maps designed for certain specific uses, such as air navigation charts, too much of the conjectural may be seriously confusing; but, in general, amplification of the sources, when done critically, is misleading only when the users of maps draw over-precise inferences from the features that have been amplified, and such danger is obviated if these features are represented by broken lines. [...]

Quantitative information on maps

Quantitative information shown on maps is peculiarly affected by the operations of amplification and simplification. By quantitative information I mean here information regarding the distribution of quantities of varying intensity. Contour lines, of course, provide such information. Statistical maps, or maps the purpose of which is to provide quantitative data, are essential tools in geography, climatology, oceanography, demography and other branches of the natural and social sciences. They are also frequently consulted by the general public in atlases, magazines and popular books. They are so widely used that it might be well if their comparatively high degree of subjectivity were more generally recognised than it is. Certain terms must first be defined (De Geer 1923: 14–23; Wright 1938). One way of setting forth quantitative data on a map is simply to copy from the sources numbers giving heights, depths, populations and so on. As this gives no visual impression of relative magnitudes, the more usual method is to show quantities by means of symbols. Three kinds of symbols may be used: point symbols, isopleths or spatial symbols. A point symbol is a dot, disk, cube, sphere or other geometrical figure of conventional form that represents a specified quantity; an isopleth is a line that represents a uniform quantity; and a spatial symbol is a colour, shading, ruling, geometrical pattern or the like that represents either a single specified quantity or a range of quantities between two specified limits.

Now it is obvious that the quantities the symbols represent – the 'mapped quantities' – seldom correspond precisely to the quantities as given in the sources. As the latter guide or control the mapmaker in selecting the appropriate symbols and in placing them on his map, they may be called 'control quantities'. The mapped quantities are almost invariably simplifications or generalisations of the control quantities. For example, a point symbol may show that a city has a population between 500 000 and 600 000, whereas the population according to the census – a control quantity – may be 578 341. From isopleths one might determine that a given point was between 20 and 40 feet above sea level, whereas a benchmark at that point would show its altitude to be 32.5 feet.

A distinction must also be made between 'locational quantities' and 'spatial quantities'. A locational control quantity is one that indicates the intensity of some condition at a 'control point', such as the altitude of the point or the temperature or barometric pressure recorded there. Locational control quantities are usually determined by instruments and may be shown by point symbols of graded sizes, though more frequently isopleths (for example, contour lines, isotherms or isobars) are used. A spatial control quantity is one that applies to the whole of a given space, the 'control space' (usually a political division); for example, the population of a county. Spatial control quantities are usually determined by census or other enumerations rather than by instruments. Point symbols are properly employed (as on 'dot maps') for showing total quantities; spatial symbols are more appropriate for showing spatial ratios (Wright 1938: 14; Raitz 1938: 246–247). On statistical maps one occasionally finds 'isopleths' that indicate roughly the distribution of spatial ratios. Spatial ratios, however, cannot be determined with respect to points, but only with respect to spaces. Hence a line that purports to pass through points at which spatial ratios are equal actually represents 'quantities' that do not exist. [...] A spatial ratio may be the ratio of a certain quantity in a space either to the area of the space (for example, density of population) or to some other quantity in the same space (for example, income per capita, birth rate per 1000 persons).

Results of generalisation of quantitative data

In deciding what quantities his symbols are to designate, the cartographer must first study carefully the general character of the distribution he intends to map. Unless he does this, he may find, for example, that he has assigned too few people to each 'dot' and consequently has to crowd so many dots into the more densely settled regions as to produce a solid black mass. Thus his map is inconsistent, since it fails to give any picture at all of the way in which the people are distributed within the black mass. Similarly, if he makes each dot represent too many people, he will meet with difficulties in placing the dots in the sparsely settled regions, where one dot may have to represent a population scattered over a very large tract. Much the same problem is encountered when spatial symbols are used. Many phenomena have a tendency toward extreme concentration in relatively small spaces; population is a notable example. If the class intervals of the densities of population represented by the spatial symbols are uniform [...] there may be too few classes to show adequately the differences in density in the regions of sparse distribution and more classes than can be legibly differentiated for those of greater concentration. Hence it is customary in mapping distributions of this type to make the class intervals narrow at the lower end of the density scale and increase the width as the upper end is approached [...].

Much the same principle also applies to the representation of relief on maps covering large tracts of country. As lowlands and plains are, on the whole far, more extensive than mountains, it is usual on such maps when relief is shown by contours and hypsometric tints to employ narrower contour intervals for the lower elevations than for the higher. [...] In this way more topographic detail can be shown for the lowlands, and there is less crowding on the highlands, than would be the case if the contour intervals were uniform. This, of course, has unfortunate effects when applied on maps that include extensive high plateaus such as those of the central Andes and Tibet, since the contour intervals are too wide to bring out adequately the often varied relief of the uplands. Thus the pattern and, with it, the adequacy of maps showing quantities are much affected by the discrimination with which the cartographer adjusts the classification of the mapped quantities shown by the symbols to the control quantities.

As the control quantities for a density map are functions of the areas of control spaces, the information provided by them will necessarily be less detailed for the parts of the map where the control spaces are relatively large. For this reason the pattern of symbols based on this information will be inconsistent within itself, unless all the control spaces are of approximately equal size, which is seldom or never the case. Furthermore, and for the same reason, it is obvious that two statistical maps of the same territory showing the same distribution will present quite different patterns if based on control data for different systems of control spaces. [...] Two statistical maps of different regions are not comparable except in a very general way if one is based on figures for control spaces with areas that are on the average larger than those of the other.

Amplification of quantitative data

Some of the inconsistencies due to lack of uniformity in the areas of the control spaces can be partly eliminated by amplifying the control quantities in the light of other evidence. Where point symbols are employed to show distributions, they can be arranged within each of the control spaces so as to conform to the probable character of the distribution therein. In placing them, other maps are consulted that make this probable character reasonably clear; for example, maps showing houses, villages, woodlands, mountains, and so on. A similar procedure can be followed where spatial symbols are used, by breaking down the control spaces into subdivisions, to each of which an estimated density is assigned with the aid of non-statistical evidence, care being taken that the estimated densities are made to conform to the known total control quantity of each space (Wright 1936). Such estimated densities might be called 'secondary control quantities', since it is they that actually govern the placing of the spatial symbols. Isopleths are drawn on a map with reference to control points. Other things being equal, the smaller the interval between the mapped quantities represented by the isopleths (for example, the contour interval), the larger the amount of detail shown. The reliability of this detail, however, depends not only on the accuracy of the control quantities but also on the geographical density of the distribution of the control points. If the control quantities are of uniform accuracy and the isopleths are located with reference to the control points merely by interpolation, the probable reliability of the information furnished by the isopleths will vary in proportion to the density of the control points.

When the isopleths are drawn by interpolation, there is a minimum of amplification. As in the case of point symbols and spatial symbols, however, the isopleths may be drawn in such a way as further to amplify the information provided by the control quantities. Thus the surveyor further amplifies his map when he 'sketches in' contours by observing the terrain in the field (or from the study of photographs) and shaping the lines accordingly. Formerly contours were nearly always 'sketched in'. [For example] Debenham (1936: 98) reports that 'The original method [of contouring] used by the Ordnance Survey was to send a levelling party to fix each of the contours on the ground, and put in lines of pegs as they levelled round each slope. The pegs were afterwards surveyed by traverse by another party. This excellent but very expensive way is now rarely followed, and the rigorous use of the word 'contour line' has largely disappeared, for the majority of maps showing contours depend on nothing more accurate than interpolation between spot heights or some such method'. Today the use of photogrammetrical plotting instruments is largely eliminating this subjective factor so far as contour mapping is concerned, though it still influences the plotting of isopleths of other types: isotherms, isobars and so on. The mapmaker must first decide whether to interpolate arbitrarily or to try to draw the lines more realistically on the basis of guided conjecture. In either case the user of the map has no way of distinguishing the parts that are probably more reliable from those that are probably less so, unless the positions of the control points are shown on the map itself or on an accompanying diagram. Unfortunately, this is seldom done on the isopleth maps in most common use, though it has been frequently recommended by climatologists and others. An interesting illustration of control on isopleth maps is given on the Veatch and Smith's (1939) 1: 20 000 charts of the submarine topography off the north-eastern United States. [...]

Although some maps present no quantitative information of the types that have just been discussed, every map is quantitative in the sense that it provides information regarding distances, areas and directions. Such quantities can be accurately determined by direct measurements on maps covering small spaces if the maps themselves are reliable. The larger the space covered, however, the less can reliance be placed on direct measurements, on account of distortions introduced by the projection. [. . .] If some of your distances are in correct proportion, others will be misleading; if your relative areas are consistent, your shapes and distances will be badly out; and so forth. The plotting of a map on a particular projection is a mechanical operation with little that is subjective about it, but the selection of the most suitable projection for any given purpose is highly subjective, requiring good judgment guided by technical knowledge. These remarks should have made it clear that the quantitative information furnished by maps is much affected by subjective influences. The most marked effects of all, perhaps, are felt on maps showing densities and other ratios. Because of the different practices employed by their makers, with different degrees of skill and judgment, such maps are seldom strictly comparable with one another or even consistent within themselves.

Synthetic information and generalisation

Up to this point we have been dealing primarily with subjective influences on the mapping of the distribution of individual phenomena, such as coastlines, populations and temperatures. One of the most important purposes that maps accomplish, however, is to show relationships of different phenomena to one another. Such relationships may be brought out by the use of different symbols on the same map to show different kinds of things: blue lines for rivers, red lines for roads, dots for cities and so on – 'mixtures' of information, in other words. Many maps present 'compounds' rather than mere mixtures – that is, the symbols themselves indicate either quantitative relationships between two phenomena or more or the coexistence of two phenomena or more at particular locations or within particular regions. Such 'synthetic information' may be the result of the compounding of only two or three elements, as on a map showing a relationship between rainfall and temperature. [. . .] Or a much larger number of elements may be compounded, as on maps showing various types of 'region' – climatic regions, soil regions, economic regions, land-use regions, 'natural regions', 'cultural regions' and so on.

The influence of subjective factors: judgment, discrimination, critical acumen and so on, is of paramount importance in mapping of this type. Whether a particular relationship between two wholly objective elements is of significance or not depends on the quality of the mapmaker's judgment. [. . .]

It is usually difficult, if not impossible, for the purposes of mapping to combine in the form of ratios, coefficients and the like more than a limited number of statistical quantities. Consequently, the more general natural and human regions that geographers have marked out on many maps either are based on the assumption that the presence of certain individual phenomena or simple combinations of phenomena within a region is a fairly reliable indication that more complex combinations exist there, or else are arbitrarily marked out on the basis of the mapmaker's general fund of information. Frequently such regions are delimited to provide a framework for teaching or for the arrangement of material in textbooks – for convenience, that is, rather than to reflect absolute realities on the earth's surface. Maps of this type may be useful as pedagogical devices. They give the student at least a rough idea of how the parts of the world differ from one another in certain large respects, but next to maps of imaginary countries they represent almost the ultimate in subjective cartography. The regions they show are the equivalent of the well defined 'periods' into which history teachers divide the course of human events, and they have both as much and as little reality as such periods.

Harmony and taste

We have seen how judgment and other subjective factors affect the reliability of the information presented by the symbols on a map. A symbol may, however, present a given phenomenon correctly as regards its distribution, character and quality and yet be altogether unsuitable. Whether or not it is suitable depends on the mapmaker's taste and sense of harmony. Where quantities such as altitude and density of population are shown by means of gradations of colour or shading, the symbolism will not be suitable unless there is some harmonious relationship between the variations in the tonal intensities of the symbols and the varying intensities of the quantities indicated by them. Although experiments have been made with a view to grading tonal intensities so that they may bear a definite mathematical relationship to the mapped quantities, for the vast majority of maps the tonal grades are established by rule of thumb [. . .]. In matters of colour and tone the rule of thumb must be applied with the skill of the artist if the results are to be good.

Medieval maps are adorned with castles, towers, sea monsters, ships and such things, and the employment of pictorial symbols is coming back today. It has certain advantages, especially on maps for popular use. By employing as point symbols conventionalised pictures [. . .] one can crowd a good deal of miscellaneous 'mixed' information into a small space, but for any single distribution one cannot present as much precise and detailed distributional information as by the use of dots or other point symbols, since each little picture is likely to occupy more space than a single point symbol. Experiments have been made in the use of pictorial patterns as spatial symbols – for example, patterns made up of tiny men, or ears of corn, or cows – but such patterns tend to look 'fuzzy' around the edges and do not lend themselves to the representation of gradations in quantity. Although in a sense they may be in better 'harmony' with the things that they represent than flat colours or shading would be, they may also be out of harmony with the purpose of the map if that purpose is to give a clear and clean-cut concept. The quality of a map is also in part an aesthetic matter. Maps should have harmony within themselves. An ugly map, with crude colours, careless line work, and disagreeable, poorly arranged lettering may be intrinsically as accurate as a beautiful map, but it is less likely to inspire confidence.

Progressiveness and conservatism

Whether a maker of maps is ever seeking and finding new things to map and developing new ways of mapping them or is a blind follower of tradition and precedent is of course partly a matter of individual character, but it is also a result of outside influences. Advances in cartography are due largely to the stimuli and opportunities that social needs give to the inventiveness of cartographers. Where a need arises, as for automobile road maps or air navigation maps, cartographers respond. War provides a powerful stimulus, as the feverish mapping activities in and out of the government today bear witness.

Conservatism in cartography is not inherently an evil when it means adherence to conventions and standards that have been tested by time and found good. The users of maps have become used to certain conventions in symbolism, and too radical departures from these may be needlessly confusing. Many of the conventions have their origin in attempts to make symbolisms conform at least roughly to certain aspects of the things mapped; for example, that of showing water in blue instead of, say, pink (on medieval maps the Red Sea was shown in red), that of showing the relative importance of places by point symbols and lettering of graded sizes or of showing railroads by crosshatched lines. The convention of grading hypsometric tints more or less according to the spectrum, from greens for the lower altitudes, through yellows for intermediate levels, to reds and violets for mountain ridges, has been found by experiment and experience to give a graphic visual impression of relative altitudes. That there are different techniques altogether for representing the character of the surface configuration of the land is hardly a sufficient reason for abandoning a well established and, on the whole, satisfactory method of representing relief. Conservatism is an evil when it means failure to keep abreast of changing social needs and technical improvements. There is a certain danger of this in peacetime, especially in large mapmaking establishments where maps are produced in immense quantities according to uniform specifications. Under such conditions changes are costly and difficult, involving the training of personnel in new methods and the purchase of new equipment. This is why so many maps in atlases as well as in series follow traditional patterns or even repeat old errors long after the need for something better has made itself felt.

Map users are human

If mapmakers are human, so too are map users. The qualities of integrity, judgment, critical acumen and the like are as much required in the interpretation of maps as in the preparation of them. Like carpenters' tools, maps should not be misused. More should not be expected of them than they can perform. Sometimes when a critic damns a compiled map because he has found errors on it in regions that he has visited, his condemnation may reveal ignorance of the nature of cartography on his part rather than carelessness on the part of the mapmaker. Not all maps can be based on new surveys. Errors that originated in the sources of a compiled map frequently could have been avoided only by not making the map. It is a misuse of maps, also, to draw unwarranted conclusions from them. To compare the area of Greenland with that of New Guinea on a map drawn on the Mercator projection is like trying to cut down an oak tree with a jig saw – it simply cannot be done; and we have seen that statistical maps are liable to similar if less obvious misinterpretation. Particularly dangerous are unwarranted conclusions as to the meaning of the facts that maps actually do disclose. Bring together all the maps there are showing languages, religions, densities of population, resources, economic regions and the like; add to them twice as many more; compare and correlate them with all the ingenuity of which you are capable – and you will not, from such study alone, get very far toward the solution of vital international and national problems. Conditions and motives that no man can map must also be considered. As was pointed out at the beginning of this paper, maps are indispensable tools in human affairs. That you cannot navigate a ship without charts, however, does not mean that you can navigate it by charts alone. Rudders and helmsmen are also necessary.

References

Debenham, F. (1936) *Map Making*, Blackie and Sons, London.
De Geer, S. (1923) On the definition, method and classification of geography. *Geografiska Annaler*, **5**, 1–37.

Eckert, M. (1908) On the nature of maps and map logic. *Bulletin of the American Geographical Society*, **40**, 344–351.
Eckert, M. (1925) *Die Kartenwissenschaft: Forschungen und Grundlagen zu einer Kartographie als Wissenschaft*, de Gruyter, Berlin.
Patton, R.S. (1927) The physiographic interpretation of the nautical chart. *The Geographical Review*, **17**, 115–117.
Raitz, E. (1938) *General Cartography*, McGraw Hill, New York.
Speier, H. (1941) Magic geography. *Social Research*, **8**, 310–330.
Veatch, A.C. and Smith, P.A. (1939) Atlantic submarine valleys of the United States and the Congo submarine valley. Geological Society of America Special Paper 7, pp. 54–55.
Wright, J.K. (1936) A method of mapping densities of population, with Cape Cod as an example. *The Geographical Review*, **26**, 103–110.
Wright, J.K. (1938) Problems in population mapping, in notes on statistical mapping, with special reference to the mapping of population phenomena, in *The Peruvian Expedition Of 1912 Under The Auspices Of Yale University & The National Geographic Society* (ed. J.K. Wright), American Geographical Society and Population Association of America, pp. 1–18.

Further reading

Herb, G. (1997) *Under the Map of Germany: Nationalism and Propaganda 1918–1945*, Routledge, London. [This monograph fleshes out the contextual detail of some of the early examples of propaganda mapping cited by Wright in the above excerpt.]
Keates, J. (1984) The cartographic art. *Cartographica*, **21** (1), 37–43. [This insightful article explores the author's attempts to rationalise the need for an aesthetic approach to map design.]
MacEachren, A.M. (1994) *Some Truth with Maps: A Primer on Symbolization and Design*, Association of American Geographers, Washington, DC. [A best practice guide for cartographic design which take on board many of the subjective aspects of mapmaking highlighted in Wright's article.]
Monmonier, M.S. (1996) *How to Lie with Maps*, 2nd edn, Chicago University Press, Chicago. [In many ways an updated version of Wright's arguments with a wide array contemporary examples and a strong argument for the integrity of 'proper' cartographic design.]

See also

- Chapter 1.7: Design on Signs/Myth and Meaning in Maps
- Chapter 1.8: Deconstructing the Map
- Chapter 1.12: The Agency of Mapping
- Chapter 3.2: Interplay of Elements
- Chapter 3.3: Cartography as a Visual Technique
- Chapter 3.4: Generalisation in Statistical Mapping
- Chapter 3.6: The Roles of Maps
- Chapter 3.9: Mapping, Modernity: Art and Cartography in the Twentieth Century
- Chapter 3.10: Affective Geovisualisations
- Chapter 5.3: Texts, Hermeneutics and Propaganda Maps

Chapter 4.3

Cognitive Maps and Spatial Behaviour: Process and Products

Roger M. Downs and David Stea

Editors' overview

At the start of the 1970s, it was intellectually fashionable amongst behavioural geographers to investigate the significance of cognitive maps, and their impacts on people's spatial behaviour. Downs and Stea's book was probably the most influential overview of the field and brought together papers from almost all of the leading exponents of this kind of research. We have excerpted Chapter 1, which explores the dimensions of cognitive mapping, distinguishing between cartographic images and the cognitive constructs that are the focus of their attention. This conceptual piece is informed by a communications model of information transmission and explores processes and defines concepts underpinning research. The authors define the concepts of perception, cognition, attitude and preference, before explaining the differences between what people need to know and what they actually know. Amongst other concepts they focus on differences between locational and attribute information, the role of incomplete, distorted, schematised, and augmented cognitive maps, and some of the behavioural reasons for the mismatch between theory and practice. They conclude by urging further experimental investigation of behavioural evidence of cognitive mapping.

Originally published in 1973: Chapter 1 in Roger M. Downs and David Stea (eds) *Image and Environment: Cognitive Mapping and Spatial Behavior,* Aldine Press, Chicago, 8–26.

Introduction

A surprising fact is associated with studies of cognitive mapping: although the emergence of this vigorously developing research area has been recent, we are not discussing something newly discovered [...]. Instead, we are concerned with phenomena so much part of our everyday lives and normal behaviour that we naturally overlook them and take them for granted.

[...]

We find that planners try to alter cognitive maps, astronauts need them, the news media use them, and advertisers tempt us with them; they are part of our everyday lives. [...] We offer a formal definition: cognitive mapping is a process composed of a series of psychological transformations by which an individual acquires, codes, stores, recalls and decodes information about the relative locations and attributes of phenomena in his everyday spatial environment.

In this paper we will expand this definition and examine the conceptual frameworks which are subsumed within it.

An analysis of cognitive mapping processes

Cognitive maps and adaptive behaviour

Underlying our definition is a view of behaviour which, [...] can be reduced to the statement that *human spatial behaviour is dependent on the individual's cognitive map of the spatial environment.* That this formulation is necessary is indicated by a comparison of the characteristics of the individual with those of the spatial environment.

The environment is a large-scale surface, complex in both the categories of information present and in the

The Map Reader: Theories of Mapping Practice and Cartographic Representation, First Edition. Edited by Martin Dodge, Rob Kitchin and Chris Perkins.

number of instances of each category. Things are nether uniformly distributed, nor ubiquitous; they have a 'whereness' quality. In contrast the individual is a relatively small organism with limited mobility, stimulus seeking capabilities, information processing ability, storage capacity and available time. The individual receives information from a complex, uncertain, changing and unpredictable source via a series of imperfect sensory modalities, operating over varying time spans and intervals between time spans. From such diversity the individual must aggregate information to form a comprehensive representation of the environment. This process of acquisition, amalgamation and storage is cognitive mapping, and the product of this process at any point of time can be considered a cognitive map.

Given a cognitive map, the individual can formulate the basis for a strategy of environmental behaviour. We view cognitive mapping as a basic component in human adaptation, and the cognitive map as a requisite both for human survival and for everyday environmental behaviour. It is a coping mechanism through which the individual answers two basic questions quickly and efficiently: where certain valued things are, and how to get to where they are from where he is.

Cognitive maps and spatial behaviour

[...] We believe that a cognitive map exists if an individual behaves as if a cognitive map exists (Stea and Downs 1970). [...] Normal everyday behaviour such as a journey to work [...] would be impossible without some form of cognitive map. [...] Admittedly, much spatial behaviour is repetitious and habitual [...] but even this apparent stimulus response sequence is not so simple: you must be ready for the cue that tells you to 'turn here' [...]. You are thinking ahead and using your cognitive map. In human spatial behaviour we consider even a series of stimulus–response connections as a 'simple' (or impoverished) form of a cognitive map, in which the general aspects of spatial relationship implicit in cognitive mapping play a minimal role. [...] The person knows that an object is valued and one way of getting to it, but knowledge of the whereness in relation to the location of other objects is absent. [...] Thus someone who knows only one route knows more about that route than just the appropriate responses at certain choice points and, because he thinks ahead, is also engaging in cognitive mapping. We are postulating the cognitive map as the basis for deciding upon and implementing *any* strategy of spatial behaviour.

However, we must make it perfectly clear that a cognitive map is not necessarily a 'map'. [...] We are using the term 'map' to designate a functional analogue. The focus of attention is on a cognitive representation, which has the functions of the familiar cartographic map but not necessarily the physical properties of such a pictorial graphic model (Blaut *et al.* 1970). [...] The cartographic map has a profound effect on our concept of a cognitive map.

Spatial information can be represented in a variety of ways. [...] All media share the same function not structure, and thus cognitive maps are derived from analogies of process, not product.

Cognitive mapping signatures and cognitive representations

[...] All of the media rely upon the same sort of spatial information, and all are employed in the same sorts of spatial behaviour: thus the inputs and outputs are specified, while the intervening storage system (the black box) is not. The way in which spatial information is encoded (mapmaking) and decoded (map reading or interpreting) gives rise to a set of operations called the signature of a given mapping code. Thus a cartographic map signature is dependent upon three operations: rotation of point of view to a vertical perspective, change in scale, and abstraction to a set of symbols [...]. Many other signatures are feasible; we have no reason to anticipate that cognitive maps should necessarily have the same form of signature as cartographic maps. Above all, we should avoid getting locked into a form of thinking through which we as investigators force a subject to produce a cartographic cognitive map and which we then verify against an 'objective' cartographic map. [...]

The issue of mapping signatures involves some fundamental theoretical and methodological issues. [...] Underlying the whole approach is the basic question: How is information derived from the absolute space of the environment in which we live, transformed into the relative spaces that determine our behaviour? The transformation can be viewed [...] as involving any or all of three fundamental operations: change in scale, rotation of perspective and a two-stage operation of abstraction and symbolisation, all of which result in a representation of relative space.

[...]

Thus, we should be interested in developing theoretical statements about the cognitive signatures that are employed in dealing with information from the spatial environment. [...] The only differences between Lynch's (1960) 'images' and city maps of cartographers lie in the degree of abstraction employed and the types of symbols chosen to depict information. [...] We should be concerned with the nature or signature of relative space as it is construed and constructed by the individual. Only if we do this can we ask how relative and absolute spaces compare and differ. [...] Some aspects of our composite cognitive maps may resemble a cartographic map; others will depend upon linguistic signatures (in which scale and rotation operations are irrelevant), and still others upon visual imagery signatures derived from eye-level viewpoints (in which the scale transformation may be disjointed or convoluted). [...]

Before considering the nature and functions of cognitive maps in more detail, we must discuss some basic definitions and attempt to clarify a few misconceptions which currently prevail.

The concepts of perception, cognition, attitude and preference

Perception and cognition: distinctions

Unfortunately, perception and cognition have been employed in a confusing variety of contexts by psychologists and other social scientists. [...] It is difficult to determine whether the process of perceiving is being discussed or [...] the product of the perception process. [...] Perception has been used in a variety of ways: to experimental psychologists it involves the awareness of stimuli through the physiological excitation of sensory receptors; to some social psychologists it implies both the recognition of social objects present in ones immediate sensory field and the impressions formed of persons or groups experienced at an earlier time. To many geographers perception is an all-encompassing term for the sum total of perceptions, memories, attitudes, preferences and other psychological factors which contribute to the formation of what might better be called environmental cognition. [...]

Given the varied uses of the terms it is difficult to distinguish between perception and cognition. [...] We reserve the term perception for the process that occurs because of the presence of an object, and that results in the immediate apprehension of that object by one or more of the senses. Temporally it is closely connected with events in the immediate surroundings and (in general) linked with immediate behaviour. [...] Cognition need not be linked with immediate behaviour and, therefore, need not be directly related to anything occurring in the proximate environment. [...]

However this distinction falls short of establishing a clear dichotomy [...]. Both refer to inferred processes responsible for the organisation and interpretation of information [...]. Cognition is the more general term and includes perception as well as thinking, problem solving and the organisation of information and ideas. A more useful definition from a spatial point of view is offered by Stea (1969). He suggests that cognition occurs in a spatial context when the spaces of interest are so extensive that they cannot be perceived of apprehended at once [...]. This scale dependent distinction [...] also suggests that we are concerned with the nature and formation of environmental cognitions rather than with briefer spatial perceptions.

Attitudes, predictions, preferences and cognitive maps

[...] The parallels between the concepts of cognitive map and attitude are marked. [...] Fishbein (1967) replaces the holistic concept of an attitude with a formation containing three components: cognitions or beliefs, affect or attitude and conations or behavioural intentions. Fishbein claims that the fact that affect, cognition and action are not always highly correlated necessitates this more complex typology. [...] Fishbein points out that attitudes, beliefs and expressed behavioural intentions are frequently brought into line with actual behaviour. [...] In other words, if the behaviour can be specified an attitude can usually be postdicted.

Finally, we must distinguish among attitudes, preferences and traits. [...] Preferences are usually considered to be: less global [...]; and less enduring over time. [....] When a given attitude pervades a wide variety of objects over a considerable period, it becomes a personality trait. [...]

Hypothetically, one could construct a scale from preference through attitude to trait, increasing in both inclusiveness and duration of the cognitive, connotative and effective components. [...]

The nature and function of cognitive maps

What do people need to know?

[...] There are two basic and complementary types of information that we must have for survival and everyday spatial behaviour: the locations and attributes of phenomena. Cognitive maps consist of a mixture of both. [...] We must also know what an object is.

Locational information is designed to answer the question, 'where are these phenomena?' and leads to a subjective geometry of space. There are two major components of this geometry, distance and direction. Distance can be measured in a variety of ways and we are surprisingly sensitive to distance in our everyday behaviour. [...] Knowledge of distance [...] is essential for planning any strategy of spatial behaviour. [...] Direction is no less important in the geometry of space, although we are less conscious of directional information. [...]

By combining distance and direction we can arrive at locational information about phenomena, but not necessarily the same as that provided by Cartesian co ordinates of cartographic map. [...] Thus, locational information is not as simple as it might appear. We must store many bits of distance and direction data to operate efficiently in a spatial environment, a process involving relatively accurate encoding, storage and decoding. Use of locational information [...], however, requires a second type of information: that concerning the attributes of phenomena.

Attributive information tells us what kinds of phenomena are out there and is complementary to locational information. [...] An attribute is derived from a characteristic pattern of stimulation regularly associated with a particular phenomenon, which, in combination with other attributes, signals the presence of the phenomenon. [...]

We can divide attributes of phenomena into two major classes: descriptive, quasi-objective or denotative; and evaluative or connotative. [...] Here we are separating attributes which are affectively neutral (descriptive) from those that are affectively charged (evaluative).

[...] An object is identified and defined by a set of attributes and bits of locational information. However, what is an object at one spatial scale can become an attribute at another [...]. The scale of analysis of the problem at hand defines what is an object and what is attributive and locational information.

What do people know?

If we compare a cognitive map with a base map of the real world [...] we find that cognitive mapping does not lead to a duplicative photographic process [...], nor does it give an elaborately filed series of conventional cartographic maps at varying spatial scale. Instead cognitive maps are complex, highly selective, abstract, generalised representations in various forms. [...] We can characterise cognitive maps as incomplete, distorted, schematised and augmented, and we find that both group similarities and idiosyncratic individual differences exist.

The incompleteness of cognitive maps

The physical space of the real world is a continuous surface which we have come to understand through a classic geometrical framework: that of Euclid. [...] There are no gaps or bottomless voids. [...] Yet all cognitive maps are discontinuous surfaces. Seemingly some areas of the earth's surface do not exist when their existence is defined by the presence of phenomena in the subject's cognitive representation. [...] [However] we must be careful in interpreting the absence of phenomena from cognitive maps as reflecting discontinuity of space.

Distortion and schematisation

By the distortion of cognitive maps we mean the cognitive transformations of both distance and direction, such that an individual's subjective geometry deviates from the Euclidian view of the real world. Such deviations can have major effects on the patterns of spatial use of the environment. [...] If people are sensitive to distance, consequent spatial behaviour patterns will be dependent on such distance distortions.

Far more significant and as yet little understood are the results of schematisation (the use of cognitive categories into which we code environmental information). We are, as Carr (1970: 518) suggests, victims of conventionality. This conventionality can be expressed in two ways. The first involves the use of those spatial symbols to which we all subscribe and which we use both as denotative and connotative shorthand ways of coping with the spatial environment. [...] However, there are other symbols dealing with geographic entities [...] which owe their cogency and importance to their mere existence. [...] Such entities have been termed the invisible landscape. As images these elements are perhaps the most purely symbolic. [...] A second aspect of schematisation or conventionality involves the very limited set of cognitive categories or concepts that we have developed in order to cope with information derived from the spatial environment. [...] Our understanding of the semantics of cognitive maps is remarkably limited.

The controversy over linguistic relativity suggests that there are cross-cultural differences in the ways in which spatial information is coded. Such differences are not only cross-cultural. [...] Downs (1970) assumed that a neighbourhood shopping centre would be clearly defined and commonly agreed upon spatial unit, with the edge of the commercial area defining the shopping centre boundary. However, residents of the area recognised four distinct subcentres.

Augmentation

[...] There is some indication that cognitive maps have non-existent phenomena added as embroidery. [...] Such distortions may be highly significant, but we know little about their causes, and nothing about their eradication.

Inter-group and individual differences in cognitive maps and mapping

[...] Underlying group perspectives are the result of three factors. Firstly, the spatial environment contains many regular and recurrent features. Secondly, people share common information processing capabilities and strategies [...]. Thirdly, spatial behaviour patterns display similar origins, destinations and frequencies. These factors in combination yield inter-group differences in cognitive maps.

The individual differences among cognitive maps emerge primarily from subtle variations in spatial activity patterns. [...] Such idiosyncrasies are particularly notable in verbal descriptions of cognitive maps – the choice of visual details shows tremendous variation from subject to subject.

In answer, therefore, to the question, 'What do we know?', we can conclude that we see the world in the way that we do because it pays us to see it in that way. [...] People behave in a world 'as they see it' – whatever the flaws and imperfections of cognitive maps, they are the basis for spatial behaviour.

How do people get their knowledge?

We have postulated a set of basic characteristics that our knowledge of the spatial environment should possess, and we have indicated the characteristics that our knowledge (or cognitive map) actually possesses. Some of the differences [...] can be attributed to the ways we acquire spatial information.

Sensory modalities

In our studies of cognitive maps we have overlooked the range and number of sensory modalities through which spatial information is acquired, and have ignored the imaginative nature of cognitive processes [...]. The visual, tactile, olfactory and kinaesthetic sense modalities combine to give an integrated representation of any spatial environment. The modalities are complementary despite our intuitive belief [...] that visual information is predominant. [...] The quality of distinctiveness or memorableness is not solely the result of the way the environment looks. [...]

Direct and vicarious sources of information

Sources have a different degree of validity, reliability, utility and flexibility. Direct sources involve face-to-face contact between the individual and, for example, a city and information literally floods the person from all his sensory modes. He must be selective in what he attends to. [...] Reinforcement and checking are continuous: erroneous beliefs about locational and attribute information are rapidly corrected by feedback from spatial behaviour.

Vicarious information about the city is by definition second-hand. [...] This is true of a verbal description, a cartographic street map, a TV film, a written description, a colour photograph or a painting. In the mapping context these modes of representation, though similar in function, are different in form because they display different signatures. [...] The result of this filtering is a complete representation, which varies with both the individual and his group membership [...]. We may not be able to translate from the signature of the street map back to the spatial environment. [...] Both active and passive information processing are tied to the spatial environment, and result from symbolic elaboration, embroidery and augmentation.

[...]

Thus we have three sorts of information available to us at any point in time. Each has distinct characteristics, validity and utility.

A terminology for change

To this point our whole discussion of cognitive mapping has been static – concepts of learning time and change have been omitted. [...] We acquire the ability to know things about our environment through a process of development. [...] Development clearly includes change; taking place over a considerable period, such change is assumed to be irreversible and [...] is also regarded as progressive. Development encompasses both growth [...] and maturation. [...] What effects or learned changes can spatial information induce? Boulding (1956) suggest three possibilities: no effect, simple accretion and complete reorganisation. The no effect case is the most frequent in the normal adult: the information simply confirms what he already knows [...]. Most of the spatial information that we receive, although essential for the successful use of the environment of any point in time, has no effect on the stored knowledge or cognitive map.

A typology of change: accretion, diminution, reorganization

The simple accretion case relates to minor changes to the cognitive map. [...] Both locational and attribute information are added to the cognitive map: a simple additive change has occurred through learning. [...] Diminution develops directly from deletion. There is no need to assume that cognitive maps undergo only progressive change [. . .] Either through the passage of time or through maturation we forget – the amount of information available through the cognitive mapping diminishes. [. . .] All stored knowledge is subject to this time decay: we need to repeat a spatial experience in order to remember the route in the future. [. . .] Diminution may also be an adaptive process. [. . .] Given our limited capacity to store and handle information, diminution maybe [. . .] ensuring that excess information is lost but important information retained. [. . .]

The most dramatic changes in cognitive maps are the result of total reorganisation. Boulding (1956) suggests that images are relatively resistant to change in their overall nature. It requires an accumulation of contrary evidence before complete reorganisation can occur. [. . .] The most frequent spatial example of such a complete reorganisation is to be found in long-distance human migration and subsequent residential site selection. [. . .]

We have examined some aspects of our cognitive maps and how they came to be. We know they are modes of structuring the physical environment [. . .]. Much of the support in contentions concerning their existence is behavioural, stemming from introspection and anecdotal evidence, but the harder experimental data are beginning to emerge. [. . .] Thus, the face of cognitive mapping is growing clearer – only the features have yet to be fully filled in.

References

Blaut, J.M., McCleary, G.F. and Blaut, A.S. (1970) Environmental mapping in young children. *Environment and Behavior*, **2**, 335–349.

Boulding, K. (1956) *The Image*, University of Michigan Press, Ann Arbor, MI.

Carr, S. (1970) The city of the mind, in *Environmental Psychology: Man and His Physical Setting* (eds H.M. Proshansky, W.H. Ittelson and L.G. Rivlin), Holt Rinehart and Winston, New York, pp. 518–533.

Downs, R.M. (1970) The cognitive structure of an urban shopping centre. *Environment and Behavior*, **2**, 13–39.

Fishbein, M. (1967) Attitude and the prediction of behaviour, in *Readings in Attitude Theory and Measurement* (ed. M. Fishbein), John Wiley & Sons, Inc., New York, pp. 477–492.

Lynch, K. (1960) *The Image of the City*, MIT Press, Cambridge, MA.

Stea, D. (1969) The measurement of mental maps: an experimental model for studying conceptual spaces, in *Behavioral Problems in Geography: A Symposium* (eds K.R. Cox and R. G. Golledge), Northwestern University Press, Evanston, IL, pp. 228–253.

Stea, D. and Downs, R.M. (1970) From the outside looking in at the inside looking out, *Environment and Behavior*, **2**, 3–12.

Further reading

Blaut, J.M., Stea, D., Spencer, C. and Blades, M. (2003) Mapping as a cultural and cognitive universal. *Annals of the Association of American Geographers*, **93** (1), 165–185. [A comprehensive meta study reviewing the cross cultural evidence for cognitive mapping as a common human trait.]

Fabrikant, S.I. and Lobben, A. (2009) Cognitive issues in geographic information visualization. *Cartographica*, **44** (3). [This themed issue includes a number of useful articles focusing upon the application of cognitive approaches to geovizualisation some forty years after the Downs and Stea book.]

Kitchin, R. (1994) Cognitive maps: what they are and why study them. *Journal of Environmental Psychology*, **14**, 1-19. [A tightly focused and critical introduction to the nature of cognitive mapping.]

Kitchin, R. and Freundschuh, S. (2000) *Cognitive Mapping: Past, Present and Future*, Routledge, London. [An edited research monograph reviewing research progress thirty years after the heyday of behavioural geography.]

Liben, L.S. (2009) The road to understanding maps. *Current Directions in Psychological Science*, **18**, 310-315. [Reviews different psychological approaches to map understanding, with a rich emphasis upon contextual differences and their impact on environmental cognition.]

See also

- Chapter 1.3: On Maps and Mapping
- Chapter 1.6: Cartographic Communication
- Chapter 1.11: Exploratory Cartographic Visualisation: Advancing the Agenda
- Chapter 3.3: Cartography as a Visual Technique
- Chapter 3.6: The Roles of Maps
- Chapter 4.9: Understanding and Learning Maps
- Chapter 4.11: Usability Evaluation of Web Mapping Sites

Chapter 4.4
Natural Mapping

James M. Blaut

Editors' overview

Building upon the behavioural geography tradition and linking with wider scholarship grounded in a cognitive approach to cartography, there has been considerable research devoted to the acquisition of mapping skills in children and indeed to cross-cultural studies exploring the variation in map use. There has also been significant research investigating changes in mapping through the time span of human history. Jim Blaut was a notable and leading figure in all of this work and his 1991 paper, excerpted here, describes evidence for map use and sets out reasons for this ubiquity. Blaut develops his theory that mapping is a shared human characteristic, probably 'hard wired' into the brain and analogous to Chomsky's notion of a language acquisition device. He suggests that the mapping potential grew, as a behavioural support mechanism, through the need to interact with macro-environments and argues that treating mapping as a natural and always present human potential ought to allow a more creative and effective curriculum for teaching map skills to be developed.

Originally published in 1991: *Transactions of the Institute of British Geographers*, **16** (1), 55–74.

The problem for theory and practice

[...] In the present essay I will sketch in the lineaments of a theory of mapping behaviour and propose a number of concrete techniques, grounded in this theory, to improve the way we teach young children about mapping and about the macro-environment. The essential theoretical argument is this: mapping behaviour is carried out by all normal human beings of all ages in all cultures; it is therefore a natural ability, or habit, or faculty, 'natural' in a sense very close to the way language acquisition is 'natural'. The practical argument is this: if we teach formal map skills to young (elementary school and preschool) children in ways that develop their own natural mapping behaviour, and if we teach geography and other macro-environmental subjects in the same ways, we will significantly improve young children's learning. [...]

Macro-environmental behaviour

As a starting-point for theorising about mapping behaviour it is useful to notice that mapping has some connection with [...] macro-environmental behaviour [...]. [...] The human individual interacts with an environment that poses [...] three dissimilar situations of action. [...] To begin with there is a definite difference between situations which present themselves immediately as social, as involving interaction with other human beings, and those which present themselves as material or, broadly speaking, as environmental. It is of course true that the categories 'social' and 'non-social' are not fully separable [...]: humans confront and solve different problems, when they are interacting with other humans and when they are interacting with material things, large or small. No less fundamental is the difference between material situations of action which are macro-environmental and those which are micro-environmental. [...] Macro-environments, at this crude level of description, are generally larger than people; micro-environments are generally smaller. The prototype of a macro-environment is a 'place'. The prototype of a

The Map Reader: Theories of Mapping Practice and Cartographic Representation, First Edition. Edited by Martin Dodge, Rob Kitchin and Chris Perkins.

micro-environment is an 'object'. [...] In sum, the three situations are: social (or transactional), micro-environmental and macro-environmental. [...]

[...] Each modality, moreover, is deployed in a partly specific way. For instance, binocular parallax is crucial in visual sensing of small and nearby things, whereas distance vision, applied to places, tends to rely more on secondary evidence, on signs and inference, in judgements of identification, size and distance. In object learning, the hands – and manipulation – play a major role; in place learning the feet play a role that is perhaps more central than that of the hands. Most generally, the range of sensory modalities available for perception tends in general to decline proportionally as the environmental setting becomes larger, more distant and more inaccessible to the person (Stea and Blaut 1973a). [...] Very young children learn about objects by touching, tasting and smelling them, but about places [...] by gazing at them, often with limited comprehension [...]. And in general: sensory perception of, and motor behaviour toward, a macro-environment [...] is fundamentally different from sensory perception of and motor behaviour toward a micro-environment, an object or object-situation. [...]

The specificity of place behaviour emerges from cultural evolution and is a cross-cultural universal. For the Palaeolithic we can infer that the development of tool use was mainly object behaviour. Locational movement [...] was primarily place behaviour. Hunting, gathering, fishing and shell fishing combined close-in object behaviour and larger-scale place behaviour. Sheltering was a mixture of both. [...] And there were elements of material culture which related most directly to object behaviour (hand tools), others most directly to place behaviour (boats, paths), others to both (shelters etc.). We can speculate that place behaviour has a longer, or at least stronger, ancestry than does object behaviour in pre-hominid evolution. [...]

Mapping behaviour as a specific adaptation for macro-environmental behaviour

[...] Cognitive processes are inseparable from behaviour; are learned in behaviour – even processes which have an innate component, like walking, also necessarily have a learned component – either in childhood enculturation or in adult interaction. The learning and teaching of macro-environmental knowledge, coping techniques and values takes place largely (not entirely) through sign communication. [...] Thus linguistic analysis gives us an important window into geographic cognition. But there are important kinds of geographic knowledge which cannot be communicated – except very awkwardly – in the natural language. Consider [...] how to describe a place. For a number of fundamental reasons the communication of information about a place is more easily done, in any cultural context, by means of mapping than by means of the ordinary language, written or spoken. Places, unlike small objects, cannot ordinarily be perceived as a whole from one vantage point. In a flat landscape, like open ocean, very distant features cannot be seen, semi-distant features can be seen only in distortion, and all features can be seen only from one side. In most terrestrial landscapes rather little can be seen except the nearest features and only their near-facing elevation. This problem [...] is immensely aggravated in young children whose eye-level is only a couple of feet above the surface of the world; what is for adults a 'horizon' problem is for little children a 'tip-toe' problem.

[...] Humans cannot ordinarily perceive a landscape as a whole from an earthbound perspective. [...] This calls for the use of a cognitive map, in which the landscape is conceptualised [...] as a whole. This involves an integration of geographic features in various ways. Most crucially, it must simultaneously present all parts of the landscape and this requires a rotation or projection. We can infer that there is normal rotation to a virtual point [...] high above the landscape and roughly orthogonal to it, that is, overhead (Mead 1938). [...]

Each place-feature must be described with at least a triad of information: (1) the nature of the feature (its semantic meaning), (2) its distance and (3) direction from some reference position [...] (Blaut *et al.* 1970) [...] In even the simplest of landscape descriptions, the number of meaning–distance–direction triads is very great. Furthermore, features distant from the observer–describer frequently are very imperfectly known, as to meaning and location, and qualified judgements [...] must be included. [...] Landscapes are more complex than objects [...]: there is simply more material reality to be dealt with in a place than in an object. [...] A landscape for this reason is more difficult to describe.

But there is another form of complexity which has received hardly any attention by geographers. Objects generally are labelled with categorical nouns and noun phrases. Landscapes, or their major parts, typically do not have such generic labels. [...] Landscapes [...] are huge relative to the perceiving person, they are vastly variable, and they do not offer countless real instances of the same thing, each clearly an example of the named entity, as we are accustomed to encountering in the case of small objects. There is, moreover, a tendency in ordinary thought and language to chop up continuous space–time experience into atomistic 'objects' in 'space' (Blaut 1971). This 'object' language surrounds us with 'things'. [...] The learning and teaching of macro-environmental knowledge, coping techniques and values takes place largely (not entirely) through sign communication. Much of this is ordinary linguistic behaviour in the natural language [...]. Thus linguistic analysis gives us an important window into geographic cognition. But there are important kinds of geographic knowledge which cannot be communicated – except very awkwardly – in the natural language. [...]

Ordinary language can describe landscapes, of course, but it must do so with complex utterances and with highly creative language. In spoken discourse such descriptions can indeed be conveyed but only slowly and sometimes with great difficulty, and with some risk that the listener will miss part of the message. [...]

Mapping is generally much more efficient than natural language as a way to describe a macro-environment. [...] A mapmaker communicates a landscape description by persuading the map reader to accept the convention that both are thinking about a landscape as though high above it, looking down, and then displays, usually on a surface, signs representing its features [...]. This activity simplifies the problems of macro-environmental description by conveying information about [...] regional processes [...].

In this view the way we deal with the complexity of landscape is to portray its features pictorially or iconically, and the way we deal with the rotation problem is to generate pictures which present an image of the landscape, which utilise the cyclopean eye in the sky. The problem with this reasoning is that the communication of knowledge [...] must convey an immense amount of information. Therefore it must [...] utilise the crucial characteristic of a sign system, namely, that each symbol stands for an infinite number of cases of the thing denoted and that the system overall is capable of generating an infinite series of descriptions, of describing an infinitely expandable body of knowledge within the field it covers (Chomsky 1965).

Maps have certain image-like qualities but only one of these qualities is of primary importance in ordinary maps. That is the rule that the map surface bears a certain projective and scale relation to the landscape surface. None of the figures on the map itself need have pictorial (self-evident) meaning; even in the simplest maps most of the figures do not have such meaning. The map, therefore, does not have to be image-like except in a very restricted sense [...]. Basically, mapping is sign communication; it is literally a language, though a limited-purpose language (Head 1984; Blaut 1969, 1987b, 1987c). It has evolved as a component of cultural evolution with the specific function of communicating and recording macro-environmental knowledge, that is, geographic knowledge. [...]. The detailed semantics of the system, [...] and the elaborated transformational or map-syntax rules [...] are cultural variables, although the possibility of cultural universals here cannot be excluded.

But [...] the classical problem of imagery in genera, and imagery in cognitive maps and physical maps, cannot be so easily exorcised (Lloyd 1982; Peuquet 1988). I suspect (and will argue) that the origins of mapping were pictorial representations of landscapes, incised in or painted on various kinds of surfaces. I will also argue that a young child imagines (or images) large landscapes that he or she cannot travel through or even see as a whole from near ground level, and this cognitive mapping, early projected into toy-play modelling, is the ontogenetic beginning of natural mapping. [...] The basic argument, then, is that imagery is central to natural mapping, but every physical map [...] is an evolved semiotic system, using language-like signs and rules of syntax (rotation and scale change), and mapping behaviour is linguistic behaviour.

Empirical evidence of early mapping in children

David Stea and I, working with several colleagues, found that young children, aged three to six years, can do creditable mapping, make things that look very much like maps, and use these maps in [...] make-believe navigation, without prior instruction. We obtained cross-cultural evidence that suggests – nothing more – that the ability may be present in children of all cultures. [...]

Our first finding was that children aged five to six years in three cultures (USA, Puerto Rico, St Vincent) can read vertical black and white aerial photographs: recognise them to be pictures of the landscape, identify on them essentially the same micro-features that would be identified in a landscape seen from the familiar horizontal perspective, and to a limited extent identify macro-features or simple landscape-gestalts like street corner, town, woods and so on.[...] (Blaut *et al.* 1970). We found also that 5- and 6-year-olds can trace maps from vertical air photos and, after the photo has been withdrawn, use the 'maps' to solve simulated navigation problems (Blaut 1969). [...] Some evidence was obtained suggesting that the air-photo identification skill, well developed at six, may not improve much (instruction aside) thereafter (Stea and Blaut 1973b). [...] Roger Hart (1971) taught third-graders (average age about eight) some rather complex elements of location theory by abstracting from aerial photographs.

[...] Toy-play behaviour was studied in three, four and five-year-olds under testing conditions which controlled for the influence of language [...] and made no use of drawing; three-year-olds were able to construct realistic though simple macro-environmental maps out of toys and the performance by the four and five-year-olds was so similar to that of the three-year-olds that we could not see clear evidence of developmental change after the age of three (Blaut and Stea 1974). More complex macro-environmental modelling was observed in older children (Stea 1976, 1982). We were unable to devise comparable methodologies for children younger than three. [...] Therefore, we obtained only informal and qualitative evidence that the mapping abilities uncovered are also present below the age of three, although this appeared to be highly likely.

A number of studies have been carried out in which this work was replicated, formally or informally. [...] Spencer *et al.* (1989) review much of this research and concur with us that young children can indeed engage in sophisticated mapping behaviour. [...] The position now has considerable support in geographic education (Natoli *et al.* 1984; King 1989). Also of importance in this context are the very many studies carried out in recent years in which it has been shown that infants and very young children acquire spatial abilities (perceptual and cognitive) at earlier ages than had been known previously to be the case. [...]

Downs and Liben (1986, 1991) question many of the Blaut-Stea findings. [...] They argue from a rather traditional theoretical position, and from an as yet unpublished empirical study, that young children do not have the high level of competence which we believe that they possess. Downs and Liben have not, as far as I know, attempted to replicate these studies, and I am not aware of other critical views of this sort. [...]

The basic protomapping skills

It appears, then, that children at and beneath the age of school entrance can and do engage in various kinds of mapping behaviour: primitive sorts of map reading, map using and mapmaking. The abilities do not result from training or from incidental exposure to map-like stimuli in a child's home environment. The problem, now, is to explain these findings.

The explanation seems to be that young children possess mapping or protomapping abilities, which permit them to represent macro-environments in language-like sign behaviour incorporating three basic skills: a semantic skill of using material sign vehicles to represent landscapes features and complexes of features; a syntactic skill of rotating (projecting) macro-environments to an overhead virtual perspective; and a second syntactic skill of scale reduction of macro-environment to map or model scale. The protomapping skills are manifested in toy play, in some forms of motor behaviour (dance, games etc.) and in the ability to read aerial photographs without prior exposure to this form of macro-environmental imagery.

[...] Three-year-olds are not cartographers. Nor can they read maps on which signs are completely non-iconic [...]. Nor can they draw maps when they cannot yet manipulate drawing or writing tools. Nor can they coordinate spatial perspectives at a level needed to perform the Piagetian tasks associated with formal decentring and abstract spatial reasoning. But they can indeed make, read and use a simple form of map, and this fact has important implications. It helps toward an understanding of mapping behaviour in all of human culture. And it provides a theoretical foundation for the teaching of mapping and basic geography, a foundation grounded in the child's own skills and experience [...]. To explain the protomapping ability, Stea and I propose a model consisting mainly of three theoretical arguments:

(1) In the most immediate casual sense, it appears that the behaviour which we observed emerges directly [...] out of toy-play behaviour, the latter having an unexpected function. Erikson and many others have shown that young children's toy play is among other things experimentation with reality, primarily having to do with social interaction but also with the manipulation and testing of material things (Erikson 1977; Piaget 1962). It seems to us that toy play [...] has a special function with regard to macro-environmental learning [...]. The act of reducing reality in scale, observing it from an overhead perspective and infusing it with small objects which, to the child, are signs representing larger objects of all sorts (including people) is, in a literal sense, model thinking [...].

But we believe that toy-play behaviour has a special and important function in relation specifically to place learning and experimentation. The macro-environment to a child of two or three is inaccessible in a proximal sense [...]. Moreover, the child cannot achieve the eye-in-the-sky perspective needed to see a macro-environment as a whole. And, finally, even the smaller pieces of a place, like cars, houses and trees, are too large and too inaccessible for a child to acquire direct experience with them under most circumstances. [...] A child has relatively little opportunity to observe and try to understand changes in the full-scale world. For all of these reasons we think that creation of a map-like model, in toy-play, [...] has the specific function of permitting experimentation with and learning about places at the ages before children are mobile enough to perceive places from various coordinated perspectives [...]. So, the first theoretical argument is that a mapping-like activity, playing with miniature landscapes consisting of toy signs and seen from above, is crucial in macro-environmental learning and explains, in an immediate sense, the precocious ability of three-to-five-year-old children to read air photos and solve mapping problems presented in toy and air-photo form.

(2) Toy-play mapping is one strategy for coping with the problem of dealing with macro-environments. Another notable one is storytelling. [. . .] We hypothesise, as our second theoretical proposition, that place learning becomes a crucial problem [. . .] at the age when the child begins to become fully mobile, able to toddle and free to move around to some extent inside and outside the dwelling (Bremner and Bryant 1985). At this level of experience, locational problems become both salient and soluble, among them problems of point-to-point navigation and problems of determining the positions, distances and (if one is lucky enough to get there) the proximal characteristics of landscape features, including their textures and what they look like from the far side. Younger children have already, we assume (following Pick and others), developed diffuse or global maps of the macro-environment as a reference frame, but full mobility suddenly sharpens these global maps, giving them detail as to features and locations (Pick 1976; Acredolo 1976). [. . .] Mapping then becomes a favoured way of dealing with these sorts of problems when direct locomotion to distant points is difficult or impossible. [. . .]

(3) It remains necessary to explain the skill itself: the fact that three-to-five-year-olds can read air photos and can engage in mapping behaviour involving the use of material signs and the rotation and reduction of landscapes to map scale and perspective. [. . .] The model which seems most plausible is a conception of a mapping or protomapping ability which emerges naturally, that is, without teaching, in infants.

It appears that an infant is ready to understand the geographical structure of the world around it and to model that world. [. . .] Shape and size constancies develop very early; a direct extension of shape constancy is the ability to project or rotate the macro-environment to the orientation of the simple map (or toy assemblage, or photo); a direct extension of size constancy is the mapping operation of scale reduction, from macro-environment to map. [. . .] Recognising not only shape- and size-identity in objects, or discriminated pieces of the environment, but also identity of meaning under many sorts of permutation, including colour and movement, seems to generalise to the use of material signs for mapped features. [. . .] We know that the child is born with equipment to establish primitive spatial coordinates for positioning things outside of itself. [. . .] This positioning includes recognition of visual regions, awareness of verticality (up and down), tactile awareness of different body surfaces, and more (Pick 1976; Presson and Somerville 1985).

Do we have 'mads' and 'masses'?

[. . .] There seems to be an intellectual structuring ability, probably in part inborn, and certainly present in every infant of our species and ours alone, which emerges as basal linguistic competence, an ability to acquire the grammatical structure of any natural language. Perhaps this same structuring ability, which Chomsky and some others call the 'language acquisition device' or 'LAD' (Chomsky 1965), also emerges as a mapping acquisition device: hence a 'MAD' (Blaut 1969: 16–17; Blaut and Stea 1974: 8–9; Blaut 1987c: 29–30, 32). The LAD structure permits an infant to begin to tie signs (phonemes plus facial expressions, gestures etc.) together by means of basic syntactic rules which are common to all languages [. . .]. Is it not reasonable to infer the existence at birth of an awareness of a device (the MAD) which gives the infant a readiness to assign primitive and tentative directions, distances and meanings to parts of the world, to orient itself crudely to a global reference frame, to display primitive locative abilities (e.g. pointing and finding hidden objects) and to map the world into both cognitive maps and material map-like models such as toy assemblages? The two abilities have common behavioural elements in the earliest infancy. [. . .] It seems very likely that the human child's competence at birth to deploy basal grammar and to learn a natural language must be connected to that child's competence to begin ordering the macro-environment. [. . .]

There is, however, considerable debate as to whether a language acquisition device exists. Chomsky sees it as a structure or faculty specific to language. [. . .] Others give much less weight to the device and much more to learning in infancy and, in particular, to social interaction. [. . .] Interestingly, from our point of view, one of Chomsky's defences of his position has been the argument that the language acquisition device may be, after all, only a special case of a broader structuring ability, a 'generalised learning mechanism', but he finds no other special case important enough to deserve comparison with the LAD (Chomsky 1985: 10). [. . .]

Cognitive mapping is certainly involved. [. . .] Primitive cognitive mapping is a phylogenetic ability. [. . .] It is displayed at all levels as behaviour involving an ability to move around in and predict (anticipate) features of the macro-environment in situations where simple stimulus–response learning cannot have produced the behaviour observed; hence a cognitive process which structures the macro-environment, a cognitive map, must be involved

(Tolman 1948). There are two basic problems, however. Firstly: is there any important connection between the phylogenetically primitive structuring involved in pre-hominid cognitive maps and the sophisticated structuring found at the level of human linguistic behaviour, human pro-positional reasoning, and the like? Secondly, it is difficult to discuss cognitive mapping without reference to the problem of imagery and image formation [...].

The connection between mapping behaviour and linguistic behaviour is much clearer when we look at written language. The connecting point is the natural skill of young children in the realm of graphicity or graphicacy: drawing, painting, writing, mapping and so on (Balchin 1972; Boardman 1989; Phillips 1989). Humans produce and communicate with two-dimensional and three-dimensional graphic signs – some highly iconic, some not – and have done so since Palaeolithic times. Both writing and environmental mapping are emergents from this graphicising behaviour, with some common underlying rules of ordering and interpretation. [...]

What I have described here as natural mapping in young children is obviously complex, diffuse and in some respects [...] obscure. We should perhaps focus a lot of attention on its most concrete and visible manifestation: the proto-mapping skills by which children rotate landscapes, reduce them in scale and semantically represent them using material sign vehicles [...]. This concatenation of behaviour, thought and materials has the coherence of a genuine culture complex. It can be looked for in children of all cultures and is transmitted by enculturation and by the mechanisms of diffusion within children's culture. It is a good source of evidence that mapping is natural in all of human culture; is a cultural universal.

Mapping behaviour as a cultural universal

A basic map is a material representation of a landscape, something which is made, read and used according to rules about scale reduction, perspective rotation and sign meaning. [...] I think it will prove true that maps are produced, read and used in all human cultures, and have been produced, read and used since the Upper Palaeolithic. I suspect that all cultures, throughout this long period, have not only made such maps but made them as tangible (non-ephemeral, 'hard') pieces of material culture (Blaut 1969, 1987c).

The proposition that mapping behaviour is a cultural universal is grounded in theoretical conceptions of the function of mapping in human culture and of the homology of mapping and natural language. But it has empirical support as well. [...]

Maps, and the behaviour involved in making, reading and using them, have basic cultural functions involving the communication, recording and heuristics of geographic information for long-term deposit or immediate use in practice, play or ritual. [...] In all cultures, to engage successfully in that behaviour an individual or group of individuals must have access to [...] a map. The map itself may serve as a permanent record; or it may be a transient way-station in communication, a medium used in order to convey information during discourse (including ritual); or it may have an essentially heuristic function [...]; or it may be generated in play among children.

The evidence that mapping is a cultural universal comes from several sources. Most persuasive, for me, is the evidence that protomapping skills are present in very young children and mapping occurs naturally at an age when children have not yet learned to read. [...] I am suggesting that any ability that is so important as to be fixed in behaviour at such an early age is likely to be extremely important in human culture. [...] Moreover, somewhat paradoxically, the fact that children engage in mapping at an age [...] when a general ability to deal with space at the level of formal operations is still clearly not present, in itself suggests that this special, limited spatial ability, relating to the map perspective and to the production, reading and use of simple maps, may be a cultural universal.

[...] Visual art is made in all cultures, and has been since the Palaeolithic. Increasingly we realise that ancient and 'primitive' art is, in addition to its many other attributes, sign communication [...]. So the boundary between visual art and language – oral–aural language – is indistinct and permeable. [...] It seems very likely that mapping emerged in the Upper Palaeolithic through a marriage of visual representation skills, linguistic rules, the exigencies of dealing with the macro-environment and art. The rules of rotation, scale and meaning probably were not adhered to strictly until much later. Certainly the function of mapping with these rules was important in the earliest epochs, and the evidence we have, for example from Mousterian cave paintings and Eastern European symbolic figuring in various media, is at least suggestive of a step toward mapping in the uses of perspective, figure size and stylisation or rudimentary signs [...]. (See the suggestive interpretations in Marshak 1977, 1979 [...]).

All of this antedates by thousands of years the earliest known writing system. The Catal Huyuk map of about 6200 BC is perhaps 2000 years older than the oldest known writing system and 4000 or more years older than the oldest known alphabetical writing system (Mellaert 1967). The record is incomplete as to the origins of both writing and mapping [...] but it seems highly probable that real maps, [...] will prove to be much older than writing. [...]

Additional evidence comes from ethnography. Here we have to make some clear distinctions. Firstly, we have to segregate the case [. . .] of communities which are believed to be distinctly mobile, perhaps semi-nomadic, are not known to possess a large stock of permanent material culture, do not have or use written languages, and do not practice much if any farming. If cultures within this category maintain non-ephemeral ('hard') maps, or make maps of whatever sort very regularly, then the case for universality would seem to be quite strong. But there are complications. Almost all cultures in this category endured intense disruption, depopulation, spatial displacement [. . .] before detailed ethnographic description was carried out, so the evidence regarding map use is at best very limited. Moreover, few ethnographers seem to have pursued this question. [. . .] Under these conditions, which prevail across the entire ethnographic world, the fact that we do not have recorded evidence of 'hard' artefactual maps from most of these cultures does not tell us very much, and there is good reason to generalise from the cases where such maps are known to have been used, and to suggest that they were indeed used elsewhere, either escaping ethnographic notice or preceding ethnographic description.

Maps are made and used in a number of such cultures, including Eskimos, Native Australians and – a somewhat special case – Marshall Islanders [. . .] (an impressive array of evidence is presented and analysed in Blakemore 1981.) Among Native Australians a permanent map record is kept [. . .] though the subject matter is usually more social–structural than conventionally geographical. Other ethnographic evidence, for instance from visual art and from toy use, exists for many cultures but has not been brought into the discussion of mapping. And there is, in addition, a great deal of not very reliable anecdotal evidence obtained at the point of contact [. . .]. Certainly today we should no longer take seriously the old argument [. . .] that people of such cultures do not have the rationality to think in abstract terms about space (Blaut 1987a). [. . .] We cannot discard the possibility that even the most mobile cultures, with little in the way of material baggage, make or made 'hard' and relatively permanent maps, everywhere.

Even if mapping behaviour is universal, the presence in a given culture of the practice of making maps that are non-ephemeral may depend [. . .] on the concrete need for such a hard record. It is often assumed that people whose subsistence activities involve significant movement [. . .] will be most likely to be mapmakers in this concrete sense. Agricultural societies, under this assumption, would be less likely to be mapmakers. But there is good reason to be more optimistic. Toy play is universal and toy-play mapping should be observable everywhere [. . .]; the same may be true of maps made for heuristic purposes. These aside, the great difficulty with the ethnographic record for sedentary agricultural peoples is, again, the colonial experience. Written records were probably kept in most societies of this type by members of socio-political superstructures. [. . .] Superstructures of this sort sometimes crumbled under colonialism. Cadastral maps were lost; indeed, the myth that colonised peoples had no concept of land ownership was propagated by colonial authorities and early ethnographers. [. . .] Probably most sedentary agricultural societies, prior to colonialism, were nested within political, urban and religious hierarchies, and presumably had access to maps. [. . .]

Cadastral maps of property and productive land, special purpose maps of irrigation and drainage systems, maps of political boundaries, village and town plans and so on should be omnipresent in class-stratified agricultural societies. Indeed, the earliest explicitly cartographic maps, embodying the formal syntactic transformations [. . .] and elaborate sign inventories, should, in principle, have their origin [. . .] with the origin of class society, perhaps during the Neolithic, on all continents, because the questions of ownership of land and rights to land (and water) [. . .] became matters of profound importance for the first time in these societies.

[. . .] It seems likely that the evidence will accumulate rapidly when we begin looking for it. [. . .] But there is an additional reason for optimism. Traditional research has tended to suffer from Euro-centric diffusionism, and has tended to accept axiomatically the assumption that the higher intellectual processes associated with modem adults [. . .] and the products of these intellectual processes, are not to be found – and not even to be sought – among (1) ancient peoples, (2) non-Western peoples and (3) children everywhere. Savages equal children equal our own ancestors. This was a cardinal principle in much of social science and psychology. [. . .] The present day survivals of this kind of thinking, which I will refer to as 'dualist' because it categorises humans into the two groupings, rational and non-rational (or abstract thinkers and practical thinkers), are still to be found among some psychologists and among a few geographers who remain influenced by this kind of division of humankind into spatial sophisticates and spatial primitives (Sack 1980; Lewis 1987; a critique of Sack is given in Blaut 1987a). [. . .]

As we pursue the search for evidence of mapping behaviour in cultures past and present, and in individual people, it will be helpful to have in mind a concrete idea of what we mean by 'mapping' and 'map'. [. . .] We should keep an eye out for behaviour which represents a macro-environment on some surface by reducing it in scale, rotating it in perspective, [. . .] and using figures or objects as signs denoting its parts. We should also keep an eye out for behaviour which communicates essentially the same meanings without the use of material artefacts. [. . .]

Conclusion

I like to think that geography as a whole is 'natural'. A young child tries to make sense out of the macro-environment, the spatial environment, and learn how to cope with it. This seems to be roughly what we big geographers do, although of course in a much more sophisticated way. Perhaps we should think of the child as a 'protogeographer' (a term I borrow from Ben Wisner). This would not lead us to undervalue the fine theoretical thinking that is done by professional geographers, any more than the fact of protomapping in children leads us to undervalue the superb product of the professional cartographer. The acorn and the oak.

References

Acredolo, L. (1976) Frames of reference used by children for orientation in unfamiliar spaces, in *Environmental Knowing: Theory, Research, and Methods* (eds G. Moore and R. Golledge), Dowden, Hutchinson & Ross, Stroudsburg, PA, pp. 165–173.

Balchin, W. (1972) Graphicacy. *Geography*, **57**, 185–195.

Blakemore, M. (1981) From way-finding to map-making: the spatial information fields of aboriginal peoples. *Progress in Human Geography*, **5**, 1–24.

Blaut, J. (1969) Studies in developmental geography. Place Perception Research Report No. 1, Clark University, Iowa.

Blaut, J. (1971) Space, structure, and maps. *Tijdschrift voor Economische en Sociale Geografie*, **62**, 1–4.

Blaut, J. (1987a) Diffusionism: a uniformitarian critique. *Annals of the Assocation of American Geographers*, **77**, 30–47.

Blaut, J. (1987b) Place perception in perspective. *Journal of Environmental Psychology*, **7**, 297–305.

Blaut, J. (1987c) Notes toward a theory of mapping behavior. *Children's Environments Quarterly*, **4**, 27–34.

Blaut, J. and Stea, D. (1974) Mapping at the age of three. *Geography*, **73**, 5–9.

Blaut, J., McCleary, G. and Blaut, A. (1970) Environmental mapping in young children. *Environment and Behavior*, **2** (3), 335–349.

Boardman, D. (1989) The development of graphicacy: children's understanding of maps. *Geography*, **74**, 321–331.

Bremner, J. and Bryant, P. (1985) Active movement and development of spatial abilities in infancy, in *Children's Searching: The Development of Search Skills and Spatial Representation* (ed. H. Wellman), Erlbaum, Hillsdale, NJ, pp. 53–72.

Chomsky, N. (1965) *Aspects of the Theory of Syntax*, MIT Press, Cambridge, MA.

Chomsky, N. (1985) *The Logical Structure of Linguistic Theory*, University of Chicago Press, Chicago.

Chomsky, N. (1988) *Language and Problems of Knowledge*, MIT Press, Cambridge, MA.

Downs, R. and Liben, L. (1986) Children's understanding of maps, in *Cognitive Processes and Spatial Orientation Animal and Man* (eds P. Ellen and C. Thinus-Blanc), Nijhoff, Dordrecht, The Netherlands.

Downs, R. and Liben, L. (1991) Understanding maps as symbols: the development of map concepts in children, in *Advances in Child Development and Behavior* (ed. H.W. Reese), Academic Press, New York.

Erikson, E. (1977) *Toys and Reasons*, Norton, New York.

Hart, R. (1971) Aerial geography: an experiment in elementary education. Unpublished MA thesis. Department of Geography, Clark University, Iowa.

Head, C.G. (1984) The map as natural language: a paradigm for understanding, in *New Insights in Cartographic Communication* (ed. C. Board), *Cartographica* Monograph 31, pp. 1–32.

King, R. (1989) Geography in the school curriculum: a battle won but not yet over. *Area*, **21**, 127–136.

Lewis, M. (1987) The origins of cartography, in *The History of Cartography, vol. 1: Cartography in Prehistoric, Ancient, and Medieval Europe and the Mediterranean* (eds J. Harley and D. Woodward), University of Chicago Press, Chicago, pp. 50–53.

Lloyd, R. (1982) A look at images. *Annals of the Association of American Geographers*, **72**, 532–548.

Marshak, A. (1977) The meander as a system: the analysis and recognition of iconographic units in Upper Palaeolithic compositions, in *Form in Indigenous Art: Schematisation in the Art of Aboriginal Australia and Prehistoric Europe* (ed. P. Ucko), Duckworth, London, pp. 286–317.

Marshak, A. (1979) Upper Palaeolithic symbol systems of the Russian plain: cognitive and comparative analysis. *Current Anthropology*, **20**, 271–295, 303–311.

Mead, G.H. (1938) *Philosophy of the Act*, University of Chicago Press, Chicago.

Mellaert, J. (1967) *Catal Huyuk a Neolithic Town in Anatolia*, Thames and Hudson, New York.

Natoli, S., Boehm, R., Kracht, J. *et al.* (1984) *Guidelines for Geographic Education: Elementary and Secondary Schools*, Association of American Geographers, Washington, DC.

Peuquet, D. (1988) Representations of geographic space: toward a conceptual synthesis. *Annals of the Association of American Geographers*, **78**, 375–394.

Phillips, R. (1989) Are maps different from other kinds of graphic information? *The Cartographic Journal*, **26**, 24–25.

Piaget, J. (1962) *Play, Dreams and Imitation in Childhood*, Norton, New York.

Pick, H. (1976) Transactional-constructivist approach to environmental knowing: a commentary, in *Environmental Knowing: Theory, Research, and Methods* (eds G. Moore and R. Golledge), Dowden, Hutchinson & Ross, Stroudsburg, PA, pp. 165–173.

Presson, C. and Somerville, S. (1985) Beyond ego-centrism: a new look at the beginnings of spatial representation, in

Children's Searching: The Development of Search Skills and Spatial Representation (ed. H. Wellman), Erlbaum, Hillsdale, NJ, pp. 1–27.

Sack, R. (1980) *Conceptions of Space in Social Thought*, Macmillan, London.

Spencer, C., Blades, M. and Morsley, K. (1989) *The Child in the Physical Environment: The Development of Spatial Knowledge and Cognition*, John Wiley & Sons, Ltd, Chichester, UK.

Stea, D. (1976) *Environmental Mapping*, Open University, Milton Keynes, UK.

Stea, D. (1982) Cross-cultural environmental modelling, in *Mind, Child, and Architecture* (eds A. Lutkus and J. Baird), University Press of New England, Hanover.

Stea, D. and Blaut, J. (1973a) Notes toward a developmental theory of spatial learning, in *Image and Environment* (eds R. Downs and D. Stea), Aldine, Chicago, pp. 51–62.

Stea, D. and Blaut, J. (1973b) Some preliminary observations on spatial learning in Puerto Rican school children, in *Image and Environment* (eds R. Downs and D. Stea), Aldine, Chicago, IL, pp. 226–234.

Tolman, E.C. (1948) Cognitive maps in rats and men. *Psychological Review*, **55**, 189–208.

Further reading

Blaut, J.M., Stea, D., Spencer, C. and Blades, M. (2003) Mapping as a cultural and cognitive universal. *Annals of the Association of American Geographers*, **93** (1), 165–185. [A meta study reviewing the cross-cultural evidence for cognitive mapping as a common human trait, developing upon many of Blaut's original ideas in this excerpt.]

Harley, J. and Woodward, D. (1987) *The History of Cartography*, Chicago University Press, Chicago. [The definitive multi-volume overview describing the history of mapping. Volumes 1 and 2 are most relevant to the issues covered by Blaut and focus on prehistoric mapping and extra European mapping traditions.]

Kitchin, R. and Freundschuh, S. (2000) *Cognitive Mapping: Past, Present and Future*, Routledge, London. [An edited research monograph reviewing research progress thirty years after the heyday of behavioural geography.]

Liben, L.S. (2009) The road to understanding maps. *Current Directions in Psychological Science*, **18**, 310–315. [Reviews different psychological approaches to map understanding, with a rich emphasis upon contextual differences and their impact on environmental cognition.]

Wiegand, P. (2006) *Learning and Teaching with Maps*, Routledge, London. [A recent overview explaining how children learn map skills and developing best education practice.]

See also

- Chapter 1.3: On Maps and Mapping
- Chapter 1.6: Cartographic Communication
- Chapter 1.10: Cartography Without 'Progress': Reinterpreting the Nature and Historical Development of Mapmaking
- Chapter 1.11: Exploratory Cartographic Visualisation: Advancing the Agenda
- Chapter 3.3: Cartography as a Visual Technique
- Chapter 3.6: The Roles of Maps
- Chapter 4.3: Cognitive Maps and Spatial Behaviour: Process and Products
- Chapter 4.9: Understanding and Learning Maps

Chapter 4.5

The Map as Biography: Thoughts on Ordnance Survey Map, Six-Inch Sheet Devonshire CIX, SE, Newton Abbot

J.B. Harley

Editors' overview

Buried in a now defunct map collecting magazine and hard to track down this short descriptive article from 1987 signals many of the themes that have emerged as significant in cartographic scholarship over subsequent decades. It describes Brian Harley's favourite map, a prosaic and commonplace Ordnance Survey Six-Inch County Series sheet centred upon the Devon, UK, town of Newton Abbot that was the author's home for seventeen years. Harley explores four ways in which this map might be read as biographical, with a power to tell particular stories. On one level the image itself has a story – a cartographic, artefactual biography. It speaks of the biographies and experiences of those enrolled in its construction. The map also relates a story of the places it depicts, which Harley interprets from the perspective of a local historian looking back at late Victorian-era Newton Abbot. In a final sense the very personal biography of Harley's life and experiences can be read through the map. So the view of cartography presented here is mutable, personal and emotionally laden, a cultural and human interpretation of the social life of the map.

Originally published in 1987: *The Map Collector,* **41**, 18–20.

I am not a collector in the normal sense of that word, though I sometimes buy and treasure maps for very personal reasons. The Ordnance Survey six inch to one mile sheet which I will describe is such a map. Like a familiar book or an album of family photographs, I am able to read it as a text whose image has meaning because it brings to the mind's eye landscapes, events and people from my own past. Personal identity is always implicated in the maps we collect. Sometimes their value is more emotional than monetary and if pleasure in collecting is also aesthetic and intellectual, it is because maps can draw from the roots of our own experience. We read them as transcriptions of ourselves (Figure 4.5.1).

My map – like any map – is biography in four senses. Firstly, the map sheet itself has a biography as a physical object designed, crafted and used in a different age. Secondly, the map serves to link us to the biographies of its makers – draughtsmen, labourers, printers and surveyors who worked to reproduce its image. Thirdly, the map is a biography of the landscape it portrays; a biography, moreover – as F. W. Maitland put it – 'more eloquent than would be many paragraphs of written discourse'. Fourthly, and of most value to me as collector, the map reciprocates my own biography. It is a rich vein of personal history, and it gives a set of coordinates for the map of memory. Let me say a little about these four *personae* of my map.

Devonshire sheet CIX, SE is a very ordinary map, and I do not seek to justify it as a work of art, ponder its aesthetics, nor to check a price index to gauge its rarity or value. Indeed, it is like hundreds of thousands of other paper maps produced in the industrial revolution of large-scale mapping in Great Britain. Measuring twelve by eighteen inches, it was printed in black and white at

The Map Reader: Theories of Mapping Practice and Cartographic Representation, First Edition. Edited by Martin Dodge, Rob Kitchin and Chris Perkins.

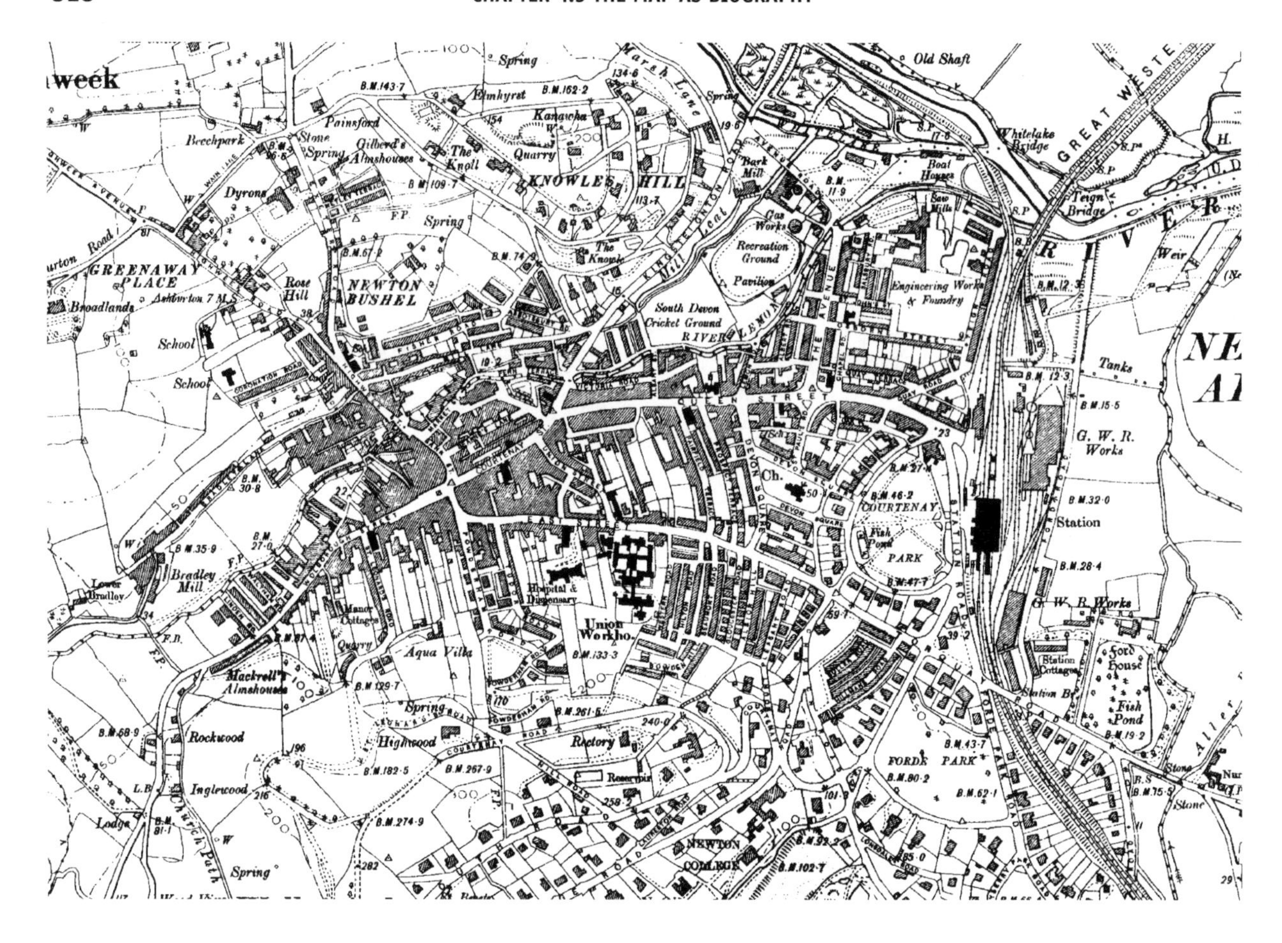

Figure 4.5.1 Much of the social history of Newton Abbot in Devon can be traced from the Ordnance Survey sheet of the area – its medieval roots, industries, the coming of Brunel's railway, the institutional buildings and recreational areas. The author's own knowledge and many personal memories of the area give the map a second, and deeper, perspective.

the Ordnance Survey Office in Southampton by a technique known as heliozincography. A few genealogical facts are provided in carefully lettered imprints at the bottom of the map. These tell us that it was surveyed in 1885–1886, at a larger scale of twenty-five inches to one mile and then reduced, to be further revised in 1904. Sometimes I wonder why carto-bibliographers fuss so much about seemingly identical maps but for the pernickety it can be said that it was reprinted on seven occasions in the 1920s and 1930s, in batches of fifty, seventy five and one hundred and fifty.

So my own copy can hardly claim to be the only extant example of this sheet. Many of those reprinted copies will now have been discarded. Others will have found a home in a national, county or local library, in an office of a land agent or solicitor or, perhaps, in a private collection. The uniqueness of my particular sheet lies, thus, not in its rarity, but in the history of how it has been used, understood and acted upon. To judge from its mint condition, it was a late developer in the world of action. It must have spent most of the half century since its final printing in 1935 in some Ordnance Survey depot, waiting long for the moment when it would be handled, read, traced and understood. Perhaps it was a duplicate, trapped at the bottom of a pile. Then it was made redundant, like doubtless many of the men and women who originally made such maps. Conversion of the large-scale series to metric scales and, later, to the even greater indignity of mere digital coordinates, led to quantities of these maps being given away by the Ordnance Survey to worthy recipients. It was after the trauma of this cartographic diaspora that I acquired Sheet CIX, SE. Now freed from the steel prison of some map chest, it adorns the wall of a lived-in room, where it has a gilded frame. It is next to the supply of cocktails so that the only member of the Charles Close Society living in Milwaukee can toast the centenary of its original birth.

So much for the map's own biography. It is the collective biographies of many such maps, suitably generalised, that gives substance to the history of cartography. Of the biographies of the makers of this particular sheet, however, there is less to say. Unlike many earlier Ordnance Survey maps, Sheet CIX, SE presents an anonymous face. We know little, in a personal sense, about the draughtsmen, photographers and printers who routinely translated the discomforts of field work into such an elegant map image. Equally, not much can be said about the surveyors into whose notebooks the details of the Devon landscape, together with its place names and the boundaries, were so meticulously entered. On the sheet itself no clues are offered as to the identity of the men, a small party of military engineers and their civilian assistants whose arrival in a Devonshire town towards the end of Queen Victoria's reign would have aroused local curiosity. We can speculate how they were gazed upon – 'foreigners' to a Devonshire mind – as they set up their theodolite at street corners, dragged the chain along pavements and stony lanes, levelled the relief of the two hills that overlook the little town, and chiselled Bench Marks for altitudes on the cornerstones of buildings or on limestone walls. We can speculate, too, how – out in the countryside and despite the printed forms that gave the surveyors legal right of entry to private property – a confrontation with an occasional irate farmer might have been sparked off as much by general rumours of a growing bureaucratic threat from Whitehall as by the immediate act of trespass by the government mappers (Figure 4.5.2).

As a biography of some six square miles of Devonshire countryside where I lived from 1969 to 1986, however, Sheet CIX, SE is much less reticent. Here the map is a transcription of culture and of individual endeavour as well as a datum line in the social history of a landscape made by ordinary men and women. Its scale and modest size convey well the intimate character of the hand-made world of the nineteenth century. By good fortune in the lottery of sheet lines, the small town of Newton Abbot stands squarely in the centre of my map frame. Founded as a market borough in the thirteenth century and set between the older parishes of Highweek and Wolborough, Newton Abbot's medieval core

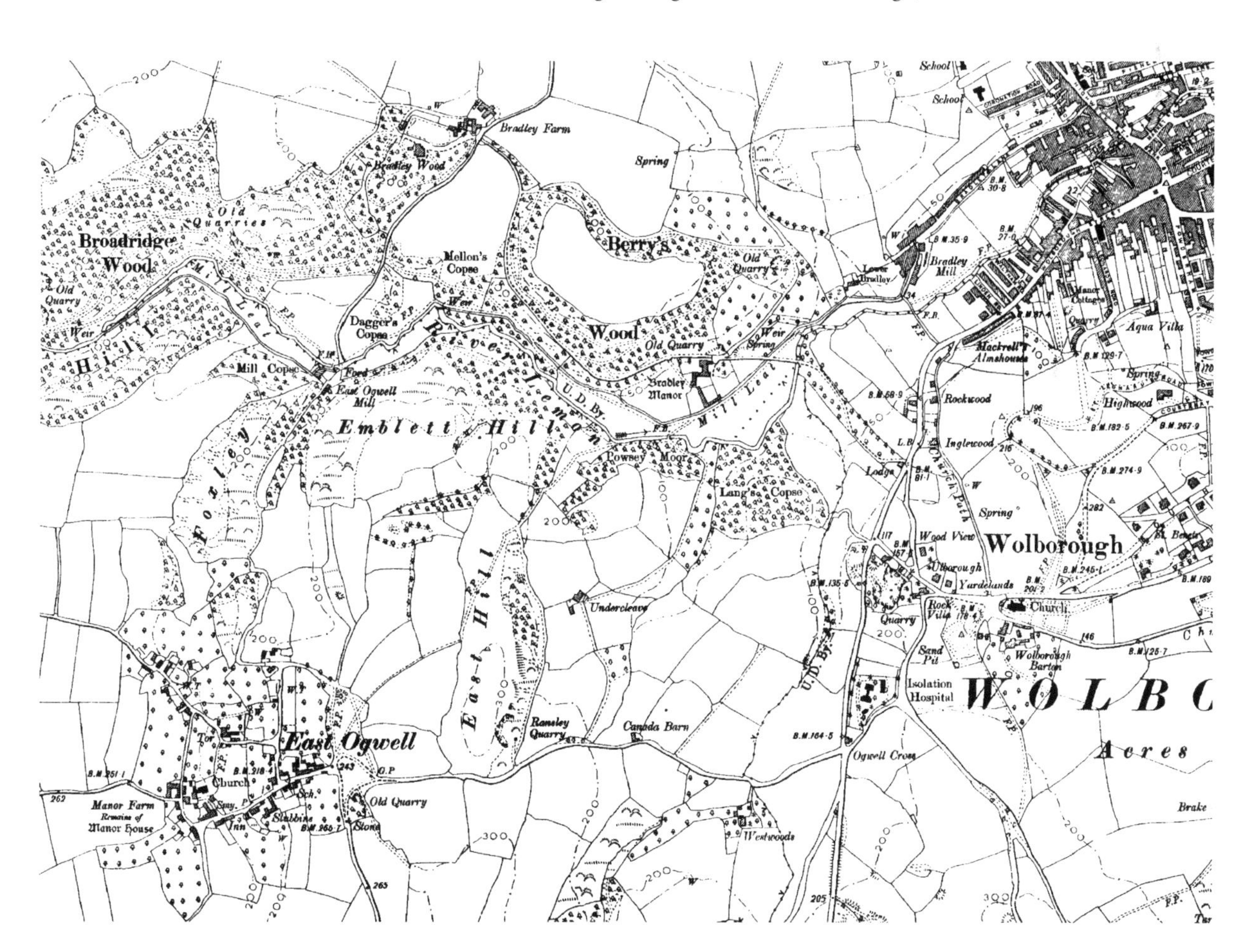

Figure 4.5.2 The 'rural' area surrounding Newton Abbot reveals how industry, in the form of quarries, has encroached into the old pattern of small villages, the domain of the manor house and church.

can be detected in the layout of its streets. In the rural areas beyond – and every field, fence, trackway, farm and cottage, wood and waste is shown – the map reveals the traces of an even older England. But by 1886, the landscape was increasingly hostage to Victorian values. Like some cinematographic still, Sheet CIX, SE reproduces images of work, of the railway age, of the doctrine of *laissez faire*, and of the consequences for local communities of a society divided by class. So, the largest public buildings on the map, shown in black, are the railway station of Brunel's Great Western line, where Bradshaw served as the guide, and the Union Workhouse, where the poor had no need of a six-inch map. Yet even if these social divisions are muted in the map, with its help we can experience the rapid changes that were transforming the Victorian town. Spartan lines of 'decent' terraces, jerry-built to house the influx of railway workers, and the scatter of gracious, gardened villas that accommodated a more prosperous middle class stand out clearly in its social geography. Yet in this sense cartography also deceives. The monotones of the map, with a false egalitarianism, give the same weight to poor and rich alike, and to both industry and the topography of pleasure. We should not forget that in the mills, foundries and engineering works, and in the clay pits and quarries, the labouring men and artisans of Newton Abbot were creating wealth for the few. But there is perhaps a softer image too. A cricket ground and its pavilion, parks, footpaths, a river and its estuary, hint at enjoyments in common on long summer Sunday afternoons when India was still the jewel of the crown and Newton Abbot was almost part of the workshop of the world.

Then, finally, there is the personal biography that lies hidden in the map. Sheet CIX, SE triggers for its present owner the memory of events lived in that place. Personal experiences and cumulative associations give to its austere lines and measured alphabets yet another set of unique meanings. Even its white spaces are crowded with thoughts as I whimsically reflect on its silences. Hung in a room of novels, poetry and music, the map ceases to be solely a document of social relevance or the utilitarian product of government policy: it is there to be read as a personal history, an affirmation that I still belong. To touch these English roots through my map, I have no need of recourse to characteristic sheets, to mathematical grids and graticules or representative fractions, nor do I require an opisometer to re-pace the pathway across the hill. Sheet CIX, SE is now transformed into a subjective symbol of place, scanned without the artifice of geometry, measured by eye without questioning its accuracy, and understood without awareness of its technical pedigree. The map is interpreted through the private code of memory.

Living for so long in such a small town allowed me to walk over much of the space shown on the map. Every square inch of its paper landscape remains so familiar that it can be read at random, and almost sensed in sleep. Its place names are not just a roll call of neighbourhoods, but of people, some now dead, others still crossing and re-crossing the town's pavements and squares and the fields of the countryside. In such a way, the map has become a graphic autobiography; it restores time to memory and it recreates for the inner eye the fabric and seasons of a former life.

The associations are often quite specific. Two of the schools on the map were attended by my children. Streets are not just the thoroughfares of busy market days but are punctuated with public houses where we can debate whether beer was stronger in Victorian England than it is today. Here is the field, the same as in 1886, with a spring and an old quarry, where a dog romped as children played. Here, too, is the lane where – not long ago – I met a woman on a summer evening: the overgrown wall of her orchard is marked on the map. And there is the trackway that led to All Saints' Church in Highweek Village and now to thoughts of my daughter's wedding. But this is also the place of sadness. The ashes of my wife and son lie buried against a north wall of that churchyard:

> 'So, the map revives her words, the spot, the time,
> And the thing we found we had to face before
> The next year's prime'

In these lines of Thomas Hardy, I feel the losses of my own family, and I have also remembered them through a map.

It is thus possible to commune with the maps we collect. I have brought with me to the shore of Lake Michigan a few square miles of English life and landscape: with this talisman I can set foot on the Devonshire soil beneath sheet CIX, SE whenever I choose. The map encompasses not so much a topography as an autobiography. No price can be put on this image of a familiar landscape and the river of life that runs through it. Yet to retrace my steps across the map is far more than a sentimental journey. To rediscover one's own past from afar is to know it better. And, finally, I begin to understand how T.S. Eliot could write:

> '. . . and the end of all our exploring
> Will be to arrive where we started
> And know the place for the first time.'

Till other landscapes and their maps crowd it out, this will remain my favourite map.

Further reading

Forrest, D. (2003) The top ten maps of the twentieth century: a personal view. *The Cartographic Journal*, **40** (1), 5–15. [A personal, yet intellectually justified, listing of 'cartographic classics', inspirational designs and mapping with an impact.]

Harley, J.B. (2001) *The New Nature of Maps: Essays in the History of Cartography*, The Johns Hopkins University Press, Baltimore, MD. [A posthumous collection of Harley's later theoretical writings, along with a useful introductory chapter and an obituary situating his work.]

Harmon, K. (2003) *You Are Here: Personal Geographies and Other Maps of the Imagination*, Princeton Architectural Press, Princeton, NJ. [A visually stimulating and thought provoking compendium of subjective and personal mappings.]

Oliver, R. (2005) *Ordnance Survey Maps: A Concise Guide for Historians*, 2nd edn, Charles Close Society, London. [The most useful guide to the interpretation of local history in Great Britain as it is depicted on Ordnance Survey topographic maps.]

Pearce, M.W. (2008) Framing the days: place and narrative in cartography. *Cartography and Geographic Information Science*, **35** (1), 17–32. [A practical design of a map that embodies narrative, but also a persuasive justification for a more personal link between experience and mapping.]

See also

- Chapter 1.8: Deconstructing the Map
- Chapter 3.10: Affective Geovisualisations
- Chapter 4.2: Map Makers are Human: Comments on the Subjective in Maps
- Chapter 4.8: Refiguring Geography: Parish Maps of Common Ground
- Chapter 5.9: Affecting Geospatial Technologies: Toward a Feminist Politics of Emotion

Chapter 4.6

Reading Maps

Eileen Reeves

Editors' overview

Over the last quarter of the twentieth century several other academic disciplines explored the changing relationships between maps and aspects of culture, in ways that were often richer and more complex than cartographic science's fixation on cognitive aspects of map reading. Reeves article, excerpted here, is a complex analysis of changing meanings attached to maps as cultural objects; it explores aspects of the shift from a textual view of certain maps to their status from the nineteenth century as part of visual culture. The author tracks from Mercator, to the metaphysical poets, to the Carte de Tendre and a Mercier painting to unpack different cartographic traditions, focusing upon the gendering of mapping, and the ways in which female map reading from the Renaissance through to around 1700 was associated with a detailed chorographical tradition, as against the mathematical and global reach of masculine ways of reading maps. Reeves argues these traditions began to break down in the eighteenth century and interrogates the anxieties occasioned by this cultural shift.

Originally published in 1993: *Word and Image,* **9** (1), 51–65.

The relationship between the sister arts of writing and painting is perhaps nowhere so close as within the confines of cartography, the science of mapmaking. Though the current tendency is to treat maps as a kind of specialized picture (Woodward 1987; Harley 1983; Clarke 1988; Nuti 1988; Tyacke 1988) and thus as an odd but entirely authentic representative of the visual arts, the very language we use to discuss cartography denotes its early and stronger alliance with the printed word. While we look at maps as we do images, we say nonetheless that we 'read' them, and it was not until the eighteenth century, when cartography's association with the written word was called into question, that a small section of the map was marked off with legends, or 'things to be read'. Until that time, every feature of the map would have been considered a legible part of a text, albeit one not meant for all readers.

The focus of the present essay will be the reasons for the consistent correlation between mapmaking and the written word, given that such parallels were invariably drawn at the expense of the visual arts. The period that I will consider most fully stretches from the Renaissance through the 1700s. [...] Moreover, the connection between certain types of maps and literacy also included a gender bias – roughly, men 'read' the most abstract of maps, while women merely looked at painted landscapes – and this was most apparent precisely when developments in cartography rendered the original dichotomy between large- and small-scale maps less than valid. The vestiges of the association between certain aspects of cartography and the written word (though perhaps not the insistence on the gender of the map reader) are present even as late as the middle of the nineteenth century, when the first systematic treatises of geography were being integrated into public education (Harley 1987: 12).

Oblique lines

It is within the context of certain Renaissance translations of Ptolemy's *Geographia* that cartography is first explicitly regarded as a text (Dilke 1987a, 1987b). The spectacular results of early modern voyages of discovery excited a new interest in Ptolemy's work, for it remained through the sixteenth century the most technically advanced treatise on the business of mapmaking. [...] In the first pages of the *Geographia*, the great Alexandrian scientist had distinguished between the tasks of the two different types of cartographers:

The Map Reader: Theories of Mapping Practice and Cartographic Representation, First Edition. Edited by Martin Dodge, Rob Kitchin and Chris Perkins.

that of the geographer, who in imitating drawings of the whole earth constructed large maps, and that of the chorographer, who depicted areas as great (or as small) as that occupied by one nation (Skelton 1969). This difference lay not only in the scale of the task but also in the skills that each required [. . .].

The distinction between the undertakings of the two types of mapmaker was based on relative, not absolute, differences. [. . .] The work of the geographer was associated with the larger field of the cosmographer, while the smaller project of the chorographer was described in terms appropriate to the still more detailed art of topography, or landscape depiction. Some of Ptolemy's Renaissance translators and commentators, eager to exploit what they saw as a fundamental difference, and to establish a distinction, based on something other than scale, tended to portray geography as a literary art, chorography a visual one. [. . .] One read a map, but merely looked at a landscape.

It was to some extent natural that maps of the whole world were regarded as texts during the Renaissance, for the similarity between geography and chorography was becoming less and less evident (Cormack 1991) In the first place, the familiarity and relative coherence of a landscape had little to do with the novelty and intricacy of the new mapmaking techniques employed by those who depicted the entire globe. Moreover, some of the associations between maps and texts depended, no doubt, on the questionable but pervasive supposition that anyone can 'see' images, but not everyone can 'read' writing. The connections drawn between large maps and texts, in other words, are an index of the interpretive difficulty which they posed for most early modern men, a difficulty that could only be compared to that of reading.

Other features of Renaissance cartography lent themselves to such interpretation. [. . .] The man who introduced the greatest change in cartography, [. . .] Gerard Mercator, [. . .] produced the first important treatise on realising that literacy and the ability to read maps were closely allied (Osley 1969). The codification of the kinds of script used on maps – capital letters for large cities and big forests, Roman type for towns and smaller wooded areas, italic letters for villages, a slanted script to show the direction in which a river flows (de Dainville 1964) – is a practice that relies not on images or icons, but rather on conventions associated with the written word to indicate the nature, size or civic status of anything on the face of the Earth. But it is the distortion developed by Mercator that provides the most striking example of the Renaissance perception of the map as a text. This innovation was an interesting revision of previous mapmaking techniques, and one that insisted on the two dimensional plane in which writing lies, rather than on the three-dimensional one in which images are placed.

While earlier maps generally used foreshortening and other perspective devices in order to acknowledge that they were merely crude approximations of the 'real' three-dimensional space, Mercator's invention [. . .] subscribed to no such spatial hierarchy. The spherical earth had to conform to the page-like map, and not vice versa. In order to make all meridians parallel – which they would be if the Earth were two-dimensional – Mercator distorted the size of regions far from the equator [. . .]. His delineation of the globe was further than ever from both the drawings which one might make of its landscape, and – more importantly – from any previous map. Here the two-dimensional status of the medium is [. . .] the thing represented, as if Mercator were seeking to copy the plane of a sheet of paper, as well as distant continents. The result was a map that all could see but few could read.

The seventeenth century abounds in the association of maps and texts and, [. . .] by this period the theme was a frequent feature of more canonical works of literature. It is perhaps within the poetry and the sermons of John Donne that references to the literary dimension of mapmaking are the most common and most learned. In his *Valediction: Forbidding Mourning*, his best-known poem, the enigmatic 'mee, who must like th'other foot, obliquely runne', is a translation of contemporary issue in both mapmaking and navigation, the cursus obliquus or loxodrome. [. . .] The movement implied by the cursus obliquus is [. . .] the spiral path traced by a compass that both 'leans' and 'hearkens', the figure for a love neither divine nor bestial, is also that of the loxodrome (Freccero 1963). That the *Valediction: Forbidding Mourning* involves a voyage is in fact crucial, for it is on Donne's journey to the Continent that the metaphor of the oblique course – and indeed the whole poem – is based. Navigation was [. . .] of three distinct types: for short distances in familiar waters one sailed along a straight line, for immense distances one sailed along a Great Circle route, and for intermediate distances one followed an oblique or loxodromic course (Waters 1958). While the first method was primitive and of limited application, the second [. . .] was little more than an elegant theory, because it involved considerable mathematical and astronomical knowledge. The third possibility, loxodromic navigation, was generally favoured by the Elizabethan mariner because of its relative ease and practicality. Following an oblique course meant crossing the net of longitudes and latitudes at a constant angle, and maps generally showed seven such fixed angles of approach, broken down into increments of 11 degrees and 15 minutes, between the baseline of the latitude and the perpendicular longitude. The constant path which the ship took was called its 'course', and the captain's representation of it on his flat map was known as a 'rhumb line' or a 'loxodrome'.

As cartographers had lately recognised, however, a constant diagonal path on a curved Earth would become that of an infinite spiral about the pole, because of the converging meridians. The loxodrome like the oblique course taken by the rational lovers of *Valediction: Forbidding Mourning*, was a compromise between the impractical routes of limited linear movement and circularity [. . .]. As should be evident, then, the cursus obliquus of the ship that bears the poet away is the visible counterpart of the unseen spiral suggested by the poem's metaphysical tendencies; what is less apparent [. . .] is that Donne's first readers would have seen the serpentine lines traced in this voyage as the analogue of the difficult lines of which the poem itself was composed. [. . .] Donne superimposed the paradoxical, oblique and winding lines of his poetry on those of the cursus obliquus, offering what is certainly one of the closest associations of literature with cartography (Reeves 1993).

Donne's overlay of verse and map inspired Andrew Marvell's *Definition of Love.* The poet complains that he and his beloved are as far apart as the two poles, and that they have no hope unless the heavens fall and force the entire globe into a single plane: 'and us to join, the world should all be cramped into a planisphere' (Marvell 1987: 50). [. . .] The term 'planisphere' serves to confuse the issue, for it normally indicates a round map on which half the globe's face is imprinted; one would need two of them to represent accurately the entire world (de Dainville 1964: 39–42). [. . .]

There were several other kinds of maps indicated by the word 'planisphere', and these usually involved some notion of the Earth as a globe whose surface might be peeled off and rearranged onto a plane. There was, for instance, the relatively recent cordiform map, one in which almost all of the polar region and a good part of the northern hemisphere lay in a heart-shaped pattern [. . .]. There was also an elliptical figure, one that sacrificed the poles but showed whatever lay between the winter and summer tropics on both sides of the Earth (de Dainville 1964: 41) [. . .]. Both [. . .] were understood to be wildly inaccurate, and neither resembles the configuration suggested in *The Definition of Love.*

Marvell was, however, drawing on still another type of planisphere: [. . .] the astrolabe, a metallic disk offering a two-dimensional picture of the celestial vault above a particular horizon. [. . .] Ptolemy himself offered a projection that reduced the armillary sphere to a more traditional planisphere, and the map that he drew featured not the polar regions but the *oikoumene* – the known inhabited part of the Earth – surrounded by the rings of the armillary sphere. [. . .] Marvell's map is a crude, almost comic, alternative to Ptolemy's elegant model. The Alexandrian scientist mapped nothing beyond the *oikoumene* while Marvell was concerned only with the polar regions. Ptolemy used nuanced colour and perspective to reduce the metal rings of the armillary sphere to a plane figure or something resembling a traditional planisphere (Neugebauer 1959); the poet proposed a thoroughly 'unreasonable use' of the projection to convert the armillary sphere to the rare planisphere. [. . .]

What sense Marvell's beloved would have made of the 'oblique lines' written in her honour is only a matter of conjecture; few women in this era would have been familiar with either the refined Ptolemaic projection [. . .]. The poem's emphasis on the explicitly sexual is [however] nicely replicated in the presumed gender-based division of its audience.

Geography and gender

It is the relationship of geography to gender, repeatedly developed in several seventeenth and eighteenth century works, that I would now like to consider. [. . .] Mademoiselle de Scudéry's famous Carte de Tendre, charts an allegorical map showing different routes towards a tender rapport with the précieuse. I have noted that the distinction between large maps – where some reference is made to the Earth's position in the heavens – and smaller ones – which were closer to landscapes – was usually presented in terms of the difference between writing and painting, geography and the correlative science of cosmography being associated with literate practices, and chorography and the ancillary discipline of topography with the visual arts. Not only does Mademoiselle de Scudéry's Carte de Tendre preserve the distinction, the novel in which it appears, Clélie, thematises the difference and aligns it with other gender-specific issues. The 'masculine' way of looking at the world [. . .] would be more global, more abstract, and less concerned with the niceties of detail: in brief, closer to what was considered the more abstruse and difficult of the sister arts of writing and painting (Schor 1987). [. . .]

When [. . .] the heroine is asked to account for her various friendships, she explains that these different relationships are by no means equal [. . .] (de Scudéry 1973). She also admonishes the eager young men who surround her that the route from the new friendships that they presumably enjoy with her to something more 'tender' involves a particularly long and arduous journey. [. . .]

Clélie's friends are intrigued [. . .] and they ask for a 'Carte de Tendre' in which these and other rules might find codification. [. . .] Clélie surprises them by producing instead the famous (or infamous) landscape known as the Carte de Tendre, a detailed map of the inroads one might attempt in establishing a tender friendship with her. [. . .] Its radical circumscription is difficult to overlook; beyond the River of Inclination is a sea which Clélie characterises as 'particularly dangerous for women', and beyond that various 'Unknown Countries', the perils of which les précieuses cannot even imagine (Filteau 1979 gives an excellent description of the technical and symbolic aspects of the map). Though Clélie presents the map as an index of the immense distances that lie between new and tender friendships, it is also clear that the kingdom portrayed there is meant to seem restrained and even minuscule to the roving masculine eye, for greater geographical knowledge is associated with the more ample sexual experience of men.

[. . .] The novel [. . .] envisages this 'alternative' way of reading the Carte de Tendre, and portrays it, moreover, in terms of two traditional distinctions, that commonly drawn between chorography and geography, and that opposing visual and literary arts. The fact that cartography seemed to have its feminine and masculine aspects – the former generally featuring defence and the latter territorial conquest – tells us more, I think, about the development of the discipline than it does about the ongoing war of the sexes. It was because small and large maps of the Earth were seen as qualitatively not quantitatively dissimilar that the distinction was first stated in terms of gender. [...] What I would like to indicate in the following discussion is the presence of a map that rivals the Carte de Tendre, and one in which the masculine and textual associations of geography are explicitly recognised.

[. . .]

The map of Clélie's heart enjoyed a circulation which that young lady surely did not (de Scudéry 1973: 408). The discrepancy between the worldwide distribution of the map and the restrained landscape which the Carte de Tendre actually represents is a mirror, of course, of the paradoxical disjunction between geography and chorography. [. . .] Though Clélie intended the map for her friends alone, the novelist notes that 'a certain constellation' caused it to be exposed throughout the world (de Scudéry 1973: 406). This appeal to astronomy, and specifically to a group of stars in the Zodiac [. . .] is a way of pointing out 'fatal forces' presumed beyond her control, and regions as remote to her as the lands lying on the far side of the 'Dangerous Sea'; for the masculine reader, however, the point is less the dainty map provided by Clélie than its widespread and even global dissemination. [. . .]

That Clélie and the other précieuses would have remained largely ignorant of the world map [. . .] goes without saying. Their girlish innocence, of course, precludes such knowledge, just as their status as chorographers implies their unfamiliarity with geography. This particular detail is not surprising, because chorography, [. . .] was traditionally defined in terms of its technical shortcomings. What is noteworthy is the unusual suggestion that those skilled in the more difficult discipline might not be competent in what was generally seen as the less demanding one. When Clélie fears that certain people, [. . .] would end up discussing it as their whim or vulgarity permitted, she insists upon the value of her map, [. . .] and, by implication, the discipline of chorography of which she is the representative. At the same time, her reference to the vulgarity [. . .] of her potential detractors serves as a subtle criticism of the vaster horizons of geography. [. . .]

This emphasis on the relationship of gender to map reading has its echoes in the eighteenth century as well. The primitive distinction between geography and chorography began to break down, almost certainly as a result of the increased use of large-scale maps in the many foreign military campaigns that animated the period (Brizzi 1976). Clearly, the long and exclusive association of pictorial detail with the feminine was no longer entirely valid, but there were nevertheless ways of designating these new map reading skills that managed to imply that the abstraction and interpretation involved were distinctly masculine arts. A striking example of this traditional bias is Philippe Mercier's *Sense of Sight*, painted between 1744 and 1748. This work [. . .] shows various forms of vision: in the background, one young woman stares into a mirror and another looks through a telescope, while three individuals in the foreground examine two different maps. [. . .] Mercier's painting is doubtless an allusion to the support which France gave Spain in 1744 in her ongoing war with the British (McKay and Scott 1983: 162–171). [. . .] The recent and initially popular military alliance would also explain the presence of the dark young man in the cape who appears absorbed in the second map, for he is surely a Spanish 'friend' of the other personages in the painting.

What is nor explained, however, is the manner in which the young woman in the foreground inspects the map before her. [. . .] There is, in the first place, an odd replication of cartographical features in her own attributes: the pearls on her head, that symmetrical and right-angled arrangement of dark and light dots, mirror, both in their position and in their tonality, the border that surrounds this problematic map, the device by which the longitude and latitude of these land masses might be known. A second instance of this pattern of replication lies in the similarity between the depiction of Port Mahon and the right sleeve of the woman's dress. This Minorcan port, [. . .] featured an extended and easily defended inlet, the length of which is [. . .] exaggerated on this map. The key to taking the island was of course this port, and the optimism of the French painter is conveyed by the suggestively unbuttoned sleeve of the young map reader. [. . .] The lighted portion of the sleeve, moreover, resembles the outline of the island of Minorca itself. [. . .]

While this latter detail is not without its erotic dimension, I do not wish to insist upon the inevitable similarities between the conquest of women and of territories. [. . .] The fact that it is not just another land mass but an abstract portion of the map – the borderline marking showing latitude and longitude – that is mirrored in the arrangement of the woman's hair implies less, I think, about masculine notions of conquest than about their prevailing sense of women's ability to read maps. It is significant that the woman appears to be gauging the distance of the island of Majorca from the Spanish mainland with her right hand, for this particular manoeuvre suggests that she is not using the coordinate system that appears at the border of the map, [. . .] but resorting to a rather more primitive, and almost child-like, system of reckoning. The device that would help her evaluate the position and relative size of the two islands is transferred – and thus trivialised – to her ornate hairstyle, where points of light and darkness form the same right-angled pattern. [. . .]

There is also the issue of the woman's gesture. As I have suggested, it seems to be a way of measuring distances, though one not nearly as sophisticated as the map itself might require. A typical early modern guide to hand signals [. . .] does in fact describe a gesture used 'to denote amplitude', (Philochirosophus 1644: 76) and it corresponds to that depicted, with one crucial difference: it is the thumb, and not the index finger, that would be extended by one who wanted to indicate width. More interesting still is the meaning attributed [. . .] to the sign that the woman is actually making [. . .] [which signifies] reproach. The gesture on which the *Sense of Sight* centres would have, therefore, a moral significance; the lady in question offers a commentary on the military ventures which her countrymen and the Spanish allies have undertaken. Whether she condemns the enterprise itself or, as is more likely, the English enemy [. . .], it would appear that she has little to do with the technical aspects of the map that lies before her. Interpretation [. . .] suggests, then, that whatever it is that the woman sees in the *Sense of Sight*, it does not involve accurate map reading, [. . .], but a moral vision at some remove from the tools she handles.

What is most pertinent, then, about Mercier's depiction of the woman with the map is its insistence on her inability to read the document that she displays to all onlookers. The *Sense of Sight* also suggests that male observers, whether within or without the picture frame, can read both the map and the various indices of the woman's ignorance [. . .]. The question that I would pose at this point is about the woman's own attitude toward her incompetence. [. . .] If she happily accepts her ignorance as a natural attribute of femininity, then the gesture retains its moral significance: women may not be able to read maps, but they can interpret and convey the ethical lessons derived from these wartime documents. It might, however, be argued that the woman's gesture is the clumsy first step of a beginner, and that she is genuinely interested in learning a discipline normally mastered by men alone. It is on this more complex (and attractive) possibility that I will focus. [. . .]

While women were not generally encouraged to take up cartography, related studies were at times recommended to them. In 1763, for instance, Demarville would publish a manual of geography which he dedicated to Queen Charlotte and expressly designed for the education of the weaker sex. [. . .] After a brief introduction to the technical aspects of geography, [. . .] this primer devoted itself to verbal descriptions of one country after another. [. . .] The author recognised, however, that even this kind of information was not always suitable for his impressionable readers. One instance of his censorship is particularly pertinent to our discussion of Minorca: Port Mahon, he wrote, 'was taken from [the English] by the French at the beginning of the late war; but in such a manner, as is better to be forgotten than repeated here' (Demarville 1763: 24).

[. . .] What is significant is the reticence that Demarville shows before his audience is the way it mirrors yet another contemporary reference to Minorca. In 1752 John Armstrong decided to tell an uninformed world about the island around which so many military ventures had been planned. Minorcans were generally uneducated, he acknowledged, and especially so the women, but there were reasons for their ignorance:

> 'There is scarce a woman in the country that writes or reads, which does not proceed from their want of capacity, but is the consequence of the jealous nature of the Men, who are not willing to furnish them with the means of intriguing, to which the heat of the climate does not a little incline them, in which however they are extremely cautious and secret' (Armstrong 1752).

[. . .] Literate Minorcan women, like a young lady who knew too much about the cowardly behaviour of otherwise worthy [British commanders], would be more than ever likely to give in to their passions. [. . .] However dissimilar the Minorcan matron and Demarville's maiden, both groups of women, if privy to the wrong kind of information, would yield themselves entirely to men for whom they were not destined.

[. . .] The Minorcan woman is already so uncontrollable that literacy would destroy what little tendency toward marital fidelity she has: even without the art of letters she manages regularly to betray her husband. Demarville's young lady, presumably a more civilized creature, can learn to read without necessarily putting her virtue into danger [. . .] The story of masculine inadequacy, however, would drive her into the arms of someone other than the husband for whom she is being schooled. [. . .] The French woman of Mercier's painting is more sophisticated still: it is probable that she can read, possible that she can read what she likes, and conceivable that she wants to learn to read a map. [. . .] No matter what it is that she is doing with her hand, it lies well within that dangerous area of the globe, as if she can at least locate the theatre of such passion and learning.

Finally, there is another meaning assigned to her gesture [. . .]: that of cuckold. This is not to suggest that the *Sense of Sight* is an actual depiction of marital infidelity, for there is no indication that the young woman is [. . .] betraying anyone with anyone else. But it does seem likely [. . .] that map literacy, emblematised by Majorca, would be the last in a series of implied threats. The young woman's open sleeve would then denote less about the anticipated victory of the French and Spanish at Port Mahon than about the burgeoning passions attributed to the novice map reader [. . .].

I would not like to suggest that all references to women's inability to read maps were in fact men's elaborately disguised fears of sexual inadequacy [. . .]. I would argue, however, that the reductive association of men with map literacy and women with ignorance of the same was most frequent precisely when those notions were imperilled. By way of example I refer to an incident in Laurence Sterne's Tristram Shandy, published about 25 years after Mercier completed his painting (Sterne 1980). Again, cartography is portrayed as a [. . .] science somewhat outside of the realm of feminine talents. When the lustful Widow Wadman decides to pursue the wounded veteran Captain Toby Shandy, she persuades him to show his various military maps to her. It is he who reads the maps, of course, while she is chiefly engaged in following his forefinger with hers over terrains that interest her not at all. [. . .] The greatest preoccupation of that worthy lady being the nature of the wound from which Captain Shandy is still recovering, she sets about making relevant inquiries. [. . .]

What might have been a illuminating – not to say emasculating – situation is happily averted, for Toby has no trouble showing the place where he was wounded: this portion of the long disused map, in fact, is so large that it has to be measured off with the inquisitive woman's scissors. At issue here, of course, is the masculinity of Captain Toby Shandy, and though he fails to assert it in the fashion that Mrs Wadman desires, he does so in a more oblique manner by his interpretation of her question, that is, by his manly tendency towards abstraction, his inclination to rend maps rather than to follow her glance towards his red plush breeches and merely look at the supposed site of his wound. [. . .]

It should be apparent, then, that even the detailed and picture-like maps used in military campaigns constituted a kind of text, whose abstract meaning women preferred to ignore when faced with more tangible possibilities. [. . .] When the distinction between geography and chorography, or large and small maps, threatened to break down, as indeed it did in the face of eighteenth century military campaigns, what was formerly the feminine domain of cartography developed its 'masculine' aspects, and that these involved precisely the same skills that were normally associated with reading.

Consider, finally, a last example from the eighteenth century [. . .]. In *Jacques le Fataliste*, Diderot (1986) tells the story of a daughter of Jean Pigeon [. . .]. 'Mademoiselle Pigeon used to go [to the Academy of Sciences] every morning with her briefcase under her arm and her box of mathematical instruments in her muff.' Unfortunately, her teacher, a certain Pierre-André Prémonval, 'fell in love with his student, and somehow by way of propositions concerning solids inscribed in spheres, a child was begotten'. Diderot recounts that the lovers were fearful of M. Pigeon's anger, and that they fled the country. [. . .]

However amusing Diderot's presentation of the story of Prémonval and Mlle Pigeon, it is disturbing in its resemblance to an older tale, the infamous encounter of two Baroque artists, Artemisia Gentileschi and Agostino Tassi. Tassi, an associate of Artemisia's father Orazio Gentileschi [. . .] had come to the Gentileschi household to give the accomplished 18-year-old girl lessons in perspective, since she would not have been permitted to study this discipline in an academy for painters. Early in 1612 Orazio Gentileschi denounced Tassi for raping Artemisia and misleading her with promises of marriage; the celebrated trial that followed forced the accused to spend eight months in prison, and firmly established Artemisia's reputation as a sexual adventuress for the next few centuries (Garrard 1989).

In both the story of Artemisia Gentileschi and of Mlle Pigeon, the desire to study some aspect of geometry is the occasion for illicit sexual activity. Artemisia, though unable to preserve her good name, [. . .] soon managed to overshadow the paintings of both Tassi and her father. [. . .] Mlle Pigeon, on the other hand, may or may not have returned to the study of perspective after her marriage, but she is known today for her biography and edition of her father's work [. . .]

In summary, given that various elements of the Prémonval–Pigeon episode were Diderot's own invention, and that their story bears a generic resemblance to the true and less pleasant incidents between Gentileschi and Tassi, it seems likely that the author saw it as a kind of cautionary tale. His audience would have understood it as yet another injunction against female map readers, those women foolish enough to go about with cartographical features inscribed in their hairdress and 'mathematical instruments' enclosed in their garments.

Orthography

By the nineteenth century the integration of small- and large-scale maps was nearly complete, a development due to more accurate methods of surveying, ones in which the curvature of even the smallest part of the Earth was reflected in chorographical and topographical maps. This transfer of techniques meant that the differences between the two areas of cartography might be minimised, and the need for the old oppositions of writing and painting, or of masculine and feminine interpretations, might be finally abandoned. The association of maps and literary activity is something curious and outmoded, and it is neglected in contemporaneous treatises on geography itself. [. . .]

Conrad Malte-Brun's (1834) well-known System of Universal Geography, does not so actively oppose the two kinds of reading. The author appears embarrassed, rather than intrigued, by activities once considered so similar. [. . .] It is the treatment of the word 'orthography' that is most significant. Malte-Brun offers a discussion of the orthographical projection in an early chapter, noting that it might also be called a 'planetary projection, since [its] principal object is to show the direct image of half the globe' (Malte-Brun 1834: 41). Several pages later he turns to the question of 'orthography' itself, which was understood in that era to mean, as I have noted, 'correct writing' and, by extension, proper spelling and penmanship. He deplores the disregard among mapmakers for standardised spelling and labelling techniques, and emphasises the illegibility of many of the names on contemporary maps. Yet the connections between the two sorts of orthography are not drawn, and the common denominator of text-like objects is no longer evoked, for the relationship which the early modern world considered so evident was no longer apparent. The irony of the association of maps and writing, of course, was that it was bound to break down and give way to the more obvious correlation with the visual arts, because once we had all learned to read maps, they looked more like pictures.

References

Armstrong, J. (1752) *History of the Island of Minorca*, C. Davis, London.

Brizzi, G.P. (1976) *La Formazione dello Classe Dirigente nel Sei-Settecento*, Il Mulino, Bologna.

Clarke, G.N.G. (1988) Taking possession: the cartouche as cultural text in eighteenth-century American maps. *Word and Image*, **4** (2), 455–74.

Cormack, L. (1991) Good fences make good neighbors: geography as self definition in early modern England. *Isis*, **82**, 639–661.

de Dainville, S.J. (1964) *Le Langage des géographes*, Editions A. et J. Picard & Cie, Paris.

Demarville (1763) *The Young Lady's Geography*, printed for R. Baldwin and T. Lounds, London.

de Scudéry, M. (1660/ 1973) *Clélie: Histoire Romaine*, Slatkine Reprints, Geneva.

Diderot, D. (1986) *Jacques the Fatalist* (trans. M. Henry), Penguin, London.

Dilke, O.A.W. (1987a) The culmination of Greek cartography in Ptolemy, in *The History of Cartography*, vol. 1 (eds J.B. Harley and D. Woodward), University of Chicago Press, Chicago, pp. 177–199.

Dilke, O.A.W. (1987b) Cartography in the Byzantine Empire, in *The History of Cartography*, vol. 1 (eds J.B. Harley and D. Woodward), University of Chicago Press, Chicago, pp. 267–274.

Filteau, C. (1979) Le Pays de Tendre: l'enjeu d'une carte. *Littérature*, **36**, 37–60.

Freccero, J. (1963) Donnes Valediction: Forbidding Mourning. *ELH*, **30**, 333–376.

Garrard, M.D. (1989) *Artemisia Gentileschi: The Image of the Female Hero in Italian Baroque Art*, Princeton University Press, Princeton, NJ.

Harley, J.B. (1983) Meaning and ambiguity in Tudor cartography, in *English Map Making 1500–1650* (ed. S. Tyacke), British Library, London, pp. 22–45.

Harley, J.B. (1987) The map and the development of the history of cartography, in *The History of Cartography*, vol. 1 (eds J.B. Harley and D. Woodward), University of Chicago Press, Chicago, pp. 1–42.

Malte-Brun, C. (1834) *A System of Universal Geography*, Samual Walker, Boston.

Marvell, A. (1987) *The Complete Poems*, Penguin, London.

McKay, D. and Scott, H.M. (1983) *The Rise of the Great Powers: 1648–1815*, Longman, New York.

Neugebauer, O. (1959) Ptolemy's Geography, Book VII, Chs 6 and 7. *Isis*, **50** (1), 22–29.

Nuti, L. (1988) The mapped views by Georg Hoefnagel: the merchant's eye, the humanist's eye. *Word and Image*, **4** (2), 545–570.

Osley, A.S. (1969) *Mercator: A Monograph on the Lettering of Maps, etc., in the 16th Century Netherlands, with a Facsimile and Translation of His Treatise on the Italic Hand and a Translation of Ghim's Vita Mercatoris*, Faber, London.

Philochirosophus, J.B. (1644) *Chironomia: Or the Art of Mannuall Rhetorique*, Thomas Harper, London.

Reeves, E. (1993) John Donne and the oblique course. *Renaissance Studies*, **7** (2), 168–183.

Schor, N. (1987) *Reading in Detail: Aesthetics and the Feminine*, Methuen, New York.

Skelton, R. (1969) *Ptolemy's Geographia*, Theattrum Orbis Terrarum, Amsterdam.

Sterne, L. (1980) *Tristram Shandy*, WW Norton, New York.

Tyacke, S. (1988) Intersections or disputed territory. *Word and Image*, **4** (2), 571–579.

Waters, D.W. (1958) *The Art of Navigation in Elizabethan and Early Stuart Times*, Hollis and Carter, London.

Woodward, D. (1987) *Art and Cartography*, Chicago University Press, Chicago.

Further reading

Klein, B. (2002) *Maps and the Writing of Space in Early Modern England and Ireland*, Palgrave, Basingstoke, UK. [A detailed analysis of the local emergence of early modern mapping metaphors in literature and culture, and in relation to cartographic publishing.]

Smith, D.K. (2008) *The Cartographic Imagination in Early Modern England: Re-writing the World in Marlowe, Spenser, Raleigh and Marvell*, Ashgate, Aldershot, UK. [The effects of mapping technologies on the geographical imagination of Tudor and Stuart England and Wales, read through some of the most important creative work of the period.]

Van den Hoonard, W.C. (1999) Map worlds: a conceptual framework for the study of gender and cartography. *International Cartographic Association Conference*, Ottawa, pp. 387–400. [A conceptual overview for changing gender relations in mapping throughout the history of cartography.]

Woodward, D. (2007) *Cartography in the European Renaissance*, Chicago University Press, Chicago. [Two definitive volumes from the History of Cartography project, charting the development of mapping across Europe in the early part of the period covered by Reeves.]

See also

- Chapter 1.8: Deconstructing the Map
- Chapter 1.10: Cartography Without 'Progress': Reinterpreting the Nature and Historical Development of Mapmaking
- Chapter 3.9: Maps, Mapping, Modernity: Art and Cartography in the Twentieth Century
- Chapter 5.2: The Time and Space of the Enlightenment Project
- Chapter 5.9: Affecting Geospatial Technologies: Toward a Feminist Politics of Emotion

Chapter 4.7

Mapping Reeds and Reading Maps: The Politics of Representation in Lake Titicaca

Benjamin S. Orlove

Editors' overview

By the 1990s anthropologists were beginning to consider maps in cultural contexts and to focus upon the mapping as a process, as well as upon an analysis of the mapped artefact. Orlove's paper excerpted here focuses upon a dispute between government ministries and peasant communities over control of the reed beds in Lake Titicaca, Peru, in the 1970s. This dispute came to a resolution with each side believing that the conflict had been resolved in its favour. In this article, maps drawn by both sides are examined in order to analyse different understandings of the conflict and to discuss the lack of 'coming together'. Orlove suggests that a critical analysis of the map as an inscription is a useful and necessary start in any critical consideration of how maps are deployed, but that an ethnographic focus upon practices is also needed in order to understand the social and political work that mapping achieves. He suggests that neither a Foucauldian nor a Gramscian reading of power is adequate as an explanation of this case and argues that postcolonial theories might usefully be deployed in analyses of the role of mapping in political encounters. So Orlove's paper is important both because of its methodological focus and because it prefigures a shift towards more participatory ways of mapping.

Originally published in 1991: *American Ethnologist*, **18** (1), 3–38.

This article examines a common sort of conflict, one between peasants and the state, in an uncommon setting, the beds of reeds [. . .] that occupy hundreds of square kilometres in Lake Titicaca, high in the Peruvian Andes. [. . .] In 1978 the Peruvian government declared these reed beds to be an ecological reserve and attempted to regulate and limit their harvest. The peasants, who had customary use rights of long standing, opposed this reserve, and they succeeded in their efforts to retain control of the reed beds. By the mid-1980s, the dispute had calmed down; the state retained weak authority in one corner of one section of the reserve, and none elsewhere. The case might seem to have some bearing on conventionally posed questions of political economy, since it suggests that one class, [. . .] gained an unusual victory over a powerful opponent. This article, however, takes a different direction [. . .] by studying the conflict through a set of maps that were drawn between 1977 and 1984, some by peasants and some by government officials. A careful examination of these visual representations [. . .] shows a surprising lack of communication between the peasants and the state. Each side views itself as being in control of the reeds, and it sees the other as accepting, rather than contesting, its position. [. . .] The absence of a shared understanding of the outcome is as striking a feature of the conflict as is its having been won by a subordinate class. This article focuses on [. . .] how can two parties, both of whom made frequent use of bureaucratic channels of communication, hold such different understanding of a situation in which they are both involved? [. . .] In this instance, two social bodies, peasants and government officials, believe not only that they each have a legitimate claim to this space, but also that they are

The Map Reader: Theories of Mapping Practice and Cartographic Representation, First Edition. Edited by Martin Dodge, Rob Kitchin and Chris Perkins.

effectively exercising these claims and that other parties acknowledge the legitimacy of their claims and accept their control of the territory.

[...] The case raises questions about the role of communication and discourse in social and political life. [...] This article draws on some work on the question of discourse and politics to address the issue of state and local power. In particular, it borrows selectively from recent work on the nature of representations in order to examine the work of two writers: Antonio Gramsci, who sought to use the notion of hegemony in order to explain the rule of specific classes, and Michel Foucault, whose view of systems of domination included not only the formal apparatus of the state but also academic and applied fields of knowledge. [...] Their writings, though pertinent, appear less applicable to Latin America, and by extension to the postcolonial world, than to Europe. They suggest that representations serve to communicate between states and local populations in a relatively straightforward manner. This case offers an example in which representations fail, quite spectacularly, to communicate in such a manner. It allows us to examine the ways in which peasants and state officials in Peru act differently from their counterparts in Europe. Most notably, the state and peasants appear isolated from one another. State officials address one another, rather than the wider citizenry, of whom the peasants form the most populous sector, and peasants seem more eager to withdraw from the state than to influence it. To analyse these issues, this article considers [...] maps in two ways.

The first examines how people draw maps, that is, the ways in which maps portray notions of the relations that social groups, categories and institutions have with one another and with specific territories. The second examines how people draw on maps, that is, the ways in which social actors use maps in social interactions, especially conflicts. The former approach could be called the study of the production of maps, the latter the study of the exchange and consumption of maps. Taken together, they compose the cartographic equivalent of 'the social life of things' (Appadurai 1986). [...] Though maps are often drawn for a specific purpose and, therefore, for a specific set of viewers, they are not restricted to those viewers. Their nature as drawings distinguishes them from more transient forms of representation such as speech. This permanence makes their content public and subject to multiple interpretations, since different people can view the same map on different occasions. In addition, maps claim to represent an external reality. [...] This claim enhances their qualities of permanence and accessibility. It also renders them comparable with one another: a set of maps of the same territory immediately lends itself to an examination of likenesses and differences. [...] Maps represent external reality as a set of features divided into a small number of recurring types [...] that are related to one another by contiguity and distance. The analysis of form consists of an examination of the representation of these features along three major dimensions: inclusion and exclusion of features, classification of features, and relations between features. [...] Where the analysis of form examines a map in relation to a particular landscape, the analysis of practice also includes the viewers, with their culturally specific ways of looking at maps. It draws on the notion that people often turn to a map for a specific purpose [...] The analysis of practice includes two components, the first of which I term 'the ethnography of viewing' and the second of which is the study of the categorisation of maps into different classes. [...]

There is a rough correspondence between the two types of analysis of maps and the two ways they will be used to examine the conflict in question here. The analysis of form helps uncover the ways in which the peasants and the state understand their relations to one another and to the landscape; the analysis of practice shows how the striking differences between their understandings remain unchallenged, allowing the situation to continue without resolution. These two types of analysis resemble the approach of Bruno Latour. He studies in detail what he calls 'inscriptions'. This term [...] appears in his ethnographic account of a biochemistry laboratory: a number of ... pieces of apparatus, which we shall call 'inscription devices', transform pieces of matter into written documents (Latour and Woolgar 1986: 88). More exactly, an inscription device is any item of apparatus or particular configuration of such items which can transform a material substance into a figure or diagram which is directly usable. [...] (Latour and Woolgar 1986: 511) [...] Latour (1987: 215–219) discusses several examples of maps, notably the relation between eighteenth century European and Chinese maps of the island of Sakhalin; of greater significance to this article, however, are his more general discussion [...] of inscriptions as concrete objects with specific histories, his analysis [...] of the use of inscriptions by social actors who try to convince others of the correctness of their interpretations, and his examination [...] of the broad political and cultural consequences and correlates of sets of inscriptions (Latour and Woolgar 1986; Latour 1987, 1988).

The following section presents contextual material on the region and summarises the conflict. Subsequent sections discuss the maps and examine them closely from the perspectives of analysis of form and analysis of practice. The final section discusses the more general questions that the case raises for the relations between politics and representation as viewed by Foucault and Gramsci, and examines the implications of the case for the study of peasants and the state.

The conflict

Lake Titicaca is quite high (3808 metres above sea level) and of a good size (8128 square kilometres); it is located in a large flat basin known as the altiplano, divided between Peru and Bolivia (Figure 4.7.1). The people who live in the densely settled portion of the altiplano closest to the lake are organised into communities of several hundred households each. The people have a strong commitment to agriculture, producing most of the food that they consume. They earn additional income through migratory wage labour, small-scale commerce, crafts and fishing. [. . .]

The tall reed that grows in the lake is called totora in Spanish [. . .]. Local peasants attach knives to long poles

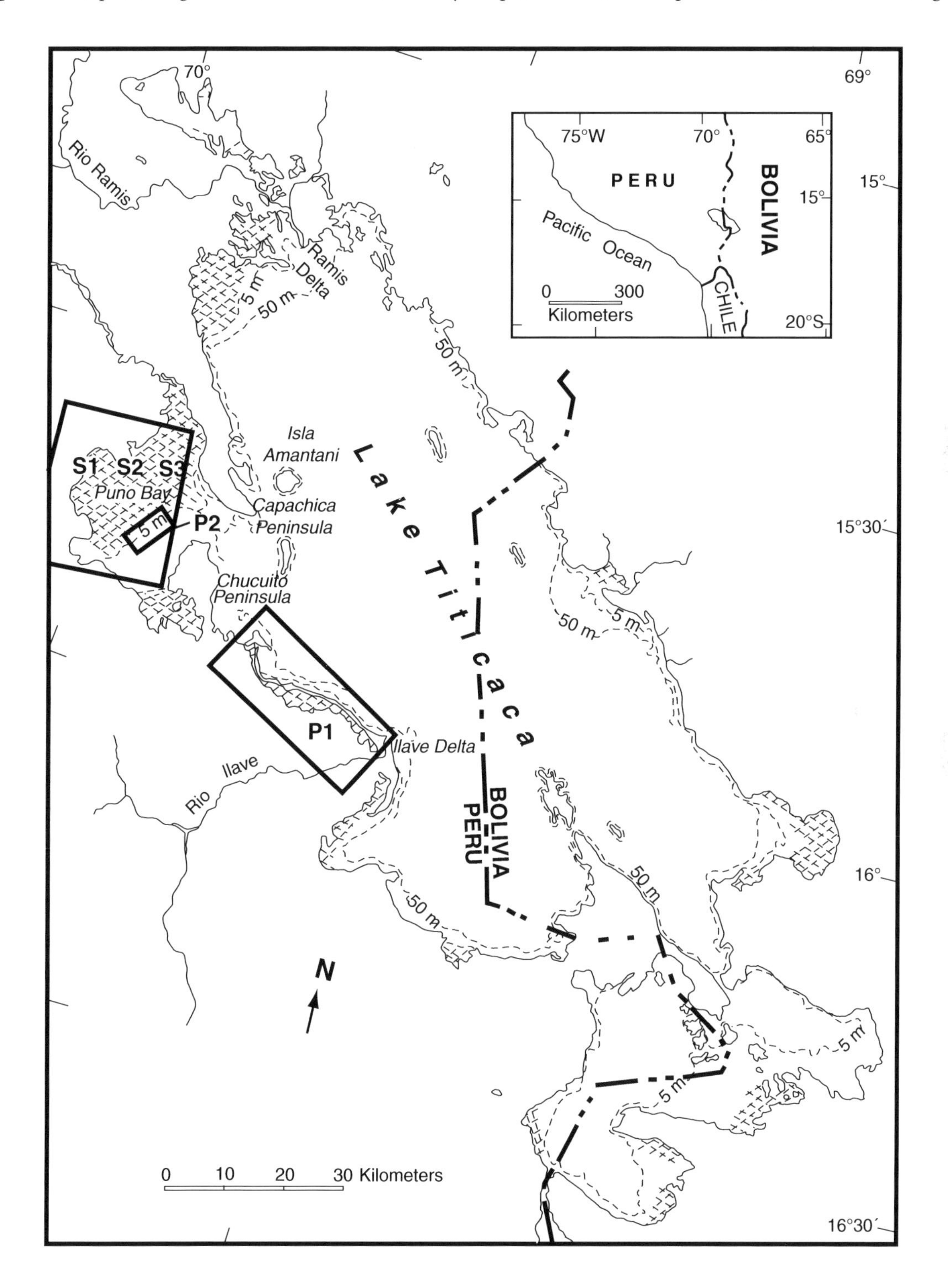

Figure 4.7.1 The Lake Titicaca region: major physical features, national boundaries and index map.

and cut the reed a foot or two below the surface of the water, leaving the base of the stem and the roots intact in the muck at the bottom of the lake so that the plant can grow again. They use the stem of the totora for many purposes, including thatching houses, making rafts and mats, and feeding cattle. [. . .] Communities manage their own totora beds. [. . .] Most communities limit the cutting of totora to certain months of the year and certain days of the week. [. . .] There is great variability in the width of the bands of water in which the reeds grow. [. . .] Some communities are self-sufficient in totora, others are lacking, and still others have a surplus. Individuals in regions deficient in totora often travel to communities with more abundant supplies and pay the community members for the right to harvest totora. [. . .]

Three major events divide the conflict over the reed beds into distinct periods. In 1975, the Peruvian government established the National Forestry Centre (Centro Nacional Forestal), commonly known by its acronym CENFOR, a subministerial branch within the Department of Forestry and Wildlife of the Ministry of Agriculture and Food; CENFOR was charged with regulating natural plant and wildlife resources through a system of national parks and reserves.

In 1978, the Peruvian government established the Titicaca National Reserve (Reserva Nacional Titicaca). In 1980, CENFOR began to issue permits for totora cutting within the Reserve. My discussion centres on three regions within the lake, shown on Figure 4.7.1: Puno Bay [. . .]; the Ramis Delta [. . .]; and the western llave Delta [. . .]. In the western llave Delta in the mid-l970s, peasant activism led to that region being excluded from the Reserve altogether; in the Ramis Delta in the late 1970s and early 1980s, it prevented the government from attempting to administer that region; in Puno Bay in the early 1980s, it reduced the area under official control. [. . .]

Word of the establishment of CENFOR had spread from official circles to the countryside, creating great concern on the part of the peasants. [. . .] They presented documents (mernoriales) which argued that CENFOR should not administer the totora beds, but should leave them in the hands of the communities instead. Representatives [. . .] met with the Minister of Agriculture and Food during his visit to the department in September 1976, a month of growing political tensions throughout the Peruvian countryside and more specifically in the department of Puno (Taylor 1987).

Six months later, in March 1977, this ministry issued interim regulations that guaranteed the peasants in the western llave Delta the right to continue harvesting totora as they always had. [. . .] The Supreme Decree of 1978 states that the government has decided to act on the basis of 'studies' that CENFOR and the Ministry of Agriculture and Food offices in Puno have conducted, and it decrees the formation of the Titicaca National Reserve [. . .] (Republica 1978: l). [. . .]

CENFOR met with differing responses from the peasants in Puno Bay and those in the Ramis Delta during the late 1970s. In the former, CENFOR had some apparent success in carrying out the requirements of the law. [. . .] Events were quite different in the Sector Ramis. [. . .] In this area, the peasants opposed the Reserve more forcefully than in the Sector Puno. Beginning in the late 1970s and extending into the early 1980s, their opposition took two forms. The first was the establishment of an organisation that pressured the government to annul the Reserve. [. . .] The second form of opposition has been a consistently antagonistic stance toward the presence of government agencies. [. . .] To date, the government agencies have made no formal response to the memoriales or to the tense confrontations. [. . .]

By 1980, Puno Bay was the only portion of the Reserve in which CENFOR exercised control. In that year, CENFOR began issuing annual permits (contratos) for totora cutting in the Sector Puno. [. . .] The government also required them to present a sketch map (croquis) of the areas in which they wished to cut reeds. These contracts offered a set of tricky possibilities. Firstly, they gave many people a chance, at least in theory, to cut totora directly instead of purchasing it from intermediaries. Formerly, only the inhabitants of lakeshore communities had cut totora, but the law extended this right to all Peruvian citizens. [. . .] Secondly, since the law asserted that the state, rather than the communities, had sovereignty over the totora beds, the law challenged one of the fundamental principles of the social and political life of peasant communities: their right to manage their territories. Finally, the contracts raised concerns about the fees that CENFOR would charge and the sanctions that forest rangers (guardias forestales) based in Puno would place on individuals or communities that did not comply with the law.

These complex possibilities touched off a series of conflicts between communities. In particular, members of communities that did not seek to obtain contracts attacked those that did. [. . .] An examination of the spatial distribution of the contracts shows a narrowing from 1980, when they were widely spread among the lakeshore districts in the province of Puno, to 1984, when they were concentrated in those areas that the government could supervise most easily [. . .] A long list of interrelated factors might explain this shift: a sense among peasants that failing to obtain contracts would entail few negative consequences, the steady peasant opposition to the Reserve, a tolerance on the part of senior government officials for reports from Reserve managers that listed relatively few contracts, a lack of coordination between Reserve officials and other government personnel (especially in the Navy and the Ministry of Fisheries) who might have assisted in the enforcement of the law, the development of informal mechanisms allowing peasants to bypass the contract requirement, and the consolidation of personal ties between Reserve officials and peasants in a few communities that did obtain contracts. [...] By the mid-1980s, then, the area of effective CENFOR control was reduced to a nucleus of the Reserve in the area closest to the city of Puno. [...] In the other lakeshore districts, peasant communities maintained customary control of the totora beds that bordered their lands, and obtained totora in the large beds in Puno Bay and the Ramis Delta through customary 'rental' arrangements.

[...] After decades, if not centuries, of managing totora on their own, peasant communities were faced with a challenge from a small state agency that sought to regulate this resource, of major economic importance to the peasants and of minor administrative significance to the state. Because of their concerns and large numbers, the peasants mobilised themselves effectively. Their organisation was largely spontaneous, since the rather limited involvement of left-wing groups followed rather than preceded the community assemblies and the establishment of the Totora League. The state took a number of steps in response to their pressures, although it did not withdraw completely from the area.

The maps

Here, attention shifts from an ethnographic reconstruction of the events in the conflict over the totora beds to an examination of representations of the totora beds. Both sides in the conflict prepared maps at different points. I include here five maps, drawn between the mid-1970s and the early 1980s I refer to the first three as 'state maps' and the other two as 'peasant maps', and I label them S1, S2, S3, PI, and P2, respectively [Figures 4.7.2–4.7.6].

These maps all include some reed beds and some portions of open water; with one exception, they also contain some lands on the lake's shores. Maps P1 [Figure 4.7.5] and S1 [Figure 4.7.2] both depict totora beds. The Totora Defense League included Map P1 [Figure 4.7.5] in a memorial that it presented to the regional office of the Ministry of Agriculture and Food in Puno in 1976 or 1977. The roughness of the drawing and the lettering indicates that it was almost certainly prepared by peasants. Map S1 [Figure 4.7.2] was published in an eight-page document, 'Reserva Nacional del Titicaca', issued in Puno by CENFOR in 1979, which summarises the decree that established the Reserve and includes a list of aquatic plants, waterfowl and tourist attractions.

[...]

Map S2 [Figure 4.7.3], the second CENFOR map in this set, was made sometime between 1979 and 1981, with1980 as its most probable date. It was included in reports that the CENFOR staff in Puno submitted to their superiors in Lima.

[...]

Maps P2 [Figure 4.7.6] and S3 [Figure 4.7.4], which both date to 1984, depict portions of the totora beds within the Reserve. They bear a much closer relation to one another than P1 [Figure 4.7.5] and S1 [Figure 4.7.2] do. Map P2 [Figure 4.7.6] is the sketch (croquis) that the lakeshore community of Quipata included with its request (solicitud) to CENFOR to be granted an extraction contract. CENFOR prepared Map S3 [Figure 4.7.4] as part of its efforts to summarise its activities for the year 1984.

Comparing the maps

[...] Peasant and state maps depict Lake Titicaca and its shores in quite different terms. I first examine these differences through the previously described analysis of form: a study of the inclusion and exclusion of features, classification of features, and relations between features. [...] Such a comparison would comment on the absence of towns in the peasant maps and the paucity of communities in the state maps and on the differing depictions of roads in order to suggest that the peasant maps present the altiplano as rural and self-contained, the state maps as urbanised and linked to other regions. The comparison, however, rests principally on the third aspect, the relations between features. It suggests that the difference between the two sets of maps is more basic than a mere question of emphasis, because they are organised differently, both as individual representations and as series of representations. They reveal different assumptions about the making and reading of maps. [...]

The peasant maps depict a series of communities, each of which consists of a focal place defined by a building or set of buildings, located at the centre of a territory enclosed by well-marked boundaries. There is a suggestion that the focal places contain plazas surrounded by buildings, since they are depicted as open triangles in Map P1 [Figure 4.7.5] and roughly circular clusters of houses in Map P2 [Figure 4.7.6]. All of the communities that are shown border on Lake Titicaca; their other boundaries consist of high points (mountain peaks that surround a valley, peninsulas that enclose a bay) or channels of water (streams, inlets that border a promontory). These communities are depicted as equivalent, both in the size and composition of their centres and in their areas, because the maps present as uniform the variable distances between them. [...] In addition, the communities are contiguous, so that no unoccupied space is left between them. Both of these maps suggest that each community controls a specific territory and that the communities jointly control the entire region.

The state maps emphasise a different dimension, that of time. [...] The first separates the Reserve (or, more strictly

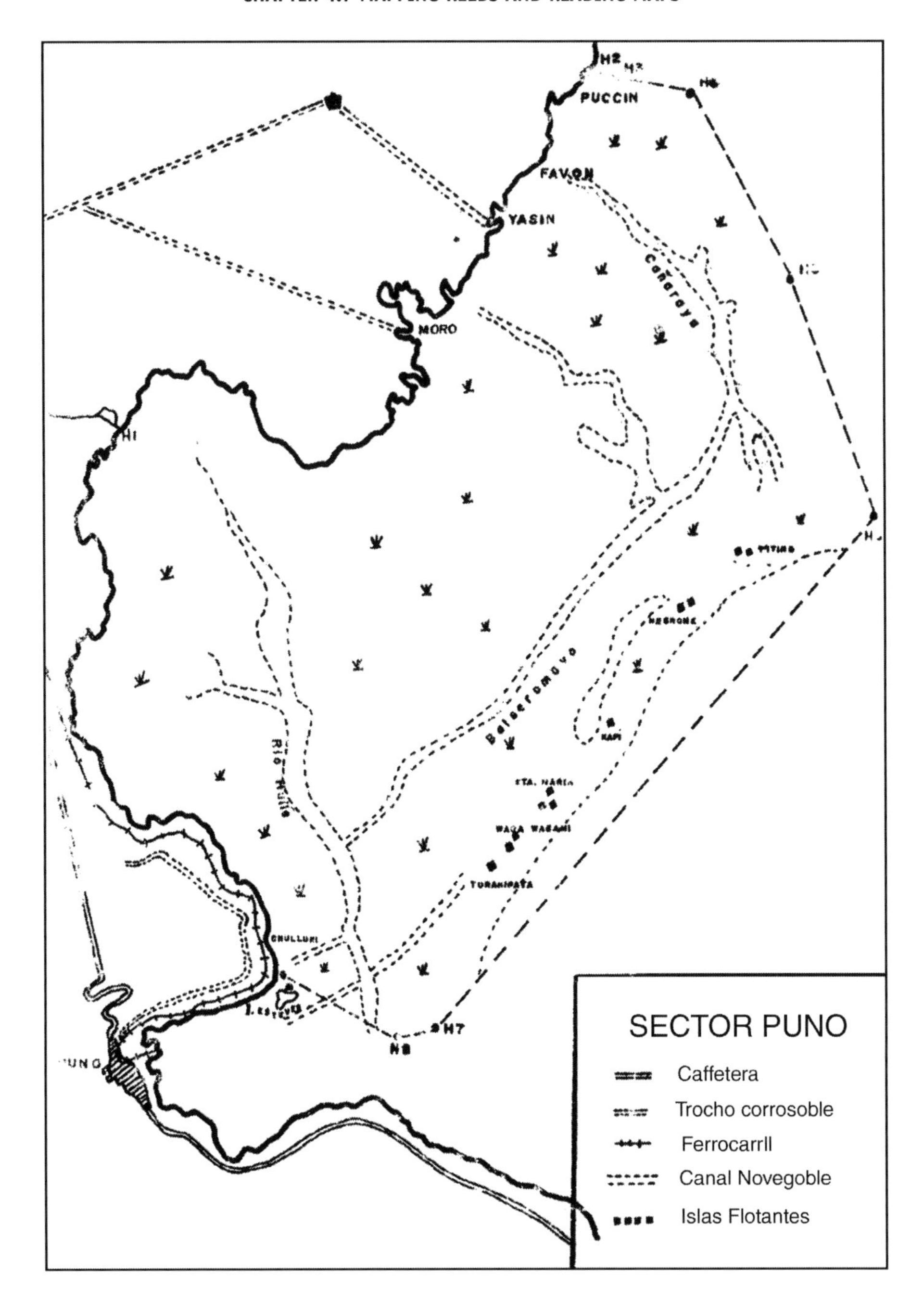

Figure 4.7.2 Map S1: Sector Puno of the Titicaca National Reserve.

speaking, Sector Puno within the Reserve) from the rest of the lake; the second divides the Reserve into different types of zones; the third allocates portions of certain zones to users under year-long extraction contracts. These three maps, then, correspond to the three activities mentioned in the law that created the Reserve: the 'studies' to determine its border, the 'development' of the Reserve, and its 'administration'. [...] The administrative decisions added in each map are duplicated in the later maps. The boundary of the Reserve remains unchanged, as a series of straight lines that connect a portion of the shore with a set of points within the lake. The gradual loss of any distinct marking of these points is the only erosion of detail in the series of three maps. [...] These maps thus constitute a narrative: they tell a story

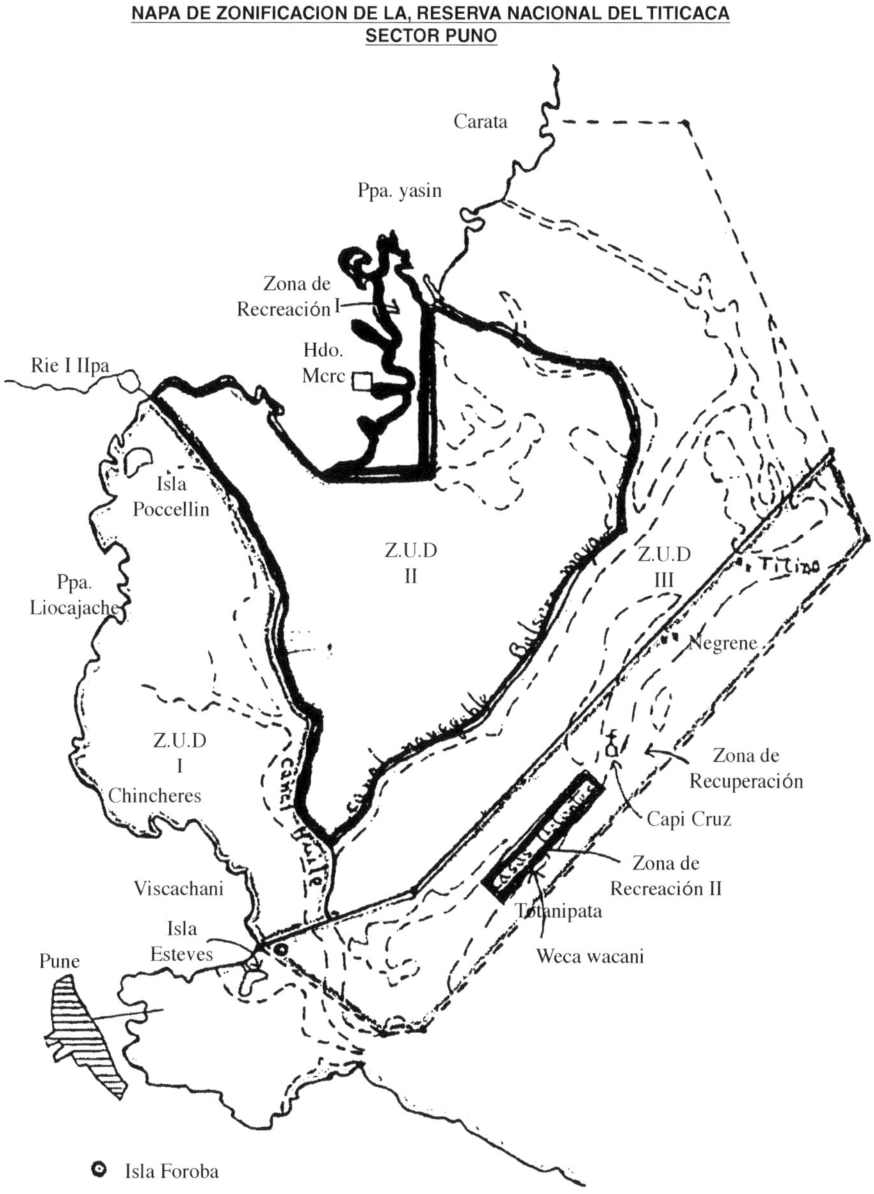

Figure 4.7.3 Map S2: Zones within the Sector Puno of the Titicaca National Reserve.

of an actor with a set of purposes carrying out a series of activities that take place over time. The structure of the narrative, to borrow a term from the study of rhetoric, is that of prolepsis: the assumption that future acts are preordained. [...] In this case, the maps depict in visual form the notion [...] that the state anticipates its own future actions. The state moves from general to specific, adding detail to its previous actions but not undoing any of them. [...] The later maps show that CENFOR has carried out all the aims assigned to it in the Supreme Decree without adding any new ones on its own: the division into 'direct utilisation zones' and 'recuperation zones' demonstrates the 'conservation of natural resources' and, in conjunction with the 'areas of totora extraction', the 'development of neighbouring populations through the rational utilisation of the flora'; the 'recreation zones' correspond to the 'promotion of local tourism'.

By contrast, maps P1 [Figure 4.7.5] and P2 [Figure 4.7.6] have different makers, as do the other maps submitted with extraction contracts. These maps [...] do not form a

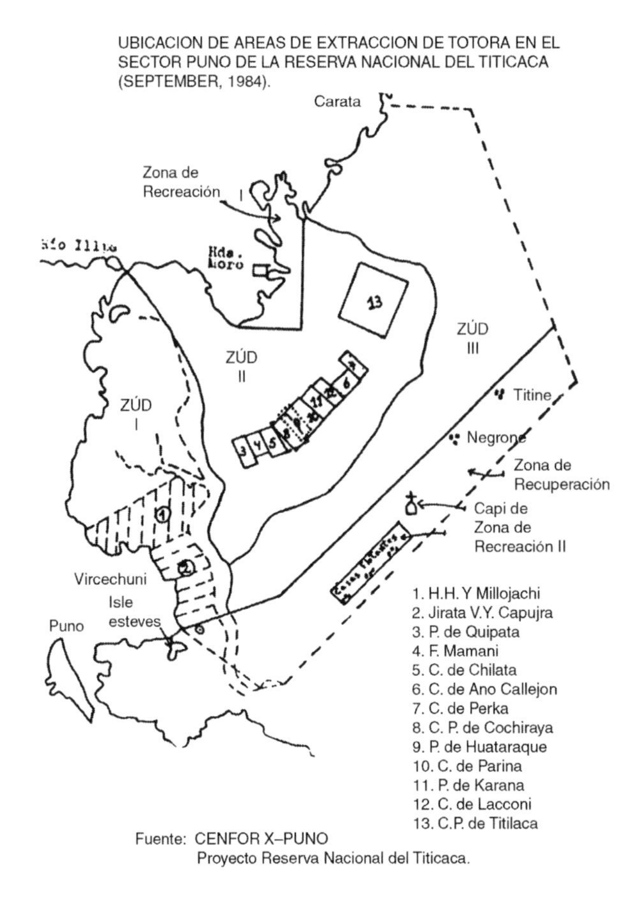

Figure 4.7.4 Map S3: Location of totora extraction areas in the Sector Puno of the Titicaca National Reserve.

temporal sequence. Although the peasant maps include no towns and the state maps omit the Sector Ramis, the peasants and the state do not erase each other from their maps.

[...]

They indicate their recognition of the other through structured and limited inclusions, carried out in ways that do not undercut their claim to sovereignty. In the peasant maps, the sole sign of the state are the flags in Map P1 [Figure 4.7.5] marking the centres of communities that held assemblies to join the Totora Defense League. [...] It seems likely that the peasants viewed these flags as indicating that the state ratified their autonomous rights to own and manage their territories, rather than as suggesting that the state delegated these rights to them and could therefore remove them (Orlove 1982). [...]

The CENFOR maps only recognise peasants when the state charters them into existence. [...] The maps thus include some of the distinctions by which the state divides peasants and which the peasants themselves do not admit: distinctions between Uros, shore dwellers and others; between recognised and unrecognised communities; between qualified and unqualified communities. The maps indicate no movement toward resolution of a fundamental difference between the peasants and the state, that each believes that it controls the reed beds. The disparity between their views remains as great in 1984 as it was in the mid-1970s, despite the changes in the intervening years. The peasant communities do not acknowledge that the state has gained some measure of control over the reed beds, at least in portions of Puno Bay. [...]

The maps also contain minor hints of a theme expressed more clearly in oral accounts of the history of the reed beds: the peasants understand their control to be ancient and unchanging and to derive legitimacy from several equally ancient and unchanging sources, in the realms of agriculture (the maps indicate the use of lands at different elevations), politics (the maps show the locations of assemblies and possibly of plazas, denoting authority), and ritual (Map P2 [Figure 4.7.6] includes boundary mountains at which ceremonies are held on certain occasions of the year).

By contrast, the state maps represent CENFOR as controlling the reed beds of the lake, without acknowledging

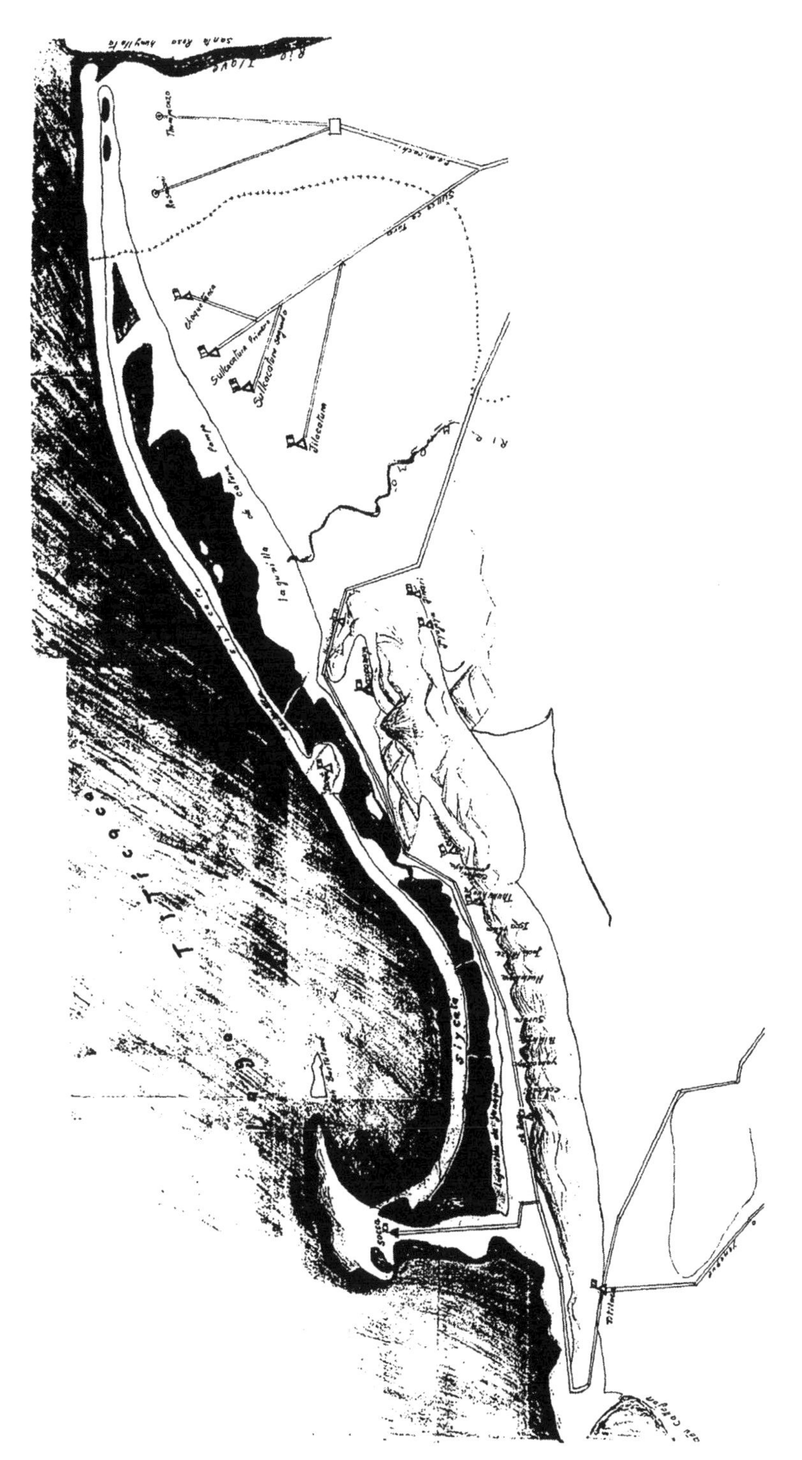

Figure 4.7.5 Map P1: The Totora Defense League.

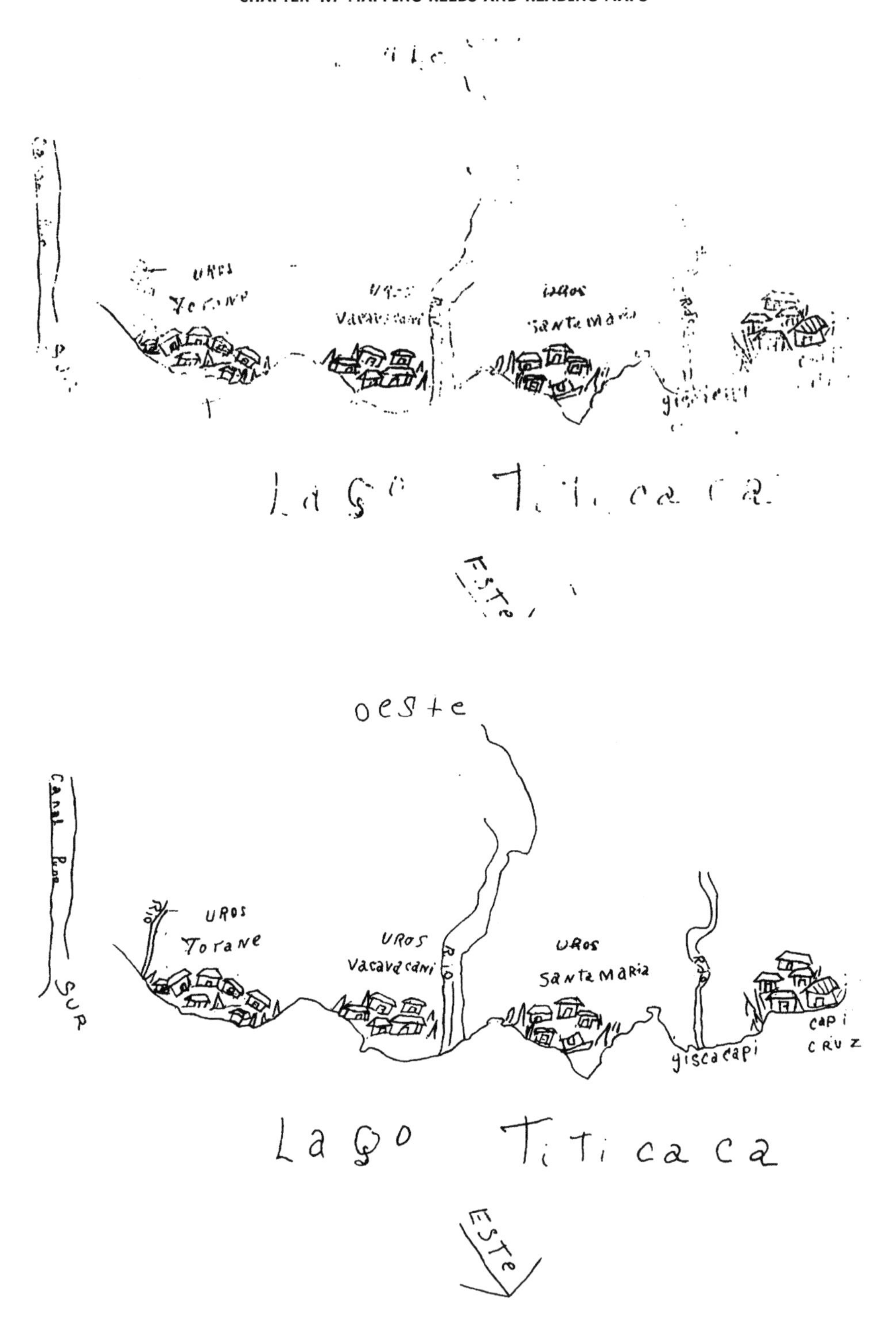

Figure 4.7.6 Map P2: Map accompanying the totora extraction contract request from the community Quipata (top) and the same map with certain details rendered more clearly (bottom).

the decline of CENFOR's power. [...] The proleptic narrative of the state, by requiring compromises by the state to be presented as prefigured in laws, makes them not real compromises at all. These differences in the spatial, temporal and social patterning of the maps suggest that the irresolution of the two accounts is not merely the result of deliberate obstinance, but rather that it stems in part from a more fundamental incapacity of each to understand the other's views because the views are incommensurable, in two senses of the word. Firstly, and more literally, there is no ratio that can convert a distance on a state map into the equivalent distance on a peasant map or vice versa. Secondly, and more metaphorically, the two sets of maps portray the relations between [...] social groups and the natural environment-in ways that are so radically different as to be incompatible. Each group presented a set of maps that was congruent with its claims for legitimate control of the reed beds.

This lack of resolution is troubling, perhaps because it violates the expectation of closure at the end of narratives [...]. It raises the question of how this incommensurability has been preserved, and thus entails an examination of the social activities and cultural conceptions that surround the circumstances in which the maps pass from the individuals who drew them to those who look at them – in other words, an analysis of practice as well as of form. [...]

The analysis of practice consists largely of two parts, the first of which might be termed an 'ethnography of viewing' and the second of which is an examination of the social and cultural categories into which peasants and officials place maps. Before taking them up in turn, however, it is worthwhile considering how an analysis of form can lead to an analysis of practice. The key links in the case are the concepts of 'implied viewer' and 'actual viewer' (Iser 1974, 1978) [...] The peasant maps clearly have state agencies as implied viewers, whether the Ministry of Agriculture and Food, to whom Map P1 was submitted, or CENFOR, which required Map P2. Nonetheless, [...] it seems that the peasants were the actual viewers who studied the maps with attention. [...] The implied viewers of the state maps seem to be the general public. [...] The state maps do not directly address the issue of their implied audience, but they do contain some clues, such as their inclusion of certain details that refer to legal terms and categories – the abbreviation 'H' for hito ('boundary point') and the phrase 'zona de recuperacion', for example. The inclusion of details unfamiliar to most Peruvians emphasises the position of the state as the source of the discourse and reduces the importance of the recipients of this discourse. If all Peruvians are the implied viewers, the actual viewers are other officials of state agencies, to whom the CENFOR employees in Puno sent reports.

These identifications of implied and actual viewers are borne out by [...] an ethnography of viewing. On some occasions, both peasant and state maps were open for viewing at the same time in the same place. When delegations from peasant communities came to the CENFOR offices they [...] entered a large room that had been the living room of a private residence before the owners rented the building to the government. [...] Taped on the walls were maps of Lake Titicaca on which were indicated the borders of the Reserve and the zones into which it was divided. These maps, quite similar to maps S2 and S3 but larger in scale [...] were in sight when the peasants stood in front of a desk and presented, to the official seated behind it, the various documents, including the maps of the proposed totora cutting sites, that were required to request a totora extraction permit.

[...] The two sets of maps appeared quite different, in straightforward visual terms, to the individuals who viewed them. The wall maps, though large, had lettering comparable in size to that of the smaller maps S1, S2 and S3, so that the names of places and physical features could not be read from the distance at which the peasant delegations viewed them; the peasants were too cautious [...] to examine the material on the office walls in order to compare it with the maps that they had drawn. The CENFOR officials also sat at some distance from the wall maps and would have had to walk over to collate the two. This ethnography of viewing shows that analysis of form may rest in part on analysis of practice; in this case, the legibility, and hence the intelligibility, of place names depends on the distance from which they are viewed.

The social activities of viewing draw on the cultural categories of maps. [...] The state has two well defined categories of maps, with established relations between the two; peasants have three relatively loosely defined categories of maps, without specified relations among them. These categories rest on the use of the maps as well as on their appearance. Visual representations that social scientists would not ordinarily classify as maps are included in all three peasant categories, and are excluded from both state categories. CENFOR officials reserved the term mapa (map) for the state maps, employing for the peasant maps the term croquis, a word that is often translated as 'sketch' but also includes the sense of 'rough draft'. This conceptualisation allowed them to present any difference between peasant maps and state maps as the result of peasant errors. [...] The term croquis [...] suggests that the peasant maps will serve as the basis for a later corrected map, drawn by state officials [...] The contrast between mapa and croquis is supported in this case by the government's use of the latter term in the 1975 law that established CENFOR, which requires the provision of such a croquis for the issuing of extraction contracts for [...] totora.

For peasants, cultural categories also blocked the comparison of the maps. Peasants have a loose set of types of maps rather than two simple named categories of maps

[...]: peasants are accustomed to several types of maps. Some could be called vernacular maps: peasants may use a stick to scratch a few lines in the earth, usually to indicate the location of some particular place. [...] Peasants occasionally draw others, which might be termed bureaucratic maps, for official proceedings like lawsuits or loan applications. [...] These are drawn only infrequently: [...] when they submit documents to officials, a written list of boundary points often substitutes for a hand-drawn map. A third type is official maps: they typically depict a whole composed of parts. The conventional political maps of Peru that adorn classroom walls are a good example. [...] In the minds of the peasant children who copy such maps as school exercises, and in the minds of the adults who see the copies, these maps may bear little more direct relation to a natural and social landscape than do other emblems contained in pedagogical assignments. [...] The peasants probably placed the maps in the CENFOR building in the category of official maps because of their size and location on a wall.

In his discussion of state-sponsored nationalism in Southeast Asia, Benedict Anderson develops the term 'logo-isation' (Anderson n.d.: 12) to refer to the way in which national maps can be used 'for banners, letterheads, official gazettes, magazines, airport decors, tablecloths, and so on' (Anderson n.d.: 13). The peasant category that I have termed 'official maps' comes close to Anderson's notion of logo, but in his view the state creates and diffuses this logo-isation of the map. His discussion omits two aspects of state-drawn maps that are of importance in this case: the possibility of multiple readings of maps and the formal categories that the state uses in its discussion of images. The Peruvian state has an official discourse regarding representations of the nation. It applies the term simbolo patrio (homeland symbol) to only four things: the flag, the shield, the national anthem and the escarapela or cockade, a small ornamental rosette made of ribbons. [...] Like the contrast between mapa and croquis, the state's definition of simbolo patrio is precise, formal, encoded in documents. The peasants' association of the national maps with the flag and the shield may simply reflect the spatial proximity of these things to one another and to government representatives in schools, police stations and administrative offices, or it may be a tactic, in de Certeau's (1984) sense of the term, that the peasants use to evade the state's claim to power.

The imputation that the maps, flags and shields are alike is potentially a serious one. As Anderson (n.d.: 7–12) points out, the precise delimitation of national borders is an important basis for state legitimacy and centralised rule, and Peru is no exception. The erasure of the difference between purportedly objective maps and more arbitrary symbols undercuts the state's claim to govern by denying the success of its closely related topographic, cadastral, administrative and policing missions (de Certeau 1984).

This analysis of practice can lead back to an analysis of form. There are two additional and related ways in which the state maps differ from the peasant maps: the former adopt more fully than the latter what Anderson (n.d.: 8) calls the 'bird's-eye convention' of depicting all features as they would appear from the air, and the former all connect spatially with one another and with other state maps. [...] The existence of only one state category of mapas is congruent with this connectedness of all state maps and this perspective in space, from which only the state sees Peru as a unity. The peasant maps contain several repeated violations of the bird's-eye convention: the flags and the mountains in P1 [Figure 4.7.5], the houses and, possibly, the viewing platforms in P2 [Figure 4.7.6]. These are all depicted as they would be seen from the ground, the houses and flags from up close and the mountains as peaks on the horizon. [...]

[...] Since the implied viewers of the peasant maps are state officials, and those of the state maps the entire Peruvian citizenry, including the peasants, the maps jointly imply an exchange between peasants and the state, a conversation in which two parties alternate the roles of speaker and listener. On closer inspection, this conversation is far less important than the two distinct, incompletely linked conversations among peasants and among state officials. [...] The separation of these two conversations corresponds to the irresolution of their differences. The lack of conversational exchanges between the two presumed interlocutors allows their highly discrepant positions to be compatible, [...] existing together without disagreement, discord or disharmony as well as an absence of conflict that would make association impossible or incongruous (Merriam 1973: 181). [...]

Viewing maps

The five maps in question here all depict similar areas, shallow portions of Lake Titicaca and, in most cases, the adjacent shores. However, they represent them in different terms. The state maps lay out a narrative of the conception and enactment of a detailed plan for a specific territory in which state officials control the natural features and human populations. The ones made by peasants tell a different story: of the control of the territory by a set of communities, all equal to one other; of the continuity of this control; of the ratification of this control by a remote state. Analyses of form and practice have explained the fact that neither peasants nor state officials have challenged the discrepancies between the sets of maps or even commented on their incommensurability.

The lack of attention to difference seems odd, granted much of the current discussion of representation and power in societies with class divisions and with centralised states. Foucault's writings, for example, emphasise the unity of systems of domination. [. . .] Foucault's interpretive frameworks seem appropriate to the case of the Lake Titicaca reed beds. [. . .] In particular, the field of cartography is reminiscent of other fields, such as medicine and sexuality, about which Foucault wrote extensively. Cartography treats its subject matter, the landscape, by categorisation and depiction, much as those other fields treat theirs, the human body. Important parallels include the implicit claim to offer accurate representations (in this case, of the world), the association with scientific fields (geography), the dispersion of the making of representations (maps) beyond the state to professionals and other social groups, and the links of the field to systems of power and domination. The links between cartography and those other fields lie in the supervision and control of the movements and activities of persons. [. . .]

Another link between the maps and Foucault's work is his emphasis on vision. This emphasis is evident in much of his work, such as his discussion [. . .] (1977) of the panopticon, and a discussion of Velazquez' painting Las meninas, in which the steady outward look of the artist depicted in the painting serves to indicate general issues of perception and representation (1970). The word 'gaze' occurs throughout his work. [. . .] This word in its sense of fixed and prolonged attention (Merriam 1973: 370) suggests control over an immobile and possibly unaware object, and thus fits closely with Foucault's analysis of discourse and power. [. . .]

Latour's account (1988) of the growth of microbiology and hygiene in late nineteenth century France, with attendant shifts in such realms as medicine, agriculture, military organisation and colonial policy, offers the notion of medical inscriptions rather than that of the medical gaze. [. . .] Latour shows fluidity, change and debate in the midst of what Foucault took to be a single coherent medical episteme.

In the instance of the conflict over the reeds, however, the state glances, rather than gazes, at peasant maps. Having requested that the peasants provide maps, in the sense of visual representations of natural and social features of a specific territory, the state turns them from maps into rough drafts; [. . .] the features of the peasant maps that propose a view different from that of the state are transformed into inaccuracies or irrelevancies [. . .] Foucault's discussions rarely include cases such as this, in which systems of power attempt neither to locate nor to control differences. The peasants in the altiplano do not have to escape the state's gaze, since they find its glances so inconsequential.

This case of viewing raises a second difficulty with the application of Foucault's work here. His analysis of discourse and practice rests on the premise that basic understandings are widely shared and deeply internalised, yet the peasants and state officials produce and interpret maps in very different ways. They would seem to lie on opposite sides of a discontinuity that separates two epistemes, a possibility that Foucault does not discuss.

If Foucault's ideas help us to understand the perceptions of the state, Gramsci's writings, [. . .] may be used to study those of the peasants. [. . .] Gramsci's focus on the associational character of political life draws attention to the community as a point at which peasant and state ideologies would compete. [. . .] Most apposite, though, is Gramsci's notion of hegemony, since it proposes that elites exercise political domination not only through direct coercion and control of resources but also through the establishment of ideologies that legitimate their rule. His ideas suggest that the dominant groups, whether social classes or the state, seek to impose their own ideologies (in this instance, a construct of state control over territory) and that a point of difference (in this instance, the incommensurability of the maps) would be the locus of struggle in which subordinate classes challenge domination through their counter-hegemonic ideologies and activities.

Although the possibility of resistance is of theoretical importance to Foucault, he devotes very little space to a discussion of practices and discourses of resistance, in contrast to his lengthy treatment of domination (Dreyfus and Rabinow 1982: 206–207, 224–226). [. . .] The term 'resistance' suggests two parties in a dominant and a subordinate position, in this case with the powerful party pushing the weaker one in some direction; unlike the term 'gaze', which attributes passivity and possibly lack of awareness to the subordinate party, the term 'resistance' implies that the subordinate party overtly recognises the hostile or threatening force and actively withstands, counteracts or alters it. Gramsci (1971: 272) draws on a similar notion in his discussion of the idea that a class or set of classes defines itself in opposition to its enemies. In the instance of the reed beds of Lake Titicaca, however, the term 'resistance' is less appropriate, suggesting a certain Eurocentric bias in Gramsci's work as well as Foucault's, because the peasants do not so much oppose the state as remove themselves from it, or at least displace or subordinate it in their accounts.

The term 'resistance' implies an engagement between two parties [. . .]. Though such engagements may take many forms, including withdrawal and retreat as well as conflict and combat, it seems that the peasants' sheer lack of attention to the state is much further from the sense of the term 'resistance' than are most other activities to which that label is attached. The peasants claim that they continue to manage their affairs in the fashion that they always have, and that the state, having withdrawn its challenges to them, once again acknowledges their right to do so; their view is one not so much of peasant resistance as of state desistance. [. . .]

A number of South Asian scholars (Guha 1983; Guha and Spivak 1988) have used a related term of Gramsci's, 'subaltern' in the name of their group, and have adopted much of his approach in their examination of the ways in which popular culture has supported, and been shaped by, resistance to hegemonic domination, especially during and after colonial rule. [...] Such a history from below has much in common with the approach taken here to the conflict over the Titicaca reeds. One recent study (Guha 1989) [...] offers a number of parallels in its emphasis on peasant notions of state legitimacy and on customary land-use practices and village authorities, but contains an important difference as well: an emphasis on direct confrontation and indirect resistance where I have described disengagement. This contrast might be attributed to the ways in which India differs from Latin America (in scale, cultural diversity, colonial history and the role of intellectuals (Kahn 1985) among others) as well to the analytical traditions from which that study has borrowed.

At the level of specific political actors, two striking features emerge from a comparison of these maps. Firstly, the state is turned inward, preparing representations for itself rather than for its control over its territory and citizenry. [...] Latin American governments are often seen as maintaining an administrative presence more effective than that found in most African and some Asian nations (Eckstein 1989). Secondly, the peasants seem autonomous rather than subordinated. [...] This view makes the rural inhabitants of the altiplano [...] something like a sovereign people, and hence a nation. [...] (Anderson 1983). It could be argued that the unusual features of this case derive from the unique ecological conditions, historical background, and cultural diversity of the altiplano. However, [...] the irresolution that characterises this conflict is created by the state's view that the peasants are passive citizens and the peasants' belief that the state is a remote, aloof supporter of their own local political institutions. [...] The peasants' view of the state as ratifying their own affairs from a distance rests on features of corporate land-holding villages, such as community assemblies and leaders with ritual and political functions, that have parallels in many other parts of the world. More broadly, a general tendency of organisations to attribute power to themselves may be reflected both in the state's perception of itself as the only source of law and of planning and in the peasants' view that their long-standing use and management of their lands legitimise their control over them.

Our discussion of the maps also raises some questions about the relation between communication and action. Firstly, it shows that the concept of 'interest', [...] has become much more complex in recent anthropological thought (Latour's work is a notable example) and, in some cases, has been abandoned altogether. In this case, it would be difficult to define, let alone to measure, the interests of peasants and government officials. [...] Each side holds distinct, complex notions of use, control and ownership of resources. The differences in these views make it difficult to establish a single authoritative account of the conflict and its outcome; this lack of a single account, in turn, challenges conventional understandings of political conflicts as having, like most Western narratives, beginnings, middles and, most important, ends. [...]

Secondly, this questioning of the notion of interests, presumably material, has turned attention to the issue of meaning, presumably mental. [...] This move shows the growing attention paid to the importance of imagination in politics [...]. Anderson's (1983) view of nations as 'imagined communities' might be complemented by the spatial idea of imagined countrysides. In the instance of the reed beds, two very different imagined countrysides coexist: one in the minds of the peasants, who rarely make maps but whose frequent travels through the area render it visible to them, and whose notions of territory, social group, ownership and authority make it comprehensible; another in the minds of government officials, who rarely leave the cities, towns and roads to enter the lake but whose maps, based on very different notions of territory, social group, ownership and authority, make the area both visible and comprehensible to them as well.

References

Anderson, B. (1983) *Imagined Communities: Reflections on the Origin and Spread of Nationalism*, Verso, London.

Anderson, B. (n.d.) *On Official Nationalism and the Colonial State MS*, unpublished mimeo

Appadurai, A. (1986) *The Social Life of Things: Commodities in Cultural Perspective*, Cambridge University Press, Cambridge.

de Certeau, M. (1984) *The Practice of Everyday Life*, University of California Press, Berkeley, CA.

Dreyfus, H.L. and Rabinow, P. (1982) *Michel Foucault: Beyond Structuralism and Hermeneutics*, University of Chicago Press, Chicago.

Eckstein, S. (1989) *Power and Popular Protest: Latin American Social Movements*, University of California Press, Berkeley, CA.

Foucault, M. (1977) *Discipline and Punish: The Birth of the Prison*, Pantheon Books, New York.

Gramsci, A. (1971) *Selections from the Prison Notebooks of Antonio Gramsci*, International Publishers, New York.

Guha, R. (1983) *Elementary Aspects of Peasant Insurgency in Colonial India*, Oxford University Press, Delhi.

Guha, R. (1989) *The Unquiet Woods: Ecological Change and Peasant Resistance in the Himalaya*, University of California Press, Berkeley, CA.

Guha, R. and Spivak, G.C. (eds) (1988) *Selected Subaltern Studies*, Oxford University Press, New York.

Iser, W. (1974) *The Implied Reader: Patterns of Communication in Prose Fiction from Bunyan to Beckett*, The Johns Hopkins University Press, Baltimore, MD.

Iser, W. (1978) *The Act of Reading: A Theory of Aesthetic Response*, The Johns Hopkins University Press, Baltimore, MD.

Kahn, J. (1985) Peasant ideologies in the Third World. *Annual Review of Anthropology*, **14**, 49–75.

Latour, B. (1987) *Science in Action: How to Follow Scientists and Engineers through Society*, Open University Press, Milton Keynes, UK.

Latour, B. (1988) *The Pasteurization of France*, Harvard University Press, Cambridge MA.

Latour, B. and Woolgar, S. (1986) *Laboratory Life: The Construction of Scientific Facts*, Princeton University Press, Princeton, NJ.

Merriam, C. (1973) *Webster's New Dictionary of Synonyms*, G & C Merriam Company, Springfield, MA.

Orlove, B.S. (1982) Tomar la Bandera: politics and punch in Southern Peru. *Ethnos*, **47** (3/4), 249–261.

Republica del Peru (1978) *Decreto Supremo No 185-78-AA*, Ministerio de Agricultura y Alimentacion, Lima, Republica del Peru.

Taylor, L. (1987) Agrarian unrest and political conflict in Puno, 1985–1987. *Bulletin of Latin American Research*, **6** (2), 135–162.

Further reading

Dodge, M., Kitchin, R. and Perkins, C. (2009) *Rethinking Maps: New Frontiers of Cartographic Theory*, Routledge, London. [An edited overview with several chapters focusing upon the need to consider the contexts in which cartography is enrolled, but also the practices that are part of everyday mapping.]

Grasseni, C. (2004) Skilled landscapes: mapping practices on locality. *Environment and Planning D: Society and Space*, **22**, 699–717. [An anthropological approach to community mapping in the Italian Alps.]

Orlove, B. (2002) *Lines in the Water: Nature and Culture at Lake Titicaca*, University of California Press, Berkeley, CA. [An evocative consideration of the wider issues in Orlove's paper that situates the mapping case study in an impassioned, rich and rigorous exploration of life around the lake.]

Sletto, B. (2009) We drew what we imagined: participatory mapping, performance and the art of landscape making. *Cultural Anthropology*, **50** (4), 443–476. [A current exploration of the empowering potential of an anthropological approach to participatory mapping in Venezuela and Trinidad.]

Wood, D. (2010) *Rethinking the Power of Maps*, Guilford, New York. [Wood uses the indigenous mapping as an example of counter-mapping and explores its complex re-negotiations of power in the twenty years after Orlove's research.]

See also

- Chapter 1.7: Design on Signs: Myth and Meaning in Maps
- Chapter 1.8: Deconstructing the Map
- Chapter 1.9: Drawing Things Together
- Chapter 1.12: The Agency of Mapping
- Chapter 1.14: Rethinking Maps
- Chapter 5.6: Whose Woods Are These? Counter Mapping Forest Territories in Kalimantan, Indonesia
- Chapter 5.7: A Map that Roared and an Original Atlas: Canada, Cartography, and the Narration of Nation
- Chapter 5.9: Affecting Geospatial Technologies: Toward a Feminist Politics of Emotion

Chapter 4.8

Refiguring Geography: Parish Maps of Common Ground

David Crouch and David Matless

Editors' overview

By the mid 1990s the social constructivist critique of cartography was widely accepted, and mapping as a form of elite knowledge was open to challenge. Few papers had, however, explored the potential offered by the form in alternative re-imaginings of place or focused on the cultural practices and processes of everyday mapping. The paper excerpted here, by Crouch and Matless, focuses upon the Parish Maps Project, initiated by the environment and arts group Common Ground in the United Kingdom, as an aesthetic intervention, but also as a political alternative form of local knowledge. Crouch and Matless consider maps in the project commissioned from artists, as well as maps produced by local communities. The paper represents a cultural weaving together of iconography and ethnography, that highlights the contradictory potential of different articulations of places offered by protagonists involved in the mapping process. It also relates the local mapping to the ethos of Common Ground and in relation to debates about place in Geography as an academic discipline.

Originally published in 1996: *Transactions of the Institute of British Geographers,* **21**, 236–255.

Introduction: Common Ground and the Parish Maps Project

The environment and arts group, Common Ground, was formed in 1983 by Sue Clifford and Angela King to promote the common, local and everyday cultural heritage, and to link conservation and the arts. Its projects have drawn upon, yet reacted against, its founders' backgrounds in social scientific geography and planning. Common Ground argued that the common was overlooked in an environmentalism placing value on the rare or exotic and that, in its desire for scientific legitimacy conservation was ignoring the very cultural values which drove everyday concerns (Clifford and King 1987; Crouch and Ward 1994; Matless 1994a). [. . .]

The Parish Maps Project, launched in 1987, was Common Ground's first major public initiative. Artists' maps were commissioned and exhibited as a catalyst for community initiatives. The project aimed to encourage communities to chart the familiar things which they value in their own surroundings, and give active expression to their affection for the everyday and commonplace, whether in town or country (Common Ground 1987; Greeves 1987a). [. . .]

[. . .] Clifford and King's experience in local protest groups had brought home the political potential of maps, yet also convinced them that people would more readily become involved in less directly political projects by emphasising place, people and personal values. [. . .] Mapping is presented as a process of self-alerting, putting people on their toes against unwanted change and producing an active sense of community. [. . .] The paradox is that the assertion of everydayness can self-consciously undermine the supposedly unreflective field it upholds.

The contradictory politics of place

The emphasis on sharing community may be strained in the process of map production. In highlighting a

The Map Reader: Theories of Mapping Practice and Cartographic Representation, First Edition. Edited by Martin Dodge, Rob Kitchin and Chris Perkins.

dialectic of the human and environmental, the individual and social, the Parish Map can bring out the different claims made to the geographical collectivities, which are places. Community maps may be made through a shared experience, rooted in everyday cultural practice (Crouch 1990; Crouch and Tomlinson 1994) but may not necessarily emerge as documents of communal harmony. [. . .] We can draw out such tensions of place by focusing on four themes: the reworking of map form; the refiguring of place; the emphasis on 'parish'; and the place of the aesthetic.

Reworking the map form

Common Ground's work forms part of a broader reworking of mapping emerging without as well as within the academy (Nash 1993). The map is deployed as the classic locational genre, yet its form and authority are unsettled through a use of older or non-western mapping traditions and by a loosening of cartographic definition to include other media. Parish Maps have appeared in ceramics, textiles, photography, paint, sculpture, collage and video as well as pen-line on paper. The standard of the Ordnance Survey (OS), which gave financial support to the Parish Maps Project, becomes one valued genre among many.

In criticising the standard of conventional cartography, Common Ground echoes Harley's linkage of maps, knowledge and power (Harley 1988). [. . .] However, Common Ground's valuing of the Ordnance Survey as one enabling cartographic language among many goes beyond any clear political or aesthetic distinction between oppressive or liberatory cartography. There is no simple distinction of hegemony and counter-hegemony here in relation to aesthetic form or production process (Rose 1994). The embrace of a multiple performativity of mapping echoes Deleuze and Guattari's (1988: 12) vision of a cartography rhizomatically inhabiting society:

> 'The map is open and connectable in all of its dimensions; it is detachable, reversible, susceptible to constant modification. It can be torn, reversed, adapted to any kind of mounting, reworked by an individual, group or social formation. It can be drawn on a Wall, conceived as a work of art, constructed as a political action or as a meditation . . . it always has multiple entryways'.

For Deleuze and Guattari, the map departs from the reproductive realism of a 'tracing'. The map has to do with performance, whereas the tracing always involves an alleged 'competence' (Deleuze and Guattari 1988: 12–13). The map contains a different relation to the matter of a place, being 'entirely oriented toward an experimentation in contact with the real' (Deleuze and Guattari 1988: 12). Such performative experimental contact is strong throughout the Parish Maps Project, although [. . .] a tension arises between Common Ground's fostering of consciously artistic experiment and a tendency, strong in many community maps, to hold to a tracing of the real. There are tensions, too, around the degree and purposes of cartographic openness and closure. [. . .] While some Parish Maps perform to a rhizomatic ideal, others seek proudly to trace, frame and root.

Refiguring place

The Parish Maps Project is one of Common Ground's several interventions in 'local distinctiveness', feeding of and into a emerging cultural politics of the local (Bird *et al.* 1993). Again, however, there is a conservative–radical tension [. . .] The map may seek place as a secure rooting ground (King 1991) or affect to unfix settlement [. . .]. Recent work has stressed place as a contested site of representation (Shields 1991) where different spatialities and temporalities conjoin. [. . .] As many different places as there are individuals and groups will effectively cohabit an area. Any sense that a map may easily trace one place-bound community is problematised. [. . .]

These tensions of place derive in part from different cartographies of tradition, modernity and difference. In one, modernity erases tradition and in so doing erases difference; in another, modernity simultaneously erases and produces difference. For one, capitalism standardises pre-capitalist variation; the other critically recognises capitalism's own manipulation and production of difference. History becomes a tale not of difference eroded but of competing differences produced. Common Ground's work veers between historically and geographically essential and dialectical versions of the local.

Parish

In the Parish Maps Project such tensions arise in part from the term 'Parish', with its overtones of tradition, spirituality and rurality An ecclesiastical as well as a secular administrative unit, parish connects settlement and surrounding land, connecting Common Ground to a long English cultural tradition of presenting place, especially rural place, in reverential, ritual, sacred terms (Matless 1994a): 'Parish' is a very laden concept. [. . .] For most of us, it is the indefinable territory to which we feel we belong which we have the measure of. Its boundaries are more the limits of our intimate allegiances than lines on a map (Mabey 1980: 36).

Can such a culturally charged banner be carried into the urban? Common Ground taps into a villaging of the city, mapping belonging at a scale below the civic, seeking the topography under the city walls. Cities are envisaged less as centralised unities, their residents looking to a civic heart, than as localised metropoli. Nevertheless, Clifford and King acknowledge a failure of Parish Mapping to root in the city:

> 'We wanted to get Parish Maps going in towns and villages, but it has been mainly in villages that they have taken off, and this has been disappointing. [. . .] '

Clifford and King acknowledge how the 'Parish' label may have frustrated the project's intentions:

> 'We wanted another title. There's no word in English that actually holds things together like Heimat in German; . . . territory . . . sounds hard; neighbourhood is not 'cuddly enough' and is American . . . Community seemed too fixed, maybe bureaucratic'.

Searching for a gentle English vernacular, Common Ground wase tugged away from the city, tugged into a standard cultural lexicon of Englishness.

Placing the aesthetic

Connecting landscape imagery to everyday environmental practices [. . .], the Parish Map places aesthetic questions at the core of environmental debates. At the same time, however, it asks questions as to what constitutes 'the aesthetic'. The Parish Map may broaden the sense of the aesthetic away from aestheticisation, away from making an art object of a place, towards a sense of the crucial presence of the aesthetic in the everyday. [. . .] Miller (1987) argues that intended symbolic meanings may be recontextualised through local processes of active consumption. [. . .]

The Parish Map can form an intervention in what Foucault (1986: 11) terms the 'aesthetics of existence' entailing less aestheticisation than a problematisation of the place of the aesthetic in social and political life, a breaking of its categorical aura [. . .] (Matless 1994b; Thacker 1993). As we will see from both the community maps and the commissioned maps, there is a tension between this Foucauldian sense of a map inhabiting and critically amplifying elements of an aesthetics of existence and the production of a discrete aesthetic object catching a place's best side. [. . .]

Parish maps by artists

The main aim of the Parish Maps Project was to promote community mapping but it is crucial first to address the role played by the initial artists' maps. Clifford and King distinguish Common Ground's work from community arts projects, feeling that the latter depend upon outside impetus, whereas their own work encourages people to set their own agenda [. . .] There remains, however, a sense that artistic example was necessary to open up different ways of seeing. [. . .] Eighteen artists were commissioned to map places they felt a particular attachment to; the maps toured in the 'Knowing your place' exhibition between March 1987 and March 1988 (Common Ground 1987; Creative Camera 1987). [. . .] These were not exercises in 'community art' and did not necessarily seek any local communal role or audience (Matless and Revill 1995). Common Ground promoted an agenda of critical celebration and we turn below to four maps which seem best to express this ethos. Firstly, however, we consider those which, in stressing either celebration or critique, highlight some of the aesthetic and political tensions in Common Ground's project.

Some maps turned away from a conventional Ordnance Surveying style to celebrate a poetic magic of place [. . .]. Pat Johns' tapestry of Topsham in Devon, Topsham Observance, surrounds an outline plan of the village and the Exe estuary with birds, flowers, a poem, a wood spirit. [. . .] Mysterious beauty and conventional cartography are parallel worlds, co-present in Topsham yet set apart. Other artists stress critique more than celebration [. . .] Atkinson's map of the iron ore mining area of Cleator Moor in west Cumbria writes graffiti of waste, radiation and health risk over the OS map, suggesting the map to be part of the language of both power and potential resistance. [. . .] Atkinson's map is a document of angry attachment, a lament rather than a celebration.

Willats also inhabits the conventional plan view to depart from place celebration. A photographic montage of the view down from a Hayes tower block, [. . .] to the debris on the ground, displays the downside of the city [. . .]. Willats makes an ironic contrast to: 'the historical idea of parish and its well ordered, homogenous community, . . . both are social territories . . . authoritatively defined' (Common Ground 1987).

The parish becomes a scale of authority to be subverted. [. . .] Other Parish Map artists, however, have no qualms about their art being a skilled presentation of beauty. In Pat Johns' Topsham, the notion that art was a bourgeois category peddling the picturesque would seem distinctly out of place.

We conclude discussion of the artists' maps by considering in more detail those which seem to echo Common Ground's ethos of critical celebration. [. . .] The maps by Balraj Khanna and Simon Lewty present places where history and memory weave a density of personal and social meaning. Khanna's *The real centre of the universe* offers a Mappa Mundi cum Hindu epic picture of Maida Vale and St John's Wood; the canals of the world in west London. [. . .] A parish is made in the city only by shadowing the surround. The edges of the picture seem curled, stopping the place seeping out into the city beyond. There is the odd conventional symbol. An MCC tie dangles by Lords cricket ground, the only place Khanna knew of on arriving in England over twenty years ago. A few permanent features stand out – church, park, cricket ground, canals – but there is no sense of any group or use dominating the field. [...] This is a place a person could move into and mark; personal names are inscribed graffitied signatures to the landscape. The surface is teeming but there are unscribed spots; the space could accommodate incoming signatures with ease. [...]

Simon Lewty's *Parish Map Old Milverton* presents a place of stories [...] A long strip, its width nine times its height, the map makes a journey through a piece of Warwickshire country on the edge of Leamington Spa [...]. The map acts as permit of entry to a geographical art of memory, where Lewty walked, swam, picked a souvenir. [...] At the centre of the reproduced detail is a trunk of a body, naively twisted with the forms of the land, turned, as on a lathe, through the process of remembering. [...] Features are dotted without a key. [...]

Common Ground's sense of art feeding on the substance of a place is given literal expression in the maps of Helen Chadwick and David Nash. [...] David Nash's map – *Parish earth, Parish stone* – of Blaenau Ffestiniog, Gwynedd, is made of place. [...] There is a sense of contact, of corporeality emphasised in the manufacture; Nash works to diminish the distance between image and substance. He also produced *A Personal Parish*, a map of names of features and substances, laid out to form a plan view in words: path, crag, granite, grass, quarry and so on. Words placed on the page to conjure the flesh of the land, a flesh Nash claims as his own corporeal space. [...] In its cartographic deposition of the place, *Parish earth, Parish stone* carries geological form and formation. [...] The colour of the map materialises this in reproduction, though the material texture is evened out beyond the original. The land is daubed as a symbol of itself, as its own colour code. The map becomes a microcosm of its source.

Helen Chadwick mapped Littleheath Woods near Croydon by merging colour photocopies of objects from the place into a map of leaves, limbs, a girl's head, animals, wires, pylons, housing. [...] The map evokes attachment and binding, a dialectic of local belonging. The photocopy method heightens the sense of place as a site of simultaneous care and containment. [...] Photocopying floats the object. [...] The photocopy catches the object but loses its texture; light touches the object but the viewer cannot. A medium of contact sends the object into a beyond. [...] Chadwick's objects carry no obvious trace of memory, cannot be touched, cannot emerge. [...] Chadwick produces a pickled simulacra of Litteheath Woods.

Each of these artists offers a distinct refiguring of place. While their maps were set up as possible examples for community groups to follow, Common Ground's work is premised on a sense that invention and creativity are not simply the province of professional artists. We now consider the production of community maps which, if not always drawing specific inspiration from the artists' maps, creatively refigure place. As we shall see, however, some of the tensions in community map production derive precisely from the sense in which these maps are held to be works of art.

Community maps

The Parish Maps made by local groups emerge from a complex collective process of geographical knowledge. Parish Mapping offers a study in the social negotiation and contestation of what matters; each map sought to realise a form of identity through local knowledge, deploying diverse strategies of production and representation, drawing on wider metaphors of heritage and styles of art and craft. [...] A common ground becomes less a common denominator than a space where different practices commingle to nation by district, county and rural community councils, and voluntary organisations such as Women's Institutes.

At the time of writing, there are over 1500 completed community maps; here we discuss four in detail to give a sense of their aesthetic and social variety. Most maps are produced on two-dimensional surfaces, paper on board, a few feet in dimension. The most common form shows a central plan fringed with motifs of local events, views, objects, buildings and stories. The layout conveys a shared yet intimate place, recognisable and familiarly encapsulated, a whole through which internal difference can be celebrated. Maps have also been produced in video, tapestry and ceramic; in one case, the map ended up as a three-sided, eight-foot high pillar, permanently sited in the street. The Project has also become a site of exchange between places, through informal meetings and regional coordination nation by district, county and rural community councils, and voluntary organisations such as Women's Institutes.

Four maps

Charlbury in Oxfordshire shows a typical layout, an A1-sized poster with a central illustrated plan and a fringe filled with details of buildings, clubs and societies – a group singing, a chess board, children playing. Historical references are inscribed over the surface. This is a consciously factual map, with carefully drawn and painted images naturalistically presenting events and things. The artist-instigator expressed a concern for accuracy, especially in relation to footpaths, where accurate record might help in their preservation. The map was produced as an item of local cultural capital, a thing of traditional beauty showing the best side of Charlbury. [...]

Standlake in Oxfordshire uses the core–fringe layout but in a very different way, seeking to open up an imaginal space for the viewer. The core street layout is a sketch, while the fringe has diverse colour sketches positioned in 70 six-inch squares making an uneven border; a 'no' sign to a local gravel pit, a gravel company's view of wild geese and JCB in harmony: the whole collage can [...] be read like a novel. The story is geographical. All the squares were painted by a different person or group; no offers were rejected. [...] A child's bedroom window view marks a private image on a public space for future memory. The squares line up in utilitarian format, with none given priority. The badges of four land-owning Oxford colleges corner the image. The purpose of the map shifted during production:

> 'it was intended as something to celebrate the opening of the new village hall; but in the process we discovered a real concern for the extension of gravel workings proposed near the village, and the Map soon developed into a means for the whole village to express concern over this'.

The sense that the map purposefully combines the objective and subjective is expressed in both form and content.

Lockwood's nine-foot square textile collage negotiates the overlapping experiences and meanings of different people across the several small settlements that make up a parish on the northern edge of the North York Moors. Places are held together by a rich brown colour over the predominant part of the map depicting the surrounding moorland. Individual sections display specific features: a children's playground, a terrace of houses, a spoil heap, hills. Stimulated by a Common Ground circular, a Labour councillor instigated meetings with the Parish Council and other bodies. Thirty people formed a working group which circulated every household and met each local club. [...] The components and relative positions of features were negotiated in a series of meetings in every hamlet. [...] Rather than promote a consciously unified or fragmented identity, different identities are inscribed over an integrated yet variegated surface.

Westbury Park's suburban Bristol map remakes the place through a game of Monopoly. Cartoons, photographs and sketches, mostly in colour, and diverse lettering and language appear all over the map surface. The map was an immediate response to an invitation at a fair [...]. The result is a kaleidoscope of images and references to objects – buildings, front gardens, shops and to the events of everyday life. [...] The map holds together and does not seek to adjudicate between differing opinions on specific sites. Produced as a poster, it contains instructions on how to cut and fold in order to make the centre pop up into a game of the place. The process of mapmaking produced a local community association and a changing of the area's recognised boundaries.

Unity and diversity

In each case, the map coordinators sought a regulated expression of demographic diversity. Even Standlake's eclecticism was given a binding consistency [...]. Charlbury's map drew on the designs of many but was finally painted by one artist with a careful and consistent organisation. [...] Standlake's use of looser sketches, exactly as the individual or group did them, has less of a sense of control, permitting a more plural, less consciously 'artistic' expression. Westbury Park sought an eclectic expression of a place at one moment in time, a loose play of graphic content. The Lockwood map was made both to bring the place together and to open up discussion of its future, with images interlocked through a thickly textured surface and richness of colour [...]. The form was collectively devised: 'it was through our meetings that we came to the idea of textile for the map'. The map instigated ongoing processes of local knowledge, it became a useful focus to get people to talk about what mattered to them ... but we are going on with other meetings now the map is done'.

The shared process of map production is central to the social aims of the project. Some Parish Maps, particularly those drawing on texture techniques of tapestry and quilting, tap into specifically gendered traditions of artistic production, whose art has indeed often been put down precisely for combining collectivity and femininity: 'Embroidery has provided a source of pleasure and power for women, while being indissolubly linked to their powerlessness' (Parker 1984: 11). When interviewed Clifford and King did not regard gender as an issue in parish mapping and saw no evidence that maps had been produced predominantly by either women or men. More detailed investigations of map production might, however, focus on the ways in which certain gendered modes of artistry are drawn upon, resisted or transgressed by those men and women making maps. [...]

This is not to suggest that a village or suburban map must necessarily inhabit this cultural space; indeed, that space is itself politically and aesthetically complex (Light 1991). Different settlement forms are nevertheless culturally charged objects whose specific ideologies might be conjured unawares in parish mapping. It would seem in the spirit of the project to attend to the cultural charge of any particular geography.

Content

The linked politics and aesthetics of parish mapping are evident in content. The Charlbury map appears as an exercise in comprehensive realism but its imagery is carefully selected. A particular iconography of the place is set up: older buildings, a flora and fauna denoting a settlement in harmony. The making of a map 'like an old painting' is also bound to a particular social aesthetic. One third of Charlbury housing is council-owned and yet nothing of the large estate appears on the map. The image of the map as a place's 'wedding photograph' would seem to entail cropping-off part of the family. The map also excludes some public buildings, notably the modernist former school, now a community and health centre, a site of significant memory and experience for much of the population. [. . .] The Charlbury map presents a kind of two-dimensional best-kept village. Such picturesqueness is, however, bittersweet. Unwilling to register a very visible architectural and social presence, placing part of the village out of cartographic sight, the mapmakers undercut their desired holistic vision of place and community.

The organisers of the Standlake map had expected most of the submitted images to be 'attractive'. Instead, the open agenda of the map both opened up an aesthetic criticism of features such as the gravel workings and widened the sense of what might be given aesthetic value. The map finds a valued beauty in humdrum everyday matrix where place is geared around a family diversity.

Local value

Often the map has a conscious political purpose, being deployed as a pragmatic language which people outside a place – planners, tourists, developers – might notice and listen to. Lockwood produced a map with a view to influencing the economic future and recognising a plural heritage, placing classic touristic images of castle and lake alongside a terrace of architecturally nondescript but locally valued houses. Charlbury also sought to highlight heritage but in a manner which suggests change held in deference to heritage rather than offering evolutionary possibility. Change in Charlbury appears to intrude on a complete scene.

Maps such as Lockwood and Standlake echo the strategies of local imaginative empowerment described by Rose (1994) in relation to the production of a community film in east London (Crouch 1994b). [. . .] Supralocal icons are called up and the maps incorporate images not only of settlement but of movement, even of potential escape; the view from a bedroom, the train heading off the Charlbury map to London. The maps continue, however, to insist that a located community remains a significant form of belonging in the contemporary world, existing alongside cross-local forms of community

Icon

While the aims of the Parish Maps Project relate in part to shared processes of production, the map as finished product is also central. The map is to be valued as both process and object. [. . .] In Charlbury, the original is on the wall in the village hall, as it is in Standlake. Both places also produced map postcards. Westbury Park's map can be bought as a poster and Lockwood's has hung in all the hamlet halls in the parish. It is expected to stay in the new visitor centre that has emerged as one idea from making the map.

Different forms of reproduction and exhibition produce very different kinds of aesthetic object and make for different spaces of readership and recontextualisation (Crouch 1994a; Miller 1987). [. . .] The map offers images which are local allegories, contains within it several spaces in different perspectives, iconically ties the local viewer who stands before it, or looks upon it as a poster or postcard, into its gaze. Hanging in or being handed around a local space, it makes that space one of reflection on what it is to be local. In its display, however, the map remains an object produced through and productive of relations of power. Such an icon of place may become confined as an object of reverence, entrusted to select guardians, a place banner to be brought out on occasion, while mechanical reproduction and dissemination may disrupt the aura of the original (Benjamin 1973), a local enshrinement of an original map (of which the postcard is a 'mere copy') in a special community site may add to a sense of the map as an object passed over into a discrete aesthetic realm, existing only for contemplation, and definitive. The map gets frozen; the implication may be that the place it shows also wants strict preservation.

Artistry

It is worth reflecting upon possible links between community map production and the artistic example set up by Common Ground in the 'Knowing your place' exhibition. The question centres on the role of individuals in map production. [. . .] Many groups have resisted the map being the work of an individual to avoid one style and/or voice becoming dominant. There also seems a resistance to the conventional ways of picturing which an appeal to the skills of a competent individual might produce. Lockwood, Standlake and Westbury Park all deploy organising devices through which different senses of place can be voiced. Some communities, however, as in Charlbury, have used local artists, professional or amateur, to produce the final maps. These rarely aspire to the aesthetic experiment of some commissioned artists' maps but rather aim for a stylishly accurate image. Different senses of artistry are at work here, the one consciously inventive, the other allowing deviation only within the bounds of naturalism. Paradoxically those community maps which engage the services of an artist may tend to be aesthetically and, perhaps by implication, socially conservative. Conservative aesthetic technique may constrain the social content and complexity of a map, fixing the locality rather than letting place flow.

The contours of place

The Parish Map bridges two forms of cultural geographical work: the ethnographic and the iconographic. Cultural geographers have tended to pursue one or the other; to make or study a Parish Map demands both. [. . .] The map is an investigation in time as well as space. It demands an opening up of a place's historical geography, encompassing a regard for the historical geography of the imagination of a place's people. This historical sense of the map can serve to unsettle environmental identity. While the multiplicity of local social identity is often stressed (Bird *et al.* 1993), environment is more commonly seen as a realm of continuity, a geographical 'base' on which cultures play.

The Parish Map [. . .] may bring out local environmental fluidity. A place does not build through time in neat sedimentary strata. Times flow, grounds rupture. Local history, far from being a secure base or bolstering heritage, can be more sodden and metamorphic. Common Ground taps into and reworks a tradition of local historical inquiry. [. . .] As older-established ways of mapping local significance – chronicles of great families, church histories – have lost their hegemony the local has become a more common historical site, a place of vernacular genealogies. Common Ground's Parish Maps work in this genre as local cultural practice.

The Parish Map also taps into tensions within recent geographical debates on place and identity. Massey (1993a, 1993b) has argued for a non-essential and 'non-parochial' view of place, with multiple and changing identities produced through a dialogue, not always happy and equal, of place, people and other places. Particularity is produced through wider interdependencies. Similarly, Parish Maps proceed from a sense that the local is semi-detached, distinct yet connected (Crouch 1994a, 1994b), yet place a different emphasis on the parochial, calling upon 'positive parochialism'. [. . .] Massey's work carries an important warning against what can at times seem to be a naive faith in the inherent virtues of community and the local but such parochial faith does suggest a need for an initially respectful encounter with all cultures of a place, even when we might not like some forms of local cultural assertion. [. . .]

In their marking out of local space, Parish Maps can be seen as exercises in the production of scale (Smith 1993). Crucial here is the nature of a place's boundary. What edges the parish? In parallel to the Parish Maps Project, Common Ground sought to revalue 'The Parish Boundary' (Greeves 1987b), urging a communal beating-of-the-bounds, walking the edge in order to confirm that within. [. . .] Maps, by contrast, allow a more fluid sense of scale; the frame may mark out a special field of care but with little sense of this being a rigid spatial barrier. [. . .] In such a context, the Parish Map serves, even in its most benign and picturesque forms, as an intervention in the social and political production of scale [. . .]. The map can become a site for struggle as well as celebration, bringing out social difference by providing one public imaginative space on which to work. Even the most outwardly conciliatory and harmonious of Parish Maps might act as a blanket thrown over difference which, in covering it, keeps it warm and stewing. The map exercises conflicting senses of the dimensions and character of a place, with the scope and nature of this geography itself becoming an active cultural figure in local politics, confirming or reshaping the contours of place. A geography refigured through a Parish Map becomes active on the local common ground.

[. . .]

References

Benjamin, W. (1973) The work of art in the age of mechanical reproduction, in *Illuminations* (ed. W. Benjamin), Fontana, London, pp. 219–253.

Bird, J., Curtis, B., Putnam, T. *et al.* (1993) *Mapping the Futures: Local Cultures, Global Change*, Routledge, London.

Clifford, S. and King, A. (1987) *Holding Your Ground: An Action Guide to Local Conservation*, Wildwood House, Aldershot, UK.

Common Ground (1987) *Knowing Your Place Exhibition Leaflet*, Common Ground, London.

Creative Camera (1987) *Location*, **4**, 10–32.

Crouch, D. (1990) The cultural experience of landscape. *Landscape Research*, **15**, 11–15.

Crouch, D. (1994a) Signs, places, lives. Working Paper, University of Odense, Odense.

Crouch, D. (1994b) Space and place in cultural identity, in *Territoriality* (ed. J. Tonboe), University of Odense, Odense.

Crouch, D. and Tomlinson, A. (1994) Collective self generated consumption; leisure, space and cultural identity in late modernity, in *Modernity, Postmodernity and Lifestyle* (ed. I. Henry), Leisure Studies Association, Brighton, UK.

Crouch, D. and Ward, C. (1994) *The Allotment: Its Landscape and Culture*, Mushroom Books, Nottingham, UK.

Deleuze, G. and Guattari, F. (1988) *A Thousand Plateaus: Capitalism and Schizophrenia*, Athlone, London.

Foucault, M. (1986) *The Use of Pleasure, vol. 2: The History of Sexuality*, Penguin, Harmondsworth.

Greeves, T. (1987a) *Parish Maps*, Common Ground, London.

Greeves, T. (1987b) *The Parish Boundary*, Common Ground, London.

Harley, J.B. (1988) Maps, knowledge and power, in *The Iconography of Landscape* (eds D. Cosgrove and S. Daniels), Cambridge University Press, Cambridge, pp. 277–312.

King, A. (1991) Mapping your roots. *Geographical Magazine*, May, 40–43.

Light, A. (1991) *Forever England: Femininity, Literature and Conservatism Between the Wars*, Routledge, London.

Mabey, R. (1980) *The Common Ground*, Hutchinson, London.

Massey, D. (1993a) Power-geometry and a progressive sense of place, in *Mapping the Futures: Local Cultures, Global Change*

(eds J. Bird, B. Curtis, T. Putnam *et al.*), Routledge, London, pp. 59–69.
Massey, D. (1993b) Questions of locality. *Geography*, **78** (2), 142–149.
Matless, D. (1994a) Doing the English village, 1945–1990: an essay in imaginative geography, in *Writing the Rural: Five Cultural Geographies* (eds P. Cloke, M. Doel, D. Matless *et al.*), Paul Chapman, London, pp. 7–88.
Matless, D. (1994b) Moral geography in Broadland. *Ecumene*, **1** (2), 127–154.
Matless, D. and Revill, G. (1995) A solo ecology: the erratic art of Andy Goldsworthy. *Ecumene*, **2** (4), 423–448.
Miller, D. (1987) *Mass Consumption and Consumer Culture*, Cambridge University Press, Cambridge.
Nash, C. (1993) Remapping and renaming: new cartographies of gender, landscape and identity in Ireland. *Feminist Review*, **44**, 39–45.
Parker, R. (1984) *The Subversive Stitch: Embroidery and the Making of the Feminine*, The Women's Press, London.
Rose, G. (1994) The cultural politics of place: local representation and oppositional discourse in two films. *Transactions of the Institute of British Geographers NS*, **19** (1), 46–60.
Shields, R. (1991) *Places on the Margin*, Routledge, London.
Smith, N. (1993) Homeless/global: scaling places, in *Mapping the Futures: Local Cultures, Global Change* (eds J. Bird, B. Curtis, T. Putnam *et al.*), Routledge, London, pp. 87–119.
Thacker, A. (1993) Foucault's aesthetics of existence. *Radical Philosophy*, **63**, 13–21.

Further reading

Clifford, S. and King, A. (2006) *England in Particular: A Celebration of the Commonplace, the Local, the Vernacular and the Distinctive*, Hodder and Stoughton, London. [An encyclopaedic and quirky campaigning call from Common Ground, this volume details the survival of the unique and everyday in English culture and landscape, and narrating the 'local' in the face of cultural homogenisation.]
Craig, W.J., Harris, T.M. and Weiner, D. (2002) *Community Participation and Geographic Information Systems*, Taylor & Francis, London. [A useful edited collection on work with PPGIS, which has some parallels to the Parish Maps Project.]
Grasseni, C. (2004) Skilled landscapes: mapping practices on locality. *Environment and Planning D: Society and Space*, **22**, 699–717. [A richly anthropological approach to community mapping in the Italian Alps.]
Perkins, C. (2007) Community mapping. *The Cartographic Journal*, **44** (2), 127–137. [An overview of different kinds of community mapping that argues for a consideration of cultural context and the mapping process.]
Wood, D. (2010) *Rethinking the Power of Maps*, Guilford, New York. [Wood uses the Parish Map Project as an example of counter-mapping and draws, in particular, on developments after the publication of the Crouch and Matless paper.]

See also

- Chapter 1.7: Designs on Signs/Myth and Meaning in Maps
- Chapter 1.8: Deconstructing the Map
- Chapter 1.12: The Agency of Mapping: Speculation, Critique and Invention
- Chapter 1.13: Beyond the 'Binaries': A Methodological Intervention for Interrogating Maps as Representational Practices
- Chapter 1.14: Rethinking Maps
- Chapter 3.9: Mapping, Modernity: Art and Cartography in the Twentieth Century
- Chapter 3.10: Affective Geovisualisations
- Chapter 3.12: The Geographic Beauty of a Photographic Archive
- Chapter 4.7: Mapping Reeds and Reading Maps: The Politics of Representation in Lake Titicaca
- Chapter 5.6: Whose Woods Are These? Counter Mapping Forest Territories in Kalimantan Indonesia
- Chapter 5.9: Affecting Geospatial Technologies: Toward a Feminist Politics of Emotion

Chapter 4.9

Understanding and Learning Maps

Robert Lloyd

Editors' overview

For many researchers map understanding and learning continues to depend on an appreciation of the thinking processes involved in map reading tasks. Lloyd's chapter, from an edited book reviewing cognitive mapping, traces the changing trajectories of this strand of work. He explores the behavioural history of what was termed psychophysics, and charts its development into more nuanced controlled experimental work grounded in cognitive approaches to map reading, exploring the perceptual processes of visual search, and reviewing in the processing of visual information. Researchers have also investigated how people remember mapped information, with continuing work focused on the operation of mental models of perceptions. Lloyd argues for a continuing search for theory based upon ongoing and carefully controlled but simplified experiments, instead of using complex maps. He concludes by examining connectionist theory linking design to use, and the role of learning and prior knowledge, with case studies focusing on animated mapping and studies of cartograms.

Originally published in 2000: Chapter 6 in *Cognitive Mapping: Past, Present and Future* (eds Rob Kitchin and Scott Freundschuh), Routledge, London, pp. 84–107.

Introduction

[...] The cartographer (mapmaker) displays the information and the map reader acquires the information. The spatial information on the map becomes spatial knowledge when patterns in the information are learned by the map reader. [...] Cartographers refer to this process as the cartographic communication model (Board 1967) [...]. A map can be thought of as an expression of a cartographer's ideas, a device for storing spatial information and a source of knowledge for the map reader. Some fundamental changes occur when the cartographer transforms spatial data into spatial information and other changes occur when the map reader transforms spatial information into spatial knowledge [...]. In a sense, cognitive maps that are learned from cartographic maps are second derivatives of the environment. Cartographers categorise, generalise and symbolise to enhance important information and eliminate any non-essential information. Cognitive mappers continue this simplification process when they encode information from the cartographer's map. Most cognitive maps not only fail to reflect all the details of the environment they represent, but also have systematic errors caused by the processes that encode them into memory (Tversky 1981, 1992).

[...]

The initial spark

Before the latter half of this century cartographers focused most of their attention on the production of maps (McCleary 1970). Maps were judged on their artistic merits as objects to be appreciated. The ability of the map to communicate information and the cognitive processes used by map readers were not a major consideration. Arthur Robinson, however, is usually credited as the first to link map design to human cognitive abilities (Robinson 1952; MacEachren 1995). Robinson's goal for cartography was to provide a scientific basis for map design. He asserted that *function* should provide the basis for design. He argued for the 'development of design principles based on objective visual tests, experience, and logic' (Robinson 1952: 13). He further pointed out that 'research in the physiological and

The Map Reader: Theories of Mapping Practice and Cartographic Representation, First Edition. Edited by Martin Dodge, Rob Kitchin and Chris Perkins.

psychological effects of colour; and investigations in perceptibility and readability in typography are being carried out in other fields' and that 'cartography cannot continue to ignore these developments' (Robinson 1952: 13–14).

Robinson's (1967) later essay on the psychological effects of colour in cartography gave the makers of maps much to consider about the role of colour in graphical communication. He argued that colour is a great simplifying and clarifying element. It can be used as a unifying agent as when it defines a figure–ground relationship. People have an emotional reaction to colours that is separate from any rational thinking. [...] He also argued that colour can effect the 'general perceptibility' of a map by affecting the legibility of the lettering and enhancing the map reader's ability to distinguish fine details. He went on to discuss a number of problems colour creates for cartographers and map readers. [...]

Power functions

Some of the earliest efforts to follow Robinson's directions adopted methods used in psychophysics that considered the relationship between a stimulus and the sensed effect of the stimulus (Stevens 1975). Using regression techniques that were becoming popular in geography, cartographers could fit a power function that connected a relevant stimulus and response (Williams 1956). The independent variables objectively measured characteristics of map symbols. The dependent variables were judgements of the characteristics made by human subjects.

Ekman *et al.* (1961) had subjects judge the volume of perspective drawings of cubes and spheres. Although power functions could be fitted for the data, it was concluded that the subject's responses reflected the perceived areas of the symbol rather than the intended volumes. This clearly indicated that the best intentions of the cartographer might be misdirected if they are based on false assumptions regarding the map reader's cognitive abilities.

Meihoefer (1969) expressed a concern about traditional methods used to construct graduated symbol maps. He argued maps with symbols that were not visually distinguishable were difficult to interpret. He also indicated that [...] 'only a small percentage of the subjects tested could distinguish subtle increases in circle size' (Meihoefer 1969: 115). [...]

Kimmerling (1975) had subjects select grey circles to partition a scale from white to black in equal steps. He then compared the functional relationship between mean estimates of grey experienced by the subjects and percentage reflectance for equally spaced grey values between 0 and 100%. He reported significant differences when estimates were done with a white and black background. He also suggested that the method used to acquire estimates from subjects could influence the results and that performing the same experiment in a map context would be difficult because simultaneous contrast effects could not be held constant.

About the time Kimmerling's paper appeared critics began to point out limitations of the psychophysical studies. Petchenik (1975) pointed out the paradigm shift that was taking place in psychology away from strict behaviourism toward what is now called cognitive psychology. The early power function studies [...] provided cartographers with no theoretical understanding of the interactions taking place between a map and a map reader.

Cox (1976) soon pointed out that the reference symbol used to make judgements of circle sizes had a significant effect on those judgements. [...] Psychological studies of the *symbolic distance effect* and *semantic congruity effect* provided a general theoretical explanation of this phenomenon (Holyoak and Mah 1982). Using reaction times to determine how quickly objects could be compared, it was consistently shown that reaction times decrease with the separation of objects on a scale such as size. [...] Reaction times for the question of which is larger or which is smaller depend on the relative sizes of the pair of objects in question. When the objects in question are relatively large (an aeroplane and a truck), deciding which is larger produces faster times. [...]

Chang (1977) argued that the methods used to estimate the sizes of circles affected the estimated parameters of the power functions relating these estimates to actual sizes. Underestimation of circle sizes was found when the ratio of the size of a circle and a referent was estimated but was only a special case when the direct magnitude of a circle was estimated with a given a referent. [...]

Castner (1983) reviewed the literature on cartographic design related to map reading and communication. He expressed a number of concerns about translating the results of map reading studies into valid map design practices. [...] The experimenter must allow some factors in the experiment to vary and control others. This causes simplified displays to be used in most map reading experiments. The results of experiments on simplified maps tell us something about the map readers' responses to the experimental stimuli, but are these findings directly applicable to designing a complex map? Although this is a valid concern, the answer is clearly not doing uncontrolled experiments with complex maps. [...] Studying complex maps that are more realistic, but have many specific characteristics that are not controlled in the experiment, may only indicate how subjects responded to a specific set of maps. [...]

Gilmartin (1981: 9) argued 'there is a need for both psychophysical and cognitive research in cartography'. [...] Medyckyj-Scott and Board's (1991) essay on cognitive cartography explained these dual needs. [...] The process of mapmaking is supported by knowledge that is, for the most part, implicit and informal. [...] The testing of individual maps as effective communication devices is generally not possible, so the effectiveness of the conventions of mapmaking cannot be guaranteed. Since craft knowledge is difficult to test, it is also difficult to generalise. [...]

Cartographers practising the craft were seeking from the psychophysical studies a simple heuristic, a rule, a tool, a simple power function that could be used to make maps. The focus of psychophysical studies was clearly on the making of the map rather than an understanding of the cognitive processes used in map reading. The focus of cognitive studies in map reading is clearly on how map reader's process and determine the meaning of information acquired from maps. [...] The goals of the craftsman and the goals of the cognitive scientist are not the same. The map stands in the middle waiting to be constructed and used.

Present

People generally read maps because they need information. They may want to know *where* a particular place is located in absolute terms [...]. A reference map can usually provide this type of information. At other times the map reader wants to know *what* a place is like. Specialised thematic maps provide this type of information [...]. These two examples assume the person knows the name of the place of interest. After identifying the name on the map, the absolute or relative location and the population size can be easily determined.

Map reading is an integration and synthesis of knowledge. Even these simple examples require both bottom-up information (the map) and top-down information (information stored in the map reader's memory). The bottom-up information is contained in the lines, colours, shapes, words and so on the cartographer has put on the map. The top-down information is prior knowledge previously acquired by the map reader [...]. It might be general factual knowledge learned in another context, but applicable to map reading, for example the meaning of words or the common names for colours. It might be cognitive abilities to process information that was learned and practised in other contexts, but applicable to map reading. [...] Other top-down information is prior knowledge about maps and the conventions used by cartographers to produce maps. [...]

Perceptual processes

Visual search

The viewer of a typical map should experience objects with information on a number of separable dimensions, for example colour, shape, or size (Shortridge 1982; Dobson 1983). Theories of visual search [...] consider how information is acquired from a visual display during an initial parallel stage of a search process and used in a subsequent serial stage to identify the target of the search. Objects located on a cartographic map would activate feature modules in memory during the parallel stage. The activation for particular features, for example colour, size or shape, is dependent on both bottom-up information (in the display) and top-down information (in the viewer's memory). Targets with unique features can be identified early in the process and are said to pop out of the display. Targets that share features with other object on the map are more difficult to find and usually require a serial search before they can be identified.

A number of visual search studies have been conducted with maps that have focused on map symbols (Lloyd 1988, 1997a), locations with particular colours (Nelson 1994), and boundaries with particular colours (Brennan and Lloyd 1993; Bunch 1998). Lloyd (1988) investigated the search for pairs of pictographic map symbols on cartographic and cognitive maps. His objective was to determine if searching for symbols previously encoded in memory was the same as searching for symbols currently being viewed on a map. [...] It was concluded that search processes used by memory and perception were not functionally equivalent.

Lloyd (1997a) considered variations in visual search for targets on maps and focused on differences between searching for targets with unique features and targets that share features with other objects. [...] The efficiency of spatial search was considered to be the time needed to find a target in a display. The best 'pop-out' effect occurred when colour was used to produce unique differences between targets and other symbols. Targets defined by orientation differences produced the slowest search times. A 'pop-out' effect also occurred when colour was combined with shape to define the target, even when the target shared features with other symbols. The location of the target on the map also had a significant effect on reaction times. Targets in the centre of the map were found faster because most subjects starting searching in the centre of the display.

Nelson (1994) has conducted experiments that considered searching for colour targets on choropleth maps. Results also demonstrated differences between the target and distractors were the most important factor affecting the search process. [...]

In a study conducted by Brennan and Lloyd (1993) subjects searched choropleth maps for a target boundary among other boundaries. [...] Their results indicated that both parallel and serial processes were important components of visual search. [...] Bunch (1998) performed a similar study using boundaries defined by opponent colours that also varied in both hue and luminance. [...] Results generally indicated that faster times occurred when the boundary colours were separated in the CIELab space. Faster times were also recorded for boundary colours that had the same hue but differed in luminance. This combination created a target with a unique hue that would pop out of the map.

Object files

[. . .] Kahneman and Treisman (1984) were the first to discuss object files as temporary episodic representations of real world objects. [. . .] The more specific and temporary object files were called *tokens*, and the more generalised and permanent structures used to label the object's identity were called *types*. [. . .] In a later paper Kahneman *et al.* (1992) argued that focusing attention on a particular object enhanced the salience of its current properties and also reactivates a recent history of the object. The process that causes a current object to evoke an item previewed in a previous visual field was called *reviewing*. As an object moves, changes characteristics or momentarily disappears from sight the reviewing process maintains the perceived continuity of the object by relating its current state to its previous states. [. . .]

An object file is opened for use by the perceptual system once we have focused attention on the object. Information is put into the file to record the changes occurring to the object over time. This is necessary to maintain the continuity of a changing object. [. . .] While viewing a map animation one may need to determine if there have been two objects on the map or one object that has changed it characteristics. The perceptual system can usually maintain the continuity of a single file and either keep it open or close the first file and create a second file. Our experiences with objects enable use to predict expected motions and behaviours that can be used to make commonsense decisions about objects we are viewing. [. . .] An object could change all of its characteristic and its label over time and still be considered the same object under some circumstances. [. . .] If an apparent change is detected in an object, we must decide if the current information related to the change, for example a new location or new colour, needs to be assigned to the existing object file or a new object file needs to be created.

When multiple objects are in a display, *correspondence* must be determined before apparent motion can be experienced. [. . .] Geographers should find it interesting that spatial and temporal constraints rather than other characteristics, for example colour, size or labels, primarily control this operation. Apparent motion is only perceived when spatial and temporal proximity is maintained. Other properties of the object, however, can change without a new object file being created. *Reviewing* of the object file must be completed to retrieve the characteristics for the object in the initial display that are no longer visible. [. . .] Apparent motion occurs because this earlier information provides the context for the change. Cammack (1995) investigate the role of object files in map reading. His results indicated no consistent reaction time advantage when objects on target maps were related to objects on preview maps by shape or colour. This supported the notion that changes in characteristics other than location are not of primary importance in maintaining object files. [. . .]

Memory processes

Mental models

Once we have acquired information from a map we need to integrate it with information acquired from other sources and store it in some organised structure in our memory [. . .] A person's mental model of an environment represents that person's learned knowledge of the environment at a point in time that can be updated as new information is acquired from additional experiences with the environment. Studies have shown that mental models can be encoded by viewing maps or navigating in an environment (McNamara *et al.* 1992; Taylor and Tversky 1992a, 1992b) or by reading verbal descriptions (Bryant *et al.* 1992; Ferguson and Hegarty 1994; Franklin *et al.* 1992; Taylor and Tversky 1992a, 1992b). [. . .] Johnson-Laird (1983) argued that images are created to provide particular views of the model from specific perspectives, but the model cannot be visualised as a whole. Our ability to visualise environments from multiple perspectives supports the contention that images can be created from mental models (Bryant *et al.* 1992; Franklin *et al.* 1992).

Researchers have suggested that mental models integrate spatial and verbal information into a single structure (McNamara *et al.* 1992). It is also thought that comprehension is increased if mental models are created while a text is being read (Waddill and McDaniel 1992). It has been reported that a text comprehension was better for subjects studying route texts over survey texts (Perrig and Kintsch 1985) and that subjects who read route texts constructed more accurate sketch maps than subjects who read survey texts (Ferguson and Hegerty 1994). This latter effect was eliminated when a map accompanied the texts.

The flexibility of mental models makes them effective representations of spatial environments. They can be used for a variety of tasks because the models integrate visual and verbal information. They could be used to plan the route between two locations in an environment (Taylor and Tversky 1992a) or to create multiple images to view an environment from different perspectives (Franklin *et al.* 1992).

Taylor and Tversky (1992b) argued that mental models of environments do not have a single perspective. [. . .] The authors argued that subjects used a mental model to verify inference statements because they could not compare inference questions directly to a representation of the text. Since inference statements were responded to with equal efficiency when a statement's perspective did not correspond to the perspective represented by the text, it was argued that mental models have no perspective. It was also reported that subjects who studied maps of the environments also created mental model with no perspective.

Lloyd *et al.* (1995) have shown that perspective free mental models of a city can be encoded by reading both route and survey description of how to go from one landmark to another, by seeing the routes traced on a vertical perspective map, or by watching a video simulations of driving the routes along streets.

Future

[...]

Predicting the future of cartographic research is not a simple task. It should be safe to say that understanding the cognitive processes used by map readers will continue to be an important goal for cartography. As those interested in spatial cognition advance toward this goal, those who practice the craft of making maps should benefit from the insights provided by cognitive research. This section of the paper discusses one broad issue related to a research perspective that could have a significant impact on cartographic research (connectionist models) and another, more specific, issue related to learning with maps (prior knowledge).

Connectionist theory

If cartographers are to develop a theory of map reading, they need to develop a perspective that allows them to (1) observe processes that directly connect map readers with their maps, (2) form hypotheses that explain these processes, and (3) develop theories that connect these processes and ultimately provide an explanation of how people read maps. Two major obstacles have made it difficult to develop a theory of map reading. Firstly, the cognitive processes are not easily observed. They take place in the brain of a map reader who, for the most part, is unaware of how he is processing information. Secondly, most map reading tasks are complex and involve multiple cognitive processes that need to be understood.

Models such as the cartographic communication model [...] are very general expressions of the elements involved, but do not tell us what we need to know. Cartographers have generally relied on information from laboratory experiments that provided surrogate measures of activities going on in the brain such as reaction times and accuracy to test hypotheses on map reading. Although these 'black box' studies have been very productive, cognitive scientists from a variety of disciplines have argued that models based on biologically correct principles have a better chance of evolving into credible theories (Kohonen 1989; Lloyd 1997b).

Such theories have been referred to as connectionist theories because they are based on models of how neurons connected as a network in the brain process information. Such studies frequently use artificial neural networks that learn the same information that humans are learning with their biological equivalents (Hewitson and Crane 1994). [...] The patterns of connection weights between neurons in artificial neural networks can provide previously unavailable information for interpreting how people process information to complete a task (Lloyd and Carbone 1995).

Lloyd (1994) conducted a map learning experiment that had both human subjects and *auto-associative* neural networks learn spatial prototypes. This type of neural network is capable of learning a prototype from a series of input patterns [...] The prototypes were for categories of maps that were supposed to be made by a number of European countries during the age of discovery. Critical information on the map related to the location, size and orientation of an island on each of the maps. An analysis based on typicality ratings indicated the human subjects and the neural networks had learned similar prototypes for the categories. [...] Another interesting finding was that the neural networks did a better job in learning some prototypes because they were not as influenced as humans were by the order in which they experienced the individual maps.

Cammack and Lloyd (1993) constructed an *Interaction Activation Competition* neural network that simulated the storage and retrieval of information from associative memory (McClellend and Rumelhart 1989). [...] The information encoded into the artificial cognitive map related to population, employment, race, politics, income and regional location. [...] By initially activating a particular neuron representing a characteristic, for example high income, one could spread activation throughout the network to activate states with this characteristic or other characteristics, for example voting Republican, associated with this characteristic. If a collection of states representing a region, for example the South, were initially activated, then activation would spread throughout the network to activate generalised characteristics of the region. This type of network can simulate the storage and retrieval of information in a cognitive map related to *where* places are and *what* they are like. By encoding a similar model for human subjects who had recently learned the same information presented on maps one could use the *Interaction Activation Competition* model to make inferences about what was actually learned during the map reading experience.

Lloyd (1998) simulated the learning of an outline map of Texas using a Kohonen neural network (Kohonen 1989). A Kohonen network consists of neurons in an input layer that provide the information to be learned. [...] The model is called self-organising because it does not learning by correcting activation errors. Instead, it learns by recognising patterns in the information being presented to it without the benefit of comparing its output to the truth. [...]

It is difficult to judge how well the Kohonen network models the process used by humans to acquire boundary information for Texas. A comparison can be made with the representation of Texas aggregated from 15 geography graduate student's sketch maps [...]. The students drew the boundaries from memory without any particular training to know the location of the 42 boundary points learned by the neural network. The even more simplified aggregate map recalled by the students may be partially due to their inability to express their knowledge through drawings or because they never actually encoded the details of the Texas boundary.

Learning and prior knowledge

Understanding how spatial knowledge is learned from maps and stored in long-term memory is a challenging research topic. It would seem reasonable to assume that things we already know might help us learn new things more effectively. Someone who has practised reading maps and who has become familiar with the conventions used by cartographers to make maps should be able to learn more effectively from a map than an inexperienced and uninformed person.

Dochy's (1991) review of the literature suggested the term 'prior knowledge' has been defined and used many different ways by researchers. In a map reading context it might refer to declarative knowledge such as the meaning of symbols on a map or the convention that certain colours are consistently used to symbolise types of objects on maps. It might also refer to procedural knowledge such as how to estimate distances between objects or how to determine the elevation of a location. At times episodic knowledge may be important if one needs to recall information from a 1950 map to compare it with information from a 1990 map. Semantic knowledge is also important in that one must know what the category lake or the category dune means if you read it on a map. Finally, strategic knowledge can make one's map reading experience more efficient. [...]

Dochy (1988) suggested a number of theoretical reasons why prior knowledge would be an important aid for learning. Map readers with previous related experiences might be able to structure the new information in long-term memory more efficiently than map readers without such experiences. The development of connections between newly learned information and previously learned information can result in multiple and redundant retrieval paths in the cognitive structure storing the information. Access to prior knowledge during the learning of new information can reduce the load on working memory and result in more information being processed in a fixed amount of time. Prior knowledge may be used to direct attention in a more selective way so that more useful information is processes and less useful information is ignored. If prior knowledge increases the total amount if information available during the learning process, then more total information is likely to be retained. The activation of prior knowledge on a specific topic generally aids the retrieval of information from memory. [...] Prior knowledge saves encoding time because fewer propositions about the new information need to be encoded if they already exist. [...]

Animation

Animated maps have been considered as a new improved approach to map communication, but studies of this potential have been inconsistent (Patton and Cammack 1996). Cutler (1998: 20) considered whether 'children would be able to comprehend and recall more information from an animated isoline map than from traditional paper isoline maps'. [...] Results indicated that a subject's prior knowledge of isoline maps and reading level significantly affected both the times it took to answer questions and the accuracy of the answers. [...] It would appear that the characteristics of the map readers were more important in predicting successful learning than the characteristics of the map presentations. [...] Students who were reading at a higher level were able to respond faster if they had experienced the paper maps, but students who were reading at a lower level responded faster if they had experienced the map animation. This suggested that students who were reading at a lower level could gain some advantage from viewing animations. Learning from paper maps consistently resulted in more accurate test scores for students at all reading levels.

Cartograms

Breding (1998) had some seventh-grade students study maps and learn the information presented on them. [...] Results indicated that familiarity with the regions significantly affected response times but did not significantly affect accuracy. It would appear that prior knowledge of the familiar regions helped students store the information in a more accessible structure, but this had no effect on the accuracy of the learned information. Students who studied familiar graduated circle maps were able to respond faster to the test questions, but they had the lowest accuracy. One explanation of this was that the simple shape of circles made it difficult to remember which circle was in what location. Since the cartograms showed population sizes with unique shapes this type of confusion is less likely. [...]

Patton (1997) considered the effect of prior knowledge on the learning of category information from maps. [...]

When the categories were coherent the information was learned significantly faster than when the categories were incoherent. This expected result suggested that the prior knowledge about categories such as *Urban* and *Rural* made it easier to learn which map belonged in which category. When subjects were required to identify maps as *Old* maps that they had learned or *New* maps they had not seen before, they were more accurate at identifying *Old* maps if they had learned coherent categories and less accurate with incoherent categories. There was also evidence that prior knowledge caused more frequent errors when considering *New* maps. [...] The prior knowledge made it much easier to learn the maps for coherent categories, but they apparently did not know as many specific details as subjects who learned the maps for incoherent categories.

A unified theory of map reading may not easily be written because map reading involves many complex processes. The path to that theory, however, will undoubtedly involve studies on how visual information is processed, integrated with prior knowledge, and used to make judgements and decisions.

References

Board, C. (1967) Maps as models, in *Models in Geography* (eds R. Chorley and P. Haggett), Methuen, London, pp. 671–725.

Breding, P. (1998) The effect of prior knowledge on eighth graders' abilities to understand cartograms. Unpublished MA thesis. Department of Geography, University of South Carolina.

Brennan, N. and Lloyd, R. (1993) Searching for boundaries on maps: the cognitive process. *Cartography and Geographic Information System*, **20**, 222–236.

Bryant, D., Tversky, B. and Franklin, N. (1992) Internal and external spatial frameworks for representing described scenes. *Journal of Memory and Language*, **31**, 74–98.

Bunch, R. (1998) Searching for bi-polar color boundaries on maps. Unpublished Masters thesis. Department of Geography, University of South Carolina.

Cammack, R. (1995) The cognition of cartographic animation with object files. Unpublished PhD dissertation. Department of Geography, University of South Carolina.

Cammack, R. and Lloyd, R. (1993) Connected space: regional neural networks. Paper presented at the Annual Meeting of the Association of American Geographers, Atlanta, Georgia

Castner, H. (1983) Research questions and cartographic design, in *Graphic Communication and Design in Contemporary Cartography* (ed. D. Taylor), John Wiley & Sons, Ltd, Chichester, UK, pp. 87–114.

Chang, K. (1977) Visual estimation of graduated circles. *The Canadian Cartographer*, **14**, 130–138.

Cox, C. (1976) Anchor effects and the estimation of graduated circles and squares. *The American Cartographer*, **3**, 65–74.

Cutler, M. (1998) The effects of prior knowledge on children's abilities to read static and animated maps. Unpublished MA thesis. Department of Geography, University of South Carolina.

Dobson, M. (1983) Visual information processing and cartographic communication: the utility of redundant stimulus dimensions, in *Graphic Communication and Design in Contemporary Cartography* (ed. D. Taylor), John Wiley & Sons, Ltd, Chichester, UK, pp. 149–175.

Dochy, F. (1988) The prior knowledge state of students and its facilitating effect on learning. OTIC Research Report 1-2.

Dochy, F. (1991) Mapping 'prior knowledge' or 'expertise': a tentative outline. OTIC Research Report 28.

Ekman, G., Lindman, R. and William-Olsson, W. (1961) A pyschophysical study of cartographic symbols. *Perceptual and Motor Skills*, **13**, 355–368.

Ferguson, E. and Hegarty, M. (1994) Properties of cognitive maps constructed from texts. *Memory and Cognition*, **22**, 455–473.

Franklin, N., Tversky, B. and Coon, V. (1992) Switching points of view in spatial mental models. *Memory and Cognition*, **20**, 507–518.

Gilmartin, P. (1981) The interface of cognitive and psychological research in cartography. *Cartographica*, **18**, 9–20.

Hewitson, B.C. and Crane, R.G. (eds) (1994) *Neural Nets: Applications in Geography*, Kluwer Academic Publishers, Boston.

Holyoak, K. and Mah, W. (1982) Cognitive reference points in judgments of symbolic magnitude. *Cognitive Psychology*, **14**, 328–352.

Johnson-Laird, P. (1983) *Mental Models: Towards a Cognitive Science of Language, Inference, and Consciousness*, Cambridge University Press, Cambridge.

Kahneman, D. and Treisman, A. (1984) Changing views of attention and automaticity, in *Varieties of Attention* (eds R. Parasuraman and D. Davies), Academic Press, New York, pp. 29–61.

Kahneman, D., Treisman, A. and Gibbs, B. (1992) The reviewing of object files: object-specific integration of information. *Cognitive Psychology*, **24**, 175–219.

Kimmerling, J. (1975) A cartographic study of equal value gray scales for use with screened gray areas. *The American Cartographer*, **2**, 119–127.

Kohonen, T. (1989) *Self-Organization and Associative Memory*, Springer, Berlin.

Lloyd, R. (1988) Searching for map symbols: the cognitive process. *The American Cartographer*, **15**, 363–378.

Lloyd, R. (1994) Learning spatial prototypes. *Annals of the Association of American Geographers*, **84**, 418–440.

Lloyd, R. (1997a) Visual search processes used in map reading. *Cartographica*, **34**, 11–32.

Lloyd, R. (1997b) *Spatial Cognition: Geographic Environments*, Kluwer Academic, Dordrecht, The Netherlands.

Lloyd, R. (1998) Learning geographic shapes. Mimeo.

Lloyd, R. and Carbone, G. (1995) Comparing human and neural network learning of climate categories. *The Professional Geographer*, **47**, 237–250.

Lloyd, R., Cammack, R. and Holliday, W. (1995) Learning environments and switching perspectives. *Cartographica*, **32**, 5–17.

MacEachren, A. (1995) *How Maps Work: Representation, Visualization, and Design*, Guilford, New York.

McCleary, G. (1970) Beyond simple psychophysics: approaches to the understanding of map perception. *Proceedings of the American Congress of Surveying and Mapping*, pp. 189–209.

McClelland, J. and Rumelhart, D. (1989) *Explorations in Parallel Distributed Processing: A Handbook of Models, Programs, and Exercises*, MIT Press, Cambridge, MA.

McNamara, T., Halpin, J. and Hardy, J. (1992) The representation and integration in memory of spatial and nonspatial information. *Memory and Cognition*, **20**, 519–532.

Medychyj-Scott, D. and Board, C. (1991) Cognitive cartography: a new heart for a lost soul, in *Advances in Cartography* (ed. J. Mueller), Elsevier Applied Science, London, pp. 201–230.

Meihoeffer, H. (1969) The utility of the circle as an effective cartographic symbol. *The Canadian Cartographer*, **6**, 105–117.

Nelson, E. (1994) Colour detection on bivariate choropleth maps: the visual search process. *Cartographica*, **31**, 33–43.

Patton, D. (1997) The effects of prior knowledge on the learning of categories of maps. *The Professional Geographer*, **49**, 126–136.

Patton, D. and Cammack, R. (1996) An examination of the effects of task type and map complexity on sequenced and static choropleth maps, in *Cartographic Design: Theoretical and Practical Perspectives* (eds C. Wood and P. Keller), John Wiley & Sons, Ltd, Chichester, UK, pp. 237–252.

Perrig, W. and Kintsch, W. (1985) Propositional and situational representations of text. *Journal of Memory and Language*, **24**, 503–518.

Petchenik, B. (1975) Cognition in cartography. *Proceedings of the Symposium on Computer Assisted Cartography*, pp. 183–193.

Robinson, A. (1952) *The Look of Maps: An Examination of Cartographic Design*, University of Wisconsin Press, Madison, WI.

Robinson, A. (1967) Psychological aspects of colour in cartography. *The International Yearbook of Cartography*, **7**, 50–59.

Shortridge, B. (1982) Stimulus processing models from psychology: can we use them in cartography. *The American Cartographer*, **9**, 69–80.

Stevens, S. (1975) *Psychophysics: Introduction to Its Perceptual, Neural, and Social Prospects*, John Wiley & Sons, Inc., New York.

Taylor, H. and Tversky, B. (1992a) Descriptions and depictions of environments. *Memory and Cognition*, **20**, 483–496.

Taylor, H. and Tversky, B. (1992b) Spatial mental models derived from survey and route descriptions. *Journal of Memory and Language*, **31**, 261–292.

Tversky, B. (1981) Distortions in memory for maps. *Cognitive Psychology*, **13**, 407–433.

Tversky, B. (1992) Distortions in cognitive maps. *Geoforum*, **23**, 131–138.

Waddill, P. and McDaniel, M. (1992) Pictorial enhancement of text memory: limitations imposed by picture type and comprehension skill. *Memory and Cognition*, **20**, 472–482.

Williams, R. (1956) Statistical symbols for maps: their design and relative values. Yale University Map Laboratory Report, New Haven, CT.

Further reading

Blaut, J.M., Stea, D., Spencer, C. and Blades, M. (2003) Mapping as a cultural and cognitive universal. *Annals of the Association of American Geographers*, **93** (1), 165–185. [An important and wide ranging meta study reviewing the cross cultural evidence for map use as common human trait.]

Fabrikant, S.I. and Lobben, A. (2009) Cognitive issues in geographic information visualization. *Cartographica*, **44** (3). [Theme issue which includes a range of relevant articles focusing upon the application of cognitive approaches to geovizualisation a decade after Lloyd's paper.]

Liben, L.S. (2009) The road to understanding maps. *Current Directions in Psychological Science*, **18**, 310–315. [Reviews different psychological approaches to map understanding, with a much greater focus on wider contexts of use than is apparent in Lloyd's paper.]

Montello, D.R. (2002) Cognitive map design research in the twentieth century. *Cartography and Geographic Information Science*, **29** (3), 283–304. [Traces the history of cognitive research in cartography, exploring its development in the USA, with a particular focus on the aftermath of Arthur Robinson's early work.]

Van Elzakker, C.P.J.M., Delikostidis, I. and van Oosterom, P.J.M. (2008) Field-based usability evaluation methodology for mobile geo-applications. *The Cartographic Journal*, **45** (2), 139–149. [A recent exploration of real-world usability testing methods grounded in the cognitive mapping tradition.]

See also

- Chapter 1.3: On Maps and Mapping
- Chapter 1.6: Cartographic Communication
- Chapter 1.11: Exploratory Cartographic Visualisation: Advancing the Agenda
- Chapter 3.3: Cartography as a Visual Technique
- Chapter 3.6: The Roles of Maps
- Chapter 4.3: Cognitive Maps and Spatial Behaviour: Process and Products
- Chapter 4.11: Usability Evaluation of Web Mapping Sites

Chapter 4.10

Citizens as Sensors: The World of Volunteered Geography

Michael F. Goodchild

Editors' overview

In the middle of the first decade of the new century a radically new way of creating, assembling and disseminating information began to gain in popularity as part of a suite of new technological solutions to mapping. This so-called 'crowd-sourcing' mode of information production has already impacted on many facets of life, and in geography and in the context of mapping has been termed Volunteered Geographic Information (VGI). Goodchild's paper was one of the first to describe its growing significance for the social production of mapping. He defines VGI highlighting the explosion of interest in using the Web to collect, share and collaboratively map, often at a global level. Goodchild exemplifies key VGI sites and assesses the technologies that facilitate this process, notably Web 2.0, georeferencing, GPS, high resolution graphics and available visualisation software. He also draws attention to the potential economic benefits of citizen cartography in an age of declining national mapping currency, and highlights the valuable contribution of citizen networks for science and emergency response of 'bottom-up' mapping initiatives. Amongst the substantive research issues flagged by Goodchild are the motivations of people engaged with the process, the impacts on standards accuracy and the authority of the map, the uneven nature of access to these technologies and the implications for privacy.

Originally published in 2007: *GeoJournal*, **69** (4), 211–221.

Introduction

In 1507, in St-Dié-des-Vosges, Martin Waldseemüller drew an outline of a new continent and labelled it *America* [. . .]. It appears that he was influenced by new books being circulated in Europe at the time, and particularly by the Soderini Letter and its purported author, Amerigo Vespucci, and the latter's claims to the continent's discovery. Although Waldseemüller withdrew the name on a later map, and although many scholars and a new biography by Felipe Fernández-Armesto (2006) cast doubt on the authenticity of the Letter, the feminine form of Vespucci's first name stuck, and was eventually adopted as the authoritative name of not one but two continents.

By today's standards this act of naming by an obscure cartographer would attract little or no attention. Modern naming in developed countries is closely regulated by a hierarchy of committees that in the United States extend from the local to the national level (Monmonier 2006). [. . .] Geographic naming has been centralised and standardised, and assigns no role to obscure individuals like Waldseemüller, who would certainly be amazed to learn that his map was recently acquired by the US Library of Congress for $10 million.

Nevertheless, the events of 1507 provide an early echo of a remarkable phenomenon that has become evident in recent months: the widespread engagement of large numbers of private citizens, often with little in the way of formal qualifications, in the creation of geographic information, a function that for centuries has been reserved to official agencies. They are largely untrained and their actions are almost always voluntary, and the results may or may not be accurate. But collectively, they represent a dramatic innovation that will certainly have profound impacts on geographic information systems (GIS) and, more generally, on

The Map Reader: Theories of Mapping Practice and Cartographic Representation, First Edition. Edited by Martin Dodge, Rob Kitchin and Chris Perkins.

the discipline of geography and its relationship to the general public. I term this *volunteered geographic information* (VGI), a special case of the more general Web phenomenon of *user generated content*, and it is the subject of this paper.

The evolving world of VGI

One of the more compelling examples of VGI is Wikimapia, which adapts some of the procedures that have been so successful in the creation of the Wikipedia encyclopaedia and applies them to the creation of a gazetteer. Anyone with an Internet connection can select an area on the Earth's surface and provide it with a description, including links to other sources. Anyone can edit entries, and volunteer reviewers monitor the results, checking for accuracy and significance. At time of writing Wikimapia had 4.8 million entries compared to Wikipedia's 7 million, describing features ranging in size from entire cities to individual buildings (each entry's geographic extent is defined by ranges of latitude and longitude). Some descriptions are extensive and include hyperlinks; for example, the entry for Madinah (Saudi Arabia) includes a picture of the Masjid-e-Nabawi and a link to the city's Wikipedia entry. Other entries describe features within the city or in the surrounding area.

Similar in some respects is the Flickr site, which allows users to upload and locate photographs on the Earth's surface by latitude and longitude. At the time of writing roughly 2.8 million photographs were being contributed each month to the site. Figure 4.10.1 shows one of the more than 2500 volunteered photographs of Uluru (Ayer's Rock) in central Australia.

At a rather different level of sophistication is MissPronouncer, a site created by Jackie Johnson to help

Figure 4.10.1 Information from the Flickr site for the area of Uluru (Ayer's Rock) in central Australia. Each symbol denotes the availability of a photograph; at the time of writing there were more than 2500 available for the area shown. Descriptive information is displayed for one such entry.

people pronounce some of the more distinctive Wisconsin place names. A full-time radio broadcaster, Ms Johnson developed the site in her spare time, and offers audio recordings of the correct pronunciation of almost 2000 places in the state. Phonic representations of place names have the advantage that they are not subject to problems over differences of alphabet (Beijing versus 北京, Baghdad versus بغداد) though the phonic rendering of common place names may vary from one language to another (e.g. Paris, Moscow).

Other VGI activities focus on the creation of more elaborate representations of the Earth's surface. OpenStreetMap is an international effort to create a free source of map data through volunteer effort. Figure 4.10.2 shows the map for part of Dublin at the time of writing. Note the incomplete nature of the map, with major streets, railways, and parks shown but with minor street detail in some areas but not others, and some streets named but not others. Dublin famously lacks a cheap, readily available digital street map, as do many other cities around the world, so this volunteer effort can potentially fill a yawning gap in the availability of digital geographic information.

When Google acquired the software previously known as Earthviewer, rebranded it, redesigned the user interface, and published an application program interface, it created a service that had immediate appeal to millions. I have described the Google Earth phenomenon as the 'democratisation of GIS' (Butler 2006), because it has opened some of the more straightforward capabilities of GIS to the general public. Whereas the creation of a 'fly-by' was previously one of the more sophisticated GIS tasks, it is now possible for a child of ten to create one in ten minutes. Google Earth and Google Maps popularised the term *mash-up*, the ability to superimpose geographic information from sources distributed over the Web, many of them created by amateurs. For example, Figure 4.10.3 shows a Google Earth mash-up of the Soho area of London during the 1854 cholera outbreak made famous by Dr John Snow (Johnson 2006). It combines a

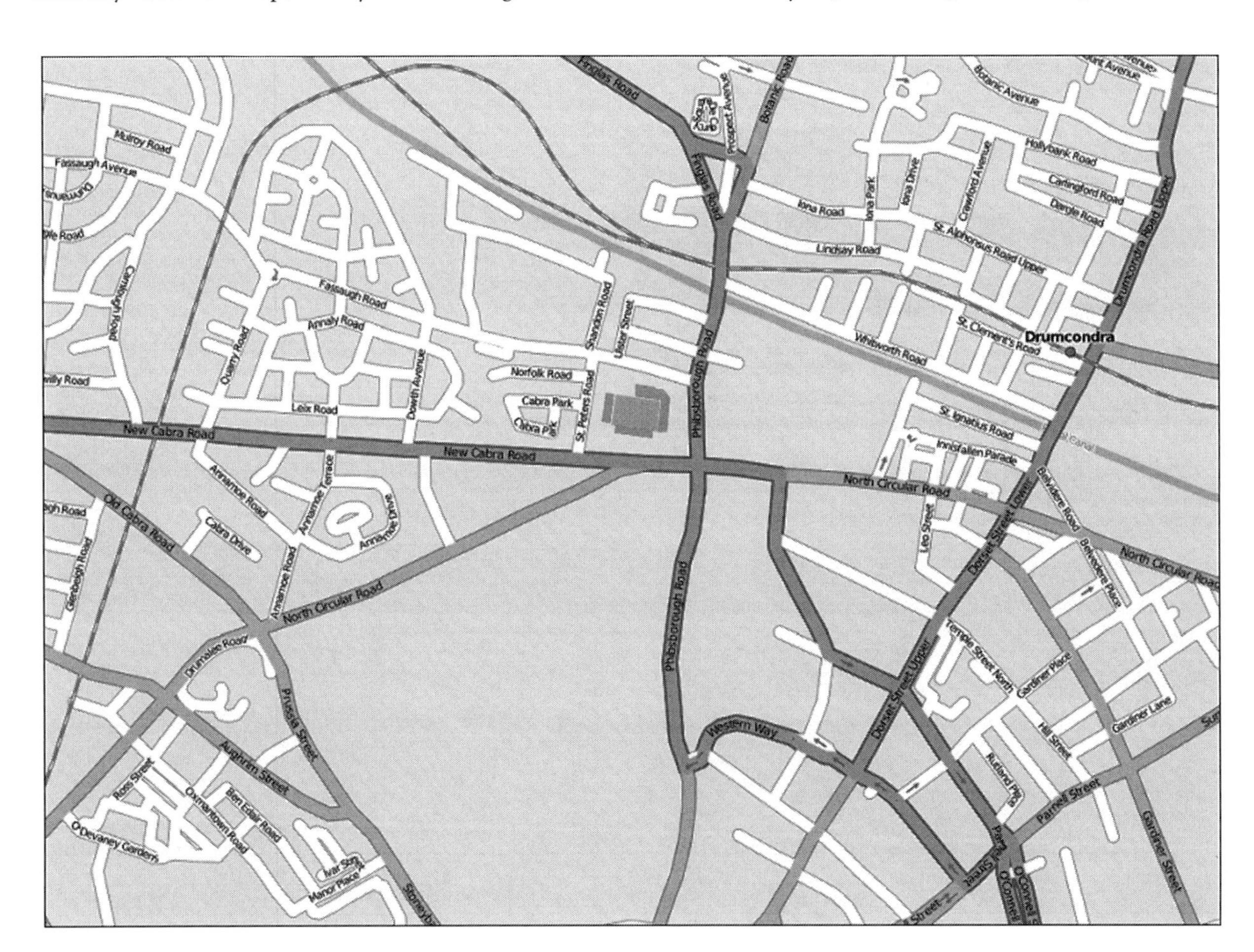

Figure 4.10.2 Part of the OpenStreetMap coverage of Dublin, Ireland. Note the missing street names and areas where no coverage is available.

street map of London from 1843 (from the online private collection of David Rumsey, a San Francisco map collector) with online data on the water sources and cholera deaths from my own Web site. Readily available software makes this kind of mash-up remarkably easy (Brown 2006) and well within the capabilities of the general public. As a result, the number of available mash-ups has reportedly reached the hundreds of thousands, and the number of downloads of the Google Earth software exceeds a 100 million.

These are just a few examples of a phenomenon that has taken the world of geographic information by storm and has the potential to redefine the traditional roles of mapping agencies and companies. In the next section I examine some of the technologies that have combined to make this possible. This is then followed by a discussion of relevant concepts and issues, and then by an analysis of the usefulness of VGI.

Enabling technologies

Web 2.0

To understand VGI, we must first ask about the technologies that make it possible. Early concepts of the Web stressed the ability of users to access remote sites through simple interfaces known as *browsers* (Mosaic, launched in 1992, was the first widely available browser). One could *surf* the Web by following *hyperlinks*, typically highlighted words that when *clicked* would initiate a download from another page or site. Web pages consisted primarily of text, but graphic images could also be included, taking advantage of the recently expanded graphics capabilities of personal computers. In all of this, however, the relationship between user (*client*) and Web page (located on a *server*)

Figure 4.10.3 A Google Earth mash-up of the area of Soho, London. The contemporary imagery base has been obscured by an 1843 map from the David Rumsey collection. Superimposed on this are the deaths (green) from cholera in the outbreak of 1854, and the water sources (red). Dr John Snow showed that the central source was responsible for the outbreak.

was essentially one-way; the user's only role was to initiate the downloading of content.

In time it became possible for the user's role to extend somewhat. Protocols were developed that allowed users to access information stored in a server's databases, and even to add records to such databases by completing forms. Airline reservation sites (e.g. Expedia), eBay and Craig's List all exploit this capability. By the early 2000s this ability of users to supply content to Web sites had grown in sophistication to the point where it became possible to construct sites that were almost entirely populated by user-generated content, with very little moderation or control by the site's owners and very little restriction on the nature of content. In some cases users could even edit the content created by others. *Blogs* and *Wikis* fall into this category, as do the sites reviewed in the previous section. Collectively, they have been termed *Web 2.0*. First and foremost, then, VGI is a result of the growing range of interactions enabled by the evolving Web.

Georeferencing

GIS relies on the ability to specify location on the Earth's surface using a small number of well defined and interoperable systems, of which latitude and longitude are by far the most universal. Most countries have some form of national grid that provides an alternative local coordinate system, and the Universal Transverse Mercator (UTM) system has been adopted for the geographic coordinates needed by many military agencies. All of these are specialised, however, and in normal human discourse it is place names that provide the basis of geographic referencing. Very few people know the latitude and longitude of their home, let alone its UTM coordinates. To enable the creation of geographic data by the general public, therefore, it is necessary to have a range of readily available tools for identifying the coordinates of locations on the Earth's surface.

Several tools now supply this need, and collectively enable VGI. The Global Positioning System (GPS) can be accessed by a wide range of consumer products, allowing location to be measured in many standard coordinate systems. Cameras can be enabled with GPS, so that digital photographs can be automatically tagged with coordinates. Some GPS receivers store entire tracks that can later be uploaded in digital form, and similar capabilities can be built into mobile phones. Coordinates can also be obtained through a process known as *geocoding*. Any recognised street address can be matched to a digital street file in a service available in most GIS software as well as on the Web.

A technically simpler option is to use the imagery available through Google Earth, Google Maps or similar services to select a location visually, and to record its coordinates by clicking. Several services allow this approach to be used to create digital records of entire streets and other features by following (*digitising*) the features on the screen; the results are then uploaded and compiled into composite digital maps. OpenStreetMap has already been cited as an example of this approach.

Geotags

A *geotag* is a standardised code that can be inserted into information in order to note its appropriate geographic location. Geotags have been inserted into many Wikipedia entries, when the contents relate to a specific location on the Earth's surface, and several sites allow such entries to be accessed from maps. [...] At the time of writing there were over 60 000 geotagged entries in the Wikipedia French-language resource alone. [...]

GPS

The Global Positioning System is arguably the first system in human history to allow direct measurement of position on the Earth's surface. GPS receivers are easy to use and provide virtually instantaneous estimates of location, often to better than 10 m accuracy. Incorporated in in-car navigation systems, GPS allows the current location of the vehicle to be compared to the contents of a digital street map. As a stand-alone device, a receiver is the basis of the popular sport of *geocaching*, which engages participants in finding hidden destinations based only on their coordinates. GPS has sparked a number of interesting VGI activities, such as the creation of maps by walking, cycling or driving. [...]

Graphics

It is easy to forget that high quality graphics are a comparatively recent innovation in the history of computing. Dynamic visualisation of three-dimensional objects, such as occurs with Google Earth, required a highly sophisticated and expensive computer as recently as 1995, and when Earthviewer appeared in 2000 only a few personal computers had the powerful graphics hardware needed to run it. Today, of course, lowly household computers have sufficient power, though devices built for video games, such as Wii, often have even greater power.

Broadband communication

Finally, VGI would be impossible without widespread access to the Internet, preferably via a high capacity connection. Many households in developed countries now have such broadband connections, using a range of satellite, cable and phone-line technologies.

Concepts

Spatial data infrastructure patchworks

It is easy to believe that the world is well mapped. Most countries have national mapping agencies that produce and update cartographic representations of their surfaces, and remote sensing satellites provide regularly updated images. But in reality world mapping has been in decline for several decades (Estes and Mooneyhan 1994). The US Geological Survey no longer attempts to update its maps on a regular basis, and many developing countries no longer sustain national mapping enterprises. The decline of mapping has many causes (Goodchild *et al.* 2007). Governments are no longer willing to pay the increasing costs of mapping, and often look to map users as sources of income. Remote sensing has replaced mapping for many purposes, but satellites are unable to sense many of the phenomena traditionally represented on maps, including the names of places. In the early 1990s, the Mapping Science Committee of the US National Research Council issued a report describing the concept of spatial data infrastructure (NRC 1993), which it defined as the aggregate of agencies, technologies, people and data that together constituted a nation's mapping enterprise.

Among the many concepts introduced in the report was that of *patchwork*, the notion that national mapping agencies should no longer attempt to provide uniform coverage of the entire extent of the country, but instead should provide the standards and protocols under which numerous groups and individuals might create a composite coverage that would vary in scale and currency depending on need. The creation of the National Spatial Data Infrastructure (NSDI) was authorised by President Clinton under Executive Order 12906 in 1994, and has provided the policy umbrella for geographic information in the United States for the past 13 years.

VGI clearly fits the model of NSDI. A collection of individuals acting independently, and responding to the needs of local communities, can together create a patchwork coverage. Given a server with appropriate tools, the various pieces of the patchwork can be fitted together, removing any obvious inconsistencies, and distributed over the Web. The accuracy of each piece of the patchwork, and the frequency with which it is updated, can be determined by local need.

Humans as sensors

Recently a great deal of attention has been devoted to the concept of sensor networks. The observational objectives of Earth science, as well as the objectives of security and surveillance, can be addressed at least in part by the installation of networks of sensors across the geographic landscape. Commonly cited examples include the network of video monitors in many major cities, proposals to instrument the ocean and seabed with sensors in the interests of science and early warning of tsunamis, and networks of traffic sensors that can provide useful information to planners, as well as real-time pictures of congestion.

It is useful to distinguish three types of sensor networks. Most examples fit the first, a network of static, inert sensors designed to capture specific measurements of their local environments. Less commonly cited are sensors carried by humans, vehicles or animals. For example, much useful research is emerging from projects that have equipped children with sensors of air pollution, in an effort to understand the factors affecting asthma. A third type of sensor network, and in many ways the most interesting, consists of humans themselves, each equipped with some working subset of the five senses and with the intelligence to compile and interpret what they sense, and each free to rove the surface of the planet.

This network of human sensors has over six billion components, each an intelligent synthesiser and interpreter of local information. One can see VGI as an effective use of this network, enabled by Web 2.0 and the technology of broadband communication.

Citizen science

The term *citizen science* is often used to describe communities or networks of citizens who act as observers in some domain of science. A perfect United States example is the Christmas Bird Count, an effort to enlist amateur ornithologists in conducting a mid-winter census of bird populations. Participants require a fairly high level of skill, and over the years a number of protocols have been established to ensure that the resulting data have high quality. An international example is Project GLOBE, an effort to enlist school children and their teachers in providing a worldwide source of high quality atmospheric observations. As with the Christmas Bird Count, a number of protocols and training programs have been established to ensure quality, and to collect, synthesise and re-distribute the results.

Both of these projects require a fair degree of training and expertise. This need for expertise would be a limiting factor in any effort to extend VGI to such comparatively sophisticated mapping themes as land use, land cover or soil class. Other forms of VGI are much less demanding, however, particularly those associated with place names, streets and other well defined geographic features.

Participant populations

Sites such as Wikimapia are open to all, as are many other VGI efforts. The Christmas Bird Count and Project GLOBE, on the other hand, place restrictions on participation in order to ensure adequate expertise. The question of *who* may volunteer has much to do with the quality of the resulting information, and a range of possibilities exist. For many years companies producing digital street maps have relied on networks of local observers to provide rapid notice of new streets, changes of street names and so on, paying them as part-time workers. Inrix is collecting tracks from hundreds of thousands of trucks and other fleets, processing and compiling the results as a source of real-time information on the state of congestion and other short-term factors affecting travel on road networks. Military personnel are important potential sources of geographic information about local battlefield conditions that can be used to augment what is available from central mapping and imagery sources. Many farmers now have elaborate systems for mapping and monitoring their fields and crops (*precision agriculture*), and constitute a potential source of data that is in many cases much more detailed and current than that available from central agricultural agencies. In essence, such developments contribute to a growing reversal of the traditional *top-down* approach to the creation and dissemination of geographic information.

Early warning

Recent events, such as the Indian Ocean tsunami or Hurricane Katrina, have drawn attention to the importance of geographic information in all aspects of emergency management, and to the problems that arise in the immediate aftermath of the event before adequate overhead imagery becomes available for damage assessment and response planning (NRC 2007). Earth-observing satellites may not pass over the affected area for several days. Images from satellites and aircraft may be obscured by clouds and smoke. Conditions on the ground may prevent the rapid downloading of digital imagery because of a lack of power, Internet connections or computer hardware and software.

On the other hand the human population in the affected area is intelligent, familiar with the area, and increasingly able to report conditions through mobile phones, using voice, text or pictures. To date there has been very little use of VGI in these situations, in part because of an almost complete lack of the tools needed to collect, synthesise, verify and redistribute the information. However, the potential to obtain almost immediate reports from geographically distributed observers on the ground will surely drive increased efforts to overcome these problems in the next few years.

Issues

Why do people do this?

In the mid 1990s the US Federal Geographic Data Committee published its Content Standards for Digital Geospatial Metadata, a format for the description of geographic data sets. The project was very timely, given the rapid increase in the availability of geographic information via the Internet that occurred at that time. Metadata were seen as the key to effective processes of search, evaluation and use of geographic information. Nevertheless, and despite numerous efforts and inducements, it remains very difficult to persuade those responsible for creating geographic data sets to provide adequate documentation. Even such a popular service as Google Earth has no way of informing its users of the quality of its various data layers, and it is virtually impossible to determine the date when any part of its image base was obtained. A recent news report concerned the apparent replacement of its coverage of New Orleans with pre-Katrina imagery, though its coverage of the Darfur region is updated almost daily.

Given this evident reluctance to provide documentation, it is perhaps surprising that the opportunity to create and publish VGI has engaged the interests of so many individuals. Why is it that citizens who have no obvious incentive are nevertheless willing to spend large amounts of time creating the content of VGI sites? What kinds of people are more likely to participate, and what drives them to be accurate (or inaccurate)?

Self-promotion is clearly an important motivator of Internet activity, and in its extreme form can lead to the exhibitionism of personal web-cams. Despite the vast resources of the Web, it is still possible to believe that *someone* will be interested in ones personal site. The popularity of some blogs can be misread as suggesting that an audience exists for *any* blog.

At a different level many users volunteer information to Web 2.0 sites as a convenient way of making it available to friends and relations, irrespective of the fact that it becomes available to all. This may underlie the popularity of sites such as Picasa, which allow contributors of personal photographs to point others to them, but it scarcely explains the popularity of Flickr or Wikimapia, where content is comparatively anonymous. Contributors to OpenStreetMap may derive a certain personal satisfaction from seeing their own contributions appear in the patchwork, and from watching the patchwork grow in coverage and detail, but there can be no question of self-promotion in this essentially anonymous project.

Authority and assertion

The traditional mapping agencies have elaborate standards and specifications to govern the production of geographic information, and employ cartographers with documented qualifications. Over the years their products have acquired an authority that derives from each agency's reputation for quality. Google, on the other hand, has no such reputation in the geographic domain. Nevertheless, users appear willing to ascribe authority to its products, perhaps because computerisation carries authority *per se* and perhaps because of the company's success in other areas, particularly its search engine.

At time of writing Google Earth's imagery over the campus of the University of California, Santa Barbara, CA, was mis-registered by approximately 20 m east–west. Further to the east in the City of Santa Barbara, the mis-registration was approximately 40 m east–west in the opposite direction, and a swath approximately 60 m wide running north–south was missing from the coverage. Any locations georeferenced from this imagery and incorporated into VGI will inherit these positional errors, and if Google re-registers the imagery at a future date that VGI will be clearly misplaced. In essence, Google has created a new *datum* or horizontal reference system that is substantially different from the current North American datum, but which is widely accepted because of the authority of Google. The shift is comparable in magnitude to that created when North American mapping agencies replaced the North American Datum of 1927 (NAD27) with the current NAD83.

VGI is sometimes termed *asserted* geographic information, in that its content is asserted by its creator without citation, reference or other authority. The early days of the Internet were characterised by a certain altruism, a belief in the essential goodness of users, and there was little anticipation of the subversive phenomena of spam, virusess, and denial-of-service attacks that now pervade the network. Similarly, many VGI efforts are driven by the kinds of altruism inherent in any kind of voluntary community effort. Can we expect, then, a similar pattern of disillusionment as antisocial elements recognise and exploit the inevitable vulnerabilities? Will there be efforts to create fictitious landscapes, or to attack and bring down VGI servers? VGI is currently a somewhat exotic domain, but if and when users begin to rely on its services a growing pattern of efforts to undermine it seems inevitable.

The digital divide

Despite the apparent openness of VGI, it remains largely the preserve of those fortunate to have access to the Internet – and broadband access in particular. While a growing fraction of citizens in developed countries have such access, it is largely unavailable to the majority of the world's population who live in developing countries. Moreover, issues of language and alphabet also affect access even for those with broadband connections, since many VGI servers support only the Roman alphabet and English. In principle, much could be achieved through mobile phones, which often have the ability to connect to the Internet and to capture images, but the tools needed to exploit this limited environment as a source for VGI do not yet exist. So while I argued above that such limited tools were potentially significant in early warning and emergency management, significant work still needs to be done to realise the potential.

The value of VGI

As I hope the examples in this paper illustrate, VGI has the potential to be a significant source of geographers' understanding of the surface of the Earth. It can be timely, a property that was particularly stressed in the discussion of early warning. By motivating individuals to act voluntarily, it is far cheaper than any alternative, and its products are almost invariably available to all (but see the earlier discussion of the digital divide).

In earlier sections I discussed why people might be motivated to create VGI, but not why they might want to use it. With sites such as Wikimapia one can learn a great deal about remote places, acquiring the kinds of information needed for planned tourist visits, or to provide background to travelogs. Sites such as OpenStreetMap often provide the cheapest source of geographic information, and sometimes the *only* source, particularly in areas where access to geographic information is regarded as an issue of national security.

It is already clear in many fields that such informal sources as blogs and VGI can act as very useful sources of military and commercial intelligence. The tools already exist to scan Web text searching for references to geographic places, and to geocode the results. Thus the most important value of VGI may lie in what it can tell about local activities in various geographic locations that go unnoticed by the world's media, and about life at a local level. It is in that area that VGI may offer the most interesting, lasting and compelling value to geographers.

References

Brown, M.C. (2006) *Hacking Google Maps and Google Earth*, John Wiley & Sons, Inc., New York.

Butler, D. (2006) Virtual globes: the web-wide world. *Nature*, **439**, 776–778.

Estes, J.E. and Mooneyhan, W. (1994) Of maps and myths. *Photogrammetric Engineering & Remote Sensing*, **60**, 517–524.

Fernández-Armesto, F. (2006) *Amerigo: The Man Who Gave His Name to America*, Weidenfeld and Nicolson, London.

Goodchild, M.F., Fu, P. and Rich, P. (2007) Sharing geographic information: an assessment of the geospatial one-stop. *Annals of the Association of American Geographers*, **97** (2), 249–265.

Johnson, S. (2006) *The Ghost Map: The Story of London's Most Terrifying Epidemic, and How it Changed Science, Cities, and the Modern World*, Riverhead, New York.

Monmonier, M. (2006) *From Squaw Tit to Whorehouse Meadow*, University of Chicago Press, Chicago.

NRC (National Research Council) (1993) *Toward a Coordinated Spatial Data Infrastructure for the Nation*, National Academies Press, Washington, DC.

NRC (National Research Council) (2007) *Successful Response Starts with a Map: Improving Geospatial Support for Disaster Management*, National Academies Press, Washington, DC.

Further reading

Elwood, S. (2008) Volunteered geographic information: future research directions motivated by critical, participatory, and feminist GIS. *GeoJournal*, **72** (3/4), 173–183. [An insightful discussion and development of the VGI concept.]

Haklay, M. (2010) How good is volunteered geographical information? A comparative study of OpenStreetMap and Ordnance Survey datasets. *Environment and Planning B: Planning and Design*, **37** (4), 682–703. [One of the first rigorous comparisons of data quality between crowd-sourced map data and professional digital cartography from established state mapping agency.]

Haklay, M., Singleton, A.D. and Parker, C. (2008) Web mapping 2.0: the neogeography of the geoweb. *Geography Compass*, **2** (6), 2011–2039. [A useful overview discussion of many dimensions of 'crowd-sourced' mapping.]

Liu, S.B. and Palen, L. (2010) The new cartographers: crisis map mashups and the emergence of neogeographic practice. *Cartography and Geographic Information Science*, **37** (1), 69–90. [Examines novel 'crowd-sourcing' practices in the context of emergency planning where participatory mapping are blended with professional expertise.]

Surowiecki, J. (2004) *The Wisdom of Crowds*, Little Brown, London. [One of the earliest social analyses of the implications of 'crowd-sourcing'.]

See also

- Chapter 1.14: Rethinking Maps
- Chapter 2.11: Extending the Map Metaphor Using Web Delivered Multimedia
- Chapter 2.12: Imaging the World: The State of Online Mapping
- Chapter 3.12: The Geographic Beauty of a Photographic Archive
- Chapter 4.11: Usability Evaluation of Web Mapping Sites
- Chapter 5.9: Affecting Geospatial Technologies: Toward a Feminist Politics of Emotion
- Chapter 5.11: Mapping the Digital Empire: Google Earth and the Process of Postmodern Cartography

Chapter 4.11

Usability Evaluation of Web Mapping Sites

Annu-Maaria Nivala, Stephen Brewster and L. Tiina Sarjakoski

Editors' overview

Shifts towards online geovisualisation entail radically different map designs, with much more control in the hands of the user. These present significant challenges for the largely commercial designers of web mapping sites. Nivala, Brewster and Sarjakoski's paper offers an empirical illustration of the potential of usability engineering as an evaluative tool to test how fit to purpose these sites might be. They deploy multiple methods, including think aloud protocols and video, to investigate user responses to the four leading mapping portal interfaces in 2006 (Google Maps, MSN, Multimap and MapQuest) and discover a strikingly significant inadequate consideration of user needs in designs. Searching, user interfaces, cartographic design and map tools were found wanting in expert and novice user tests in relation to specified assessment criteria. The paper concludes with design recommendations for this kind of mapping site.

Originally published in 2008: *The Cartographic Journal*, **45** (2), 129–138.

Introduction

Recent technological developments have provided new tools and techniques for designing interfaces and interacting with Websites. Many people use these sites for locating places and businesses, and for planning visits to unfamiliar places. Figures gathered from Web mapping sites' own Web pages give an indication of their popularity; one site states that it has over 40 million unique visitors each month (MapQuest 2007), while another maintains a unique user base of over 10 million, ranking consistently in the top 10 Websites by traffic in the United Kingdom (Multimap 2007). Web maps are often freely available and not only provide the map, but different map tools and map-related services.

However, the use of Web maps is not always straightforward. [...] Traditional map design and evaluation methods may no longer always be valid. Koua and Kraak (2004) crystallised this problem by stating that map use studies that have long been carried out in the field of cartography are not fully compatible with new interactive visualisations, which can have new representational spaces and user interfaces. So how can it be guaranteed that today's maps using different (new) technologies will fulfil user requirements?

Usability engineering – a term used to describe methods for analysing and enhancing the usability of software – is an approach to help design products that take into account the new technical environments and user requirements (Nielsen 1993). Usability is defined in the ISO 9241 standard as 'the effectiveness, efficiency and satisfaction with which specified users achieve specified goals in particular environments' (ISO 1997). The ISO 13407 standard gives instructions to achieve user needs by utilising the User-Centred Design (UCD) approach throughout the entire life cycle of a system (ISO 1999). Making systems more usable may have noticeable benefits for users by guaranteeing easy-to-use systems. [...]

The Map Reader: Theories of Mapping Practice and Cartographic Representation, First Edition. Edited by Martin Dodge, Rob Kitchin and Chris Perkins.

A user-centred design can provide financial benefits for the system developer in reduced production costs, reduced support costs, reduced costs in use, and improved product quality (Earthy 1996).

Several researchers have observed the lack of thorough usability engineering in cartographic visualisation and geo-visualisation, for instance, MacEachren and Kraak (2001), Fuhrmann *et al.* (2005), van Elzakker (2005) and Nivala *et al.* (2007). The aim of the present study was to identify potential usability problems of Web mapping sites in order to provide guidance for the future design of such services.

Previous usability evaluations of on-screen maps

Previous research on the usability of Web mapping sites seems to be rare. However, several usability evaluations have been carried out in relation to other on-screen maps.

[...]

A usability evaluation of Web maps, similar to the study presented here, was carried out by Skarlatidou and Haklay (2006), who arranged workshops for assessing the usability of seven public Web mapping sites. In their method, users carried out six to seven tasks with the sites. Qualitative data was gathered through the 'thinking aloud protocol' and questionnaires and quantitative data by measuring the total time each user was performing each task, as well as the total number of clicks. Through measuring the users' performance, Skarlatidou and Haklay drew conclusions on which sites were the most and least usable and discussed the qualitative findings of their evaluation.

Method

The aim of this study was to identify potential usability problems with Web mapping sites and gather qualitative information to suggest guidelines for the design of future sites. Four different Web mapping sites were evaluated in this study: Google Maps (abbreviated in this paper as GM, available at http://maps.google.com), MSN Maps & Directions (MD, http://maps.msn.com), MapQuest (MQ, http://www.mapquest.com) and Multimap (MM, http://www.multimap.com). [...] They all consisted of an interactive 2D map application with zooming and panning options. Additionally, users were able to search for different locations and directions for routes.

Procedure

Several experiments were carried out in order to identify as many potential usability problems [...] as possible. Firstly, a typical scenario for [...] was drawn up: 'A tourist is planning to visit London and uses a Web mapping site for planning the trip beforehand'. Part of the evaluation was conducted as a series of user tests (with eight 'general' users), with the other part involving the evaluation of the maps by experts (eight cartographers plus eight usability engineers). Altogether, [...] 32 different evaluations were carried out. [...] The experiments were run in a Windows environment using either desktop or laptop PCs. Evaluations were carried out from August–September 2006 and the results presented here are based on the content of the Web mapping sites at that time.

User tests

[...] Eight test users were involved in the evaluation (five males, three females), with ages ranging from 19 to 35. With the exception of one person, all users had previous experience of using several different types of maps [...]. All of the users regarded their map reading skills to be fairly good or excellent.

The use scenario was described to the users at the beginning of the test. Following this, the test instructor gave the users one pre-defined task at a time, which they would try to complete by using the Web map (Table 4.11.1). The participants were given a Web map site that they had not used before. The users were then encouraged to 'think aloud' and describe the reasoning behind their actions. During the tests, the computer screens were recorded with a video camera to support subsequent data analysis.

Expert evaluations

Sixteen experts (eight cartographers and eight usability engineers) were involved in the evaluation (eight males, eight females), with ages ranging from 23 to 45. The term 'expert' here means a postgraduate student in cartography or usability engineering or a person who has already worked as a cartographer or usability specialist. The expert evaluators were given the use scenario and a list of typical user tasks (Table 4.11.1) and asked to go through the Web maps

Table 4.11.1 Usability evaluation tasks

	Task Description
1	You are planning to visit London during a weekend. Identify the most ideal location for a hotel by using the map site. Describe the reasons behind your choice.
2	Show the same place you chose during the previous task (the screen view returned back to the start page).
3	Find Roupell Street in London and point it out on the map to the test instructor.
4	Find the most northerly street in London with 'smith' included in its name.
5	What is the distance between Buckingham Palace and Piccadilly Circus?
6	Show the route that you would use if you were to walk from Sumner Road to Gresham Street.
7	Find London Bridge.

carefully, and [...] write down all the problems they encounter [...]. The experts were asked to list all the usability problems found [...].

Analysis

The video data from the user tests were analysed by writing down everything that the users had problems with and/or commented as a problem in some way. The same was done with the expert evaluations and all the negative findings were picked up from the evaluation reports. [...] Usability problems were grouped under four different categories (1–4) according to the severity of the problem (categories modified from Nielsen 1993) [...]. To make the rating more objective, a conductor of the experiment judged the severity of each problem together with one cartographer and one usability expert.

Results

Altogether, 403 usability problems were found with different evaluation methods. In total, 343 unique problems were identified: 69 in Google Maps, 83 in MSN Maps & Directions, 92 in MapQuest and 99 in Multimap (Table 4.11.2).

Severity of the problems

In total, 33 catastrophic problems were identified (severity category 1), in addition to 138 other major problems (category 2), 127 minor problems (category 3) and 44 cosmetic problems (category 4). From GM only one catastrophic problem was found, whereas MD and MM generated the same number of the most serious problems (13). GM also had the smallest amount of major problems (21).

Usability problems and design guideline suggestions

[...] Preliminary suggestions for design guidelines are given at the end of each category. While some of the guidelines may be 'self-evident' among map designers, the fact that problems emerged during this evaluation suggests that some aspects were not as evident for the designers of these specific Web mapping sites.

The user interface

First impressions are important when entering Web mapping sites. Despite this, there were a lot of problems relating to 'start pages' and user interfaces (UIs).

Layout

In many cases, the home pages of the Web maps appeared to be overloaded with different types of information (advertisements, links, images), and the users commented that these looked messy and prevented them from finding relevant information (MD, MQ, MM). Some of the links (e.g. school and insurance sites) were considered irrelevant for ordinary tourist users (MQ, MM). Users were also confused about inactive image links that did not seem to have any purpose (MD). Distractive animations were considered very annoying (MM). Some home pages were criticised for not indicating that they actually were about maps at all, that is there was no image or preview of a map (MD, MQ). [...]

The overall layout of the UI was also criticised. For instance, the search box was considered too small and its location wrong because it was not in the centre of the screen (MQ). Some users did not notice the search box during the first 5–10 minutes of trying to find something on the map (GM, MM). The grouping of the map tools, search boxes and general UI tools was also criticised because function buttons were distributed all over the screen (MM) and advertisements placed disturbingly between some of the function buttons and the map window (MM).

Functionality

Where links in the UI opened in the same browser window as the map (MM), which was then easily lost, this was considered to be a problem. In the home page of one Web map there were three different maps that looked like links, but when the user clicked on the UI nothing happened, and

Table 4.11.2 The number of usability problems found from different evaluation methods

	User tests	Cartographic experts	Usability experts	No. of problems	No. of unique problems
Google Maps (GM)	38	17	25	80	69
MSN Maps and Directions (MD)	57	21	18	96	83
MapQuest (MQ)	50	26	32	108	92
Multimap (MM)	71	32	16	119	99
Total	*216*	*96*	*91*	*403*	*343*

there was no clear feedback as to why not (MM). In some cases there was no quick way back to the home page [...] (MD). With one Web mapping site, taking the bigger map view caused all the other functions to be pushed away from the screen [...] (MD). From the analysis of feedback from this part of the experiment, several design guidelines are recommended.

Design guideline 1: the user interface

Layout

- The home page should be clear and simple.
- Intuition is important; the user should be able to start using the map immediately when entering the page.
- There should be a modest number of adverts and animations and these should be located in such a way that they do not disturb the user.
- Information presented on the UIs should be placed logically; attention should be paid to the grouping of the various tools.
- The search box should be given a principal role in the layout.

Functionality

- Links in the UI should not be opened in the same browser window as the map.
- There should always be a short cut back to the home page.

The map

In sites where the actual map took quite a small amount of the space on the Web page (MD, MM), this was criticised, as it made it difficult to get an overall picture of a location. [...]

Map visualisation

Maps were criticised for looking like they were designed to appear as paper maps instead of Web maps, because their visualisation was messy, confusing, restless and awful to look at on a computer screen (MD, MM). Some maps were regarded as being quite sketch-like (MM), old fashioned (MD), or the map projection looked weird to the participants (MM).

The use of colours was also criticised. For instance, the background colour of built-up areas was considered unreadable and the text was not optimal in contrast (MD); similar colours made it difficult to distinguish between shopping areas and hospitals (MQ); and some colours were considered to be unintuitive (MQ: built-up area). [...]

Some maps were overloaded with information and/or colours at certain scales (GM: overview map of London; MM), while others just looked unpleasant [...] (MQ). The categorisation of streets for different scales was also thought inappropriate in some cases (MQ). On the other end of the scale were comments that the map was too general (MD, MQ) and, for instance, that the information on the overview map was not sufficient to support decision making because street names and so on would be needed (MM, MD [...]). There were also problems with the text on the maps; the placement of the text was poor (GM, MD), the text was not legible (GM: hybrid map; MD), or the font size was too small for a Web map (MM). Some street and district names were also messy at the biggest scale, so the user could not read them (MD).

Some symbols caused problems because they 'stood out' in relation to the other symbols, especially if it was not clear what they were and why they were emphasised (MD, MQ, MM). [...]. On a small-scale map, the names of towns and so on looked like links from users' point of view, but did not work as such (GM, MM). Many symbols were also misinterpreted (red squares on MM; train tracks on MD).

Some maps also generated problems at different zoom levels. For instance, some symbols (MD) or text (MQ) appeared and disappeared randomly with different scales; the step between map scales was too large (MQ); and in many cases, the visualisation between different scales was distinctively different (MD, MQ, MM). [...]

The information included in the maps was criticised as being insufficient, especially regarding public transportation (railway stations, airports, timetables etc.) and different types of tourist attraction, points of interest and landmarks (GM, MD, MQ, MM). In terms of completeness, the data were considered to be inconsistent; some airports and hotels were shown on the map from a specific location, while others were not (GM, MM). [...] Data accuracy seemed to be insufficient [...] (MM). Sometimes it was impossible to find information about where the map data was from and when was it gathered (MQ).

Map tools

There was either no legend for the maps, or the participants were not able to find it (GM, MQ, MM, MD). Some users had problems realising that they could actually perform searches on the map (GM, MM). Estimating distances was also difficult, mainly because some of the users did not realise that there was a scale bar (MD, MM). One scale bar only showed miles, while some of the users only understood the metric system (MM). It was also criticised that the scale bar could only be used for a rough estimation of distance (MD, MM). Some users wanted a grid in the map for comparing different locations and estimating the distances between them (GM, MQ).

Mistakes in design were also observed. For example, the map-size buttons did not work if a route was shown on the screen, although they appeared to be active buttons (MD). At times, parts of the map were covered by zoom buttons and scale bars (GM). The scale bar was also considered to disappear on the map window because it was so tiny (MQ, MM) and/or poorly designed (MM). Some participants criticised the lack of an option to [. . .] show or hide different data layers or symbols on the map (GM, MD), especially because some of the maps were overloaded with so many different objects (MM). An option to highlight various classes of object [. . .] was also called for (MQ).

In addition, a link to print the map was missing (MQ, MM), as was an option to save a search [. . .] (MQ). A route direction tool [. . .] was also required (MM), as was an indication of north (MM). Users also wanted to add markers to the map in order to make re-finding a certain location easier (MQ). Some sites provided an option to change the map area, but either users did not realise this or did not understand how it worked (MD). It was also annoying for the participants that the setting for the map size was not retained for the next query (MD).

Panning was sometimes considered problematic and too slow (MQ, MM). If there was no feedback, users often thought that they had missed the button [. . .]. Participants were also confused about the different types of zoom setting and their relationship with each other (MM). [. . .] Sometimes the zoom function was criticised as being old fashioned (with scale numbers) and confusing for ordinary users (MD, MM). Zooming was also considered problematic when there were neither steps nor animation when switching between different zoom scales (MD, MQ), because users lost the location that they were looking at earlier. With one site, zooming moved the search result out of the map window because the search result was not centred on the map when starting off (MM). Some participants would have liked to point at the area into which they were interested in zooming (MD). It was found confusing that the map could be zoomed by clicking on it, as the cursor did not change when it was pointed at the map (MD). It was also considered annoying that clicking on the map did not just centre the view [. . .] (MM). It was surprising to the users that clicking on the map re-focused and re-centred it, when they only wanted to point on it (GM). Accidental zooming also occurred when participants used the scroll wheel of the mouse when they wanted instead to scroll down the search results window (GM).

From these results, a series of design guidelines were developed that relate to the map.

[. . .]

Design guideline 2: the map

Visualisation

- The map should be visualised according to the properties of the computer screen.
- The map should be optimised for viewing on a computer screen.
- Maps should be simple and intuitive and pleasant to use.
- Colours should be in harmony.
- Each map scale should be considered separately: what information should be included and how it should be visualised at each of the scales.
- Information about data accuracy and validity should be provided.

Map tools

- Map tools should be distinctive, but not obscure too much information on the map.
- A route-measuring tool would be beneficial (in addition to a scale bar).
- New tools would be beneficial for users: an option to add markers on the map; to click on different objects in order to get more information about them; to customise the map by checking 'boxes' to show or hide different data layers or symbols on the map (e.g. tourist attractions, hotels, restaurants); and incorporate an easy way to print and e-mail the map.
- The scale bar (and other) units should be customisable.
- A continuous click-and-drag option would be best for panning.
- Scale increments should not be too great, allowing users to follow a specific location while zooming in and out.
- Scale numbers (ratios or representative fractions) should not be used. Instead, scale should be indicated by more commonly used terms (such as street level, city level, country level etc.).

Search operations

A significant number of usability problems were found relating to queries and searches. [. . .]

Search criteria/logic

One site was different from the others in that it supported a 'free search', whereby the user could type their search criteria more liberally in one or two search boxes (GM) The other sites provided users with different search boxes, each requiring a certain kind of text, for example country, address, place name.

[...] The free search was liked because it is the way people normally find information when using search engines. However, it was also considered to be confusing. [...] For example, one user typed in the search box 'road1 to road2' and then pressed 'get directions', but got no results (GM). It was also commented upon that [...] users may like to have access to at least some shortcut buttons (instead of always having to search). The positive elements of the other search type (MD, MQ, MM) were that people are more used to having separate search boxes for 'location', 'directions' and 'businesses' with Web maps, and most of the time people also know what to type in each search box. On the other hand, the boxes were not very flexible and often required the data to be typed exactly in the correct way. [...] It was observed that the users wanted to make [...] several separate searches simultaneously (multisearches), so that the different objects would appear on the same map at the same time (GM, MD, MM). Moreover, people did not know whether or not the search they carried out was only going to include the area currently shown on the map (GM, MM). [...] The only way to search for addresses or directions was via entering text, whereas it would be helpful to have the map as an interface as well [...] (MD). Searching for addresses was also not always easy [...] The user needs to know that in the United Kingdom the house number is placed first for an address (MD). Users were also frustrated by not being able to search for anything else other than addresses (i.e. places, MQ).

Default settings

The severe usability problems encountered most often related to the default settings of the Web mapping sites, which, in the worst case, prevented some of the participants from using the sites. For example, if the user typed 'London Bridge', the site would only give results from the USA (GM), because the participant did not notice the USA default, or did not know how to change it (MD, MQ). It was considered especially frustrating that the search box always went back to the default settings [...] (MD, MQ). [...]

Search results

Often, the participants did not know how the search results matched their search criteria. [...] If only one result matches the search criteria, the result is shown on the map without any explanation of how it matched the search (GM). On another occasion, the user got a list of 'Londons' in the USA and did not realise that these were not in the desired country (MQ). The participants also tried to use two or more search criteria at the same time and often got a map with the result displayed (MM). However, this was not always the correct result, since it sometimes returned only one search criterion. The users did not always realise this. [...]

The users had to be sure about what they were looking for because some searches gave a number of incorrect results, even though the search was very well defined (e.g. 'Big Ben' gave results everywhere else except London) (MM). Search results were even more confusing when they were based on similar sounding place names; 'London tower' gave the result 'Lake Teterower' (a lake in Germany) and 'Longbridge' in Birmingham (MD). One user got 25 results for a simple search, because everything that included the searched name or even sounded the same was included ('Tussaud' resulted in 'Tosside' and 'Thickwood' [...]) (MM).

Performing route queries became problematic when users did not know which of the search results was the start or end points on a map of their route (GM). It was also criticised that users could not easily change the start or finish of a route already shown on the screen (GM, MD). For 'directions search' the participants would have liked to have all the possible choices (results) for the end and start to be shown on the map to help choose between them (MM).

[...] One site made the route suggestion automatically based on the previous road search (MQ). It was also confusing that [...] they got the result as a text description, not as a map as they had expected (MQ) [...]. Some participants also wanted multistop route searches, enabling them to search for routes from A to C via B (MM, MQ). More choice for customising routes was also required: quickest, shortest and with different transportation modes (GM, MD). It was also noted that it would be good to be able to search for businesses (e.g. restaurants) along a specific road or route (GM, MD).

Criticism of the visualisation of search results, for instance, when a street result was visualised with a pin instead of linear highlighting (GM, MD) was also present. Comparing different search results was considered difficult, because they were not shown on the map at the same time (GM, MD) or because they were shown in different scales (MD, MM). Sometimes, the users had to open another map window to compare a distance between two locations (MD).

The search results were occasionally shown on the map on top of each other, so that the users were not able to see them all (GM, MM). One site centred the map according to the result without any visual emphasis of its location (MD). The users did not realise that this was a result. [...] (MD, MM). The same problem occurred when searching for routes (MD) and roads (MQ) [...]. The search results were also easily lost on a map (GM, MM). The users commented that there should be an option allowing a quick return to the search result instead of constantly having to click back to the search page (MQ, MM). With one site, the zooming did not work when clicking the search result [...] (MM). With one Web mapping site there was no route shown after performing a directions search, only points indicating where to turn, and these were considered difficult to read (MM). More dramatically, the route visualisation changed between map scales (MM). Sometimes the search results were given on the same scale as the map preceding the search, therefore the whole route was not always shown on the map (MM). The lack of an option to print out the visualisation of the route was criticised (MM) [...].

A number of guidelines related to the search operations were proposed as an outcome of these evaluation results and are provided in the following section.

Design guideline 3: the search operations

Functionality

- Different types of searches should be supported.
- Users should know with what type of criteria the search is carried out.
- A list of users' previous searches should be saved and provided to them.
- It should be made clear to users what the search results are based on and how they relate to the query.

Visualisation

- The results should be centred on the map and distinctively visualised, taking into account the symbols that are already in use on the map.
- The result symbols should not cover the map too much and be on top of each other.
- The defaulting map scale should give enough information for the user to check whether or not the result is correct.
- It would be beneficial to show all the possible results on the map, so that the user can choose the correct option among them.
- Street and route search results should be visualised with a line.
- Route search results should be displayed on a tailored map scale so that the user sees the entire route.

Help and guidance in an error situation

Error situations are often inevitable with map sites [...]. It was observed that, in some situations, there was no proper help available. Instructions on how to start using the Web map were missing from some sites, or the existing instructions were not considered useful (MQ, MM).

Some error messages did not look like a message and users did not notice them appearing on the screen (MM). If the error message was given clearly, it was not always informative (GM, MM). Some of the sites did not provide any help (or the users did not find it) for using the map (GM) or for looking for streets and directions (MQ). Some sites gave examples for help in using the searches, but they were also confusing [...] (GM). Sometimes the 'help' was not what the user expected; the user needed help for finding locations but only got a legend for tools (MM).

Design guideline 4: help and guidance

- The user should be provided with help in map use and in other functions in the site.
- Error messages should be clear, informative and distinctive.
- Users should be informed of current default settings and how they can be changed.

Discussion and conclusions

By identifying pitfalls in existing Web maps, it is possible to offer recommendations on how to design Web mapping sites that are easier to use and attractive to different groups of users. A possible bias in this study, however, may be drawn from the fact that the Web maps included in the evaluation were well known and widely used. It might therefore be expected that such sites have fewer problems than more unfamiliar applications as a result of their popularity. The evaluation of the Web mapping sites nevertheless identified a considerable number of severe usability problems. If these were typical for Web maps that are in use every day and by large numbers of people, it would be interesting to investigate the usability of smaller, less familiar map applications. [...]

Even though many usability problems were identified, some of the problems may have been exacerbated by the tasks chosen for this study. Different sites can have different objectives, and the use-scenario may not have corresponded exactly to that for which these sites were originally designed. In fact, some of these Web maps may not have been designed for use by tourists. [...]

As map sites are unquestionably visual in their nature, distractive advertisements and messy user interfaces were criticised. A map that frustrates the user from the very beginning may cause very negative feelings towards it. Some of the users actually stated that in a real-life situation they would have given up trying to complete the tasks with some of these sites and tried another Web map. This is important, as it may be that the product developer who can design the most usable application will win the battle for market dominance.

Some of the Web maps have been in existence longer than others, which may have biased the results of this study. Some users may have been attracted by newer ideas and these might have received more positive comments because of that. On the other hand, some of the users valued traditional types of services because they are used to them. This was especially obvious when the different search criteria of the sites were discussed. Some people have been used to making Web searches with search engines, and they also wanted to carry out map searches in the same 'free' manner. Others needed more structured or guided searches. The challenge remains to design sites that different types of people can use without getting frustrated or without facing a lot of problems in using them.

Another challenge is that some of the participants had hardly used any types of maps at all and, for them, the use of these sites was especially difficult; some of the users did not even realise that the map scale could be changed or that searches could be carried out for different objects. This is understandable, since Web maps deal with complicated spatial data and may allow a high degree of interactivity between the user and the site. How can we help ordinary Internet users to realise the variety of map sites and their functionality and benefit from their use? The observed lack of guidance within these sites does not help in this situation.

The Web maps often offer links to different additional services (such as hotels and tourist attractions), which either have their own map interface or no map at all. If a user wanted information on how to use the underground rail network to get from one tourist attraction to a hotel, at least three different maps and services had to be opened at the same time: an underground route map, a map with hotels on it, a map with tourist attractions on it, and perhaps even a base map for combining all this information. If all of these have their own maps with different scales and visualisations, users will find it difficult to combine the information. The best solution would be to have all these embedded within the same map service, or to have harmonised maps between different services. [...]

References

Earthy, J. (1996) *Development of the Usability Maturity Model*, INUSE Deliverable D511, Lloyd's Register, London.

Fuhrmann, S., Ahonen-Rainio, P., Edsall, R.M., *et al.* (2005) Making useful and useable geovisualisation: design, and evaluation issues, in *Exploring Geovisualisation* (eds J. Dykes, A.M. MacEachren and M.J. Kraak), Elsevier, New York, pp. 553–566.

ISO 9241-1 (1997) *Ergonomic Requirements for Office Work with Visual Display Terminals (VDTS) – Part 1: General Introduction*, International Organisation for Standardisation, Geneva.

ISO 13407 (1999) *Human-Centered Design for Interactive Systems*, International Organisation for Standardisation, Geneva, Switzerland.

Koua, E.L. and Kraak, M.J. (2004) A usability framework for the design and evaluation of an exploratory geovisualisation environment. *Proceedings of the 8th International Conference on Information Visualisation, IV'04*, IEEE Computer Society Press.

MacEachren, A.M. and Kraak, M.J. (2001) Research challenges in geovisualisation. *Cartography and Geographic Information Science*, **28**, 3–12.

Nielsen, J. (1993) *Usability Engineering*, Academic Press, San Diego, CA.

Nivala, A.-M., Sarjakoski, L.T. and Sarjakoski, T. (2007) Usability methods' familiarity among map application developers. *International Journal of Human–Computer Studies*, **65** (9), 784–795.

Skarlatidou, A. and Haklay, M. (2006) Public web mapping: preliminary usability evaluation. *GIS Research UK 2005*, Nottingham.

van Elzakker, C.P.J.M. (2005) From map use research to usability research in geo-information processing. *Proceedings of the 22nd International Cartographic Conference*, Coruna, Spain.

Further reading

Alaçam, Ö. and Dalci, M. (2009) A usability study of webmaps with eye tracking tool: the effects of iconic representation of information human-computer interaction. *New Trends Lecture Notes in Computer Science*, **5610**, 12–21. [Explores the functionality of web map interfaces using usability engineering techniques and extends the work of Nivala *et al.* through the use of different methodologies.]

Dykes, J.A., MacEachren, A.M. and Kraak, M.J. (2005) *Exploring Geovisualization*, Elsevier, Amsterdam. [A large edited volume showing the possibilities opened out by a wider appreciation of the links between design and use in geovisualisation.]

Fabrikant, S.I. and Lobben, A. (2009) Cognitive issues in geographic information visualization. *Cartographica*, **44** (3). [Theme issue of the journal including articles on the application of cognitive approaches to geovizualisation charting theories implicit in the work of Nivala *et al.*]

MacEachren, A.M. (1995) *How Maps Work*, Guilford, New York. [A systematic evaluation linking semiotic and cognitive approaches to mapping with strong links between design and use.]

Peterson, M.P. (2008) *International Perspectives on Maps and the Internet*, Springer, New York. [A useful overview volume documenting the very great diversity of mapping served over the Internet.]

See also

- Chapter 1.11: Exploratory Cartographic Visualisation: Advancing the Agenda
- Chapter 2.11: Extending the Map Metaphor Using Web Delivered Multimedia
- Chapter 2.12: Imaging the World: The State of Online Mapping
- Chapter 3.5: Strategies for the Visualisation of Geographic Time-Series Data
- Chapter 3.10: Affective Geovisualisations
- Chapter 4.9: Understanding and Learning Maps
- Chapter 5.11: Mapping the Digital Empire: Google Earth and the Process of Postmodern Cartography

SECTION FIVE

Power and Politics of Mapping

Chapter 5.1

Introductory Essay: Power and Politics of Mapping

Rob Kitchin, Martin Dodge and Chris Perkins

Introduction

There is a long tradition of historical analysis that examines the production of maps, their development over time and their role in society. Such analysis implicitly concerns the power of mapping to influence social and economic relations in particular places and times. More recently, research has focused specifically on the politics and power of mapping; how power is captured in and communicated through maps to assert command and control of territory and socio-spatial relations; how power is bound up in the very creation and use of maps; and how mapping practices are used to resist and contest the exercise of power over space. Much of this research is framed within what has been termed critical cartography (Harley 1989; Crampton and Krygier 2005) and critical GIS (Pickles 1995; Curry 1998; Schuurman 1999; O'Sullivan 2006). Critical cartography is post-positivist in its approach, drawing on a range of social theory to re-examine cartographic representations and the wider milieu of mapping processes. It is often avowedly political in its analysis of mapping praxis, seeking to deconstruct the work of maps and the science that produces them, often undertaking to produce alternative maps that are more sensitive to the power relations at play. On the one hand, this has led to an examination of the power of maps and the work they do in the world, and on the other to new forms of collaborative and counter-mapping that seek to produce empowering and emancipatory cartographies, which subvert the status quo. In both cases, there is an explicit recognition that maps are a product of power at work and that they are powerful tools in struggles of domination and resistance. In this section excerpts from a number of key readings that seek to document and theorize the power of maps are provided.

Cartographic power, nation building and colonial conquest

> 'As much as guns and warships, maps have been the weapons of imperialism'. Brian Harley, *Maps, Knowledge and Power*, 1988.

Mapping has been, and remains, a key device in the formation of nation states, colonial projections of power and the control of distant imperial lands. This is achieved in part because of the unique properties of maps to project a coherent representation of territorial continuity and the unity of people to a common cause (be it monarch, religion or national ideology).

Maps then have been important devices in forming national identity and nation building. Anderson's (1991: 175) thesis of nationalism as imagined community, for example, highlights the extensive symbolic power of 'map-as-logo', deployed in an 'infinitely reproducible series, available for transfer to posters, official seals, letterheads, magazine and textbook covers, tablecloths and hotel walls.

The Map Reader: Theories of Mapping Practice and Cartographic Representation, First Edition. Edited by Martin Dodge, Rob Kitchin and Chris Perkins.

Instantly recognisable, everywhere visible.' Maps showing space divided according to political authority are a powerful assertion of state sovereignty and have become so ingrained as a 'natural' template that such borders are present even in maps which are not explicitly political (e.g. weather maps). The symbolic power of cartography to make borders is endlessly exploited in the 'grand games' of geopolitics between states, where the 'maps provided the master image of the nation's superiority and centrality in global affairs' (Vujakovic 2002: 198), such as Halford Mackinder's cartographic articulation of the 'Eurasian heartlands' thesis at the height of British imperial power.

The instrumental role of Western mapping in imperial exploitation through the erasure of indigenous peoples from the colonisers' maps provides perhaps the strongest evidence of the malignant power of cartography. In the partition of India, the annexation of Palestinian land, or the '*terra nullius*' of Australia, cartography has been integral to colonial practices, providing both spatial justification and a rationalising tool for colonisers, past and present. For example, Bassett's (1994: 333) analysis of maps made by European imperial powers at the end of the nineteenth century demonstrates how effectively they 'promoted the appropriation of African space under the rhetoric of commerce and civilisation.'

Winichakul (1994, excerpted as Chapter 5.4) provides a detailed example of how mapping was a key instrument in the formation of a nation, charting the tensions between the Siam royal court and the struggle between French and British colonial interests in South East Asia. Competition in surveying and a small number of cartographic artefacts at the end of the nineteenth century reveal the constructive power of mapping. Up to this point, Siam was largely unmapped, in terms of formalised Western representational science, and its territorial borders were tacitly known by local knowledge and observed tribal customs. Through the process of surveying and mapping Siam underwent a cultural re-imagining to produce a new 'geo-body' (a socio-geographical understanding of the country). Winichakul discusses how the cartographic representations produced did not simply reveal the geography of Siam, but also brought forth a new sense of what Siam was and could become; they anticipated a shared vision of a nation, rather than depicting one already established. Moreover, maps enabled monarchical power to assert its authority over territory and to enforce new forms of administrative control, significantly enhancing their power to influence local communities and shape social life. In a similar vein, Herb (1997) on Germany, Ramaswamy (2010) on India and Schulten (2001) on the United States of America analyse the power of mapping to shape national consciousness in the service to certain interests.

Along comparable lines, Sparke's article (1998, excerpted as Chapter 5.7), *The Map that Roared*, documents the ways in which such large-scale, centrally organised and administered Statist cartographic programmes produced a 'geo-body' that had the power to undermine the validity of local knowledges and obliterate the legitimacy of indigenous mapping traditions. By carefully tracing out how First Nations maps, territorial claims and knowledge were treated during a long court trial, Sparke reveals the subtle ways in which Western cartographic practice built up and maintained its hegemonic status as the only legitimate form of spatial representation, and thus the arbiter of property claims and disputes. The select set of map artefacts of the Canadian government thus enjoyed a particular sovereign status that worked for the interests of the state and settler populations at the expense of indigenous First Nations peoples. This kind of cartographic power is evident in many colonial and postcolonial struggles including contemporary geopolitical situations (e.g. Gregory's 2004 analysis of cartographic logics underpinning imperial moves in Palestinian land, Iraq and Afghanistan; all areas were well mapped by earlier rounds British colonial cartographers and surveyors.)

That maps have this power is, for a large part, due to the fact that they have certain, universal qualities. As Harvey (1989, excerpted as Chapter 5.2) notes, Western European cartography was transformed during the Renaissance, adopting perspectivism and Cartesian rationality to seek to produce a universal system for mapping the whole of the known world. For Latour (1992, excerpted as Chapter 1.9) this new scientific approach enabled maps to become 'immutable mobiles'; that is, mechanisms used to generate and circulate cartographic information which fixed particular meanings. The form maps took (in terms of scale, legend, symbols, projection etc.) became familiar and standardised through established protocols so that cartography became a stable, combinable and transferable form of knowledge, portable across space and time. As such, maps produced in distinct political and cultural contexts, say in the royal courts of France, Germany, Portugal, Spain, The Netherlands and so on, became decipherable and applicable to someone from another country because they shared a body of common principles and standards that rendered them easily legible. Moreover, cartographic data transported from around the planet in the form of latitude and longitude observations and measured surveys could be reliably interpreted and meaningfully applied to update charts of an area, or be combined with other information, despite the fact that the cartographer was unlikely to have ever visited the area they were mapping. As such, the media of maps became increasingly important because they were mobile, immutable, flat and foldable (and therefore easily carried), modifiable in scale, reproducable, capable of being recombined and layered, but also optically consistent and amenable to insertion into other texts. The results were significant, one can argue, because they contributed to the efficiency and effectiveness of small European nations projecting their military and commercial power over far distant lands and with large indigenous populations.

Like Harvey, Latour contends that these qualities allowed exploration and trade and ultimately contributed to the brutal violence of colonialism by: making territory knowable, navigable and claimable; allowing control to be exerted from afar; and enabling knowledges about new territories to be effectively transported globally. Maps became a vital part in the cycle of knowledge accumulation that allowed explorers to '*bring* the lands *back with them*' and to successfully send others in their footsteps (Latour 1987: 220, original emphasis). Latour thus argues that the European mapmakers of the Renaissance produced centres of calculation (key institutions of knowledge accumulation and cartographic practice) that came to dominate much of the world. In so doing, he contends that the maps produced did not simply represent space at a particular time, but were mappings bringing into being new space-times. Maps opened up new possibilities – such as reliable long distance trade and territorial conquest by tiny forces, operating many thousands of kilometres from home – and thus created new geographies and histories. Maps consequently served political and economic interests, enabling the demarcation of boundaries, assigning property rights, detailing rights of passage, securing transportation routes and guiding military campaigns. Such pursuits were critical for those in power, such as the sovereign or religious elites, to assert, exploit, control, maintain and extend their effective rule over people and places. As time went on, Western cartography became ever-more sophisticated in design and capacity to project power, including the effective display of statistical knowledge relating to populations (providing a spatial overview of inhabitants as well as lands) and the use of propaganda mapping explicitly aimed at creating particular views about specific places and to reinforce national and regional identities (Anderson 1991; Pickles 1991, excerpted as Chapter 5.3).

An important way that the power of the 'cartographic gaze' works, is by dehumanising the landscape, allowing powerful groups to exercise power at a distance, 'removed from the realm of face-to-face contacts' (Harley 1988: 303). Maps are foundational to modern systems of governmentality, as evidenced in the extensive use of statistical mapping by state bureaucracies. These cartographies are designed to produce a 'rationality of calculability of populations' (Crampton 2004: 43), where people can be managed *through* the map more easily because action can be taken without witness to human consequences. Indeed, maps come to symbolise the governmental processes of regimentation in which particular places, individual homes and complex lives are rendered as mere dots. This kind of de-socialisation of space through cartographic abstraction is seen most brutally in the military. Modern war making is now frighteningly like a map game in which death is played out on digital geospatial interfaces that render human landscapes into an impersonal terrain of targets and threats that can be engaged by so-called precision-guided weapons (Gregory 2010).

The meaning and power of maps

In addition to examining in broad terms how maps have been enrolled as potent instruments of state control and colonial security, there is now a significant body of literature examining in detail how power is constituted in the very design and creation of maps, and how maps are used to reproduce specific power relations. For example, Wood and Fels (1986, excerpted as Chapter 1.7), Harley (1989, excerpted as Chapter 1.8) and Pickles (1991, excerpted as Chapter 5.3) all argue that all maps are inherently ideologically loaded, vested with the interests of their creators. This is most visible in maps employed as overly propagandist displays, designed to reshape how people think about a particular area or stir up emotional response to an issue, but is inherent in even the most seemingly benign maps, such as the supposedly neutral, scientific productions of the topographic map, or school atlas. This is because all map designers have to make a whole series of decisions regarding content, presentation, scale and so on that directly affect what the map communicates and how it is read. As a consequence, maps designed by state agencies claim a particular authority and communicate selective messages and include all kinds of 'silences' about other information. Over the past two decades a number of scholars have actively critiqued such maps from a variety of perspectives, such as feminism and post-colonialism.

This analysis looks beyond the aesthetic connoisseurship of the map collector or the rules of 'good design' considered in Chapter 3.1, and focuses on the 'second text' of the map. As such, deconstructing the map means exposing the reasons behind the selectivity of what is displayed and demystifying the origins of the signs used. Everything about the look of a map is subjective and to some extent arbitrary in semiotic terms, but people usually ignore this because they read modern maps as 'natural', having been thoroughly indoctrinated into the conventions of cartographic sign systems (i.e. a blue line for a river). This has important implications because '[o]nce it is accepted that certain conventions are "natural" or "normal", the danger is that they acquire a coercive and manipulative authority' (Harley 2001: 202).

For example, feminist scholars have critiqued the Cartesian rationality of modern cartography as being a particularly masculinist way of thinking and representing the world. Such a way of thinking employs the 'god trick' of a disembodied and emotionless view from nowhere, floating some way above the Earth, wherein spatial relations can be holistically mastered and manipulated (Haraway 1991; Rose 1993). As noted by Huggan (1994, excerpted as Chapter 5.5), from a feminist perspective mapping codifies, defines, encloses and excludes, subjugating land to a male gaze and representation (also see Kwan 2007, excerpted as Chapter 5.9). Such an approach pre-supposes that it is possible to objectively and neutrally capture and process the world, and to know, dominate and master it. From a related perspective, Brown and Knopp (2008, excerpted as Chapter 5.10) detail how Seattle's gay history had been written out of the city's spatial register through past maps silences concerning the venues important to its gay citizens. Maps then are most often hetero-normative; that is, they assume and reinforce a heterosexual orthodoxy, wherein traditional maps only portray a heterosexual world.

Other work along this vein includes consideration of the potential of mapping to reinforce able-bodied stereotypes and map a world that fails to serve the interests of different groups of disabled people (see for example Matthews and Vujakovic 1995 on mapping for wheelchair users; and Gleeson 1996 on visually impaired people and their marginalisation through sighted map design). Other social categories are also 'off the map' with interests that are rarely mapped out. Research in this context has focused in particular upon ethnicity and the Othering potential of mapping that reflects largely white governing interests (Winlow 2001); but research has also focused on social class (Harley 1988) and age (Gerber 1993). The last twenty years has also seen a significant rise in the amount of 'map art', (Wood 2010), in which artists are playing with norms of cartographic representation to challenge different politics of space (Biemann 2002; Mogel and Bhagat 2007).

Cartographic power, surveillant knowledge and spatial control

As well as expressing power through their meaning, and selectivity and 'silences', maps can work explicitly as tools of the powerful for controlling territory and populations by enabling spatial surveillance and rendering people visible and identifiable to those in power. As Crampton (2003, excerpted as Chapter 5.8) and others such as Monmonier (2002) detail, maps have long been employed by states as a means to plot and track social, economic and environmental phenomena through statistical mapping. For example, during the nineteenth century a panoply of new forms of data generation, such as censuses, health and education records, housing registers, crime counts and so on were introduced as means to monitor societal changes, with much of these data represented in newly developed thematic mapping (Robinson 1982). Indeed, maps became important tools for identifying and addressing particular societial problems that were deemed significant or threatening, such as John Snow's celebrated epidemiological mapping of cholera cases in London which provided evidence that the disease was water borne. Mapping became a vital instrument for new, emerging systems of governmentality (how societies are organised and governed to fulfil certain aims) by revealing key spatial patterns and processes (Joyce 2003), and the surveillant potential of digital technologies described in Chapter 2.1 continues to grow.

The myriad ways that the state has come to rely on 'power through the map' to govern means that it is still far and away the largest patron of cartography, but mapping is also integral in capitalist accumulation by (re)ordering lived lives into markets, potential profits or obstructions to consumption. For example, geodemographic mapping profiles individuals, fitting them into idealised consumer types, fixing them into a spatial grid of quantifiable economic value and ranking them based on their 'worth' or 'risk' (Curry 1997; Goss 1995). This easily leads to the discriminatory practices of 'redlining' – the term is derived from the mapping practice where communities deemed unprofitable or high risk and are denied services (e.g. Hillier's 2005 historical analysis of mortgage loan discrimination in Philadelphia).

In recent years, improvements in surveillance systems and mapping technologies have led to marked change in the ability to track and profile people and places. As Dodge and Kitchin (2005) show, the digital age has brought with it a qualitative shift in the amounts and kinds of data that can be generated and analysed. It has now become feasible and cost effective to harvest vast sums of data, at an increased spatial granularity, to process and map this data in real time, to collate and combine data in ever more sophisticated ways, to distribute the data instantly, display it on maps against other relevant layer, and to store it in multiple forms for future use. Maps become a medium through which it is possible to spy in real time on most citizens. For example, it is possible to track the mass movements of people and vehicles through cities by mapping data automatically generated by ANPR traffic cameras, smartcard-ticketing on metros and mobile phone identifiers (Ratti *et al.* 2006). These changes raise significant concerns with respect to civil rights, equity and privacy, and yet they are supported by powerful discourses concerning security, safety and economic rationality as well as opening up profitable new opportunities for business, which inexorably encourage continued implementation for the foreseeable future.

Cartographic power, counter-maps and participatory mapping

While the potent role of cartographic power in social domination by the state and corporation is unquestioned, such hegemonic mapping is dialectical because it also opens up new ways to resist. The practical and rhetorical power of maps to articulate alternative perspectives is always available. The power of the map can be used to re-frame the world in the service of progressive interests and to challenge inequality (such was the goal of the Peters projection project), while the logo-map used to bolster the state can be re-imagined as a potent emblem in anti-colonial struggles (Huggan 1994, excerpted as Chapter 5.5). Wherever power is expressed it is met with some forms of resistance and often counter movements, yet until recently maps were underutilized to challenge authority. Given the need to access data, specialist cartographic resources and advanced cartographic skills, the limits to counter-mapping are perhaps unsurprising. However many of the same technologies that facilitate cartographic surveillance have been enrolled to create new forms of counter- and participatory mapping that seek to empower and emancipate people from specific forms of oppression (such as Paglen's 2009 use of surveillant tactics and techniques to expose the extent of the secret state; for an example of protest cartography, see Colour Plate Six).

Greater availability of mapping software, new open-source tools and online services have drastically reduced the resource base needed to produce professional looking maps and have enabled users to scrutinise official data sets in new ways and share their own data for analysis. These trends have contributed to more people being able to produce what Peluso (1995, excerpted as Chapter 5.6) has called 'counter-maps'; maps that challenge power and hegemony of state and commercial maps by representing other interests, but which maintain the same standards of production. In that sense, counter-maps appropriate the state's 'techniques and manner of representation' in order to re-territorialise the area being mapped and to make a case for a redistribution of resources. Their creation and circulation is designed to empower citizens and enable resistance and protest. Counter-maps, then, are explicitly political in ambition and seek to counterbalance the discourses of government and capital by inserting local views into the decision making process. In Peluso's case, the counter-maps were of forest areas and resources as delineated by local communities who used the maps to challenge omissions of settlement and biodiversity, the categorisation of land and management, and the placement of boundaries. In Sparke's case (1998, excerpted as Chapter 5.7), the First Nations tribes used counter-maps to challenge the territorial claims and political administration.

Cartographic power has also been exploited by environmental pressure groups and anti-globalisation activists to counter the dominant corporate discourses by using the authority of the map against itself (e.g. maps of the ozone hole over the Antarctic become potent images in the mid-1980s). This kind of counter-hegemonic cartographic potential is evident in the work of radical geographer Bunge (1975: 150), and his expeditionary geography, mapping socially-polarised urban America, to 'depict a region of super-abundance adjacent to a region of brutal poverty' (Figure 5.1.1). In many examples of counter-cartography, the actual maps themselves are not alternative in design terms, making use conventional cartographic signs (e.g. Bunge's 1975 dot maps, or Kidron and Segal's 1995 use of choropleth mapping). The distinction that marks these mapping projects as 'subversive' is that they exploit the authority of cartography to ask difficult questions by mapping the types of human phenomena (war, poverty, violence against women) and landscape features

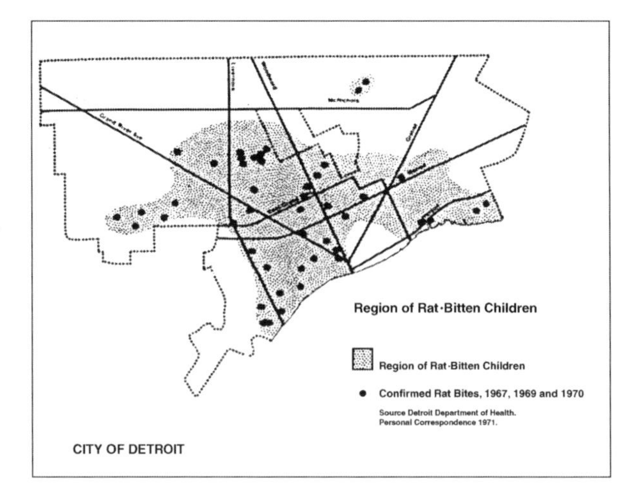

Figure 5.1.1 Example of the counter-cartography of William Bunge showing the rhetorical power of thematic maps to challenge the status-quo. (Source: Bunge 1975: 161.)

(toxic waste sites, rat bites) that are usually deemed insignificant, inappropriate or otherwise 'difficult' by mainstream government and commercial cartography and, therefore, generally left unmapped. They confront the norms of society by using the conventional signs of the society's elite. Another significant tactic in counter-cartography is changing scale and opening up authorship, for example in eco-mapping, which stresses the importance of mapping local areas by local people (Aberley 1993), and the empowering of marginalised groups, such as having physically disabled people map their experiences of hostile streetscapes (Kitchin 2002).

The inclusion of local voices is often pursued through a strategy of participation. Participatory maps are produced with and by, as opposed to for, local groups. For example, Kitchin (2002) reports a participatory mapping project with a group of disabled people to create an access map of a town to illustrate the problems of urban inaccessibility and to campaign for inclusive planning. The group worked together to devise a work plan, identify issues, create a symbol set, survey the landscape, create and distribute the final map. In so doing, the participants not only took charge of the process, but gained new skills and knowledge, and helped influence local decision making. This process of collaboration and negotiation can be very rewarding to both researcher and locals but it can also be fraught with all kinds of issues and be time consuming, as detailed by Brown and Knopp (2008, excerpted as Chapter 5.10). Such 'bottom-up' mapping is not without is own politics and partiality of representations.

More recently, internet mapping services have allowed users to access and interact with growing volumes of geographic data, such as street maps, high resolution aerial photographs and satellite imagery, by using straightforward interfaces to produce their own maps. For example, as Farman

(2010, excerpted as Chapter 5.11) and Geller (2007, excerpted as Chapter 2.12) detail, Google Earth is one such online platform that enables users to access, interact with, and update spatial data and to share related information such as overlays, photographs, video clips, artwork, notes and so on. Moreover, Google Earth is complemented with bulletin boards that allow mappers to discuss issues relating to the platform, the data it uses and the data uploaded by other users.

In this sense, Google Earth is an example of what Crampton (2009) terms 'Mapping 2.0'; mapping that is distributed, participatory and social. Mapping 2.0 offers a new form of mapping experience in which users can become authors and through which the content is built collaboratively. This collaboration is a form of so-called 'crowd-sourcing', wherein many people volunteer pertinent information usually on their local patch, as detailed by Goodchild (2007, excerpted as Chapter 4.10). Another well documented example of a collaborative mapping is OpenStreetMap, an open source project that largely uses 'crowd sourced' GPS data to provide an alternative online mapping system to commercial and state products. (See visualisation of the extent of OSM mapping in Colour Plate Five.) The resultant detailed map database is distinctive in that it is a wiki (everything is editable by everyone) and is available to be used in projects without the burden of restrictive copyright licenses that often limit how government and commercial data can be used. Similarly, there are discussion forums that encourage collaboration and debate, and data are open to be edited and updated by other users (which is not the case with commercial and state data). Mapping 2.0 therefore has political and practical ramifications, as it radically blurs the division between mapmaker and map user, and begins to expose the partiality of authorship and the ways authority of map representations has to be manufactured.

Conclusion

The chapters excerpted in this section all make the case that maps are not neutral, value-free spatial representations of the world. Rather, they contend that power is inherently bound within their very making and representation, in their design and content, to communicate spatial relations in a certain manner that seeks to assert or reproduce a particular way of thinking about the world. Maps then are ideologically loaded, representing the interests of their creators, forming part of an armoury of political instruments used to underpin claims with respect to territory, to monitor people and police the places they live. Given the power of maps, cartography has played an important role in the building of nations and national identities, the development of empires and colonies, including the waging of war and violence, and in the construction of efficient trading routes and the accumulation of capital. Maps have served, and very much continue, to extend and reproduce the power and influence of those that created them. More recently, this power has been harnessed by those who are usually subjugated by such maps through the production of counter-maps that seek to provide an alternative viewpoint and subvert dominate socio-spatial relations. Indeed, new mapping technologies, along with more access to relevant data, are significantly reshaping who can produce maps and how they are produced, in the process reconfiguring established cartographic power relations. As such, a somewhat paradoxical situation is arising – on the one hand, mapping is being evermore used by states and corporations as a medium through which to survey and control populations, and, on the other, maps are being used to provide counter-discourses to states and corporations with the aim of producing more emancipatory and empowering outcomes. There is no denying then the power of maps.

References

Aberley, D. (1993) *Boundaries of Home: Mapping for Local Empowerment*, New Society Publishers, Gabriola Island, BC.

Anderson, B. (1991) *Imagined Communities: Reflections on the Origins and Spread of Nationalism*, 2nd edn, Verso, New York.

Bassett, T.J. (1994) Cartography and empire building in nineteenth-century West Africa. *Geographical Review*, **84** (3), 316–335.

Biemann, U. (2002) Remotely sensed: a topography of the global sex trade. *Feminist Review*, **70**, 75–88.

Brown, M. and Knopp, L. (2008) Queering the map: the productive tensions of colliding epistemologies. *Annals of the Association of American Geographers*, **98** (1), 40–58. (Excerpted as Chapter 5.10.)

Bunge, W. (1975) Detroit humanly viewed: the American urban present, in *Human Geography in a Shrinking World* (eds R. Abler, D. Janelle, A. Philbrick and J. Sommer), Duxbury Press, North Scituate, MA, pp. 149–181.

Crampton, J. (2003) Cartographic rationality and the politics of geosurveillance and security. *Cartography and Geographic Information Science*, **30** (2), 135–148. (Excerpted as Chapter 5.8.)

Crampton, J.W. (2004) GIS and geographic governance: reconstructing the choropleth map. *Cartographica*, **39** (1), 41–53.

Crampton, J. (2009) Cartography: maps 2.0? *Progress in Human Geography*, **33**, 91–100.

Crampton, J. and Krygier, J. (2005) An introduction to critical cartography. *ACME: An International E-Journal for Critical Geographies*, **4** (1), 11–33.

Curry, M.R. (1997) The digital individual and the private realm. *Annals of the Association of American Geographers*, **87** (4), 681–699.

Curry, M.R. (1998) *Digital Places: Living with Geographic Information Technologies*, Routledge, London.

Dodge, M. and Kitchin, R. (2005) Codes of life: identification codes and the machine-readable world. *Environment and Planning D: Society and Space*, **23** (6), 851–881.

Farman, J. (2010) Mapping the digital empire: Google Earth and the process of postmodern cartography. *New Media & Society*, **12**, doi: 10.1177/1461444809350900. (Excerpted as Chapter 5.11.)

Geller, T. (2007) Imaging the world: the state of online mapping. *IEEE Computer Graphics and Applications*, (March/April), 8–13. (Excerpted as Chapter 2.12.)

Gerber, R. (1993) Map design for children. *The Cartographic Journal*, **30**, 154–157.

Gleeson, B. (1996) A geography for disabled people. *Transactions of the Institute of British Geographers*, **18** (1), 387–396.

Goodchild, M. (2007) Citizens as sensors: the world of volunteered geography. *GeoJournal*, **69** (4), 211–221. (Excerpted as Chapter 4.10.)

Goss, J. (1995) We know who you are and we know where you live: the instrumental rationality of geodemographics systems. *Economic Geography*, **71**, 171–198.

Gregory, D. (2004) *The Colonial Present: Afghanistan, Palestine, Iraq*, John Wiley & Sons, Ltd, Chichester, UK.

Gregory, D. (2010) War and peace. *Transactions of the Institute of British Geographers*, **35**, 154–186.

Haraway, D. (1991) *Simians, Cyborgs, and Women: The Reinvention of Nature*, Routledge, New York.

Harley, J.B. (1988) Maps, knowledge and power, in *The Iconography of Landscape* (eds D. Cosgrove and S. Daniels), Cambridge University Press, Cambridge, pp. 277–312.

Harley, J.B. (1989) Deconstructing the map. *Cartographica*, **26** (2), 1–20. (Excerpted as Chapter 1.8.)

Harley, J.B. (2001) Can there be a cartographic ethics? in *The New Nature of Maps: Essays in the History of Cartography* (ed. P. Laxton), The Johns Hopkins University Press, Baltimore, MD, pp. 197–207.

Harvey, D. (1989) *The Condition of Postmodernity*, Blackwell, London. (Excerpted as Chapter 5.2.)

Herb, G.H. (1997) *Under the Map of Germany: Nationalism and Propaganda 1918–1945*, Routledge, London.

Hillier, A.E. (2005) Residential security maps and neighborhood appraisals: the Home Owners' Loan Corporation and the case of Philadelphia. *Social Science History*, **29** (2), 207–233.

Huggan, G. (1994) *Territorial Disputes: Maps and Mapping Strategies in Contemporary Canadian and Australian Fiction*, University of Toronto Press, Toronto. (Excerpted as Chapter 5.5.)

Joyce, P. (2003) *The Rule of Freedom: Liberalism and the City in Britain*, Verso, London.

Kidron, M. and Segal, R. (1995) *The State of the World Atlas*, Penguin, London.

Kitchin, R. (2002) Participatory mapping of disabled access. *Cartographic Perspectives*, **42**, 50–62.

Kwan, M.-P. (2007) Affecting geospatial technologies: toward a feminist politics of emotion. *The Professional Geographer*, **59** (1), 22–34. (Excerpted as Chapter 5.9.)

Latour, B. (1987) *Science in Action*, Harvard University Press, Cambridge, MA.

Latour, B. (1992) Drawing things together, in *Representation in Scientific Practice* (eds M. Lynch and S. Woolgar), MIT Press, Cambridge, MA, pp. 19–68. (Excerpted as Chapter 1.9.)

Matthews, M.H. and Vujakovic, P. (1995) Private worlds and public places – mapping the environmental values of wheelchair users. *Environment and Planning A*, **27**, 1069–1083.

Mogel, L. and Bhagat, A. (2007) *An Atlas of Radical Cartography*, Journal of Aesthetics and Protest Press, Los Angeles.

Monmonier, M. (2002) *Spying with Maps*, University of Chicago Press, Chicago.

O'Sullivan, D. (2006) Geographic information science: critical GIS. *Progress in Human Geography*, **30** (6), 783–791.

Paglen, T. (2009) *Blank Spots on the Map: The Dark Geography of the Pentagon's Secret*, World Dutton, New York.

Peluso, N.L. (1995) Whose woods are these? Counter-mapping forest territories in Kalimantan, Indonesia. *Antipode*, **27** (4), 383–406. (Excerpted as Chapter 5.6.)

Pickles, J. (1991) Texts, hermeneutics and propaganda maps, in *Writing Worlds: Discourse, Text and Metaphor in the Representation of Landscape* (eds T.J. Barnes and J.T. Duncan) Routledge, London, pp. 193–230. (Excerpted as Chapter 5.3.)

Pickles, J. (1995) *Ground Truth: The Social Implications of Geographic Information Systems*, Guilford, New York.

Ramaswamy, S. (2010) *The Goddess and the Nation: Mapping Mother India*, Duke University Press, Durham, NC.

Ratti, C., Williams, S., Frenchman, D. and Pulselli, R.M. (2006) Mobile landscapes: using location data from cell phones for urban analysis. *Environment and Planning B: Planning and Design*, **33** (5), 727–748.

Robinson, A.H. (1982) *Early Thematic Mapping in the History of Cartography*, University of Chicago Press, Chicago.

Rose, G. (1993) *Feminism and Geography: The Limits of Geographical Knowledge*, Polity Press, Cambridge.

Schulten, S. (2001) *The Geographical Imagination in America, 1880–1950*, University of Chicago Press, Chicago.

Schuurman, N. (1999) Critical GIS: theorizing an emerging discipline. *Cartographica*, **36** (4), 5–21.

Sparke, M. (1998) A map that roared and an original atlas: Canada, cartography, and the narration of nation. *Annals of the Association of American Geographers*, **88** (3), 463–495. (Excerpted as Chapter 5.7.)

Vujakovic, P. (2002) Whatever happened to the 'New Cartography'? The world map and development mis-education. *Journal of Geography in Higher Education*, **26** (3), 360–380.

Winichakul, T. (1994) *Siam Mapped: A History of the Geo-Body of a Nation*, University of Hawai'i Press, Honolulu, HI. (Excerpted as Chapter 5.4.)

Winlow, H. (2001) Anthropometric cartography: constructing Scottish racial identity in the early twentieth century. *Journal of Historical Geography*, **27** (4), 507–528.

Wood, D. (2010) *Rethinking the Power of Maps*, Guilford, New York.

Wood, D. and Fels, J. (1986) Designs on signs/myth and meaning in maps. *Cartographica*, **23** (3), 54–103. (Excerpted as Chapter 1.7.)

Chapter 5.2

The Time and Space of the Enlightenment Project, from *The Condition of Postmodernity*

David Harvey

Editors' overview

The leading geographer of his generation, over the past four decades David Harvey has detailed how the dominant global mode of production – capitalism – uses spatial differences and inequalities to generate fresh rounds of investment and exploitation to further the ends of the accumulation of wealth. In this excerpted chapter from his most cited work, *The Condition of Postmodernity*, he provides an historical analysis of how changes in both space and time were conceptualised and considers the role of cartography in enabling merchants, monarchs and nations to extend their power, influence and trade across the globe. Through the adoption perspectivism and Cartesian rationality the world became knowable, navigable and claimable in new scientific ways, framed within Enlightenment discourse and mathematical rigour. Cartographic knowledge thus became an important commodity in its own right. Mapmaking, Harvey argues, was driven by notions of objectivity, practicality, functionality and ordering, underpinned by economic as well as political imperatives, such as demarcating boundaries, assigning property rights, detailing rights of passage and transport. Such pursuits were critical for those in power to assert, exploit, control, maintain and protect their assets.

Originally published in 1989: Chapter 15 in David Harvey, *The Condition of Postmodernity*, Blackwell, Oxford, 240–259.

[...] In this chapter, I shall look briefly at the long transition that prepared the way for Enlightenment thinking about space and time.

In the relatively isolated worlds (and I use the plural advisedly) of European feudalism, place assumed a definite legal, political and social meaning indicative of a relative autonomy of social relations and of community inside roughly given territorial boundaries. Within each knowable world, spatial organisation reflected a confused overlapping of economic, political and legal obligations and rights. External space was weakly grasped and generally conceptualised as a mysterious cosmology populated by some external authority, heavenly hosts or more sinister figures of myth and imagination. The finite centred qualities of place (an intricate territory of interdependence, obligation, surveillance and control) matched time-honoured routines of daily life set in the infinity and unknowability of 'enduring time' (to use Gurvitch's term). Mediaeval parochialism and superstition were paralleled by an 'easy and hedonistic psychophysiological' approach to spatial representation. The mediaeval artist 'believed that he could render what he saw before his eyes convincingly by representing what it felt like to walk about, experiencing structures, almost tactilely, from many different sides, rather than from a single overall vantage' (Edgerton 1976). Mediaeval art

The Map Reader: Theories of Mapping Practice and Cartographic Representation, First Edition. Edited by Martin Dodge, Rob Kitchin and Chris Perkins.

and cartography, interestingly, seem to match the sensibility portrayed in de Certeau's 'spatial stories'.

[...]

The Renaissance, however, saw a radical reconstruction of views of space and time in the Western world. From an ethnocentric viewpoint, the voyages of discovery produced an astounding flow of knowledge about a wider world that had somehow to be absorbed and represented. They indicated a globe that was finite and potentially knowable. Geographical knowledge became a valued commodity in a society that was becoming more and more profit conscious. The accumulation of wealth, power and capital became linked to personalised knowledge of, and individual command over, space. By the same token, each place became vulnerable to the direct influence of that wider world through trade, intraterritorial competition, military action, the inflow of new commodities, of bullion and the like. But by virtue of the piecemeal development of the processes shaping it, the revolution in conceptions of space and time was slow to unfold.

Fundamental rules of perspective – rules that broke radically with the practices of mediaeval art and architecture, and which were to dominate until the beginning of the twentieth century – were elaborated in mid-fifteenth century Florence by Brunelleschi and Alberti. This was a fundamental achievement of the Renaissance; it shaped ways of seeing for four centuries. The fixed viewpoint of perspective maps and paintings 'is elevated and distant, completely out of plastic or sensory reach.' It generates a 'coldly geometrical' and 'systematic' sense of space which, nevertheless, gives 'a sense of harmony with natural law, thereby underscoring man's moral responsibility within God's geometrically ordered universe' (Edgerton 1976: 114). A conception of infinite space allowed the globe to be grasped as a finite totality without challenging, at least in theory, the infinite wisdom of the deity. 'Infinite space is endowed with infinite quality,' wrote Giordano Bruno at the end of the Renaissance, 'and in the infinite quality is lauded the infinite act of existence' (cited in Kostof 1985: 537). The chronometer, which gave strength and measure to the idea of time's arrow, was likewise rendered theoretically compatible with God's infinite wisdom by attributing infinite qualities to time analogous to those which attached to space. The attachment was of immense importance. It meant that the idea of time as 'becoming' – a very human sense of time which is also contained in the idea of time's arrow – was separated from the analytical and 'scientific' sense of time which rested on a conception of infinity that was preferred (though not by the authorities in Rome) broadly for religious reasons. The Renaissance separated scientific and supposedly factual senses of time and space from the more fluid conceptions that might arise experientially.

Giordano Bruno's conceptions, which prefigured those of Galileo and Newton, were in practice so pantheistic that Rome burned him at the stake as a threat to centralised authority and dogma. In so doing, the Church was recognising a rather significant challenge that infinite time and space posed to hierarchically conceived systems of authority and power based in a particular place (Rome).

Perspectivism conceives of the world from the standpoint of the 'seeing eye' of the individual. It emphasises the science of optics and the ability of the individual to represent what he or she sees as in some sense 'truthful', compared to superimposed truths of mythology or religion. The connection between individualism and perspectivism is important. It provided an effective material foundation for the Cartesian principles of rationality that became integrated into the Enlightenment project. It signalled a break in artistic and architectural practice from artisan and vernacular traditions towards intellectual activity and the 'aura' of the artist, scientist or entrepreneur as a creative individual. There is also some evidence to connect the formulation of perspectivist rules with the rationalising practices emerging in commerce, banking, book-keeping, trade and agricultural production under centralised land management (Kostof 1985: 403–10).

The story of Renaissance maps, which took on entirely new qualities of objectivity, practicality and functionality, is particularly revealing. Objectivity in spatial representation became a valued attribute because accuracy of navigation, the determination of property rights in land (as opposed to the confused system of legal rights and obligations that characterised feudalism), political boundaries, rights of passage and of transportation, and the like, became economically as well as politically imperative. Many special purpose map representations, such as the portolan charts used by navigators and estate maps used by landowners, already existed, of course, but the importation of the Ptolemaic map from Alexandria to Florence around 1400 appears to have played a crucial role in the Renaissance discovery and use of perspectivism:

> 'The portolans did not furnish a geometrical framework for comprehending the whole world. The Ptolemaic grid, on the other hand, posed an immediate mathematical unity. The most far-flung places could all be precisely fixed in relation to one another by unchanging coordinates so that their proportionate distance, as well as their directional relationships, would be apparent... The Ptolemaic system gave the Florentines a perfect, expandable cartographic tool for collecting, collating and correcting geographical knowledge. Above all, it supplied to geography the same aesthetic principles of geometrical harmony which Florentines demanded of all their art'. (Edgerton 1976)

The connection with perspectivism lay in this: that in designing the grid in which to locate places, Ptolemy had imagined how the globe as a whole would look to a human eye looking at it from outside. A number of implications then follow. The first is an ability to see the globe as a knowable totality. As Ptolemy himself put it, 'the goal of chorography is to deal separately with a part of the whole', whereas 'the task of geography is to survey the whole in its just proportion'. Geography rather than chorography became a Renaissance mission. A second implication is that mathematical principles could be applied, as in optics, to the whole problem of representing the globe on a flat surface. As a result, it seemed as if space, though infinite, was conquerable and containable for purposes of human occupancy and action. It could be appropriated in imagination according to mathematical principles. And it was exactly in such a context that the revolution in natural philosophy, so brilliantly described by Koyré (1957), which went from Copernicus to Galileo and ultimately to Newton, was to occur.

[...]

If spatial and temporal experiences are primary vehicles for the coding and reproduction of social relations (as Bourdieu suggests), then a change in the way the former get represented will almost certainly generate some kind of shift in the latter. This principle helps explain the support that the Renaissance maps of England supplied to individualism, nationalism and parliamentary democracy at the expense of dynastic privilege. But, as Helgerson points out, maps could just as easily function 'in untroubled support of a strongly centralised monarchic regime,' though Philip II of Spain thought his maps sufficiently subversive to keep them under lock and key as a state secret. Colbert's plans for a rational spatial integration of the French nation state (focused as much upon the enhancement of trade and commerce as upon administrative efficiency) are typical of the deployment of the 'cold rationality' of maps used for instrumental ends in support of centralised state power. It was, after all, Colbert, in the age of French Absolutism, who encouraged the French Academy of Sciences (set up in 1666) and the first of the great mapmaking family, Jean Dominique Cassini, to produce a coherent and well ordered map of France.

The Renaissance revolution in concepts of space and time laid the conceptual foundations in many respects for the Enlightenment project. What many now look upon as the first great surge of modernist thinking took the domination of nature as a necessary condition of human emancipation. Since space is a 'fact' of nature, this meant that the conquest and rational ordering of space became an integral part of the modernising project. The difference this time was that space and time had to be organised not to reflect the glory of God, but to celebrate and facilitate the liberation of 'Man' as a free and active individual, endowed with consciousness and will. [...] Enlightenment thinkers similarly looked to command over the future through powers of scientific prediction, through social engineering and rational planning, and the institutionalisation of rational systems of social regulation and control. They in effect appropriated and pushed Renaissance conceptions of space and time to their limit in the search to construct a new, more democratic, healthier, and more affluent society. Accurate maps and chronometers were essential tools within the Enlightenment vision of how the world should be organised.

Maps, stripped of all elements of fantasy and religious belief, as well as of any sign of the experiences involved in their production, had become abstract and strictly functional systems for the factual ordering of phenomena in space. The science of map projection, and techniques of cadastral surveying, made them mathematically rigorous depictions. They defined property rights in land, territorial boundaries, domains of administration and social control, communication routes and so on with increasing accuracy. They also allowed the whole population of the earth, for the first time in human history, to be located within a single spatial frame. The grid that the Ptolemaic system had provided as a means to absorb the inflow of new information had by now been corrected and filled out, so that a long line of thinkers, from Montesquieu to Rousseau, could begin to speculate on the material and rational principles that might order the distribution of populations, ways of life and political systems on the surface of the globe. It was within the confines of such a totalising vision of the globe that environmental determinism and a certain conception of 'otherness' could be admitted, even flourish. The diversity of peoples could be appreciated and analysed in the secure knowledge that their 'place' in the spatial order was unambiguously known. In exactly the same way that Enlightenment thinkers believed that translation from one language to another was always possible without destroying the integrity of either language, so the totalising vision of the map allowed strong senses of national, local and personal identities to be constructed in the midst of geographical differences. Were not the latter after all entirely compatible with the division of labour, commerce and other forms of exchange? Were they not also explicable in terms of different environmental conditions? I do not want to idealise the qualities of thought that resulted. The environmentalist explanations of difference put forward by Montesquieu and Rousseau hardly appear enlightened, while the sordid facts of the slave trade and the subjugation of women passed Enlightenment thinkers by with hardly a murmur of protest. Nevertheless, I do want to insist that the problem with Enlightenment thought was not that it had no conception of 'the other' but that it perceived 'the other' as necessarily having (and sometimes 'keeping to') a specific *place* in a spatial order that was ethnocentrically conceived to have homogeneous and absolute qualities.

[...]

I think it useful, however, to pave the path to understanding the break into modernist ways of seeing after 1848 with a consideration of the tensions that lay within Enlightenment conceptions of space. The theoretical, representational and practical dilemmas are also instructive in interpreting the subsequent move towards postmodernism.

Consider, as a starting point, de Certeau's contemporary critique of the map as a 'totalising device'. The application of mathematical principles produces 'a formal ensemble of abstract places' and 'collates on the same plane heterogeneous places, some received from tradition and others produced by observation'. The map is, in effect, a homogenisation and reification of the rich diversity of spatial itineraries and spatial stories. It 'eliminates little by little' all traces of 'the practices that produce it'. While the tactile qualities of the mediaeval map preserved such traces, the mathematically rigorous maps of the Enlightenment were of quite different qualities. Bourdieu's arguments also apply. Since any system of representation is itself a fixed spatial construct, it automatically converts the fluid, confused, but nonetheless objective spaces and time of work and social reproduction into a fixed schema. 'Just as the map replaces the discontinuous patchy space of practical paths by the homogeneous, continuous space of geometry, so the calendar substitutes a linear, homogeneous, continuous time for practical time, which is made up of incommensurable islands of duration each with its own rhythm'. The analyst, Bourdieu continues, may win 'the privilege of totalisation' and secure 'the means for apprehending the logic of the system which a partial or discrete view would miss,' but there is also 'every likelihood that he will overlook the change in status to which he is subjecting practice and its product', and consequently 'insist on trying to answer questions which are not and cannot be questions for practice'. By treating certain idealised conceptions of space and time as real, Enlightenment thinkers ran the danger of confining the free flow of human experience and practice to rationalised configurations. It is in these terms that Foucault detects the repressive turn in Enlightenment practices towards surveillance and control.

This provides a useful insight into 'postmodernist' criticism of the 'totalising qualities' of Enlightenment thought and the 'tyranny' of perspectivism. It also highlights a recurring problem. If social life is to be rationally planned and controlled so as to promote social equality and the welfare of all, then how can production, consumption and social interaction be planned and efficiently organised except through the incorporation of the ideal abstractions of space and time as given in the map, the chronometer and the calendar? Beyond this there lies another problem. If perspectivism, for all its mathematical rigour, constructs the world from a given individual viewpoint, then from whose perspective is the physical landscape to be shaped? The architect, designer, planner could not preserve the tactile sense of mediaeval representations. Even when not directly dominated by class interests, the producer of space could only produce 'alien art' from the standpoint of its inhabitants. Insofar as the social planning of high modernism reincorporated these elements into its practical applications, it likewise stood to be accused of the 'totalising vision' of space and time to which Enlightenment thinking was heir. The mathematical unities given by Renaissance perspectivism could, from this standpoint, be regarded as just as totalising and repressive as the maps.

Let me follow this line of argument a bit further in order to capture the central dilemma of defining a proper spatial frame for social action.

The conquest and control of space, for example, firstly requires that it be conceived of as something usable, malleable and therefore capable of domination through human action. Perspectivism and mathematical mapping did this by conceiving of space as abstract, homogeneous and universal in its qualities, a framework of thought and action which was stable and knowable. Euclidean geometry provided the basic language of discourse. Builders, engineers, architects and land managers for their part showed how Euclidean representations of objective space could be converted into a spatially ordered physically landscape. Merchants and landowners used such practices for their own class purposes, while the absolutist state (with its concern for taxation of land and the definition of its own domain of domination and social control) likewise relished the capacity to define and produce spaces with fixed spatial coordinates. But these were islands of practice within a sea of social activities in which all manner of other conceptions of space and place – sacred and profane, symbolic, personal, animistic – could continue to function undisturbed. It took something more to consolidate the actual use of space as universal, homogeneous, objective and abstract in social practice. In spite of the plethora of utopian plans, the 'something more' that came to dominate was private property in land, and the buying and selling of space as a commodity.

[. . .]

Enlightenment thinkers sought a better society. In so doing they had to pay attention to the rational ordering of space and time as prerequisites to the construction of a society that would guarantee individual liberties and human welfare. The project meant the reconstruction of the spaces of power in radically new terms, but it proved impossible to specify exactly what those terms might be. State, communitarian and individualistic ideas were associated with different spatial landscapes, just as differential command over time posed crucial problems of class relations, of the rights to the fruits of one's labour, and of capital accumulation. Yet all Enlightenment projects had in common a relatively unified commonsense of what space and time were about and why their rational ordering was important. This common basis in part depended on the popular availability of watches and clocks, and on the capacity to diffuse cartographic knowledge by cheaper and more efficient printing techniques. But it also rested upon the link between Renaissance perspectivism and a conception of the individual as the ultimate source and container of social power, albeit assimilated within the nation state as a collective system of authority. The economic conditions of the European Enlightenment contributed in no uncertain measure to the sense of common objectives. Increased competition between states and other economic units created pressure to rationalise and coordinate the space and time of economic activity, be it within a national space of transport and communications, of administration and military organisation, or the more localized spaces of private estates and municipalities. All economic units were caught up in a world of increasing competition in which the stakes were ultimately economic success (measured in the bullion so dear to the mercantilists, or by the accumulation of individualised money, wealth and power as lauded by the liberals). The practical rationalisation of space and time throughout the eighteenth century – a progress marked by the rise of the Ordnance Survey or of systematic cadastral mapping in France at the end of the eighteenth century – formed the context in which Enlightenment thinkers formulated their projects. And it was against this conception that the second great turn of modernism after 1848 revolted.

References

Edgerton, S.Y. (1976) *The Renaissance Rediscovery of Linear Perspective*, Basic Books, New York.

Kostof, S. (1985) *A History of Architecture: Settings and Rituals*, Oxford University Press, Oxford.

Koyré, A. (1957) *From the Closed World to the Infinite Universe*, The John Hopkins Press, Baltimore, MD.

Further reading

Brotton, J. (1997) *Trading Territories: Mapping the Early Modern World*, Reaktion, London. [Brotton's book provides an effective discussion of the cartographic underpinning of commercial trade.]

Edney, M.H. (1997) *Mapping an Empire: The Geographical Construction of British India 1765–1843*, University of Chicago Press, Chicago. [An important study of application of Cartesian ways of ordering spatial knowledge drawing upon detailed empirical work on Indian Survey.]

Lestringant, F. (1994) *Mapping the Renaissance World: The Geographical Imagination in the Age of Discovery*, Polity Press, Cambridge. [A lively and wide ranging historical analysis of the emergence of world mapping in the West.]

Turnbull, D. (1996) Cartography and science in early modern Europe: mapping the construction of knowledge spaces. *Imago Mundi*, **48**, 5–24. [Turnbull's article offers insights into the emergence of scientific mapping practices.]

See also

- Chapter 1.8: Deconstructing the Map
- Chapter 1.9: Drawing Things Together
- Chapter 1.12: The Agency of Mapping: Speculation, Critique and Invention
- Chapter 5.4: Mapping: A New Technology of Space; and Geo-body
- Chapter 5.5: First Principles of a Literary Cartography
- Chapter 5.7: A Map that Roared and an Original Atlas: Canada, Cartography, and the Narration of Nation

Chapter 5.3

Texts, Hermeneutics and Propaganda Maps

John Pickles

Editors' overview

In this excerpted chapter, Pickles makes the case for understanding maps not simply as technical productions, but as texts that are constructed and read. He argues that the easily asserted division between the technical and political, between inaccuracies caused by the difficulties of precisely representing spatial relations and political decisions made by cartographers seeking to convey a particular message (between 'objective' and 'propaganda' maps), need to be deconstructed. For Pickles, all maps are texts, and all are ideologically loaded, asserting messages and meanings, regardless of any claims to objectivity and truth. Through an historical analysis of overtly propagandist maps Pickles argues for a hermeneutic approach to cartographic scholarship that seeks to provide a contextual, interpretive and 'nested' account of maps as texts. Such an approach recognises that the transmission and reception of a map is not a straightforward, linear process, rather analysis of maps must seek to frame the making of a map within contexts in which it is made, the other texts within which it is presented, the situation into which it is projected, and the world of the reader. In other words, the meaning and interpretation of maps cannot simply be controlled through technical means, and it is foolhardy to conceptualise them, as per the communication model, as if this is the case.

Originally published in 1991: Chapter 12 in *Writing Worlds: Discourse, Text and Metaphor in the Representation of Landscape* (eds Trevor J. Barnes and James T. Duncan), Routledge, London, 193–230.

[...]

Modern cartography and communication models of meaning

Present day cartography is a product of the Cartesian world, in which the map is the scaled representation of the real. Under this view, maps are devises to information transmission involving the basic rules of graphic communication based on a one-on-one correspondence of the world and the message sent and received (Meuhrcke 1972). Communication involves the mechanical transfer of messages from a sender (inputs) through a medium (transfer) to a receiver (outputs) (Monmonier 1975). As with any communication system the model requires that information from the sender be encoded and that the receiver decode the information (Figure 5.3.1). Information is conveyed, and, in so far as the cartographer, map and map reader all receive the same information, distortion is avoided (Robinson and Petchenik 1975) (Figure 5.3.2). The measure of communication efficiency in the mapping process is related to the amount and accuracy of information transmitted. The cartographer's task is to devise better approximations between raw data and the map image (Muehrcke 1972). Thus, the map is an objective tool for transmitting information. In so far as the technical production does not distort the data collected from the real world the 'good cartographer' is successful. By contrast, if the cartographer deliberately selects information to support his or her argument, and seeks to produce a map that has visual impact, then he or she becomes the 'propaganda cartographer' (Ager 1977).

The Map Reader: Theories of Mapping Practice and Cartographic Representation, First Edition. Edited by Martin Dodge, Rob Kitchin and Chris Perkins.

In this model the sources of error are limited and can be specifically ascribed to either a domain of technical control or a domain of ideological/political control (Figure 5.3.3). Error is introduced (1) as a result of technical error (the poor choice of symbols, inappropriate projection or incorrect scale); (2) as a result of malicious intent to deceive (as with Haushofer's (1928) cartography, where some ulterior motive lies behind the construction of the map and where technique is subsumed to the attempt to persuade); or (3) from untrained decoders (map readers) or perceptual or value differences among interpreters.

SENDER → encoding → MEDIUM → decoding → RECEIVER

Figure 5.3.1

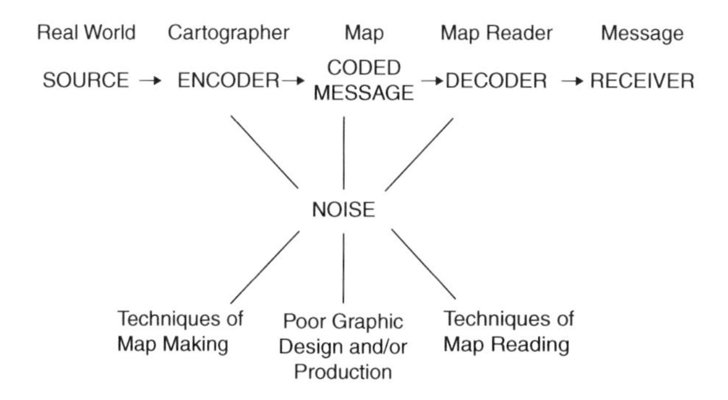

Figure 5.3.2

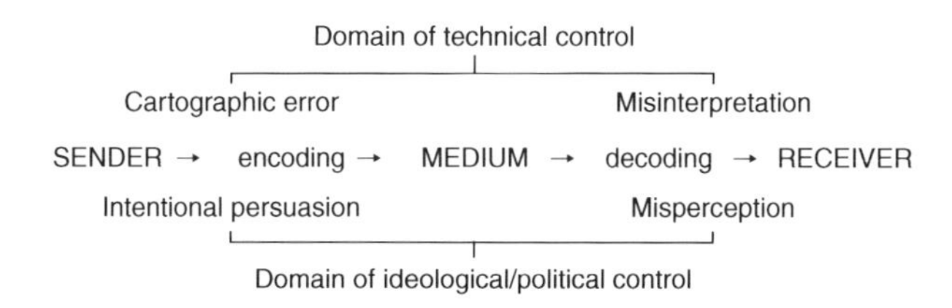

Figure 5.3.3

Such a neat division between the technical and the political produces, in turn, useful practical distinctions for the working cartographer: cartographic technique is seen as an ongoing approximation to the real, presupposing a correspondence or representational theory of truth. The distinction between fact is thereby mirrored by the separation of the good cartographer and the propaganda cartographer (the latter being banished from the halls of science). In this sense, adapting correspondence rules between world, fact and map raises no fundamental and problematic questions of interpretation or of context.

Representationalism, correspondence rules and the separation of fact and value, however, do present immense conceptual difficulties for accounts of distortion and error. The ideological is to be expunged but from a world that disavows its own ideology. The only remedies open in such a world are to retreat into technique, to reject interpretation as bias, and to accept an 'openness to alternatives', which permits claims of the sort that 'because interpretation is always a factor the cartographer must exercise care and judgement'. Not surprisingly, then, in the present century, as objective science has increasingly assumed that its methods give access to the true, the view of propaganda has altered radically. Science is seen not as a persuasive enterprise but as a claim to true knowledge. A good map is one in which the image received by the map user corresponds to that intended (inscribed) by the mapmaker and where the image inscribed (and received) is an accurate representation of the real world. Mapmaking and map reading are seen to involve the straightforward transmittal of information in a philosophically and practically unproblematic manner. In particular, cartography does not seek to persuade, to convince or to argue; it does not select techniques of representation on the basis of their visual impact; and in the choice of subject matter, what is centred on the page, what is consigned to the edge of the map, and which scale and projection shall be used, the cartographer is guided by rules of scientific procedure and convention.

Propaganda maps

While the cartographer tries to present information accurately, comprehensively, with a balanced design, and without favouring one side of an issue, the propaganda cartographer seeks 'to produce a map which has visual impact and is not only believable, but goes a stage further – is convincing' (Ager 1977: 1). Propaganda maps are to be guarded against because the cartographer has used the wrong method and has 'failed to communicate correctly with the user' (Ager 1977: 14). [. . .] 'The propagandist's primary concern is never the truth of an idea but its successful communication to a public. [. . .]' (Speier 1942: 31–33).

In what is probably the most systematic treatment in English of propaganda maps to date, Tyner (1974: 2) suggests the name 'persuasive cartography' to distinguish propaganda, suggestive, advertising, journalistic and subjective cartography from other forms. 'Persuasive cartography is a type of cartography whose main object or effect is to change or in some way influence the reader's opinion, in contrast to most cartography which strives to be objective.' Persuasive cartography thus seeks to manipulate symbols in order to influence some group about the value of some idea, opinion or action. Where some malicious intent lies behind this manipulation, propaganda is used.

[...]

[C]urrent theories of cartography, working with a distinction between objective cartography and propaganda maps, are founded on erroneous epistemological grounds. The distinction is articulated in terms of poles of objectivity and bias. When interrogated by cartographers themselves, however, the distinction breaks down, and leaves us with a series of difficult questions to answer. Are only those maps which use certain graphical techniques to create a distorted impression to be considered propaganda maps? Or are there certain categories of map use that constitute propaganda, while other uses do not? Are all maps propaganda maps?

Contextual interpretation: the sociology of propaganda maps

Solving the problem of distortion and representation by adopting the view that all maps are propaganda maps does not deal with our problem, it merely sidesteps the issue. Since all maps are constructed images, and since all images are interpretations of a particular context, we gain little by merely repeating that maps and propaganda maps are both interpretations and distortions. We remain caught within the metaphysics of presence, which presupposes some foundational object against which the distortions and interpretations can be measured: that some interpretation-free image could be produced that does not distort the world. Moreover, in arguing that all maps are propaganda maps we seem to deny the value of any theory which seeks to judge maps critically. Because all maps distort, what is important is the intention behind the construction of the map, and the use to which a map is put. [...]

Propaganda maps and hegemony

[...]

The increasingly sophisticated techniques of propaganda have often been seen as problematical for a society which protects free speech. By the late nineteenth and early twentieth centuries such techniques were seen to pose a threat to democratic practices and institutions themselves. Liberal political thought saw propaganda as a potentially dangerous form of distortion of true claims and accurate positions, which in the hands of the unscrupulous and corrupt might pose a threat to liberal democracy itself. [...]

Gramsci (1981: 80, n. 49) saw the techniques of public persuasion (of which propaganda was one) as central elements in forging the relationship between consent and coercion in the establishment of hegemony: [...] Propaganda aims at persuading large groups of people to believe something or act in a way that they would not, in the normal course of events. Propaganda techniques are, then, techniques of persuasion which fail to abide by established and accepted norms of accuracy and truth. They seek to manipulate relationships in order to persuade people about a particular claim to truth.

With such techniques institutions concerned with the process of establishing hegemony capture the discursive field and reconstitute the discourse of the age and the place. One means by which the state has attempted in the past to capture the discursive field is precisely not only through the appropriation of space (and the map) to its purposes, but by the symbolic constitution of mapped space as national space. In so doing, fledgling national territories sought to establish a national identity abroad, and to create a national ideology at home, sometimes in the face of internal disunity or rebellion. The map has been used frequently in this way to express some sense of national identity. Here the link between map and symbol becomes clear. [...]

Propaganda maps and war

[...]

[M]aping is an interpretative act not purely a technical one in which the product – a map – conveys not merely the facts but also and always the author's intention, and all the acknowledged and unacknowledged conditions and values any author (and his/her profession, time and culture) bring to a work. Thus, like all works, the map carries along with it so much more than the author intended. Also, like any text, the map takes on a life (and a context) of its own beyond the author's control. The map is a text, like any other in this regard, whose meaning and impact may go far beyond the limits of technique, *mens auctoris*, and the mere transmittal of information.

Propaganda maps and time

The perception of graphical images is not a purely psychological reception of information but a complex social play of images present and absent, in the context of other symbolic, ideological and material concerns. All cartography operates within and makes use of such unacknowledged pre-conditions and more or less accepted symbolic forms and mapping conventions. Yet the impact of these techniques and effects becomes particularly clear when we turn to obviously propagandistic texts. At a very basic level a particular iconography has been appropriated to the goals of political and commercial propaganda. The globe and the map both stand as icons, repeated as an unchallenged vocabulary of advertising, Cold Warism or national boosterism.

[...] Temporal sequencing and contextual memory are vital to the work of all graphic designers and artists. In modern cartography they are made more powerful by new techniques of design and production. Principles and techniques including scale and projection, assimilation and contrast, negative and positive, grouping and line, symmetry, double symmetry and asymmetry, reversible images, and the perceptual influence of angles, provide the graphic designer with powerful optical devices to manipulate and create the required field of visual effects. Modern cartography linked with new technologies of image production and transmittal has many new and powerful techniques for the creation of new forms of dynamic imagery. [. . .]

Textual interpretation of the propaganda map

Traditional theories of mapping and maps are of little use to us when we push them to incorporate propaganda maps. Without the foundation of an unproblematic theory of representation to fall back upon, cartographers are forced to retreat to the position that all maps are distorting and, hence, all maps function as propaganda maps. But as we have already seen this is merely to sidestep the issue, and raises other serious questions about the sorts of claims we can make and the work we can do. Both approaches fail because they do not appreciate the textual qualities of maps. In this section I will begin to unpack this textuality in the form of a 'nested' account of maps as texts. This nested account will address three issues: the world and the text, the text in a text, and the analysis of the work itself.

We encounter ambiguity the moment we ask, what is the content of a graphic image? Clearly it is the real world, the real situation, the landscape, the scene. The mapmaker reduces this object field according to established principles of objectification, abstraction, reduction and idealisation to create the map. In this sense all maps are thematic abstractions involving reduction of one form or another. In a way different from the photographic image, however, this reduction is a transformation. In order to move from the real situation to the map it is necessary to divide up this reality into units and to constitute these units as signs substantially different from the object they communicate. The map is thus a coded message whose relationship to the object world it portrays is complex.

What is the nature of the coding? The map is a message. As the previous discussion suggests, cartographers and geographers have traditionally taken this message to involve a source, a medium and a receiver. The source is the cartographer (and his or her body of received techniques and style), the medium is the map (and the often ignored immediate contexts within which the map is embedded) and the receiver is the map reader (as a public 'readership'). This view construes the map too narrowly, however. It ignores the other texts within which the map is itself embedded and with which it is co-determined. It ignores the context into which the map is projected. It one-sidedly places emphasis on the intended message and fails to consider possible unintended meanings. Finally, it has no way of accounting for the ability of graphic images to conjure up other texts (maps, photographs, books etc.) and embed them in any reading of their own codes. By way of illustration let us ask, what is the medium of the map's message? In the communication models discussed above, the medium is the map. But how can this be? The medium is the report, the article, the book, the magazine, within which the map appears. [. . .]

We are faced with layers of textuality: the map itself, the immediate context of the map (its caption, the chapter and the work of which it is a part) and the wider context of the map (the opus of the individual cartographer or school, the opus to which the text itself belongs, the socio-cultural context of the work). But although the map is an embedded figure, the map is also an object that has a structural autonomy independent of both its production and its use, and thus requires an analysis of the work itself. This will not be definitive, but will always have to be situated alongside a sociological analysis of text and context, of production and use. Even an analysis of the work itself cannot divorce the map entirely from its context, for the map is not an isolated object. It has a title, and fits within the body of a text, along with a set of other maps, or, if it is a single map, it is framed and displayed in some manner.

In terms of the internal construction of the map, the message of the map is carried by two different structures, one of which is graphic, the other of which is linguistic. Yet the consideration given to the linguistic components of the map has been mainly restricted to the design and effectiveness of the graphicality of lettering (size, print style, placement). The importance of the linguisticality of the map has been ignored. While in other graphic forms (photography, painting) the graphical and linguistic elements are complementary, in the map they operate almost uniquely as inseparable from each other. This inseparability is also typical of certain forms of advertising, poster art and modernist art forms such as Dadaism. Here the linguistic elements are embedded within the image, not incidentally but as intrinsic components of the whole picture. [. . .]

In the map the symbolic graphic image is embedded in a written text (a paper, a book, an atlas) and rarely has an existence beyond the body of the text and the discursive aims of the research of which it is a part. Moreover, the symbols and words in the map are interbedded: the names of places, features and other descriptors are integral to the visual image, and call for a special form of construction and present specific difficulties for analysis. In particular, such interbedded texts (maps, poster, commercial and Dadaist art) are correspondingly much closer to the tract: the commercial, political poster or artwork. It is in this intersection that much of what has been called propaganda mapping arises. The issue becomes clearer on closer analysis.

For Barthes all 'imitative' arts comprise two messages: a denotive message and a connotive message. The realist painter and photographer stake their reputations on their work being predominantly denotive, in the sense that the representation of the objects is a re-presentation in which the objects represented are objects from the world, without transformation. Traditional cartographic theory presents the map as a purely denotive message. But, as we have already seen in the mapping process, objects to be represented are transformed and reconstituted as signs and symbols substantially different from the objects they communicate. That is to say, the map is a coded message.

[...]

The transmission and reception of the map image are not the straightforward, linear processes presumed in the communication model. The coded image (the map, linguistic and graphic) is also connotive. Through the fusion of horizons between the reader's world and the world of the map (and the mapmaker) the map connotes a variety of meanings. Thus, the reading of the map is always historical and 'depends on the reader's "knowledge" just as though it were a matter of a real language, intelligible only if one has learned the signs' (Barthes 1987: 28). The map is a purposive cultural object with reasons behind its construction and values associated with its reading. To suggest otherwise is to fail to see its status as made object. The map is always and necessarily an expression of an idea. In mediating the transformative processes of abstraction, reduction, thematisation and idealisation, the cartographer selects, sifts and emphasises this or that aspect of the world under consideration, and articulates an image in the rebus linking graphic and linguistic codes.

[...]

Hermeneutics

[In] the second half of the twentieth century a new metaphor – the text metaphor – has arisen as the template for understanding and framing social life. In this period the text metaphor has colonised certain domains of study – painting, film, landscape and, most recently, social life.

[T]he adoption of the reading metaphor in [...] interpretive approaches has given rise to a methodology which is largely implicit, often derived from years of apprenticeship and practice, and informed by the keen eye of the fieldworker, but unguided by the detailed and explicit debates about hermeneutic methodology developed elsewhere in the social sciences and in philosophy. [...] Only recently have geographers begun to explore the necessity for, and the character of, an explicit hermeneutic canon and its associated interpretive theories and approaches. [...]

Without [...] hermeneutics our ability to develop intersubjectively accepted criteria for assessing claims must be in question. [...] Hermeneutic method has been seen as a technique appropriate only to some aspects of social enquiry, but also as a hindrance to materialist critique (Walker 1989). In this context the radical and universal nature of hermeneutic understanding has yet to be fully explicated. [...]

How, then, do we read maps, especially those in which problems of interpretation are compounded by distortion, error and lies? [...] [A]s texts the map and the landscape present innumerable problems of determining authorship, establishing a syntax and structure by which to read (and knowing what not to read), and distinguishing and relating the various levels of determination that historically constituted any particular map or landscape. [...] From its conception the propaganda map aims to be a distortion. [...] The propaganda map is, therefore, the archetypical problematical text requiring hermeneutic interpretation, and provides a potentially good starting point for elaborating the methods of interpretation; philology, hermeneutics and criticism. [...]

Philology places strong demands on the act of interpreting texts (be they poems, landscapes, maps or social actions). Is the text the one it is claimed to be: is the ascribed authorship correct, did the text fulfil the role it is claimed to have filled, and so on? Is it a coherent whole? What does the text say about its own world? What does the text now mean? What is the relationship between the meaning of a text and the intention of the author in creating it? Given that some of these texts may have been authored by people who are no longer known or who were anonymous at the time of production, that they may have originated in worlds about which we now know little or nothing, and that only fragments may now be extant, are we really able to retrieve the *mens auctoris* (the author's intention)? And if we are, then in what sense can we claim to have access to the *mens auctoris*? Does the work constitute something independent of and different from that intention? And, if we cut our interpretation loose from the author's intention, how do we avoid the danger of arbitrary and forced interpretations?

Strict concern for the *mens auctoris* would, of course, place us in an untenable position as social scientists. The antiquarian may be able to bracket his/her present world and become immersed 'fully' in the world of the other, of the past, of the author. This option is never open to the social scientist (nor, practically, even to the antiquarian). We ask questions always from the standpoint of the present, and in every case we carry out a retrieval of the author, his/her intentions, and the work, in order to make them meaningful in our present worlds (be they conceptually, temporally or geographically removed). We cannot derive the text's 'own meaning' in this way, but must recognise that in any interpretation the work has an autonomy of its own beyond the intensions of its author(s). In this process the author retains a claim on the surface details of the work: the site, the literal and symbolic content intended, the date of production, the materials and techniques used in production. But with regard to the meaning of the word as a work, the author's claim is more tenuous. The author's intention is not fully determinative of the meaning of the text. It is for this reason that all interpretations of works must also be destructive retrievals, that is they must situate the work in a context meaningful to the reader.

Consequently, where the interpreter is able to make explicit what the author left implicit or did not recognise or understand, he or she may know more about a work than its author. The philological concern thus corresponds to a lower hermeneutic, which is concerned to establish a critical edition of a text, to verify that the text is the text it is claimed to be, that is it has not been falsified, that it is (or is not) a coherent whole and is not a pastiche of several authors, that it is authentic and that it is complete. Higher hermeneutics takes as its task the proper understanding of the meaning of a text, how it related to its own world (and subsequent worlds in which it has had an existence) and how it is to be related to our present world. These claims become clearer when we recognise that symbols may be of one of two types: univocal or equivocal. Univocal signs, like symbols in symbolic logic or mathematics, have one designated meaning. Cartographic symbols (church, castle, urban area) have often been seen to be univocal symbols in this manner. Their correspondence, however, is not of the same kind as that of those symbols of logic or mathematics where the equivalence is complete. In the case of the symbol 'church' the equivalence, as with mathematical notation, is purely formal. Except in this trivial formal equivalence the symbol is actually equivocal: the age, style, denomination and size of the church remain open to interpretation from the context of the whole map and the text within which it is embedded. Equivocal symbols may have several layers of meaning and are the true focus of hermeneutics (Ricoeur 1971).

For any rigorous interpretation several conditions must pertain: the integrity of the meaning of the text must be preserved in such a way that meaning is derived from, not projected into, the text (canon l). Consequently, the interpreter has the responsibility to bring him/herself into a harmonious relationship or proper attunement with the text (canon 2). This is neither a call for a slave-like adherence to the text or the tradition to which it belongs, nor to argue that any reading can fully preserve the meaning of the text or bring the reader into full harmony with it. It is an argument that critique must be rooted in the claims, conventions and forms of the text. The interpreter must give an optimal reading of the text and of the meaning the text must have had for those for whom it was written and show what the text means for us today in the context of modern views, interests and prejudices (canon 3).

Of course, the difficulty for any interpretation is precisely where to begin and where to end in reconciling these three canons. This is the importance of the canon of the hermeneutic circle (canon 4): the whole must be understood from its parts, and all the parts must be understood from the whole. All texts have a certain anticipation of their parts from the whole, leading to a wider understanding of the whole and the parts. Part–whole relationships permeate all readings of texts at all levels of analysis and critique, and specifically include: (a) the relationship of the text to its own intrinsic parts; (b) the relationship of language– text–language; (c) the relationship of cultural context– text–cultural context; (d) the relationship of author and his/her world text as part of this world. Finally, since clear and unambiguous texts do not need interpretation (by definition), all other texts must be complemented with suitable assumptions in order for the interpreter to make things explicit which the author (and/or subsequent readings) left implicit (canon 5). In this way we can say that the interpreter understands the test better than the author. All readings of texts must use (implicitly or explicitly) these hermeneutic canons. They are standard methods of textual exegesis, general philology, linguistics and literary criticism. [. . .] It is the adoption of a theory of interpretation predicated on the notion that symbols are the mirror of the world that has caused contemporary theories of maps to be one-sidedly concerned with the *mens auctoris* and the objectivity of the map, marginalised the importance of propaganda maps (and distortion generally), and led to the current practical gulf between social theory and the theory of maps.

Conclusion: writing and theory

[A]n effective critique of the distortive and ideological nature of propaganda maps must be based on a wider conception of what constitutes propaganda. That is, the ideological and propagandistic elements of contemporary 'scientific' maps must also be assessed at those points where the cartographer shares the ideology of his/her age, where accepted practices are founded on particular ideologies, and where unchallenged interests influence the form and content of the theory and practice of mapping. Examples of the ways in which cartography shares and reproduces the values of the age are numerous and some are well known: the continued public use of the Mercator and adapted Mercator projections, the ideological fixation on 'north at the top' maps, and the ridicule which greets, for example, 'the Australian's view of the world'.

Interpreting the meaning of maps also requires that other issues be considered. Two symbolic systems are involved: graphical images and writing systems. Not only does the image exist in a reductive relationship to the world, but graphical systems always also exist as an interplay between images and linguistic texts and contexts, thus creating a multiplicity of cross-cutting structures of a writing system; maps also contain within them the spoken and written in a relationship which is never exactly correspondent (i.e. maps 'play' in two registers). For cartographers this complexity of meaning has generally been seen as a technical problem to be either dissolved by careful adherence to established mapping practices or explained in terms of the creative ability of the mapmaker. In seeing this complexity of meaning in terms resolvable by technical means reducible to individual skills, traditional theories of maps have failed to address the issue as one which is inherent to the form and always in need of unpacking. [. . .]

[. . .]

[W]e need to think through the discursive nature of texts and to clarify the canons of interpretation necessary to the critique of an embedded text. The hermeneutic canons presented in this chapter are only the necessary first canons of any rigorous interpretation. Cartographers and geographers must clarify their own interpretive frameworks if we are to understand and control our production and use of maps, and we might begin by questioning our current understanding of texts, writing and 'reading'.

References

Ager, J. (1977) Maps and propaganda. *Society of University Cartographers Bulletin*, **11**, 1–15.

Barthes, R. (1987) *Image, Music, Text*, Hill & Wang, New York.

Gramsci, A. (1981) *Selections from the Prison Notebooks*, Lawrence & Wishart, London.

Haushofer, K. (1928) Die suggestive karte. *Bausteine zur Geopolitik*, 342–348.

Monmonier, M.S. (1975) *Maps, Distortion, and Meaning*, Association of American Geographers, Washington, DC.

Muehrcke, P. (1972) *Thematic Cartography*, Association of American Geographers, Washington, DC.

Ricoeur, P. (1971) The model of the text: meaningful action considered as a text. *Social Research*, **38**, 529–562.

Robinson, A.H. and Petchenik, B.B. (1975) The map as a communication system. *The Cartographic Journal*, **12** (11), 7–15.

Speier, H. (1942) Magic geography. *Social Research*, **8**, 310–330.

Tyner, J.A. (1974) Persuasive cartography: an examination of the map as a subjective tool of communication. Unpublished PhD thesis. Department of Geography, University of California, Los Angeles, CA.

Walker, R. (1989) What's left to do? Some principles to live by. *Antipode*, **21** (2), 133–165.

Further reading

Edsall, R.M. (2007) Iconic maps in American political discourse. *Cartographica*, **42** (4), 335–347. [A recent informative research article on the propagandist work of cartographic imagery.]

Herb, G.H. (1997) *Under the Map of Germany: Nationalism and Propaganda 1918–1945*, Routledge, London. [A detailed empirical study of cartographic power and propaganda in an important historical period.]

Monmonier, M. (1996) *How to Lie with Maps*, University of Chicago Press, Chicago. [An accessible and entertaining introduction to many ways that maps can be consciously and unconsciously manipulated to sell a particular viewpoint on space.]

Pickles, J. (2004) *A History of Spaces: Cartographic Reason, Mapping, and the Geo-coded World*, Routledge, London. [An updated and extended work on the politics of cartography including the restatement of his theories on the eponymous propaganda map.]

See also

- Chapter 1.7: Designs on Signs/Myth and Meaning in Maps
- Chapter 1.8: Deconstructing the Map
- Chapter 1.9: Drawing Things Together
- Chapter 1.12: The Agency of Mapping: Speculation, Critique and Invention
- Chapter 3.9: Maps, Mapping, Modernity: Art and Cartography in the Twentieth Century
- Chapter 3.10: Affective Geovisualisations
- Chapter 5.4: Mapping: A New Technology of Space; and Geo-body
- Chapter 5.5: First Principles of a Literary Cartography
- Chapter 5.7: A Map that Roared and an Original Atlas: Canada, Cartography, and the Narration of Nation

Chapter 5.4

Mapping: A New Technology of Space; Geo-Body, from *Siam Mapped: A History of the Geo-Body of a Nation*

Thongchai Winichakul

Editors' overview

Maps have long been an important technology of nationhood, colonialism and trade. Winichakul, in his book, *Siam Mapped*, provides a detailed account of the role of cartography in creating and defining the Siam nation. Cartography did not simply reveal the geography of Siam, but through expedition and survey was a crucial endeavour in producing the modern geo-body of Siam; that is, a sense of what Siam was and could become. The maps produced anticipated a spatial reality and imagined community, rather than depicting one. In other words, Winichakul details that the initial maps of Siam were a model for, rather than a model of, what it purported to represent. Maps thus enabled the monarch to assert his authority over territory, exert territorial administration over the whole state and dissipating the role of local chiefs, and also to assert the country's place in the world of nations. In addition, surveying and cartographic projects were critical technologies of modernisation and contributed to the planning and building of modern infrastructure such as roads, railways and telegraph lines.

Originally published in 1994: Chapters 6 and 7 in Thongchai Winichakul, *Siam Mapped: A History of the Geo-Body of a Nation*, University of Hawai'i Press, Honolulu, HI, pp. 113–139.

Geography had been a powerful science inseparable from the knowledge of the East from the early days of the Europeans' journeys until nineteenth century colonialism. For the Malay peninsula, the Chinese were the customary visitors and had made several coastal maps of the area. Some of them became valuable sources for Europeans, such as Marco Polo, whose maps were made in 1292–1294. Since then, the map of this region was included in many atlases of classical, medieval and early modern Europe. Siam in particular, however, appeared rather late in the European maps of this region. It was not included in the Portuguese discoveries until the latter half of the sixteenth century. Thereafter Siam was well recognised by prominent mapmakers.

Siam in Western maps

The French and Dutch in the seventeenth century were the leaders in cartographic techniques. The French court, in particular, established a scientific society led by many generations of cartographer. Both countries were also among the leading European powers in the Oriental expeditions of the time. France's close connection with Siam in the 1680s resulted in the advance of geographical knowledge and the mapping of Siam. The French envoys and cartographers of the court of Louis XIV published many maps of Siam and passed their knowledge to other European mapmakers. [...]

Nonetheless, on all the maps before the second half of the nineteenth century, the detail of Siam was more or less limited to the coastal areas. The interior had been, until then, *terra incognita* to the European. [...]

[...]

Since Mercator invented the latitude–longitude matrix covering the entire globe, the world has been full of blank

The Map Reader: Theories of Mapping Practice and Cartographic Representation, First Edition. Edited by Martin Dodge, Rob Kitchin and Chris Perkins.

squares waiting to be filled in. The New World was 'discovered'. African Africa was found. The unexplored places were opened up and inscribed on the map. Indeed, modern mapmaking had inspired innumerable missions to fulfil its desire to plot the entire world. [...] The desire for geographical knowledge seemed to be an integral part of colonial expansion since it became a master science for colonial acquisition as well as for explorers and administrators.

Western mapping in siam

So far there is no evidence on the impact of Dutch and French mapmaking since the seventeenth century on Siamese mapping. Until the reign of Mongkut (1851–1868), surveys neither interested nor bothered the Siamese court, except in the case of potential war routes to Bangkok – such as Crawfurd's map of the Chao Phraya channel, which sparked off a protest from the court. [...]

Siam under Mongkut's regime was different. The elite were more families with Western ideas and scientific instruments. Globes and maps were among the instruments enjoyed by the Siamese. How important they were for Mongkut may be seen from the fact that among the gifts selected by Western envoys to present to him, maps of many countries and cities were often included. It is hard to say whether the giver wanted the recipient to have these maps, with a certain hidden agenda in mind, or the latter desired to obtain and possess them. It is safe to say that the maps were special enough to be given to Rex Siamensium, the name Mongkut preferred to call himself in the correspondence with foreigners. As for the Siamese elite, having witnessed envoys from so many distant countries and having had knowledge about them for some time, particularly having seen those countries on maps, could they resist imagining or desiring to have Siam be on a map just as those civilised countries were? Siam was out there, to be included on the globe. Yet it was to a considerable extent *terra incognita* in mapping terms, even to the Siamese elite. It was there; but it had yet to be fully recognised and accounted for.

The Siamese elite of Mongkut's regime was more cooperative and ready to deal with foreigners even with a traditional map. In fact, more than cooperative, the regime was active and creative in expanding the role of mapping in state affairs. In the last five years of the reign Bangkok issued a large number of letters and instructions asking local authorities about the frontiers on the Burmese and Cambodian borders. There were also many communications concerning surveys and mapping of many localities within the realm of Siam proper, such as Phitsanulok, Phimai and Prachin. Some communications concerned the survey with the French at Sisophon (then on the Thai–Cambodian border, not inside Cambodia as it is today). Perhaps this was the first sign of the new concept of territorial administration.

Around the same time, negotiations on many boundaries on both the western and eastern fronts with the British and French were proceeding. But there was no evidence that Siam had launched any attempt to draw a map of its own geo-body. Only in 1866, when he knew that a French exploring team was surveying the areas along the Mekhong, did Mongkut realise that Siam must do likewise. A Dutchman was soon appointed to head a team of surveyors to the Mekhong areas from Nan, Luang Phrabang, then eastward to Mukdahan. [...] Apart from this assignment, there is no record of any Siamese survey of its boundary until the 1880s.

Mapping and topographical surveys seemed to play a much greater role in the modernisation projects in Bangkok and certain provincial areas. This role continued and even increased in the following reign, as Siam moved rapidly toward modernisation. The growth of urbanism and construction projects in Bangkok, particularly roads, railways and telegraph lines, required the increasing role of mapping technology in terms of knowledge, technicians and facilities.

Since it was a new technology to the Siamese, however, the task of mapping in those early days was undertaken mostly by foreigners, even by those who were not technicians. [...] In 1880 the British India government requested permission from the Siamese court to conduct a survey in Siam in order to complete the triangulations for the boundary map of British India. They had completed their triangulations from India to their eastern frontier, that is, Burma. To accomplish the boundary mapping of this front, however, they needed to make connected triangulations into Siam. [...] The Siamese court was frightened. An urgent meeting of ministers and senior officers was called to consider the request because many still believed that such a survey was the first step of a foreign invasion. This reaction was understandable, since it was not a survey for a construction project in a particular locality, an undertaking with which the Siamese rulers were familiar. [...] The proposed location of marking points, both of them sacred sites, was another cause of apprehension. Indeed, the proposal tells us a lot about the imperialistic insensitivity of the British and modern geography. Besides, the initiative came from the British, to be done by the British, for the benefit of the British.

[...]

For the rulers in Bangkok, the role of mapping had been expanding rapidly. It had become as necessary to Siam as roads, electricity, the telegraph and railways. The first group of mapping officials was formed as early as 1875 by selecting about fifty men from the Royal Bodyguard, the first Western-style regiment in Siam. [...] In 1882 Damrong recommended the establishment of the first mapping school to train officials to be assistants to the European technicians. The school limited its students according to demand, and most of them were the descendants of high ranking government officials. Among the courses offered in this school were Western mathematics and astronomy as well as the use of sophisticated scientific devices. The students also learned to calculate coordinates and many other topographical measurements. In fact, this school was one of the few

Western-style schools in Siam at the time. And it was the only school run by the Siamese government which offered intensive studies of English and Western scientific knowledge, since this knowledge was necessary for the job. [...] Three years later, in 1885, the Royal Survey Department was founded. It was responsible for all surveys, planning and mapping projects of the government. Mapping was no longer a foreign technology in Siam.

[...]

For the first time the regime was attempting to know the units which comprised the realm in territorial terms. Undoubtedly, this was a consequence of the new vision created by the modern geographical discourse of mapping. Mapping was both a cognitive paradigm and a practical means of the new administration. It demanded the reorganisation and redistribution of space to suit the new exercise of administrative power on a territorial basis. The name of the new system – *thesaphiban* (protection of territory) – reflected these changes honestly.

[...]

It seems that Siam expected mapping to be the means which could determine once and for all the boundary of the realm. By mapping, that is to say, the ambiguity of margins was expected to be eradicated and the clear-cut limits of the realm of Siam would appear. Mapping technology was no longer alien or suspicious to them. Apparently they realised that in order to counter the French claim, modern geography was the only geographical language the West would hear and only a modern map could make an argument. Mapping had frightened the court in the early years of the reign. Now it became an indispensable technology to decide and establish the geo-body of Siam.

[...]

Mapping cross fire: a lethal weapon unleashed

The relation between map and military force was remarkable. The desire of the force was to make the territory exclusive and map it. [...] Mapping spearheaded the conquest. Nevertheless, since the spheres of influence of both sides [Siamese and French] had never been defined and, in fact, were overlapping, a modern boundary could be anywhere in those marginal – in every sense of the word – areas. A proposed boundary therefore was a speculation which, depending on one's point of view, was equally truer and falser than another proposal. In actual practice, the survey of an area by one side was done alongside the military advance. The military decided the extent of territorial sovereignty and provided the authority under which mapping could be executed, not vice versa. Force defined the space. Mapping vindicated it. Without military force, mapping alone was inadequate to claim a legitimate space. But the legitimation of the military presence was always substantiated by a map. Mapping and military became a single set of mutually reinforcing technology to exercise power over space in order to define and create the geo-body of Siam. [...] The geo-body was being created literally on paper. A new life for Siam was about to begin.

Geo-body

The map of bounded Siam appeared for the first time after the Paknam crisis of 1893. Ironically, it was eventually an outcome of cooperation between Britain, France and Siam. By 1893, only the boundary and map of the western front between Siam and Burma were finished. On all other fronts, except for a short boundary at Battambang and the one between Kedah and Perak, there were only topographical surveys and sketches. So all the data and work done by the Siamese and French mapping officials were gathered together with British cooperation. In 1897 two maps were produced by Siam. The first one was published in England. The other map, published in Calcutta under the title *Phaenthi phraratcha-anakhet sayam r.s. 116* (Map of the Boundary of Siam 1897) [...] Both maps stated clearly that whenever there was a gap in the survey by Siam, the maps drawn by the British and French had been copied to fill in the missing sections. Practically and symbolically, Siam had its first geo-body and its representation made, filled and shaped, at least in part, by Western powers.

[...]

The geo-body of Siam was reshaped by many treaties with Britain and France in 1893, 1899, 1902, 1904 and 1907 and by means of cartographic techniques. Siam and both superpowers set up many committees to decide the boundaries and many detailed ad hoc agreements for each section of the boundary with specific maps for each. [...]

Emergence of the geo-body: a victory of mapping

It depends on one's point of view whether the contest between Siam and France for the upper Mekhong and the entire Lao region was a loss or a gain of Siam's territory. But it certainly signalled the emergence of the geo-body of Siam. And the ultimate loser was not, in fact, Siam. The losers were those tiny Chiefdoms along the routes of both the Siamese and the French forces. Not only were they conquered – a fate by no means peculiar to them – but they were also transformed into integral parts of the new political space defined by the new notions of sovereignty and boundary. Another ultimate loser was the indigenous knowledge of political space. Modern geography displaced it, and the regime of mapping became hegemonic.

It was this triumph of modern geography that eliminated the possibility, let alone opportunity, of those tiny chiefdoms being allowed to exist as they had done for centuries. In other words, the modern discourse of mapping was the ultimate conqueror. Its power was exercised through the actions of major agents representing the contending countries. The new geographical knowledge was the force behind every stage of conceiving, projecting, and creating the new entity.

From the beginning, it was a new knowledge – a new geographical 'language' by which information originated and the new notion of the realm of Siam was conceived. It became a framework for thinking, imagining and projecting the desired realm; it became in effect the language in which Siam was to be discussed. But since the reality did not yet exist, the new geography served as the vision for the geo-body of Siam still to be created. Its requisites – the new kind of boundary, sovereignty and margin – were formed at various moments in time and place and in different fashions. When all parties involved became preoccupied by the new conception of a state, drafts and sketches of the limits of Siam were drawn even before an actual survey was carried out. Siam's geo-body was anticipated, and desired, by all parties. But the anticipated extent of the entity to be created was varied. The imperialist concept of the two European powers and the royal hegemonic ambition of Siam were in conflict.

At that point, mapping was no longer merely a conceptual tool for spatial representation. It became a lethal instrument to concretise the projected desire on the earth's surface. Not only was mapping a necessary device for new administrative mechanisms and for military purposes – functions that seemed modest and merely instrumental – but indeed the discourse of mapping was the paradigm within which both administrative and military operations worked and served. In other words, mapping turned both operations into its mechanism to realise its projection, to concretise its 'enunciation'. It transformed human beings of all nations, people whose actions were heroic or savage, honourable or demeaning, into its agents to make the mapped space come true. Siam was bounded. Its geo-body emerged. Mapping created a new Siam – a new entity whose geo-body had never existed before.

Communication theory and common sense alike persuade us that a map is a scientific abstraction of reality. A map merely represents something which already exists objectively. In the history of the geo-body, this relationship was reversed. A map anticipated a spatial reality, not vice versa. In other words, a map was a model for, rather than a model of, what it purported to represent. A map was not a transparent medium between human beings and space. It was an active mediator. In this case, all the requisites of the map of a nation had not been given in pre-modern Siam and thus had to be created to meet the demands of a map. The outcome was the result of the contending anticipation expressed on each claimant's map. Perhaps more than has been realised, the regime of mapping did not passively reflect Siam. Rather, it has actively structured 'Siam' in our minds as well as on earth.

In fact, the ambiguous relationship between a map and its anticipated object – and the potential of the reverse relationship – can be found in the ambiguity of the notion of geography itself. In English, 'geography', as well as *phumisat* in Thai, refers to the knowledge or study of a spatial object as well as to such an object itself. The ambiguous and intricate relationship between the two notions indicates that an object can be what a knowledge of it allows it to be. Of course, we may question the relationships of other branches of knowledge and their object (subject of study), especially those whose double notions are signified by the same term, such as history.

In a conventional history, the making of modern Siam was often seen as the outcome of the reform and modernisation undertaken by the Siamese elite. Siam's territory was the outcome of the 'national integration' which consolidated its formerly disintegrated units by means of 'internal' mechanisms. The West was an 'external' power which jeopardised the survival of Siam and dismembered 'parts of its body'. Siam appeared as much victimised as the West appeared cruel. In the history of the geo-body, however, the annexation of the otherwise autonomous units was executed ambitiously and aggressively by the new administrative mechanism as well as by military force. But together they were only one side of the attempt to inscribe the geo-body of Siam on the earth's surface. They were the positive identification of the realm of Our space.

The other side of the emergence of Siam's geo-body was the making of the Others' space by the imperialists. Through diplomacy and military conquest, they delimited the extremities of the domain of Siam's space by identifying the limits of domains of their colonies. The Others surrounding Siam were also concretised and delimited in the same process. What distinguished Siam from the Others was not language, culture or religion, since Siam took over many formerly 'foreign' tributaries as parts of its realm. It was simply the space that was left over from direct colonialism. Siam was the space between. This was a negative identification of the geo-body of Siam. Whether Siam lost its territories to the imperialists or simply was the loser in the expansionist contest depends on one's perspective. But the indisputable fact remains: the colonial powers helped constitute the present geo-body of Siam.

The emergence of the geo-body of Siam was not a gradual evolution from the indigenous political space to a modern one. It was a displacement of the former by the latter at various moments both by foreign powers and by the Siamese themselves. Strategically, the new discourse threatened, destabilised or simply made the existing discourse ambiguous and then displaced it. The presence of the geo-body of Siam is an effect of the hegemony of modern geography and mapping. It is a phenomenon in which a domain of human space has been inscribed in one way rather than another. This phenomenon will last as long as the knowledge that inscribes it remains hegemonic. Not only is the geo-body of a nation a modern creation; if we perceive history in a *longue durée* of the earth's surface and humankind, it is also ephemeral. There are other knowledges of space, either residual or emerging, operating to contend with the geo-body. The presence of the geo-body is always subject to challenge.

Beyond territory and geography

[...]

The geo-body actively takes part in generating new ideas, new values and new culture, even beyond its primary task of spatial definition. The role of the map of Siam has been similarly active not only in representing the territory of Siam but in conveying other meanings and values as well. A map is frequently used to represent nationhood – to arouse nationalism, patriotism or other messages about the nation. [...]

As a symbol of nationhood, the map of Siam has become one of the most popular logos for organisations, political parties, business firms and trademarks. The use of a map may become more serious if it is indispensable for conveying the proper message of a particular symbol or is designed to arouse sentimental effects. [...]

Sometimes the appearance of a map is not serious since it is not designed for any sentimental effect. In commercial usage, for example, a map may be decorated, distorted or transformed for visual appeal. It may not look like a map at all. It may be a caricature of the map of a nation used in a very casual manner. [...] Can a caricature of a map arouse ationalism, royalism or other serious sentimental responses?

A map is usually taken out of its contextual origin, that is, the earth's surface. In many cases, there is no symbol to indicate the coordinates or the surrounding countries in a geographical textbook. A map may float. Moreover, there may be no mapping symbol or any convention. Yet floating maps even without mapping conventions can communicate to anyone familiar with the map. This is because all the maps in the emblems and advertisements cited above are no longer maps. They no longer represent the nation's territoriality. Rather, they are signifiers which signify the map of a nation. They are signs of the map of a nation. They have meanings and values and can send messages because they refer to the map of such a nation, which has been loaded with the meanings and values of nationhood. In other words, the map of a nation becomes a signified. In the words of Roland Barthes, it becomes a metasign: it has become an adequately meaningful sign in itself, not necessarily with a further reference to the territoriality of that nation. By signifying the map of that nation, these map-like signs can signify other meanings and values carried by the map. And in the reverse direction, becoming a metasign, the map of a nation can generate values and meanings which have nothing to do with territory at all.

At this point, we may realise that the relationship between a map and space becomes even more complex. It is hard to confine a map to its assumed nature as a spatial representation. It has moved too far away from its technical origin to return to its creator, the cartographer. It no longer belongs to the cartographer, who has lost control over it completely. Independent of the object as well as its human creator, it becomes a common property in the discourse of a nation.

In many ways, a map contributes its share to the human knowledge of a nation. As a sign, it is an effective and active mediation which can even create a geo-body; as a metasign, it is an object of reference in itself and can create more meanings and values beyond its origin. In addition to the fact that it monopolises the means of human conceptualisation of the artificial macrospace called a nation, both roles allow it to reign comfortably over the domain of knowledge of nationhood and also bring it close to being a natural entity.

By way of example, we can perhaps point to a never-ending number of cases in which the geo-body and a map as a discourse, knowledge, a sign, a metasign, operate to generate meanings and conceptual shifts. But one of the most significant effects the geo-body and a map have on our knowledge is their power to shape our conception of the past. Here the issue is the conjuncture between the new geographical knowledge and that of the past. How did the geo-body and map generate history? In what ways must history be changed in order to come to terms with the emergence of the geo-body and its disruptive origin?

Further reading

Grasseni, C. (2004) Skilled landscapes: mapping practices on locality. *Environment and Planning D: Society and Space*, **22**, 699–717. [An anthropological approach to community mapping in the Italian Alps.]

Ramaswamya, S. (2001) Maps and mother goddesses in modern India. *Imago Mundi*, **53** (1), 97–114. [Research paper discussing the role of cartography in the construction of Indian national identity.]

Sletto, B. (2009) We drew what we imagined: participatory mapping, performance and the art of landscape making. *Cultural Anthropology*, **50** (4), 443–476. [A current exploration of the empowering potential of an anthropological approach to participatory mapping in Venezuela and Trinidad.]

Wood, D. (2010) *Rethinking the Power of Maps*, Guilford, New York. [Wood uses the indigenous mapping as an example of counter-mapping and explores its complex re-negotiations of power.]

See also

- Chapter 1.8: Deconstructing the Map
- Chapter 1.9: Drawing Things Together
- Chapter 1.12: The Agency of Mapping: Speculation, Critique and Invention
- Chapter 4.7: Mapping Reeds and Reading Maps: The Politics of Representation in Lake Titicaca
- Chapter 4.8: Refiguring Geography: Parish Maps of Common Ground
- Chapter 5.2: The Time and Space of the Enlightenment Project
- Chapter 5.7: A Map that Roared and an Original Atlas: Canada, Cartography, and the Narration of Nation

Chapter 5.5

First Principles for a Literary Cartography, from *Territorial Disputes: Maps and Mapping Strategies in Contemporary Canadian and Australian Fiction*

Graham Huggan

Editors' overview

In this piece, Huggan examines the conceptualisation of maps and their power, critiques these conceptualisations, drawing from social theory, and details an understanding of maps from a literary perspective. In the initial section he sets out an understanding of maps as models, documents and claims. Maps, he details, are both products and processes – representation of the world that do work in the world. They possess textual properties, but they may also become textual events. Maps facilitate the pursuit of profit and the struggle for territory, and act as a source of authority. The reading and use of maps, however, are open to three challenges – feminist, regionalist and ethnic– that question their silences and displacements, and highlights how maps can be critically interpreted in ways that contest their claims. Interpreting the work of maps through a literary lens, Huggan explains how they function as powerful devices in three ways: as icons, as motifs and as metaphors. He forwards a literary cartography focused on the definition, connotation and literary function of maps, and especially how maps are implicated in territorial strategies to control, not simply represent, land.

Originally published in 1994: Chapter 1 in Graham Huggan, *Territorial Disputes: Maps and Mapping Strategies in Contemporary Canadian and Australian Fiction*, University of Toronto Press, Toronto, pp. 3–33.

> Map me no maps, sir!
> my head is a map,
> a map of the whole world.
>
> *Henry Fielding (1967), Rape upon Rape*

Defining the map

The map as model

[. . .]

Cartography has never been, and never will be, an exact science. [. . .] [M]aps, by their very nature, are never more than approximations of the environments they purport to represent. In Korzybski's (1958: 58) famous phrase, 'a map is not the territory it represents,' although, 'if correct, it has a similar structure to the territory, which accounts for its usefulness'. Yet how 'correct' can maps ever be? Whatever their degree of scientific accuracy, maps are neither exact nor entirely objective; they are, after all, controlled by human interests, and at best offer 'not a copy, but a semblance of reality, filtered by the mapmaker's motives and perceptions' (Korzybski 1958: 310). The subjective aspect of maps involves more, however, than the motives and perceptions of the mapmaker; for maps constitute a communication system that engages an often complex set of transactions between mapmaker and map reader (Robinson and Petchenik 1977).

The Map Reader: Theories of Mapping Practice and Cartographic Representation, First Edition. Edited by Martin Dodge, Rob Kitchin and Chris Perkins.

Neither mapmaker nor map reader can be considered, strictly speaking as an individual. Explorers and surveyors, designers and printers, publishers and politicians can all be thought of as participating in the process of mapmaking; and although the decisions taken in reading a map may reflect individual choice, they are also influenced, directly or indirectly, by wider social relations and cultural attitudes. The meaning of a map thus emerges from a transactional process involving a number of different interest groups; in the words of Wilbur Zelinsky (1973: 3): 'The map exists and has meaning only as it connects with other aspects of an interlocking communicative structure'. The map is both product and process: it represents both an encoded document of a specific environment and a network of perpetually recoded messages passing between the various mapmakers and map readers who participate in the event of cartographic communication. The accuracy of a map obviously depends on its precision of detail and refinement of delivery, yet it also depends on explicit or tacit perceptual conventions that differ widely from culture to culture. Maps, in this sense, are the unstable products of social, historical and political circumstance.

If the map is deficient as a copy of reality, it may function efficiently as a model of reality. As Christopher Board (1967: 672) puts it, 'maps [are] representational models of the real world ... They are also conceptual models containing the essence of some generalisation about reality. In that role, maps are useful analytical tools which help investigators to see the real world in a new light, or even to allow them an entirely new view of reality'. Board (1967: 672) outlines a dialectic between the formulation of the map, during which 'the real world is concentrated in model form', and the implementation of the map, during which 'the model is tested against reality'. As a result of this dialectical interaction, Board (1967: 672) suggests, 'the cycle may begin again with [a] revised view of the real world'. Other commentators, however, do not share Board's confidence in the viability of the map as a conceptual model of reality. Philip Muehrcke (1978: 309), for example, claims that a map-like conception of reality is inherently suspect; it may well have the effect of alienating us from our environment, rather than of uniting us with it: 'The deficiency of both the "map as reality" and "reality as map" attitudes is that they fall short of uniting us with our environment in all aspects of experience. They encourage us to ignore the independent underlying structure of our existence, upon which our survival and the well-being of our world depend'. Muehrcke's humanistic sentiments are echoed by John Vernon, for whom maps encourage a geocentric point of view in which distance intervenes between the world and its perceiver. At its most extreme, the attitude fostered by the map induces a kind of schizophrenia by persuading its user to believe that the world can be transformed into an object. The conceptual model of reality provided by the map may thus contribute to the rigidly dualistic philosophy that has enabled '[Western] civilisation to confirm its absolute space of reasonableness, cleanliness, freedom and wealth, precisely by creating equally absolute but sealed-off spaces of madness, dirt, slavery and poverty' (Vernon 1973: 17). Vernon's (1973) argument is overstated, but instructive in so far as it emphasises the map's tendency towards simplification. As Vernon suggests, maps are necessarily simplified models of the environment they represent; however elaborate their 'modelling system', they remain generalised, incomplete and relativistic representations of reality (Keates 1982; Monmonier 1977).

The concept of a 'modelling system' is usually associated with the Russian semiotician Yuri Lotman (1977). All texts, says Lotman (1977: 239), are codified modelling systems; maps provide examples of plot-less texts characterised by their definite order of internal organisation, while plotted texts 'cross the forbidden border which the plot-less structure establishes'. Maps may acquire a plot, however: 'If we draw a line across the map to indicate ... the possible air or sea routes, the text then assumes a plot: an action will have been introduced which surmounts the structure' (Lotman 1977: 239). Maps possess textual properties, but they may also become textual events; the act of reading the map does not restrict itself to the decoding of the model, but also involves itself in the further recoding of the modelling system.

Two cartographic theorists informed by Lotman's semiotics are Denis Wood and John Fels. For Wood and Fels (1985: 54), 'every map is at once a synthesis of signs and a sign in itself: an instrument of depiction – of objects, events, places – and an instrument of persuasion – about these, its makers, and itself. Like any other sign, it is the product of codes: conventions that prescribe relations of content and expression in a given semiotic circumstance'. In much the same way as Lotman (1977) distinguishes between the syntagmatic (internal) and paradigmatic (external) codes that inform the literary text, Wood and Fels (1985: 54) distinguish between the 'intrasignificant' codes that 'govern the formation of the cartographic icon, the deployment of visible language, and the scheme of their joint representation' and the 'extrasignificant' codes which 'govern the appropriation of entire maps as sign vehicles for social and political expression – of values, goals, aesthetics and status – as the means of modern myth'. The map's status as a model depends, then, on the coherence of its internal structure, but also on the degree and scope of its external influence. Wood and Fels's (1985) reminder that maps may assume a mythical status through the force of their sign production does not contradict their capacity to model the 'real world'; it merely emphasises the potential for discrepancy that exists between the model (or modelling system) and the 'reality' represented by the model.

The map as document

Wood and Fels's (1985) reference to the mythmaking potential of maps serves as a reminder of the map's contingent status as historical document. Since their rudimentary beginnings, maps have often relied more on conjecture than on fact. Indeed, many ancient and medieval maps did not set out to record fact at all, but rather to reinforce belief. In the theocentric T-O (*orbis terrarum*) maps of the Middle Ages, for example, Jerusalem was placed at the centre of a spherical universe, while in other, highly schematised maps from the same period, 'perfect celestial realms were located,' with rich embellishment, 'above imperfect terrestrial worlds' (Thrower 1972: 34; Figure 5.5.1). Significant changes, however, were to follow the discoveries of the Age of Exploration and the development of cartographic instruments and techniques in sixteenth and seventeenth century Europe. A combination of scientific rigour and exploratory zeal, coinciding with the revival of Ptolemy's projections and with the geographical discoveries of America and of a sea route to India, reinforced the intellectual authority and commercial success of Renaissance Europe. The improved measuring techniques of the world maps of this period, the most famous being Mercator's (1569), paved the way for the development of specialised topographic, hydrographic and thematic mapping traditions in the seventeenth and eighteenth centuries. Despite the progress of the Scientific Revolution, late Renaissance and early Enlightenment maps were still riddled with errors and fanciful conjecture.

Mapmaking could hardly be considered a frivolous activity, however; the Spaniards, leaders in the Renaissance exploration of the New World, are known to have destroyed, or to have bought up and hidden, whole editions of books and maps because they were thought to disseminate the wrong kind of information. [. . .] Given this rigorous and in some cases ruthless censorship, it is hardly surprising that many of the maps and charts that survive from the period are miscellaneous and unsubstantiated. Maps based on the systems of Ptolemy and Mercator continued to mix fact with fable; they were also misinterpreted, or artfully doctored, by their users. By the end of the seventeenth century, the history of cartography had become a history of contractual abuse between mapmaker and map user; as a result, the documentary value of the map was eroded, and the 'evidence' it presented was distorted, wilfully altered, abridged or censored – if it had ever been accurate in the first place.

[. . .]

The new 'scientific' cartography of the seventeenth and eighteenth centuries paved the way for the imperialist expansion of the nineteenth century. The map's value consisted not only in its putative accuracy as a document, but also in its increasing desirability as a consumer good. If the New Science lent authority to the document, the desire for commercial expansion increased its value as a commodity; a growing supply of maps and charts duly materialised to meet the consumer demands of the individual or corporate buyer and to 'facilitate the reorganisation of patterns of trade and political control' (Mukerji 1983: 81). The map itself became both a facilitator of the profit-making venture and a pawn in the struggle for dominion; whereas the eighteenth century cadastral (estate) map had functioned primarily as a symbol of private ownership, the maps of the imperialist period were to become symbols of corporate gain or national conquest. Imperial maps expressed the reality of conquest while promoting and legitimising the idea of empire; moreover, as J.B. Harley (1989: 282) remarks, 'the graphic nature of the map gave its imperial users an arbitrary power that was easily divorced from the social responsibilities and consequences of its exercise. The world could be carved up on paper'. In the scramble for Africa and other overseas colonies, the map realised its potential as a formidable political weapon. In fostering the notion of a socially empty space, the blank map was fully exploited by the colonisers of the new, 'virgin' lands; blank maps proved equally valuable to the commercial and geopolitical agents of imperialism in countries such as Africa and India, which, although densely populated, could be impersonally refashioned for the purposes of political control and economic gain.

Figure 5.5.1 Medieval T-O Map, Isidore of Seville, 1475.

The map as claim

This brief history of cartography suffices to emphasise the map's considerable authority as a geopolitical claim. The authority that maps confer upon their makers – or impose upon their readers – brings with it obvious opportunities for political manipulation: opportunities seized upon with particular relish by the practitioners of the propaganda map, which represents 'a wilful exploitation of the inherent limitations of maps to distort, exaggerate or deny facts' (Quam 1943: 22). All maps deceive, but propaganda maps are designed to deceive: readily comprehensible but subtly manipulative, they use the authoritative status and alleged neutrality of the map as means of reinforcing 'the peculiar credulity with which maps are generally accepted' (Quam 1943: 32). The map's efficacy as a claim, like its impact as a political weapon, rests on the combined effect of its diverse strategies: the delineation and demarcation of territory; the location and nomination of place; the inclusion and exclusion of detail within a preset framework; and the choice of scale, format and design. Many of these strategies are obvious, but some are subliminal, reflecting the subtlety with which maps operate as forms of social knowledge or as agents of political expediency.

[...]

The hidden rules of cartography contribute to the map's status as a symbol of political authority. By de-emphasising or excluding minority interests, maps reveal themselves as 'pre-eminently a language of power, not of protest ... The ideological arrows have tended to fly largely in one direction, from the powerful to the weaker in society. The social history of maps ... appears to have few genuinely popular, alternative, or subversive modes of expression' (Harley 1989: 300–301). Although the map's authority has periodically been challenged by those who read its claim to veracity as a disguised expression of the will to power, the voices of the challengers have often gone unheard; the map has engendered a language of protest without seeming to compromise its own language of power. Three dissident groups which challenge the representational accuracy, historical authenticity and political authority of the map demonstrate further that maps are ultimately neither copies nor semblances of reality but modes of discourse which reflect and articulate the ideologies of their makers.

Challenging the map

The feminist challenge

Although explicitly authoritative in their mode of expression, maps may be read in ways that contest, rather than confirm, their discursive claims. An example of this contestatory reading is provided by feminist theorists and/or creative writers who view the ethnocentric tendencies of the map, its makers' choice to displace (or discard) what they cannot accommodate, as an analogue for the marginalisation of women in patriarchal society. 'The map is not the territory,' writes Canadian poet Betsy Warland (1985) in an ironic rejoinder to Korzybski: 'where did we originate/are we a displaced civilization? ... our country/our bodies/edge/boundaries of viciousness: each country's conviction to colonise us'. Warland's gesture of angry denial finds support in the theoretical position taken up by feminist theorists such as Hélène Cixous. Cixous (1975) equates male writing with 'marked' writing: the patriarchal discourse which authorises itself by 'marking' the female voice as 'other'. [...] To inscribe their own femininity, women must break with this tradition and the rationale which nurtures it. The strategies characteristic of mapping – strict codification, definition, enclosure, exclusion – are precisely the strategies Cixous (1975) wishes to counter; for, in her opinion, [...] [t]he map operates [...] as a dual paradigm: for the phallocentric discourse that inscribes woman as 'other', and for the rationalistic discourse that inscribes the land as 'other'. Predicated on the principle of the binary opposition, these two mutually supportive discursive systems legitimise the subservience of woman as a 'logical' counterpart to the conquest of nature: woman, like the land, becomes an enslaved object of male representation. To reclaim their own subjecthood, suggests Cixous (1975), women must challenge the paradigms that inform patriarchal representation, displacing, undermining and eventually discrediting the propositions put forward by the patriarchal system.

The connection between patriarchy and a teleological 'language of proposition', in which 'meaning, origin and forming [are] posited as the limit of any attempt at clarification', has been further explored by Julia Kristeva (1980: 280–281). Kristeva claims that the structure of language, and the viability of the propositions it puts forward, depend on the 'metaphysical solidarity' of the logos, which functions as both source (*arche*) and goal (*telos*) of linguistic activity. If the status of the logos is called into question, the structure it informs and the propositions it supports are undermined; if it is then identified as the locus of male authority, the homogenising categories of patriarchal discourse – its meaning, origin and forming, to retain Kristeva's terms – are subverted. Since the structure can no longer be perceived as unified, the propositions it encapsulates lose their authority; what remains, however, is not a total breakdown in signification (a lapse into the meaningless) but an opening up of the field of signification (a newly recognised permissiveness that affords the opportunity for alternative meanings). This strategy of displacement is counter-discursive: it both subverts established or dominant discursive modes and provides the impetus for new or previously outlawed forms of expression (Terdiman 1990). If displacement is regarded as a prerequisite for new, formerly suppressed or disallowed projections of self, women can be seen in this sense both as mapbreakers engaged in the dismantling of a patriarchal system of representation and as mapmakers involved in the plotting of new coordinates for the articulation of (female) knowledge and experience.

[...]

Several options are open to women in their challenge to the patriarchal authority invested in the map. They may choose to reject the map outright as a symbol of authority or as a mode of representation, or to accept the paradigm but alter its terms of reference, transforming the map into a vehicle for the organisation and celebratory expression of female experience. The wide range of possible responses is indicated by the adaptability of key cartographic terms such as the boundary. Boundaries may be perceived as means of exclusion, symbolic devices for the marginalisation of women in patriarchal culture, or as means of territorial delimitation which allow women to define their own space and to give shape to their own experience.

The regionalist challenge

A similar set of options is open to regionalists, who may choose either to discredit the national map or, alternatively, to realign it in accordance with the experience of their own culture group.

[...]

I shall define regionalism loosely here as that set of attitudes, perceptions, circumstances and their emotional colorations which allows one (usually larger or more powerful) group to dominate other (usually smaller or less powerful) groups or, conversely, which enables one (or more) of these latter groups to identify and validate itself. [...] The primary concern is often 'not so much with place as with power; with perceived effectiveness, with lines of connection and authority – and not so much with the nature of power as with the possession or placement of power. So place remains important but comes to exist as a metaphor of structure' (New 1972: 5).

Hence the fascination of regional writers with the figure of the map; for the map, after all, is a metaphor of structure whose own lines of connection and authority are contained within a definite, and apparently coherent, framework. The notion of literary regionalism, then, cannot be defined solely by the expression of a range of responses to the particularity of place; it must also account for the choice of spatial metaphors to evoke that particularity, and for the structural properties of those metaphors as they interact with one another and within the overall structure of the text (Adamson 1980). The use of the map as metaphor, however, is inherently problematic: firstly, because maps contain and restrict, as well as organise and orient, space; and, secondly, because maps support the notion of a total, 'closed' structure, a notion currently called into question by literary and historical theorists, social scientists and creative writers alike (Derrida 1967). So, if the 'truly regional voice,' as New (1972: 6) believes, is one that 'declares an internal political alternative', is it not also likely to be one that questions the totality of the defining structure, the immobility of the designated centre, or even the coherence and completeness of the literary text? As notions of centre and periphery are redefined, 'region' may come to denote the semantic slippage between definitions of place rather than the circumscribed assertion of local identity. This slippage may then be identified with the counter-discursive strategies that enact what Raymond Williams (1973: 335) has called 'an unlearning of the inherent dominative mode'. Thus, like that cluster of dissident or revisionary discursive positions taken up within feminism, the various alternatives put forward in the name of regionalism may be considered as 'new configurations': rhetorical spaces that disrupt or discredit the notion of a 'central' locus of authority.

The ethnic challenge

Another series of new configurations which has come to challenge the territorial imperative of the 'centre' may be loosely bracketed under the heading 'ethnicity.' Ethnicity clearly means more than the expression of a distinctive ethnic identity. [...] The unstable emotional component and shifting geographical perspective of ethnicity make it difficult to define clearly; as a result, the rhetorical space occupied by ethnicity, like those spaces occupied by feminism and regionalism, is frequently ambivalent. Ethnicity is perhaps more closely concerned with territorial than with cartographic principles: it refers to the expression of social power and to the relations between space and society rather than to the abstract representation of a geographical environment (Ardrey 1966; Dubreuil and Tarrab 1976; Sack 1986). The socially motivated, often politically manipulated, dimensions of the map indicate a connection, however, between the expression of ethnicity and the representation of ethnocentrism. The connection, once again, is by no means clear-cut. The ethnic should certainly not be equated with the ethnocentric: the former expresses a 'vision, both ethical and future-oriented' (Fischer 1986: 196), the latter a prejudice nurtured on myths of cultural superiority that are 'justified' by a real or imagined past. Ethnic minority groups often consider, and dispute, ethnocentric maps as modes of hegemonic discourse; but they also define and designate their own territory in a gesture that might itself be perceived as ethnocentric. The geographical map may, therefore, function as a catalyst for ethnic dispute or as a representational medium for the expression of ethnic status. Usually, however, the expression of ethnic territoriality involves a resistance to conventional forms of cartographic representation. This resistance, as Michel de Certeau has pointed out, does not so much involve a challenge to the terms of cartographic discourse as an implied refusal to operate within them (Boelhower 1987). In his discussion of the territorial principles of the Latin American Indians, de Certeau (1985: 229) demonstrates that 'the designation of a *locus proprius* ... enables the resistance to avoid being disseminated in the occupiers' power grid, to avoid being captured by the dominating, interpretive systems of discourse ... it maintains a difference rooted in an affiliation that is opaque and inaccessible to both violent appropriation and learned cooptation. It is the unspoken foundation of affirmations that have political meaning to the extent that they are based on a realisation of coming from a different place . . . on the part of those whom the omnipresent conquerers dominate'.

The same is true of spatial representation in ethnic writing. Like feminist or regional writers, ethnic writers may focus on the disruptive activity of mapbreaking or on the reconstitutive activity of mapmaking; but they are usually involved to some extent in both: the reconstituted map has altered its terms of reference, not to avoid being subsumed within the dominant cartographic discourse but precisely to resist that avoidance. 'Ethnicity', like 'feminism' and 'regionalism', may thus come to be considered as that set of rhetorical strategies which activates a slippage of meaning between prescribed (cartographic) definitions. The easy ethnocentric distinction between 'our' territory and 'theirs' is consequently blurred, indicating a fault line between the neat rhetorical divisions inherent in conventional (Western) cartographic discourse. The hard lines of the geopolitical boundary are as likely to betray vulnerability as to display force; as Edward Said suggests, they are perhaps less unpredictable, but not necessarily any more 'objective' or definitive, than the lines we draw in our heads to designate our own imaginative territory and to exclude others from it.

The delineation of physical geography, argues Said, is complicated by the operations of 'imaginative geography': the representation of space, it would seem, owes as much to the subtlety of cultural perception as to the putative accuracy of technical presentation. A modern map, of course, is not likely to display the same errors as, say, a medieval one, but its detailed representation of the physical environment is not entirely free from the suppositions of imaginative geography (Allen 1976; Said 1979). In this context, maps should perhaps be measured on a continuum from those belonging to the category of 'technical cartography', which might include standard topographical maps, computerised maps, and the like, to those belonging to the category of 'imaginative cartography', which is perhaps best exemplified in the various 'countries of the mind' and 'landscapes of the imagination' of creative writing.

[. . .]

The map can be seen as a symbol of centralized political authority or as the expression of a dominant cultural imperative. Ethnic writing can be considered in this context as operating counter-discursively to the discourse of 'mainstream' culture; in so doing it identifies, and resists, the map as a spatial paradigm of cultural imperialism.

[. . .]

Fictionalizing the map

The map as literary device

Maps function as literary devices in three ways: as icons, as motifs and as metaphors. The map as icon is usually situated at the frontispiece of the text, directing the reader's attention towards the importance of geographical location in the text that follows, but also supplying the reader with a referential guide to the text. The map operates as a source of information but, more importantly, it challenges the reader to match his/her experience of the text with the 'reality' represented by the map. The map, in this sense, supplies an organisational principle for the reading of the text: information gleaned from the text is referred back to the map for verification, so that the act of reading the text involves an alternation between verbal and visual codes. Maps in literary texts differ in this respect from landscapes. Like maps, landscapes are cultural images, but their function in literary texts is most often one of symbolic identification, whereas maps, more conceptual in design, invite the reader to consider (and in some cases to question) the duplicating procedures of mimetic representation. Conventional maps persuade the reader to 'go beyond the physical presence of ink on paper to the real world referents of the symbols' (Muehrcke and Muehrcke 1974: 319), a process of adjustment which often involves the recognition that maps differ from the 'reality' they represent. Maps in literary texts highlight this process, and in some cases exacerbate this difference, by juxtaposing two sets of conventions: the verbal and the visual. The process of matching map to text, or text to map, involves the reader in a comparative activity that may bring to the surface flaws or discrepancies in the process of mimetic representation; for this reason, maps are prevalent in contemporary literature, especially in those self-reflexive fictions which problematise the notion of mimesis or the referential function of language. Writers of fantasy are attracted by the pictorial extravagance of the map, which invents even as it attempts to 'document'. Writers operating in an ironic mode are also attracted by this quality of extravagance, although their emphasis is more likely to be on the distinction between the map and the 'reality' it purports to represent. An early example of a writer combining fantasy with ironic distance is Jonathan Swift. Swift's maps in *Gulliver's Travels* are a deliberately incongruous mixture of the real and the imaginary: his fictitious islands are charted with scientific precision, accentuating gaps in contemporary geographic knowledge, and embellished with irrelevant detail, poking fun at the visual extravagances of contemporary (late seventeenth/early eighteenth century) cartography (Figure 5.5.2). In his mockery of the pretensions of the New Science, Swift takes particular aim at cartographers such as Hermann Moll, who, in 1719, less than ten years before the publication of *Gulliver's Travels*, had presumed to publish 'A New and Correct Map of the Whole World.' (Case 1958; Moore 1961) Ironic reflections on the nature and scope of geographic knowledge, Swift's maps also testify to the power of his imagination: they serve at once to

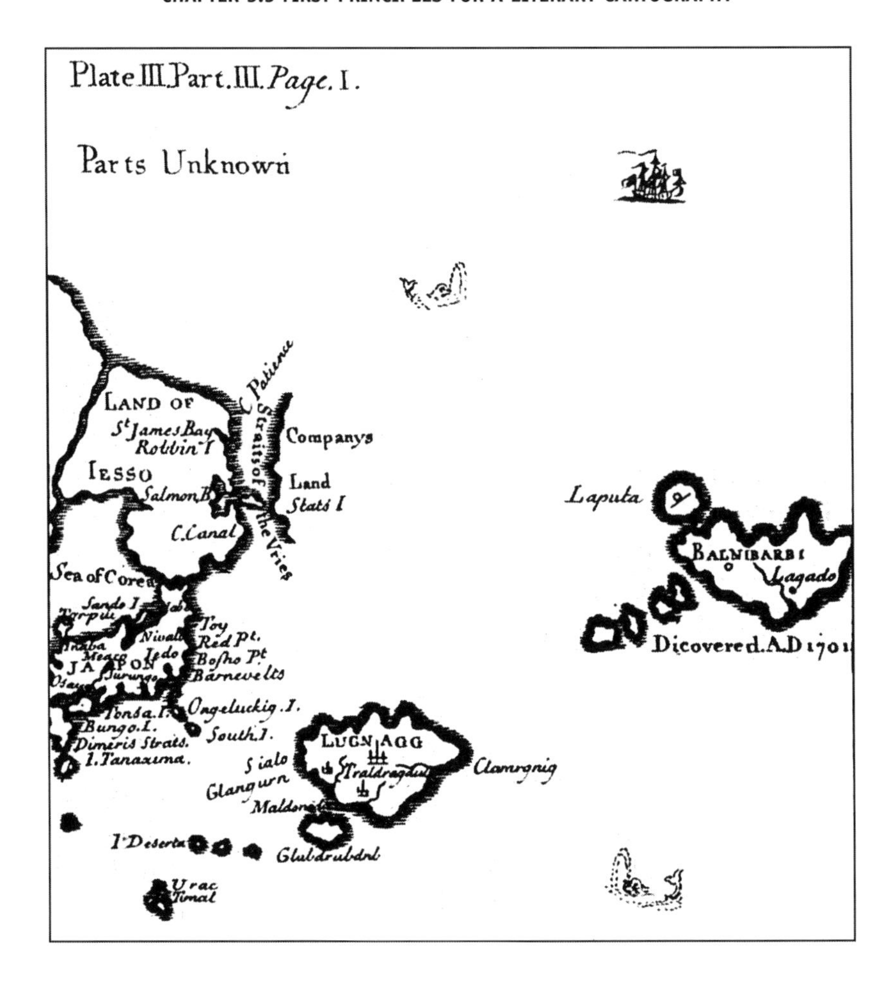

Figure 5.5.2 Detail of 'The Gulliverian Hemisphere', from Swift, 1766.

deride the conjectural worlds of cartography and to celebrate the invented worlds of fiction.

[...]

In [...] James Dickey's (1970) *Deliverance*, the map functions as a motif supporting the theme of design. When the protagonist leaves his job (as a graphic designer) to embark on a challenging journey into the interior, he believes that he has 'come out of the map'. But the holiday he had previously 'designed' fails to meet the definite standards required of it; the trip goes out of control, and he is lucky to escape with his life. The controlled abstractions of the map are shown in the process to be no match for the unpredictable events of the 'real' world. If Swift exploits the map as icon, [...] Dickey the map as motif, [both] use the map as metaphor. Maps are frequently used as metaphors in literary texts, usually of structure (arrangement, containment) or of control (organisation, coercion). Some of the best examples are provided in the work of the nineteenth century novelist Jules Verne, whose *Voyages extraordinaires* set out to define, enclose and control a whole imaginative world. Cyrus Smith's expedition in *L'Ile mystérieuse* is a microcosm of Verne's fictional project. Cast away on a remote South Pacific island, Smith and his followers proceed to explore, map and transform it. Their map, however, is not only a means of orientation around the island; it is also a prerequisite for the colonisation of the island. Geographical information is turned to political ends. The map of the newly named Lincoln Island functions as a metaphor of control, where control entails appropriation of territory, development of a hierarchical system of government, domestication of nature, establishment of a communications network, and all the other trappings of a colonial regime. [...] The map can also be considered as a metaphor of structure; the colonisers' geometrification of their environment thus sheds light on the meticulous structural organisation of the text and on the larger, 'comprehensive' schema of the *Voyages extraordinaires*. Like his geography, Verne's fiction operates on the principles of classification and conquest, the map's 'technology of possession' (McClintock 1988: 151) enabling it both to describe the immediate environment and to inscribe upon it the greater cultural imperative of colonial expansionism. (Chesneaux 1966).

[...]

But the map is not just a metaphor of artistic self-possession; it is also one of textual organisation. Alan Sillitoe (1975: 686) expresses it well: 'Just as a general needs maps upon which to plan his campaign or fight his battles, so an author requires them for his novels and stories, even if they exist only in memory, or in his imagination . . . it is better to get them down in black and white, better still in many colours . . . for they can be just as much a part of the notes for a novel as those key phrases and paragraphs with which you prepare the ground for one'. For Sillitoe (1975: 689), the consultation of existing maps, or the invention of new ones, becomes 'a stabilising factor giving play to both rigidity and fantasy . . . a way of fixing the mind, and at the same time recognising no limits to the prison in which it seem[s] that one [has] been born'. By guiding and shaping the production of fiction, claims Sillitoe, the map operates both as a visual complement to the written text and as a prerequisite for the writing of the text.

Many twentieth century writers might agree with Sillitoe's technical measures but disagree with his fictional principles; for the map, in twentieth century fiction, has often tended to function as a metaphor of the appearance of control rather than of its actual exercise. The map may operate as a powerful symbol of political control while remaining an inadequate referential guide. Its inaccuracies and lacunae, moreover, may expose ideological inconsistencies in the controlling agency, so that the map's attempt to locate and orient the individual in the 'real' world is re-identified as a disguised form of political manipulation. It is one thing, however, to expose the map as a false guide, quite another to replace it.

A contemporary writer who turns the transformative potential of the map as metaphor to his advantage is the Guyanese novelist Wilson Harris. A trained surveyor, Harris is well versed in cartographic principles and procedures. In his fiction, however, maps function primarily as metaphors. Harris distinguishes between the map as a paradigm of conquest and as a medium of perceptual transformation. In the first instance, maps are limited, even monolithic structures which are made to represent the one-sided views and self-serving ambitions of conquistadorial cultures; in the second, they are perceived as multidimensional, providing the means by which different, apparently incompatible cultures may be cross-fertilised in 'the [universal] womb of space' (Harris 1983: 137). If the map is perceived as a vehicle of imaginative reconciliation rather than as a mirror of cultural prejudice, it can serve as a metaphorical device for the unification of disparate, or warring, cultures.

[...]

Harris substitutes a unifying metaphysical construct for a divisive cultural one. The narrator is 'free[d] from material restraint and possession' (Harris 1983: 108) by recognising that the map does not have to be a touchstone for his own, narrowly defined ambitions; instead, it may provide a medium for the visualisation of new, unforeseen opportunities: opportunities for both personal and social/cultural growth. Far from confirming the authority of the dominant culture, the map allows for the possibility of a new kinship both between and within cultures: a kinship that satisfies the need for communication and cultural exchange rather than yielding to the strident demands of personal ambition. Since the map is inevitably filtered through the perceptions of its makers, it may be remade if it is perceived in a new light or from a different angle. Thus [...] Harris does not simply reject the map as metaphor but insists that, with perceptual adjustment, it may be considered as an agent of change rather than as a safeguard of existing authority.

Not all writers are as optimistic as Harris about the possibility of transforming the map. For the New Zealand writer Janet Frame (1964), maps are metaphors of claustrophobic containment which designate societies whose Procrustean distinctions between the 'normal' and the 'abnormal' thinly disguise social, racial and class prejudices. [...]

With similar sardonic emphasis, the Somali writer Nuruddin Farah (1986) writes of maps as insidious mechanisms that justify the dispossession of minority peoples. The *de facto* boundaries, generic definitions and designated no man's lands of geopolitical maps are used as metaphors for the divided loyalties of those caught up in the Horn of Africa war. The following conversation between the protagonist Askar and his uncle/mentor Hilaal is worth quoting at length because it brings into focus many of the cartographic issues I have previously discussed:

Hilaal: Do you carve out of your soul the invented truth of the maps you draw? Or does the daily truth match, for you, the reality you draw and the maps others draw?

Askar: Sometimes I identify a truth in the maps which I draw. When I identify this truth, I label it as such . . . I hope, as dreamers do, that the dreamt dream will match the dreamt reality – that is, the invented truth of one's own imagination. My maps invent nothing. They copy a given reality, they map out the roads a dreamer has walked, they identify a notional truth.

Hilaal: The question is, does truth change? Or do we? . . . Better still, who or what is more important, the truth or its finder? You look at a map of the British colonies in Africa, say . . . now compare the situation today with its ghostly past and someone may think that a great deal of change has taken place and that names of a number of countries have been altered to accommodate the nationalist wishes of the people of those areas. But has the more basic truth undergone a change? Or have we? . . . There is truth in maps. The Ogaden, as Somali, is truth. To the Ethiopian mapmaker, the Ogaden, as Somali, is untruth. (Farah 1986: 216–8)

Farah's dramatisation of the politics of cartography draws attention to the map's problematic status as metaphor. [. . .]

First principles for a literary cartography

Despite the prevalence of maps in contemporary literary texts, there has been little attempt on the part of critics or theorists to establish principles for a literary cartography. This chapter has gone some way towards compensating for the deficiency: firstly, by considering the map as a representational construct (or model), a historical document and a geopolitical claim; secondly, by outlining theoretical counter-arguments to strategies involved in or implied by the cartographic process; and,thirdly, by illustrating the map's function as icon, motif and metaphor in selected literary texts. The practice of literary cartography can thus be hypothetically based on the definition, connotation and literary function of maps. It also proceeds from a distinction between maps and other spatial metaphors or paradigms; for if a map is not equal to the territory it represents, neither is it equal to a landscape. The principles of literary cartography, like those of landscape theory, are fundamentally concerned with the process of representation; but whereas the symbolic representation of landscapes in literature is primarily directed towards the question of how the land is perceived, the metaphoric function of maps in literature is addressed first and foremost to the issue of how the land is *controlled.* Like maps, landscapes are culturally determined and susceptible to political manipulation; maps, however, draw more immediate attention than landscapes to their status, function and implication as power structures. In this sense, literary cartography not only examines the function of maps in literary texts, but also explores the operations of a series of territorial strategies that are implicitly or explicitly associated with maps. Some of these strategies are beneficial to their users, as in the attempt to order, direct and articulate personal, social or cultural experience; others involve power relations that serve to reinforce existing divisions within society or to exacerbate cultural prejudices, as in the attempt to enclose, restrict or wilfully control experience. The function of maps and mapping strategies in literary texts is therefore frequently ambivalent: maps may be simultaneously perceived as useful tools and as dangerous weapons.

Furthermore, maps may either facilitate the relation between real and represented worlds or exacerbate the distinction between them. Since, like literary texts, maps are conventional systems of representation, they may be used as paradigms for an investigation of the procedures – and of the ontological and epistemological problems – of mimesis. For this reason, maps often feature in self-reflexive texts; they also tend to feature in texts by feminist writers anxious to liberate themselves from, or to revise supremacist assumptions behind, patriarchal representation; by regional writers critical of representations that promote or protect the values of the (metropolitan) 'centre'; and by ethnic writers resistant to or avoiding circumscription within the homogenising discourses of the cultural 'mainstream.'

References

Adamson, A. (1980) Identity through metaphor: an approach to the question of regionalism in Canadian literature. *Studies in Canadian Literature,* **5** (1), 83–99.

Allen, J.L. (1976) Lands of myth, waters of wonder: the place of the imagination in the history of geographical exploration, in *Geographies of the Mind: Essays in Historical Geosophy in Honor of John Kirtland Wright* (eds D. Lowenthal and M.J. Bowden), Oxford University Press, New York, pp. 41–61.

Ardrey, R. (1966) *The Territorial Imperative,* Atheneum, New York.

Board, C. (1967) Maps as models, in *Models in Geography* (eds R. Chorley and P. Haggett), Methuen, London, pp. 671–725.

Boelhower, W.Q. (1987) The culture of the map, in *Through a Glass Darkly: Ethnic Semiosis in American Literature* (ed. W.Q. Boelhower), Oxford University Press, New York, pp. 46–49.

Case, R. (1958) The geography and chronology of *Gulliver's Travels,* in *Four Essays on Gulliver's Travels* (ed. R. Case), Peter Smith, Gloucester, MA, pp. 50–68.

Chesneaux, J. (1966) *Une Lecture Politique de Jules Verne,* Maspéro, Paris.

Cixous, H. (1975) Le rire de la méduse. *L'Arc,* **61**, 39–54.

de Certeau, M. (1985) *Heterologies: Discourse on the Other* (trans. B. Massumi) University of Minnesota Press, Minneapolis, MN.

Derrida, J. (1967) La structure, le signe et le jeu dans le discours des sciences humaines, in *L'écriture et la Différence* (ed J. Derrida), Seul, Paris, pp. 409–428.

Dickey, J. (1970) *Deliverance,* Laurel, New York.

Dubreuil, G. and Tarrab, G. (1976) *Culture, Territoire et Aménagement,* Éditions Georges Le Pape, Montréal, Canada.

Farah, N. (1986) *Maps,* Pan, London.

Fielding, H. (1967) Rape upon rape: or, the justice caught in his own trap, in *Plays and Poems,* vol. 2 (ed. H. Fielding), Barnes and Noble, New York, pp. 79–156.

Fischer, M. (1986) Ethnicity and the post-modern arts of memory, in *Writing Culture* (eds J. Clifford and G. Marcus), University of California Press, Berkeley, CA, pp. 194–233.

Frame, J. (1964) *Scented Gardens for the Blind*, Braziller, New York.

Harley, J.B. (1989) Maps, knowledge and power, in *The Iconography of Landscape* (eds D. Cosgrove and S. Daniels), Cambridge University Press, Cambridge, pp. 277–312.

Harris, W. (1983) *The Womb of Space: The Cross-Cultural Imagination*, Greenwood, Westport, CT.

Keates, J.S. (1982) *Understanding Maps*, Longman, London.

Korzybski, A. (1958) *Science and Sanity: An Introduction to Non-Aristotelian Systems*, Institute of General Semantics, Lakeville, CT.

Kristeva, J. (1980) *Desire in Language: A Semiotic Approach to Literature*, Columbia University Press, New York.

Lotman, Y. (1977) *The Structure of the Artistic Text*, University of Michigan Press, Ann Arbor, MI.

McClintock, K. (1988) Maidens, maps and mines: the reinvention patriarchy in colonial South Africa. *South Atlantic Quarterly*, **87** (1), 147–192.

Monmonier, M. (1977) *Maps, Distortion and Meaning*, Association of American Geographers, Washington, DC.

Moore, J.R. (1961) The geography of *Gulliver's Travels. Journal of English and Germanic Philology*, **40**, 214–228.

Muehrcke, P.C. and Muehrcke, J.O. (1974) Maps in literature. *The Geographical Review*, **64** (3), 317–338.

Muehrcke, P. and Muehrcke, J. (1978) *Map Use: Reading, Analysis and Interpretation*, JP Publications, Madison, WI.

Mukerji, C. (1983) *From Graven Images: Patterns of Modern Materialism*, Columbia University Press, New York.

New, W.H. (1972) *Articulating West*, New Press, Toronto.

Quam, O. (1943) The use of maps in propaganda. *Journal of Geography*, **42** (1), 21–32.

Robinson, A.H. and Petchenik, B.B. (1977) The map as a communication system. *Cartographica*, **19**, 92–110.

Sack, R. (1986) *Human Territoriality: Its Theory and History*, Cambridge University Press, Cambridge.

Said, E. (1979) *Orientalism*, Vintage, New York.

Sillitoe, A. (1975) A sense of place. *Geographical Magazine*, **47** (11), 685–689.

Swift, J. (1776/1925) *Gulliver's Travels*, Oxford University Press, London.

Terdiman, R. (1990) *Discourse/Counter-Discourse: The Theory and Practice of Symbolic Resistance in Nineteenth Century France*, Cornell University Press, Ithaca, NY.

Thrower, N.J. (1972) *Maps and Man: An Examination of Cartography in Relation to Culture and Civilization*, Prentice-Hall, Englewood Cliffs, NJ.

Vernon, J. (1973) *The Garden and the Map: Schizophrenia in Twentieth-Century Literature and Culture*, University of Illinois Press, Urbana, IL.

Warland, B. (1985) The map is not the territory. *Kunapipi*, **7** (3), 145.

Williams, R. (1973) *The Country and the City*, Chatto and Windus, London.

Wood, D. and Fels, J. (1985) Designs on signs/myth and meaning in maps. *Cartographica*, **23** (3), 54–103.

Zelinsky, W. (1973) The first and last frontier of communication: the map as mystery. *Bulletin of the Geographical and Map Division*, **94**, 2–8.

Further reading

Barbara, P. and Lorenz, H. (2009) Mapping the ontologically unreal: counterfactual spaces in literature and cartography. *The Cartographic Journal*, **46** (4), 333–342. [A useful and up to date review of literary mapping.]

Conley, T. (2006) *Cartographic Cinema*, University of Minnesota Press, Minnesota, MN. [This monograph presents a rich and comprehensive analysis of the intersections of cinematic space and map representations.]

Moretti, F. (1998) *Atlas of the European Novel 1800–1900*, Verso, London. [An inventive and insightful exercise in literary mapping.]

Padron, R. (2004) *The Spacious Word: Cartography, Literature, and Empire in Early Modern Spain*, University of Chicago Press, Chicago. [A detailed empirical case study in an important historical context for mapping.]

Pickles, J. (2004) *A History of Spaces: Cartographic Reason, Mapping, and the Geo-coded World*, Routledge, London. [An extended work on the politics of cartography including discussion on the textuality of mapping.]

See also

- Chapter 1.7: Designs on Signs/Myth and Meaning in Maps
- Chapter 1.8: Deconstructing the Map
- Chapter 1.12: The Agency of Mapping: Speculation, Critique and Invention
- Chapter 3.9: Mapping Modernity: Art and Cartography in the Twentieth Century
- Chapter 5.3: Texts, Hermeneutics and Propaganda Maps
- Chapter 5.7: A Map that Roared and an Original Atlas: Canada, Cartography, and the Narration of Nation
- Chapter 5.9: Affecting Geospatial Technologies: Toward a Feminist Politics of Emotion
- Chapter 5.10: Queering the Map: The Productive Tensions of Colliding Epistemologies

Chapter 5.6

Whose Woods are These? Counter-Mapping Forest Territories in Kalimantan, Indonesia

Nancy Lee Peluso

Editors' overview

In her analysis Peluso makes the claim that all mapping of environments is an inherently political act because it documents the location and extent of natural resources and reveals their latent wealth for possible exploitation. In the case of mapping rainforests, state cartography works against forest dwellers because it consolidates state control over territory and facilitates large-scale accumulation strategies of exploitation. She charts how states are not the only groups that can use maps to assert claims, however, detailing the ways local activists, with various kinds of help, have produced counter-mappings to delineate and formalise their claims to forest territories and resources that villages have traditionally managed. Such counter-maps challenge the authority and power of state cartography by providing alternative representations and discourse. Moreover, their creation and circulation can empower citizens and enable resistance and protest, including the building of local and international alliances.

Originally published in 1995: *Antipode*, **27** (4), 383–406.

Forests are repositories of great wealth and ecological importance; politically, they are much more than that. Forests are often located in critical spaces that states want to control: international border areas as well as zones which might be deemed 'sensitive' because of either their political–ecological importance or sociological composition. Historically, forests have also been the outposts of 'outlaws' and 'outcasts' and the base for many an opposition force to imperialistic powers – from tenth century 'China' to fourteenth century Java to twentieth century Peru and Vietnam (Menzies 1992). Forest mapping was embraced early by emerging European states, first for establishing political boundaries and later for management (Kain and Baigent 1992).

Mapping of forest resources is, therefore, an intrinsically political act: whether drawn for their protection or production, they are drawings of a nation's strategic space. Forest maps pinpoint the location of valuable and accessible timber and mineral resources (Buisseret 1992) and are used for zoning protection of fragile, steep or biologically diverse areas. Forest maps have been an important tool for state authorities trying to exclude or include people within the same spaces as forest resources; maps increase state control over spaces which are sources of social unrest and valuable resources (Menzies 1992). Mapping facilitates large-scale accumulation strategies that work to forest dwellers' disadvantage, and consolidates state control over politically sensitive areas such as border zones (Girot and Nietschmann 1993).

This paper examines the origins, implementation and implications of forest mapping in two different forms in Kalimantan, Indonesia. In Indonesia, forest maps have been an important tool of state land managers and supporting international institutions [...]. In response to two decades of intensive industrial timber exploitation and the Indonesian government's superseding of customary

The Map Reader: Theories of Mapping Practice and Cartographic Representation, First Edition. Edited by Martin Dodge, Rob Kitchin and Chris Perkins.

forest rights through official planning and mapping efforts, an alternative or 'counter-mapping' movement has begun. Local activists, with international and sometimes government assistance of various sorts, are using sketch maps to delineate and formalise claims to forest territories and resources their villages have traditionally managed. In some cases they are matching their sketch maps to points on the Global Positioning System (GPS) and the official Indonesian forest planning maps using sophisticated software (Sirait *et al.* 1994; Momberg 1994). The goal of these efforts is to appropriate the state's *techniques* and *manner of representation* to bolster the legitimacy of 'customary' claims to resources. The practical effect is far reaching: the use of maps and a highly 'territorialized' strategy redefines and reinvents customary claims to standing forest resources and harvestable products as claims to the land itself. [...]

[...]

The politics of mapping

[...]

Much of the 'politics of mapping' theory is based on local/national histories in early modern and contemporary Europe and in the colonised 'New World' (including the USA and Canada, with some attention to Europeanised Latin American localities). It accordingly fails to capture the distinctiveness of contemporary Third World mapping politics. The most intensive state mapping initiatives arrived on the 'scenes' of the Third World with global capitalism firmly entrenched and in advanced stages, particularly in the tigerish economies of East and Southeast Asia. The advanced stage of mapping technology at which both national mappers and local 'counter-mappers' have entered the game is also relevant, insofar as using the new tools both raises the stakes of resource mapping and offers new political openings for resource users. [...]

Harley (1989), Kain and Baigent (1992) and others have contended that cartography and mapping are uniquely sources of power for the powerful (cf. Wood 1992). However, if maps can be seen as one of many 'authoritative resources' that states mobilise to consolidate their own power (Giddens 1984, cited in Harley 1988), then local groups' appropriation of the technology of mapping may help to counterbalance, or at least offset, the previous monopoly of authoritative resources by the state or capital. This requires understanding the social and political contexts within which maps are used by local groups. Just as inclusion and exclusion are powerful political tools used by states and state-legitimated organisations to control and allocate resource access (Harley 1988; Menzies 1992), local groups can claim power through mapping by using not only what is on a map, but what is *not* on it. One effect of having multiple maps of a single forest, for example, could be to challenge the accuracy of a 'standard' map used for planning.

An important element of such a challenge to state authority to create maps is the re-insertion of people on resource maps. Individual homesteads, settlements and villages are routinely excluded from maps of private and state land holdings. This practice grew out of efforts of sixteenth and seventeenth century estate holders to 'know' and manage lands held within their extensive domains, to enclose or privatise land from the commons, or of states' attempts to claim jurisdiction over wildlands or resource-rich areas (Harley 1989). [...] When court authorities established legal precedent by accepting the validity of maps in land disputes [...] the role of maps as tools of the powerful was enhanced.

Not all people were excluded from forest maps at all times, however, and the inclusion of people was also a mechanism for exerting control. [...] As the types of rights to land and resources changed in importance, maps became more explicit means of controlling resource access. Thus, in Norway from the seventeenth century, maps were used to settle disputes over both individually held and common lands [...].

Contrary to the conclusion on hegemony that Harley (1989) draws from his extensive research on the politics of mapping, maps can be used to pose alternatives to the languages and images of power and become a medium of empowerment or protest. Alternative maps, or 'counter-maps' as I call them here, greatly increase the power of people living in a mapped area to control representations of themselves and their claims to resources (Orlove 1989). Local people may exert control directly by making their own maps or entrust a representative of their choice, such as a local NGO, to perform the task. Counter-maps thus have the potential for challenging the omissions of human settlements from forest maps, for contesting the homogenisation of space on political, zoning or property maps, for altering the categories of land and forest management, and for expressing social relationships in space rather than depicting abstract space in itself (Sack 1983; Lefebvre 1991; Vandergeest and Peluso 1995). Counter-mapping can be used for alternative boundary making and 'to depict strategies of resistance: where to block ... unwise development, to identify landscapes that have been damaged, to describe alternatives to the incremental destruction of sustaining habitats' (Aberley 1993: 4).

[...] If we understand maps and cartography as part of an elite language of the powerful, then could we interpret the teaching of mapping skills to local people to be a new form of empowerment? In other words, is the process of counter-mapping a 'vernacularisation' of maps of a similar calibre? Although it is difficult to imagine the spread of mapping skills as having anywhere near the impact of the spread of print and the capacity to read, there are several ways in which counter-mapping can have a major impact.

I would argue that while counter-mapping has some potential to transform the role of mapping from 'a science of princes' (Harley 1988: 281), it is unlikely to become a 'a science of the masses' simply because of the level of investment required by the kind of mapping with the potential to challenge the authority of other maps. Investment in specialised computers and software and knowledge will make the costs of mapping prohibitive for *most* local people, particularly in poor areas. This of course creates openings for new types of power relations around the control and knowledge of mapping technologies, both in local class relations and in the relationships between NGOs and local villagers. However, although there will necessarily be gatekeepers involved in the mapping enterprise, there are multiple ways that gates can be kept. What ultimately may be more important for the 'masses' is not the technology itself, but the content of the maps produced and the way the knowledge and information on the maps is distributed.

Another question must be asked in the course of re-representing claims to resources and formalising them in the terms of the state as formal property rights. Whereas abstract space on a map represents merely state claims to power rather than a state capacity to enforce its claims, local people's actual control may be enhanced by exclusion from the map. When the degree of state surveillance increases, for example because of an increased value of resources or because of a reduction in resources located elsewhere, local people's inclusion on the map is more desirable. Once mapping begins, however, a new locus of negotiation and potential conflict over resource claims takes centre stage – the allocation of resources and ostensible settlement of claims among local users by establishing boundary lines between individuals' claims. The process of mapping almost forces the reinterpretation of customary rights to resources *territorially*, thereby changing both the claim and the representation of it from rights in trees, wildlife or forest products to rights in land.

Indonesian forests and forest mapping

In Indonesia, prior to the early 1970s, when the government developed a plan for the management of Outer Island timber resources, state forest management and planning was concentrated in Java (Peluso 1992). In Kalimantan, forest land use planning effectively began with the passing of Basic Forestry Law No. 5/1967, which empowered the national government to control, manage and administer all state forest lands (Barber 1989; Zerner 1990). Until 1966, Indonesia's first president, Sukarno, had pursued economic policies oriented toward domestic self-sufficiency, shunning most foreign investment, particularly by the 'Western' (Europe, North America, Australia) capitalist countries. When Suharto took over as president in 1967, he immediately set the stage for foreign investment and capitalist development, with the passage of Foreign Investment Act No. 1/1967, representing a major reversal of economic and foreign policies from those of the previous regime. Foreign logging industries from Japan, the Philippines, the USA and Europe were granted timber concessions, called HPH (*Hak Pengusahaan Hutan* – Permit for Forest Industry) in the Outer Islands (Manning 1972). At the end of 1989, some 561 concessions were in operation, 294 of them in Kalimantan (FAO/GOI, cited in Potter 1995). Untold numbers of concessions had long since folded, after having extracted and exported hundreds of thousands of cubic meters of timber.

The first of three mapping episodes directed at 'forest management' in Kalimantan accordingly consisted of notoriously inaccurate and secretive concession locations. Anecdotal evidence indicates that these maps revealed border conflicts, multiple permitting of territories, and illegal entry of one concession operator onto the concessions granted another. They ignored the physical conditions of the forest itself in designating these concession areas for timber production and whether competing claims and forms of management were already in place. [. . .]

These maps were replaced between 1981 and 1985 when provincial foresters collaborated with colleagues in agriculture, public works and agrarian affairs to develop plans and maps. This second set of state forest maps was called the Consensus Forest Land Use Plan (TGHK, or Tata Guna Hutan Kesepakatan) (Moniaga 1993; Potter 1995). [. . .] Once again, no account was taken of local people's previous claims to these lands, nor of existing vegetative cover (Potter 1995). [. . .]

The third and most recent state attempt to map forests in Kalimantan and other 'outer' islands of Indonesia is the Regional Physical Planning Programme for Transmigration (generally known as RePPProt), a collaborative effort between the GOI's Ministry of Transmigration and the Land Resources Department of the Overseas Development Administration (ODA) in London. The maps are part of a larger regional planning effort, which in the case of Kalimantan is to involve the resettlement of millions of people from Java, Bali and Lombok and the creation of agricultural estates – principally for palm oil and rubber. The labour for these estates will be drawn from both immigrant and locally born populations. Using Landsat data and aerial photographs, actual land use cover is being mapped and the areas included in different forest land use categories are being reconsidered. The discrepancies between the earlier TGHK maps and the RePPProt maps are striking.

These latest planning maps also include settlement areas around urban areas and villages, cultivated fields outside of these settlements and planned forest areas. The maps underestimate, indeed, lack knowledge of, forest-based populations' claims to and management of forest territories, as well as their actual patterns of forest and agricultural land use. For example, shifting cultivation is considered as a 'non-permanent' use of the land, although it is recognised that this may not agree with the views expressed under local customary rights. The villages associated solely with this extensive form of land use are not permanent in the long term, although some may remain on the same site for 10–20 years or more before moving elsewhere (RePPProt West Kalimantan Executive Summary).

[...]

The notion that villages practicing shifting cultivation inevitably move is outdated and historically correct for only a subset of the groups labelled shifting cultivators in Kalimantan. Recent research has shown that many groups in West Kalimantan have remained settled for several hundred years. Moreover, their land management techniques include not only protecting forest but also creating it (Peluso 1993; Padoch 1994). The RePPProt planners have not completely ignored their lack of knowledge of customary systems, but they have neither emphasised their importance in the executive summary, nor made recommendations about what to do for the purposes of their mapmaking exercise. [...]

[...]

Both in Indonesian law and by verbal consensus, Indonesian planners recognise that extensive systems of customary law and practice *(hukum adat* and *hak ulayat)* exist throughout Indonesia, and often overlap with forest territories and resources claimed by the state, though they have no maps or other documents formally indicating their extent. Forest Law No. 5 states that the rights of indigenous peoples to land and resources covered by *adat* should be respected, except when these conflict with national or the (undefined) 'public interest.' Essentially this means that by law national development initiatives and planning, such as that represented by the RePPProt, can override customary practices, laws, and claims, in the interests of the 'public' represented by the state. [...]

Territorial claims and counter-mapping

Mapping by government land use planners focuses on the land itself. In other words, maps are part of a larger resource management strategy with a strong territorial component (Sack 1983; Peluso 1992; Vandergeest and Peluso 1995). This represents a shift from an emphasis on the control of the resources on the land (in the case of forests) and of the labourers needed to extract those resources (Peluso 1992) to a territorialized strategy emphasising the control of land itself.

State land use planners recognise the following categories of local people's (i.e. not state) land and forest management: shifting cultivation (which they never call swidden [households]), permanent cultivation (defined as continuous cultivation of at least one crop per year on wet or dry fields), thatch/brush/secondary growth. However, they only recognise local people's *territorial* rights to areas they define as 'permanent cultivation' (GOI, n.d., Executive Summary: 30). Forest planners recognise people's *adat* claims to certain forest trees and plants producing products such as rattan, fruit, honey, illipe nuts, resins (damar) and rubber, even when these occur in state-claimed forest territories. State recognition of individual trees in the forest, however, does not translate into recognition of villagers' claims to portions of the forest *as territorial entities.* In fact, exactly the opposite is true: certain species and individual claims to them are recognised in part to allow the state to claim the forest as territory and to allocate exploitation rights (to corporations, not to villages, as a general rule) as it sees fit. Such rights include rights to harvest timber (through concessions) and rights to convert the forest to plantation tree cropping, whether oil palm, rubber, or pulpwood species (in the case of lands categorised 'Convertible Forest'). [...]

Local counter-mapping initiatives and territoriality

Two different counter-mapping strategies have been developed in response to this situation. The first is through efforts of outsiders working for international organisations. They have suggested mapping as a way of clearly depicting and protecting local claims to territory and resources to a government that in the past ignored them. The second has been initiated by Indonesian NGOs which request or contract the services of key international groups to learn the uses of counter-mapping strategies to document forest uses, claims and population distribution. Both strategies involve (1) the uses of low and high technology mapping techniques necessitating villagers' formation of political alliances with international NGOs and foreign experts, and (2) the assertion of specific and permanent territorial claims to resources. The key theoretical questions about the impacts of counter-mapping on resource control are to what degree new notions of territoriality reflect older ones; how the reinvention of these traditions benefits or works to the detriment of customary practice, law and resource distribution; and how the intervention of NGOs (whether locally, nationally or internationally based) affects the villagers' access to and control over forest resources.

[...]

Counter-mapping as a joint forest management strategy

One of the two counter-mapping strategies described here has been applied in the Kayan Mentarang Reserve. The Kayan Mentarang Reserve, set within the mountainous territory along the East Kalimantan–Sarawak border, was gazetted in 1980. Culturally and biologically diverse, it contains potentially important archaeological remains and is home to 12 distinct ethno-linguistic groups. [...] Since approximately 1990, the Worldwide Fund for Nature, The Indonesian Department of Forest and Nature Conservation (PHPA) and the Indonesian Institute of Sciences have been cooperating to develop a long-term conservation program in this 1.6 million hectare reserve, one of the largest in Asia. Their activities include an inventory of the reserve's extensive human and natural resources, documentation of local knowledge and resource management systems, and, most recently, efforts to record this information on maps. The maps are intended to form the basis of talks for identifying customary forest tenure boundaries, in order to assess how indigenous ways of organising and allocating space might support or conflict with the objectives of forest protection, for evaluating different means of coordinating indigenous resource management systems with government-instituted systems of management, and as a basis for formal legal recognition and protection of customary forest tenure arrangements (Sirait *et al.* 1994).

With funding from the Ford Foundation, a subproject within the reserve area was established, called the 'Culture and Conservation' project. The goal of the project was to record oral histories, indigenous knowledge and village dynamics related to resource management. The mapping component was added at the suggestion of a colleague at the Environment and Policy Institute of the East–West Center. Using a method developed by Fox (1990), sketch maps of local land use and resource territories were constructed. Sketch maps reflecting local people's ways of talking about resources and their claims to them were combined with points on the GPS. A geographic information system was used to match field data with data on official land use and topographic maps. In this way, the counter-mapping agencies hoped to identify territorial conflicts, establish resource use boundaries, and better understand the ways local people conceptualise their resources.

[...]

Prior to the gazetting of the reserve and the allocation of other uses by the Forest Department, the villagers kept a majority of their lands (66% or 11 844 ha) in two protected forests, one to be used only by the village council when wood and other products are needed for village development projects, the other to be kept for product extraction by widows and orphans (Sirait *et al.* 1994). Other human use of these lands is relatively rare. Another 31% (5419 ha) of the village land is standing forest, used for collecting firewood, construction wood, resins, fruits and other non-timber forest products by all the villagers (Sirait *et al.* 1994). Only 4% (631 ha) of the village land is under swidden cultivation, but this figure includes swidden fallows under various management regimes including fruit tree groves and rattan gardens (which are typically interspersed with a variety of planted and self-sown species in swidden fallow forests).

In sharp contrast to the locally produced maps, the land uses on the TGHK maps show no regard for current village uses or claims. On the basis of these maps, and with no ground checks, government forest planners allocated more than 50% of the village's land – mostly its standing forest – to two external users: the Kayan Mentarang Reserve and a timber concession. The reserve includes land classified as protection forest and convertible forest; the timber concession includes land classified as convertible forest and limited production forest. More than half the village's protected forest falls within the reserve or within another adjacent protected forest area. 25%, however, falls either in the concession, in convertible forest not yet allocated, or in limited production forest (Sirait *et al.* 1994: 415, Table 1). All of the village's cultivated land is in either the reserve or the concession area (Sirait *et al.* 1994). Note that such an aberration does not benefit the timber concession or the reserve because lands under village agriculture contain neither mature hardwood species for logging nor the species intended for protection. Moreover, were the villagers to prevail in a decision over whose maps to use, the outcome would be more standing forest than the government has presently planned.

The question raised by these discrepancies is whether the counter-map has a chance of recognition by the government. This requires thinking about the changes that would have to be made in the government's current uses. Two major things need to happen to give the villagers' total jurisdiction over their forest. Firstly, the status of the Kayan Mentarang Nature Reserve would need to be changed to a National Park or a Biosphere Reserve, in order to allow some 'traditional' uses of the forest by local people. This would in effect allow the recognition of local people's customary rights. However, since the counter-map was made, a request to change the reserve's status was put forth. A concurrent request by the logging company for permission to build a road through the proposed park to the timber concession led the Minister to turn down both.

The second change required would be to alter the forest concession agreement. This would entail changing the boundaries of the concession, a much more expensive and contestable task than changing from one conservation status to another. If villagers were allowed to control the whole conversion forest area, but none of the production forest, they would only regain a quarter of the territory usurped by the concession. The other alternative is for the villagers, the concessionaire and the Forest Department to work out a management plan. A recent government ruling has placed the burden on the concessionaire 'to recognise the existence of customary land and reach a consensus with the villagers about its management' (Sirait *et al.* 1994: 416). In practice, very few timber companies have actually carried out this new requirement. In virtually all cases, conflicts over territory, resource rights and road building continue apace. [...]

Despite initial difficulty in surmounting such obstacles, the 'Culture and Conservation' mapping project has several factors operating in its favour. Firstly, as it is one of the biggest contiguous reserved parks in Asia, developments within it are likely to have an important impact regionally, particularly if it successfully integrates people into the planning process and the majority of local people feel they have benefited after implementation. Secondly is the participation of international institutions with a history of involvement in and influence on resource management policy in Indonesia. Some of these programmes have emphasised taking the needs of local people into account. [...] A third element in the mapping project's favour is the appropriation of the government's own mapping methods and planning tools, including the topographic map series and the GPS. Indonesia has invested considerable funds in GIS technologies, satellite technology and computerised resource management tools; acquisitions that now make the state somewhat vulnerable to counter-mapping strategies. Moreover, when peasant groups meet government mappers on their own ground, as it were, their efforts have greater legitimacy than if the maps were simple sketches. Finally, counter-mappers have allies within the Indonesian state itself. The Ministry of Forestry has been involved in reserve planning and oversight since the beginning, including at least some discussions concerning the roles and status of indigenous peoples living in or adjacent to the park. [...]

Counter-mapping strategies initiated by local NGOs

Both structurally and in terms of goals, mapping projects initiated by local NGOs unfold somewhat differently. In Kalimantan, as in parts of eastern Indonesia, several local NGOs have requested the services of mapping experts to teach and aid them in mapping village land use. The relationships between the NGOs and their international supporters differ from those engaged in the co-mapping strategy described above. These NGOs work autonomously: they do not share management of the project with government agencies, or with internationally-based NGOs like WWF. The NGOs hire the experts (sometimes the same individuals working in the project described above) with funding that they apply for on their own (but often from some of the same sources as those mentioned in previous section). [...]

Some of these NGOs' goals in mapping include documenting current and historical land uses and claims as well as locating and counting forest-dependent populations by ethnicity. In doing so, they intend to legitimate claims to areas that have not already been 'converted' into production forests or plantations. They also hope to counter the impact of the national census, which inadequately represents the diversity of local populations and therefore works against local claims formerly protected by customary law *(adat)*. [...]

The nationalistic thrust of Indonesian policy over the past 50 years has emphasiSed the homogeniSing aspects of national unity, at the expense of the country's rich ethnic diversity, although both phrases are part of the nation's motto ('Unity in Diversity'). The number of people heir to particular ethnic identities, therefore, has become an important and scarce bit of information. When the central state's mechanism for counting citizens does not differentiate them by ethnicity, the geographic extent of local claims remains unclear. Ethnic diversity and identity, expressed among other ways, through resource management and control strategies, and codified by *adat,* is an important aspect of what these local NGOs wish to document. Relating population figures to forest maps is thus a first step in understanding where conflict might arise between claimants with aboriginal or historical claims and newcomers to the local scene, including both newly settled migrants and government-sponsored resource exploitation projects. In a less formal, but no less territorialized manner, the NGOS want to help local people document their claims to the resources within particular lands and the rights to convert forest to other land uses, as they did for centuries before the nationalisation of forest land.

[...]

Both mapping strategies described above necessarily involve more educated, often urbanised members of these subethnic groups, representing 'local' situations of which they may no longer be a permanent part. The technology being used necessitates this – at least in these early stages. Moreover, they are providing a voice from these localities which has been missing from previous representations of these forested spaces. The more detailed these maps become, however, the more important will become the question of which local voices are represented.

Discussion and conclusion

Counter-mapping is a uniquely late-twentieth century phenomenon, made possible in part by both technological developments and the last decade's push toward participatory politics and management strategies. This paper presents two means by which local people are gaining access to the tools of the powerful – maps and mapping technologies developed by and for state international resource planners and managers – and shows how they are using them to legitimise their claims to land and resources. Regardless of their future success or failure in changing state policy and state maps, however, the cases raise several critical theoretical issues. Most critical, perhaps, is the potential maps have for 'freezing' the dynamic social processes which are referred to as 'customary law.'

Secondly, will an independent strategy to map and claim resources fare better than an inclusive one that works with government forest agencies and international environmental groups with a strong presence in Indonesia? As Foucault, Anderson, Giddens and others have discussed, the use of a new medium of expression, in this case maps, to express social relations has transformative power. The fear of 'freezing' custom is not a new argument for Indonesia (or the former Dutch East Indies). Many writers have argued that the codification of customary law, the writing down of oral traditions, the legalising of flexible law codes, generally resulted in 'freezing' these traditions, taking away their characteristic flexibility, and therefore changing their very nature (Lev 1985; cf. van Vollenhoven in Holleman 1981). [. . .]

Since mapping is the visual or representational aspect of the 'writing' of custom, it too can be accused of affecting the flexibility of land use and claims to resources. [. . .] The question is whether maps will *preclude* future changes that ignore the information on the map. I think the answer to this question in terms of land use is no: maps may or may not be a covenant, despite the current fascination with them as a planning tool. [. . .] Once a group's map is empowered by both state recognition and local acceptance, the map can become a tool for negotiation of local land use controls – separating protection forest from agricultural land, for example. But empowerment should also bring the ability to change the map, to renegotiate its terms, and to alter the contents of what may remain somewhat abstract space at a larger scale. In addition, many of the boundaries on the ground are unlikely to remain as strict and clear as they will appear on maps. Maps may influence the direction and impact of change, but change, like flexibility, is an important part of customary practice or law. Like customary rules transmitted orally, or even like written customary or statutory laws, maps can be changed as practice, use and values change, or as rights are transferred between generations or out of the hands of the original holders.

[. . .]

The main purpose of the maps described here is to document and establish boundaries between forest villagers and external claimants, from the local point of view, and to re-claim for local people some of the territory being appropriated by state and international forest mapping projects. Local notions of territoriality have had to change as extensive land-based projects have threatened them; they will change further with mapping. Yet, given the alternate futures – of not being on the map, as it were, being obscured from view and having local claims obscured – there almost seems to be no choice. Both in forest mapping and generally in Indonesia's natural resource politics, local people's views and claims have not been adequately recognised, and even more rarely accepted on their own terms. Some translation is needed into the terms of those who would claim them. Maps give local people the power to do so.

References

Aberley, D. (1993) *Boundaries of Home: Mapping for Local Empowerment*, New Society Publishers, Gabriola Island, BC.

Barber, C.V. (1989) The state, the environment, and development: the genesis and transformation of social forest policy in New Order Indonesia. Unpublished PhD thesis. University of California, Berkeley, CA.

Buisseret, D. (1992) *Monarchs, Ministers, and Maps: The Emergence of Cartography as a Tool of Government in Early Modern Europe*, University of Chicago Press, Chicago.

Fox, J. (1990) Sketch mapping as a diagnostic tool in forest management, in *Keepers of the Forest: Land Management Alternatives for Southeast Asia* (ed. M. Poffenberger), Kumarian Press, Westport, pp. 119–133.

Giddens, A. (1984) *The Constitution of Society*, University of California Press, Berkeley, CA.

Girot, I. and Nietschmann, B. (1993) The geopolitics and ecopolitics of the Rio San Juan. *National Geographic Research and Exploration*, **8** (1), 52–63.

GOI (Government of Indonesia). Regional Physical Planning Programme for Transmigration (RePPProt), ODA and Departemen Transmigrasi, Jakarta.

Harley, J.B. (1988) Maps, knowledge, and power, in *The Iconography of Landscape* (eds D. Cosgrove and S. Daniels), Cambridge University Press, New York, pp. 277–312.

Harley, J.B. (1989) Deconstructing the map. *Cartographica*, **26**, 1–20.

Holleman, J.F. (1918/1981) *Van Vollenhoven on Indonesian Adat Law: Selections from Het Adatrecht van Nederlandsch-Indie*, vol. 1, Martinus Nijhoff, The Hague, The Netherlands.

Kain, R.J. and Baigent, E. (1992) *The Cadastral Map in the Service of the State: A History of Property Mapping*, University of Chicago Press, Chicago.

Lefebvre, H. (1991) *The Production of Space*, Blackwell, Oxford.

Lev, D.S. (1985) Colonial law and the genesis of the Indonesian State. *Indonesia*, **40**, 57–74.

Manning, C. (1972) The timber boom in East Kalimantan. *Bulletin of Indonesian Economic Studies*, **7**, 30–61.

Menzies, N.K. (1992) Strategies of inclusion and exclusion in China's forest management. *Modern Asian Studies*, **26**, 719–733.

Momberg, F. (1994) Participatory tools for community forest profiling and zonation of conservation areas: experiences from the Kayang-Mentarang Nature Reserve, East Kalimantan, Indonesia. Paper presented at the Borneo Research Council Third Biennial International Conference, Pentavank, 1–14 July.

Moniaga, S. (1993) Toward community-based forestry and recognition of adat property rights in the Outer Islands of Indonesia, in *Legal Frameworks for Forest Management in Asia: Case Studies of Community – State Relations* (ed. J. Fox), East–West Center Occasional Papers of the Program on Environment, Honolulu, HI.

Orlove, B.S. (1989). Maps of Lake Titicaca: the politics of representation in encounters between peasants and the state in Peru. Unpublished manuscript.

Padoch, C. (1994) The woodlands of Tae: traditional forest management in Kalimantan, in *Forest Resources and Wood-Based Biomass Energy* (ed. W. Bentley), Science Publishers, New York, pp. 131–150.

Peluso, N.L. (1992) *Rich Forests, Poor People: Resource Control and Resistance in Java*, University of California Press, Berkeley, CA.

Peluso, N.L. (1993) Coercing conservation: the politics of state resource control. *Global Environmental Change*, **4** (2), 199–217.

Potter, L. (1995) Forest degradation, deforestation, and reforestation in Kalimantan: towards a sustainable land use? in *Borneo in Transition: People, Forests, Conservation and Development* (eds C. Padoch and N.L. Peluso), Oxford University Press, Oxford.

Sack, R.D. (1983) *Human Territoriality: Its Theory and History*, Cambridge University Press, Cambridge.

Sirait, M., Prasodjo, S., Podger, N. *et al.* (1994) Mapping customary land in East Kalimantan, Indonesia. *Ambio*, **23** (7), 411–417.

Vandergeest, P. and Peluso, N.L. (1995) Territorialization and the Thai State. *Theory and Society*, **24** (3), 385–426.

Wood, D. (1992) *The Power of Maps*, Guilford, New York.

Zerner, C. (1990) *Legal Options for the Indonesian Forestry Sector*, United Nations Food and Agriculture Organization, Jakarta, Indonesia.

Further reading

Harris, L.M. and Hazen, D. (2006) Power of maps: (counter) mapping for conservation. *ACME: An International E-Journal for Critical Geographies*, **4** (1), 99–130. [A useful review paper on counter-mapping.]

Hodgson, D. and Schroeder, R.A. (2002) Dilemmas of counter-mapping community resources in Tanzania. *Development and Change*, **33**, 79–100. [An early research article on the issue of counter-mapping.]

Wainwright, J. (2009) Cartography, territory, property: postcolonial reflections on indigenous counter-mapping in Nicaragua and Belize. *Cultural Geographies*, **16** (2), 153–178. [Contemporary discussion of counter-mapping.]

See also

- Chapter 1.8: Deconstructing the Map
- Chapter 4.7: Mapping Reeds and Reading Maps: The Politics of Representation in Lake Titicaca
- Chapter 4.8: Refiguring Geography: Parish Maps of Common Ground
- Chapter 5.4: Mapping: A New Technology of Space; and Geo-body
- Chapter 5.5: First Principles of a Literary Cartography
- Chapter 5.7: A Map that Roared and an Original Atlas: Canada, Cartography, and the Narration of Nation
- Chapter 5.10: Queering the Map: The Productive Tensions of Colliding Epistemologies

Chapter 5.7

A Map that Roared and an Original Atlas: Canada, Cartography, and the Narration of Nation

Matthew Sparke

Editors' overview

Sparke's paper examines in detail what occurs when two different concepts of space, time and territorial jurisdiction – that of two First Nations tribes and the Canadian government – come into conflict. His analysis traces a court case concerning sovereignty and land rights in British Columbia, wherein the First Nations tribes of Gitxsan and Wet'suwet'en challenged the legitimacy and accuracy of state maps, presenting their own maps of their nations and their claim to space. Sparke details the political, cartographic and postcolonial discourses through which the maps were created, framed, interpreted and used, and how the Canadian state sought to assert the hegemony of its cartography as superior, and how the First Nations sought to destabilise such claims through, what he terms, 'contrapuntal cartographies': a strategic re-voicing of the subdominant position to reveal the power relations at play and to counter-balance them. Such a re-voicing reveals the contingency of the epistemology underpinning Western cartography.

Originally published in 1998: *Annals of the Association of American Geographers*, **88** (3), 463–495.

On May 11, 1987 a trial over sovereignty and land rights began in the Supreme Court of British Columbia. The case had been brought by two First Nations – the Wet'suwet'en and the Gitxsan – against the federal government of Canada and the provincial government of British Columbia. Ken Muldoe, the chief whose Gitxsan name Delgamuukw

Exhibit 102 Traditional Boundaries of the Gitxsan and Wet'suwet'en Territories

---

25: The Court: We'll call it the map that roared.
26: Mr. Plant: I beg your pardon.
27: The Court: We'll call it the map that roared.

(*Delgamuukw v. the Queen*, Trial Transcripts, vol. 108: 6871)

served as the official abbreviation for all the plaintiffs making the case, concluded his opening address to the court by summarising the view of two First Nations that the trial was not a simple appeal to the law of the land but rather a political negotiation within the Canadian legal system. 'The purpose of this case, then, is to find a process to place Gitxsan and Wet'suwet'en ownership and jurisdiction within the context of Canada. We do not seek a decision as to whether our system might continue or not. It will continue.' (Quoted in Monet and Skanu'u 1992: 23). As the 318 days of evidence and 56 days of closing argument proceeded, the Wet'swuet'en and Gitxsan continued to affirm their traditions of self-government, educating the court and those members of the Canadian public who followed the immense trial [...] about their understandings of space, time and territorial jurisdiction.

Later in the same year, another rather different affirmation of nation emerged before the Canadian public in the

The Map Reader: Theories of Mapping Practice and Cartographic Representation, First Edition. Edited by Martin Dodge, Rob Kitchin and Chris Perkins.

more muted but institutionally privileged form of the first volume of a new historical atlas of Canada. More than an academic exercise prosecuted solely for the benefit of historical and geographical research, the *Historical Atlas of Canada* was also a national project that narrated an origin story of Canada. National in production and funding as well as in organisation and scope, the *Atlas* as a whole received more than Can$6 million in public funds. [...] Initially published just in time for the Christmas commercial season, the attractive coffee-table sized volume – with 27 000 copies of the English edition sold by November 1993 (Piternick 1993: 21) – found a much bigger audience than the one that was packed into the small Smithers, BC, courtroom on the opening day of *Delgamuukw v. the Queen.*

There are many other obvious differences between these two examples of national negotiation. The trial took place in a legal setting, whereas the *Atlas* was produced in a grant-maintained academic context and disseminated through commercial networks. More significantly, the court case involved a direct and necessarily adversarial conflict between colonial and anti-colonial forces, whereas in the *Atlas*, these antagonistic power relations found joint expression in a more interwoven and ambivalent national text. In addition, notwithstanding the editors' remarkable attempts to represent the historical geography of everyday life – home life, home building and the quotidian routines of survival on the frontier of colonial contact – the literally textual character of the *Atlas* meant that it was less immediately related than the court case to the actual experiences of people. The court case, like the *Atlas*, involved a whole series of politicised representations of experience, but it also directly coordinated and controlled such experience within the confines of the court. Such differences notwithstanding, this article focuses on how, as coinheritors of overlapping historical geographies, both these examples of national negotiation had much in common as graphic and, indeed, cartographic negotiations of the meanings of space, territory and state jurisdiction. Critical to both were maps of national space, and clearly evident in both was the paradoxical capacity of such cartography to function variously for and against the exercise of modern state power. Indeed, while scholars such as Benedict Anderson (1991) have discussed the general *hegemonic* effect of national mapping, they have rarely addressed the *counter-hegemonic* effect of cartographic negotiations. It is the tensions between the hegemonic and the counter-hegemonic, or, more precisely, the inter-articulation of dominant and oppositional forms of hegemony, that the two case studies considered here help clarify. In juxtaposition, they illustrate an ambivalence in cartography that in turn points to a profound ambivalence in the very narration of the nation state itself.

[...]

Geo-graphic supplements

The political

Thanks in part to an engagement with the work of Michel Foucault (especially 1979, 1980), political geographers are increasingly coming to terms with a considerably widened understanding of what counts as 'the political'. Once power is understood in a non-sovereigntist and more relational way as something that is exercised in social relations rather than held in the hands of individuals and states, then, following Foucault, power relations, politics and, in turn, political geographies can be found at work across all scales of social life. [...]

[...] In a field in which sovereigntist conceptions of 'the state' holding power have chronically framed scholarship, Mitchell (1991) has offered instead an alternative Foucauldian account of how congeries of power relations ranging from national planning to the disciplining of modern armies create the net effects we subsequently homogenise and call 'the state.' Mitchell's ascending analysis of state power is valuable for geographers, I think, because it simultaneously centres the question of spatial ordering. It is this question that informs my own inquiries, and in these I follow Mitchell's (1988) longer elaboration of the processes that, as he describes it, 'enframed' the ordered space of the state effect that was colonial Egypt. [...]

[E]nframing can be said to involve a process of spatial abstraction that simultaneously occults the process through which abstract space becomes represented as discontinuous from the lived spaces of everyday life. This thesis clearly resonates with Lefebvre's now-famous argument about the production of abstract space in modern capitalism: abstract space that is abstracted and organised as decorporealised, bureaucratised and commodified (1991). [...] Considered thus as condensation points for struggles over the enframing of space, law and cartography provide useful case studies not only because they are exemplary institutional arenas in which abstract state-space is reproduced and reworked, but also, as Blomley (1994) has shown so well, because they are *constitutively* interlinked. [...]

The cartographic

Blomley's research into the history of Western property law led him to the valuable work of Helgerson (1992) on the emergence of state cartography in Elizabethan England. Attending to the recursive proleptic effects of mapping – the way maps contribute to the construction of spaces that later they seem only to represent – Helgerson highlights how the new national maps of Saxton, Camden and others were part and parcel of a civil and conceptual revolution in which it was made to seem as if the land itself spoke of the kingdom as a single state. [. . .]

Much like Mitchell's analysis of the dualisms inaugurated by colonial modernity as an exhibitionary order in Egypt, Helgerson (1993) shows that the appearance of this conceptual gap was crucial to the development of the fledgling state as something seemingly separate from society.

The constitutive role of cartography in the production of some early modern state effects has of course been further documented and analysed by historians of cartography. Most notably it was one of Brian Harley's key foci of analysis in his works on the power relations underpinning cartography (1988, 1992). What Harley's debunking of the myths of cartographic objectivity also showed, of course, was that this critical contextualisation of cartography could be further extended to examine the central role mapping played in the organisation and consolidation of imperial rule by European states overseas. [. . .]

[. . .] Showing how cartography can operate both for and against colonialism not only deepens the scholarly work of critical cartography, it also counters the too-speedy denunciation of maps and mapping as metaphors of domination (Pile and Rose 1992). It simultaneously serves as a corrective to the trend toward 'imperial nostalgia' – a febrile fascination with the glory days and travels of colonialists – among certain celebrants of postcolonialism. [. . .]

The postcolonial

As Bruce Willems-Braun (1997: 704) has carefully explained [. . .] the notion of 'postcoloniality' as a historic period is problematic insofar as it 'can easily be taken to assume a historical rupture between past and present'. In a country like Canada, as Willems-Braun also emphasises, such an assumption can have the effect of actually concealing the ongoing effects of colonialism as an active force in day-to-day life. By contrast, as Gyan Prakash (1992: 8) has argued, postcolonial theory at its best can instead 'force a radical rethinking and reformulation of forms of knowledge and social identities authorised by colonialism and western domination'. [. . .] Jane Jacobs's work is a particularly significant reference point here because her study of aboriginal politics in urban Australia examines cartography as an ambivalent (post)colonial mode of spatial representation.

She notes that the 'emphasis on the hegemonic effect of the map may well overstate the power of the cartographic imagination' (Jacobs 1996:150), and proceeds to show how some contemporary aboriginal remappings produce a destabilising doubling of the colonial cartography drawn for tourists, a doubling, she suggests, that effectively 'detours' the map (p. 151). To be sure, Jacobs (1996: 154) also notes that the decolonising political potential of this detouring of the tourist landscape is minimal: '[it] probably will not result in Aborigines gaining significant or meaningful land rights in relation to Brisbane'. Nevertheless, her keen attention to cartographic destabilisation through remapping illustrates and introduces the concept-metaphor I call 'contrapuntal cartographies'.

The notion of a contrapuntal *reading* of cartography is drawn from Edward Said (1993), who uses the musical metaphor to break down singularised and unidirectional understandings of the culture of imperialism. Reading contrapuntally, argues Said (1993: 51), involves 'a simultaneous awareness both of the metropolitan history that is narrated and of those other histories against which (and together with which) the dominating discourse acts'. Said thus reworks the formal musical meaning of the term, suggesting that a contrapuntal interpretation involves a strategic re-voicing of the subdominant to make it equal to the dominant and thus to orchestrate a balance that can potentially edify and educate an audience about the power relations of culture. In the two cases examined here, this strategic contrapuntal orchestration provides political purchase on what are, at once, more state-related and overtly geographical forms of national narration. As case studies, they in turn highlight [. . .] how in a national struggle 'to reclaim, rename and re-inhabit the land,' the impulse is indeed 'cartographic' (Said 1993: 226). More than this, though, these examples reveal the intrinsic articulation between the colonial and meaningfully *post*colonial that runs deeper than any act of interpretation by the single composer/reader evoked by Said's metaphor. In examining these points of real articulation, and therefore ambivalence, in the contrapuntal cartographies presented by the Delgamuukw trial and the *Historical Atlas*, Bhabha's conceptual claims about the constitution of the modern nation become valuable.

The imagination and narration of nation, Bhabha (1994: 146–151) argues, is deeply marked by ambivalence. On the one hand, there is what he describes as the self-certain *pedagogy* of national discourse. Not unlike the maps described by Helgerson, such pedagogy teaches, among other things, that the spaces of everyday life along with all 'the people' can be abstracted into the nation-space, all territory transformed into a new national tradition. On the other hand, Bhabha argues that such national pedagogy always has to come to realisation through supplementary *performances*, the actual putting into practice and place of the teachings. It is in such unavoidable performance, he suggests, that the political unity of the national narrative falls apart (split, in Bhabha's metaphorics, by the same displacement that Derrida elaborated in the supplementary structure of *writing*). The pedagogy has to be performed and put in place, and yet each time it is thereby supplemented, it transforms homogeneity into heterogeneity. As a result, 'the very act of the narrative performance interpellates a growing circle of national subjects' (Bhabha 1994: 145).

This canny articulation of national pedagogy displaced in performance is invaluable for coming to terms with the contrapuntal cartographies of postcolonial national negotiation. [. . .]

[. . .] Bhabha offers no account of the historical production of abstract space. Consequently, he ignores the possibility of how space can be produced and thereby performed pedagogically in powerful ways that convene and thereby potentially co-opt plural traditions and histories into the abstraction of the single territorial collectivity constituting the state. Bhabha's treatment of space, and what other scholars have critiqued as his bracketing of the political (Xie 1996; Mitchell 1997), would thus seem to be related. In order to draw attention to the implications of this weakness, I am seeking here to rework and thus, as it were, re-perform his argument geographically through the examination of the contrapuntal cartographies of these two case studies.

Contrapuntal cartographies

Jean-Pierre Wallot, the Dominion Archivist who wrote the second of two forewords to volume 1 of the *Atlas*, compared the volume to a 'musical score', primarily as a way of coming to terms with how it represented demographic, economic, cultural, governmental and social relationships altogether in the space of the nation-state. As a result of such amalgamation, he wrote, 'Canadians will have a better understanding of themselves, and, it is hoped, will be inspired to extend the frontiers of knowledge even further' (Wallot 1987). Beyond this achievement, I argue that the more radical and creative aspect of the *Atlas* has been to provide a cartographic 'musical score' which, once given contrapuntal voicing, can enable its national Canadian audience to rethink the colonial frontiers of national knowledge itself.

By scrupulously mapping the supposed beginnings of the nation, the *Atlas* subverts any punctual notion of a singular national origin, displacing it with an invitation to readers to re-evaluate the ways in which the template of contemporary Canada is imposed proleptically on a heterogeneous past. Most particularly, it enables Canadians to reconsider the discontinuous positions of native peoples – their positions quite literally in diverse geographies, on the continent, before the arrival of the English and French – as a disjunctive series of national traditions at the ends of the frontier of Canada as nation. As such, its contrapuntal aspects exemplify what Bhabha (1994: 148) calls 'a liminal signifying space that is *internally* marked by the discourses of minorities, the heterogeneous histories of contending peoples, antagonistic authorities and tense locations of cultural difference'.

Still more tense were the differences in and over location evidenced in the court case. Like the *Atlas*, the contrapuntal dualities of *Delgamuukw v. the Queen* made the *location* of national discourse a contentious question through a repeated return to maps. Not only were cartographic tools and arguments used by the defense (the BC and federal governments), they were also a key component of the Wet'suwet'en and Gitxsan people's own attempts to outline their sovereignty in a way the Canadian court might understand. It was one such map of Gitxsan and Wet'suwet'en territory (Exhibit 102) that Chief Justice Allan McEachern was beginning to unfold when he declared, 'We'll call it the map that roared.' In the immediate context of trying to open up a huge paper reproduction of the First Nations' map, his words appeared to refer to the colloquial notion of a 'paper tiger' (*Delgamuukw v. the Queen*, Trial Transcripts 1988: 6871). They also may have been a reference to the 1959 Peter Sellers movie satirising Cold War geopolitics, 'The Mouse that Roared.' As such the comments might be interpreted as a derisory scripting of the plaintiffs as a ramshackled, anachronistic nation. But as Don Monet, a cartoonist working for the Gitxsan and Wet'suwet'en, made clear, the Chief Justice's reference to a roaring map simultaneously evoked the resistance in the First Nations' remapping of the land: the cartography's roaring refusal of the orientation systems, the trap lines, the property lines, the electricity lines, the pipelines, the logging roads, the clear-cuts and all the other accoutrements of Canadian colonialism on native land.

McEachern (1991) ultimately dismissed the Gitxsan and Wet'swuet'en's claims with a remarkably absolutist set of colonialist claims about the extinguishment of aboriginal rights. In its original format, his judgment spanned almost 400 pages and, in arguments ranging widely from the Chief Justice's view of First Nations societies to his understanding of Canadian history, he systematically dismissed Gitxsan and Wet'suwet'en claims to ownership, jurisdiction and damages for the loss of lands and resources since the establishment of the colony. Nevertheless, his comments on Exhibit 102, the map showing the 'Traditional Boundaries of the Gitxsan and Wet'suwet'en Territories,' would also appear to betray, albeit unconsciously, a real recognition of Gitxsan and Wet'suwet'en agency and territorial survival. That very same agency recently recorded far more fulsome vindication when the Supreme Court of Canada handed down its decision overturning McEachern's judgment and opening up the possibility of a new trial, or at least greater bargaining leverage for the two First Nations with the provincial and federal governments (Appeal Transcripts 1997). This was a massive turnaround in native rights litigation more generally in Canada, and, as well as making local news (Bell 1997), was reported on the front page of the *New York Times* (DePalma 1988). But before all this, the clash of 'antagonistic authorities' referred to by

Bhabha as the internal mark of the nation-space was already clear in the original courtroom discourse over cartography, and, for the same reason, it will serve as the first of my two case studies.

The trial: pedagogy performing the policing of performance

[...]

By entering into the Canadian legal process of the trial in the Supreme Court of British Columbia, the Wet'suwet'en and Gitxsan peoples were insinuating their claims into the terms of reference of the dominant discourse. Theirs was a contested and compromising entry, and while the claim to 58 000 square kilometres of land did have a crucial antagonistic effect, it did not come with the same rhetorical ease with which Bhabha's account of 'the supplementary' seems to spirit its way towards almost ontological conclusions. In fact, so dubious an ally was the Canadian law seen to be that many sympathetic commentators, both from within First Nations communities and without, criticised the chiefs for embarking on such a compromising strategy. [...] It was not simply that the Gitxsan and Wet'suwet'en had to deal with the adversarial protocols and far from sly spatiality of the courtroom. It was also that the actual arguments made by the two First Nations had to work within the framework of Canadian colonial law. [...]

Coming into the Canadian court, the two First Nations had to attempt to insert their voices and speak their claims in a way that would successfully communicate their primarily oral knowledge and understanding of territorial jurisdiction to a white judge trained in the abstractions and textual formalities of the modern western state. [...] Central to all these courtroom abstractions was the removed, bureaucratised and disembodied conception of abstract space that, Mitchell argues, is constitutive of modern state effects. [...]

[...]

Negotiating with maps

In addition to the formal spatiality of the court and its subversion, Monet's cartoon also highlights another more directly cartographic theme. On the one side, he pictures Antgulilibix singing the *limx'ooy*. On the other is the Chief Justice, surrounded by his written records and maps. The latter, with their Cartesian grid base and their orientation system organised by the North Pole arrow, stand here as paradigmatic of the proper pedagogy of the courtroom. Monet uses them as both an indication of the rationalisations used by the government lawyers and examples of the 'proper way to approach the problem', according to the Chief Justice. It was precisely the rules of this game played in abstract space that the two First Nations had to negotiate. Given that ultimately they had to communicate their territorial knowledge to this judge in this court, they translated their oral knowledge into a series of maps.

This produced, I think, one of the clearest examples of Satsan's point about playing the game in order to change it. Through the medium of modern mapping, they articulated their claim to their territories in a way the judge might understand. In the process, they were effectively cartographing their lands as First Nations within the abstract state-space of Cartesian cartography. Simultaneously, they were supplementing the provincial and federal mapping of the land with maps based on Gitxsan and Wet'suwet'en oral knowledge. As such a repetition with a difference, a performance of the pedagogy of the place of the people, the maps served at once both to communicate in and disrupt the cartographic conventions of the court.

[...]

For the Gitxsan and Wet'suwet'en, these questions of mapping and tradition were [...] complexly interrelated, insofar as the point of producing their maps as trial affidavits related to the translation of the oral histories of the Houses into modern maps. Thus, unlike the land claims of the Labrador Inuit, the northern Ontario Cree and Ojibwa, and the northern Saskatchewan Chipewyan, the Gitxsan and Wet'suwet'en were not using map biographies of their occupancy and resource utilisation in order to claim title on a resource-tenure basis (Usher *et al.* 1992). Instead, they presented the court with a series of maps that mapped their Houses and thus their territory over, or more accurately perhaps *under*, British Columbian provincial maps [...]. Because such provincial maps had historically been imposed over the territory in a way that almost erased its pre-colonial spatiality, this cartographic representation of the Houses also served to chart the sheer density of the palimpsest produced by the whole series of pre-colonial, colonial and postcolonial inscriptions.

The process restored some meaning to spaces more usually covered by apparent emptiness on modern Canadian representations of the province. Inscribing anti-colonial names and places in the middle of the colonial coverage, it also addressed head-on what Bhabha (1994: 217) describes as the 'problem of signifying the interstitial passages and processes of cultural difference that are inscribed in the 'in-between''.

[...]

Yet the maps, while following these abstract rules of proper and stately court procedure, nevertheless also transformed them by establishing a toponymy for the subdivisions with the names of the First Nations' Houses. In effect, they depict the First Nations territories as seemingly independent nation-states with their own internal boundaries.

One way of looking at the maps, therefore, is as illustrations of how the counter-hegemonic can rearticulate the hegemonic in its own oppositional terms. In this way, the First Nations' territorial jurisdiction was successfully communicated all the while the audience in the theatre of nation-state pedagogy witnessed a performance with a difference. Evidently the maps in general had disruptive effects in court: they did indeed 'roar.' But beyond the temporary disruptions to the orderliness of the court, the unfolding of specific First Nations' maps, such as those charting the internal boundaries, also evidenced a systematic re-codification of the land. The territories repeatedly presented in the government maps as so many square miles of resources were thereby actively represented as a landscape rich with the historical geographies of Wet'suwet'en and Gitxsan names and meanings.

[...]

There were many other moments when the accuracy of the Gitxsan and Wet'suwet'en map boundaries was challenged. [. . .] The fact that the maps were translations of songs, and that such House songs of territory and history were unlikely to match up perfectly with discretely delineated blocks of territory in abstract space, was of little concern. Instead, the Chief Justice simply judged the cartographic affidavits by the standards of the colonial state. As a result, the evidence of House ownership presented by the Gitxsan and Wet'suwet'en was ultimately dismissed.

[. . .]

Whatever names and meaning the Gitxsan and Wet'suwet'en had sought to map back on to the landscape with their cartography was in such moments lost on the judge. For him, instead, the landscape was enframed as 'beautiful, vast and almost empty.' It might seem that these are just terms of innocent, almost sublime reflection. And yet they are also terms of colonial conquest, terms which, with easy appeals to the abstractions of European aestheticism, serve to empty the landscape. Combined with the court' abstract cartographic conventions in the 'Reasons for Judgement,' they explain a great deal of the Chief Justice's 'proper way to approach the problem,' and why, ultimately, he ruled to dismiss the suit. As Harley describes the enframing effect of colonial cartography, the Chief Justice's overall view might also be said to have dispossessed the Gitxsan and Wet'suwet'en (1992: 531) 'by engulfing them with blank spaces'.

[. . .]

The *Atlas:* from the pedagogic root to performative routes

[. . .]

[I]t might seem that an atlas of Canada would offer little relief from the pattern of nationalist hegemony identified by Bhabha as the transformation of territory into tradition. Such a presumption would be highly premature, however, and, in the case of the *Historical Atlas*, rather inaccurate. Certainly the *Atlas* can be read as a classic example of national pedagogy. Through its powerful social status as a teaching tool, its traditional evolutionary narrative, and its imposition of modern Canadian names and shapes on the pre-colonial past, it might indeed seem to transform the nation's modern territoriality 'into the archaic, atavistic temporality of Traditionalism' (Bhabha 1994: 149). And yet, I will argue, the *Historical Atlas* does not so much 'displace' anxiety about Canada's irredeemably plural modern space as actively celebrate it as the very stuff of Canadian tradition. This, I suggest, has a number of critical implications about Bhabha's arguments concerning space, and the links between space, performativity and disruption. At a more practical level, though, the example of the *Atlas* shows how a seemingly hegemonic narrative of the nation can also function through its very rigour and ideals of comprehensiveness to open up spaces for counter-hegemonic questioning. Overall, my account of the *Atlas* testifies to the wisdom in José Rabasa's words about Mercator's original atlas. Such an atlas, he argued, must be understood: as simultaneously constituting a stock of information for a collection of memory and instituting a signalling tool for scrambling previous territorializations. Memory and systematic forgetfulness, fantastic allegories and geometric reason coexist in the Atlas with an apparent disparity (Rabasa 1985: 3).

It is this simultaneity and coexistence of pedagogic stockpiling with performative scrambling that the court's strict divisions dichotomised in *Delgamuukw v. the Queen.* Performance was policed to severely curtail any chance of turning the map into a signalling tool that might further First Nations' re-territorializations. The court and counsel for the defence pontificated with strident pedagogy while the Wet'swuet'en and Gitxsan struggled to have their cartographic performance even recognised as such. In the *Atlas*, by comparison, the hegemonic and counter-hegemonic were far more closely intertwined. As such, they can usefully be compared with two different sets of entangled roots and routes. One set is the pedagogic national-genesis story with its *singular root* marked by the subtitle given to the English edition of volume 1 by the University of Toronto Press: *From the Beginning to 1800.* The other, less sacred-sounding set, the set made up of the *multiple routes* charted performatively in space in the *Atlas*, instead found its more plural reflection in the doubled-up sensitivity of the French edition's subtitle: *Des Origines à 1800* (Harris and Dechêne 1987). Below, I track back and forth between these two root systems to show how the plural routes of travel, contact and interaction, mapped out in the *Atlas*, displace its chronological narrative's transcendental truth claims to a national root in the soil of North America.

Teaching and reading

Perhaps the most obvious illustration of the *Atlas*'s pedagogic status was the way it was planned, packaged and disseminated as a teaching and research tool. National atlases have long been regarded as having a crucial educational function, and even cartographers who have pondered the supposed 'Mathematical Basis of National Atlases' acknowledge that the final role of such cartography is 'cultural and educational' (Fremlin and Sebert 1972: 30). The *Historical Atlas* was no exception to this pattern. From its inception, it was advanced before the national funding agency – the Social Sciences and Humanities Research Council of Canada – as addressing some of the nation's pressing teaching needs. [. . .]

[. . .]

Understood as a vital element in such national educational endeavours, it is hard to imagine a more 'national-pedagogical' positioning of the *Atlas*. Acclaimed in the national press as a symbol of Canada, celebrated by academics as an empirically rich cartographic essay, and widely purchased by the public who could afford it, the *Atlas* as both medium and message seemed to reach almost anthemic as well as hegemonic status. Yet even as it did so, the very lengths to which its creators went in order to chart the complex regional dynamics of Canadian history and to simultaneously make the work speak to audiences across the country, invited questions.

The French edition, which made the *Atlas* accessible to francophone readers (who quickly purchased all the available copies), notably led to criticism. To be sure, the *Atlas* was widely praised in Québec for making a united Canada visible: '[u]n plaisir pour l'oeil et pour l'intelligence,' declared Yvon Lamonde (1987). But at the same time, some French-speaking readers attentive to the *Atlas*'s ideals of comprehensiveness and national inclusiveness worried over whether the documentation of development dynamics had not been skewed by English Canadian interests. [...] what I prefer to highlight here is how the wide address of the *Atlas* inevitably led some readers to question the adequacy of its coverage. The calls for comprehensiveness and inclusiveness that it simultaneously issued and answered with every massively detailed map, also called forth demands for more specific details from particular constituencies. This form of invitation to critique and rethinking is one of the more interesting achievements of the *Atlas*, and it serves as such as a good example of Bhabha's notion of displacement through narrative performance. In order to bring the radical implications of such displacement more clearly into focus, I next turn to the question of how the *Atlas*'s Eurocentric chronology was itself displaced in volume 1 by the details demanded by the actual work of cartography. As scholarly maps, the *Atlas* plates were, as Dean's foreword emphasises, 'inexorably tyrannical taskmasters'. That same tyranny, I argue, helped introduce a multiplicity of routes where there might otherwise only have been a simple historicist root.

Chronology and cartography

The clearest expression of the singularising root system in the *Atlas* is the way its narrative evolution anchors the 'beginning' of Canada in time, or, to be more precise, in European historicist chronology. Following the temporal logos of this chronology, Plate 1 begins the so-called 'Prehistory' section of the *Atlas* with a map of 'The Last Ice Sheets, 18 000–10 000 BC.' Here, as it were, is Canada's ice-bound garden of Eden, a picture of the glacial past dated geologically, labelled with anachronistic but also seemingly objective geographic terms like 'New Québec Ice,' and mapped in such a way as to present, under the gentle purple hue of the glaciers, an apparently unified and non-American space of collective Canadian experience. There follows, in linear evolutionary sequence, a careful charting of the so-called 'Indian' arrival on the continent.

Despite the pre-European context, the land is, nevertheless, still named the 'New World' in another anachronistic application of modern labels. Plates drawn up by paleobotanists, glacial geologists and archaeologists proceed to map the most recent scientific findings concerning native peoples, plotting their positions like so many specimens in the translucent and icily anaemic geography of transparent state-space. Positioned as early arrivals in this New World space, the first peoples are collectively (re) described by the *Atlas* as 'Canada's first immigrants', which is to say they are reduced to early arrivals in a national pageant of immigration. As such early immigrants, these peoples – ranging from those identified as the 'Fluted Point' people and the 'Northern Interior Microblade' to the 'Late Palaeo-Eskimo' – are brought in turn into the national narrative without compromising its unified, if icy, starting point 18 000 years BC. [. . .]

[. . .] Aboriginal cultures are definitely assigned a complex set of spatial positions on the maps. But because those positions are organised according to the historicist logic of archaeological chronology, there is what Fabida Jara and Edmundo Magana (1982: 117) call an 'evolutionist taxonomy' at work in the map series, a taxonomy whose disciplined and repeated reference to 'diagnostic artifacts', 'cultural sequences' and academic debates over 'poor data', turns this first part of the *Atlas* into the cartographic equivalent of a state-managed archaeological museum.

[. . .]

[. . .] Most immediately and powerfully, the chronology is disrupted by the maps in the subsequent sections that represent the changing geographies of aboriginal societies *after* the arrival of the Europeans. With the advent of Western History, it seems, aboriginal people are not at all banished from the scene. Instead, there are maps of trade and warfare in the St Lawrence (Plate 33), depictions of Iroquois and Algonquian seasonal movements (Plate 34), and brilliantly detailed maps – better described perhaps as cartographic monographs – by Conrad Heidenreich, displaying the complex spatial histories of native groups in relation to the developing fur trade in the Great Lakes region (Plates 35, 37, 38, 39 and 40). In addition to all this, the development of the Northwest is not told as the traditional heroic tale of colonial discovery, but rather as a haphazard history of imperial competition, error and negotiation: a spatial history that is itself punctured by Plate 59, showing the maps of a Chipewyan and two Blackfoot guides. Admittedly these latter maps were drawn at the request of explorers and the plate itself is entitled 'Indian Maps', a terminological homogenisation of first peoples that is also symptomatic of European epistemological imperialism. Still the reader of the *Atlas* can find in such moments a vivid representation of the complexity of different aboriginal geographies and knowledges (Plate 66, which maps language groups and trading relations among West Coast groups around 1800, is another prime example). Certainly, Trigger's complaint that aboriginal histories are commonly confined by chronology to Canada's archaeological prehistory finds a substantive rebuttal in the form of these later plates.

[. . .]

Territory and traditions

One of the reasons why the final plate in the *Atlas* carries authoritative weight as a rebuttal to national narratives of extinguishment is that it establishes an aboriginal presence in terms of a conventional nation-state cartography. Plate 69's title is 'Native Canada, ca. 1820,' and as such claims comprehensive coverage of 'Native' movements from sea to sea. The plate, in other words, negotiates with the abstractions of the modern Canadian nation-state. This is what bestows so much authority on the resulting cartographic product, and yet it comes at a cost. [. . .]

In relation to Plate 69, we might thus note that the vision of 'Native Canada, ca 1820' could never have been glimpsed through the divers native eyes of the time. The people so carefully placed on the map did not see this coherent vision. It is a *post hoc* and indeed abstract reconstruction based largely on European records. Certainly it is a *re*vision, insofar as prior atlases of Canada rarely marked any place, let alone such a 'full place' for native peoples. But, as a revision that abstracts the heterogeneous diversity of native geographies into one comprehensive map, the plate returns us to what Bhabha calls 'the question of social visibility, the power of the eye to naturalise the rhetoric of national affiliation and its forms of collective expression.' Considered in these terms, the collective vision of the plate seems to pre-emptively nationalise what were, and what sometimes remain, the non-national or, at least, non-Canadian realities of native life. The map as a technology of vision does indeed seem to naturalise this nationalisation, concealing its abstracting work even as it turns the diversity of native geographies into a unifying common denominator for the whole of the territory of what is now Canada.

[. . .]

Like Plate 69, the earlier maps of the Prehistory section accomplish the same abstraction of aboriginal geographies onto a collective national stage. [. . .] [It] also recoded them as 'immigrants' following the conventional narrative of Canada as a multicultural New World nation-state.

[. . .]

Disseminating conclusions

My critical point about Bhabha's thesis should now be clear. If national pedagogy is always linked with the timing of historicism, while performance is always affiliated with the disruptive putting in place of such traditional teaching, where is there room for a critical account of the nation-enframing effects of spatial abstraction? To be sure, such abstraction is always performed, but, as such, it may have immediately obvious homogenising effects like those found in both the Chief Justice's comments in the trial and in the nationalising collective vision of 'Native Canada, ca 1820' in the *Atlas.* Alternatively, this abstraction may follow a more complex trajectory through the Gitxsan and Wet'suwet'en negotiations in the trial and the cartographies of spatial heterogeneity, put to work in interstate comparisons of Canada with the United States in the *Atlas.* Bhabha's account provided some purchase on how such mapping can lead to disruptive performances, but his argument seems to remain hard-pressed to explain whether such displacement can actually achieve very much: whether it can be used as a lever for resistance by those marginalised in the modern nation-state, or whether it can be simply internalised by a more geographically open-ended, yet still hegemonic, narrative of nation. Where is resistance located exactly, who articulates it, and what are its limits if it is always already found in the locution of location? These questions seem unanswerable in the terms of Bhabha's 'Dissemi-Nation.' [. . .]

[. . .]

References

Anderson, B. (1991) *Imagined Communities: Reflections on the Origin and Spread of Nationalism*, Verso, London.

Appeal Transcripts (1997) *Delgamuukw, also known as Earl Muldoe, suing on his own behalf and on behalf of the members of the Houses of Delgamuukw and Haaxw (and others suing on their own behalf) and on behalf of thirty eight Gitksan Houses and twelve Wet'suwet'en Houses v. Her Majesty the Queen in Right of the Province of British Columbia*, Supreme Court of Canada, Ottawa.

Bell, S. (1997) In historic judgment, top court strengthens Indian land claims. *The Vancouver Sun*, 12 December, A1–A2.

Bhabha, H. (1994) Dissemination: time, narrative and the margins of the modern nation, in *The Location of Culture* (ed. H. Bhabha), Routledge, New York, pp. 139–170.

Blomley, N. (1994) *Law, Space and the Geographies of Power*, Guilford, New York.

DePalma, A. (1988) Canadian Indians celebrate vindication of their history. *New York Times*, 9 February, A1/A8.

Foucault, M. (1979) *Discipline and Punish: The Birth of the Prison* (trans. A. Sheridan), Vintage Books, New York.

Foucault, M. (1980) *Power/Knowledge: Selected Interviews and Other Writings 1972–1979*, Pantheon, New York.

Fremlin, G. and Sebert, L.M. (1972) Mathematical basis of national atlases. *Cartographica*, **9** (1), 30–45.

Harley, J.B. (1988) Silences and secrecy: the hidden agenda of cartography in early modern Europe. *Imago Mundi*, **40**, 57–76.

Harley, J.B. (1992) Deconstructing the map, in *Writing Worlds: Discourse, Text and Metaphor in the Representation of Landscape* (eds T.J. Barnes and J.S. Duncan), Routledge, New York, pp. 231–247.

Harris, R.C. and Dechêne, L. (1987) *Atlas Historique Du Canada, 1: Des origines à 1800* (trans. M. Paré; cartographer G.J. Matthews), Les Presses De L'Université De Montréal, Montréal, Canada.

Helgerson, R. (1992) *Forms of Nationhood: The Elizabethan Writing of England*, University of Chicago Press, Chicago.

Helgerson, R. (1993) Nation or estate? Ideological conflict in the early modern mapping of England. *Cartographica*, **30** (1), 68–74.

Jacobs, M.J. (1996) *Edge of Empire: Postcolonialism and the City*, Routledge, New York.

Jara, F. and Magana, E. (1982) Rules of imperialist method. *Dialectical Anthropology*, **7**, 115–136.

Lamonde, Y. (1987) Le temps dans l'espace. *Le Devoir*, 17 October.

Lefebvre, H. (1991) *The Production of Space* (trans. D. Nicholson-Smith), Blackwell, Cambridge, MA.

McEachern, A. (1991) Reasons for judgement, *Delgamuukw et al. v. The Queen in Right of British Columbia et al.*, Dominion Law Reports, 4th series, 79, pp. 185–640.

Mitchell, T. (1988) *Colonising Egypt*, Cambridge University Press, Cambridge.

Mitchell, T. (1991) The limits of the state: beyond statist approaches and their critics. *American Political Science Review*, **85**, 77–96.

Mitchell, K. (1997) Different diasporas and the hype of hybridity! *Environment and Planning D: Society and Space*, **15**, 533–553.

Monet, D. and Skanu'u (Wilson A.) (1992) *Colonialism on Trial: Indigenous Land Rights and the Gitxsan and Wet'suwet'en Sovereignty Case*, New Society Publishers, Philadelphia, PA.

Pile, S. and Rose, G. (1992) All or nothing? Politics and critique in the modernism/postmodernism debate. *Environment and Planning D: Society and Space*, **10**, 123–136.

Piternick, A. (1993) The historical atlas of Canada: the project behind the product. *Cartographica*, **30** (4), 21–31.

Prakash, G. (1992) Postcolonial criticism and Indian historiography. *Social Text*, **31/32**, 8–28.

Rabasa, J. (1985) Allegories of the Atlas, in *Europe and Its Others*, vol. 2 (ed. F. Barker), University of Essex, Colchester, UK, pp. 1–16.

Said, E. (1993) *Culture and Imperialism*, Alfred A. Knopf, New York.

Trial Transcripts. *Delgamuukw v. the Queen* in the Supreme Court of British Columbia, Action No. 0843, Smithers Registry.

Usher, P., Tough, F. and Galois, R. (1992) Reclaiming the land: aboriginal title, treaty rights and land claims in Canada. *Applied Geography*, **12**, 109–132.

Wallot, J.P. (1987) Foreword, in *The Historical Atlas of Canada, vol. 1: From the Beginning to 1800* (ed. Harris), University of Toronto Press, Toronto, Canada.

Willems-Braun, B. (1997) Reply: On cultural politics, Sauer, and the politics of citation. *Annals of the Association of American Geographers*, **87**, 703–708.

Xie, S. (1996) Writing on boundaries: Homi Bhabha's recent essays. *Ariel*, **27** (4), 155–166.

Further reading

Blomley, N. (1994) *Law, Space and the Geographies of Power*, Guilford, New York. [An important study of territorial control and the role of cartographic knowledge.]

Craib, R.B. (2004) *Cartographic Mexico: A History of State Fixations and Fugitive Landscapes*, Duke University Press, Durham, NC. [This insightful empirical analysis demonstrates the significance of mapping projects in the national story of Mexico.]

Ramaswamya, S. (2001) Maps and mother goddesses in modern India. *Imago Mundi*, **53** (1), 97–114. [Another empirical analysis considering how popular map imagery was enrolled in the nationalistic construction of India.]

Said, E. (1993) *Culture and Imperialism*, Alfred A. Knopf, New York. [A key work on the role of knowledge, such as cartography, in the production on colonial space.]

See also

- Chapter 1.8: Deconstructing the Map
- Chapter 1.9: Drawing Things Together
- Chapter 1.12: The Agency of Mapping: Speculation, Critique and Invention
- Chapter 4.8: Refiguring Geography: Parish Maps of Common Ground
- Chapter 5.2: The Time and Space of the Enlightenment Project
- Chapter 5.4: Mapping: A New Technology of Space; and Geo-body
- Chapter 5.6: Whose Woods Are These? Counter Mapping Forest Territories in Kalimantan, Indonesia

Chapter 5.8

Cartographic Rationality and the Politics of Geosurveillance and Security

Jeremy W. Crampton

Editors' overview

Cartography has long played a role as device to manage populations. In this excerpted paper, Crampton provides an historical overview of the mapping in the production of state security and surveillance, tracing through the maps role in providing a mode of spatial analysis and overview of people, social phenomena, and potential threats and risks. Maps, he argues, are a critical technology in the nexus between government, knowledge and power, enabling states to practice certain forms of governmentality through the monitoring and disciplining of populations. Advances in surveillance and GIS technologies are broadening and deepening capabilities to track and profile subjects, supported by an ever stronger rationale for their deployment. Such advances raise significant questions about the nature of governmentality, privacy and civil rights.

Originally published in 2003: *Cartography and Geographic Information Science,* **30** (2), 135–148.

Introduction

Security is now the dominant framework for understanding the modern world in the United States. It is enshrined as both law (e.g. the USA Patriot Act 2001) and official policy (i.e. the National Security Strategy of the USA 2002). Significant resources are being dedicated to maintaining and increasing security, as well as the deployment of technologies to identify and pre-empt possible threats to security. Many of these developments represent important shifts in policy and mark a new political era that abandons the policy of containment and deterrence of the Cold War for one of threat assessment and unilateral action. Threat assessment requires successful surveillance in order to achieve security. [...]

[...]

My purpose in this paper is to critically examine the role played by mapping and GIS in the production of security and surveillance, along with its historical genesis. An increasingly significant component of security discourse comes from a spatial or geographic standpoint, and in particular from cartography and GIS. Indeed, the relevance of mapping and GIS was demonstrated soon after the terrorist attacks on 11 September [2001], when the *New York Times* and NBC published several powerful LIDAR (light detection and ranging) images by The Center for the Analysis and Research of Spatial Information (CARSI) at Hunter College. The LIDAR images showed a 'before' and 'after' view of the World Trade Centers in 3D (Clarke 2003). [...]

[...] The Association of American Geographers (AAG) meanwhile launched a workshop funded by the National Sciences Foundation (NSF) on 'geographical dimensions to terrorism' and established a list of priority action and research items. The first priority action item listed is to '[e]stablish a distributed national geospatial infrastructure as a foundation for homeland security' (Cutter *et al.* 2002: 2), which would include geospatial databases and GIS analysis. This effort is based on the DHS 'Information

The Map Reader: Theories of Mapping Practice and Cartographic Representation, First Edition. Edited by Martin Dodge, Rob Kitchin and Chris Perkins.

Analysis and Infrastructure Protection' (IAIP) Directorate's budget request for 'development and maintenance of a *complete and accurate mapping* of the Nation's critical infrastructure and key assets' (US Government 2003: 472, emphasis added). [...]

In recent years, the increase of surveillance (particularly electronic surveillance) has given rise to the sobriquet 'surveillance society' (Pickles 1991; Lyon 1994), to capture the idea that surveillance has become an 'institutionally central and pervasive feature of social life' (Lyon 1994: 24). Although surveillance can be dated to classical times, writers on surveillance associate modern industrial society with its institutionalisation. [...]

In an early paper, Pickles warned of the potential for GIS to be used for surveillance (Pickles 1991), and during the early 1990s, John Pickles and Brian Harley worked on a book about the ideology of maps. Harley's death in 1991 cut this project short, but the legacy appeared as the book *Ground Truth* (Pickles 1995). Smith went even further by equating GIS with military conquest in the Gulf War (Smith 1992). Following these early critiques, GIS users and their critics entered a long period of rapprochement (see Schuurman 2000 for a full account) [...]

Approach

[...] I argue that the renewed emphasis on security and surveillance is part of a long-standing series of historical linkages between government, knowledge and technologies of power. These historical linkages were forged during the rise of modern industrial societies in the eighteenth century. Although technologies change over time and the relationship between public and private life has become increasingly more problematised by technological developments (Alderman and Kennedy 1997; Curry 1997; Lyon 1994), there are important historical continuities that can still be traced. I argue that these continuities are constituted as *governmental rationalities* (motivating discourses) that inform practices of government.

The perspective adopted here is a broadly Foucauldian one, utilising work on 'governmentality,' or how people have governed themselves and others (Foucault 1991). Governmentality is the exercise of government 'beyond the state' (Rose and Miller 1992), including government of the family, of the economy, and techniques of self-government (or 'ethics', Foucault 1997). Foucault's analysis of government was concerned with how individuals and populations were divided and grouped according to norms. This occurred at either the individual level (what he called 'discipline'; Foucault 1977) or at the group or population level (an aspect of government he called 'biopower'; Foucault 1978). Although Foucault looked at particular practices in their time and place, he understood them as constitutive of larger ways of thinking, or rationalities. These rationalities are historically bound, that is, they come into being and reach dominance at certain moments, but they do not exist by necessity or in the same way over time. [...] Foucault's goal was to trace these rationalities and to seek out the counter-acting tendencies that may be in a position to resist them.

It would be a mistake to locate such resistance in the form of an alternative GIS (Curry 1998). Governmental reason has the effect of both increasing capacities (power) and establishing increased control (knowledge) about subjects. It would not be possible to retain new capacities if GIS or surveillance are rejected (not least because they can be manifested in any number of ways, ranging from neighbourhood activism against anti-gay hate crimes to defence mapping for war). It is, therefore, better to work from within to improve our relationship to technology. [...]

As part of the effort to enrich our understanding of mapping and GIS, we can use Foucault's historical method to study how mapping and GIS are used in contemporary surveillance and security. In particular, two parallels from early nineteenth century cartography are informative, as they cast light on our shared problems of threat and risk. Firstly, security and risk were used to think of space and people as *resources* that required management and protection. Secondly, space and individuals were understood through *normalising surveillance*. Surveillance (including 'geosurveillance' specifically concerned with locations and distributions across spatial territories) was therefore an important technology of control tied to discourses of resource management and normalisation. We can conclude from this historical comparison that it is not technologies of surveillance – mapping or GIS *per se* – that are problematic, but rather the underlying political rationality of normalisation which constituted people and the environment as threatened resources under risk of hazard.

[...]

Maps as government: moral statistics in early nineteenth century Europe

In early nineteenth century France, fear over the threat of crime had reached such heights that when a map was published that appeared to deny a relationship between crime rates and education levels there was an immediate outcry. Education was commonly thought to be an effective preventative measure against crime. Areas with higher educational levels would have lower crime rates. Crime was an activity of the uneducated lower classes; they had a '*penchant au crime*' was a famous phrase of the time

(Robinson 1982: 161). However, the 1829 maps, which employed the latest techniques of 'comparative statistics,' showed the precise opposite – areas with high education levels had high crime levels. [...]

If education was not the cause of crime, then what was? The startling possibility arose that crime could occur anywhere. These crime maps were published by Italian and French statisticians Adriano Balbi and André-Michel Guerry, who had deep interest in 'moral statistics' or social problems (e.g. crime, education, birth rates, suicide). The maps were remarkable for another reason too; they were one of the first examples of the choropleth technique which had been invented by Charles Dupin just three years earlier (Robinson 1982; Crampton 2003). Dupin's choropleth maps were exceptionally popular methods for revealing the moral statistics of his day, and they were extensively emulated. After Balbi and Guerry (who was awarded a special prize in 1864 by the Academy of Sciences for his work) came D'Angeville with health and wealth choropleths in 1836, Charles Joseph Minard, who popularised proportional symbol maps in the mid-nineteenth century, and many others. So once social problems could be grasped in their distribution across territories, policies could be implemented to address them. [...]

Maps have long been associated with this effort because they provide a picture of where things are so that there can be a 'right disposition' of resources and people over the territory (Foucault 1991: 93). This idea of a rightful distribution is important because it requires comparison to some norm. Territorial mapping has occurred for thousands of years to assist in inventories and taxation, and it is perhaps surprising that it was only in the early nineteenth century that thematic maps were invented. Why were they not deployed previously? In fact, it turns out that thematic or statistical maps were part of a more general effort to govern by means of statistical analysis. It was only with the development of descriptive and probabilistic statistics, and the formulation of society in terms of likelihoods and norms, that thematic maps could emerge. Thematic statistical maps appeared at precisely the same moment that society came to understand itself in statistical terms for purposes of regulation (policing in the larger sense) and management. A few examples will illustrate how this occurred.

In the 1820s the Belgian statistician Adolphe Quetelet derived the new analytics of probability theory and the normal distribution curve. These advances were keyed to societal problems that were thought to be amenable to governmental intervention. Quetelet was concerned about the social upheavals in Europe during the 1830s and centred his analysis of social variation around *l'homme moyen*, or the average man (his needs and typical actions and the nature of error or deviation away from this norm). Total human variation could thus be justifiably reduced to divergence around a norm. If these norms could be properly and reliably determined, then this would be extremely useful in dealing with the 'great masses of registered facts' about populations (Sir John Herschel (1857), quoted in Atkins and Jarrett (1979)). The positivist conception of science that emerged at the end of the seventeenth century gave epistemological primacy to observable data that was value-free, a primacy that is underpinned by statistics. As Atkins and Jarrett show, statistical inference and significance tests on samples also permitted populations to be compared and known (how much they vary around a mean, for example, in their susceptibility to infant mortality). In sum, the sciences were founded around the governmental concerns of knowledge, statistics and population. During the nineteenth century great strides were made in the sciences of statistics, probability and statistical mapping. These did not occur in isolation from one another, nor more interestingly, from the question of politics – indeed, they were stimulated and put into the service of 'political' problems. Thematic mapping was part and parcel of this political problematic. [...] The ability to identify one's resources and thus to exploit them was necessary for the secure governing of the state.

Perhaps the most visible and influential practice of using statistics to help govern the state occurs during the great decennial censuses of many European countries (from 1790 in the United States). Although in Europe these censuses were depicted in maps in the early nineteenth century, in the USA it was not until the ninth census in 1870 that results were shown cartographically. These maps appeared in 1874 in America's first statistical atlas (Walker 1874; Hannah 2000).

Hannah's excellent analysis of the 1870 census atlas using Foucault's work on governmentality sheds considerable light on the spatial politics of knowledge at this time. The atlas had a tremendous impact on cartographic representations of space in the following decades. In particular, it introduced thematic mapping to the United States in a concerted manner (although several maps from the 1860 census had appeared; Schwartz and Ehrenberg 2001: plate 177). Maps from the census were first presented at the American Geographical Society (AGS) in 1871, where, according to J.B. Jackson, they received so much attention that the Secretary of the Interior 'was persuaded to authorise a special atlas' [...]

Security: discipline and biopower

[...] [B]oth discipline and biopower are ways of dividing and grouping either individuals or populations according to norms (biopower is so-called because it deals with biological factors of birth and death rates, fecundity, issues affecting the health of a population, and so on). In order to understand how governmentality arose we can examine discipline and biopower in the context of historical changes in juridicality and criminality. Prior to the legal reforms of the eighteenth and early nineteenth centuries, Foucault argued the law focused on the nature of the crime committed, the evidence of guilt or innocence, and the system of penalties to be applied. In other words: crime and punishment. The person of the criminal was important only insofar as he or she was the individual to which the crime would be attributed. With the reforms, this hierarchy was reversed, the crime was merely an indicator of something more significant – the 'dangerous individual' (Foucault 1977: 252).

The law was now interested in the potential danger of the individual: 'The idea of *dangerousness* meant that the individual must be considered by society at the level of his potentialities, and not at the level of his actions; not at the level of the actual violations of an actual law, but *at the level of the behavioural potentialities they represented*' (Foucault 2000: 57, original emphasis). Punitive responses thus had to be appropriately tailored to perceived threat. [...]

We can understand the emergence of thematic mapping in the early nineteenth century as a similar preventative measure, to get a better description of where potential threats to the health of a population, such as crime, poor education and high birth rates, were occurring. From this knowledge it became possible to differentiate neighbourhoods of the city and to classify space in terms of dangerousness. These maps produced a picture of normality and abnormality. It also became paramount to identify and locate dangerous people or places, based on the risk they posed.

How was this dangerousness determined and how was it prevented? Dangerousness and security threats are measured against a set of norms. Only by establishing normal behaviour as a baseline could deviations from that behaviour be detected. It was through mapping that a society understood the spatial disposition of its resources, where the healthy areas of the country were, which places had 'abnormally' high rates of infant mortality or early marriages or where crimes against property were above average (all these were the subject of early nineteenth century choropleth maps). In other words, politics was becoming more and more concerned with the problems of normalising space and territory.

Then, as today, security was achieved through surveillance, supervision and management. The national censuses were not carried out from mere curiosity but, rather, to assist the state to take measure of the population, to provide numbers for the emerging discipline of moral statistics, and to lay out what could be considered normal and abnormal (Hannah 2000). Foucault's reference to the 'panopticon,' the all-seeing architectural device of Jeremy Bentham (1748–1832), philosopher of law and sometime prison reformer, is well known (Foucault 1977; Hannah 1997). [...] The importance of the panopticon is that it provided multiple orderings of space which permitted and encouraged observation (e.g. in the schoolroom, the army, the monastery or the factory; see also Philo 1992; Elden 2001: 133–150) and embedded people in a structure of power, knowledge and normalisation.

Foucault pointed out that to administer an individual's dangerousness required a new field of *expertise*. In crime mapping, for instance, these experts are usually part of the judicial system itself (typically, GIS researchers or staff in police departments across the United States), but there are also extra-judicial experts. The latter include purveyors of criminal statistical GIS software such as 'Crimestat,' organisers of crime mapping conferences, academics, providers of monitoring anklets and security systems for surveillance, web sites of parolee maps, listservs discussing problems of crime mapping, and neighbourhood associations agitating for offenders to be registered. These extra-judicial experts are an important part of power–knowledge relations for assessing spatial dangerousness, indicating that the state is not always the source of governing rationalities. This is why concepts of politics are kept deliberately broad – the strategies of governing go far beyond the state.

Experts need new *techniques*. Here too, crime mapping plays an important role, by providing the ability to perform 'geoprofiling.' Geoprofiling is a disciplinary technique for determining the typical spatial patterns of an individual with the goal of predicting that person's behaviour or targeting them for surveillance. Graham (1998) discussed implications of regulating space by what he calls 'surveillant simulation' (i.e. digital surveillance or 'control at a distance', such as electronic tagging; Bloomfield 2001), which acts in this disciplinary manner. Graham highlighted four cases of surveillance: as social control especially of criminality; in and around consumption; differential deployment over space (transport informatics); and the utility industry. To this list we could add others, such as surveillance in the workplace (a practice established in the nineteenth century), or in the pursuit of leisure activities (e.g. the Visionics face recognition software and CCTVs at stadium sporting events), or walking down the street [...]

Graham's discussion is very useful in identifying some concerns with surveillance technology. However, we might register a point of caution concerning social control. As far as Foucault's work is concerned, 'social control' should not be interpreted as a condition of total domination over life. Rather, Foucault emphasised *discipline* and the governmental *management* of a problem. Historically, since the late eighteenth and early nineteenth centuries, with the demographic expansion of the populace and increasing industrialisation, governments have realised that crime, poverty and madness cannot be eliminated. With this realisation, societies emerged from *sovereignty* into *discipline*. Whereas under sovereignty the ruler or sovereign exercised a system of total control based in violence, in disciplinary societies power relations are distributed 'without recourse, in principle at least, to excess, force or violence' (Foucault 1977: 177). Thus, '[o]ne might say that the ancient right to *take* life or *let* live was replaced by a power to *foster* life or *disallow* it to the point of death [in capital punishment]' (Foucault 1978: 138, original emphasis).

Discipline is a question of 'correcting' toward a norm, of 'reducing gaps' between actual and normal behaviour (Foucault 1977: 179). In a disciplinary society we are complicit in our acquiescence to technologies of power because we exist in a relation of freedom. Management and discipline of a problem give rise to normalisation and technologies of government, but also leave space, in Foucault's well known phrase, for resistance and the 'insurrection of subjugated knowledges' (Foucault 2003: 7). Local knowledges are not anti-science or anti-GIS, but rather are 'an insurrection against the centralising power effects that are bound up with the institutionalisation and workings of any scientific discourse organised in a society such as ours' (2003: 9). The target to be analysed is not science, its methods or techniques, but the rationality which informs it and gives it its truth status. Reason has a history, in this case the cartographic reason in which people are subjectified as calculable, at-risk resources to be managed.

[...]

There are two levels in which Foucault argues we can understand normalising rationalities. One of these is at the level of the individual and is characterised by discipline (Foucault 1977, see also the course summaries of his lectures from 1971–1972 and 1972–1973 in Rabinow 1997). For example, bodies are trained and regularised (e.g. drilling the troops), fixed into appropriate places (e.g. children in the classroom or patients in the hospital), observed or surveilled (e.g. offenders in prison), and recorded via bookkeeping, inspections and reports. This 'disciplinary' society emerged at the end of the seventeenth and early eighteenth centuries. Foucault argued that a second system of power emerged in the early eighteenth century that regulated, counted and surveilled the mass of people as a population. Foucault called this 'biopolitics of the population' (Foucault 1978: 139) or, more simply, 'biopower', and that is why the census, thematic mapping and statistics were used to measure and record birth and death rates, crime, disease, and so on. The target of surveillance in the disciplinary society is the individual, whereas it is the distribution ('disposition') of the population over its territory that constitutes biopower.

Discipline of the individual and regulation of the population are characteristic functions of power in modernity. Firstly, at both the individual and population levels, surveillance is critical to ensure the optimisation of the capabilities of the individual and to measure the health of the population. Once measured, the results constitute a set of norms against which further behaviour can be evaluated and corrective mechanisms applied (such as birth control incentives).

Secondly, norms are almost inevitably understood in a calculative, statistical manner. Foucault points to the emergence of demography and 'the evaluation of the relationship between resources and inhabitants, the constructing of tables analysing wealth and its circulation' (1978: 140). Demographers, geographers and cartographers began to observe and quantitatively investigate birth and death rates, hygiene and the need 'to medicalise the population' (Foucault 2003: 244), accidents and various anomalies that might need to be addressed through such things as 'insurance, individual and collective savings, safety measures' (Foucault 2003, p. 244), and the establishment of actuary tables and life expectancies. [...]

Thirdly, population and territories across which they are distributed are understood as resources. This is a political issue in the sense that the population that occupies it must be governed, managed and harboured from harm (Hannah 2000). [...] Although Foucault does not use the word *geosurveillance*, the necessity for it is firmly built into his descriptions of modern society.

[...]

Geosurveillance expertise and techniques in contemporary crime mapping

It was argued above that dangerousness, risk and norms were established through specific techniques and expertise. The crime map is an important means of constructing knowledge about the city and its inhabitants and for implementing policies to manage a crime situation. Crime maps are tied to the rise of social statistics such as the FBI Uniform Crime Report or UCR (collected since the 1930s), but also local police reports, victim reports and corporate loss reports. These maps help to construct a discourse of risk. [...]

Today we see a hyper-extension of these developments in the surveillant systems deployed by the police to monitor and check residents as they go about their daily business. These include closed circuit TVs [...], and the FBI's new DNA database, the Combined DNA Index System (CODIS) [...].

Crime mapping employs many techniques of geosurveillance. One set of techniques revolves around those who have already offended and are in the criminal system. There are about 6.5 million people classified as offenders in the USA; incarcerated offenders numbered 1.96 million at the end of 2001. However, there are also about 4.66 million people on probation or parole, which usually involves some degree of self-reporting (Bureau of Justice Statistics 2002). In addition, a small number of people not in local or state jails are supervised by other techniques, including community service, work release, weekend reporting, electronic monitoring and other programmes. Jail populations have steadily increased during the 1990s (by about 30% between 1995 and 2001), prompting increased efforts to supervise this population outside the facility (Aungles and Cook 1994). During the same period, the percentage of people supervised outside jail rose from 6.4 to 10.0% (34 869 to 70 804). The number of people supervised by electronic monitoring has, however, remained the same for several years, at about 10 000. This number is less than 15% of those monitored outside jails, and a tiny 1.4% of the total jail population.

Several technologies have been developed to monitor out-of-jail offenders. A common technology is an ankle bracelet or tag, which emits an RF radio signal that can be detected by a device in the home linked to the phone system. A more advanced approach is to use GPS. It, too, is often based on an anklet worn by the offender which can receive GPS signals and transmit its location (through the cell phone system) to the company's monitoring centre. [...] This monitoring is geographically flexible: 'Each map is tailored for a specific parolee. A map can show, for instance, areas where a paroled paedophile must remain clear of – such as a school – when going to and from an off-site counselling session' (Chabrow 2002). [...]

In addition to tracking known offenders there are efforts by the crime mapping community to develop geoprofiling. The theory of geoprofiling was developed by Kim Rossmo in 1995 and has since been implemented in software that can make a predictive surface of a criminal's location. Geoprofiling is 'an investigative methodology that uses the locations of a connected series of crime to determine the most probable area of offender residence' (Rossmo 2000: 1). Geoprofiling is based on the concept of offender behavioural profiling of the offender's characteristics, in this case his or her home location. Rossmo uses well known principles of geography to show that crime is committed by people near (but not too near) their own homes. Rossmo claims that with five to six incidents traceable to one person, his software can reduce the search area by up to 90%. Crime maps enable geoprofiling to isolate behaviour which does not conform to the norm. [...]

The need for blanket monitoring and surveillance arises because of the perceived ubiquity of the threat. Often we do not know in advance where the risk or the danger will be. [...] [W]hen places such as the World Trade Center are attacked it confirms that we are already 'inside' the threat zone and therefore must constantly map everything. Thus the new reality of the threat: it is everywhere and so must be the surveillance. The same reasoning applies to the Bush administration's controversial plans for TIPS (Terrorism Information and Prevention System), which was proposed in early 2002 but has since been dropped from Homeland Security. In this plan, citizens and workers who often go into residential neighbourhoods (postal workers, cable TV installers, truck drivers) would be recruited to call a government hotline if they saw suspicious activity. The idea was to benefit from as many as a million sources of surveillance in ten pilot cities (these cities were never specified).

Other terrorist information hotlines, such as the FBI's Terrorist Tipline, remain in operation (as well as Amber Alerts about child abductions, Coast Watch, Highway Watch and River Watch, for reporting chemical or biological spills, the ATF Hotline for reporting suspicious firearm activity, and even the Treasury Department's FinCEN for reporting financial crimes). The 'Uniting and Strengthening America by Providing Appropriate Tools Required to Intercept and Obstruct Terrorism' Act of 2001 (the USA Patriot Act or USAPA) also widely broadened existing statutes permitting electronic surveillance. According to an analysis by the Electronic Freedom Foundation (EFF), USAPA 'expands all four traditional tools of surveillance – wiretaps, search warrants, pen/trap orders and subpoenas' (EFF 2001) – and made it legal to search homes without a warrant or install surveillance devices in people's homes without notification until long afterwards (HR 3162, §§202, 210, 213). We have thus reached an analogous situation to that faced by the citizens of Paris in 1829 when they were presented with the Balbi and Guerry crime maps: we fear crime and threats to our security from everywhere, and it is no surprise that normative governmental rationality gives rise to widespread geosurveillance in order to manage these threats.

Conclusion: the risks of security

The events of 11 September 2001 have caused many people to ask whether there has been a fundamental shift in the balance between surveillance and privacy. While mapping and GIS offered critical assistance in the immediate aftermath of the attacks, the more general role of geographic technologies in constituting geosurveillance and security has yet to be determined. I have argued in this paper that a fruitful understanding of mapping and GIS in the context of security can be gained by tracing the underlying motivations and rationalities of geosurveillance to its origins at the beginning of the nineteenth century. These rationalities were directly concerned with governing (e.g. counting, measuring and establishing norms) individuals and populations in their distributions across territories. Working within a broadly Foucauldian perspective, I have especially highlighted how a *rationality of security* is constructed in which geosurveillance is deployed as a response to dangerousness, and in which the environment and people are constructed as *at-risk resources* subject to normalisation and management.

The issue of security is often contrasted against issues of privacy or civil rights. The two are seen in balance, sometimes moving more toward one side (more surveillance in times of threat) and sometimes to the other (reassertion of civil rights in times of peace). As Curry (1997) has argued, however, while privacy has long been a phenomenon, its coding as a right is new, and the nature of that right has changed alongside changes in technology. The fact is, the battle has largely been won in favour of a surveillant rationality even before 11 September 2001. Foucault's insight is that we should understand the rationality *itself* 'behind' security, geosurveillance and rights; that is to say what justifies it and gives it its status as truth. Opposing surveillance by appealing to civil liberties is problematic because the latter are easily constructed in different ways.

[...]

The object of these government activities is not to impose an all-powerful system of domination, but rather one in which problems are managed by establishing spatial norms. Analysing the choice of problems at any given time reveals what is at issue for that society. Likewise, the strategies adopted to address these problems, such as a discourse of hazard, threat or dangerousness, guide our interpretation of the underlying rationality.

[...] [I]ntervention can occur by those subject to normalisation. It is not a question of being 'anti-GIS' or anti-geosurveillance, but rather one of critically understanding the relationship between technology and rationalities, and how rationalities are integral to policing, policy making and politics. This is for two reasons; firstly, GIS and cartography are important technologies in the production of knowledge for governmentality (i.e. geosurveillance, discipline and biopower), and, secondly, that knowledge is subject to normalisation that casts people and space as at-risk resources. [...]

References

Alderman, E. and Kennedy, C. (1997) *The Right to Privacy*, Vintage Books, New York.

Atkins, L. and Jarrett, D. (1979) The significance of 'significance tests', in *Demystifying Social Statistics* (eds J. Irvine, I. Miles and J. Evans), Pluto Press, London, pp. 87–109.

Aungles, A. and Cook, D. (1994) Information technology and the family: electronic surveillance and home imprisonment. *Information Technology and People*, **7**, 69–80.

Bloomfield, B. (2001) The right place at the right time: electronic tagging and the problems of social order/disorder. *The Sociological Review*, **49**, 174–201.

Bureau of Justice Statistics (2002) Prisoners in 2001. http://www.ojp.usdoj.gov/bjs/pub/pdf/p01.pdf.

Chabrow, E. (2002) Every move you make, every breath you take... *Information Week*, 30 August. www.informationweek.com/story/IWK20020830S0027.

Clarke, K.C. (2003) *Getting Started with Geographic Information Systems*, Pearson Education, New Jersey.

Crampton, J.W. (2003) Are choropleth maps good for geography? *GeoWorld*, **16** (1), 58.

Curry, M.R. (1997) The digital individual and the private realm. *Annals of the Association of American Geographers*, **87**, 681–699.

Curry, M.R. (1998) *Digital Places: Living with Geographic Information Technologies*, Routledge, London.

Cutter, S.L., Richardson, D. and Wilbanks, T. (2002) *The Geographical Dimensions of Terrorism: Action Items and Research Priorities*, Association of American Geographers, Washington, DC.

EFF (2001) Electronic Freedom Foundation Analysis of the Provisions of the USA PATRIOT Act. www.eff.org/.

Elden, S. (2001) *Mapping the Present: Heidegger, Foucault, and the Project of a Spatial History*, Continuum, London.

Foucault, M. (1977) *Discipline and Punish*, Vintage Books, New York.

Foucault, M. (1978) *History of Sexuality: An Introduction*, Vintage Books, New York.

Foucault, M. (1991) Governmentality, in *The Foucault Effect: Studies in Governmentality* (eds G. Burchell, C. Gordon and P. Miller), University of Chicago Press, Chicago, pp. 87–104.

Foucault, M. (1997) The ethics of the concern for self as a practice of freedom, in *Ethics Subjectivity and Truth. Essential Works of Foucault 1954–1984*, vol. 1 (ed. P. Rabinow), The New Press, New York, pp. 281–301.

Foucault, M. (2000) Truth and juridical forms, in *Power: Essential Works of Foucault 1954–1984*, vol. 3 (ed. J.D. Faubion), The New Press, New York, pp. 1–89.

Foucault, M. (2003) *Society Must be Defended*, Picador, New York.

Graham, S. (1998) Spaces of surveillant simulation: new technologies, digital representations, and material geographies. *Environment and Planning: Society and Space*, **16**, 483–504.

Hannah, M.G. (1997) Imperfect panopticism: envisioning the construction of normal lives, in *Space and Social Theory* (eds G. Benko and U. Strohmayer), Blackwell, Oxford, pp. 344–359.

Hannah, M.G. (2000) *Governmentality and the Mastery of Territory in Nineteenth Century America*, Cambridge University Press, Cambridge.

Lyon, D. (1994) *The Electronic Eye. The Rise of Surveillance Society*, University of Minneapolis Press, Minneapolis, MN, p. 270.

Philo, C. (1992) Foucault's geography. *Environment and Planning D: Society and Space*, **10** (2), 137–161.
Pickles, J. (1991) Geography, GIS and the surveillant society, in *Papers and Proceedings of Applied Geography Conferences*, **14** (eds J.W. Frazier, B.J. Epstein, F.A. Schoolmaster and H. Moon), pp. 80–91.
Pickles, J. (1995) *Ground Truth: The Social Implications of Geographic Information Systems*, Guilford, New York.
Rabinow, P. (1997) *Ethics, Subjectivity and Truth: Essential Works of Foucault 1954–1984*, vol. 1, The New Press, New York.
Robinson, A.H. (1982) *Early Thematic Mapping in the History of Cartography*, University of Chicago, Chicago, IL.
Rose, N. and Miller, P. (1992) Political power beyond the state: problematics of government. *British Journal of Sociology*, **43** (2), 173–205.
Rossmo, D.K. (2000) *Geographic Profiling*, CRC Press, Boca Raton, FL.
Schuurman, N. (2000) Trouble in the heartland: GIS and its critics in the 1990s. *Progress in Human Geography*, **24**, 560–572.
Schwartz, S.I. and Ehrenberg, R.E. (2001) *The Mapping of America*, 2nd edn, Wellfleet Press, Edison, NJ.
Smith, N. (1992) Real wars, theory wars. *Progress in Human Geography*, **16** (2), 257–271.
US Government, Office of the President (2003) *Budget of the United States Government, Fiscal Year 2004 – Appendix*, US Printing Office, Washington, DC.
Walker, F.A. (1874) *Statistical Atlas of the United States*, J. Bien, New York.

Further reading

Crampton, J. (2007) The biopolitical justification for geosurveillance. *Geographical Review*, **9** (3), 389-403. [A more recent extension of Crampton thinking in the governmental implications of mapping.]
Dobson, J. and Fisher, P. (2003) Geoslavery. *IEEE Technology and Society Magazine*, Spring, 47-52. [An interesting polemical essay on the risks of geospatial technologies to constrain personal freedoms in new ways.]
Klinkenberg, B. (2007) Geospatial technologies and the geographies of hope and fear. *Annals of the Association of American Geographers*, **97** (2), 350-360. [This paper provides a balancing anecdote to the thesis that cartographies are all about controlling people and highlights the real empowering potential of mapping for social activism and environmental justice.]
Monmonier, M. (2002) *Spying with Maps: Surveillance Technologies and the Future of Privacy*, University of Chicago Press, Chicago. [An informative and accessible monograph considering the ways that maps are deployed for monitoring and control.]

See also

- Chapter 2.3: Manufacturing Metaphors: Public Cartography, the Market, and Democracy
- Chapter 2.4: Maps and Mapping Technologies of the Persian Gulf War
- Chapter 2.6: Cartographic Futures on a Digital Earth
- Chapter 2.10: Mobile Mapping: An Emerging Technology for Spatial Data Acquisition
- Chapter 2.11: Extending the Map Metaphor Using Web Delivered Multimedia
- Chapter 4.2: Map Makers are Human: Comments on the Subjective in Maps
- Chapter 4.10: Citizens as Sensors: The World of Volunteered Geography
- Chapter 5.11: Mapping the Digital Empire: Google Earth and the Process of Postmodern Cartography

Chapter 5.9

Affecting Geospatial Technologies: Toward a Feminist Politics of Emotion

Mei-Po Kwan

Editors' overview

Drawing on a range of literatures – critical cartography and GIS, feminist geographies and affective geographies – in this excerpted paper Kwan argues that cartography has been dominated by a masculinist, Cartesian rationality that has led to a disembodied, emotionless conceptualisation of maps. In contrast, she forwards a vision of mapping that combines representational and non-representational ideas. In so doing, she emphasises the corporeal, affective and unwritable dimensions of maps and their creation and use, suggesting that they be understood through their practices and performances, rather than as representations which merely communicate spatial relations. The production and use of maps is embodied and our understanding of maps needs to be attentive to emotions, feelings, values and ethics that are an integral aspect of geospatial practices. Through such an affective turn, she argues that more moral geospatial practices become possible.

Originally published in 2007: *The Professional Geographer,* **59** (1), 22–34.

Geospatial technologies include a broad range of technologies for collecting, storing, displaying or analaysing geographical information (e.g. geographical information systems [GIS], global positioning systems and remote sensing). Much has been written about the limitations and social implications of geospatial technologies (GT) since the early 1990s (Sheppard 1993; Curry 1994; Pickles 1995). Critiques have focused largely on issues of epistemology, representation, power, ethics, privacy violation and the non-civilian deployment of these technologies. With contributions by critical geographers from diverse perspectives, considerable progress has been made in the nascent subfields of critical GIS and critical cartography to date (Schuurman 1999; Kyem and Kwaku 2004; Sheppard 2005; Crampton and Krygier 2006; Del Casino and Hanna 2006; Elwood 2006; Knigge and Cope 2006; Kwan and Knigge 2006; Pavlovskaya 2006; Propen 2006; Sieber 2006; Ghose 2007; Kwan and Aitken 2009).

Among recent critical perspectives on GT, feminist geographers have provided new insights since the early 2000s (Nightingale 2003; Gilbert *et al.* 2005; McLafferty 2005b). Sara McLafferty (2002, 2005a), for instance, examines the role of GIS in feminist activism and explores how GIS-based power/knowledge may empower or marginalise women activists as spatio-political scale shifts. Marianna Pavlovskaya (2002, 2004) examines the link between urban restructuring and the micro-geographies of women's everyday lives in Moscow through a grounded story composed with GIS. Marie Cieri's (2003) study of queer tourism highlights how GT can be used to explore the gendered and sexualised geographies of urban space. I renegotiate the meanings of GIS at the intersection of science, art and subjectivities (Kwan 2002a). I have also argued that GIS can be a site for deconstructing the binary understanding of geographical method and have called for a recovery of the critical agency of GIS users or researchers (Kwan 2002b, 2004). [...]

In this article I seek to develop feminist perspectives on GT along new directions. [...] Since emotions affect

The Map Reader: Theories of Mapping Practice and Cartographic Representation, First Edition. Edited by Martin Dodge, Rob Kitchin and Chris Perkins.

research processes and findings (Bennett 2004) and are highly political but rarely an important consideration in public policy (Kwan and Aitken 2009), bringing emotions back to bear upon GT practices may offer new insights about ways of using GT that contest the dominant understanding and meanings of GT and their relationships with the social and political world (e.g. using GT as a means of resistance or political protest).

[...] Non-representational theories challenge 'the epistemological priority of representations as the grounds of sense-making' or as the means for acquiring knowledge about the world (McCormack 2003: 488). They emphasise the importance of the corporeal, affective and unwritable dimensions of existence and turn our attention from representations to practices and performances (Nash 2000; Kwan and Aitken 2009). Critical reflections that non-representational thinking inspired have pushed our understanding of maps, cartography and GIS from conventional notions of representations toward feminist notions of performance and performativity (e.g. Del Casino and Hanna 2006). Artists and scholars in cultural studies have recently drawn on these feminist notions to explore the use of GT as locative media for self-expression and articulation of emotional geographies (Parks 2001). These experimentations hint at new geospatial practices (or performances) that contest our understanding of GT as representational or communicative media.

[...] I argue that an attention to the importance of affect (feelings and emotions) in social life and research and the performative nature of GT practices offers a 'distinctive critical edge' to feminist work on GT (Jenkins *et al.* 2003: 59), and that GT can be a fruitful analytic project for feminist geographers. [...] Drawing on feminist conceptualisation of affect (Thien 2005), I argue that geospatial practices need to be embodied and attentive to the effects of emotions, which mediate the social and political processes through which our subjectivities are reproduced (Harding and Pribram 2002; Bennett 2004). This not only involves reintroducing long-lost subjectivities of the researcher, the researched and those affected by GT back to geospatial practices, but also involves making emotions, feelings, values and ethics an integral aspect of geospatial practices. Only then will moral geospatial practices become possible, and only then can we hope that the use and application of GT will lead to a less violent and more just world.

Bodies and emotions matter

Geospatial technologies are designed, created and used by humans, and a large proportion of their application is for understanding or solving problems of individuals and social groups. Bodies, however, are often absent or rendered irrelevant in contemporary practices of GT. This 'omission of the body' occurs in two different but related senses (Johnson 1990: 18). Firstly, although bodies are involved in the development and use of GT, there is little room in these technologies to allow for any role of the practitioner's subjectivities, emotions, feelings, passion, values and ethics. Secondly, despite the fact that a large number of bodies are affected by the application of GT (e.g. people profiled by geodemographic application, and civilians who were annihilated as 'collateral damage' by GPS-guided smart bombs that missed their targets), bodies are often treated merely as things, as dots on maps, or even as if they do not exist (Gregory 2004; Hyndman 2005).

The dominant disembodied practices of GT, however, are contestable as they are largely the result of a particular understanding of science and objectivity (Kwan 2002a). This historically specific and socially constructed notion of science, as Donna Haraway (1991) argues, is predicated on the positionality of a disembodied master subject with transcendent vision. With such disembodied and infinite vision, the knower is capable of achieving a detached view into a separate, completely knowable world. The kind of knowledge produced with such disembodied positionality denies the partiality of the knower, erases subjectivities and ignores the power relations involved in all forms of knowledge production (Foucault 1977). Haraway (1991: 189) calls this decorporealised vision 'the god-trick of seeing everything from nowhere.'

Closely associated with this view of science is a gendered notion of knowledge production and academic scholarship that privileges rational thought over 'irrational' emotionality (Bennett 2004). This 'marginalisation of emotion,' as Anderson and Smith (2001: 7) put it, 'has been part of a gender politics of research in which detachment, objectivity and rationality have been valued, and implicitly masculinised, while engagement, subjectivity, passion and desire have been devalued, and frequently feminized.' Geography in particular has tended to 'deny, avoid, suppress or downplay its emotional entanglements' (Bondi *et al.* 2005: 1). Yet, to paraphrase Anderson and Smith, there are times and places where lives are explicitly lived through pain, love, hate, anger, hope, fear and passion. If the world is imbued with complex emotional geographies, GT practices are more relevant to real lives if they allow us to take the spatial, temporal and social effects of feelings into account. To neglect how our research and social life are mediated by feelings and emotions is to exclude a key set of relations through which lives are lived, societies made and knowledge produced (Anderson and Smith 2001).

As GT practitioners, our decisions to adopt particular research agendas and engage with particular issues (e.g. emergency response) are often motivated by the emotions provoked by events such as wars, environmental problems and 'natural' disasters (Bennett 2004). In fieldwork involving interaction between GT researchers and research participants, emotions expressed by the researched may provide insights about their relationships with others and their social worlds. In social and political contexts involving interaction among diverse groups of stakeholders, there are inevitably underlying feelings and tensions that cannot be clearly articulated or communicated – like 'the thrown-chairs, the put-downs, the red faces and the hugs' and 'the anger, the frustrations, the sadness and the joys' in planning meetings that involve the data or results generated by GT (Kwan and Aitken 2009). Contemporary life is also imbued with emotionally intense encounters brought about by real time media coverage of events around the globe (e.g. planes crashing into buildings, dead bodies of tsunami victims, and violent encounters in anti-globalisation protests). Exploring and developing new GT practices that are attentive to bodies and emotions is therefore an important and fruitful feminist project.

The critical project that aims to bring bodies and emotions back in GT practices entails several important elements. As feminist GT practitioners, we can appropriate the power of GT, contest the dominant uses of these technologies, and reconfigure the dominant visual practices to counter their objectifying vision. We can experiment with new geospatial practices that better articulate the complex realities of gendered, classed, raced and sexualised spaces and experiences of individuals. These new practices should help us understand emotions in terms of their 'socio-spatial mediation and articulation rather than as entirely interiorised subjective mental states' (Bondi *et al.* 2005: 3). While being attentive to how emotions, subjectivities and spaces are mutually constitutive in particular places and at particular times, these new practices should also take into account the existence of different kinds of bodies (e.g. pregnant, disabled, old, mutilated, dead) and their socially encoded meanings in relation to specific spatial, temporal and cultural contexts (Rose 1993; Laws 1997; Domosh and Seager 2001; Longhurst 2001). [...]

As feminist GT practitioners we deeply care about the subject(s) of our research and are 'emotionally committed to our work', and our geospatial practices should be infused with a sense of 'emotional involvement with people and places' (Bondi *et al.* 2005: 2). We can develop GT practices that entail this emotional involvement and help express meanings, memories, feelings and emotions for our subjects. We can draw on the emotional power of moving images and the techniques in narrative cinema to create GIS movies or visualisations that tell stories about the lives of marginalised people, highlight social injustice, and – we hope – effect social change (Aitken 1991; Aitken and Craine 2006).

[...]

Subject(ive) mapping with Global Positioning Systems

Lisa Parks (2001: 209), a cultural critic and video artist, contests the meanings of Global Positioning System (GPS) by using it as an interactive technology for 'plotting the personal.' She explores whether the GPS can be used to document human movement and everyday experiences in a way similar to that of photography, home videos and travelogues. She highlights the paradoxical nature of GPS and argues the need for critical strategies that struggle over the meanings of satellite technologies. [...] Parks contends that what state-sponsored and commercial digital mapping projects share is their quest for total vision and total knowledge of the planet. She argues that the personal plot (personal map) she explored works against this centralisation of vision and knowledge by insisting that GPS need not be used to articulate the agendas of the state or business. Instead it can be used as a means of storytelling and a technology for self-expression.

Parks explored GPS as a means of articulating the politics of location through linking and interpreting an individual's global position (location data produced through satellites) with her subject position (historically and socially constituted identities). Through a discussion of GPS tracks of her movements that she recorded in two recent trips, one in California and the other in Alice Springs, Australia, she illustrates how GPS maps might produce such politics. As Parks (2001: 216) puts it, 'At each juncture I entered a waypoint, ensuring that each moving trace would be remembered. I was reminded here of my own mobility relative to theirs – and that my GPS map of California would look quite different from that of a migrant worker, a Chinese pharmacist, a high-tech executive or a groaning seal for that matter'. She suggests that GPS maps (or personal plots) offer 'new ways of visualising social difference that are based on human movement rather than physiognomy or pigmentation' (Parks 2001: 211). These visualisations, she argues, enable us to conceptualise more precisely how identities are constituted through material, bodily movements.

[...] Parks argues, the GPS map combines the objective and omniscient discourse of cartography with the subjective, grounded experience of the user. Visual representation of the moving body by GPS introduces the possibility of subject(ive) mapping. Although represented as a series of lines and dots, the body's movement transforms the map from an omniscient view of territory into an individualised expression. By plotting the personal, GPS inscribes embodied practices into the discourse of mapping and allows the user to call into question the objective status of the map by inflecting it with personal movement. The producer of the GPS map is none other than the body that travelled, walked or moved along a certain trajectory carrying a GPS receiver. The practice of plotting the personal, then, figures the user as subject, produced through a series of movements and encounters. [...]

[...] The GPS maps [...] represent the possibility of a mediated experience, as they often necessitate storytelling and narration because what they reveal is seen and experienced from very specific and personal points of view. When used as a technology of self-reflection, GPS invites the user to see herself as a subject in motion, as an author and a reader, reflexively inscribing personal trajectories onto the text of the social world of her everyday life. In this light, GPS receivers can be used as technologies of self-expression, creating spatial interpretation and social understanding as much as they can be used as tracking and monitoring devices.

The Amsterdam RealTime Project, as Amy Propen (2006) describes it, shares critical intent similar to Parks's personal plot. In the Amsterdam project, real time location data from the GPS-enabled personal digital assistants (PDAs) of the participants were sent to a central server via wireless Internet connection. As the GPS tracks were visualised against a black background without showing any information about the city (e.g. streets or parcel boundaries), the participants' movements in real time construct their own maps and representations of the city. Through creating personalised maps of the city with the performances of their own bodies (recorded and visualised as GPS tracks), the project participants (who were all volunteers) were the authors (subjects) of these plots and at the same time were being portrayed in these maps (objects). The project, therefore, contests the conventional distinction between author and reader, subject and object, performance and representation (Del Casino and Hanna 2006).

Collaborative 3D GIS videography

[...] [F]eminist GT practitioners can draw on the emotional power of moving images and techniques in narrative cinema to create GIS movies that tell emotionally provocative stories or that highlight social injustice (Deleuze 1986, 1998; Aitken 1991; Aitken and Craine 2006). Cinema, in Gilles Deleuze's (1998: 15) view, tells 'stories with blocks of movements/duration'. As Stuart Aitken (1991: 105) argues, the frame sequence in a motion picture 'portrays the dynamic interaction between people and their social and physical environments', and the foundations of successful narrative cinema lie in a unique portrayal of this dynamic interaction.

In a recent project, I explored ways of using moving images generated by GIS for articulating emotional geographies and contesting the objectifying vision of GIS-based 3D geovisualisation. Drawing on the methods in visual ethnography, visual sociology and film studies (e.g. Banks 2001; Pink 2001; Rose 2001; Buckland 2003), I created a 3D GIS movie that is more an artistic and expressive visual narrative than an objective recording generated with the aid of scientific visualisation. As Sarah Pink (2001) suggests, video materials should not be treated merely as visual facts but rather as representations in which the collaborations and strategies of self-representation of those involved are part of their making. For visual ethnographers, video is not simply a data collecting tool but a technology that participates in the negotiation of social relationships and a medium through which ethnographic knowledge is produced. Participatory video has been used by feminist geographers in action research that seeks to encourage communities to 'analyse their social world and to explore the construction of meaning' (Kindon 2003: 143). The collaborative use of video, as Kindon (2003: 143) suggests, has 'considerable transformative potential in terms of the action it may generate'.

Based on these notions of participatory video and narrative cinema, I developed 'collaborative 3D GIS videography', a method of creating videos using moving images rendered by a 3D GIS for articulating the personal experience and story of a particular research participant. I produced a video based on the oral history of a Muslim woman in Columbus, Ohio (who was a key informant of the study), about her feelings when travelling and undertaking activities outside her home shortly after 11 September 2001 (hereafter 9/11). The purpose of the study was to understand the impact of post-9/11 anti-Muslim hate crimes on the perception of safety and use of public space of the Muslim women in Columbus, Ohio, study. Several months after 9/11, I travelled with her for one day as she drove her minivan to undertake her normal out-of-home activities. As we passed through various routes, she recalled her feelings and fear when she saw particular buildings or stores (and her oral narrative was recorded). Using the textual transcripts of such audio recordings, the field notes I took on that day, and the activity diary and map sketches she completed during an in-depth interview, I portrayed her body's space–time trajectory and her emotions as she moved around the study area with a 3D GIS.

Contrary to the high-angle perspective commonly used in 3D geovisualisation, the video that I produced adopts her point of view (in the literal sense) as the vantage point. The moving images of the video show what she saw (rendered by 3D GIS) as she drove through various routes in the study area on a particular day after 9/11; her movement is portrayed as a personalised space–time trajectory that is colour coded to reflect the level of fear and perceived danger she experienced, and the buildings along the road were also colour coded to indicate the level of perceived danger she experienced as she passed them. Audio clips from her oral narrative were also incorporated, resulting in a video that not only shows the routes and the spaces her body moved through, but also tells her story through the images and her oral narrative as she recalls what happened to her life and how she negotiated the hostile urban spaces after 9/11. It shows what she saw and experienced from her personal point of view (i.e. from the position of a driver who was travelling along various roads in the study area). It is a powerful form of individualised storytelling based on her personal movements, memories, feelings and emotions.

The 3D GIS video I produced seeks to 'present its subject matter in a subjective, expressive, stylised, evocative and visceral manner' (Buckland 2003: 145). It is not an 'objective' or impartial video recording of anything that can be captured by a conventional video camera. Its scenes have many physical elements that are considered to be parts of the objective reality and scientifically visible 'facts' of the study area (e.g. buildings and roads), but they are rendered from the GIS database with symbolic and artistic techniques, which helped to create an expressive visual narrative that was produced collaboratively with the informant. For instance, a green line was used to represent the tiny comfort zone that she experienced as she drove her minivan through a major road in the study area, and the oppressive effect of the hostile urban environment was symbolically represented by colouring the surrounding buildings as red blocks. Further, instead of being filmed, represented as a protagonist, and being watched by spectators, the informant does not appear in the video. She is the person who saw and acted, and mainly her emotions, feelings, memories and experiences find expression in the video. The video produced is, therefore, not only about her but also for her – she is situated at the centre of its production. It portrays her emotional geographies in terms of the dynamic interaction between her feelings and the post-9/11 urban environment of the study area. Through this shift from a spectator's viewpoint to the protagonist's (subject's) viewpoint, the video contests the objectifying gaze of conventional 3D geovisualisation practices through a particular spatial and visual organisation of its elements.

GT art practices as politics of resistance

As Parks's and my own work have shown, GT can be appropriated as media for self-expression and articulation of emotional geographies. These experimentations contest the detachment, rationality and objectifying vision entailed in conventional GT practices. Map artists and art activists have long created art maps that contest the authority and content of official maps – witness the maps produced by the Surrealists and the Situationists (Krygier 2006; Varanka 2006; Wood 2006). Art maps are often created by extensively reworking pre-existing maps, 'redrawing, digitally altering, painting over, and reorienting the original images' (Wood 2006: 10). They point toward worlds other than those mapped in official maps and seek to 'produce new configurations of space, subjectivity and power' (kanarinka 2006). Each art map is therefore not only a 'work of art' but also a 'political action' (Deleuze and Guattari 1987: 12). Similarly, GT can be appropriated as a digital art medium and used to create artworks that protest against social injustice and violence. GT art practices can be undertaken or performed as a form of resistance (Deleuze 1998; Kaufman 1998; Klebesadel 2003).

Based on these notions of art practices as politics of resistance, I have explored GIS as an artistic medium for generating digital artwork using GIS software and data. As GIS was not developed and designed for artistic work, my GIS art project intends to challenge the understanding of GT as scientific apparatus for producing objective knowledge or as an instrument of domination. I seek to destabilise the fixed meanings of GT that have precluded their use in novel and creative ways. Through my GIS art I also articulate my discontent with the use of GT in wars and international conflicts that result in large numbers of civilian casualties (Gregory 2004; Hyndman 2005, 2007). I also protest against the use of these technologies in any applications that violate personal rights and privacy, as in geodemographic and surveillance applications.

I have explored the aesthetic potential of GIS by experimenting with various artistic styles and techniques. [...] My GIS art project was undertaken out of my sadness in light of the human casualties resulted from the attacks at the World Trade Center and the Pentagon on 9/11 as well as the ensuing wars and violent conflicts in the Middle East. In the project, GIS was used to create digital images that are aesthetically pleasing, but none of the visual elements in these images corresponds to any particular object in the world. [...] Through this abstract and non-representational GIS art practice, GIS is momentarily dissociated from any precepts of science, objectivity, transcendent vision, exploitation, surveillance or control. I thus participated in the cultural politics of contending the meanings of GT (albeit at a personal level), as cultural politics 'are contestations over meanings, over borders and boundaries, over the ways we make sense of our worlds, and the ways we live our lives' (Mitchell 2000; 159). Through this geospatial aesthetics grounded on my concern about the role of GT in global violence, I insist that GT should be used primarily for creating a more just and peaceful world, as when the technologies are used in research on environmental justice or for empowering marginalised social groups (Mennis 2002; McLafferty 2005a). In the project, GIS was used as a medium of passionate politics for countering the dominant practices. It is in this sense that my GIS art project can be understood as part of a broader counter-hegemonic struggle over GT, as a form of questioning, and a form of protest and resistance.

My GIS art project and Parks's (2001) personal plot, however, are largely personal endeavours. In order to influence public policy and to effect broader social change, politics of resistance at the individual level needs to be scaled up and connected to collectively practiced politics. The recent trend of increasing collaboration between researchers, artists and community groups in projects that seek to understand people's feelings and concerns may be indicative of how this connection can be made (Rose 1997). For instance, the Greenwich Emotion Map Project engaged art activists and local residents to reflect on the social change taking place on the Greenwich Peninsula (Nold 2005). It was a mapping project that aimed at understanding how local residents feel about the area based on their personal exploration and journeys. In the project, biomapping devices worn by participants recorded their emotional response (their body's level of stimulation) to and interaction with their immediate environment, and a GPS tracked the routes they took. On returning to the studio, the information and photos taken along the way were uploaded and interpreted by participants to create a personal visual narrative. The resulting emotion maps encouraged participants' personal reflection on the complex relationship between them, their local environment and their fellow citizens. The project allowed local residents of the Greenwich Peninsula to visualise where they feel stressed and excited, to articulate their concerns, and to engage with wider community issues (Nold 2005).

[...]

Toward embodied practices and passionate politics

The wars following 9/11 have taken an enormous human toll, sometimes with the assistance of GT such as GPS and remote sensing. The failures that Hurricane Katrina revealed, which many had hoped to be able to avoid through the help of GT, are also disconcerting. As feminist GT practitioners, we need to think carefully about the kinds of geospatial practices that are truly relevant to the contemporary world. We should engage in the development of GT practices that help to create a less violent and more just world. I have argued in this article that embodied practices and passionate politics of GT that are attentive to bodies, emotions and subjectivities will help us move beyond software and data to focus on real people and real lives. Drawing on recent developments in feminist thinking, I suggest that attention to the importance of affect and possibilities of performing (practicing) GT as resistance would lead to distinctively feminist contribution to research and practice on GT.

[...]

In order to effect broader social change, however, it is important to scale our care or concern from the personal/local level up to larger contexts. Although most of the projects I describe in this article were undertaken as personal endeavours, our personal politics of resistance needs to be scaled up to the level of collectively practiced feminist politics. Collaborative projects undertaken by GT researchers, feminist/art activists and community groups throughout the world offer important inspiration for how this may be accomplished (McLafferty 2002, 2005a; Kanarinka 2006). As feminist GT practitioners, we should develop innovative means to protest against the use of GT for violence and to engage in political activism that turn violence and fear into hope. Only when emotions, feelings, values and ethics as well as a commitment to social justice become integral elements of our geospatial practices will moral geospatial practices become possible. Only then can GT help create a less violent and more just world.

References

Aitken, S. (1991) A transactional geography of the image-event: the films of Scottish director, Bill Forsyth. *Transactions of the British Institute of Geographers*, **16**, 105–118.

Aitken, S. and Craine, J. (2006) Guest editorial: Affective geovisualizations. *Directions Magazine*, 7 February. http://www.directionsmag.com/article.php?article_id=2097&trv=1.

Anderson, K. and Smith, S.J. (2001) Editorial: Emotional geographies. *Transactions of the Institute of British Geographers*, **26**, 7–10.

Banks, M. (2001) *Visual Methods in Social Research*, Sage, London.

Bennett, K. (2004) Emotionally intelligent research. *Area*, **36** (4), 414–422.

Bondi, L., Davidson, J. and Smith, M. (2005) Introduction: geography's emotional turn, in *Emotional Geographies* (eds L. Bondi, J. Davidson and M. Smith), Ashgate, Aldershot, UK, pp. 1–18.

Buckland, W. (2003) *Film Studies*, McGraw-Hill, Chicago.

Cieri, M. (2003) Between being and looking: queer tourism promotion and lesbian social space in Greater Philadelphia. *ACME: An International E-Journal for Critical Geographies*, **2**, 147–166.

Crampton, J.W. and Krygier, J. (2006) An introduction to critical cartography. *ACME: An International E-Journal for Critical Geographies*, **4** (1), 11–33.

Curry, M. (1994) Image, practice, and the unintended impact of geographical information systems. *Progress in Human Geography*, **18**, 441–459.

Del Casino, V.J. and Hanna, S.P. (2006) Beyond the binaries: a methodological intervention for interrogating maps as representational practices. *ACME: An International E-Journal for Critical Geographies*, **4** (1), 34–56.

Deleuze, G. (1986) *Cinema 1: The Movement-Image* (trans. H. Tomlinson and B. Habberjam), University of Minnesota Press, Minneapolis, MN.

Deleuze, G. (1998) Having an idea in cinema, in *New Mappings in Politics, Philosophy, and Culture* (eds E. Kaufman and K.J. Heller), University of Minnesota Press, Minneapolis, MN, pp. 14–19.

Deleuze, G. and Guattari, F. (1987) *A Thousand Plateaus: Capitalism and Schizophrenia* (trans. B. Massumi), University of Minnesota Press, Minneapolis, MN.

Domosh, M. and Seager, J. (2001) *Putting Women in Place: Feminist Geographers Make Sense of the World*, Guilford, New York.

Elwood, S. (2006) Negotiating knowledge production: the everyday inclusions, exclusions, and contradictions of participatory GIS research. *The Professional Geographer*, **58**, 197–208.

Foucault, M. (1977) *Power/Knowledge: Selected Interviews and Other Writings 1972–1977*, Pantheon, New York.

Ghose, R. (2007) Politics of scale and networks of association in public participation GIS. *Environment and Planning A*, **39** (8), 1961–1980.

Gilbert, A., Melissa, R. and Masucci, M. (2005) Moving beyond 'Gender and GIS' to a feminist perspective on information technologies: the impact of welfare reform on women's IT needs, in *A Companion to Feminist Geography* (eds L. Nelson and J. Seager), Blackwell, Oxford, pp. 305–321.

Gregory, D. (2004) *The Colonial Present: Afghanistan, Palestine, Iraq*, Blackwell, Oxford.

Haraway, D. (1991) *Simians, Cyborgs, and Women: The Reinvention of Nature*, Routledge, New York.

Harding, J. and Pribram, E.D. (2002) The power of feeling: locating emotions in culture. *European Journal of Cultural Studies*, **5**, 407–426.

Hyndman, J. (2005) Feminist geopolitics and September 11, in *A Companion to Feminist Geography* (eds L. Nelson and J. Seager), Blackwell, Oxford, pp. 565–577.

Hyndman, J. (2007) Feminist geopolitics revisited: body counts in Iraq. *The Professional Geographer*, **59**, 35–46.

Jenkins, S., Jones, V. and Dixon, D. (2003) Thinking/doing the F' word: on power in feminist methodologies. *ACME: An International E-Journal for Critical Geographies*, **2** (1), 57–63.

Johnson, L. (1990) New courses for a gendered geography. *Australian Geographical Studies*, **28** (1), 16–28.

Kanarinka (2006) Art-machines, body-ovens and map recipes: entries for a psychogeographic dictionary. *Cartographic Perspectives*, **53**, 24–40.

Kaufman, E. (1998) Introduction, in *New Mappings in Politics, Philosophy, and Culture* (eds E. Kaufman and K.J. Heller), University of Minnesota Press, Minneapolis, MN, pp. 3–13.

Kindon, S. (2003) Participatory video in geographic research: a feminist practice of looking? *Area*, **35** (2), 142–153.

Klebesadel, H. (2003) Feminist activist art in action. *Newsletter of Wisconsin Women's Network*. www.wiwomensnetwork.org.

Knigge, L. and Cope, M. (2006) Grounded visualization: integrating the analysis of qualitative and quantitative data through grounded theory and visualization. *Environment and Planning* A, **38** (11), 2021–2037.

Krygier, J. (2006) Jake Barton's performance maps: an essay. *Cartographic Perspectives*, **53**, 41–50.

Kwan, M.P. (2002a) Feminist visualization: re-envisioning GIS as a method in feminist geographic research. *Annals of the Association of American Geographers*, **92**, 645–661.

Kwan, M.P. (2002b) Is GIS for women? Reflections on the critical discourse in the 1990s. *Gender, Place and Culture*, **9** (3), 271–279.

Kwan, M.P. (2004) Beyond difference: from canonical geography to hybrid geographies. *Annals of the Association of American Geographers*, **94**, 756–763.

Kwan, M.P. and Aitken, S. (2009) GIS and qualitative research: geographical knowledge, participatory politics, and cartographies of affect, in *Handbook of Qualitative Research in Human Geography* (eds D. DeLyser, S. Aitken, M. Crang *et al.*), Sage, London, pp. 286–303.

Kwan, M.P. and Knigge, L. (2006) Doing qualitative research using GIS: an oxymoronic endeavor? *Environment and Planning* A, **38** (11), 1999–2002.

Kyem, P. and Kwaku, A. (2004) Of intractable conflicts and participatory GIS applications: the search for consensus amidst competing claims and institutional demands. *Annals of the Association of American Geographers*, **94**, 37–57.

Laws, G. (1997) Women's life courses, spatial mobility, and state policies, in *Thresholds in Feminist Geography: Difference, Methodology, Representation* (eds J.P. Jones, H.J. Nast and S.M. Roberts), Rowman and Littlefield, Oxford, pp. 47–64.

Longhurst, R. (2001) *Bodies: Exploring Fluid Boundaries*, Routledge, London.

McCormack, D.P. (2003) An event of geographical ethics in spaces of affect. *Transactions of the Institute of British Geographers*, **28**, 488–507.

McLafferty, S. (2002) Mapping women's worlds: knowledge, power and the bounds of GIS. *Gender, Place and Culture*, **9** (3), 263–269.

McLafferty, S. (2005a) Geographic information and women's empowerment: a breast cancer example, in *A Companion to*

Feminist Geography (eds L. Nelson and J. Seager), Blackwell, Oxford, pp. 486–495.

McLafferty, S. (2005b) Women and GIS: geospatial technologies and feminist geographies. *Cartographica*, **40** (4), 37–45.

Mennis, J. (2002) Using geographic information systems to create and analyze statistical surfaces of population and risk for environmental justice analysis. *Social Science Quarterly*, **83**, 281–297.

Mitchell, D. (2000) *Cultural Geography: A Critical Introduction*, Blackwell, Oxford.

Nash, C. (2000) Performativity in practice: some recent work in cultural geography. *Progress in Human Geography*, **24** (4), 653–664.

Nightingale, A. (2003) A feminist in the forest: situated knowledges and mixing methods in natural resource management. *ACME: An International E-Journal for Critical Geographies*, **2** (1), 77–90.

Nold, C. (2005) Greenwich Emotion Map. http://www.emotionmap.net.

Parks, L. (2001) Plotting the personal: global positioning satellites and interactive media. *Ecumene*, **8** (2), 209–222.

Pavlovskaya, M.E. (2002) Mapping urban change and changing GIS: other views of economic restructuring. *Gender, Place and Culture*, **9** (3), 281–289.

Pavlovskaya, M.E. (2004) Other transitions: multiple economies of Moscow households in the 1990s. *Annals of the Association of American Geographers*, **94**, 329–351.

'Pavlovskaya, M.E. (2006) Theorizing with GIS: a tool for critical geographies? *Environment and Planning* A, **38** (11), 2003–2020.

Pickles, J. (1995) *Ground Truth: The Social Implications of Geographic Information Systems*, Guilford, New York.

Pink, S. (2001) *Doing Visual Ethnography*, Sage, London.

Propen, A.D. (2006) Critical GPS: toward a new politics of location. *ACME: An International E-Journal for Critical Geographies*, **4** (1), 131–144.

Rose, G. (1993) *Feminism and Geography: The Limits of Geographical Knowledge*, University of Minnesota Press, Minneapolis, MN.

Rose, G. (1997) Performing inoperative community: the space and the resistance of some community arts projects, in *Geographies of Resistance* (eds S. Pile and M. Keith), Routledge, New York, pp. 184–202.

Rose, G. (2001) *Visual Methodologies*, Sage, London.

Schuurman, N. (1999) Critical GIS: theorizing an emerging science. *Cartographica* Monograph 53.

Sheppard, E. (1993) Automated geography: what kind of geography for what kind of society. *The Professional Geographer*, **45**, 457–460.

Sheppard, E. (2005) Knowledge production through critical GIS: genealogy and prospects. *Cartographica*, **40** (4), 5–21.

Sieber, R. (2006) Public participation geographic information systems: a literature review and framework. *Annals of the Association of American Geographers*, **96**, 491–507.

Thien, D. (2005) After or beyond feeling? A consideration of affect and emotion in geography. *Area*, **37** (4), 450–456.

Varanka, D. (2006) Interpreting map art with perspective learned from J.M. Blaut. *Cartographic Perspectives*, **53**, 15–23.

Wood, D. (2006) Map art. *Cartographic Perspectives*, **53**, 5–14.

See also

- Chapter 1.11: Exploratory Cartographic Visualisation: Advancing the Agenda
- Chapter 1.13: Beyond the 'Binaries': A Methodological Intervention for Interrogating Maps as Representational Practices
- Chapter 1.14: Rethinking Maps
- Chapter 2.6: Cartographic Futures on a Digital Earth
- Chapter 3.9: Maps, Mapping, Modernity: Art and Cartography in the Twentieth Century
- Chapter 3.10: Affective Geovisualisations
- Chapter 4.8: Refiguring Geography: Parish Maps of Common Ground
- Chapter 5.5: First Principles of a Literary Cartography
- Chapter 5.10: Queering the Map: The Productive Tensions of Colliding Epistemologies

Chapter 5.10

Queering the Map: The Productive Tensions of Colliding Epistemologies

Michael Brown and Larry Knopp

Editors' overview

Ontologically and epistemologically, queer geographies seem antithetical to the Cartesian rationality of cartographic representation and GIS technologies. Queer theory, rooted in poststructuralism, rejects the totalising certainties and essentialising tendencies of traditional cartography. Instead, it recognises the contingent, relational, contextual, contradictory and paradoxical nature of everyday life. How then to produce a map of gay and lesbian life in Seattle over several decades that at once conforms to the conventions and aesthetics of modern maps, and to the ideas and ideals of queer theory? This is the task that Michael Brown and Larry Knopp set themselves and which is detailed in this excerpted paper. Their solution is one that embraces the notion of colliding epistemologies – a tacking back and forth between a queer theoretical framework and the more conventionally scientific strictures of cartography to produce an artefact that looks like a conventional map, but was authored through a contested, negotiated, multilayered process. The paper provides an ethnography of this process, detailing the intricacies of the various steps through which they progressed to produce a map that was sensitive to the politics of the community it seeks to represent.

Originally published in 2008: *Annals of the Association of American Geographers*, **98** (1), 40–58.

Queer geography has often positioned itself – and been positioned by mainstream geography – as antithetical to the realms of more traditional, orthodox disciplinary anchors like cartography and GIScience (Binnie 1995; Brown 1995; Browne 2006). Emerging as it has from the discipline's critical theoretical turn, queer geography's intellectual ken in queer theory and poststructuralism has certainly contributed to a thoroughgoing interrogation of Cartesian rationality, Euclidean spatial ontologies and the often norming fixity inherent in cartographic representation. Queer geography has thus been quite distanced from conventional GIScience, often seeking alternate modes of spatial representation (Cieri 2003). Meanwhile, GIScience has neglected queer subjects and topics, even when it has broached related areas of feminism and postcolonialism (Rundstrom 1995; Kwan 2002a; McLafferty 2002; Schuurman and Pratt 2002) and incorporated a sophisticated critical awareness of the social constructedness of spatial data, the vexing relations between epistemology and representation, and the political dimensions of geographic information systems (GIS) (Harvey 2000; Aitken 2002; Crampton 2004; Kyem 2004; Pickles 2004; Schuurman 2004). [...]

Amid this dissonance and centrifuge, it is arguably remarkable that two queer geographers might employ GIS techniques in an effort to execute an action research project epistemologically and politically anchored in queer theory. Our project entailed work with an all-volunteer non-profit organisation in Seattle, Washington (the Northwest Lesbian and Gay History Museum Project), dedicated to recording, saving and disseminating the twentieth century history of lesbians, gays and other sexual dissidents in the US Pacific Northwest. Our work involved, quite simply, making a map using GIS and visual design software. [...]

By chronicling our research experience ethnographically, our purpose here is to highlight the productive tensions of

The Map Reader: Theories of Mapping Practice and Cartographic Representation, First Edition. Edited by Martin Dodge, Rob Kitchin and Chris Perkins.

what we call colliding epistemologies in our use of GIS. In the course of our work, we experienced a number of problems and challenges arising from our attempts to tack back and forth between a queer theoretical framework and the more conventionally scientific strictures of cartography and GIS. Many of these problems have been anticipated by feminist and other critical geographers writing about GIS (Sieber 2000; Kwan 2002a; McLafferty 2002; Pavlovskaya 2002; Schuurman 2002b; Schuurman and Pratt 2002; Kyem 2004); but others resonate especially with queer geography, and have more to do with a broader set of intellectual paradoxes that emerge when the relentless and insistent poststructuralist drive to critically deconstruct (i.e. to queer) in academia engages with more pragmatic forms of activism and voluntarism (e.g. to make an essentialising map; Fyfe and Milligan 2003; Cameron and Gibson 2005), or when different literatures are brought into conversation with one another through empirical praxis.

[...]

Between multiple literatures

Queer urban history

[...]

Although queer urban history in the United States has been sensitive to space, place and environment, then, it has been relatively devoid of explicit spatial theorisations and follow-through (Hornsey 2002; Houlbrook 2005; Nash 2006). Maps are rarities, references to geographers' work are scant, scales tend to be unexamined, and temporality still tends to be privileged over spatiality. We aimed, through a map informed by multiple literatures within geography, to extend this literature (and Atkins's work on Seattle specifically) by providing a chorological perspective (Harris 1992; Baker 2003; Holdsworth 2003).

Sexuality and space and queer geography

From the sexuality and space literature we took several key assumptions that not only foregrounded a geographical imagination to the project, but also theoretically energised it with cautions about the complexity of the relations among identity, space, and place. Three points were key. The first is that urban space is heteronormatively structured and performed (Valentine 1993; Hubbard 1998, 2001; Brickell 2000; Podmore 2001, 2006). Secondly is that queer space is characterised by duality, fluidity and simultaneity (Valentine and Skelton 2003; Knopp 2004). Thirdly, there are wide arrays of both institutional and individual resistances in the city that are both intellectually and politically important for geographers to appreciate, but they are never completely emancipatory. [...]

In this broad intellectual framework, we recognise cartography and mapping as key interventions in disrupting the heteronormativity of space, at the very least. [...]

Yet despite the fact that an explicit spatial perspective or geographical imagination pervades the urban sexuality and space literature, relatively few geographers interested in sexuality and space have used cartographic techniques in their research (see, however, Elder 2003; Brown *et al.* 2005; O'Reilly and Crutcher 2006). We speculate that this absence is at least in part due to the insights from queer theory discussed earlier, which raise profound and problematic epistemological and ontological challenges to cartography and GIS and to representation more generally (Pickles 2004). Furthermore, because queer theory has stressed that sexuality and desire are central to understanding all human phenomena, that sexual subjectivities are often fluid rather than fixed, and that space is multidimensional, socially constructed and discursive as well as material, fixing sexual subjectivities on a map inevitably foregrounds some queer lives and experiences at the expense of others (Knopp 2004). Finally, like so much poststructural theory, queer theory stresses that knowledge is always produced in the context of power relationships, and that representation is always mediated, partial and political. Accordingly, queer geography is suspicious of pre-given or universal frameworks for understanding and seeks ways of knowing and representing that are more inclusive than exclusive (Binnie 1995). The pre-given areal units, boundaries and scales of censuses and other bureaucratically-produced data therefore tend a priori to be treated very critically, as are data themselves and mappings thereof. [...] The sexuality and space literature thus impels us to contest heteronormativity by mapping gay and lesbian Seattle, but prevents us from ignoring or dismissing the profound epistemological and ontological challenges inherent in representing queer identities in space (Hubbard 2002).

Feminist, critical and participatory GIS

Maps obviously exemplify a spatial perspective, and as such the tools of cartography and GIScience seemed the clearest way to augment the historicity of gay urban studies with a geographical imagination. So we turned to the critical literature on cartography and GIS. Maps and, in particular, GIS constitute powerful tools of visualisation, which has clear political resonance with the Seattle history project. They are powerful, of course, because of the visual nature of their output (the map as artefact), but also because of their flexibility, efficiency, expandability and increasing affordability, all of which mean that more different kinds of spatial information, and more different types of citizens, can make use of the techniques and technologies.

[...]

Concerns about resources and expertise, as well as the deeper epistemological and ontological preoccupations of queer theory, familiarly echo throughout debates in critical GIS (Perkins 2003; Schuurman 2004, 2006). Kwan (2002b: 276) and Schuurman (2002a) in particular have argued for critical – one might say queer – forms of visualisation that employ GIS as a 'subversive practice'. These critical forms include both a more reflexive and imaginative interpretation of the GIS-produced images themselves and the use of more sophisticated GIS techniques to visualise unconventional topics and processes, such as what Kwan (2002a: 654) calls 'the closeted spatiality' of African-American women and what Schuurman (2004: 143) refers to as 'a wider range of ontologies'.

We are very much inspired, then, by a desire to espond to these calls within critical feminist GIS for more 'translation tales' (Perkins 2003: 342), 'connections' (Hanson 2002: 301), 'hybridity' (Kwan 2004: 756), 'reconciliation' (Schuurman 2002a: 73) and 'writing the cyborg' (Schuurman 2002b: 261) around GIS and social geography, and in particular for more ethnographic accounts of critical GIS practices (Schuurman 2002b; Matthews *et al.* 2005). As a specific means to this end, we followed Matthews, Detwiler and Burton's (2005) approach in combining ethnographic research with GIS. Ethnography can explicate the points of tension between colliding epistemologies that often prove the most difficult moments in mixing methods or different kinds of data.

We were also inspired by the democratic possibilities of participatory GIS, although we do not share its often behaviouralist epistemology or naive pluralist urban political imagination. Participatory GIS programmes have long been heralded as empowering local community organisations in their dealings with state bureaucracies, pluralising the forms of local knowledge and spatial information at stake in decision making, and democratising input into public decision making (Obermeyer 1998; Carver 2003). An especially productive move around this literature is the so-called counter-mapping efforts of community organisations that use GIS (Wood 1992; Sparke 1998).

Geographers have chronicled the ways that marginalised groups have used 'the master's tools' toward their own ends. The critical turn in GIS, however, has simultaneously produced several careful and revealing studies that illustrate just how difficult it can be to reach that promise. Like all political and decision making processes, those around participatory GIS are shot through with multidimensional power relations of class, race and gender, structural forces of capitalism, unequal access to spatial data, and unanticipated and antidemocratic outcomes (Aitken and Michel 1995; Harris and Weiner 1998; Sieber 2000; Aitken 2002; Elwood 2002; McLafferty 2002; Crampton 2004; Esnard *et al.* 2004; Grasseni 2004; Kyem 2004; Norheim 2004).

To make a map: the (not so) mundane story

The process behind the research and production of the map is outlined in this section, although we hasten to add the steps were not as discrete or as ordinal as this discussion implies. The History Project produced a hand-drawn, photocopied, black-and-white version of 'Claiming Space' in 1996. By the spring of 2003, the map had outlived its usefulness. There were simple errors on it, and several new locations had been discovered that the group felt needed to be included on the map. When we became volunteers, the group felt that with two geographers on board it had the opportunity to produce a higher quality version of the map. Somewhat taken aback by being interpellated as cartographers, because we were 'the geographers,' we took to the project as a new and creative challenge, with a novel opportunity to work between cartography/GIS and queer geography. We set ourselves the goal of having a new, expanded and more professional-quality map that more fully conveyed the richness, complexity, plurality and even fluidity of Seattle's queer historical geography by the time of the city's Gay Pride Festival in June 2004.

[. . .] As a group, we began by amassing a single, expandable database of the eighty-one major sites of significance from the 1996 map. The database included variables on the address, zip code, current census tract, decade with which the venue was most associated, start date, end date, type of site, relevance to the community, and comments.

[. . .]

We then turned to historical archives to add important locations in the post-war era that needed to be added to the map. These archival materials included the History Project's own oral history collection, its scattered files containing various ephemera and memorabilia, the Seattle Gay Community Center News (1974–1975) and, most helpfully, the Seattle Gay News, which was available on microfiche from its origins in 1976 to the present. [. . .]

Next, we tapped long-time, elder Seattle residents and activists to confirm, triangulate and add to the data found in the newspaper archive. [. . .]

Members reviewed the new entries in an attempt to triangulate the new locations. [. . .] Following the design of the 1996 map, we allocated the venues by decade, which were colour coded on the map. We also decided at this point to link virtually every site on the map to a colour coded descriptive annotation, using a numeric identifier that was also colour coded by decade. The annotation (usually one to three sentences), it was felt, turned locations into places, and therefore added context and polyvocality to the cartography.

Once locations were plotted on the Seattle street grid, smaller neighbourhood-scale maps were printed and circulated among members for validation. [...]

To make a map: ethnography

The productive tensions of colliding epistemologies

Given that the map project was guided by a queer epistemology and ontology, the project's constitutive politics necessarily reflected a process of negotiation and compromise with almost life-like forms of positivism, realism, pragmatism and Cartesian rationality that insinuated themselves into the algorithms, hardware and ongoing interpretation of our map production. In other words, a distinctly non-queer epistemology infused the thinking of even ourselves and our non-academic queer colleagues. Of course, the literature anticipated these forces (Aitken 2002; Schuurman 2002a, 2006), but we were nevertheless struck by how vexing this problem was, and the degree to which it manifested itself even in discussions and decision making among group members ourselves.

Tensions appeared early. For example, early discussions among group members revolved around epistemological debates over deductive versus inductive orientations toward mapmaking. Initially, some members asserted that to be intelligible to the community, a single, a priori grand narrative was essential, and that our discussions should be focused on which story we wanted to adopt before any data collection could take place. The movement of gay space from Pioneer Square to Capitol Hill was one. Another was, 'early on there was a small cluster of bars that constituted gay space, but over time sites queered, expanded and diffused throughout the city as queer folk became more accepted' (i.e. Atkins 2003 'from exile to belonging').

Others, ourselves included, tried to exemplify the queer theory and politics around us, and questioned the need for a predetermined narrative on the grounds that we did not wish to preclude new and multiple stories from emerging out of the map production process itself. We were at pains, for example, to include residential space from the pre-Stonewall era, on the grounds that private space was important in gay and lesbian lives, and because it was so hard to get from the archive. The importance of sites that were queered by sexual dissidents (like The Green Parrot, a second-run movie theatre, or the Ben Paris lunch counter downtown) seemed impossible to ignore given queer historical geography's lessons, but they did not fit the teleological narrative. Our inductive position was driven by our motivation to uncloset queer space empirically and visually, at the same time retaining as much as possible a sense of the queerness of what we were representing (Schuurman 2002b).

Amicable as these debates were, they were relentless, exhaustive and time consuming. The problem, of course, is that data never only speak for themselves. Although arguably more multiple and diverse, the stories that emerged from the 'winning' more inductive approach were still incomplete, imperfect and somewhat chaotic. Still, we argued that our insistence on a more inductive practice would produce real time reflexivity on both our parts and those of future map consumers about the many possible stories that the map might tell (or inspire). [...]

There were also epistemological tensions around what was to constitute a site of significance, worthy of a dot on our map. Some members (not always the deductivists) argued for consistent and rigid criteria, on the grounds that such a practice would resolve in a more 'professional' way the crisis of representation that we were confronting. Others (again, including ourselves) argued for more flexible and situational decision making. In still another axis of difference, the basis of significance for some members had to be public and documented, whereas for others it was more personal and affective. In the end our decisions were necessarily flexible and inconsistent, largely due to the free-wheeling, democratic, all-volunteer ethos of the History Project. [...]

Other issues pitting our queer theoretical orientation against a more orthodox scientific mind-set presented themselves at subsequent stages of the process. Data collection, cleaning and coding were particularly complicated moments. They were shaped not only by pragmatic considerations but also issues of memory, negotiation, trust and serendipity. For example, preliminary map drafts for internal consumption precipitated numerous discussions about point patterns that were emerging, and the kinds of locations that were and were not making their way onto the map. [...] The point here is not just that members had different and changing epistemological approaches to cartography, but that each of us 'knew' the data differently (or even multiply). Epistemological tensions such as these are crucial to appreciate when understanding how the map was made.

Representing the unrepresentable

A key issue in the queer historiography is the concern over imposing present-day categorisations of sexualities anachronistically on historical subjects who had no or different language for their positionality (Halperin 2002). For us, the visceral experience of representing the historical geography of such subjects became a recurrent exercise in representing the unrepresentable. We were also reminded throughout the project of how difficult it can be to represent connectivity, fluidity, multiplicity and multiple scales when making a simple two-dimensional map, especially given our rudimentary skills set. Although Schuurman (2006) insists that both critical theorists and GISers deal equivalently with these vexations, our very uneven familiarity with both cannot be discounted here. We were struck by how productive these representational challenges turned out to be. This dilemma was a function not only of the queer subject matter, but also because of our own limits and ignorance about the technology and cartography. [. . .] Some non-point phenomena, such as gay pride parades and migration patterns, were left off the map altogether, due to the practical difficulty of accommodating them technologically (given our skills set at the time). These decisions, however imperfect, were part of the colliding epistemologies in the map's production.

In our attempt to represent the places behind (within?) the dotted locations, we used annotations and illustrations. They were our imperfect attempts to convey a sense of multidimensionality and contexts, as well as change and movement in sites, such as multiple private residences that functioned periodically as public meeting places. [. . .]

[. . .] some locations confounded easy categorisation, as in the case of a 1970s gay radio programme that was broadcast from a home on north Capitol Hill. Furthermore, we wanted to avoid introducing more divisions and hierarchies into a set of phenomena that were in fact fluid and hybrid. Experiments with this option seemed to us to leave the map cluttered and overwritten. Many locations contained multiple and different sites over the years, and in fact evolved from one to the other rather than having hard edges or distinct moments of transformation. The inevitable essentialising that was done to fix these points in space, we hold, has led to a deeper appreciation among History Project volunteers in narrating such moving or contested sites on our simple map.

In trying to represent queer space fully on a two-dimensional map, we found ourselves forced to imagine the closet visually through different epistemologies simultaneously. In fact, as we discuss later, we and other group members were constantly frustrated by the fact that actual lived experiences included processes, practices and experiences that were quite variable and unpredictable in both time and space. A good example here would be the Crescent Tavern, a bar known as either a lesbian or a gay bar, a bar known as homophobic or homosocial, and a space that was not simply one of many Capitol Hill drinking establishments, but the organising site for the Lesbian Mothers National Defense Fund. [. . .]

For the data we were mapping to be culturally resonant, we agreed to associate each site with only one decade in the annotation. This decision was a hugely non-queer move, in that many sites actually spanned multiple decades and the construction of 'decade' itself is quite obviously an artificial and particular reification of time. Yet that was how people recalled most sites. We defend this move on the grounds that it allowed us to highlight the site's symbolic power, which is an exemplar of the constitutive politics underlying the project. In these ways, colliding epistemologies were not smoothly reconciled nor solved, and although they might be common problems in cartography, they productively resonated with our attempts to keep this a queer project.

Productive pragmatics

One of the things we find interesting about how all of these tensions were negotiated is the crucial and ironic role that pragmatics played in approaching, if not achieving, some of our more queer objectives. In turning 'soft' data into 'hard,' and acquiescing to our prior decisions about scope, scale, boundaries and temporal categories (decades), we not only claimed space (literally) that was otherwise unclaimed or deemed homophobic, thereby securing a historical archive in the context of a community whose history is only just beginning to take material form, but actually helped facilitate the constitution of a queer political subjectivity. [. . .] Thus a key political outcome of the map was that it constituted a visible, historical and tangible 'we' that, ironically, included many different and evolving identities over time, precisely the point made by Nestle (1983).

[. . .]

A final instance of the productive nature of our pragmatics is the will to actually produce something, despite limited resources, imperfect knowledge and a very informal and all-volunteer working environment. Making the map itself was important to the very ontology of the group. It gave us a clear focus, and its completion would bolster the group's sense of purpose and efficacy. Now to be sure, this claim for GIS's potential certainly does not inoculate us or the map from queer critique, perpetual and relentless as it always is. Nevertheless, for that critique to occur, the map had to be drawn in the first place. Yet we must also stress the counterfactual here: the alternative was to forego an important and practical form of visualisation and representation (i.e. to not make the map). At this juncture, then, one might ask, is it irresponsible to be only queer? Should we also offer representations that are useful? And isn't the map itself a critique of the closeting of urban history, and of the heteronormative presumptions of urban geography more generally?

The contingent nature of facts and truth

Something that is hinted at on the map, but that is much better appreciated through the telling of this story about the making of the map, is the contingent and negotiated nature of facts and truth, and the productive work that such 'facts' and 'truths' enable the map to do.

[. . .]

More to the point, though, certain sites made their way onto the map – thereby becoming reified – whereas others did not, based solely, at times, on the fact that one or more of our members could confirm their existence (or their significance). Obviously this contingency implied another, namely that of who among us was present and who was not at a particular meeting. Compounding the issue was the fact that different members had different sentimental attachments or other personal interests in seeing certain sites on the map. The result was a process of discussion and negotiation that yielded highly contingent decisions, at times, about what ended up being represented and what did not. Although this partiality might be seen, according to traditional social scientific rules and conventions, as little more than a flaw, the fact is that these same memories, discussions and negotiations almost certainly stimulated memories and imaginations as much as occluding them. The effect of that, once map viewers get involved, is cumulative. Yes, certain sites did not find their way onto the map, due to lack of memory or advocacy on the parts of group members. [...]

Still another contingent basis for facts and truth emerged because we are an all-volunteer organisation. We cannot ignore the simple fact that reminiscing, memory work is a vital source of data (Grasseni 2004) but it was always contingent on who was a member of the History Project, and whether or not they showed up to meetings at which discussions and decisions took place that year. Structurations of class, race and gender among all of us volunteers also contingently affected the information on the map. Very little on the map speaks to the experiences of non-white and other ethnic minority queer people. In Seattle, there remains much to be uncovered about the historical geographies of queer African Americans, Asian Americans, and native people, in particular. Clearly the map's relative silence on these groups had a great deal to do with the fact that our group was overwhelmingly white, and a suspicion on the part of elderly queers of colour toward such a white organisation. [...]

[...]

Power and its discontents

Finally, with respect to colliding epistemologies and the productive tensions associated with them, we became very aware of the importance of various forms of power in this knowledge production process. Again, this is less immediately visible on the map than it is a product of the map production process. But it is still both.

Our own authority in the History Project as 'the geographers', and Michael's as 'the technician', were particularly important. We were interpellated by our compatriots in a way that gave us a certain amount of freedom that others did not enjoy. Naturally this was uncomfortable for us, but in certain ways it facilitated production of the map. In one obvious and profoundly consequential way it allowed us to use our academic credentials and institutional resources to secure funding and research assistant support, as well as access to software. [...]

Still another form of structural power we acknowledge is that of the market relations within which software developers, printers and we were embedded. Although it is difficult to see these structures of power as productive, given the emancipatory purpose of the map, we recognise that our relatively privileged positions within these structures – both in terms of economic class and professional status – has contributed mightily to our ability to get the job done. [...]

Conclusion: ethnography and uncloseting the queerness of map production

The key point about the productive nature of all of the tensions and problems involved in the map's production, as we see it, is not that they resulted in some kind of consensus (Kyem 2004) or reconciliation (Schuurman 2002a). They were not simply technical problems, nor were they just opportunities for queer rumination. Rather it is that they opened the process of map production and consumption to multiple forms of representation, multiple ways of knowing, and multiple interpretations. No single notion of 'queer' made its way onto the map, nor does the map exclude all but a narrow range of readings. It is far from all things to all people, but it is most certainly not a singular rendering of Seattle's queer historical geography. Still, absent this ethnographic reflection on its production, the map, due to its own physical, visual, technical and technological reifications, closets the queerness of that production.

[...] GIS can be an integral part of a politics of uncloseting urban (and other) spaces that are otherwise heteronormatively represented and imagined. By fixing and making visible queer spaces and places – particularly from the past – a constitutive politics of individual and collective identity, community, history and belonging is made possible. Moreover, seemingly fixed visual representations, such as those on our map, might in fact be both derivative and productive of much more queer spatialities and knowledges. The intellectual, political and cultural work that making the map entailed, and that its existence continues to inspire, makes this clear. Despite their seemingly built-in epistemologies and ontologies, cartography and GIS did not dictate our own epistemologies and ontologies. As Sieber (2000) would predict, there were some conforming properties to our GIS, but there were also resistances. Our multiple subject positions demonstrated this point, as does the fact that we used GIS technologies and algorithms productively to do much more than just map points in space. Moreover, the reflexivity brought out in this ethnography is a key step in achieving the cyborg hybridity that feminist GIS envisions. Surely the map's own colliding epistemologies evince a queer sensibility, no matter how Cartesian the artefact!

[...]

[T]he political valence of participatory GIS can be extended beyond distributional issues. 'Claiming Space' was part of a constitutive, not just distributional or oppositional, politics. The visual artefact itself (the map) and what happened around it were both important. The map not only relied on but produced – and continues to produce – new and different kinds of knowledge, by affirming identities, sparking imaginations and inspiring activism.

We also note that collisions between the epistemologies underlying our critical and ethnographic methods, and epistemologies more easily (and typically) anticipated by GIS technologies, were absolutely crucial moments in our process. These moments facilitated the production of multiple and hybrid forms of data that were then translated into fixed points on the map. So, in this case, GIS, rather than necessarily privileging certain kinds of knowledge and data over others, in fact mediated between epistemologically and ontologically dissonant ways of knowing.

[. . .]

References

Aitken, S. (2002) Public participation, technological discourses and the scale of GIS, in *Community Participation and Geographic Information Systems* (eds W. Craig, T. Harris and D. Weiner), Taylor & Francis, New York, pp. 357–366.

Aitken, S. and Michel, S. (1995) Who contrives the real in GIS? Geographic information, planning, and critical theory. *Cartography and Geographic Information Systems*, **22**, 17–29.

Atkins, G. (2003) *Gay Seattle: Stories of Exile and Belonging*, University of Washington Press, Seattle.

Baker, A. (2003) *Geography and History: Bridging the Divide*, Cambridge University Press, Cambridge.

Binnie, J. (1995) Coming out of geography: towards a queer epistemology. *Environment and Planning D: Society and Space*, **15**, 223–237.

Brickell, C. (2000) Heroes and invaders: gay and lesbian pride parades and the public/private distinction in New Zealand media accounts. *Gender, Place and Culture*, **7**, 163–178.

Brown, M. (1995) Ironies of distance: an ongoing critique of the geographies of AIDS. *Environment and Planning D: Society and Space*, **13**, 159–183.

Brown, M., Knopp, L. and Morrill, R. (2005) The culture wars and urban electoral politics: sexuality, race, and class in Tacoma, Washington. *Political Geography*, **24**, 267–291.

Browne, K. (2006) Challenging queer geographies. *Antipode*, **38**, 885–893.

Cameron, J. and Gibson, K. (2005) Participatory action research in a poststructuralist vein. *Geoforum*, **36**, 315–331.

Carver, S. (2003) The future of participatory approaches using geographic information: developing a research agenda for the 21st century. *URISA Journal*, **15**, 61–71.

Cieri, M. (2003) Between being and looking: queer tourism promotion and lesbian social space in greater Philadelphia. *ACME: An International E-Journal for Critical Geographies*, **2**, 147–166.

Crampton, J. (2004) GIS and geographic governance: reconstructing the choropleth map. *Cartographica*, **39**, 41–53.

Elder, G. (2003) *Hostels, Sexuality, and the Apartheid Legacy*, Ohio University Press, Athens, OH.

Elwood, S. (2002) GIS use in community planning: a multidimensional analysis. *Environment and Planning* A, **34**, 905–922.

Esnard, A., Gelobter, M. and Morales, X. (2004) Environmental justice, GIS, and pedagogy. *Cartographica*, **38**, 53–61.

Fyfe, N. and Milligan, C. (2003) Out of the shadows: exploring contemporary geographies of voluntarism. *Progress in Human Geography*, **27**, 397–413.

Grasseni, C. (2004) Skilled landscapes: mapping practices of locality. *Environment and Planning D: Society and Space*, **22**, 699–717.

Halperin, D. (2002) *How to do the History of Homosexuality*, University of Chicago Press, Chicago.

Hanson, S. (2002) Connections. *Gender, Place and Culture*, **9**, 301–303.

Harris, C. (1992) Power, modernity and historical geography. *Annals of the Association of American Geographers*, **81**, 671–683.

Harris, T. and Weiner, D. (1998) Empowerment, marginalization, and community-integrated GIS. *Cartography and Geographic Information Systems*, **25** (2), 67–76.

Harvey, F. (2000) The social construction of geographical information systems. *International Journal of Geographical Information Science*, **14**, 711–713.

Holdsworth, D. (2003) Historical geography: new ways of imagining and seeing the past. *Progress in Human Geography*, **27**, 486–493.

Hornsey, R. (2002) The sexual geographies of reading in postwar London. *Gender, Place and Culture*, **9**, 371–384.

Houlbrook, M. (2005) *Queer London*, University of Chicago Press, Chicago.

Hubbard, P. (1998) Sexuality, immorality, and the city. *Gender, Place and Culture*, **5**, 55–72.

Hubbard, P. (2001) Sex zones: intimacy, citizenship, and public space. *Sexualities*, **4**, 51–71.

Hubbard, P. (2002) Sexing the self: geographies of engagement and encounter. *Social and Cultural Geography*, **3**, 365–381.

Knopp, L. (2004) Ontologies of place, placelessness, and movement: queer quests for identity and their impacts on contemporary geographic thought. *Gender, Place and Culture*, **11**, 121–134.

Kwan, M.-P. (2002a) Feminist visualization: re-envisioning GIS as a method in feminist geographic research. *Annals of the Association of American Geographers*, **92**, 645–661.

Kwan, M.-P. (2002b) Is GIS for women? Reflections on the critical discourse in the 1990s. *Gender, Place and Culture*, **9**, 271–279.

Kwan, M.-P. (2004) Beyond difference: from canonical geography to hybrid geographies. *Annals of the Association of American Geographers*, **94**, 756–763.
Kyem, P. (2004) Of intractable conflicts and participatory GIS applications: the search for consensus amidst competing claims and institutional demands. *Annals of the Association of American Geographers*, **94**, 37–57.
Matthews, S., Detwiler, J. and Burton, L. (2005) Geoethnography: coupling geographic information analysis techniques with ethnographic methods in urban research. *Cartographica*, **40**, 75–90.
McLafferty, S. (2002) Mapping women's worlds: knowledge, power and the bounds of GIS. *Gender, Place and Culture*, **9**, 263–269.
Nash, C.J. (2006) Toronto's gay village (1969–1982): plotting the politics of gay identity. *The Canadian Geographer*, **50**, 1–16.
Nestle, J. (1983) Voices from lesbian herstory. *The Body Politic*, **96**, 35–36.
Norheim, R. (2004) How institutional culture affects results: comparing two old-growth forest mapping projects. *Cartographica*, **38**, 35–52.
Obermeyer, N. (1998) The evolution of public participatory GIS. *Cartography and Geographic Information Systems*, **25**, 65–66.
O'Reilly, K. and Crutcher, M. (2006) Parallel politics: the spatial power of New Orleans's Labor Day parades. *Social and Cultural Geography*, **7**, 245–265.
Pavlovskaya, M. (2002) Mapping urban change and changing GIS: other views of economic restructuring. *Gender, Place and Culture*, **9**, 281–289.
Perkins, C. (2003) Cartography: mapping theory. *Progress in Human Geography*, **27**, 341–351.
Pickles, J. (2004) *A History of Spaces: Cartographic Reason, Mapping, and the Geo-coded World*, Routledge, London.
Podmore, J. (2001) Lesbians in the crowd: gender, sexuality, and visibility along Montreal's Boul. St-Laurent. *Gender, Place and Culture*, **8**, 333–455.
Podmore, J. (2006) Gone underground? Lesbian visibility and the consolidation of queer space in Montreal. *Social and Cultural Geography*, **7**, 595–625.
Rundstrom, R. (1995) GIS, indigenous peoples, and epistemological diversity. *Cartography and Geographic Information Systems*, **22**, 45–57.
Schuurman, N. (2002a) Reconciling social constructivism and realism in GIS. *ACME: An International E-Journal for Critical Geographies*, **1**, 73–90.
Schuurman, N. (2002b) Women and technology in geography: a cyborg manifesto. *The Canadian Geographer*, **46**, 258–265.
Schuurman, N. (2004) *GIS: A Short Introduction*, Blackwell, Oxford.
Schuurman, N. (2006) Formalization matters. *Annals of the Association of American Geographers*, **96**, 726–739.
Schuurman, N. and Pratt, G. (2002) Care of the subject: feminism and critiques of GIS. *Gender, Place and Culture*, **9**, 291–299.
Sieber, R.E. (2000) Conforming (to) the opposition: the social construction of geographical information systems in social movements. *International Journal of Geographical Information Science*, **14**, 775–793.
Sparke, M. (1998) Mapped bodies and disembodied maps, in *Places Through the Body* (eds H. Nast and S. Pile), Routledge, London, pp. 305–337.
Valentine, G. (1993) (Hetero)sexing space: lesbian perceptions and experiences of everyday spaces. *Environment and Planning D: Society and Space*, **11**, 395–413.
Valentine, G. and Skelton, T. (2003) Finding oneself, losing oneself: the lesbian and gay scene as a paradoxical space. *International Journal of Urban and Regional Research*, **27**, 849–866.
Wood, D. (1992) *The Power of Maps*, Guilford, New York.

Further reading

Crampton, J.W. (2010) Cartography: performative, participatory, political. *Progress in Human Geography*, **33** (6), 840-848. [A current review of diverse work in participatory mapping and consideration of its wider significance to cartographic praxis.]
Elwood, S. (2008) Volunteered geographic information: future research directions motivated by critical, participatory, and feminist GIS. *GeoJournal*, **72** (3/4), 173-183. [This paper details developments in open and user-generated mapping projects and how this might link to more critical agendas in empowering communities and bringing new voices into cartographic production.]
Parker, B. (2006) Constructing community through maps? Power and praxis in community mapping. *The Professional Geographer*, **58**, 470-484. [A useful discussion of myriad issues underlying participatory mapping methods.]

See also

- Chapter 1.13: Beyond the 'Binaries': A Methodological Intervention for Interrogating Maps as Representational Practices
- Chapter 1.14: Rethinking Maps
- Chapter 3.10: Affective Geovisualisations
- Chapter 4.6: Reading Maps
- Chapter 5.5: First Principles of a Literary Cartography
- Chapter 5.9: Affecting Geospatial Technologies: Toward a Feminist Politics of Emotion

Chapter 5.11

Mapping the Digital Empire: Google Earth and the Process of Postmodern Cartography

Jason Farman

Editors' overview

In this paper, Farman examines how the technologies underpinning Google Earth have enabled new Web 2.0 forms of mapping that can be characterised as being more distributed, participatory and social than convention cartography. Rather than maps that people simply view and use, Google Earth invites its users to contribute content and to actively complement and subvert existing annotations and to participate in interactive dialogue through bulletin boards. As such, Google Earth offers a new form of map experience, according to Farman, in which all users can become authors, but one in which the Google corporation nonetheless holds a powerful position as the ultimate arbiter and gatherer of content. Farman draws on the critical cartography and GIS literature to examine Google Earth, but also details its emancipatory and empowering qualities, arguing that it embodies a postmodern cartography.

Originally published in 2010: *New Media & Society*, **12**, doi: 10.1177/1461444809350900.

Introduction

[. . .] [W]hat type of colonialism could be present in the seemingly 'neutral' technology of Google Earth? By connecting this popular GIS to the colonial history of cartography, this article analyses the cultural implications of this software program and the potential dangers that are often attributed to GIS. I also seek to counter these critiques by showing how Google Earth uniquely engages its users, not as disembodied voyeurs, but as participants in global dialogue, represented spatially on the digital map. Ultimately, this study seeks to find a way in which recontextualisation and subversion from the 'master representations' of maps can be achieved within the authorial structure of the digital map rather than re-authoring the existing software.

Digital mapping and Google Earth

[. . .] While many school-aged children around the world are presented with the Mercator map in the classroom, the ability to access a wider variety of maps in an online realm offers the possibility to visualise the space of the earth in a different way. [. . .]

While the consequences of accessing and comparing an unprecedented number of maps is an important step forward for cartography, comparing several maps with one another is not a 'new' method. What is new are the advancements made by emerging GIS programs such as Google Earth that allow for spatial debate of maps within maps, new levels of interactivity and user agency with maps, and the ability for non-professionals to engage in these activities. These options have instigated a massive step forward for how users interact with maps. [. . .]

The Map Reader: Theories of Mapping Practice and Cartographic Representation, First Edition. Edited by Martin Dodge, Rob Kitchin and Chris Perkins.

[Google Earth] falls under the category of GIS and has made this once-specialised software available and usable for the mass market. It compiles satellite imagery and aerial photographs into a 3D virtual globe that can be interacted with in a wide variety of ways. Once started, the program situates viewers from roughly the same distance to Earth as some of the Apollo 8 whole-earth photographs – about 16 000 miles – and then zooms in on (or 'flys to' in Google Earth terminology) the user's region. The baseline resolution [...] can be as good as 0.15 metres and up to one metre in largely populated areas of Europe or North America. [...] An historical timeline was added to version 5.0 in early 2009, which allows users to scroll through archived imagery of an area. [...]

One of the most important contributions that Google Earth makes in the field of cartography is the social network that has developed around the program called the 'Google Earth Community'. This network, which is essentially a spatial Bulletin Board System (BBS), was integrated into the early versions of the program. Members of the community can post placemarks that relate information about a specific location for any user to see. Many in the Google Earth Community also create 'overlays' that offer a literal replacement or augmentation of the existing map, such as a detail of the path of Cyclone Nargis and the affected areas in Myanmar. These overlays can be downloaded and implemented by any user of the program. Thus, users can spatially debate the very tool they are using while simultaneously augmenting the borders in Google Earth to offer a different map altogether.

Critiques of geographic information systems

[...]

One reason that mapping technologies such as Google Earth often avoid critique is their use of satellite and aerial photography. Though the photograph has undergone intense scrutiny in the digital age in regards to its status as an index of reality, the photograph still holds a connection to material space that is unmatched by handcrafted maps. Peirce (1998a: 322), who famously wrote that 'representations have power to cause real facts', brought notions of indexicality in visual representations to the forefront of semiotics. His studies posit the index (under which the photograph can be categorised) as being 'in contrast to the icon's relatively straightforward resemblance and the symbol's conventionality or arbitrariness' (Doane 2007: 2). Instead, the index 'stands for its object by virtue of a real connection with it, or because it forces the mind to attend to that object' (Peirce 1998b: 14). Photography's indexical nature prompts an evaluation of it as, according to Barthes (1981: 77), an index of an 'absolute [...], irrefutabl[e] present'. Thus, as users of Google Earth engage with the historical timeline function, the satellite or aerial photograph serves as an index of a specific moment in time and a representation of that ontological materiality captured by the photographic technology. Since the science of cartography has historically overshadowed the art of mapmaking (Harley 2001: 35), hand-drawn maps close the ambiguous gap between product and authorship. Harley (2001: 38) notes that the move from 'the manuscript age to the age of printing' caused an accentuation of the division of labour in the production of maps and, as maps become more reproducible, the sense of a single creator with a singular purpose becomes less obvious. This accentuation is accelerated in the photographic age of mapping. While photographs are often associated with a photographer (the 'witness' snapping the shutter in a specific moment in time), satellite and aerial photographs used in programs like Google Earth are more commonly associated with the machinery that produces them than the person or organisation capturing or compiling them. This association between machine and product distances maps like Google Earth from a sense of subjectivity and instead emphasises the objective nature of photographic representations of earth. The result, as Sontag (1977: 154) argues, is that the 'photograph is not only an image (as a painting is an image), an interpretation of the real; it is also a trace, something directly stencilled off the real, like a footprint or a death mask'. While early maps, created through drawing, painting, etching or other methods, often attempted to distance the creator from the representation, they still functioned less as an index than as an icon (in Peirce's terms). [...]

Though cartographic methods that precede the photographic era sought standardisation and to be 'factual statements written in the language of mathematics' (Harley 2001: 36–37), these media forms were more readily associated with subjectivity (the hand of the hand-drawn map) than is associated with satellite and aerial photography. [...] [S]atellite and aerial photographs' link to machinic production from orbital locations instead point toward disembodiment, the dislocation of the subject, and objectivity.

[...] As Google Earth zooms in to the earth from a distance, the 'disembodied master subject' as Donna Haraway theorised is 'seeing everything from nowhere' (1991: 189). These representations are believed to be objective; they are simply images of reality and outside the realm of cultural interpretation. The problem with positioning GIS as software that simply gathers empirical data and presents it as fact is that such 'scientific objectivity' is typically situated and privileges those in power. The reading of objective space is indeed a 'reading', an interpretation that is never outside of the culture that produced such a reading. [...]

The relationship between technological gazing and being 'owned' by the gaze is particularly apt to the cartographic technology of GIS as seen in the Google Earth software. Maps have been, as previously noted by Edney, a way for empires to intimately know the territory they have conquered and controlled. The tools associated with GIS technology have many ties to militaristic uses, such as the implementation of aerial photography and satellite imagery. [...] Satellites were immediately understood to have a significant military function. [...] Thus, the technological gaze of aerial and satellite imagery – the essence of the interactive maps presented in Google Earth – has a long history with war and imperialism and subsequently has a historical relationship in the ways maps delineate 'us' versus 'them' as well as defining 'our territory'.

The digital empire

Here I return to the question I presented at the beginning of this article: if Google Earth's ancestry is colonial cartography and the tools it utilises (aerial and satellite imagery) are rooted in militaristic uses, what, if anything, is the empire mapped by this GIS? I want to argue that Google Earth's charting of the globe onto an interactive, web-based GIS is inherently connected to the desire to map out a new territory: the digital empire. Here I draw from Hardt and Negri's (2000: xi) redefinition of the term empire, in which they argue that imperialism, as it was known, no longer exists but has been transformed. The role of the nation state in acts of oppression and domination has undergone a progressive decline and has been replaced by a 'new form of sovereignty'. They continue by noting, 'Empire is the political subject that effectively regulates these global exchanges [of economies and cultures], the sovereign power that governs the world' (xi). [...] [S]uch a redefinition of empire is useful in identifying how corporations that control the flows of information and the infrastructure behind those flows now wield powerful global control. Google, currently one of the key corporations dominating information flows, is thoroughly invested in its role in 'modulating networks of command' (Hardt and Negri 2000: xiii). As Givler (quoted in Stripling 2008), executive director of the Association of American University Presses, recently said, 'I'm worried that Google is fast becoming our sole access point for information seeking [...] and I think that's a dangerous and unhealthy situation'. One such 'network of command' that reiterates Google's dominance of information is the data visualisation technology of GIS.

Since cartography, the delineation of borders, and the naming of territories have such historical intimacy to the control that empires wield, Google's sustained interest in digital maps have made them a key node of command over the 'information empire' [...] While mapping Hardt and Negri's new empire is a task that has not been very successful in a traditional cartographic sense (since it would require mapping the flows of information rather than the geographical borders of nation states), connecting the flows of information to the geographic map actually is one means of visualising McLuhan's 'global village' (which has indeed been actualised in the digital age). Though the World Wide Web continues to be a mostly unmapped territory for most internet users, there is still a desire to locate oneself spatially within cyberspace. One possibility for beginning to chart this new global and distributed power is to replicate the visual connectivity that was initiated by the 'Whole-Earth' photographs of the Apollo space missions. By representing the new global village as a virtual globe that can be navigated and interacted with, Google has taken the steps to chart out visually the territory that it has sought to command: an interconnected global village.

While the relationship between Google and the nation state is quite different than the imperial relationship between Britain and the East India Company, it is important to note that Google's corporate concerns (even the positioning of the company as persistently developing technologies that advance human knowledge and interaction) are fundamentally linked to political concerns. From disputes of the proper labelling of Taiwan to the disappearance of Tibet from the program, the creation or erasure of national borders has caused worldwide debates that demonstrate the indelible link between this technology and political concerns. National governments, such as Chile, have demanded that Google change the borders on its program to accurately reflect the borders that have been previously established. Google responded to Chile's demands, correcting the border near the town of Villa O'Higgins (named after a national hero who fought for independence) to reside in Chile instead of Argentina (Haines 2007). However, Google has remained silent to the requests of Taiwan to be labelled as its own country instead of a province of China.

Another historically problematic issue with Google Earth that inherently ties into political issues is the map projection used. Wood (1992: 57) points out that, while there are quite possibly an infinite number of map projections through which we can turn the spherical globe into a planar representation, each projection works toward certain purposes to the detriment of others. Rather than the traditional cartographic problem of transforming a globe into a planar representation, Google Earth instead faces the opposite problem. This GIS is made up of various flat photographs that need to be altered into a 3D sphere and, as with any map projection, distortion occurs. The effects of this distortion and its political consequences are determined by the mathematical projection used. [. . .] Though the projection Google Earth uses (an equirectangular projection) is well suited for a spherical representation of Earth, any decision regarding which projection to use is far more politically loaded than simply choosing the projection that best represents 'reality'. [. . .]

But as we have already seen, the attention to 'propaganda' is an alibi. It does nothing but deflect attention from the fact that the selection of any map projection is always to choose among competing interests, is inescapably to take – that is to promote, to embody in the map – a point of view. (Woods 1992: 60, emphasis original). [. . .] These decisions (the delineation of borders and the choice of map projection) reiterate the authorial control Google has over the representation it presents to its users. Since maps are, by and large, accepted as representing some ontological reality that exists beyond the limited subjectivity of its viewers, a transference of the power of the gaze is placed upon the viewer rather than the cartographer. By accepting the map as reality, the viewer enters into partnership with the map's author over the hegemonic assumptions such a visual representation makes. Acceptance of the map without question to the authorial nature of its design shifts ownership of the gaze onto the map user. Approaching the world around them with the assumptions of objective empiricism, their gaze into the world becomes a scientific one, outside of the realm of critique. However, as Wood (1992: 19) argues, if the map were acknowledged as creating the boundaries rather than representing them, it would no longer function as the tool that embodied reality.

The social network intervenes

Google Earth functions to trouble this transference of the gaze by including a crucial element to the map's own deconstruction: the fundamental component of a participatory culture. One major draw to the Google Earth program is the interactive nature it offers with a social network, the 'Google Earth Community'. By integrating a social network with GIS technology, the authorial nature of the map can be brought into public debate and reconfigured by the user-generated content created by the community. [. . .] [T]he Google Earth Community is a BBS that is spatially related to particular locations on the map. Users post forum comments that relate to particular pinpoints users stick onto the map. For example, in July 2007, a Slovenian member of the Google Earth Community, in his first post to the BBS, noted that the border between Italy and Slovenia was incorrect at the city of Nova Gorica. The other members of the community responded, compared maps, and linked to the site through which users can report errors to Google.

The border was then changed by Google to include Nova Gorica in Slovenia. (However, the label still reads 'Nova Gorica, Italy' as of this writing.) In another example, one user placed a pinpoint (or a 'placemark') on Lhasa, Tibet, that said, 'No Human Rights Here'. As users clicked on the placemark, the community member's post opened up to discuss the human rights violations committed by the Chinese government in Tibet. Various users responded, asynchronously in forum style, to the post, debating the current situation in Tibet and sharing the latest news about the location's border disputes.

Utilising [. . .] online social networking, [. . .] Google Earth is able to connect people across borders in the discussion of those borders. [. . .] Google Earth is able to present these debates spatially, associating the community dialog with the visual representation of the space being discussed.

Users can take dialog about the map one step further: they can actually replace or alter the map through the use of 'overlays'. Overlays function as a way for users to augment the map by offering a different visual representation of a specific area and can range from the simple – such as a user replacing the low resolution imagery of Bora Bora in French Polynesia with a higher resolution aerial photograph – to the complex – such as an animated overlay that shows the shrinking Artic icecaps. The overlays highlight the fact that the maps are not simply static visual facts to be received, but instead flexible signs that can be engaged in free play. In the history of mapping, the notion of the overlay is not new. [. . .] However, incorporating the overlay into the social network – in which the overlay can operate as a piece of the larger bricolage – is what is truly revolutionary about the Google Earth Community's overlays. Upon entering Google Earth and engaging the Google Earth Community, it becomes quickly obvious that there is not a 'central' map of authority that will dominate user interactions; instead, the map users are initially presented with is acted upon, changed and replaced. This is a very different experience of maps than in other eras of cartography. The user-generated content of the Google Earth community brings this symbol, which has enjoyed the status of being a grounded sign, into a relationship with the users that allows them to engage in free play. Such levels of interactivity with maps have historically been reserved for those in positions of cartographic skill or authoritarian power. Since maps are 'inherently rhetorical images' (Harley, 2001: 37), rhetorical devices can be utilised to convey significant meaning across the information visualisation tool of Google Earth overlays. [. . .] [T]he sheer volume of user-generated content in conjunction with the spatial dialogs that develop around these overlays give this online community potential for a radical reinvention of the way we read maps.

The problems of interactivity and agency

Does the inclusion of a social network that is able to interact and alter the maps within Google Earth solve the fundamental problems posited by cultural cartographers and theorists? Some may argue that there is nothing neither new nor revolutionary about the Google Earth Community's overlays, since they rely on a level of skill to produce them and simply utilise the tools made available by Google. However, what one person has termed to me as the 'empire of technological skill' in the creation of overlays is very far from reserved for the specialist. In Google Earth, the creation of overlays is done in a way that is familiar to anyone who has uploaded a picture to a social networking site like MySpace or Facebook [. . .] Though the skill level to contribute an overlay to the Google Earth map is not necessarily a barrier to many computer users, there are still many barriers that people take issue with. There are, after all, the cartographic, design and coding decisions made by Google that necessarily structure and limit the ways users interact with the maps and with each other. After all, Google is the one that made the option of overlays available to users in the first place. [. . .] Also, as with almost all BBSs, there is a forum moderator, who ultimately decides what content is appropriate for the bulletin board and what content or users will not be allowed past the gates.

An even graver issue is the problem of access, as Google Earth is a broadband-intense program. While many cannot contribute to the spatial debates played out in Google Earth because they do not have access to a computer, even those who have access to a computer may not be able to participate due to the intense graphic and bit rate requirements of the program. [. . .] Programs like Google Earth are designed with a very specific user in mind, one who has broadband access and a computer that can handle the graphics requirements of the software. Thus, the question needs to be raised that, while dialog and debate over maps can take place within the map itself of Google Earth, do the users who are able to engage these debates represent a diverse range of perspectives?

Aren't we forced to read Google Earth as simply reiterating Western dominance over information distribution and adhering to centralised power over user interactions as laid out by the Google corporation? My response is, no, we do not have to read Google Earth as remaining within the static authorial control of its authors/programmers and system requirements. Drawing from the rich debates that have surrounded the term 'interactivity' in such fields as electronic literature or game studies (Ryan 1991), I argue that resistance to master narratives can come through a recontextualisation from within the existing structures. [. . .] Interactivity . . . tends to function as a normative term – either fetishised as the ultimate pleasure or demonised as a deceptive fiction' (Kinder 2002: 4). For my analysis of Google Earth, I find it vital to locate the user in a relationship to the software that neither overemphasises dominance over the program (through fetishising interactivity) nor situates the user as always constrained by the limits of the program (thus demonising interactivity). Instead, by engaging issues of interactivity and agency within the very structure that potentially limits interactivity and agency, the social network as a community is positioned to enact agency. This potential for agency comes through the implementation of the very tools that limit them through a repurposing, reimagining, and reconfiguring of master representations in conjunction with user-generated content. [. . .]

While influential and inspiring feminist authors, such as Lorde (1983: 94–101), argue that 'the master's tools will never dismantle the master's house', I believe that any level of interactivity that leads to social reform comes from a recontextualisation of the existing master narratives – a refiguring that ultimately works to deconstruct the grounded signification demanded by any master narrative. Arguments which claim that interactivity and agency are impossible within Google Earth, because Google provides the tools of interactivity, go against our experience of navigating through everyday life and the authorial structures that bound us on every side. Despite the fact that boundaries exist according to authorial structures, we have the ability to 'freely' navigate the space and ultimately recontextualise the spaces that we inhabit. [. . .]

Such a reading of interacting with the existing structures to formulate a path of resistance resonates strongly with the work of Debord, particularly in the ways that his ideas of derive and détournement correspond to notions of bricolage. Theories built around the derive, defined as a wandering through the urban landscape that allows the drifter to reconfigure the sign and map systems of the city, and détournement, understood as an alteration of existing semiotic structures via a 'reuse of pre-existing artistic elements in a new ensemble' (Debord 1959), work well with the ability to reconfigure existing structures to ultimately subvert master representations. By navigating/wandering the 'psychogeography' of Google Earth (to use Debord's term), the user is embodied as he or she engages the sign systems and begins to reconfigure them through a bricolage of user-generated content. As Debord and Wolman (1956) write in their 'A User's Guide to Détournement', 'Détournement not only leads to the discovery of new aspects of talent; in addition, clashing head-on with all social and legal conventions, it cannot fail to be a powerful weapon in the service of a real class struggle. The cheapness of its products is the heavy artillery'. [. . .]

Combining Derrida's notions of bricolage with Debord's theory of détournement, we have a method for recontextualisation of master representations that corresponds with the potentials present in and through the Google Earth Community's utilisation of user generated content. Though some argue that utilising tools that are outside of master representations in order to subvert these dominating structures would be ideal and even necessary, such an approach is a produced myth. While it can be argued that 'all discourse is bricoleur', it must also be noted that there is no 'subject who supposedly would be the absolute origin of his own discourse and supposedly would construct it "out of nothing", "out of whole cloth"', since this subject 'would be the creator of the verb, the verb itself' (Derrida 1978: 285). Such notions of discourse outside existing structures tend to return to metaphysical and theological ideas, for which, Lévi-Strauss noted, also do not exist outside bricolage. Again I return to the notion that we are indeed bound at every side, yet we are importantly bound by bricolage with which we may become interactors. By engaging the bricolage – the 'heavy' and 'cheap' artillery Debord spoke of and Derrida defined as the instruments at our disposal – users of Google Earth engage in the process of rhetorical and flexible nature of maps rather than simply relying on their static authorship. [...] [U]sers should begin to engage software such as Google Earth as a tool that can radically recontextualise master representations and discursive structures through the bricolage of user-generated content. This user-generated content disseminated in Google Earth by the social network is a tool that ultimately reimagines the status of the map presented by Google and the viewer's relationship to that map. Through spatial discussions and map overlays, users become interactors involved in the representation of the social space of the global village. Though it is often argued that the age of the internet is a borderless space, borders are constantly reiterating their presence. From the disputes over borders within the Google Earth program to the borders established by the software and its system operators to limit the types of interactions users can have with this GIS, many feel so bounded by these borders to argue that such authorities need to be replaced by a complete re-authoring of the software.

Such perspectives unfortunately do not take advantage of the potential that bricolage has for major social change of re-evaluating the static nature of maps and cartography. As Google Earth and digital mapping programs continue to be growing objects of study in the field of new media, theorists and designers will need to analyse the ways that the software's interface fosters or discourages user debate and dialog about the very interface users employ. Programs like Google Earth alter the ways users inhabit mixed reality spaces that encourage seamless collaboration between material landscape and digital interface. As such, studies must interrogate the ways that these interfaces (digital and material) rhetorically situate their own methodology in order to promote or discourage critical dialog. Further studies need to also analyse the audience of this critical dialog, especially since many of the emerging devices and interfaces are available and usable by a very specific demographic. Can programs designed for a broadband-only audience actually be used to confront issues of the digital divide rather than reiterate the distance between those who have access to the necessary tools and those who do not? Studies should continue to analyse the consequence of broadband-intense programs and consequences on shifting definitions of the digital divide.

References

Barthes, R. (1981) *Camera Lucida: Reflections on Photography* (trans. R. Howard), Hill and Wang, New York.

Debord, G. (1959) Détournement as Negation and Prelude. Bureau of Public Secrets. www.bopsecrets.org/SI/3.detourn.htm.

Debord, G. and Wolman, G.J. (1956) A User's Guide to Détournement. Bureau of Public Secrets. www.bopsecrets.org/SI/detourn.htm.

Derrida, J. (1978) *Writing and Difference* (trans. A. Bass), University of Chicago Press, Chicago.

Doane, M.A. (2007) Indexicality: trace and sign introduction. *Differences: A Journal of Feminist Cultural Studies*, **18** (1), 1–6.

Haines, L. (2007) Google cedes Chilean village to Argentina. *The Register*, 30 April 2007. http://www.theregister.co.uk/2007/04/30/google_cedes_village/ (accessed October 2008).

Haraway, D. (1991) *Simians, Cyborgs, and Women: The Reinvention of Nature*, Routledge, New York.

Hardt, M. and Negri, A. (2000) *Empire*, Harvard University Press, Cambridge, MA.

Harley, J.B. (2001) Text and contexts in the interpretation of early maps, in *The New Nature of Maps: Essays in the History of Cartography* (ed. J.H. Andrews), The Johns Hopkins University Press, Baltimore, MD, pp. 31–49.

Kinder, M. (2002) Hot spots, avatars, and narrative fields forever: Bunuel's legacy for new digital media and interactive database narrative. *Film Quarterly*, **55** (4), 2–15.

Lorde, A. (1983) The master's tools will never dismantle the master's house, in *This Bridge Called My Back: Writings By Radical Women of Color* (eds C. Moraga and G. Anzaldúa), Kitchen Table, New York.

Peirce, C.S. (1998a) New elements, in *The Essential Peirce: Selected Philosophical Writings*, vol. 2 (ed. The Pierce Edition Project), Indiana University Press, Bloomington, IN, pp. 300–324.

Peirce, C.S. (1998b) Of reasoning in general, in *The Essential Peirce: Selected Philosophical Writings*, vol. 2 (ed. The Pierce

Edition Project), Indiana University Press, Bloomington, IN, pp. 11–26.

Ryan, M. (1991) *Possible Worlds, Artificial Intelligence, and Narrative Theory*, Indiana University Press, Bloomington, IN.

Sontag, S. (1977) *On Photography*, Picador, New York.

Stripling, J. (2008) Still searching. *Inside Higher Education*, 29 October. http://insidehighered.com/news/2008/10/29/google.

Wood, D. (1992) *The Power of Maps*, Guilford, New York.

Further reading

Cloud, J. (2001) Imaging the world in a barrel: the CORONA reconnaissance satellite programme and the clandestine cold war convergence of the Earth sciences. *Social Studies of Science*, **31** (2), 231–251. [An insightful analysis of the development of US spy satellites systems and the ways the new images from space circulated, including in civilian cartographic production.]

Cosgrove, D. (2001) *Apollo's Eye: A Cartographic Genealogy of the Earth in the Western Imagination*, The Johns Hopkins University Press, Baltimore, MD. [Cosgrove's monograph provides a definitive discussion of the multiple meanings of the earth image and the power that surrounds such global cartographic vision in different historical contexts.]

Goodchild, M.F. (2008) What does Google Earth mean for the social sciences, in *Geographic Visualization: Concepts, Tools and Applications* (eds M. Dodge, M. McDerby and M. Turner), John Wiley & Sons, Ltd, Chichester, UK, pp. 11–23. [An informed discussion of the potential and pitfalls of Google Earth, and allied virtual globes, as tools for cartographic display and spatial analysis tasks of academic researchers.]

Pickles, J. (1995) *Ground Truth: The Social Implications of Geographic Information Systems*, Guilford, New York. [This edited collection contains a range of important critical essays, and whilst they predate the release of Google Earth by a decade, they have much relevance in terms of highlighting the many social issues and cultural contexts underlying this powerful cartographic technology.]

Zook, M. and Graham, M. (2007) The creative reconstruction of the Internet: Google and the privatization of cyberspace and digiplace. *GeoForum*, **38** (6), 1322–1343. [A useful critical analysis of the role of search engines and how their arcane algorithms for processing geographic queries and generating automatic mapping have significant social implications for representation of the world.]

See also

- Chapter 1.13: Beyond the 'Binaries': A Methodological Intervention for Interrogating Maps as Representational Practices
- Chapter 1.14: Rethinking Maps
- Chapter 2.4: Maps and Mapping Technologies of the Persian Gulf War
- Chapter 2.5: Automation and Cartography
- Chapter 2.6: Cartographic Futures on a Digital Earth
- Chapter 2.12: Imaging the Word: The State of Online Mapping
- Chapter 4.10: Citizens as Sensors: The World of Volunteered Geography
- Chapter 5.8: Cartographic Rationality and the Politics of Geosurveillance and Security

Index

The Map Reader: Theories of Mapping Practice and Cartographic Representation, First Edition. Edited by Martin Dodge, Rob Kitchin and Chris Perkins.